Representation Theory

For further volumes:
http://www.springer.com/series/6253

Algebra and Applications

Volume 19

Algebra and Applications aims to publish well written and carefully refereed monographs with up-to-date information about progress in all fields of algebra, its classical impact on commutative and noncommutative algebraic and differential geometry, K-theory and algebraic topology, as well as applications in related domains, such as number theory, homotopy and (co)homology theory, physics and discrete mathematics.

Particular emphasis will be put on state-of-the-art topics such as rings of differential operators, Lie algebras and super-algebras, group rings and algebras, C^*-algebras, Kac-Moody theory, arithmetic algebraic geometry, Hopf algebras and quantum groups, as well as their applications. In addition, Algebra and Applications will also publish monographs dedicated to computational aspects of these topics as well as algebraic and geometric methods in computer science.

Alexander Zimmermann

Representation Theory

A Homological Algebra Point of View

Springer

Alexander Zimmermann
Département de Mathématiques LAMFA
Université de Picardie
Amiens
France

ISSN 1572-5553 ISSN 2192-2950 (electronic)
ISBN 978-3-319-07967-7 ISBN 978-3-319-07968-4 (eBook)
DOI 10.1007/978-3-319-07968-4
Springer Cham Heidelberg New York Dordrecht London

Library of Congress Control Number: 2014942058

Primary: 16E30, 16S34, 18E30, 20C05, 20C20
Secondary: 16D90, 16E20, 16G10, 16P20, 16P40, 16S34, 16T10, 18E35, 18G05, 18G10, 18G15, 20J05

Printed on acid-free paper

Springer is part of Springer Science+Business Media (www.springer.com)

Für Susanne

Preface

This book gives an introduction to the representation theory of finite groups and finite dimensional algebras via homological algebra. In particular, it emphasises common techniques and unifying themes between the two subjects.

Until the 1990s, the representation theory of finite groups and the representation theory of finite dimensional algebras had developed quite independently of one another, with few connections between them. Then, starting at about 1990 a unifying theme was discovered in form of methods from homological algebra. In the representation theory of algebras, homological approaches had already been accepted and used for quite a long time. This was certainly a consequence of the most influential work of Auslander and Reiten who preferred this abstract approach. The techniques of triangulated categories were brought to the representation theory of algebras by Happel in [1]. However, before this time, methods from homological algebra were not yet fully accepted in the representation theory of groups. The relevance of stable categories was known, but no systematic approach was really undertaken. The situation was different for the representation theory of Lie algebras and of algebraic groups. Since that part of representation theory is very close to algebraic geometry, homological algebra methods and in particular derived categories had long been established as an important tool.

The bridge from the representation theory of algebras to the representation theory of groups via homological algebra was fully established in 1989, when Rickard proved a Morita theory for derived categories and Broué pronounced his most famous abelian defect conjecture. The most appealing abelian defect conjecture uses very deep knowledge of the structure of algebras and its homological algebra. It cannot be approached by dealing only with finite groups, progress can be achieved only by taking into consideration further algebraic structures and their homological algebra. Most interestingly, the conjecture and the methods to approach it offer a bridge to questions close to algebraic topology, and is at once a very challenging field of applications for the representation theory of algebras. Now, more than two decades after the abelian defect group conjecture was stated, categories are everywhere in the representation theory of groups and homological algebra methods as well as methods from the representation theory of algebras are omnipresent. In the reverse direction, motivations and methods from the representation theory of finite groups have become central and important in the representation theory of finite dimensional algebras.

While Broué's abelian defect conjecture was in a certain sense, the crystallisation point for the use of common techniques in the representation theory of groups and of algebras, it is now commonly admitted that these modern techniques are useful when studying group representations, even if one is not really interested in Broué's abelian defect conjecture. This is the point of view we will present. We shall present the homological algebra methods that are useful for all kinds of questions in the representation theory of algebras and of groups, and thus provide a solid background for further studies. We will not, or better to say almost not, consider the abelian defect conjecture, but the reader will acquire knowledge of the necessary techniques which will enable him to pursue developments in the direction of this and related conjectures.

At the present time, no textbook is available which gives an introduction to representations of groups and algebras at the same time, focussing on methods from homological algebra. In my opinion, a student learning the representation theory of either groups or algebras should master techniques from both sides and the unifying homological algebra needed. This book is meant to give an introduction with this goal in mind.

More explicitly, the general principle is the following. Classical representation theory deals with modules over an algebra A or over a group algebra $A = KG$. We emphasise here the category $\mathcal{C} = A\text{-}mod$ of these modules. It soon becomes clear that the module category of these algebras or groups encodes the relevant phenomena, but these cannot easily be obtained from the latter. To circumvent this problem, another category $\mathcal{D}_{\mathcal{C}}$ obtained from the category $\mathcal{C}$ is introduced. The choice of the category $\mathcal{D}_{\mathcal{C}}$ depends on the kind of question one wants to study. The main kind of problem is then to characterise in terms of $\mathcal{C}_1$ and $\mathcal{C}_2$ when $\mathcal{D}_{\mathcal{C}_1}$ is equivalent to $\mathcal{D}_{\mathcal{C}_2}$. Moreover, it is obviously interesting to find properties of $\mathcal{C}_1$ that are invariant under an equivalence between $\mathcal{D}_{\mathcal{C}_1}$ and $\mathcal{D}_{\mathcal{C}_2}$, so that $\mathcal{C}_2$ also has this property whenever there is such an equivalence and $\mathcal{C}_1$ has the requested property.

In our case we first study the module category $\mathcal{C} = A\text{-}mod$, and choose $\mathcal{D}_{\mathcal{C}} = \mathcal{C}$. The relevant equivalence then is called Morita equivalence. Then, in a second attempt, we choose $\mathcal{D}_{\mathcal{C}}$ to be the stable category, to be introduced later. A rich structure is available for this choice of $\mathcal{D}_{\mathcal{C}}$, at least when A is self-injective. There are two possible equivalences in this case, in order to strengthen possible implications if necessary. Third, we choose $\mathcal{D}_{\mathcal{C}}$ to be the derived category of complexes of A-modules. In this case, many classical results are known, and in some sense the structure is aesthetically the most appealing. Recently, an equivalence weaker than equivalences between derived categories, but more general than stable equivalences, coming from so-called singular categories has been examined intensively, and we shall present some of the known results. In addition to this, structure results from very classical theory also heavily depend on equivalence between these subsequent categories. For example, the structure theorem of blocks with cyclic defect group is intimately linked with questions about equivalences in stable categories. We shall present the details in the relevant Chap. 5. Of course, for all this, some preparations are necessary. Not all of the necessary mathematics is part of the standard material in algebra classes, so we need to close this gap. In the introductory chapters, all the necessary background is presented in a uniform way.

Where is it appropriate to start and where to end? In my opinion, the classical representation theory of finite groups and semisimple algebras is still a very good starting point. The subject is explicit, provides plenty of nice examples, and is appealing for most people entering the subject. Moreover, when introducing the basic objects it is natural to pave the way in a manner that the methods needed to understand the bridge between representations of algebras and of groups become familiar naturally. The student I have in mind has taken a basic linear algebra course and knows some basics on rings, fields and groups, as I believe is done nowadays at all universities around the world. Normally, the student is at the end of his fourth year of university and wants to learn representation theory. In this sense, the book is self-contained. The very few results that are not proved concern number theoretic statements on completions or set theoretical details, where a proof would have led to distant foreign grounds. In this case references are given, so that the reader willing to explore these points can find accessible sources easily. I also believe that a mathematician wanting to see what this theory is about and trying to learn the main results should be able to profit from the book. This is the goal I tried to achieve.

The book is meant to give the material for lectures starting at the level of the end of the first year of a master's degree in the Bologna treaty scheme of the European Union from 1999, then specializing in the first half of the second year and finally choosing a more advanced topic in the second half of the last year. The first chapter has a level of difficulty appropriate to the first year of the master's degree, the second chapter is an introduction to group representations, and should be adapted to the first semester of the second year. The following chapters can be studied in parts only, and offer many different ways to approach more specialised topics. Chapter 1 is compulsory for the rest of the book, in the sense that the basic methods and main motivations from ring and module theory are provided there. Of course, the advised reader can skip this chapter. Chapter 2 gives an introduction to the representation theory of groups, which frequently provides examples and applications for more general concepts throughout the book. Again, most of the content can also be obtained from other text books, but some concepts are original here and cannot easily be found elsewhere.

The second part of the book deals with more advanced subjects. The homological algebra used in representation theory is developed and the main focus is given to equivalences between categories of representations. Chapter 3 is the entrance gate to the second part. Chapters 4–6 give specialisations in different directions, and depending on how the reader wants to continue, different parts of Chap. 3 are necessary. Sections 3.1–3.3 are necessary for each of the remaining chapters, whereas Chap. 4 does not need any further information. Section 3.4 is the technical framework for all that is studied in Chaps. 5 and 6. Section 3.5 is crucial and it prepares Sects. 3.6 and 3.7. In particular, Sects. 3.5 and 3.7 are both used in an essential way in Chaps. 5 and 6. Section 3.9 is technical and is used for a few details in proofs appearing in Chap. 6.

Chapter 4 presents the classical Morita theory from the 1950s in such a way that the more recent developments of the subsequent chapters become natural. As main applications, we give Külshammer's proof of Puig's structure theorem for nilpotent blocks in Sect. 4.4.2 and Gabriel's theorem on algebras presented by quivers and relations in Sect. 4.5.1.

Chapter 5 describes stable categories and the equivalences between them. As an application the structure of blocks with cyclic defect groups as Brauer tree algebras is proved in Sect. 5.10. Section 5.11 gives a very nice and not widely known result due to Reiten, and the method of proof is interesting in its own right. As far as I know, this is the first documentation of this result in book form.

Chapter 6 then introduces the reader to derived equivalences, and at least from Sect. 6.9 onwards parts of Chap. 5 are also used. For Rickard's Morita theory I choose a proof which combines results from Rickard and from Keller. The proof of the basic fact may be slightly more involved than in Rickard's original approach, but it has the advantage that it answers questions left open in Rickard's original approach concerning the actual constructibility of the objects. Readers interested in getting the most direct proof for Rickard's Morita theory for derived categories may like to consult the treatment in [2]. However, only an existence proof is given there and no construction of the basic object is shown, except in a very specific situation.

All chapters arose from lectures I gave at the Université de Picardie over the years, and most chapters served as class notes for lectures I gave at the appropriate level. I owe much to the students and their questions and remarks during the lectures, alerting me when the presentation of preliminary versions had to be improved.

Representation theory is vast, and even the topics covered in the book can be developed much further. In order to keep the book to a reasonable size, I decided to leave out many important topics, such as Rickard's result on splendid equivalences. Initially I planned to include much more of the ordinary representation theory as well so that it could be used as manuscript for an entire one-semester course. However, this project would have increased the book further to an unreasonable size, and since the subsequent chapters do not use ordinary representation theory, I had to refrain from developing the first chapter further in this direction. The representation theory of algebras is in some sense unthinkable without an introduction to Auslander-Reiten quivers and almost split sequences. I have not given an introduction to this most important theory here. Many textbooks, such as [3] or [4] have appeared recently and cover Auslander-Reiten theory in a very nice and an accessible way. Almost split sequences have not been used throughout the book, although many properties in Chap. 5 in particular can be shown with only a little use of this technique. The book [4] gives some examples, and [5] also uses them without hesitation and with breathtaking speed. Another omission is the rapidly developing theory of cluster categories and cluster algebras. I feel that the theory is not yet settled enough to be given a definitive treatment. A very nice and useful topic which I omitted, is geometric representation theory, in the sense of methods from algebraic geometry used in the representation theory of algebras. It would have made a great deal of sense to include this powerful tool, however, it needs quite a few prerequisites in algebraic geometry at various levels. To introduce the algebraic geometry needed for even the most elementary purposes would have increased the size of the book considerably.

There are many colleagues I am indebted for help, comments, remarks and ideas. Mentioning some colleagues implies that many others will be forgotten. I apologise to those not explicitly mentioned. Serge Bouc spent hundreds of hours with me in discussions during our weekly reading group over the past 16 years. I thank him, in particular, for his permission to include his generalisation of Hattori-Stallings traces to Hochschild homology. Further, I thank Thorsten Holm, Bernhard Keller, Henning Krause, Jeremy Rickard, Klaus Roggenkamp, Zhengfang Wang and Guodong Zhou for numerous discussions which either directly or indirectly influenced this work. I thank Yann Palu and Guodong Zhou for careful reading of parts of the manuscript and for alerting me on those occasions where I was about to write nonsense. Finally, I want to mention that the very pleasant atmosphere at the university in Amiens made this project possible.

Amiens, August 2013 Alexander Zimmermann

References

1. Happel, D.: Triangulated categories in the representation theory of finite-dimensional algebras, London Mathematical Society Lecture Note Series, vol. 119. Cambridge University Press, Cambridge (1988)
2. König, S., Zimmermann, A.: Derived Equivalences for Group Rings (with contributions by Keller, B., Linckelmann, M., Rickard, J., Rouquier, R.) Lecture Notes in Mathematics, vol. 1685. Springer, Berlin (1998)
3. Assem, I., Simson, D., Skowroński, A.: Elements of the representation theory of associative algebras. Techniques of representation theory, vol. 1. Cambridge University Press, New York (2006)
4. Auslander, M., Reiten, I., Smaló, S.: Representation theory of artin algebras. Cambridge studies in advanced mathematics. Cambridge University Press, Cambridge (1995)
5. Benson, D.: Representations and cohomology, vol. I and II. Cambridge University Press, Cambridge (1991)

Contents

1 Rings, Algebras and Modules 1
1.1 Basic Definitions 1
1.1.1 Algebras 1
1.1.2 Modules 3
1.2 Group Representations 7
1.2.1 Group Algebras and Their Modules 8
1.2.2 Maschke's Theorem 11
1.3 Noetherian and Artinian Objects 14
1.4 Wedderburn and Krull-Schmidt 16
1.4.1 The Krull-Schmidt Theorem 16
1.4.2 Wedderburn's Structure Theorem 22
1.4.3 Base Field Extension and Splitting Fields 31
1.5 Group Rings as Semisimple Algebras 33
1.6 Radicals and Jordan–Hölder Conditions 34
1.6.1 Jacobson- and Nil-Radical 35
1.6.2 The Jordan–Hölder Theorem 41
1.7 Tensor Products 48
1.7.1 The Definition and Elementary Properties 48
1.7.2 Immediate Applications for Group Rings 59
1.7.3 Mackey's Formula 66
1.8 Some First Steps in Homological Algebra; Extension Groups 72
1.8.1 Projective and Injective Modules 72
1.8.2 *Ext*-groups 80
1.8.3 Low Dimensional Group Cohomology 101
1.8.4 Short Exact Sequences of Groups and Group Cohomology 107
1.9 More on Projective Modules 116
1.9.1 The Module Approach 116
1.9.2 The Idempotent Approach 121
1.10 Injectives and Projectives May Coincide: Two Classes of Such Algebras 125
1.10.1 Self-Injective Algebras and Frobenius Algebras 125
1.10.2 Symmetric Algebras 134

1.11 Algebras Defined by Quivers and Relations: Two Classes of Algebras 140
1.11.1 Hereditary Algebras 140
1.11.2 Special Biserial Algebras 145
References 152

2 Modular Representations of Finite Groups 155
2.1 Relatively Projective Modules 155
2.1.1 Relatively Projective Modules for Subalgebras 155
2.1.2 Vertex and Source 161
2.1.3 Green Correspondence 165
2.2 Clifford Theory 169
2.2.1 Clifford's Main Theorem 169
2.2.2 Group Graded Rings and Green's Indecomposability Theorem 173
2.3 Brauer Correspondence 176
2.4 Properties of Defect Groups, Brauer and Green Correspondence 181
2.5 Orders and Lattices 185
2.5.1 Discrete Valuation Rings 185
2.5.2 Classical Orders and Their Lattices 189
2.6 The Cartan-Brauer Triangle 197
2.6.1 Grothendieck Groups 198
2.6.2 Cartan and Decomposition Maps 199
2.6.3 The Cartan-Brauer Triangle 202
2.7 Defining Blocks by Non-vanishing Ext^1 Between Simple Modules 204
2.8 The Structure of Serial Symmetric Algebras 209
2.9 Külshammer Ideals 214
2.9.1 Definitions of Külshammer Spaces 214
2.9.2 The Number of Simple Modules 217
2.9.3 Külshammer Ideals of Symmetric Algebras 219
2.9.4 Further Properties of Külshammer Ideals of Group Algebras 224
2.10 Brauer Constructions and p-Subgroups 228
2.10.1 The Brauer Homomorphism and Osima's Theorem . . . 228
2.10.2 Higman's Criterion, Traces and the Brauer Quotient 231
2.10.3 Brauer Pairs 234
2.11 Blocks, Subgroups and Clifford Theory of Blocks 240
2.11.1 Brauer Correspondence for Bigger Groups 240
2.11.2 Blocks and Normal Subgroups 244

2.12 Representation Type of Blocks, Cyclic Defect 251
2.12.1 The Structure of Blocks with Normal Cyclic Defect Group 251
2.12.2 Blocks of Finite Representation Type 252
References 256

3 Abelian and Triangulated Categories 259
3.1 Definitions 259
3.1.1 Categories 259
3.1.2 Functors 263
3.1.3 Natural Transformations 266
3.2 Adjoint Functors 273
3.3 Additive and Abelian Categories 279
3.3.1 The Basic Definition 279
3.3.2 Useful Tools in Abelian Categories 281
3.4 Triangulated Categories 288
3.5 Complexes 299
3.5.1 The Category of Complexes 300
3.5.2 Homotopy Category of Complexes 304
3.5.3 The Derived Category of Complexes 319
3.6 An Application and a Useful Tool: Hochschild Resolutions . . . 340
3.6.1 The Bar Resolution 340
3.6.2 Second Hochschild Cohomology, Extensions and The Wedderburn-Malcev Theorem 343
3.6.3 Unicity of the Form on Frobenius Algebras 348
3.7 Hypercohomology and Derived Functors 349
3.7.1 Left and Right Derived Functors 350
3.7.2 Derived Tensor Product, Derived Hom Space, Hyper Tor and Hyper Ext 352
3.8 Applications: Flat and Related Objects 360
3.9 Spectral Sequences 371
3.9.1 Spectral Sequence of a Filtered Differential Module 372
3.9.2 Spectral Sequences of Filtered Complexes 376
3.9.3 Spectral Sequence of a Double Complex 378
3.9.4 Short Exact Sequences Associated to Spectral Sequences 381
3.9.5 Application: Homomorphisms in the Derived Category 382
References 384

4 Morita Theory 387
4.1 Progenerators 387
4.2 The Morita Theorem 393

4.3 Properties Invariant Under Morita Equivalences 400
4.4 Group Theoretical Examples 402
4.4.1 Elementary Examples 402
4.4.2 Nilpotent Blocks 406
4.5 Some Basic Algebras 412
4.5.1 Basic Hereditary Algebras and Gabriel's Theorem 412
4.5.2 Frobenius Algebras and Self-injective Algebras II 416
4.6 Picard Groups 419
References 424

5 Stable Module Categories 427
5.1 Definitions, Elementary Properties, Examples 427
5.1.1 Definition of Stable Categories 427
5.1.2 Syzygies 428
5.1.3 Representation Type and Equivalences of Stable Categories 430
5.1.4 The Case of Self-Injective Algebras; Triangulated Structure 431
5.2 Some Examples of Stable Categories 440
5.3 Stable Equivalences of Morita Type: The Definition 443
5.3.1 Some Auxiliary Results on Functors Between Stable Categories 443
5.3.2 Broué's Original Definition 446
5.3.3 Stable Equivalences Induced by Exact Functors...... 448
5.4 Stable Equivalences of Morita Type Between Indecomposable Algebras 455
5.5 Stable Equivalences of Morita Type: Symmetricity and Further Properties 463
5.6 Image of Simple Modules Under Stable Equivalences of Morita Type 466
5.7 The Stable Grothendieck Group 469
5.7.1 The Definition 469
5.7.2 Application to Nakayama Algebras 472
5.8 Stable Equivalences and Hochschild (Co-)Homology 474
5.8.1 Stable Equivalence of Morita Type and Hochschild Homology Using Bouc's Trace 474
5.8.2 Stable Equivalences of Morita Type and Hochschild Cohomology 485
5.9 Degree Zero Hochschild (Co-)Homology 486
5.9.1 Stable Centre 486
5.9.2 Stable Cocentre 497
5.9.3 Külshammer Ideals and Stable Equivalences of Morita Type 500

5.10 Brauer Tree Algebras and the Structure of Blocks with Cyclic Defect Groups 504
5.10.1 Brauer Tree Algebras 505
5.10.2 Representation Finite Local Blocks 513
5.10.3 Algebras Which Are Stably Equivalent to a Symmetric Serial Algebra. 515
5.10.4 Blocks with Cyclic Defect Group. 528
5.11 Stable Equivalences and Self-Injectivity 534
5.11.1 Functor Categories 535
5.11.2 The Proof of Reiten's Theorem 546
References 554

6 Derived Equivalences 557
6.1 Tilting Complexes 557
6.1.1 Historical and Methodological Remarks 557
6.1.2 An Equivalence Gives a Tilting Complex 558
6.2 Some Background on Co-algebras 563
6.2.1 Co-algebras, Bi-algebras, Hopf-Algebras. 564
6.2.2 Derivations, Coderivations 568
6.3 Strong Homotopy Action 574
6.3.1 Strong Homotopy Actions for Complexes Without Self-Extensions 574
6.3.2 Morphisms Between Strong Homotopy Actions. 577
6.3.3 Constructing an Algebra Action Out of a Strong Homotopy Action 580
6.4 Constructing a Functor from a Tilting Complex; Keller's Theorem. 587
6.5 Rickard's Morita Theorem for Derived Equivalences 589
6.5.1 The Theorem and Its Proof 589
6.5.2 Derived Equivalences of Standard Type 594
6.6 The Case of Two-Term Complexes 596
6.6.1 The Two-Sided Tilting Complex of a Two-Term Tilting Complex. 597
6.6.2 Endomorphism Rings of Two-Term Tilting Complexes 600
6.7 First Properties of Derived Equivalences. 605
6.7.1 A First Non-trivial Example 605
6.7.2 Local Rings and Commutative Rings 608
6.7.3 Hochschild (Co-)Homology and Derived Equivalences 612
6.7.4 Symmetric Algebras. 616

6.8 Grothendieck Groups and Cartan-Brauer Triangle 618
6.8.1 The Grothendieck Group of a Triangulated Category 619
6.8.2 Derived Equivalences and Cartan-Brauer Triangles . . . 622
6.9 Singularity and Stable Categories as Quotients of Derived Categories 627
6.9.1 Definition and First Properties of the Singularity Category 627
6.9.2 Singularity Categories and Self-Injective Algebras; Link to the Stable Case 631
6.9.3 Homomorphisms in Singularity Categories as Limits 637
6.9.4 Algebras with Radical Squared 0; an Example 639
6.10 Examples 645
6.10.1 Blocks of Cyclic Defect 645
6.10.2 Okuyama's Method 651
6.10.3 Derived Equivalences Defined by Simple Modules . . . 657
6.11 Broué's Abelian Defect Group Conjecture; Historical Remarks 665
6.12 Picard Groups of Derived Module Categories 668
6.12.1 The General Definition 668
6.12.2 Fröhlich's Localisation Sequence 672
6.12.3 An Example: The Brauer Tree Algebras 677
References 692

Index 697

Notations

$\hookrightarrow, \twoheadrightarrow$	Monomorphism, epimorphism
$\simeq$	Is isomorphic to
$\trianglelefteq$	Is a normal subgroup of
$\overset{!}{=}$	Suppose equality
$- \otimes_A^{\mathbb{L}} -$	Left derived tensor product
$[A, A]$	Commutator space
A^{op}	Opposite algebra
$A^{\times}$	Unit group of the algebra A
$\mathcal{A}b$	Category of abelian groups
A-*Mod*	Category of all A-modules
A-*mod*	Category of finitely generated A-modules
A-*Proj*	Category of projective A-modules
A-*proj*	Category of finitely generated projective A-modules
A-$\underline{Mod}$	Stable category of A-modules
A-$\underline{mod}$	Stable category of finitely generated A –modules
$\mathbb{C}$	Field of complex numbers
$C_G(Q)$	Centraliser of Q in G
$\mathcal{C}^{op}$	Opposite category of the category $\mathcal{C}$
$C(\mathcal{A})$	Category of complexes in $\mathcal{A}$
$C^+(\mathcal{A})$	Category of left bounded complexes in $\mathcal{A}$
$C^-(\mathcal{A})$	Category of right bounded complexes in $\mathcal{A}$
$C^b(\mathcal{A})$	Category of bounded complexes in $\mathcal{A}$
$C^{\emptyset,+}(\mathcal{A})$	Category of complexes in $\mathcal{A}$ with left bounded homology
$C^{\emptyset,-}(\mathcal{A})$	Category of complexes in $\mathcal{A}$ with right bounded homology
$C^{\emptyset,b}(\mathcal{A})$	Category of complexes in $\mathcal{A}$ with bounded homology
$C^{-,b}(\mathcal{A})$	Category of left bounded complexes in $\mathcal{A}$ with bounded homology
$C^{+,b}(\mathcal{A})$	Category of right bounded complexes in $\mathcal{A}$ with bounded homology
$D(\mathcal{A})$	Derived category of complexes in $\mathcal{A}$
$D^-(\mathcal{A})$	Derived category of right bounded complexes in $\mathcal{A}$
$D^+(\mathcal{A})$	Derived category of left bounded complexes in $\mathcal{A}$
$D^b(\mathcal{A})$	Derived category of bounded complexes in $\mathcal{A}$

$D_{\mathcal{A}'}(\mathcal{A})$	Derived category of $\mathcal{A}$ with homology in $\mathcal{A}'$
$D_{sg}(A)$	Singularity category
$DPic_k(A)$	Picard group of the derived category
$\mathcal{E}ns$	Category of sets
$Ext^*(M, M)$	*Ext*-algebra Proposition
$GL_n(K)$	Group of $n \times n$ invertible matrices over K
$H(M)$	Homology of a complex
$HH_n(A)$	Hochschild homology of degree n
$HH^n(A)$	Hochschild cohomology of degree n
$Inn(A)$	Group of inner automorphisms of A
$K(\mathcal{A})$	Homotopy category of complexes in $\mathcal{A}$
$K^-(\mathcal{A})$	Homotopy category of right bounded complexes in $\mathcal{A}$
$K^+(\mathcal{A})$	Homotopy category of left bounded complexes in $\mathcal{A}$
$K^b(\mathcal{A})$	Homotopy category of bounded complexes in $\mathcal{A}$
$K^{-,b}(\mathcal{A})$	Homotopy category of right bounded complexes in $\mathcal{A}$ with bounded homology
$K\backslash G/H$	Double classes of G modulo H and K
$M\uparrow_H^G$	Induced module from the subgroup H to G
${}^{\alpha}M$	Module twisted by automorphism α
${}_{\alpha}M_{\beta}$	Twisted bimodule
$Nil(A)$	Nil radical of the algebra A
$\mathbb{N}$	Set of natural numbers $\{0, 1, 2, \ldots\}$
$N_G(Q)$	Normaliser of Q in G
$\Omega^n_M, \Omega^n(M)$	n-th syzygy of M
$Out(A)$	Automorphisms modulo inner automorphisms
$Pic_K(A)$	Picard group
$Picent(A)$	Central Picard group
$\mathbb{Q}$ and $\mathbb{R}$	Field of rational and real numbers
$\hat{\mathbb{Q}}_p$	Field of p-adic numbers
$\mathbb{R}Hom(-,?)$	Right derived *Hom*
$\mathrm{rad}(M)$	Jacobson radical of the module M
$\mathrm{soc}(M)$	Socle of the module M
$\mathfrak{S}_n$	Symmetric group of degree n
$\tau_{\leq n}, \tau_{\geq m}$	Truncation of a complex
$T_n(A)$	Külshammer spaces
$Tor_n^A(M,N)$	Torsion groups
$U^{\perp}$	Orthogonal of the set U
$X[1], X[n]$	Shifted complex by n degrees
$Z(A)$	Centre of an algebra
$\mathbb{Z}$	Ring of all integers
$\hat{\mathbb{Z}}_p$	Ring of p-adic integers

Chapter 1
Rings, Algebras and Modules

In this first chapter we provide the necessary facts in elementary module theory, we define the concept of a representation, and give elementary applications to representations of groups. We also provide a short introduction to the basic concepts leading to homological algebra, as far as it is necessary to understand the elementary modular representation theory of finite groups as it is developed in Chap. 2. We restrict ourselves to a selection of those properties that are going to be used in the sequel and avoid developing the theory in directions which are not explicitly used in later chapters. This way the book remains completely self-contained, without being encyclopedic, and the choice will also allows us to fix a coherent notation throughout.

1.1 Basic Definitions

This book deals with groups and algebras. In order to fix the notation, to recall the basic facts and to keep the book as self-contained as possible we shall briefly give the necessary definitions of the most important objects and some of their properties.

1.1.1 Algebras

A field in this book is always commutative, a ring is always associative and has a unit, but may be non-commutative. The centre of a ring A is denoted by $Z(A)$ and is defined to be

$$Z(A) := \{b \in A \mid a \cdot b = b \cdot a \ \forall a \in A\}.$$

It is clear that $Z(A)$ is a commutative subring of A. Ring homomorphisms are always assumed to preserve the unit.

A. Zimmermann, *Representation Theory*,
Algebra and Applications 19, DOI: 10.1007/978-3-319-07968-4_1,

Definition 1.1.1 Let K be a commutative ring and let A be a ring. Then a K-*algebra* is the structure of the ring A together with a ring homomorphism $\epsilon_A : K \longrightarrow Z(A)$.

Remark 1.1.2

- Observe that a $\mathbb{Z}$-algebra is nothing else than a ring.
- Very often we are interested in the case of a field K and then mostly the case where A is of finite dimension over K. However, sometimes we need to use a broader concept, and it will be important to be able to pass to general commutative rings K.
- For all $\lambda \in K$ and all $a \in A$ we simply write $\lambda \cdot a := \epsilon_A(\lambda) \cdot a$.

Example 1.1.3 A few constructions are used frequently in the sequel.

1. Let K be a field and let $A = End_K(K^n)$ be the set of square n by n matrices over K. Then this is a K-algebra of finite dimension n^2 as a K-vector space, with additive ring structure being the sum of matrices defined by adding each coefficient separately and multiplicative structure the matrix multiplication. The homomorphism of K to the centre of A is given by sending $k \in K$ to the diagonal matrix with diagonals entries all being equal to k.
2. If A_1 and A_2 are both K-algebras, then $A_1 \times A_2$ is a K-algebra as well. Indeed, if $\lambda_1 : K \longrightarrow Z(A_1)$ is the homomorphism defining the algebra structure of A_1, and if $\lambda_2 : K \longrightarrow Z(A_2)$ is the homomorphism defining the algebra structure of A_2, then
$$\lambda_1 \times \lambda_2 : K \ni k \mapsto (\lambda_1(k), \lambda_2(k)) \in A_1 \times A_2$$
defines an algebra structure on $A_1 \times A_2$.
3. If L is a field extension of K, then L is a K-algebra. If D is a skew field with centre L, then D is an L-algebra. Moreover, if A is an L-algebra, then by restricting the mapping $L \longrightarrow Z(A)$ to a subfield K of L, one sees that A is also a K-algebra.

As usual, once we have defined the objects, we are interested in structure preserving maps.

Definition 1.1.4 Let K be a commutative ring and let A and B be K-algebras with structure morphisms $\epsilon_A : K \longrightarrow A$ and $\epsilon_B : K \longrightarrow B$. A ring homomorphism $\varphi : A \longrightarrow B$ is an *algebra homomorphism* if $\epsilon_B = \varphi \circ \epsilon_A$ as mappings $K \longrightarrow B$. In other words, for all $\lambda \in K$ and all $a \in A$, we have $\varphi(\lambda \cdot a) = \lambda \cdot \varphi(a)$.

A K-algebra homomorphism φ is an *algebra monomorphism* (*epimorphism, isomorphism*) if φ is a ring monomorphism (epimorphism, isomorphism).

Example 1.1.5 A few easy observations illustrate the definition.

- The ring of complex numbers $\mathbb{C}$ forms a $\mathbb{C}$-algebra A_1 by the structural map $\lambda_{\mathbb{C}} = id_{\mathbb{C}} : \mathbb{C} \longrightarrow Z(A_1)$. However, the complex numbers $\mathbb{C}$ also forms another $\mathbb{C}$-algebra A_2 via the structural map 'complex conjugation' $\mu_{\mathbb{C}}(z) = \overline{z} : \mathbb{C} \longrightarrow A_2$. These two algebra structures are isomorphic by putting $\varphi(a) = \overline{a} \in A_2$ for any $a \in A_1$.

- The composition of two K-algebra homomorphisms is a K-algebra homomorphism.
- An algebra homomorphism which is invertible as a ring homomorphism is also invertible as an algebra homomorphism. This follows immediately from

$$(\varphi \circ \lambda_A = \lambda_B) \Rightarrow \left(\lambda_A = \varphi^{-1} \circ \varphi \circ \lambda_A = \varphi^{-1} \circ \lambda_B\right).$$

1.1.2 Modules

We shall be interested mainly in the action of algebras. The corresponding concept is a module, and this will be the central object studied in this book.

Definition 1.1.6 Let K be a commutative ring and let A be a K-algebra. Then a *left A-module M* is an abelian group and a mapping $\mu : A \times M \longrightarrow M$ (we shall write $\mu(a, m) =: a \cdot m$) such that

- $1_A \cdot m = m \; \forall m \in M$
- $(a_1 + a_2) \cdot m = (a_1 \cdot m) + (a_2 \cdot m)$
- $(a_1 \cdot a_2) \cdot m = a_1 \cdot (a_2 \cdot m)$
- $a \cdot (m_1 + m_2) = (a \cdot m_1) + (a \cdot m_2)$

for all $a, a_1, a_2 \in A$ and $m, m_1, m_2 \in M$.

Definition 1.1.7 Let K be a commutative ring and let A be an K-algebra. Then the *opposite algebra* $(A^{op}, +, \cdot^{op})$ (or A^{op} for short) is A as a K-module, with multiplication $\cdot^{op}$ defined by $a \cdot^{op} b := b \cdot a$ for all $a, b \in A$.

Definition 1.1.8 For an algebra Aa *right A-module* is a left A^{op}-module.

Example 1.1.9 Let us illustrate this concept by some easy examples.

- For a field K, a K-module is nothing other than a K-vector space.
- For any K-algebra A, the abelian group $\{0\}$ with only one element is an A-module, which will be denoted by 0.
- A $\mathbb{Z}$-module is just an abelian group.
- For a field K a $K[X]$-module M is the action of a K-linear endomorphism ψ on a K-vector space M. The endomorphism ψ is given by $\mu(X, -)$.
- Let $\psi : A \longrightarrow B$ be an algebra homomorphism and let M be a B-module with structure map $\mu : B \times M \longrightarrow M$. Then M is an A-module if we define $A \times M \longrightarrow M$ as the composition $\mu \circ (\psi \times id_M)$.
- For any $a \in A$ the set $A \cdot a := \{b \cdot a |\; b \in A\}$ is a (left) A-module by multiplication on the left. If $a = 1$, then we call this module the *regular A-module*.

We would like to compare modules.

Definition 1.1.10 Let K be a commutative ring, let A be a K-algebra and let M and N be A-modules. Then an *A-module homomorphism* $M \longrightarrow N$ is a group homomorphism $\alpha : M \longrightarrow N$ such that for all $a \in A$ one has $\alpha(a \cdot m) = a \cdot \alpha(m)$ for all $m \in M$ and $a \in A$.

Of course, for a K-algebra A and two A-modules M and N an A-module homomorphism $\alpha : M \longrightarrow N$ is K-linear. Instead of saying that a map α between A-modules is a homomorphism of A-modules, we sometimes say that *α is A-linear.*

Definition 1.1.11 For a K-algebra A and A-modules M and N, a homomorphism $\alpha : M \longrightarrow N$ of A-modules is

- an *epimorphism* if α is surjective and we write $M \twoheadrightarrow N$,
- a *monomorphism* if α is injective and we write $M \hookrightarrow N$,
- an *isomorphism* if α is bijective.

An *endomorphism* of M is a homomorphism $M \longrightarrow M$ and an *automorphism* is a bijective endomorphism.

A *submodule* of an A-module M is a subset N of M which is again an A-module, and such that the inclusion mapping $N \longrightarrow M$ is a monomorphism of A-modules. The submodule N is a *proper submodule* of M if $N \neq M$. We write $N \leq M$ (resp. $N < M$) to express that N is a (resp. proper) submodule of M. Two A-modules M and N are *isomorphic* if there is an isomorphism $\alpha : M \longrightarrow N$. One writes in this case $M \simeq N$. Denote by $Hom_A(M, N)$ the set of homomorphisms $M \longrightarrow N$, by $End_A(M)$ the set of endomorphisms of M and by $Aut_A(M)$ the set of automorphisms of M.

Trivially, the intersection $\bigcap_{i \in I} M_i$ of a family of submodules M_i, $i \in I$, of an A-module M is again a submodule of M.

The A-module $\sum_{i \in I} N_i$ is the intersection of all submodules S of M so that N_i is a submodule of S for all $i \in I$. If $I = \{1, \ldots, t\}$ is finite, then we write $\sum_{i \in I} N_i = N_1 + \cdots + N_t$.

Remark 1.1.12 Observe that for a commutative ring K, a K-algebra A and an A-module M the set $End_A(M)$ is a K-algebra. The set $Aut_A(M)$ is its unit group.

Lemma 1.1.13 *Let $\alpha : M \longrightarrow N$ be a homomorphism of A-modules.*

1. *The* image $\mathrm{im}(\alpha) := \alpha(M) := \{\alpha(m) \mid m \in M\}$ *is an A-submodule of N.*
2. *The* kernel $\ker(\alpha) := \alpha^{-1}(0) := \{m \in M \mid \alpha(m) = 0\}$ *is an A-submodule of M.*
3. *If M has an A-submodule L, then M/L is an A-module as well.*
4. *The* cokernel $\mathrm{coker}(\alpha) := N/\mathrm{im}(\alpha)$ *is an A-module.*
5. $\mathrm{im}(\alpha) \simeq M/\ker(\alpha)$.
6. *If α is an isomorphism of A-modules, then α^{-1} is an isomorphism of A-modules as well.*

Proof

1. As for vector spaces, it is sufficient to show that for $a, b \in A$ and $y_1, y_2 \in \mathrm{im}(\alpha)$ one gets $ay_1 + by_2 \in \mathrm{im}(\alpha)$. Since $y_1, y_2 \in \mathrm{im}(\alpha)$, there are elements $x_1, x_2 \in M$ with $\alpha(x_1) = y_1$ and $\alpha(x_2) = y_2$. Hence,

$$ay_1 + by_2 = a\alpha(x_1) + b\alpha(x_2) = \alpha(ax_1 + bx_2) \in \mathrm{im}(\alpha).$$

2. Let $x_1, x_2 \in \ker(\alpha)$ and $a, b \in A$. Then

$$\alpha(ax_1 + bx_2) = a\alpha(x_1) + b\alpha(x_2) = 0$$

and therefore $ax_1 + bx_2 \in \ker(\alpha)$.
3. Define for any $a \in A$ and $m \in M$ the expression $a \cdot (m + L) := am + L$. This is well-defined since L is an A-module and gives an A-module structure since M is an A-module.
4. This follows from 1 and 3.
5. Define $\overline{\alpha} : M/\ker(\alpha) \longrightarrow N$ by $\overline{\alpha}(m + \ker(\alpha)) := \alpha(m)$ for all $m \in M$. Since $\alpha(\ker(\alpha)) = 0$, this is well-defined, and of course the image is equal to $\mathrm{im}(\alpha)$. The mapping $\overline{\alpha}$ is a monomorphism, since it is A-linear, α being A-linear, and since $\overline{\alpha}(m + \ker(\alpha)) = 0$ implies $\alpha(m) = 0$ which is equivalent to $m + \ker(\alpha) = 0$.
6. We need to show $\alpha^{-1}(ax_1 + bx_2) = a\alpha^{-1}(x_1) + b\alpha^{-1}(x_2)$ for $x_1, x_2 \in N$ and $a, b \in A$. Now, $x_1 = \alpha(y_1)$ and $x_2 = \alpha(y_2)$ for $y_1, y_2 \in M$. Using the fact that α is an A-module homomorphism, one obtains the required property.

This proves the lemma. □

The following property is sometimes useful.

Lemma 1.1.14 *Let K be a commutative ring and let A be a K-algebra. Let M be a K-module. If M is an A-module, then we obtain a homomorphism of K-algebras $A \xrightarrow{\lambda} End_K(M)$ given by left multiplication. If we have a homomorphism of K-algebras $A \xrightarrow{\nu} End_K(M)$, then there is a unique structure of an A-module on M so that the homomorphism $A \xrightarrow{\lambda} End_K(M)$ given by left multiplication is ν.*

Proof Suppose that M is an A-module with structure map $\mu : A \times M \longrightarrow M$. Then define for each $a \in A$ a mapping $M \longrightarrow M$ by $\lambda(a) : m \mapsto \mu(a, m) =: a \cdot m$. Since $\mu(1, m) = m$ for all $m \in M$, we get $\lambda(1) = id_M$. Since for all $m \in M$ and $a_1, a_2 \in A$ we get $(a_1 \cdot a_2) \cdot m = a_1 \cdot (a_2 \cdot m)$ and $(a_1 + a_2) \cdot m = (a_1 \cdot m) + (a_2 \cdot m)$, we obtain $\lambda(a_1 \cdot a_2) = \lambda(a_1) \circ \lambda(a_2)$ and $\lambda(a_1) + \lambda(a_2) = \lambda(a_1 + a_2)$. Since for all $m_1, m_2 \in M$ and $a \in A$ we get $a \cdot (m_1 + m_2) = (a \cdot m_1) + (a \cdot m_2)$ and, using that A is a K-algebra, we have a ring homomorphism $K \longrightarrow Z(A)$ which implies that $\lambda(a)$ is K-linear. Hence λ is a well-defined algebra homomorphism.

Conversely, let $A \xrightarrow{\nu} End_K(M)$ be an algebra homomorphism, then define $\mu : A \times M \longrightarrow M$ by $\mu(a, m) := \nu(a)(m)$. This shows at once that the homomorphism given by multiplication from the left is ν. The axioms for an algebra homomorphism translate as in the previous step to the axioms for M to be an A-module. □

Another very useful concept generalises from vector spaces to modules. Recall that when we have a finite dimensional K-vector space V, then one may choose a K-basis and V is isomorphic to K^n for some $n \in \mathbb{N}$, or in a different notation

$$V \simeq \underbrace{K \oplus K \oplus \cdots \oplus K}_{n \text{ terms}} \simeq \bigoplus_{j=1}^{n} K_j$$

where $K_j = K$ is a one-dimensional K-vector space.

Let now A be an R-algebra and let M_1 and M_2 be two A-modules with homomorphisms $\mu_1 : A \longrightarrow End_R(M_1)$ and $\mu_2 : A \longrightarrow End_R(M_2)$. Then the set theoretical product $M_1 \times M_2$ is an A-module as well by setting $\mu_{M_1 \times M_2} := \mu_1 \times \mu_2$; or in more explicit terms for all $a \in A$ and $m_1 \in M_1, m_2 \in M_2$ one defines $a \cdot (m_1, m_2) := (a \cdot m_1, a \cdot m_2)$. Of course this generalises to arbitrary direct products: Let I be a non-empty set and suppose given an A-module M_i for each $i \in I$. Then the set theoretical direct product $\prod_{i \in I} M_i$ is an A-module by setting

$$a \cdot (m_i)_{i \in I} := (a \cdot m_i)_{i \in I}$$

for every $(m_i)_{i \in I} \in \prod_{i \in I} M_i$ and $a \in A$. This construction is the *direct product construction of a module*. For finite sets I there is no difference to the direct sum construction. Indeed, we encounter the following problem. Given an A-module N, an infinite index set I and for each $i \in I$ an A-module M_i, for every $i_0 \in I$ the projection

$$\begin{array}{rcl} \prod_{i \in I} M_i & \xrightarrow{\pi_{i_0}} & M_{i_0} \\ (m_i)_{i \in I} & \mapsto & m_{i_0} \end{array}$$

is a homomorphism of A-modules since

$$\begin{aligned} \pi_{i_0}((m_i)_{i \in I} + (m'_i)_{i \in I}) &= \pi_{i_0}((m_i + m'_i)_{i \in I}) = m_{i_0} + m'_{i_0} \\ &= \pi_{i_0}((m_i)_{i \in I}) + \pi_{i_0}((m'_i)_{i \in I}) \end{aligned}$$

and

$$\pi_{i_0}(a(m_i)_{i \in I}) = \pi_{i_0}((am_i)_{i \in I}) = am_{i_0} = a\pi_{i_0}((m_i)_{i \in I}).$$

Suppose, moreover, we have a homomorphism $\varphi_i : M_i \longrightarrow N$ for each $i \in I$. There is no way the homomorphisms φ_i can induce a homomorphism of A-modules $\varphi : \prod_{i \in I} M_i \longrightarrow N$. One would like to define "$\varphi = \sum_{i \in I} \varphi_i \circ \pi_i$" as in the case where I is finite. But infinite sums are not defined in general for A-modules and so there is no way for such a φ to exist. Observe, however, that given homomorphisms $\psi_i : N \longrightarrow M_i$ for each $i \in I$, then

$$N \xrightarrow{\psi} \prod_{i \in I} M_i$$
$$n \mapsto (\psi_i(n))_{i \in I}$$

is a homomorphism of A-modules satisfying $\pi_i \circ \psi = \psi_i$ for all $i \in I$.

The solution for this problem is the *direct sum construction of modules*. Let I be a set and suppose given for each $i \in I$ an A-module M_i. Then define the coproduct of the M_i; $i \in I$ to be those elements in the product which are 0 for all but a finite number of coordinates:

$$\coprod_{i \in I} M_i := \bigoplus_{i \in I} M_i := \left\{ (m_i)_{i \in I} \in \prod_{i \in I} M_i \middle| \quad |\{i \in I \mid m_i \neq 0\}| < \infty \right\}.$$

Then for all $i_0 \in I$ there is a map

$$M_{i_0} \xrightarrow{\iota_{i_0}} \coprod_{i \in I} M_i$$
$$m_{i_0} \mapsto (m'_i)_{i \in I}$$

defined by putting $m'_i = 0$ if $i \neq i_0$ and $m'_{i_0} = m_{i_0}$. Suppose now we have a homomorphism $\varphi_i : M_i \longrightarrow N$ for each $i \in I$. Then

$$\coprod_{i \in I} M_i \xrightarrow{\varphi} N$$
$$(m_i)_{i \in I} \mapsto \sum_{i \in I} \varphi(m_i)$$

is a homomorphism of A-modules, since only a finite number of the m_i are non-zero and so the sum is well defined. Moreover, this is the only homomorphism of A-modules satisfying

$$\iota_i \circ \varphi = \varphi_i \ \forall i \in I$$

as is immediately verified.

1.2 Group Representations

By Lemma 1.1.14 a module over a K-algebra A is given by an algebra homomorphism $A \longrightarrow End_K(M)$. In the case of a group, we want to study representations as well. Here, staying in the philosophy expressed previously, we understand a representation to be a concrete realisation of an abstract group by an action on a "real world object", in our case a K-vector space where K is a field.

1.2.1 Group Algebras and Their Modules

Since group elements have inverses, the same should be true for the representation of a group. So, a representation of a group G is a homomorphism of G to the group of invertible matrices.

Definition 1.2.1 Let G be a group and K be a field. A *representation of G over K* of degree $n \in \mathbb{N}$ is a group homomorphism $G \longrightarrow GL_n(K)$. If the kernel of this homomorphism is all of G, then we say that G *acts trivially*, and if in addition $n = 1$, we call the representation the *trivial representation*.

There are at least two obvious drawbacks to this concept. First, it would be nice if we could phrase this idea in our setting of modules over an algebra. Second, we see that we would like to use at least some of the additional information the matrices provide. One of the most obvious ones is the fact that $GL_n(K)$ is the unit group of an algebra, the ring $Mat_{n\times n}(K)$ of degree n square matrices. Just as we can form linear combinations of invertible matrices of the same size to get a (not necessarily invertible) square matrix, we can *formally* form K-linear combinations of group elements and get a K-algebra. This is the concept of a group ring.

Definition 1.2.2 Let G be a group and let R be a commutative ring. The *group ring* RG as an R-module is the R-module $\oplus_{g\in G} R \cdot e_g$, where $R \cdot e_g = R$ for all $g \in G$. This becomes an R-algebra when one sets

$$\left(\sum_{h_1\in G} r_{h_1} e_{h_1}\right) \cdot \left(\sum_{h_2\in G} s_{h_2} e_{h_2}\right) := \sum_{g\in G}\left(\sum_{h\in G} r_h s_{h^{-1}g}\right) e_g.$$

Occasionally we shall also call a group ring a *group algebra* if we want to stress the algebra structure.

We should explicitly mention that if G is infinite, the fact that RG is defined as a *direct sum* $\coprod_{g\in G} R$ of copies of R as R-modules implies that an element $x = \sum_{g\in G} r_g e_g$ has only a finite number of non-zero coefficients r_g. In other words, for any $x = \sum_{g\in G} r_g e_g \in RG$ we have

$$|\{g \in G \mid r_g \neq 0\}| < \infty.$$

Now, the group of invertible elements $(RG)^\times$ of RG contains G as a subgroup. Indeed, the mapping

$$\begin{aligned} G &\longrightarrow (RG)^\times \\ g &\mapsto 1 \cdot e_g \end{aligned}$$

is a group monomorphism. The image of this homomorphism will again be denoted by G.

Using this we can formulate an important property of group rings, namely that RG is universal for having G as a subgroup of its unit group.

Lemma 1.2.3 *Let R be a commutative ring and let A be an R-algebra. Then for every group homomorphism $\alpha : G \longrightarrow A^\times$ there is a unique homomorphism of R-algebras $\beta : RG \longrightarrow A$ with $\alpha = \beta|_G$.*

Proof We have to define $\beta(\sum_{g\in G} r_g e_g) := \sum_{g\in G} r_g\alpha(g)$ in order to get an R-linear mapping $RG \longrightarrow A$ with $\alpha = \beta|_G$. Moreover,

$$\begin{aligned}\beta\left(\left(\sum_{g\in G} r_g e_g\right)\cdot\left(\sum_{g\in G} s_g e_g\right)\right) &= \sum_{g\in G;h\in G}\left(\sum_{gh=k} r_g s_h\right)\alpha(k)\\ &= \sum_{g\in G;h\in G}\left(\sum_{gh=k} r_g s_h\right)\beta(e_k)\\ &= \beta\left(\sum_{g\in G} r_g e_g\right)\cdot\beta\left(\sum_{g\in G} s_g e_g\right).\end{aligned}$$

This proves the lemma. □

Lemma 1.2.4 *Let R be a commutative ring. For any integer n and any group homomorphism $\varphi : G \longrightarrow GL_n(R)$ there is a unique RG-module structure $\hat{\varphi}$ on R^n such that $\hat{\varphi}|_G = \varphi$. Moreover, if K is a field, then the restriction of the structure map $\hat{\varphi} : KG \longrightarrow End_K(M)$ of a KG-module M induces a representation of G over K of dimension* $\dim_K(M)$.

Proof This follows immediately from Lemma 1.2.3. Indeed, by definition $GL_n(R) = Mat_{n\times n}(R)^\times$ and so, by this lemma, there is a unique ring homomorphism $RG \longrightarrow Mat_{n\times n}(R)$ restricting to φ. This proves the lemma. □

In elementary terms the previous lemma constructs the module structure on R^n in the following way. Suppose φ is a group homomorphism $\varphi : G \longrightarrow GL_n(R)$. Then

$$\hat{\varphi} : RG \ni \sum_{g\in G} r_g e_g \mapsto \sum_{g\in G} r_g\varphi(g) \in End_R(R^n)$$

is an R-algebra homomorphism.

Lemma 1.2.4 implies that there is no need to distinguish between modules over the group ring KG and K-representations of the group G. Since we defined the concept of a group representation only over fields, the concept of an RG-module for a commutative ring R is more appropriate as a definition of a representation of a group over a commutative ring R. If φ is a representation of G over R, then we call the RG-module obtained by Lemma 1.2.4 the *module afforded by* φ.

Lemma 1.2.5 *Let ϕ and ψ be two representations of the same dimension over a field K of a group G and suppose that there is an invertible matrix T so that for all $g \in G$ one has $T \cdot \phi(g) = \psi(g) \cdot T$. Then the modules M_ϕ and M_ψ which are induced by ϕ and ψ are isomorphic. If the modules M_ϕ and M_ψ afforded by ϕ and ψ are isomorphic as KG-modules, then there is an invertible matrix T such that $T \cdot \phi(g) = \psi(g) \cdot T$.*

Proof Suppose there is a matrix T which conjugates $\phi(g)$ to $\psi(g)$ for all $g \in G$. Then T yields a K-linear endomorphism $K^n \longrightarrow K^n$ which induces a KG-isomorphism between M_ϕ and M_ψ. Indeed, $T\phi(g)m = \psi(g)Tm$ for all $m \in K^n$ is exactly the equation needed to show that T is KG-linear.

Suppose that there is an isomorphism $T : M_\phi \longrightarrow M_\psi$. This isomorphism is K-linear, hence given by an invertible matrix T, and since it is KG-linear, $T\phi(g)m = \psi(g)Tm$ holds for all $m \in K^n$ and $g \in G$. Hence, $T\phi(g) = \psi(g)T$ for all $g \in G$. This proves the lemma. □

We would like to see some examples of how this concept is going to work.

Example 1.2.6 Let $K = \mathbb{C}$ and $G := C_n$ be the cyclic group of order n generated by c, say.

1. Let us study the 1-dimensional $\mathbb{C}C_n$-modules up to isomorphism. It is necessary and sufficient to define a group homomorphism $C_n \longrightarrow GL_1(\mathbb{C}) = \mathbb{C}^\times$. Since $c^n = 1$ in C_n, in order to be able to find images of c we need elements $x \in \mathbb{C}$ satisfying the same relation $x^n = 1$. Fix a primitive n^{th} root of unity ζ_n in $\mathbb{C}$. We get n possibilities, namely $\varphi_m(c^\ell) = (\zeta_n^m)^\ell = \zeta_n^{m\ell}$. By Lemma 1.2.5 we see that φ_{m_1} and φ_{m_2} lead to isomorphic representations if and only if $m_1 - m_2 \in n\mathbb{Z}$.

   For example, if $n = 3$ and $\zeta_3 =: j$, we have the following three possibilities.

   (a) The first is $\varphi_1(c) = 1$ and therefore $\varphi_1(c^n) = 1^n = 1$ for all $n \in \mathbb{N}$.
   (b) The second is $\varphi_j(c) = j$ and therefore $\varphi_j(c^2) = j^2$ as well as $\varphi_j(1) = 1$ of course.
   (c) The third is $\varphi_{j^2}(c) = j^2$ and therefore $\varphi_{j^2}(c^2) = j^4 = j$ and, of course, we have $\varphi_{j^2}(1) = 1$.

   We observe that φ_j and φ_{j^2} are conjugate complex representations.
2. Let us study representations of dimension d. As before, for any representation φ we have that $\varphi(c) =: M_d$ is an invertible matrix of size d of order n; i.e. $M_d^n = 1$. Since M_d can be conjugate into Jordan normal form, and the power of a conjugate matrix is the conjugated of the power, one sees that M_d is diagonalisable with diagonal coefficients $\zeta_n^{u_1}, \zeta_n^{u_2}, \cdots, \zeta_n^{u_d}$ for a vector $(u_1, u_2, \dots, u_d) \in \mathbb{Z}^d$. Since $\varphi(c^\ell) = M_d^\ell$, the representation is fixed by the choice of this single matrix M_d. By Lemma 1.2.5 the $\mathbb{C}C_n$-module M_φ which corresponds to the representation φ is isomorphic to the direct sum $\bigoplus_{j=1}^d M_{\hat{\varphi}_{u_j}}$.

We have classified all finite dimensional complex representations of cyclic groups.

The phenomenon that a finite dimensional $\mathbb{C}G$-module M is a direct sum of smaller modules is quite typical. An A-module M is *finitely generated* if there is a finite subset S of M such that M is the smallest submodule of M containing S.

Definition 1.2.7 Let K be a commutative ring, A be a K-algebra and M be an A-module.

- M is said to be *simple* if any non-zero submodule N of M is equal to M.
- If M is finitely generated, then M is *semisimple* if there are simple A-modules $S_1, S_2, \ldots, S_n$ such that $M \simeq S_1 \oplus S_2 \oplus \cdots \oplus S_n$.
- If K is a field, then we say that A is *semisimple* if and only if every finite dimensional A-module is semisimple.

Note that the definition implies that the module 0 is not simple. This is just a technical exclusion which simplifies the statements later on. Consider a vector space of infinite dimension over a field K. This K-module should be called semisimple as well, but the above definition does not include this case. A possibly infinitely generated module is called semisimple if each submodule is a direct factor. In a second step one shows that this implies that this definition implies that the module is the direct sum of simple submodules. When we speak of semisimple modules, we will always assume that they are finitely generated. The interested reader may consult Lam [1] for the general case.

1.2.2 Maschke's Theorem

We now prove one of the most important results in the representation theory of finite groups. The result actually states that if G is a finite group of order n and K is a field with $n \cdot K = K$, then any module of KG is semisimple.

Theorem 1.2.8 (Maschke 1905) *Let G be a finite group and let K be a field. If the order of G is invertible in K, then KG is semisimple.*

Proof We shall first show that for any KG-module M and any submodule N of M there is another submodule L of M such that $M \simeq N \oplus L$.

Once this is done we proceed by induction on the dimension of M. The statement is clear for one-dimensional modules. Suppose we have shown that any module of dimension up to n is semisimple, and that we have given an $n+1$-dimensional module M. If M does not have a proper non-zero submodule, then M is simple. If this is not the case, then there is a proper non-zero submodule N of M. By the argument which we are about to develop below, there is another submodule L of M such that $M \simeq L \oplus N$. Since the dimensions of L and of N are both at most n, these two modules are semisimple, and hence M is semisimple.

Suppose that N is a proper non-zero submodule of M. Then the quotient $\tilde{L} := M/N$ is a KG-module as well. Denote by $\pi : M \longrightarrow \tilde{L}$ the mapping $\pi(m) = m+N$.

We may choose a K-basis $\mathcal{N}$ of N which we may complete by vectors $\mathcal{L}$ to a K basis $\mathcal{M} = \mathcal{N} \cup \mathcal{L}$ of M. Then $\{l + N|\ l \in \mathcal{L}\}$ is a K-basis of $\tilde{L}$. Denote by $\tilde{\rho}$ the K-linear mapping $\tilde{L} \longrightarrow M$ defined by $\tilde{\rho}(l + N) := l$ for any $l \in \mathcal{L}$. By definition, $\pi \circ \tilde{\rho} = id_{\tilde{L}}$. Moreover, π is a KG-module homomorphism, but there is however no reason why $\tilde{\rho}$ should be a KG-module homomorphism.

Define

$$\begin{aligned} \rho : \tilde{L} &\longrightarrow M \\ l + N &\mapsto \frac{1}{|G|} \sum_{g \in G} e_g^{-1} \tilde{\rho}(e_g \cdot (l + N)) \end{aligned}$$

Remark 1.2.9 Observe that we assumed that $|G|$ is invertible in K, and so the term $\frac{1}{|G|} \in K$ makes sense. This is the only occasion when we shall use this hypothesis.

Since $\tilde{\rho}$ is a well-defined homomorphism of K-vector spaces, this is true for ρ as well. Moreover, for any $l + N \in \tilde{L}$ one has

$$\begin{aligned} \pi \circ \rho(l + N) &= \pi(\frac{1}{|G|} \sum_{g \in G} e_g^{-1} \tilde{\rho}(e_g \cdot (l + N))) \\ &= \frac{1}{|G|} \sum_{g \in G} e_g^{-1} \pi(\tilde{\rho}(e_g \cdot (l + N))) \\ &= \frac{1}{|G|} \sum_{g \in G} e_g^{-1} e_g \cdot (l + N) \\ &= l + N \end{aligned}$$

and so, $\pi \circ \rho = id_{\tilde{L}}$.

Furthermore, ρ is a KG-module homomorphism. We need to show that for any $l + N \in \tilde{L}$ and any $h \in G$ one has $\rho(e_h \cdot (l + N)) = e_h \cdot \rho(l + N)$. This is the case:

$$\begin{aligned} \rho(e_h \cdot (l + N)) &= \frac{1}{|G|} \sum_{g \in G} e_g^{-1} \tilde{\rho}(e_g \cdot (e_h \cdot (l + N))) \\ &= \frac{1}{|G|} \sum_{g \in G} e_g^{-1} \tilde{\rho}(e_{gh} \cdot (l + N)) \\ &= e_h \cdot \frac{1}{|G|} \sum_{g \in G} e_h^{-1} e_g^{-1} \tilde{\rho}(e_{gh} \cdot (l + N)) \\ &= e_h \cdot \frac{1}{|G|} \sum_{g \in G} e_{gh}^{-1} \tilde{\rho}(e_{gh} \cdot (l + N)) \\ &= e_h \cdot \frac{1}{|G|} \sum_{gh \in G} e_{gh}^{-1} \tilde{\rho}(e_{gh} \cdot (l + N)) \end{aligned}$$

$$= e_h \cdot \frac{1}{|G|} \sum_{g \in G} e_g^{-1} \tilde{\rho}(e_g \cdot (l + N))$$
$$= e_h \cdot \rho(l + N).$$

Therefore, ρ is a homomorphism of KG-modules, and $L := \text{im}(\rho)$ is a KG-submodule of M.

We need the following Lemma.

Lemma 1.2.10 *Let A be an R-algebra for a commutative ring R and let M be an A-module. Suppose there is an A-module endomorphism φ of M such that $\varphi \circ \varphi = \varphi$. Then $M \simeq \text{im}(\varphi) \oplus \ker(\varphi)$.*

Remark 1.2.11 An element φ in a ring with $\varphi \circ \varphi = \varphi$ is called an *idempotent element*.

Proof $\text{im}(\varphi)$ and $\ker(\varphi)$ are both A-submodules of M. Let $m \in \text{im}(\varphi) \cap \ker(\varphi)$. Then there is an $m' \in M$ such that $m = \varphi(m')$. Hence,

$$m = \varphi(m') = \varphi \circ \varphi(m') = \varphi(m) = 0.$$

Let $m \in M$. Then

$$\varphi(m - \varphi(m)) = \varphi(m) - \varphi \circ \varphi(m) = \varphi(m) - \varphi(m) = 0$$

and therefore $m - \varphi(m) \in \ker(\varphi)$. As a consequence,

$$m = (m - \varphi(m)) + \varphi(m) \in \ker(\varphi) + \text{im}(\varphi).$$

This proves the lemma. □

We continue with the proof of Maschke's theorem. Since $\pi \circ \rho = id_{\tilde{L}}$, we have

$$(\rho \circ \pi) \circ (\rho \circ \pi) = \rho \circ (\pi \circ \rho) \circ \pi = \rho \circ id_{\tilde{L}} \circ \pi = \rho \circ \pi.$$

Moreover, since $\pi \circ \rho = id_{\tilde{L}}$, the mapping ρ is injective, $\ker(\rho \circ \pi) = \ker(\pi) = N$ and $\text{im}(\rho \circ \pi) \simeq \text{im}(\pi) = \tilde{L}$. Hence, $M \simeq N \oplus M/N$. This finishes the proof. □

We first illustrate this result with a small example.

Example 1.2.12 Let $G = \mathfrak{S}_n$ be the symmetric group on n letters which we may assume to be $\{1, 2, \ldots, n\}$ and let K be any field of characteristic 0 or at least $n + 1$. Let N_n be an n-dimensional K-vector space, and fix a basis $\{b_1, b_2, \ldots, b_n\}$ in N_n. Then N_n becomes a $K\mathfrak{S}_n$-module if one defines $\sigma \cdot b_i := b_{\sigma(i)}$ for any $\sigma \in \mathfrak{S}_n$ and any $i \in \{1, 2, \ldots, n\}$. Of course, the K-sub vector space generated by $\sum_{i=1}^n b_i$ is invariant under the action of any $\sigma \in \mathfrak{S}_n$. Hence, this one-dimensional subspace T is a $K\mathfrak{S}_n$-subspace of N_n. Since any $\sigma \in \mathfrak{S}_n$ acts as the identity, T is the trivial $K\mathfrak{S}_n$-module. By Maschke's Theorem 1.2.8 we know that $N_n \simeq T \oplus N_n/T$. It is an easy exercise to show that this direct sum decomposition holds if and only if $nK = K$. We shall continue to work on this example in Example 1.7.30.

1.3 Noetherian and Artinian Objects

Being a finite dimensional algebra over a field often is too strong a condition. The appropriate concept is that of a Noetherian or artinian module, which we will introduce now. These concepts are by far less restrictive, but are sufficiently strong to allow the most important results for finite dimensional algebras, at least those properties that are interesting to us.

Definition 1.3.1 Let K be a commutative ring, let A be a K-algebra and let M be an A-module.

- M is said to be *Noetherian* if, whenever there is a sequence
$$M_1 \subseteq M_2 \subseteq M_3 \subseteq \cdots \subseteq M$$
of A-submodules of M, then there is an $n_0 \in \mathbb{N}$ such that $M_n = M_{n_0}$ for all $n \geq n_0$.
- M is said to be *artinian* if, whenever there is a sequence
$$M \supseteq M_1 \supseteq M_2 \supseteq M_3 \supseteq \dots$$
of A-submodules of M, then there is an $n_0 \in \mathbb{N}$ such that $M_n = M_{n_0}$ for all $n \geq n_0$.
- A is said to be *left (right) Noetherian* if the left (right) regular module is Noetherian.
- A is said to be *left (right) artinian* if the left (right) regular module is artinian.

Of course, if A is an algebra over a field K, then any A-module of finite dimension as a K-vector space is Noetherian and artinian.

Example 1.3.2 The ring of integers $\mathbb{Z}$ is Noetherian since for any ideal $I = n\mathbb{Z}$ of $\mathbb{Z}$, an ideal $J = m\mathbb{Z}$ contains I if and only if m divides n. There are only finitely many divisors of n, and the statement is proven.

The ring of integers is not artinian, since the sequence of ideals
$$\mathbb{Z} \supseteq 2\mathbb{Z} \supseteq 4\mathbb{Z} \supseteq 8\mathbb{Z} \supseteq \dots$$
obviously is not finite.

The properties in the following lemmas are essential and are the reason for the importance of the notions Noetherian and artinian.

Lemma 1.3.3 *Let A be a ring and let M be a left module over A. Then the following statements are equivalent.*

1. *M is Noetherian.*
2. *Every submodule of M is finitely generated.*
3. *For any submodule N of M, both N and M/N are Noetherian.*

Proof **1 implies 2**: Let N be a submodule of M and let $\{m_i \mid i \in I\}$ be a generating set of N, so that every proper subset of $\{m_i \mid i \in I\}$ does not generate N. Then if I is not finite we may find inductively for every $n \in \mathbb{N}$ an element $m(n) \in \{m_i \mid i \in I\}$ so that $m(n) \notin \sum_{j=1}^{n-1} A \cdot m(j)$. Hence we have constructed the A-submodules $M(n) := \sum_{j=1}^{n} A \cdot m(j)$ of N, and hence of M, with the property that $M(n-1) \subset M(n)$ for all $n \in \mathbb{N}$ and $M(n-1) \neq M(n)$ for all $n \in \mathbb{N}$. This contradicts the property of M being Noetherian.

2 implies 3: We first prove that every submodule of M is Noetherian.

Let $N_1 \subset N_2 \subset N_3 \subset \cdots \subset N$ be an ascending sequence of A-submodules of N. Then $N_1 \cup N_2 \cup N_3 \cup \cdots =: \tilde{N}$ is an A-submodule of N, and hence of M. Therefore $\tilde{N}$ is finitely generated, by 2. This shows that $\tilde{N}$ is generated by $x_1, \ldots, x_s$ for some elements $x_1, \ldots, x_s$ of N. But each x_i is in $\bigcup_j N_j$, and hence there is an integer t so that $x_j \in N_t$ for all $j \in \{1, \ldots, s\}$. Therefore $\bigcup_j N_j = N_t$ and $N_u = N_t$ for all $u \geq t$. This shows that N is Noetherian.

Let $\overline{N}_1 \subset \overline{N}_2 \subset \overline{N}_3 \subset \cdots \subset M/N$ be a strictly increasing sequence of A-submodules of M/N. The pre-images $N_i := \{n \in M \mid n + N \in \overline{N}_i\}$ form a strictly increasing sequence of submodules of M. By the previous step this sequence is finite. Hence M/N is Noetherian. This contradiction proves the statement.

3 implies 1: Suppose N and M/N are both Noetherian. Let $L_1 \subseteq L_2 \subseteq \cdots \subseteq M$ be an increasing sequence of submodules. Then $(L_1 + N)/N \subseteq (L_2 + N)/N \subseteq \cdots \subseteq M/N$ is an increasing sequence of submodules of M/N and since M/N is Noetherian this sequence is constant from n_0 onwards: $(L_{n_0} + N)/N = (L_n + N)/N$ for all $n \geq n_0$. Now, the sequence $L_{n_0} \cap N \subseteq L_{n_0+1} \cap N \subseteq \cdots \subseteq N$ is a sequence of submodules of N. Since N is assumed to be Noetherian, there is an $n_1 \in \mathbb{N}$ such that $L_{n_1} \cap N = L_n \cap N$ for all $n \geq n_1$. Since $(L_n + N)/N \simeq L_n/(L_n \cap N)$ we obtain from the fact that $(L_{n_0} + N)/N = (L_n + N)/N$ for all $n \geq n_0$ that the sequence $(L_n)_{n \in \mathbb{N}}$ is stationary from n_1 onwards. □

Lemma 1.3.4 *Let A be a ring and let M be a left module over A. Then the following statements are equivalent.*

1. *M is artinian.*
2. *For any submodule N of M, both N and M/N are artinian.*

Proof Let $N \supset N_1 \supset N_2 \ldots$ be a strictly decreasing sequence of submodules of N. Then $M \supset N \supset N_1 \supset \ldots$ is a strictly decreasing sequence of submodules of M. Since M is artinian this sequence is finite. Let $M/N \supset \overline{N}_1 \supset \overline{N}_2 \ldots$ be a strictly decreasing sequence of submodules of M/N. Then the pre-images $N_i := \{n \in M \mid n + N \in \overline{N}_i\}$ form a strictly decreasing sequence of submodules of M. Hence the sequence is finite.

Conversely suppose N and M/N are artinian. The proof then follows exactly the lines of the last step of Lemma 1.3.3. It should be no problem to leave the details to the reader as an exercise. □

Proposition 1.3.5 *Let A be an algebra and let M be an A-module. Then*

- *M is Noetherian if and only if every non-empty set of submodules of M has a maximal element.*
- *M is artinian if and only if every non-empty set of submodules of M has a minimal element.*

Proof Suppose M is Noetherian and let $\mathcal{X}$ be a set of submodules of M. Then let $\mathcal{Y}$ be a non-empty ascending chain of submodules in $\mathcal{X}$. Since M is Noetherian, this ascending chain of submodules is actually finite. Let $\mathcal{Y} = \{M_0, M_1, \dots, M_n\}$ with $M_n < M_{n-1} < \cdots < M_0$. Hence M_0 is a maximal element of $\mathcal{Y}$ and by Zorn's lemma $\mathcal{X}$ has maximal elements.

Suppose conversely that every non-empty subset of submodules contains a maximal element. Then take an increasing chain of submodules $M_0 \leq M_1 \leq \cdots \leq M$ of M. Since $\mathcal{X} = \{M_0, M_1, \dots\}$ is a non-empty set of submodules of M, it contains a maximal element, say M_n. Hence, $M_m = M_n$ for all $m \geq n$.

The proof for artinian modules is completely analogous. □

A nice and useful property of Noetherian modules is that surjective endomorphisms are automorphisms.

Proposition 1.3.6 *Let A be an algebra and let M be a Noetherian A-module. Then any surjective endomorphism of M is injective.*

Proof Let $\varphi : M \longrightarrow M$ be a surjective endomorphism of the A-module M. As usual we put $\varphi^1 := \varphi$ and $\varphi^n := \varphi \circ \varphi^{n-1}$ for all $n \geq 2$. Then the sequence $(\ker(\varphi^n))_{n\in\mathbb{N}}$ of submodules of M is increasing, and hence, since M is Noetherian, becomes stationary. Let n_0 be an integer so that $\ker(\varphi^n) = \ker(\varphi^{n_0})$ for all $n \geq n_0$. Let $m \in \ker(\varphi)$. Since φ is surjective, φ^k is also surjective for all k. Hence there is an $x \in M$ such that $\varphi^{n_0}(x) = m$. Then $0 = \varphi(m) = \varphi^{n_0+1}(x)$ and so $x \in \ker(\varphi^{n_0+1}) = \ker(\varphi^{n_0})$. Therefore $m = \varphi^{n_0}(x) = 0$ and φ is injective. □

1.4 Wedderburn and Krull-Schmidt

The main reason why Maschke's result Theorem 1.2.8 is one of the most fundamental in the representation theory of finite groups is that the structure theory of finite dimensional semisimple algebras is known in great detail. We shall now develop this structure.

1.4.1 The Krull-Schmidt Theorem

For the moment we do not know anything about the unicity of a decomposition of a semisimple module into its factors. This is the very important Krull-Schmidt theorem.

It holds more generally, replacing simple modules by indecomposable modules as factors and semisimple algebras by general finite dimensional algebras.

Definition 1.4.1 A non-zero A-module M is called *indecomposable* if whenever $M \simeq N \oplus L$, then either $N = 0$ or $L = 0$. Modules which are not indecomposable are *decomposable.*

We have seen that (by definition) semisimple indecomposable modules are simple. In general, however, indecomposable modules need not be simple.

Example 1.4.2 Let K be a field and let $A = K[X]/(X^2)$ be the so-called ring of dual numbers. A is a K-algebra of dimension 2 over K, and the regular module is indecomposable but not semisimple. Indeed, the only non-zero proper ideal of A is $X \cdot K[X]/(X^2)$.

Theorem 1.4.3 (Krull-Schmidt) *Let K be a field and let A be a K-algebra. Let M be a Noetherian and artinian A-module. Suppose there are indecomposable A-modules $M_1, M_2, \ldots, M_m$ and $N_1, N_2, \ldots, N_n$ such that*

$$\bigoplus_{j=1}^{m} M_j = M = \bigoplus_{i=1}^{n} N_i.$$

Then $m = n$ and there is a permutation $\sigma \in \mathfrak{S}_n$ such that $M_j \simeq N_{\sigma(j)}$ for all $j \in \{1, 2, \ldots, n\}$.

We divide the proof of the Krull-Schmidt theorem into three lemmas. Each lemma is quite simple, but actually the lemmas are interesting in their own right. An element a in a ring A is *nilpotent* if there is an $n \in \mathbb{N}$ such that $a^n = 0$. This applies in particular to the endomorphism ring A of a module.

Lemma 1.4.4 (Fitting 1935) *Let K be a commutative ring, let A be a K-algebra and let M be a Noetherian and artinian A-module. Then for any endomorphism u of M there is a decomposition $M \simeq N \oplus S$ such that the restriction $u|_N$ of u to N is a nilpotent endomorphism of N and the restriction $u|_S$ of u to S is an automorphism of S. One may take $N = \ker(u^m)$ and $S = \operatorname{im}(u^m)$ for a large enough integer m.*

Proof Define $u^k := u^{k-1} \circ u$ and $u^1 = u$ for any positive integer k.

- We get a sequence

$$\ker(u) \subseteq \ker(u^2) \subseteq \ker(u^3) \subseteq \cdots \subseteq M.$$

  Since M is Noetherian, there is an integer k_0 such that $\ker(u^{k_0}) = \ker(u^{k_0+1})$. Hence, for any $k \geq k_0$ one has $\ker(u^{k_0}) = \ker(u^k) =: N$.
- We get another sequence of submodules

$$M \supseteq \operatorname{im}(u) \supseteq \operatorname{im}(u^2) \supseteq \operatorname{im}(u^3) \supseteq \cdots \supseteq 0$$

which has to be finite, since M is artinian. So, there is an integer k_1 such that $\mathrm{im}(u^{k_1}) = \mathrm{im}(u^k) =: S$ for any integer $k \geq k_1$.

- Put $k_2 := \max(k_1, k_0)$.
- We shall show that $u|_S$ is an automorphism of S. Indeed,

$$u(S) = u(\mathrm{im}(u^{k_2})) = \mathrm{im}(u^{k_2+1}) = \mathrm{im}(u^{k_2}) = S$$

  and so, $u|_S$ is an epimorphism.
- We shall show that $u|_N$ is a nilpotent endomorphism of N. Indeed, if $n \in N$, then $n \in \ker(u^{k_2}) = \ker(u^{k_2+1})$. As a consequence,

$$0 = u^{k_2+1}(n) = u^{k_2}(u(n))$$

  and so $u(n) \in \ker(u^{k_2}) = N$. By definition $(u|_N)^{k_0} = 0$. Hence, $u|_N$ is a nilpotent endomorphism of N.
- If $m \in \mathrm{im}(u^{k_2}) \cap \ker(u^{k_2}) = S \cap N$, then there is an $m' \in M$ such that $m = u^{k_2}(m')$. Since $m \in \ker(u^{k_2})$ one has

$$0 = u^{k_2}(m) = u^{2k_2}(m').$$

  Hence, $m' \in \ker(u^{2k_2}) = \ker(u^{k_2})$. So, $m = u^{k_2}(m') = 0$.
- The previous statement gives that $N \cap S = 0$. Since $\ker(u) \subseteq N$, we also obtain that $\ker(u|_S) = \ker(u) \cap S \subseteq N \cap S = 0$. Therefore $u|_S$ is injective. By the definition of S we get that $u|_S$ is surjective, and hence an automorphism.
- Let $m \in M$. Then, since $\mathrm{im}(u^{k_2}) = \mathrm{im}(u^{2k_2})$, there is an $m' \in \mathrm{im}(u^{k_2})$ with $u^{k_2}(m) = u^{k_2}(m')$. Hence, $m = m' + (m - m')$ and $m - m' \in \ker(u^{k_2})$ whereas $m' \in \mathrm{im}(u^{k_2})$.

This finishes the proof of Fitting's lemma. □

In order to formulate the second lemma in a concise way we need another notion.

Definition 1.4.5 A ring A is *local* if the set of non-invertible elements of A form a two-sided ideal.

Lemma 1.4.6 *Let A be a K-algebra and let M be a non-zero indecomposable A-module so that Fitting's lemma holds for all endomorphisms of M. Then $End_A(M)$ is a local ring. Each endomorphism u of M is either nilpotent or bijective.*

Proof Given an endomorphism u of A, then Fitting's Lemma 1.4.4 implies that there is a decomposition $M \simeq N \oplus S$ into submodules where $u|_N$ is a nilpotent endomorphism of N and $u|_S$ is an automorphism of S. Since M is indecomposable, either N or S is 0. If $N = 0$, then u is an automorphism. If $S = 0$, then u is nilpotent, and therefore not invertible.

Suppose now that u is not invertible. We need to show that for any endomorphism v of M, $u \circ v$ and $v \circ u$ are also not invertible. But since u is not invertible, and since M is indecomposable, $M = N$ and u is nilpotent. If $\ker(u) = 0$, then u is injective. Injective endomorphisms cannot be nilpotent. Hence $\ker(u) \neq 0$ and therefore $\ker(v \circ u) \supseteq \ker(u) \neq 0$. Therefore $v \circ u$ is not invertible. Since u is not invertible, u is nilpotent and there is an integer k such that $u^k = 0$. If u is surjective, $0 = u^k(M) = u^{k-1}(u(M)) = u^{k-1}(M) = \cdots = M$ and hence u cannot be surjective. Hence $u \circ v$ is not surjective, and therefore $u \circ v$ is not invertible.

We need to show that if u_1 and u_2 are not invertible endomorphisms, then $u_1 + u_2$ is not invertible. Suppose to the contrary that $u_1 + u_2$ is invertible. Define $v_1 := u_1 \circ (u_1 + u_2)^{-1}$ and $v_2 := u_2 \circ (u_1 + u_2)^{-1}$. Then

$$v_1 + v_2 = u_1 \circ (u_1 + u_2)^{-1} + u_2 \circ (u_1 + u_2)^{-1} = (u_1 + u_2) \circ (u_1 + u_2)^{-1} = id_M.$$

Since neither u_1 nor u_2 is surjective, neither v_1 nor v_2 is surjective. By the first paragraph of the proof, v_1 and v_2 are both nilpotent of order m, say. Hence,

$$\begin{aligned} id_M &= (id_M - v_2) \circ (id_M + v_2 + v_2^2 + v_2^3 + \cdots + v_2^{m-1}) \\ &= v_1 \circ (id_M + v_2 + v_2^2 + v_2^3 + \cdots + v_2^{m-1}) \end{aligned}$$

and therefore v_1 is surjective. Fitting's lemma implies that $M = S$ with respect to v_1 and therefore v_1 is an automorphism of M. This contradicts the fact that u_1 is not invertible. □

The next lemma is of a bit more technical nature.

Lemma 1.4.7 *Let M and N be two non-zero A-modules, and suppose that N is indecomposable. Let $u \in Hom_A(M, N)$ and $v \in Hom_A(N, M)$. Then*

$$v \circ u \in Aut_A(M) \Rightarrow u \text{ and } v \text{ are isomorphisms.}$$

Proof $e := u \circ (v \circ u)^{-1} \circ v \in End_A(N)$.

$$\begin{aligned} e \circ e &= (u \circ (v \circ u)^{-1} \circ v) \circ (u \circ (v \circ u)^{-1} \circ v) \\ &= u \circ (v \circ u)^{-1} \circ (v \circ u) \circ (v \circ u)^{-1} \circ v \\ &= u \circ (v \circ u)^{-1} \circ v \\ &= e. \end{aligned}$$

Since N is indecomposable, by Lemma 1.2.10 we get $e \in \{0, id_N\}$. As

$$\begin{aligned} 0 \neq id_M = id_M^2 &= (v \circ u)^{-1} \circ (v \circ u) \circ (v \circ u)^{-1} \circ (v \circ u) \\ &= (v \circ u)^{-1} \circ v \circ e \circ u \end{aligned}$$

we get $e = id_N$. Since $id_N = e = u \circ (v \circ u)^{-1} \circ v$, the morphism u is surjective. Since $v \circ u \in Aut_A(M)$, the morphism u is injective.

Thus, u is an isomorphism, and since $v \circ u$ is an automorphism, v is an isomorphism as well. □

We are now ready to prove Theorem 1.4.3. We shall use induction on m.

If $m = 1$, then M is indecomposable, and therefore $n = 1$ as well. Hence, $M_1 = M = N_1$.

Suppose the theorem is proven for $m - 1$. Let M be an A-module and suppose there are submodules $M_1, M_2, \cdots, M_m$ and $N_1, N_2, \cdots, N_n$ of M such that

$$\bigoplus_{j=1}^{m} M_j = M = \bigoplus_{i=1}^{n} N_i.$$

Hence, for any $x \in M$ there are uniquely determined elements $e_i(x) \in M_i \subseteq M$ for all $i \in \{1, 2, \ldots, m\}$ such that $x = \sum_{i=1}^{m} e_i(x)$. Each of the mappings e_i is an idempotent endomorphism of M. Likewise, for any $x \in M$ there are uniquely determined elements $u_j(x) \in N_j \subseteq M$ for all $j \in \{1, 2, \ldots, n\}$ such that $x = \sum_{i=1}^{n} u_j(x)$. Each of the mappings u_j is an idempotent endomorphism of M.

Let $v_j := e_1 \circ u_j \in Hom_A(N_j, M_1)$ and $w_j := u_j \circ e_1 \in Hom_A(M_1, N_j)$ for all $j \in \{1, 2, \ldots, n\}$. As $\sum_{j=1}^{n} u_j = id_M$,

$$\begin{aligned}\left.\left(\sum_{j=1}^{n} v_j \circ w_j\right)\right|_{M_1} &= \left.\left(\sum_{j=1}^{n} (e_1 \circ u_j \circ u_j \circ e_1)\right)\right|_{M_1} \\ &= \left. e_1 \circ \left(\sum_{j=1}^{n} u_j\right) \circ e_1 \right|_{M_1} = e_1|_{M_1} = id_{M_1}\end{aligned}$$

and so, by Lemma 1.4.6, there is a $j_0 \in \{1, 2, \ldots, n\}$ such that $v_{j_0} \circ w_{j_0} \in Aut_A(M_1)$. By Lemma 1.4.7, since N_{j_0} is indecomposable, v_{j_0} and w_{j_0} are both isomorphisms.

We still need to show that $M/M_1 \simeq M/N_{j_0}$.

By the particular structure of w_{j_0} as a composition of projection-injections with respect to the direct sum decompositions, we see that we are in the following situation. Let U, V, X, Y be A-modules such that one has an isomorphism $U \oplus V \xrightarrow{\varphi} X \oplus Y$ which is given by

$$\varphi = \begin{pmatrix} \varphi_{U,X} & \varphi_{V,X} \\ \varphi_{U,Y} & \varphi_{V,Y} \end{pmatrix}.$$

Suppose further that $\varphi_{U,X}$ is an isomorphism. Then

$$\tau = \begin{pmatrix} id_U & -\varphi_{U,X}^{-1} \circ \varphi_{V,X} \\ 0 & id_V \end{pmatrix} \in Aut_A(U \oplus V).$$

Since τ is an automorphism and since φ is an isomorphism, the composition $\varphi \circ \tau$ is an isomorphism as well. But

$$\varphi \circ \tau = \begin{pmatrix} \varphi_{U,X} & \varphi_{V,X} \\ \varphi_{U,Y} & \varphi_{V,Y} \end{pmatrix} \cdot \begin{pmatrix} id_U & -\varphi_{U,X}^{-1} \circ \varphi_{V,X} \\ 0 & id_V \end{pmatrix} = \begin{pmatrix} \varphi_{U,X} & 0 \\ \varphi_{U,Y} & \left(\varphi_{V,Y} - \varphi_{U,Y} \circ \varphi_{U,X}^{-1} \circ \varphi_{V,X}\right) \end{pmatrix}$$

and hence

$$\left(\varphi_{V,Y} - \varphi_{U,Y} \circ \varphi_{U,X}^{-1} \circ \varphi_{V,X}\right) : V \longrightarrow Y$$

is an isomorphism.

As a consequence we get an isomorphism

$$\bigoplus_{j=2}^{n} M_i \simeq M/M_1 \longrightarrow M/N_{j_0} \simeq \bigoplus_{j \neq j_0} N_j.$$

This completes the proof of the Krull-Schmidt theorem. □

Example 1.4.8 We give some examples to show that this is really a remarkable result.

1. Let $A = K$ be a field and consider the module $M = \bigoplus_{i \in \mathbb{N}} K$ of a countably infinite direct sum of copies of K. Then

$$M \simeq M \oplus K \simeq M \oplus M.$$

   Of course, M is neither Noetherian nor artinian.
2. The well-educated reader will know that for any square free positive integer d the ring of integers $\mathcal{O}_{\mathbb{Q}(\sqrt{-d})}$ of $\mathbb{Q}(\sqrt{-d})$ is not a principal ideal domain except for a few small exceptions. On the other hand, the ideal class group of each of these rings is always finite. Hence, in almost all cases there is a non-principal ideal $\mathfrak{a}$ such that there is an integer $n > 0$ for which $\mathfrak{a}^n$ is principal. Since $\mathfrak{a} \oplus \mathfrak{b} \simeq \mathfrak{a} \cdot \mathfrak{b} \oplus \mathcal{O}_{\mathbb{Q}(\sqrt{-d})}$, this gives examples of a Noetherian ring not satisfying the Krull-Schmidt theorem. An incidence of this phenomenon is the ring $\mathbb{Z}[\sqrt{-5}]$ in which one has two decompositions of 6 into prime elements

$$(1 - \sqrt{-5}) \cdot (1 + \sqrt{-5}) = 6 = 2 \cdot 3.$$

   Since both of the elements 2 and $(1+\sqrt{-5})$ are prime elements, the ideal generated by 2 and $(1 - \sqrt{-5})$ is not principal: If it was principal, then a generator a of the ideal would divide 2 and $(1 + \sqrt{-5})$. Recall that the norm of an element $x + \sqrt{-5}y$ is $(x + \sqrt{-5}y)(x - \sqrt{-5}y) = x^2 + 5y^2$. The norm is multiplicative, and takes values in $\mathbb{Z}$ if $(x, y) \in \mathbb{Z}^2$. Since the norm of 2 is 4 and the norm of $(1 + \sqrt{-5})$ is 6, the norm of a would be 2. But $2 = x^2 + 5y^2$ does not have solutions $(x, y) \in \mathbb{Z}^2$.

For more details on this subject one may consult Hasse's classical text [2] or Curtis-Reiner's comprehensive book on representation theory [3, 4].

3. R.G. Swan shows in [5] that for the generalised quaternion group Q_{32} of order 32 there is an ideal $\mathfrak{a}$ in $\mathbb{Z}Q_{32}$ which is not free as a $\mathbb{Z}Q_{32}$-module, but

$$\mathfrak{a} \oplus \mathbb{Z}Q_{32} \simeq \mathbb{Z}Q_{32} \oplus \mathbb{Z}Q_{32}.$$

1.4.2 Wedderburn's Structure Theorem

We have seen in Sect. 1.4.1 that given a finite group G and a field K in which the order of G is invertible, any module M is semisimple with an essentially unique decomposition into a direct sum of submodules. We do not know yet how to obtain a list of isomorphism classes of simple modules. This is the purpose of the present section.

Towards Wedderburn's Theorem

The first step in this direction is the following statement.

Lemma 1.4.9 (Schur) *Let K be a field and let A be a K-algebra. Let V be a simple A-module. Then $End_A(V)$ is a skew-field containing K.*

Remark 1.4.10 Of course, it might happen that $End_A(V)$ is not only a skew-field but even a field. This is the case, for example, if V is the trivial KG-module of a group G.

Proof Let $u \in End_A(V)$.

- The image $\text{im}(u)$ and the kernel $\ker(u)$ are submodules of V.
- Since V is simple, the only submodules of V are 0 and V.
- If $\text{im}(u) = 0$, then $u = 0$. Hence, u is either 0 or surjective.
- If $\ker(u) = V$, then $u = 0$. If $\ker(u) = 0$, then u is injective.

Hence, u is either 0 or an automorphism, and using Remark 1.1.12, $End_A(V)$ is a K-algebra in which any non-zero element is invertible. □

Example 1.4.11 Let Q_8 be the quaternion group of order 8. Recall that Q_8 is generated by three elements a, b, c subject to the relations $a^4 = 1, ab = c, a^2 = b^2 = c^2$. A $\mathbb{Q}Q_8$-module is given by the rational quaternion algebra $\mathbb{H}_{\mathbb{Q}}$. This is the $\mathbb{Q}$-algebra of dimension 4 with $\mathbb{Q}$-basis $\{1, i, j, k\}$. This becomes a $\mathbb{Q}$-algebra by the following multiplication rules:

- 1 is the neutral element of the multiplication;
- $i \cdot j = k$, $j \cdot k = i$, $k \cdot i = j$;
- $i^2 + 1 = j^2 + 1 = k^2 + 1 = 0$;
- extend this $\mathbb{Q}$-bilinearly.

The verification that this indeed is an associative $\mathbb{Q}$-algebra is a straightforward, but slightly lengthy calculation. The multiplicative inverse of

$$x = x_1 \cdot 1 + x_i \cdot i + x_j \cdot j + x_k \cdot k$$

is

$$x^{-1} := \frac{1}{x_1^2 + x_i^2 + x_j^2 + x_k^2} \cdot (x_1 \cdot 1 - x_i \cdot i - x_j \cdot j - x_k \cdot k)$$

as one verifies by elementary multiplication. We show that $\mathbb{H}_{\mathbb{Q}}$ is a $\mathbb{Q}Q_8$-module by declaring that
a acts on $\mathbb{H}_{\mathbb{Q}}$ as multiplication *on the left* by i
b acts on $\mathbb{H}_{\mathbb{Q}}$ as multiplication *on the left* by j
c acts on $\mathbb{H}_{\mathbb{Q}}$ as multiplication *on the left* by k.
Since the relations of Q_8 are satisfied by definition in $\mathbb{H}_{\mathbb{Q}}$, this gives $\mathbb{H}_{\mathbb{Q}}$ the structure of a $\mathbb{Q}Q_8$-module.

Moreover, since the multiplication in $\mathbb{H}_{\mathbb{Q}}$ is associative, multiplication *on the right* by elements of $\mathbb{H}_{\mathbb{Q}}$ gives a homomorphism $\mathbb{H}_{\mathbb{Q}} \longrightarrow End_{\mathbb{Q}Q_8}(\mathbb{H}_{\mathbb{Q}})$. Since $\mathbb{H}_{\mathbb{Q}}$ is a skew-field, this homomorphism is injective. Moreover, a $\mathbb{Q}$-linear endomorphism of $\mathbb{H}_{\mathbb{Q}}$ which has to commute with left multiplication with i, j and k, is actually an endomorphism of $\mathbb{H}_{\mathbb{Q}}$, which is $\mathbb{H}_{\mathbb{Q}}$-linear. Then the image of the unit element determines every endomorphism by the fact that the endomorphism has to be linear with respect to the left module structure of $\mathbb{H}_{\mathbb{Q}}$ over itself. Moreover, any image in $\mathbb{H}_{\mathbb{Q}}$ defines an endomorphism. So, $\mathbb{H}_{\mathbb{Q}} \longrightarrow End_{\mathbb{Q}Q_8}(\mathbb{H}_{\mathbb{Q}})$ is an isomorphism. We see that skew-fields may occur in Schur's lemma.

Observe that $\mathbb{H}_K$ can be defined analogously for all fields K, and then $\mathbb{H}_K$ is a skew-field as soon as the equation $x_1^2 + x_2^2 + x_3^2 + x_4^2 = 0$ has only the trivial solution $x_1 = x_2 = x_3 = x_4 = 0$ in K.

Remark 1.4.12

1. The so-called Brauer group $Br(K)$ classifies skew-fields D which are finite dimensional K-algebras and with centre K, up to a certain equivalence. The subgroup of $Br(K)$ consisting of those skew-fields D which occur as endomorphisms of simple KG-modules for some group G is called the Schur subgroup of $Br(K)$.
2. The centre Z of a skew-field is a field. Given an A left module M then M is an $End_A(M)$-module by putting $s \cdot \varphi := \varphi(s)$ for every $s \in S$ and $\varphi \in End_A(S)$. Hence, a simple A-module S is an $End_A(S)$-module. Since Z is commutative, this defines a Z-module structure on S. So, simple modules are always vector spaces over a field.

The next lemma is almost trivial.

Lemma 1.4.13 *Let K be a commutative ring and let A be a K-algebra. If V and W are two non-isomorphic simple A-modules, then $Hom_A(V, W) = 0$.*

Proof Let $u \in Hom_A(V, W)$ be a homomorphism. Then $\ker(u)$ is a submodule of V, and so either $u = 0$ or u is injective. If u is injective, then $u(V)$ is a submodule of W isomorphic to V. Since W is simple, $u(V) = W$. Since W is non-isomorphic to V we get a contradiction. Hence, $u = 0$. □

Recall the definition of the opposite algebra from Definition 1.1.7.

Lemma 1.4.14 *Let K be a commutative ring and let A be a K-algebra. Then the endomorphism ring of the regular A-module ${}_AA$ is isomorphic to A^{op}.*

Proof Let $\varphi \in End_A({}_AA)$. Then $\varphi(1_A) =: a_\varphi$ and $\varphi(b) = \varphi(b \cdot 1_A) = b \cdot \varphi(1_A) = b \cdot a_\varphi$ for all $b \in A$. Hence, every endomorphism φ of ${}_AA$ is given by multiplication on the right by an $a_\varphi \in A$. Clearly, $a_\psi = a_\varphi$ for two endomorphisms ψ and φ of A if and only if $\psi = \varphi$. Moreover, ${}_AA \ni b \mapsto ba \in {}_AA$ is in $End_A({}_AA)$. Finally, for all $b \in A$

$$ba_{\varphi\circ\psi} = (\varphi \circ \psi)(b) = \varphi(\psi(b)) = (ba_\psi)a_\varphi = b(a_\psi a_\varphi)$$

and so, $a_\psi a_\varphi = a_{\varphi\circ\psi}$. As a consequence, $End_A({}_AA) \simeq A^{op}$. □

Remark 1.4.15 The reason for this phenomenon is the fact that we write homomorphisms on the left. If we had written homomorphisms on the right, and composed them accordingly, we could have avoided the use of the opposite algebra here. However, the convention is a strong one, and many readers would find a systematic use of mappings written on the right confusing. Nevertheless, one should take note that the appearance of the opposite algebra comes only from this tradition and has no intrinsic reason.

Wedderburn's Result and First Consequences

We use the notation $Mat_{n\times n}(R)$ to denote the matrix ring of square matrices with coefficients in R. We are ready to formulate the main result of this subsection.

Theorem 1.4.16 (Wedderburn 1907 and 1925) *Let K be a field and let A be a finite dimensional semisimple K-algebra. Then there is an integer $m \in \mathbb{N}$, skew-fields $D_1, D_2, \dots, D_m$ with centres containing K, and integers $n_1, n_2, \dots, n_m$ such that there is an isomorphism of algebras*

$$A \simeq \prod_{i=1}^{m} Mat_{n_i\times n_i}(D_i).$$

Moreover, the skew-fields D_i and the integers n_i for $i = 1, \dots, m$ are completely determined by A.

Proof By Lemma 1.4.14 we know that the K-algebra A is isomorphic to the K-algebra $(End_A({}_AA))^{op}$.

Now, since A is semisimple, ${}_AA$ is isomorphic to a direct sum of simple modules

$$ {}_AA \simeq \bigoplus_{i=1}^{m} \bigoplus_{j=1}^{n_i} S_i $$

where we have numbered the modules in such a way that $S_{i_1} \simeq S_{i_2}$ if and only if $i_1 = i_2$, where there are exactly n_i different summands in the direct sum decomposition which are isomorphic to S_i, and so we have expressed possible multiplicities by the second direct sum over j.

Hence,

$$ A \simeq \left(End_A \left(\bigoplus_{i=1}^{m} \bigoplus_{j=1}^{n_i} S_i \right) \right)^{op}. $$

But, by Lemma 1.4.13 $Hom_A(S_i, S_j) = 0$ whenever $i \neq j$ and so,

$$ A \simeq \prod_{i=1}^{m} \left(End_A \left(\bigoplus_{j=1}^{n_i} S_i \right) \right)^{op}. $$

But now, for A-modules X, Y, V, W one always has

$$ Hom_A(X \oplus Y, V \oplus W) \simeq \begin{pmatrix} Hom_A(X, V) & Hom_A(Y, V) \\ Hom_A(X, W) & Hom_A(Y, W) \end{pmatrix}. $$

Therefore,

$$ End_A \left(\bigoplus_{j=1}^{n_i} S_i \right) \simeq Mat_{n_i \times n_i}(End_A(S_i)) $$

with multiplication given by the usual matrix multiplication, and composition of mappings in $End_A(S_i)$. Schur's Lemma 1.4.9 shows that $End_A(S_i) = D_i'$ is a skew-field with centre containing K. For the last step we observe that if D is a skew-field with centre containing K, then D^{op} is again a skew-field with centre containing K. Moreover, for two algebras A and B, since A and B commute in the direct product, we get $(A \times B)^{op} \simeq A^{op} \times B^{op}$. Finally, for any K-algebra B we have

$$ \begin{aligned} (Mat_{n \times n}(B))^{op} &\simeq Mat_{n \times n}(B^{op}) \\ M &\mapsto M^{tr} \end{aligned} $$

denoting by M^{tr} the transpose of a matrix. We just apply all of this to our situation, defining $D_i := D_i'^{op}$ and obtain the requested isomorphism.

The unicity of the skew-fields D_i and the dimensions n_i for $i \in \{1, \dots, m\}$ are a consequence of the fact that the skew-fields are the endomorphism algebras of the simple modules, that the Krull-Schmidt theorem implies the unicity of the decomposition of the regular module into a sum of simple submodules, and that the n_i are the multiplicities of the isomorphism classes of these simple modules as a direct factor of the regular representation. □

The Wedderburn structure theorem has a lot of consequences. Here we mention just a few in order to illustrate the importance of this result.

Corollary 1.4.17 *A skew-field that is a finite dimensional vector space over an algebraically closed field is commutative. In particular, if A is a finite dimensional semisimple K-algebra, and if K is algebraically closed, then there are integers $n_1, \dots, n_m$ such that*

$$A \simeq \prod_{j=1}^{m} Mat_{n_j \times n_j}(K).$$

Proof Let D be a finite dimensional skew-field over K and let $d \in D$. Since D is a finite dimensional K-vector space, by the Cayley-Hamilton theorem d has a minimal polynomial $P_d(X) \in K[X]$, that is a polynomial with leading coefficient 1 and smallest possible degree having d as root. Since K is algebraically closed, the degree of P is 1, and so $d \in K$. Hence $D = K$. □

This applies in particular to group rings of finite groups G over an algebraically closed field K so that the order of G is invertible in K.

Corollary 1.4.18 *Let G be a finite group. Then there is an integer m and integers $n_1, n_2, \dots, n_m$ such that $|G| = n_1^2 + n_2^2 + \dots + n_m^2$ and these integers n_i for $i \in \{1, \dots, m\}$ are the dimensions of the simple KG-modules, where K is any algebraically closed field in which the order of G is invertible.*

Proof Compare the dimensions of the algebras in Corollary 1.4.17. □

Corollary 1.4.19 *If A is semisimple, commutative, and finite dimensional over K then there are finite dimensional field extensions K_i over K such that $A \simeq \prod_{j=1}^{m} K_i$.*

Proof Indeed, A being abelian, in the Wedderburn structure theorem no matrix ring can occur, and nor can any skew-field. □

Corollary 1.4.20 *Let A be a semisimple finite dimensional K-algebra and let S be a simple A-module. Then there is an $i \in \{1, \dots, m\}$ such that*

$$S \simeq S_i \simeq \begin{pmatrix} D_i \\ \vdots \\ D_i \end{pmatrix}_{n_i}.$$

Proof Indeed, if S is a simple A-module, take $s \in S \setminus \{0\}$. Then $A \cdot s$ is a non-zero A-submodule of S, so that then $S = As$ and therefore S is a quotient of ${}_AA$. Since ${}_AA$ is semisimple, there is an $i \in \{1, \ldots, m\}$ such that $S \simeq S_i$. Since

$$S_i \simeq Hom_A({}_AA, S_i) \simeq Hom_A\left(\bigoplus_{\ell=1}^{m}\bigoplus_{j=1}^{n_i} S_\ell, S_i\right) \simeq \begin{pmatrix} D_i \\ \vdots \\ D_i \end{pmatrix}_{n_i}$$

the result follows. □

Recall from Remark 1.4.12 item 2 that every simple A-module S is a vector space over the skew field $End_A(S)$.

Corollary 1.4.21 *Let A be a finite dimensional semisimple K-algebra. Let S be a simple A-module. Then exactly* $\dim_{End_A(S)}(S)$ *copies of S are direct factors of the regular A-module ${}_AA$.*

Proof This is precisely the fact that in $Mat_{n\times n}(D)$ a simple direct factor corresponds to a column, and there are exactly as many columns as rows in a square matrix. □

Example 1.4.22 Note that $\mathbb{Z}/2\mathbb{Z} \times \mathbb{Z}/3\mathbb{Z}$ is a semisimple artinian $\mathbb{Z}$-algebra.

Wedderburn's theorem continues to be true in this more general setting, as is seen in the following corollary.

Corollary 1.4.23 *Let A be a semisimple artinian algebra. Then A is isomorphic to a finite direct product of matrix rings over skew fields.*

Proof The regular A-module is still semisimple, and therefore its endomorphism ring is a direct product of matrix rings over skew-fields. □

Definition 1.4.24 A module M which is generated by a single element is called a *cyclic module.* A *generator* of the cyclic module M is an element s such that $As = M$.

Remark 1.4.25 We see by the proof of Corollary 1.4.20 that simple modules are cyclic, with any non-zero element being a generator. If S is a simple module, and $s \in S\setminus\{0\}$, then we obtain an A-module epimorphism $A \ni a \mapsto a \cdot s \in S$ with kernel the so-called annihilator $ann(s) := \{a \in A \mid a \cdot s = 0\}$ of s. Hence $S \simeq A/ann(s)$ by Lemma 1.1.13. The regular module is also cyclic. However, in this case only units can be used as generators.

We now show that (direct products of) matrix rings over skew-fields are actually semisimple.

Definition 1.4.26 Let K be a field and let A be a K-algebra. We say that the algebra A is *simple* if for each two-sided ideal I of A we have $I = 0$ or $I = A$.

An example of an infinite dimensional simple algebra will be given in Example 1.4.32. We shall now show that finite dimensional simple algebras are actually semisimple.

Proposition 1.4.27 *Let K be a field and let A be a finite dimensional simple K-algebra. Then A is semisimple.*

Proof Let S be a simple left A-submodule of the regular module ${}_AA$, i.e. a minimal left ideal of A. Then define the left ideal $J := \sum_{a\in A} S\cdot a$ where the multiplication is taken inside A, which makes sense since $S \subseteq A$. Hence, by definition of J, we see that J is in fact a non-zero two-sided ideal of A. Therefore $J = A$ since A is simple.

However J is semisimple. Indeed, let $B_1 \subseteq A$ so that $L_1 := \sum_{b\in B_1} S\cdot b \neq A$, but so that for any $a \in A$ we have either $S\cdot a \subseteq L_1$ or $L_1 + S\cdot a = A$. Then, $S\cdot a$ is either 0 or isomorphic to the simple A-submodule S of A. Indeed, right multiplication by a gives an epimorphism $S \twoheadrightarrow S\cdot a$ of A-modules and since S is simple, the kernel can only be all of S or 0. Let $a \in A$ such that $L_1 + S\cdot a = A$. Then $S\cdot a \neq 0$ and therefore $L_1 \cap S\cdot a \leq S\cdot a$ and since $S\cdot a$ is simple, we get $L_1 \cap S\cdot a = 0$. Hence $A = L_1 \oplus S\cdot a$. Now, let $b \in B_1$ such that $S\cdot b \neq 0$, and let $B_2 \subseteq B_1$ such that $L_2 := \sum_{b\in B_2} S\cdot b$ has the property that for every $b_2 \in B_1 \setminus B_2$ we have $S\cdot b = 0$ or $L_2 + S\cdot b_2 = L_1$. By the very same argument as for L_1 we obtain that $L_2 \oplus S\cdot b_2 = L_1$. We recursively construct submodules L_i so that $L_i \oplus S\cdot b_i = L_{i-1}$ and proceeding by induction on the dimension of L_i we get that $A = J = \bigoplus_{j=1}^m S\cdot b_j$ for certain elements $b_j \in A$. Hence, the regular A-module is semisimple.

We now show that all finitely generated modules are semisimple. Let M be a finitely generated A-module. Then there is a generating set $m_1,\dots,m_t$ of M as A-module, and the map

$$({}_AA)^t \xrightarrow{\mu} M$$

$$(a_1,\dots,a_t) \mapsto \sum_{j=1}^t a_j m_j$$

is an epimorphism of A-modules. Since the regular module ${}_AA$ is semisimple, $({}_AA)^t$ is also semisimple. We need a lemma.

Lemma 1.4.28 *Each submodule of a finitely generated semisimple A-module is a direct factor. Moreover, all submodules and quotients of finitely generated semisimple modules are semisimple.*

Proof Let S be semisimple and let $S = S_1 \oplus \cdots \oplus S_t$ for simple submodules S_i, with $i \in \{1,\dots,t\}$. Let T be a submodule of S. We shall show that T is a direct factor of S. Consider the set $\mathfrak{X} := \{J \subseteq \{1,\dots,t\} \mid T \cap \sum_{j\in J} S_j = \{0\}\}$. Of course $\emptyset \in \mathfrak{X}$, and so $\mathfrak{X}$ is non-empty. Let J_0 be a maximal set in $\mathfrak{X}$ and then $M := T + \sum_{j\in J_0} S_j = T \oplus \bigoplus_{j\in J_0} S_j$. We claim that $M = S$. Suppose $S_{i_1} \not\leq M$ for $i_1 \in \{1,\dots,t\}$. Since

S_{i_1} is simple, $S_{i_1} \cap M = 0$ and so $S_{i_1} \oplus M = T \oplus S_{i_1} \oplus \bigoplus_{j \in J_0} S_j$, contradicting the maximality of J_0. This shows that $T \leq S \Rightarrow \exists_{T' \leq S} : S = T' \oplus T$.

Let S be semisimple and let T be a submodule of S. By the first step $S = T \oplus T'$ for some submodule T' of S. Since $T' \simeq S/T$ we only need to show that submodules of semisimple modules are semisimple. Let $\mathfrak{T}$ be the set of simple submodules of T and let $U := \sum_{V \in \mathfrak{T}} V$. Then U is a submodule of T, which is a submodule of S, and hence there is a submodule X of S so that $(U \oplus T') \oplus X = S$. Hence $(U \oplus X) \oplus T' = T \oplus T' = S$ and therefore $U \oplus X \simeq T$.

Suppose $X \neq 0$. We claim that X contains a simple submodule. Indeed, let $x \in X \setminus \{0\}$. Then let Y be a maximal submodule of X such that $x \notin Y$. By Zorn's lemma such a submodule exists. Since Y is also a submodule of S there is a submodule Z of X such that $U \oplus T' \oplus Y \oplus Z = S = U \oplus T' \oplus X$. Hence $Y \oplus Z \simeq X$. Let Z' be a non-zero proper submodule of Z. By the maximality of Y we get that $x \in Z' \oplus Y$. This shows that $Z' \oplus Y \simeq X \simeq Z \oplus Y$ and hence $Z = Z'$. Therefore Z is simple.

Since $Z \leq X \leq T$ with Z simple, $Z \in \mathfrak{T}$, and by the definition of U we get that $Z \subseteq U$. But this shows that the sum $U \oplus X = T$ is not direct, a contradiction. Hence $U = T$.

Let $\mathfrak{U} \subseteq \mathfrak{T}$ be maximal so that $\sum_{V \in \mathfrak{U}} V = \bigoplus_{V \in \mathfrak{U}} V =: W$. Then, as above, there is a submodule L of S such that $W \oplus L = T$. Again, if $L \neq 0$, by the same argument as above, L has a simple submodule W' and so $W' \in \mathfrak{U}$, contradicting the direct sum property. This proves the second statement. □

Lemma 1.4.28 shows that $\ker \mu$ and M are both semisimple. This proves Proposition 1.4.27. □

Remark 1.4.29 Let D be a skew field. Then $Mat_{n\times n}(D)$ is a simple algebra. Indeed, let M be a non-zero element of an ideal I of $Mat_{n\times n}(D)$. Let $d_{i,j} \neq 0$ be a non-zero coefficient of M. Then we may multiply from the right and from the left by a matrix, containing only one non zero entry, namely 1, in the i-th row and column, and likewise by a matrix with only one non-zero coefficient, being 1, in the j-th row and column. Multiply these matrices from the left and from the right on M to get an element of I with only one non-zero coefficient, namely $d_{i,j}$ in position (i, j). Then by multiplying permutation matrices from the left and from the right, we may produce matrices in I with only one non-zero coefficient, namely $d_{k,\ell}$ in position (k, ℓ) for any choice of k and ℓ. Linear combinations of these matrices will give any matrix of $Mat_{n\times n}(D)$, and so $I = Mat_{n\times n}(D)$. Therefore $Mat_{n\times n}(D)$ is simple.

A consequence is that the algebras described by Wedderburn's theorem are all semisimple.

Wedderburn's theorem suggests that simple algebras should resemble matrix rings over skew-fields. This is not true in general. We shall give an example below, which shows that simple algebras can be quite complicated if they are not finite dimensional over a field.

Definition 1.4.30 Let X be a set and let K be a commutative ring. The *free K-algebra* $K\langle X\rangle$ on X is the algebra which has a K-basis consisting of words of elements of X. The multiplication of two words is given by concatenation. Extend this multiplication K-bilinearly.

Example 1.4.31 If X has only one element, then $K\langle X\rangle = K[X]$. If X has two elements x and y, then a K-basis of $K\langle X\rangle$ contains, for example, the expressions

$$w_1 := xyyxxyxyyx = xy^2x^2yxy^2x$$

and

$$w_2 := xyyxyxxxy = xy^2xyx^3y.$$

We then get

$$w_1 \cdot w_2 = xyyxxyxyyxxyyxyxxxy = xy^2x^2yxy^2x^2y^2xyx^3y.$$

We shall now consider the following example.

Example 1.4.32 Let K be a field. Form the polynomial ring in one indeterminate $K[X]$ and attach a formal derivative $\frac{\partial}{\partial X}$ which should act just as a derivative in analysis. That is, we observe that for all $n \in \mathbb{N}$ we have

$$\left(\frac{\partial}{\partial X}X - X\frac{\partial}{\partial X}\right)(X^n) = (n+1)X^n - nX^n = X^n$$

and so $\frac{\partial}{\partial X}$ is modelled by another formal variable Y satisfying $YX - XY = 1$, where the unit 1 stands for multiplication by the identity.

We define the first *Weyl algebra* to be the quotient of the free algebra on two variables X and Y by the two-sided ideal generated by $YX - XY - 1$:

$$W_1(K) := K\langle X, Y\rangle/(YX - XY - 1).$$

Similarly for every $n \in \mathbb{N}$ we define the *Weyl algebra in n variables* by the quotient of the free algebra in $2n$ variables

$$\{X_1, X_2, \dots, X_n; Y_1, Y_2, \dots, Y_n\}$$

by the two-sided ideal

$$I_n := (Y_iX_i - X_iY_i - 1; X_iX_j - X_jX_i; Y_jY_i - Y_iY_j; X_iY_j - Y_jX_i)_{1\le i,j\le n; i\neq j} :$$

$$W_n(K) := K\langle X_i, Y_i\rangle_{1\le i\le n}/I_n.$$

The Weyl algebra has many interesting properties. A lot of information can be found in McConnell and Robson [6].

Lemma 1.4.33 *If K is a field of characteristic 0, then the algebra $W_n(K)$ is simple for all $n \in \mathbb{N}$.*

Proof We first observe that for all $c \in W_n(K)$ one has a unique expression

$$c = \sum_{\alpha,\beta} \gamma_{\alpha,\beta} X^\alpha Y^\beta$$

for multi-indices $\alpha, \beta \in \{1, \dots, n\}^{\mathbb{N}_0}$ and $\gamma_{\alpha,\beta} \in K$. The easiest way to see this is to realise the Weyl algebra concretely as an algebra of polynomials together with polynomial differential operators. Then we see that

$$X_i c - c X_i = -\frac{\partial}{\partial Y_i} c$$

for all $i \in \{1, \dots, n\}$. Actually this is clear for monomials and by the additivity of the commutator with X_i one gets the statement. Analogously,

$$Y_i c - c Y_i = \frac{\partial}{\partial X_i} c$$

for all $i \in \{1, \dots, n\}$. Now, let M be a two-sided non-zero ideal of $W_n(K)$. Then take $c \in M \setminus \{0\}$. Applying successively commutators with X_i and Y_i for all $i \in \{1, \dots, n\}$, up to the degree of Y_i and X_i, we get that there is an element of $K \setminus \{0\}$ in M. Non-zero elements of K are invertible of course, and so $M = W_n(K)$. This is what we claimed. □

1.4.3 Base Field Extension and Splitting Fields

There is a general construction to extend the defining field of an algebra and a module. Let K be a field, let L be an extension field and let A be a finite dimensional K-algebra. Then take a K-basis $B = \{b_1, b_2, \cdots, b_n\}$ of A. An algebra carries in addition a multiplicative structure, and by bilinearity of the multiplication it is sufficient to define the multiplication on the basis elements. There are elements $c_{i,j}^k \in K$ for $(i, j, k) \in \{1, 2, \dots, n\}^3$ with

$$b_i \cdot b_j = \sum_{k=1}^{n} c_{i,j}^k b_k.$$

These elements $c_{i,j}^k$ are called *structure constants* of the algebra. Being associative, having a unit, and being distributive can be expressed easily in terms of (quite nasty but elementary) equations between the structure constants. So, one may just define A_L as the space with the same basis as an L-vector space, and multiplication being defined by the same structure constants $c_{i,j}^k$ for $(i, j, k) \in \{1, 2, \ldots, n\}^3$. This is an L-algebra, since the equations valid for the structure constants assure associativity, distributivity, etc.

Let M be a finite dimensional A-module for the finite dimensional K-algebra A. Let $\{m_1, \ldots, m_s\}$ be a K-basis for M. Then there are elements $u_{i,j}^k \in K$ such that

$$b_i \cdot m_j = \sum_{k=1}^{s} u_{i,j}^k m_k \ \forall i \in \{1, \ldots, n\};\ j \in \{1, \ldots, s\}.$$

Being a module can be expressed again in terms of equations between the structure constants of the algebra and the elements $u_{i,j}^k$. Consider now the L-vector space M_L of dimension s with the basis $\{m_1, \ldots, m_s\}$. This becomes an A_L-module with the same structure constants $u_{i,j}^k$.

Definition 1.4.34 A field K is called a *splitting field* for a finite dimensional K-algebra A if the endomorphism ring of every simple A-module is isomorphic to K. The finite dimensional K-algebra A is called *split semisimple* if A is semisimple and K is a splitting field for A. An A-module M is *absolutely simple* if the module M_L is a simple A_L-module for all extension fields L over K.

Remark 1.4.35 The proof of Wedderburn's theorem shows that Corollary 1.4.17 and 1.4.18 are true under the weaker assumption that K is a splitting field for the algebra.

Do we really need to go to algebraically closed fields in order to get a splitting field for a semisimple K-algebra A? The answer is no. There is always a finite extension to achieve this goal.

Proposition 1.4.36 *Let A be a finite dimensional semisimple K-algebra, then there is an extension field L of K such that* $\dim_K(L) < \infty$ *and such that A_L is split semisimple.*

Proof We first go to the algebraic closure $\overline{K}$ of K. We know that $A_{\overline{K}}$ is split semisimple. Suppose given an algebra isomorphism $\psi : A_{\overline{K}} \longrightarrow \prod_{j=1}^m Mat_{n_j \times n_j}(\overline{K})$. Choose a K-basis

$$\mathcal{B} := \{b_1, \ldots, b_\ell\}$$

of A and take a $\overline{K}$-basis $\mathcal{C}_j := \{c_{j1}, \ldots, c_{j_{n_j^2}}\}$ of each of the matrix rings $Mat_{n_j \times n_j}$ $(\overline{K})$ comprised of elementary matrices having a single non-zero entry 1. Then $\mathcal{C} := \bigcup_{j=1}^m \mathcal{C}_j$ is a $\overline{K}$-basis of $\prod_{j=1}^m Mat_{n_j \times n_j}(\overline{K})$ and

$$\psi^{-1}(\mathcal{C}) := \bigcup_{j=1}^{m} \{\psi^{-1}(c_{j1}), \ldots, \psi^{-1}(c_{j_{n_j^2}})\}$$

is a $\overline{K}$-basis of $A_{\overline{K}}$. Hence, there is a base change matrix T from $\psi^{-1}(\mathcal{C})$ to $\mathcal{B}$. Let L be the smallest field containing all the coefficients of T. Then, since each of the sets $\psi(\psi^{-1}(C_j))$ for $j \in \{1, \ldots, m\}$ is a basis of the full matrix ring, the restriction ψ_L of ψ to A_L defines an isomorphism

$$\psi_L : A_L \longrightarrow \prod_{j=1}^{m} Mat_{n_j \times n_j}(L).$$

Since all of the elements in $\overline{K}$ are algebraic, and hence all of the coefficients of T are algebraic over K, and since we encounter only a finite number of such coefficients, the field L is of finite dimension over K. This proves the proposition. □

1.5 Group Rings as Semisimple Algebras

Let G be a finite group and let K be a field such that the order of G is invertible in K. The goal is now to determine the data of Wedderburn's theorem if possible entirely in terms of data of the group G and the field K.

We start with the number of simple KG-modules. Recall that the centre of an algebra A is given by

$$Z(A) := \{a \in A | \ \forall b \in A : ba = ab\}.$$

The centre of an R-algebra A is a commutative R-algebra. The definition implies immediately that two isomorphic algebras A and B have isomorphic centres:

$$A \simeq B \Rightarrow Z(A) \simeq Z(B).$$

The isomorphism of the centres is the restriction of the isomorphism $A \simeq B$ to the subset $Z(A)$. Moreover, if A_1 and A_2 are two K-algebras, then

$$Z(A_1 \times A_2) \simeq Z(A_1) \times Z(A_2).$$

Lemma 1.5.1 $Z(Mat_{n\times n}(A)) = \{z \cdot I_n | \ z \in Z(A)\}$ *for any algebra A, where*

$$I_n = \begin{pmatrix} 1 & 0 & & \ldots & 0 \\ 0 & & \ddots & \ddots & \vdots \\ \vdots & & \ddots & \ddots & 0 \\ 0 & \ldots & & 0 & 1 \end{pmatrix}.$$

Proof Indeed, the inclusion "$\supseteq$" is clear. The other inclusion comes from the fact that first, any element in $Z(Mat_{n\times n}(A))$ must commute with any element of the form $b \cdot I_n$ and hence

$$Z(Mat_{n\times n}(A)) \subseteq Mat_{n\times n}(Z(A)).$$

Moreover, the fact that elements in the centre of $Mat_{n\times n}(A)$ commute with elementary matrices and permutation matrices implies the statement. □

Corollary 1.5.2 *Let K be a splitting field for a finite dimensional semisimple K-algebra A. Then $\dim_K(Z(A))$ is the number of isomorphism classes of simple A-modules.*

Proof This is an immediate consequence of Lemma 1.5.1, Corollary 1.4.17 and Remark 1.4.35. □

Is it possible to express the number of isomorphism classes of simple KG-modules in purely group theoretic terms? This is actually the case. For an element $g \in G$ let $C_g := \{hgh^{-1} \mid h \in G\}$ be the conjugacy class of G.

Lemma 1.5.3 *Let R be a commutative ring and let G be a finite group. Then $\{\sum_{h\in C_g} e_h \mid g \in G\}$ is an R-basis of $Z(RG)$.*

Proof Since RG is a free R-module with basis $\{e_g \mid g \in G\}$, the set $\{\sum_{h\in C_g} e_h \mid g \in G\}$ is clearly linearly independent. Moreover,

$$x = \sum_{g\in G} x_g e_g \in Z(RG) \Leftrightarrow \forall h \in G : e_h \cdot x \cdot e_h^{-1} = x.$$

But,

$$e_h \cdot x \cdot e_{h^{-1}} = \sum_{g\in G} x_g e_h e_g e_{h^{-1}} = \sum_{g\in G} x_g e_{hgh^{-1}} = \sum_{g\in G} x_{h^{-1}gh} e_g$$

and therefore $x \in Z(RG)$ if and only if $e_h \cdot x \cdot e_h^{-1} = x$, which is equivalent to $x_g = x_{h^{-1}gh}$ for all $h \in G$. This proves the lemma. □

We come to our first result, which can be regarded as a central statement and which has considerable practical importance. The proof no longer poses any problems.

Theorem 1.5.4 *Let G be a finite group and let K be a field in which the order of G is invertible. Suppose that K is a splitting field for KG. Then the number of simple KG-modules up to isomorphism is equal to the number of conjugacy classes of G.*

Proof Lemma 1.5.3 and Corollary 1.5.2 prove the statement. □

1.6 Radicals and Jordan–Hölder Conditions

Ordinary representation theory of finite groups, i.e. over fields of characteristic 0, can be done using the following three basic theorems. First, Maschke's theorem shows that the group ring of a finite group over a ground field is semisimple whenever the group order is invertible in the ground field. Second, Wedderburn's theorem determines the structure of semisimple finite dimensional algebras over a ground field, and third Krull-Schmidt's theorem proves that a direct sum decomposition of finite dimensional modules into indecomposables is unique up to permutation and isomorphism of factors.

We have seen that the hypothesis on the group order being invertible is indeed a non-trivial hypothesis and that not all algebras are semisimple. What can we say about these more general algebras?

1.6.1 Jacobson- and Nil-Radical

Working with Noetherian modules almost always includes the use of a particularly important result, namely Nakayama's lemma, which is shown below. This involves the notion of a radical.

Definition 1.6.1 Let A be an algebra and let M be an A-module. The *Jacobson radical* $\mathrm{rad}(M)$ of M is the intersection of all maximal A-submodules of M. We sometimes call the Jacobson radical the radical if no confusion may occur. Moreover, $\mathrm{rad}(A)$ is the radical of the regular A-module.

Example 1.6.2 The radical of $\mathbb{Z}$ is 0, as it is the intersection of all prime ideals. Indeed, an element is in the radical of $\mathbb{Z}$ if it is divisible by all primes.

The radical of a semisimple artinian ring is 0 by Lemma 1.4.28.

Lemma 1.6.3 *Let A be a ring. Then* $\mathrm{rad}(A)$ *is a two-sided ideal of A.*

Proof By definition, $\mathrm{rad}(A)$ is a left ideal of A. If $\mathfrak{m}$ is a maximal ideal of A, then $A/\mathfrak{m}$ is a simple A-module. If S is a simple A-module, then by Remark 1.4.25 we get that $S = A\cdot s$ for any $s \in S\setminus\{0\}$ and $S \simeq A/ann(s)$ where $ann(s) := \{a \in A \mid a\cdot s = 0\}$. Moreover, $ann(s)$ is a maximal ideal. Indeed, if $ann(s) < \mathfrak{m}$, then $A/\mathfrak{m}$ is a quotient of $A/ann(s) \simeq S$. Now, $ann(S) := \bigcap_{s\in S\setminus\{0\}} ann(s)$ is a two-sided ideal of A since S is an A-module. Therefore $\mathrm{rad}(A)$ is included in the intersection of the two-sided ideals $ann(S)$ for all simple A-modules S. If S is a simple A-module, then $S \simeq A/\mathfrak{m}$ for some maximal ideal $\mathfrak{m}$. If $a \in ann(S)$, then $a\cdot A \subseteq \mathfrak{m}$ and hence $a \in \mathfrak{m}$. Therefore $ann(S) \le \mathfrak{m}$ and the intersection of $ann(S)$ over all simple A-modules is included in $\mathrm{rad}(A)$. Hence we get that the intersection of $ann(S)$, the intersection taken over all simple A-modules, equals $\mathrm{rad}(A)$. But these annihilators $ann(S)$ are two-sided ideals of A. □

Proposition 1.6.4 *Let A be an algebra and let M be a Noetherian A-module. Then* $\mathrm{rad}(A)M = M$ *implies* $M = 0$.

Proof Since M is Noetherian, Lemma 1.3.3 implies that M is finitely generated by $m_1, m_2, \ldots, m_n$, say. Suppose that this is a minimal generating set of M. Then there are elements $j_1, j_2, \ldots, j_n$ in $\mathrm{rad}(A)$ such that

$$m_1 = j_1 m_1 + j_2 m_2 + \cdots + j_n m_n.$$

Hence,

$$(1 - j_1)m_1 = j_2 m_2 + \cdots + j_n m_n.$$

If $1 - j_1$ is not (left-)invertible in A, it generates a principal left ideal different from A and hence it belongs to a maximal left ideal $\mathfrak{m}$ of A. But j_1 is in the radical of A, hence it also belongs to $\mathfrak{m}$. As a consequence $1 \in \mathfrak{m}$, which is absurd. Hence $1 - j_1$ is left-invertible in A and therefore $\{m_2, m_3, \ldots, m_n\}$ is a generating set of M. This contradicts the minimality of n. □

Proposition 1.6.5 (Nakayama's Lemma) *Let A be an algebra, let M be a Noetherian A-module and let N be an A-submodule of M. Then*

$$(N + \mathrm{rad}(A)M = M) \Rightarrow (N = M).$$

Proof Proposition 1.6.4 shows that $M/N = 0$ and therefore $M = N$. □

Another consequence is that for Noetherian and artinian algebras $\mathrm{rad}(A)$ is always nilpotent.

Lemma 1.6.6 *Let A be a Noetherian and artinian algebra. Then* $\mathrm{rad}(A)$ *is a nilpotent two-sided ideal of A.*

Proof Lemma 1.6.3 shows that we only need to show that $\mathrm{rad}(A)$ is nilpotent. Since A is artinian, there is an integer n such that

$$\mathrm{rad}(A)^n = \mathrm{rad}(A)^{n+1} = \mathrm{rad}(A) \cdot \mathrm{rad}(A)^n.$$

By Proposition 1.6.4 we get $\mathrm{rad}(A)^n = 0$. □

Remark 1.6.7 Hopkins' Theorem 1.6.20 below shows that artinian algebras are Noetherian.

The following characterisation of the radical is very useful.

Lemma 1.6.8 *Let A be an algebra and let M be an artinian A-module. Then* $\mathrm{rad}(M)$ *is the smallest submodule of M so that $M/\mathrm{rad}(M)$ is semisimple.*

Proof Let $\mathcal{F} := \{T \leq M \mid M/T \text{ is semisimple or } 0\}$. Since $M \in \mathcal{F}$, this set is non-empty. Since M is artinian, Proposition 1.3.5 shows that $\mathcal{F}$ contains a minimal element N. Since M/N is semisimple, projection on each simple direct factor gives a morphism $M \longrightarrow S$ for a certain simple A-module S. The kernel of this morphism is a maximal submodule of M, since S is simple. N is then the kernel of the sum of these morphisms, which in turn is the intersection of the maximal submodules which occur as the kernel of the projection onto the direct factors of M/N. Hence $\mathrm{rad}(M) \leq N$.

Since M is artinian, $\mathrm{rad}(M)$ is the intersection of a finite number of maximal submodules. Indeed, if this were not the case it would be easy to produce an infinite decreasing sequence. Let $\mathrm{rad}(M) = M_1 \cap M_2 \cap \cdots \cap M_n$ with n minimal and M_i being a maximal submodule of M for each $i \in \{1, 2, \ldots, n\}$. Then $M/\mathrm{rad}(M) \simeq \oplus_{i=1}^n M/M_i$ is semisimple. Hence $N \le \mathrm{rad}(M)$. □

Lemma 1.6.9 *Let A be an algebra, let M and N be A-modules and let $f \in Hom_A(M, N)$. Then $f(\mathrm{rad}(M)) \le \mathrm{rad}(N)$.*

Proof Let $g : N \to S$ be an epimorphism to a simple A-module S. Then $g \circ f$ is a homomorphism to a simple A-module. Hence $g \circ f$ is either 0 or surjective. If $g \circ f$ is surjective, $(g \circ f)(\mathrm{rad}(M)) = 0$, and therefore $f(\mathrm{rad}(M)) \subseteq \ker(g)$. If $g \circ f = 0$, then $f(\mathrm{rad}(M)) \subseteq \ker(g)$ as well. Hence $f(\mathrm{rad}(M)) \le \mathrm{rad}(N)$. □

Lemma 1.6.10 *Let A be an algebra and let M be an A-module. Then* $\mathrm{rad}(A) \cdot M \le \mathrm{rad}(M)$. *If $A/\mathrm{rad}(A)$ is an artinian algebra, then* $\mathrm{rad}(M) = \mathrm{rad}(A) \cdot M$.

Proof Let $m \in M$. Then we define a homomorphism $A \xrightarrow{\mu_m} M$ given by $a \mapsto a \cdot m$. Lemma 1.6.9 shows that $\mu_m(\mathrm{rad}(A)) = \mathrm{rad}(A) \cdot m \le \mathrm{rad}(M)$, and hence $\mathrm{rad}(A) \cdot M = \sum_{m \in M} \mathrm{rad}(A) \cdot m \le \mathrm{rad}(M)$.

If $A/\mathrm{rad}(A)$ is artinian, by Lemma 1.6.8, we get that $A/\mathrm{rad}(A)$ is semisimple artinian. Moreover, the module structure of A on M induces an $A/\mathrm{rad}(A)$-module structure on $M/(\mathrm{rad}(A) \cdot M)$. Since $A/\mathrm{rad}(A)$ is semisimple artinian, $M/(\mathrm{rad}(A) \cdot M)$ is also semisimple, and therefore $\mathrm{rad}(M/(\mathrm{rad}(A) \cdot M)) = 0$ by Lemma 1.6.8. Since $\mathrm{rad}(M/(\mathrm{rad}(A) \cdot M)) = \mathrm{rad}(M)/(\mathrm{rad}(A) \cdot M)$, we get $\mathrm{rad}(M) = \mathrm{rad}(A) \cdot M$. □

Definition 1.6.11 Let A be an artinian algebra and let M be an A-module. Then we call the A-module $M/\mathrm{rad}(M)$ the *head (or top)* of M.

Corollary 1.6.12 *Let A be an algebra and let M be an artinian A-module. Then* $\mathrm{rad}(M/\mathrm{rad}(M)) = 0$.

Proof Let S be a semisimple A-module. Then $\mathrm{rad}(S) = 0$, as is easily seen by considering the projections on the simple direct factors. Now, since M is artinian, Lemma 1.6.8 implies that $\mathrm{rad}(M/\mathrm{rad}(M)) = 0$. □

We can sharpen Lemma 1.6.6. If A is only artinian, then we also get that the radical is nilpotent. This follows from the definition and the study of the nil-radical.

The proof of the following proposition is essentially taken from [3].

Proposition 1.6.13 *Let A be an artinian algebra. Then a left ideal I of A either contains an idempotent element or is nilpotent.*

Proof Let $\mathcal{X}$ be the set of non-nilpotent left ideals of A,

$$\mathcal{X} := \{J \le A \mid J \text{ is not nilpotent}\}.$$

Since A is artinian, and since $A \in \mathcal{X}$, Proposition 1.3.5 implies that the set $\mathcal{X}$ contains a minimal element J_0. Since J_0 is not nilpotent, J_0^2 is also not nilpotent. Hence, $J_0^2 \in \mathcal{X}$ and $J_0^2 \leq J_0$. Therefore, by minimality of J_0, we get $J_0^2 = J_0$. Let $\mathcal{Y}$ be the set of sub-ideals of J_0 which are not annihilated by J_0:

$$\mathcal{Y} := \{L \leq J_0 \mid J_0 L \neq 0\}.$$

Again, $J_0 \in \mathcal{Y}$ and the fact that A is artinian implies by Proposition 1.3.5 that $\mathcal{Y}$ contains a minimal element L_0.

Since $J_0 L_0 \neq 0$, there is an $\ell \in L_0$ such that $J_0 \ell \neq 0$. But, $J_0 \ell \in \mathcal{Y}$. So, by minimality of $\mathcal{Y}$, we get $J_0 \ell = L_0 \ni \ell$. Hence there is an $a \in J_0$ with $a\ell = \ell$. This implies that $\ell = a^m \ell$ for all integers m. Since $L_0 \neq 0$, also $\ell \neq 0$ and so a is not nilpotent.

$ann_{J_0}(\ell) := \{b \in J_0 \mid b\ell = 0\}$ contains $n_1 := a^2 - a$ and is properly contained in J_0. Clearly, $ann_{J_0}(\ell)$ is an ideal. By minimality of J_0, $ann_{J_0}(\ell)$ is nilpotent and hence so is n_1. If $a^2 - a = 0$ we are finished.

Suppose not. Put $n_i := a_{i-1}^2 - a_{i-1}$ and $a_i := a_{i-1} + n_i - 2a_{i-1}n_i$. We claim that n_i is divisible by n_{i-1}^2, that n_i commutes with a_{i-1}, that a_i is not nilpotent and that n_i is nilpotent. Then

$$\begin{aligned} n_{i+1} = a_i^2 - a_i &= (a_{i-1} + n_i - 2a_{i-1}n_i)^2 - a_{i-1} - n_i + 2a_{i-1}n_i \\ &= a_{i-1}^2 + n_i^2 + 4a_{i-1}^2 n_i^2 + 2a_{i-1}n_i \\ &\quad -4a_{i-1}^2 n_i - 4a_{i-1}n_i^2 - a_{i-1} - n_i + 2a_{i-1}n_i \\ &= (a_{i-1}^2 - a_{i-1}) - n_i + n_i^2 + 4(a_{i-1}^2 - a_{i-1})n_i^2 - 4(a_{i-1}^2 - a_{i-1})n_i \\ &= n_i - n_i + n_i^2 + 4n_i^3 - 4n_i^2. = n_i^2(4n_i - 3) \end{aligned}$$

Hence n_{i+1} commutes with a_i, is divisible by n_i^2 and is nilpotent, since n_i is nilpotent by induction hypothesis. Since a_{i-1} is not nilpotent and since n_i commutes with a_{i-1}, the element $a_i = a_{i-1} + n_i - 2a_{i-1}n_i$ is also not nilpotent.

Since then n_i is divisible by $n_1^{2^i}$, for large enough i one has

$$a_{i-1}^2 - a_{i-1} = n_i = 0.$$

Therefore for large enough i the element a_{i-1} is idempotent, non-nilpotent and hence different from 0. □

Lemma 1.6.14 *Let A be an artinian ring, and let N_1 and N_2 be nilpotent left ideals of A. Then $N_1 + N_2$ is a nilpotent left ideal of A.*

Proof Suppose $N_1^{n_1} = 0 = N_2^{n_2}$. Then any element in $(N_1 + N_2)^{n_1+n_2}$ is a sum of products of elements of N_1 and of N_2. Since for any $x \in N_1$ and $y \in N_2^j$, $xy \in N_2^j$, and likewise exchanging N_1 and N_2, we get $(N_1 + N_2)^{n_1+n_2} = 0$. □

Definition 1.6.15 Let A be an artinian ring. Then the sum of all nilpotent left ideals is the *nil-radical* $\mathrm{Nil}(A)$.

Proposition 1.6.16 *For A an artinian ring,* $\mathrm{Nil}(A)$ *is a two-sided ideal of A. The nil-radical is the smallest two-sided ideal of A so that $A/\mathrm{Nil}(A)$ does not contain any nilpotent ideals.*

Remark 1.6.17 It should be observed that we do not claim that $A/\mathrm{Nil}(A)$ does not contain nilpotent elements. We only claim that $A/\mathrm{Nil}(A)$ does not contain nilpotent ideals. There are non-commutative artinian rings with $\mathrm{Nil}(A) = 0$ but containing nilpotent elements. An example is a 2×2 matrix ring over a field.

Proof of Lemma 1.6.16 By Lemma 1.6.14 the nil-radical is a left ideal of A. Since A is artinian there are a finite number of nilpotent left ideals $N_1, N_2, \ldots, N_k$ such that $\mathrm{Nil}(A) = N_1 + N_2 + \cdots + N_k$. Otherwise $\mathrm{Nil}(A) = \sum_{i \in \mathbb{N}} N_i$ for nilpotent ideals N_i and $\sum_{i \geq j} N_i \neq \sum_{i>j} N_i$. We obtain an infinite decreasing sequence of ideals, contradicting the fact that A is artinian.

For $a \in A$, suppose $\mathrm{Nil}(A)a$ is not nilpotent. Then, by Proposition 1.6.13, the left ideal $\mathrm{Nil}(A)a$ contains a non-zero idempotent element $e \neq 0$. Therefore $e = fa = (fa)^2 = fafa$ for $f \in \mathrm{Nil}(A)$. But now

$$e = fa = (fa)^2 = \cdots = (fa)^m = f\left((af)^{m-1}\right)a.$$

Since $f \in \mathrm{Nil}(A)$, also $(af)^{m-1} \in \mathrm{Nil}(A)^{m-1}$, which is 0 for large enough m.

Suppose $A/\mathrm{Nil}(A)$ contains a nilpotent ideal $\overline{L}$. Then $\overline{L} = L/\mathrm{Nil}(A)$ for some ideal L of A. We claim that L is nilpotent. If not, again by Proposition 1.6.13, L contains a non-zero idempotent element, and so does $\overline{L}$. This contradiction proves that $L + \mathrm{Nil}(A)$ is nilpotent and bigger than $\mathrm{Nil}(A)$. This contradicts the construction of $\mathrm{Nil}(A)$.

Since for every two-sided ideal $J < \mathrm{Nil}(A)$ one has that $\mathrm{Nil}(A)/J$ is a nilpotent ideal of A/J, the two-sided ideal $\mathrm{Nil}(A)$ is the smallest nilpotent ideal such that $A/\mathrm{Nil}(A)$ does not contain any nilpotent ideals. Any two-sided ideal J such that A/J does not contain a nilpotent ideal must itself contain all nilpotent ideals of A, and hence must contain $\mathrm{Nil}(A)$. □

Proposition 1.6.18 *Let A be an artinian ring. Then* $\mathrm{Nil}(A) = \mathrm{rad}(A)$.

Proof Let S be a simple A-module. Then $\mathrm{Nil}(A) \cdot S$ is a proper submodule of S, and hence $\mathrm{Nil}(A) \cdot S = 0$. Hence $\mathrm{Nil}(A) \supseteq \mathrm{rad}(A)$.

Conversely, by Wedderburn's theorem $A/\mathrm{rad}(A)$ is a direct product of matrix rings over fields, and hence does not contain any nilpotent ideals. Hence $\mathrm{Nil}(A) \subseteq \mathrm{rad}(A)$. □

Corollary 1.6.19 *Let A be an artinian ring. Then* $\mathrm{rad}(A)$ *is nilpotent.*

Indeed, $\mathrm{Nil}(A)$ is nilpotent, and hence so is $\mathrm{rad}(A) = \mathrm{Nil}(A)$.

A very surprising consequence of the above is that artinian rings are Noetherian.

Proposition 1.6.20 (Hopkins [7]) *Every left artinian ring A is left Noetherian.*

Proof Since A is artinian, we get from Corollary 1.6.19 that $\mathrm{rad}(A)$ is nilpotent, say $\mathrm{rad}(A)^n = 0$. Then the decreasing sequence

$$A \geq \mathrm{rad}(A) \geq \mathrm{rad}^2(A) \geq \cdots \geq \mathrm{rad}^{n-1}(A) \geq \mathrm{rad}^n(A) = 0$$

has the property that $\mathrm{rad}^{m-1}(A)/\mathrm{rad}^m(A)$ is semisimple. Semisimple artinian modules are Noetherian. We proceed by downwards induction on ℓ. For $\ell = m$ one sees that $\mathrm{rad}^{m-1}(A)$ is semisimple artinian, and hence Noetherian. Suppose $\mathrm{rad}^\ell(A)$ is Noetherian, $\mathrm{rad}^\ell(A)$ is a Noetherian submodule of $\mathrm{rad}^{\ell-1}(A)$ and the quotient $\mathrm{rad}^{\ell-1}(A)/\mathrm{rad}^\ell(A)$ is artinian semisimple, and hence Noetherian as well. Then, by Lemma 1.3.3, $\mathrm{rad}^{\ell-1}(A)$ is also Noetherian. □

The concept of successive radicals of submodules deserves a name.

Definition 1.6.21 Let A be a K-algebra and let M be an artinian A-module. The series of submodules

$$\begin{array}{ccccccccc} M & \geq & \mathrm{rad}(M) & \geq & \mathrm{rad}(\mathrm{rad}(M)) & \geq & \mathrm{rad}(\mathrm{rad}(\mathrm{rad}(M))) & \geq \cdots \geq 0 \\ \| & & \| & & \| & & \| & \\ M_0 & & M_1 & & M_2 & & M_3 & \end{array}$$

of M is the *Loewy series* of M. The successive quotients M_i/M_{i+1} are the *Loewy factors*, and the smallest integer n for which $M_n = 0$ is the *Loewy length.*

We will consider a special case, namely the case of the radical of a group algebra KG of a p-group G over a field K of characteristic $p > 0$. In many practical applications, and also theoretical considerations, this will be a most important property.

Proposition 1.6.22 *Let K be a field of characteristic $p > 0$ and let G be a (finite) p-group. Then* $\mathrm{rad}(KG)$ *is of dimension $|G| - 1$ and has a K-basis $\{g - 1 | g \in G \setminus \{1\}\}$. Moreover, the only simple KG-module is the trivial module.*

Proof We prove this fact by induction on $|G|$, using in an essential way that the centre of a p-group is non-trivial. If $|G| = p$, the statement is clear since given a generator c of G, the ideal $KG \cdot (1 - c)$ is nilpotent of degree p. Moreover, $KG/KG \cdot (1 - c)$ is one-dimensional, c acting trivially.

Let G be a p-group of order p^n and let $c \in Z(G)$ be a central element. We may assume that c is of order p, replacing if necessary c by a power of c. Denote by C the group generated by c. Then $KG \cdot (1-c)$ is nilpotent of degree p again, c being central. Hence, $KG \cdot (1 - c) \subseteq \mathrm{rad}(KG)$ by Proposition 1.6.18. But, $KG/KG \cdot (1 - c) \simeq KG/C$ by the natural residue class mapping $G \longrightarrow G/C$. Since $|G/C| < |G|$ and since G/C is a p-group again, $\mathrm{rad}(K(G/C))$ has a K-basis given by the elements $gC - 1$ for $gC \in (G/C) \setminus \{C\}$ and the trivial module is the only simple KG/C-module. Hence the only simple KG-module is the trivial module as well, since the kernel of the mapping $KG \longrightarrow KG/C$ is in $\mathrm{rad}(KG)$. Therefore,

$$KG/\text{rad}(KG) = K,$$

the trivial module. As a consequence the elements $g - 1$ are all mapped to 0, hence are in the kernel of the mapping $KG \longrightarrow KG/\text{rad}(KG)$. Therefore, they belong to $\text{rad}(KG)$, they are K-linearly independent, the elements of $G \setminus \{1\}$ being K-linearly independent in KG, and there are exactly $|G| - 1$ such elements. This proves the statement. □

We shall give a more conceptual proof in Corollary 2.9.8.

1.6.2 The Jordan–Hölder Theorem

We have already seen the fundamental role of the Krull-Schmidt theorem. In general, however, this does not give us full information. Indeed, given an algebra A, the Krull-Schmidt theorem shows that given a Noetherian and artinian A-module M then M can be decomposed in an essentially unique way into indecomposable modules. In the semisimple situation we have seen that indecomposable modules are simple, and the classification of simple modules implies the knowledge of any given module. If A is not semisimple, indecomposable modules can be constructed from two copies of one simple module in various ways to obtain a fairly complicated structure. We give a first example.

Example 1.6.23 Let K be a field and let $K\langle X, Y\rangle$ be the free algebra with two generators X and Y (cf Definition 1.4.30 for a precise definition) and let A be the quotient $K\langle e_1, e_2, X, Y\rangle/I$ where I is the smallest two-sided ideal of $K\langle X, Y\rangle$ containing the set

$$\{X^2, Y^2, XY, YX, e_1^2 - e_1, e_2^2 - e_2, e_1e_2, e_2e_1,$$
$$e_1X - Xe_2, e_1Y - Ye_2, e_1X - X, e_1Y - Y, e_2X, e_2Y, Xe_1, Xe_2\}.$$

More economically and suggestively we write A as a quiver algebra as follows.

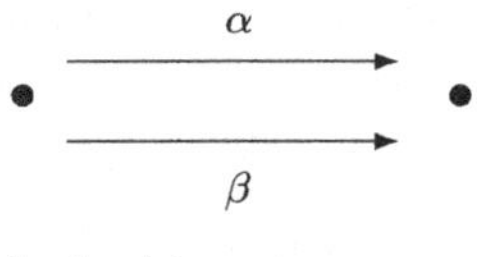

A K-basis of A is given by paths in this quiver, as well as the "lazy paths" given by the vertices. Multiplication of two paths is given by composition of these paths, and the composition of two paths is 0 if concatenation of paths does not make sense, i.e. if the ending vertex of the first path is not the starting vertex of the second. It is clear that the above quiver corresponds to the algebra described above. The lazy path on the left vertex corresponds to e_1, the lazy path on the right vertex corresponds to e_2. The arrow α is an incarnation of X and β is an incarnation of Y. The algebra A is the so-called *Kronecker quiver algebra*. For a more proper introduction and definition of quivers, quiver algebras and relations see Definition 1.11.5.

We shall consider 3-dimensional modules over this algebra. For technical and pedagogical reasons we shall look at right modules. There is the module $(K^3)_1$ on which e_1 acts as identity and the other generators as 0. On the other extreme there is the module $(K^3)_2$ on which e_2 acts as the identity and the other generators as 0.

Further, there is a module $K_{2,1,2}$. The action of A on $K_{2,1,2}$ is given by the following. Fix a K-basis v_2^+, v_1, v_2^- of $K_{2,1,2}$ and let e_i act as identity on $v_i^{(\pm)}$ and as 0 on the other basis elements. Further $v_1 \cdot X = v_2^+$ and $v_1 \cdot Y = v_2^-$. The other operations are 0. One checks with a little effort, but still at an elementary level that the relations are satisfied for this operation. For example, $v_1(Y - Ye_2) = v_2^- e_2 - v_2 = 0$.

Then there is a module $K_{1,2,1}$ with a K-basis $\{v_1^+, v_2, v_1^-\}$ and operation defined by $v_i^{(\pm)} \cdot e_i = v_i^{(\pm)}$, $v_1^+ \cdot X = v_2$, $v_1^- \cdot Y = v_2$, the remaining operations of the generators being 0. Again, one easily checks that the relations are satisfied for this operation.

It can be shown (cf [8]) that this is a complete list of indecomposable modules of dimension 3, up to isomorphism.

There are only two one-dimensional modules, up to isomorphism, namely K_1 on which e_1 acts as identity and e_2 acts as 0, and K_2 on which e_2 acts as identity, and e_1 acts as 0. Now $Hom_A(K_2, K_{2,1,2})$ is at least (in fact exactly) two-dimensional. Indeed, $1 \in K_2$ can be mapped to v_2^+ or to v_2^-. These two mappings are A-linear, since on both elements the only generator which acts non-trivially, namely as identity, is e_2. Define the mappings $\varphi_{\lambda,\mu} \in Hom_A(K_2, K_{2,1,2})$ by $\varphi_{\lambda,\mu}(1) := \lambda v_2^+ + \mu v_2^-$. Then, if $(\lambda, \mu) \neq (0, 0)$, define the module

$$B_{2,1,2}^{(\lambda,\mu)} := K_{2,1,2}/\mathrm{im}(\varphi_{\lambda,\mu}).$$

We show that $B_{2,1,2}^{(\lambda_1,\mu_1)} \simeq B_{2,1,2}^{(\lambda_2,\mu_2)}$ if and only if $[\lambda_1, \mu_1]$ and $[\lambda_2, \mu_2]$ define the same point on the projective line $\mathbf{P}(K^2)$. In other words,

$$B_{2,1,2}^{(\lambda_1,\mu_1)} \simeq B_{2,1,2}^{(\lambda_2,\mu_2)} \Leftrightarrow \exists \zeta \in K \setminus \{0\} : (\lambda_1, \mu_1) = (\zeta\lambda_2, \zeta\mu_2).$$

Proof If $\exists \zeta \in K \setminus \{0\} : (\lambda_1, \mu_1) = (\zeta\lambda_2, \zeta\mu_2)$ then an isomorphism is given by sending the class of v_1 to the class of v_1 and the class of v_2^+ to the class of ζv_2^+ whenever $\lambda_1 \neq 0$, or the same with v_2^- otherwise.

Conversely, if there is an isomorphism, say $B_{2,1,2}^{(\lambda_1,\mu_1)} \xrightarrow{\psi} B_{2,1,2}^{(\lambda_2,\mu_2)}$, then $\psi(v_1)$ must be a multiple of v_1, since $\psi(v_1) = av_1 + bv_2^-$ gives

$$av_1 + bv_2^- = \psi(v_1) = \psi(v_1 e_1) = \psi(v_1)e_1 = (av_1 + bv_2^-)e_1 = av_1.$$

Now, $a \neq 0$ since otherwise ψ would have a kernel and we get $b = 0$. We may suppose that $a = 1$ since any isomorphism can be multiplied by a non-zero scalar. Analogously, $\psi(v_2^-) = bv_2^-$ for $b \in K \setminus \{0\}$.

$$bv_2^- = \psi(v_2^-) = \psi(v_1 X) = \psi(v_1)X = v_2^+ = \frac{\mu}{\lambda} v_2^-.$$

This proves the statement. □

We conclude that there is a family of indecomposable A-modules $B_{2,1,2}^{(\lambda,\mu)}$ which are all pairwise non-isomorphic to each other when $[\lambda,\mu]$ runs through the projective line $\mathbf{P}(K^2)$. Each of the modules $B_{2,1,2}^{(\lambda,\mu)}$ has a submodule isomorphic to K_2 and quotient K_1 where K_2 and K_1 are the two one-dimensional (whence simple) modules introduced above. We say for short that $B_{2,1,2}^{(\lambda,\mu)}$ fits into an exact sequence

$$0 \longrightarrow K_2 \longrightarrow B_{2,1,2}^{(\lambda,\mu)} \longrightarrow K_1 \longrightarrow 0.$$

Hence, the Krull-Schmidt theorem is far from telling the whole story about the possible modules over a non-semisimple algebra A.

As we have seen in Example 1.6.23, the modules $B_{2,1,2}^{(\lambda,\mu)}$ all have a simple submodule K_2 such that the quotient by this simple module is the simple module K_1. Of course, this property can be generalised in such a way that a module M has a simple submodule S_1 and the quotient Q_1 need not necessarily be simple, but again has a simple submodule S_2 with quotient Q_2, and continuing in this way, for some n the module Q_n is simple.

Definition 1.6.24 Let K be a commutative ring and let A be a K-algebra. A *composition series* for an A-module M is a sequence of submodules

$$0 = M_m \le M_{m-1} \le M_{m-2} \le \dots M_1 \le M_0 = M$$

such that for each $i \in \{1, 2, \dots, m\}$ the A-module M_{i-1}/M_i is simple. The *family* $(M_{i-1}/M_i \mid i \in \{1, 2, \dots, m\})$ of simple modules are called *composition factors*. The number m is called the *composition length* of M.

We should note here that in a family we do count multiple occurrences, but we do not care about the order of occurrence. Up to now, the composition factors and the composition length have depended on the composition series and not only on the module. The Jordan–Hölder Theorem 1.6.26 below will show that these data actually do not depend on the chosen composition series.

Example 1.6.25 Let $A = K[X]$ be the polynomial ring over K and let $P(X) \in K[X]$ be a polynomial. Suppose $P = P_1 \cdot P_2 \cdot \dots \cdot P_m$ where $P_i(X) \in K[X]$ are irreducible polynomials for all $i \in \{1, 2, \dots, m\}$. Then the A-module $M := K[X]/P(X)K[X]$ has the composition factors $(K[X]/P_i(X)K[X] \mid i \in \{1, 2, \dots, m\}$. Indeed, a composition series is given by

$$0 \le \frac{P(X)}{P_1(X)} M \le \frac{P(X)}{P_1(X)P_2(X)} M \le \dots \le P_m(X)M \le M.$$

The composition length is m in this case. Multiple factors may occur for multiple irreducible factors. In particular, $K[X]/X^3K[X]$ has three composition factors, namely three occurrences of the module $K[X]/XK[X]$.

The main result of this section is the following.

Theorem 1.6.26 (Jordan–Hölder theorem) *Let K be a commutative ring and let A be a K-algebra. Then an A-module M is Noetherian and artinian if and only if M has a composition series. The composition factors of any two composition series of M are equal.*

For the proof we need a Lemma.

Lemma 1.6.27 *Let K be a commutative ring, let A be a K-algebra and let M be an artinian A-module. Then M has a simple submodule.*

Proof If M is simple there is nothing to show. If not, let

$$\mathcal{F}_M := \{N \leq M \mid N \text{is a proper non-zero submodule of } M\}.$$

This is a non-empty family of submodules of M. Since M is artinian, there is a minimal element S. If S is not simple, then it contains a proper submodule S_1, which is in $\mathcal{F}_M$, and which is smaller than the minimal element S. Hence S is simple. □

Definition 1.6.28 Let K be a commutative ring, let A be a K-algebra and let M be an artinian A-module. The maximal semisimple submodule T of M is the *socle* of M.

Remark 1.6.29 Suppose M is artinian. Then

$$\operatorname{soc}(M) = \sum_{S \leq M \text{ and } S \text{ simple}} S.$$

Just as in the proof of Lemma 1.4.28 we denote by $\mathfrak{X}$ the set of simple submodules of M. Clearly $\operatorname{soc}(M) \leq \sum_{S \in \mathfrak{X}} S$. By Lemma 1.6.27 the submodule on the left-hand side is not zero.

Let $\mathcal{Y}$ be the set of subsets Y of $\mathfrak{X}$ such that $\sum_{S \in Y} S = \bigoplus_{S \in Y} S$. This set $\mathcal{Y}$ is partially ordered by inclusion, is non-empty since it includes one-element sets, and by a standard application of Zorn's lemma there is a maximal element Y_m in $\mathcal{Y}$. Hence $N := \sum_{S \in Y_m} S = \bigoplus_{S \in Y_m} S \leq M$ and N is semisimple. We get that $N = \sum_{S \in \mathfrak{X}} S$. Indeed, otherwise let S_0 be a simple submodule of M with $S_0 \not\leq N$. Since S_0 is simple, $S_0 + N = S_0 \oplus N$, contradicting the maximality of Y_m.

Proof of Theorem 1.6.26 Suppose M is Noetherian and artinian. Since M is artinian, by Lemma 1.6.27 there is a simple submodule M_1 of M. If $M_1 = M$ we have a composition series

$$0 \leq M_1 = M.$$

If not, take a simple submodule S_2 of M/S_1 and observe that $S_2 = M_2/M_1$ for some submodule M_2 of M; i.e. M_2 is the preimage of S_2 under the projection $M \longrightarrow M/M_1$. If $M_2 = M$, we have a composition series

$$0 \leq M_1 \leq M_2 = M.$$

We will recursively construct submodules M_i of M so that M_i contains M_{i-1} for all i and so that M_i/M_{i-1} is simple.

Suppose M_i is already constructed. If $M = M_i$ we have a composition series with the sequence $(M_j)_{j\leq i}$. Otherwise take a simple submodule S_{i+1} of M/M_i and let M_{i+1} be the pre-image of S_{i+1} in M. Then, $M_{i+1}/M_i \simeq S_{i+1}$.

Since M is Noetherian, the increasing sequence $(M_i)_{i\in\mathbb{N}}$ becomes stationary. Hence there is an $n \in \mathbb{N}$ such that $M_n = M$. We obtain a composition series of M.

Claim 1.6.30 *Suppose we have a composition series*

$$0 < M_0 < M_1 < M_2 < \cdots < M_m = M$$

and another increasing sequence of submodules of M with simple successive quotients

$$0 < M_0' < M_1' < M_2' < \cdots < M.$$

We claim that then $M_m' = M$ and there is a permutation $\sigma \in \mathfrak{S}_{m+1}$ such that

$$M_i/M_{i-1} \simeq M'_{\sigma(i)}/M'_{\sigma(i-1)}.$$

Proof of Claim 1.6.30 We use induction on m.

If $m = 0$, M is simple and trivially $M = M_0 = M_0'$.

If $m > 0$ we form

$$0 < M_0' \cap M_0 < M_1' \cap M_0 < \cdots < \bigcup_{m'=0}^{\infty} M'_{m'} \cap M_0 = M \cap M_0 = M_0$$

and see that there is an i_0 such that $M'_{i_0} \cap M_0 = M_0$ and $M_i' \cap M_0 = 0$ for all $i < i_0$ since M_0 is simple. Then

$$0 < M_1/M_0 < M_2/M_0 < \cdots < M_m/M_0 = M/M_0$$

and

$$\begin{aligned} 0 < M_0' < M_1' < \cdots < M'_{i_0-1} < M'_{i_0+1}/M_0 \\ < M'_{i_0+2}/M_0 < \cdots < \bigcup_{m'=0}^{\infty} M'_{m'}/M_0 = M/M_0 \end{aligned}$$

are two series of submodules for M/M_0 with simple successive quotients since $M_i' \cap M_0 = 0$ for $i < i_0$. By the induction hypothesis we see that the series

$$0 < M_0' < M_1' < \cdots < M/M_0$$

is of length $m - 1$ and hence the set of composition factors of the two composition series coincide up to a permutation. □

Suppose M has a composition series

$$0 < M_0 < M_1 < \cdots < M_m = M.$$

We need to prove that M is Noetherian and artinian.

Let

$$0 \leq J_0 \leq J_1 \leq J_2 \leq \cdots \leq M$$

be an increasing sequence of submodules of M. Then

$$0 \leq J_0 \cap M_0 \leq J_0 \cap M_1 \leq J_0 \cap M_2 \leq \cdots \leq J_0 \cap M_m = J_0$$

is an increasing series of submodules of J_0 with successive quotients being 0 or simple. Indeed, for any $i \geq 1$ we have

$$(J_0 \cap M_i)/(J_0 \cap M_{i-1}) \subseteq M_i/M_{i-1}$$

and M_i/M_{i-1} is simple. Proceeding in this way we refine the sequence

$$0 \leq J_0 \leq J_1 \leq J_2 \leq \cdots \leq M$$

to an increasing sequence of submodules with simple successive quotients. Claim 1.6.30 shows that the refined sequence is finite, bounded by the composition length, and hence M is Noetherian.

Let

$$0 < \cdots < J_1 < J_0 = M$$

be a decreasing sequence of submodules of M. If $J_{m+2} \neq 0$, then

$$0 < J_{m+1}/J_{m+2} < J_m/J_{m+2} < J_{m-1}/J_{m+2} < \cdots < M/J_{m+1}$$

is a strictly increasing sequence of submodules of length $m + 2$ with non-zero successive quotients. This contradicts Claim 1.6.30 and hence M is artinian. □

Remark 1.6.31 The occurrence of composition factors in composition series is not unique in general. A simple example is Example 1.6.25. A module with a unique composition series is called *uniserial*.

A ring A with the property that whenever P is an indecomposable direct factor of A, then P has a unique composition series, is called *serial*.

An example of a local serial algebra is given by $K[X]/X^m$ for some integer m and some field K. Another example is $\mathbb{Z}/p^n\mathbb{Z}$ where p is a prime number. Later we will see more occurrences in Sects. 2.8 and 2.12.

We finish this section with a nice property of radicals and socles.

Proposition 1.6.32 *Let A be a K-algebra and let M and N be two artinian A-modules. Then*

$$\text{rad}(M \oplus N) = \text{rad}(M) \oplus \text{rad}(N)$$

and

$$soc(M \oplus N) = soc(M) \oplus soc(N).$$

Proof Since $\text{soc}(M)$ and $\text{soc}(N)$ are semisimple, $\text{soc}(M) \oplus \text{soc}(N)$ is also semisimple, and hence

$$\text{soc}(M) \oplus \text{soc}(N) \leq \text{soc}(M \oplus N).$$

Likewise $M/\text{rad}(M)$ and $N/\text{rad}(N)$ are both semisimple, and hence

$$M/\text{rad}(M) \oplus N/\text{rad}(N) = (M \oplus N)/(\text{rad}(M) \oplus \text{rad}(N))$$

is semisimple. Therefore

$$\text{rad}(M) \oplus \text{rad}(N) \geq \text{rad}(M \oplus N).$$

Denote by $\pi_M : M \oplus N \longrightarrow M$ and by $\pi_N : M \oplus N \longrightarrow N$ the natural projections, and by $\iota_M : M \longrightarrow M \oplus N$ and by $\iota_N : N \longrightarrow M \oplus N$ the natural injections. Let S be a simple submodule of $M \oplus N$. If $\pi_M(S) = 0$ or if $\pi_N(S) = 0$, then S is a submodule of $\text{soc}(M) \oplus \text{soc}(N)$. Since S is simple, $\pi_M(S)$ is either 0 or a submodule of M isomorphic to S. Hence, if $\pi_M(S) \neq 0$, then $\pi_M(S) \subseteq \text{soc}(M)$ and likewise for N. But

$$id_{M\oplus N} = \iota_M \circ \pi_M + \iota_N \circ \pi_N$$

and so we get for any $s \in S$ that

$$s = (\iota_M \circ \pi_M + \iota_N \circ \pi_N)(s) = (\iota_M \circ \pi_M)(s) + (\iota_N \circ \pi_N)(s) \in \text{soc}(M) \oplus \text{soc}(N).$$

Similarly, if U is a maximal submodule of $M \oplus N$, and if $\pi_M(U) = M$ or if $\pi_N(U) = N$, then U contains $\text{rad}(M) \oplus \text{rad}(N)$. Suppose otherwise that $\pi_M(U) \neq M$ and $\pi_N(U) \neq N$. Then $\pi_M(U)$ is maximal in M, since otherwise a larger module $V \neq M$ in M would produce the submodule $\pi_M^{-1}(V)$ of $M \oplus N$, which strictly contains U, and hence is equal to U by maximality of U. Therefore $U \geq \text{rad}(M)$. But then we get

$$U = (\iota_M \circ \pi_M + \iota_N \circ \pi_N)(U) = (\iota_M \circ \pi_M)(U) + (\iota_N \circ \pi_N)(U) \geq \mathrm{rad}(M) \oplus \mathrm{rad}(N).$$

This proves the proposition. □

1.7 Tensor Products

We are going to introduce a fundamental construction, the tensor product, which is absolutely necessary for higher algebra. In particular this is the abstract tool used for induced modules and Mackey's formula, which in turn are the most important tools in the representation theory of finite groups over an arbitrary field.

Furthermore, tensor products are in some respect counterparts to homomorphisms, in a sense which can be made very precise. The precise formulation is given in Lemma 1.7.9 below, which will play an extremely important role in the subsequent material.

1.7.1 The Definition and Elementary Properties

The first concept to introduce is a free abelian group generated by a set.

Definition 1.7.1 An abelian group A is *free on a subset* S of A if for every abelian group B and every mapping $\varphi_S : S \longrightarrow B$ (as a set !) there is a unique homomorphism $\varphi : A \longrightarrow B$ such that the restriction $\varphi|_S$ of φ to S equals φ_S.

Consider the abelian groups $(\mathbb{Z}^n, +)$ for any integer n. The group $(\mathbb{Z}^n, +)$ is a free abelian group on the set

$$\{(1, 0, 0, \dots, 0), (0, 1, 0, \dots, 0), \dots, (0, 0, \dots, 0, 1)\}$$

in the definition below. Indeed, the above set is a basis of $\mathbb{Z}^n$ in the sense that for any abelian group B and any n elements $b_1, b_2, \cdots, b_n$ of B there is a unique homomorphism of abelian groups $\varphi : \mathbb{Z}^n \longrightarrow B$ such that

$$\begin{aligned} \varphi(1, 0, 0, \dots, 0) &= b_1, \\ \varphi(0, 1, 0, \dots, 0) &= b_2, \\ &\cdots\cdots\cdots\cdots \\ \varphi(0, 0, \dots, 0, 1) &= b_n. \end{aligned}$$

$\varphi(a_1, a_2, \dots, a_n) = \sum_{i=1}^n a_i b_i$ is a group homomorphism and is the unique one with the above properties.

Suppose A is a free abelian group on a set S_A, and suppose B is another free abelian group on a set S_B. If S_A and S_B are of the same cardinality (i.e. there is a bijection $\beta : S_A \longrightarrow S_B$) then A and B are isomorphic. Indeed, β defines a unique homomorphism of groups $\widehat{\beta} : A \longrightarrow B$ restricting to β on S_A, and β^{-1} defines a unique homomorphism of groups $\widehat{\beta^{-1}} : A \longrightarrow B$ restricting to β on S_B. Now, the identity on S_A is the unique group homomorphism $A \longrightarrow A$ restricting to the identity on S_A. But, $\widehat{\beta^{-1}} \circ \widehat{\beta}$ is a group homomorphism restricting to $\beta^{-1} \circ \beta = id_{S_A}$ on S_A. Therefore, by the unicity, $\widehat{\beta^{-1}} \circ \widehat{\beta} = id_A$. Analogously, $\widehat{\beta} \circ \widehat{\beta^{-1}} = id_A$. Hence, $\widehat{\beta}$ is an isomorphism.

Until now, we have not shown that free abelian groups exist, except when S is a finite set. Let S be a set of any cardinality. Let

$$F_S := \{f \in Map(S, \mathbb{Z}) |\ f(s) = 0 \text{ for almost all } s \in S\}$$

be the set of those mappings (as a set) of S to the abelian group $(\mathbb{Z}, +)$ which are 0 for all but a finite number of points $s \in S$. This is clearly an abelian group by pointwise addition: $(f+g)(s) := f(s) + g(s)$ for all $f, g \in Map(S, \mathbb{Z})$ and $s \in S$. Define for every $s \in S$ the mapping $f_s \in Map(S, \mathbb{Z})$ by $f_s(t) := 0$ if $s \neq t$ and $f_s(s) = 1$. Let us prove that F_S is free on the set $\{f_s |\ s \in S\}$. Given an abelian group B and a family $(b_s)_{s \in S}$ of B, we want to define a homomorphism of abelian groups $\beta : F_S \longrightarrow B$ so that $\beta(f_s) = b_s$ for all $s \in S$. For every $f \in F_S$ put

$$\beta(f) := \sum_{s \in S,\ f(s) \neq 0} f(s) b_s.$$

Since for every $f \in F_S$ there are only a finite number of $s \in S$ such that $f(s) \neq 0$, the sum is well defined. Let $s \in S$. Then

$$(\beta(f_s))(t) = \sum_{s \in S,\ f_s(t) \neq 0} f_s(t) b_t = b_s.$$

The definition of β implies immediately that β is a group homomorphism and we just showed that $\beta(f_s) = b_s$ for all $s \in S$. Hence, F_S is free on $\{f_s |\ s \in S\}$; and this set is of the same cardinality as S.

We have just proved the following.

Lemma 1.7.2 *For every set S there is a free abelian group F that is free on a set F_S of the same cardinality as S.*

We are ready to define the tensor product.

Definition 1.7.3 Given a ring A, a left A-module M and a right A-module N, take the free abelian group $F_{N \times M}$ on $N \times M$. In this abelian group we look for the smallest subgroup $R_{N \times M; A}$ which contains all the elements

$$\begin{array}{ll}(R1) & (n_1 + n_2, m) - (n_1, m) - (n_2, m)\ ;\ \forall m \in M; n_1, n_2 \in N \\ (R2) & (n, m_1 + m_2) - (n, m_1) - (n, m_2)\ ;\ \forall m_1, m_2 \in M; n \in N \\ (R3) & (nr, m) - (n, rm)\ ;\ \forall r \in A; n \in N; m \in M\end{array}$$

The *tensor product of N and M over A* is

$$N \otimes_A M := F_{N\times M}/R_{N\times M;A}.$$

A brief look at the definition makes it clear that the tensor product is a subtle thing which is not so easy to understand. Also, the tensor product $N \otimes_A M$ is an abelian group, and nothing more in general.

We denote by $n \otimes m$, for $n \in N$ and $m \in M$, the image of (n, m) in $N \otimes_A M$ and call $n \otimes m$ an elementary tensor. The elements of $N \otimes_A M$ are sums of the form $\sum_{i=1}^{k} n_i \otimes m_i$ and there are in general many very different ways to present the same element using the relations in $R_{N\times M;A}$. In addition the quotient $N \otimes_A M = F_{N\times M}/R_{N\times M;A}$ may be 0 even when neither of the modules N or M is 0.

Example 1.7.4 If p and q are relatively prime integers, then

$$\mathbb{Z}/p\mathbb{Z} \otimes_{\mathbb{Z}} \mathbb{Z}/q\mathbb{Z} = 0.$$

Indeed, since $N \otimes_A M$ is generated by $n \otimes m$ for $n \in N$ and $m \in M$, which is the image of (n, m) in $N \times M$, we only need to show that $n \otimes m = 0$ in $\mathbb{Z}/p\mathbb{Z} \otimes_{\mathbb{Z}} \mathbb{Z}/q\mathbb{Z}$, for $n \in \mathbb{Z}/p\mathbb{Z}$ and $q \in \mathbb{Z}/q\mathbb{Z}$. Now, since p and q are relatively prime integers, there are integers x and y such that $1 = px + qy$ by Euclid's algorithm. Hence,

$$\begin{aligned}
n \otimes m &= (n \cdot 1) \otimes m \\
&= n \cdot (px + qy) \otimes m \\
&= (npx \otimes m) + (nqy \otimes m) \text{ by Definition 1.7.3 (R1)} \\
&= (npx \otimes m) + (n \otimes qym) \text{ by Definition 1.7.3 (R3)} \\
&= (0 \otimes m) + (n \otimes 0) \text{ since} n \in \mathbb{Z}/p\mathbb{Z} \text{ and } m \in \mathbb{Z}/q\mathbb{Z}.
\end{aligned}$$

Returning to the general case $N \otimes_A M$ for a moment, $n \otimes 0 = 0 \in N \otimes_A M$ and $0 \otimes m = 0 \in N \otimes_R M$ for all $n \in N$ and $m \in M$. Indeed, denote by 0_N resp. 0_M the zero element in M. Then

$$n \otimes 0_M + n \otimes 0_M = n \otimes (0_M + 0_M) = n \otimes 0_M,$$

and so $n \otimes 0_M = 0_{N\otimes_A M}$. Likewise we see that $0_N \otimes m = 0_{N\otimes_A M}$ for all $n \in N, m \in M$. In particular, we may take $n = 0$ so that $0_N \otimes 0_M = 0_{N\otimes_A M}$.

One would like to know now how it is possible to control these strange objects reasonably well. The main method is via the following universal property.

Proposition 1.7.5 *Given a ring A, a left A-module M and an right A-module N, and an abelian group B. Then for any map (as sets !)* $\varphi : M \times N \longrightarrow B$ *satisfying*

$$\varphi((n_1 + n_2, m)) = \varphi((n_1, m)) + \varphi((n_2, m)) \; ; \; \forall m \in M; n, n_1, n_2 \in N$$
$$\varphi((n, m_1 + m_2)) = \varphi((n, m_1)) + \varphi((n, m_2)) \; ; \; \forall m_1, m_2 \in M; n \in N$$

$$\varphi((nr, m)) = \varphi((n, rm)) \ ; \ \forall r \in A$$

there is a unique homomorphism of abelian groups $\hat{\varphi} : N \otimes_A M \longrightarrow B$ *with* $\hat{\varphi}(n \otimes m) = \varphi(n, m)$. *A mapping* φ *satisfying the above properties is called* A*-balanced.*

Proof Since $F_{N \times M}$ is free there is a unique homomorphism of abelian groups $F_\varphi : F_{N \times M} \longrightarrow B$ extending φ. Since we assumed the above properties of φ, one gets that $F_\varphi(R_{N \times M; A}) = 0$. Therefore, $\hat{\varphi}$ exists with the required properties.

The uniqueness follows from the fact that the elements $n \otimes m$ generate $N \otimes_A M$ and so the images of all other elements in $N \otimes_A M$ are determined. □

A first application is the following Lemma.

Lemma 1.7.6 *Let* R *be a ring, let* N_1 *and* N_2 *be right* R*-modules, and let* M_1 *and* M_2 *be left* R*-modules. Then for every homomorphism* $\alpha : N_1 \longrightarrow N_2$ *of right* R*-modules and every homomorphism* $\beta : M_1 \longrightarrow M_2$ *of left* R*-modules there is a unique homomorphism of abelian groups* $\gamma : N_1 \otimes_R M_1 \longrightarrow N_2 \otimes_R M_2$ *satisfying* $(\alpha \otimes \beta)(n \otimes m) = \alpha(n) \otimes \beta(m)$. *Let* $\gamma =: \alpha \otimes \beta$.

Proof Since in the tensor product $N_2 \otimes_R M_2$ elements satisfy the required relations, Proposition 1.7.5 shows that $\alpha \otimes \beta$ exists and is unique. □

Definition 1.7.7 Let R and S be two rings. Then an *R-S-bimodule* is an $R \times S^{op}$ left module. More explicitly, one has an action of R on the left, of S on the right and that the actions commute in the sense that for every $r \in R, s \in S, m \in M$ one has $(rm)s = r(ms)$.

Lemma 1.7.8 *Let* R*,* S*,* T *and* V *be rings. Let* M *be an* R*-*V*-bimodule, let* N *be an* S*-*R*-bimodule, and let* U *be a* T*-*R*-bimodule. Then*

1. $N \otimes_R M$ *is an* S*-*V*-bimodule satisfying* $s \cdot (n \otimes m) \cdot v = (sn) \otimes (mv)$ *for all* $s \in S, v \in V, n \in N$ *and* $m \in M$.
2. $Hom_R(N, U)$ *is a* T*-*S*-bimodule via the operation defined by* $(t \cdot f \cdot s)(n) := tf(sn)$ *for all* $s \in S, n \in N, t \in T$ *and* $f \in Hom_R(N, U)$.

Proof

1. By symmetry we only need to show the S-module structure. For every $s \in S$ the mapping $\lambda_s : N \times M \longrightarrow N \otimes_R M$ given by $\lambda_s(n, m) := (sn) \otimes m$ is obviously R-balanced just using the defining properties of a module and of the tensor product on the right of the mapping. By Proposition 1.7.5 there is a mapping $\hat{\lambda}_s : N \otimes_R M \longrightarrow N \otimes_R M$ satisfying the required property. Since N is a bimodule, $1_S \cdot (n \otimes m) = n \otimes m$, the operation is distributive and $(s_1 s_2) \cdot (n \otimes m) = s_1 \cdot (s_2 \cdot (n \otimes m))$ for all $s_1, s_2 \in S$ and $n \in N, m \in M$.

2. The fact that $(t \cdot f \cdot s)(n) := tf(sn)$ for all $s \in S$, $n \in N$, $t \in T$ and $f \in Hom_R(N, U)$ defines for all $s \in S$, $t \in T$, $f \in Hom_R(N, U)$ an element in $Hom_R(N, U)$ is clear using that f is R-linear and that N is a bimodule. The fact that this defines a bimodule structure on $Hom_R(N, U)$ is an easy exercise using only the definitions of a bimodule and an R-linear homomorphism.

This proves the lemma. □

Lemma 1.7.9 *Let R, S, T and V be four rings and let M be a T-V-bimodule, N be an S-T-bimodule, N' another S-T-bimodule, U be an R-S-bimodule, and W an R-V-bimodule. Then*

1. *$U \otimes_S (N \otimes_T M) \simeq (U \otimes_S N) \otimes_T M$ where the isomorphism maps $(u \otimes (n \otimes m))$ to $((u \otimes n) \otimes m)$ for all $u \in U$, $n \in N$, $m \in M$.*
2. $$Hom_R(U \otimes_S N, W) \simeq Hom_S(N, Hom_R(U, W))$$

   as T-V-bimodules. The isomorphism restricts to an isomorphism

   $$Hom_{R-V}(U \otimes_S N, W) \simeq Hom_{S-V}(N, Hom_R(U, W))$$

   and to an isomorphism

   $$Hom_{T-R}(U \otimes_S N, W) \simeq Hom_{T-S}(N, Hom_R(U, W)).$$

3. $$U \otimes_S (N \oplus N') \simeq (U \otimes_S N) \oplus (U \otimes_S N')$$

   and

   $$(N \oplus N') \otimes_T M \simeq (N \otimes_T M) \oplus (N' \otimes_T M).$$

Proof

1. We first define for every $u \in U$ a mapping $\lambda_u : N \times M \longrightarrow (U \otimes_S N) \otimes_T M$ by $\lambda_u(n, m) := (u \otimes n) \otimes m$. It is immediate to verify that λ_u is T-balanced, and so it induces a group homomorphism $\lambda_u : N \otimes_T M \longrightarrow (U \otimes_S N) \otimes_T M$. Now, in a second step we define a mapping $U \times (N \otimes_T M) \longrightarrow (U \otimes_S N) \otimes_T M$ by $(u, x) \mapsto \lambda_u(x)$ for every $u \in U$ and $x \in N \otimes_T M$. Here, it might be a little less clear why this map is S-balanced. A proof is as follows. Take $u_1, u_2 \in U$ and $x = \sum_{i=1}^{k} n_i \otimes m_i \in N \otimes_T M$. Then

$$\begin{aligned}\lambda_{u_1+u_2}(x) &= \sum_{i=1}^{k} ((u_1 + u_2) \otimes n_i) \otimes m_i \\ &= \sum_{i=1}^{k} ((u_1 \otimes n_i) \otimes m_i + (u_2 \otimes n_i) \otimes m_i)\end{aligned}$$

$$= \lambda_{u_1}\left(\sum_{i=1}^{k} n_i \otimes m_i\right) + \lambda_{u_2}\left(\sum_{i=1}^{k} n_i \otimes m_i\right)$$
$$= \lambda_{u_1}(x) + \lambda_{u_2}(x).$$

Similarly one shows the other conditions.

2. Define

$$\Psi : Hom_R(U \otimes_S N, W) \longrightarrow Hom_S(N, Hom_R(U, W))$$
$$f \mapsto (n \mapsto (u \mapsto f(u \otimes n)))$$

First, for any $f \in Hom_R(U \otimes_S N, M)$ and any $n \in N$, the mapping $(u \mapsto f(u \otimes n))$ is R-linear. Indeed, $(ru \mapsto f(ru \otimes n))$, but $f(ru \otimes n) = rf(u \otimes n)$. Moreover, $\Psi(f)$ is S-linear, since

$$(\Psi(f)(sn))(u) = f(u \otimes sn) = f(us \otimes n) = (\Psi(f)(n))(us)$$
$$= (s \cdot \Psi(f)(n))(u).$$

Define

$$\Phi : Hom_S(N, Hom_R(U, W)) \longrightarrow Hom_R(U \otimes_S N, W)$$
$$f \mapsto (u \otimes n \mapsto (f(n))(u))$$

We need to show that this is well-defined, i.e. $\Phi(f)$ is S-balanced. But this follows immediately from the fact that f is S-linear. Moreover, Φ is inverse to Ψ.

Now, it is easy to show that Φ maps T-linear maps to T-linear maps, and likewise Ψ. Similarly, Φ maps V-linear maps to V-linear maps, and similarly Ψ.

3. $u \otimes (n + n') \mapsto (u \otimes n) \oplus (u \otimes n')$ defines a mapping $U \otimes_S (N \oplus N') \longrightarrow (U \otimes_S N) \oplus (U \otimes_S N')$ and $(u \otimes n) \oplus (u' \otimes n') \mapsto (u \otimes n) + (u' \otimes n')$ defines a mapping in the other direction. The fact that they are mutually inverse is clear. Associativity in the first variable is proved similarly.

This proves the lemma. □

Lemma 1.7.10 *Let K be a commutative ring and let A be a K-algebra. If M is a right A-module and N is a left A module, then $M \otimes_A N$ is a K-module.*

If K is a field and M and N are finite dimensional over K, then $\dim_K(M \otimes_K N) = \dim_K(M) \cdot \dim_K(N)$. *Moreover, $A^s \otimes_A N \simeq N^s$ as an A left module. In particular $A \otimes_A N \simeq N$ via the mapping satisfying $a \otimes n \mapsto an$ for each $a \in A$ and $n \in N$.*

Proof The fact that $M \otimes_A N$ is a K-module is a consequence of Lemma 1.7.8.

We shall now prove $A^s \otimes_A N \simeq N^s$. By Lemma 1.7.9 (3) it is sufficient to show $A \otimes_A M \simeq M$, i.e. the case $s = 1$. Observe that $A \times M \ni a \times m \mapsto am \in M$ is clearly balanced. Hence it defines a mapping $A \otimes_A M \longrightarrow M$. We shall define an inverse $M \longrightarrow A \otimes_A M$ by putting $m \mapsto 1 \otimes m$ for all $m \in M$. Since $a \otimes m = 1 \otimes am$ for all $a \in A$ and $m \in M$ these two maps are left and right inverses of each other.

If K is a field, by the above one gets $K^n \otimes_K K^m \simeq K^{n \cdot m}$ as K-vector spaces for all $n, m \in \mathbb{N}$. □

Remark 1.7.11 The same proof yields that even for infinite direct sums $(\oplus_{i \in I} A) \otimes_A N \simeq \oplus_{i \in I} N$ for arbitrary index sets I. Here $\oplus_{i \in I} N$ is the set of those mappings $I \longrightarrow N$ such that the image of almost every $i \in I$ is 0.

Note that $(\prod_{i \in I} A) \otimes_A N \neq \prod_{i \in I} N$ in general. Indeed, the definition of the tensor product implies that an element of the left-hand side is a finite linear combination of elementary tensors $x \otimes n$ for $x \in \prod_{i \in I} A$ and $n \in N$. The right-hand side is a direct product of copies of N, whence an a priori infinite product.

Lemma 1.7.12 *Given a commutative ring R and R-algebras A and B, then $A \otimes_R B$ becomes an R-algebra when one puts $(a_1 \otimes b_1) \cdot (a_2 \otimes b_2) = (a_1 a_2 \otimes b_1 b_2)$ for all $a_1, a_2 \in A$ and $b_1, b_2 \in B$.*

Proof In Lemma 1.7.10 we have already seen that $A \otimes_R B$ is an R-module. The additive structure is therefore clear. We need to define the multiplicative structure.

For every $a_1 \in A$ and $b_1 \in B$ define a mapping $A \times B \longrightarrow A \otimes_R B$ by

$$(a_1, b_1) \cdot (a_2, b_2) := (a_1 a_2 \otimes b_1 b_2)$$

for all $a_2 \in A$ and $b_2 \in B$. It is immediate that this is R-balanced and so this defines a mapping

$$A \otimes_R B \longrightarrow A \otimes_R B$$

satisfying $(a_1, b_1) \cdot (a_2 \otimes b_2) = (a_1 a_2 \otimes b_1 b_2)$ for all $a_1, a_2 \in A$ and $b_1, b_2 \in B$. Moreover, it is clear that $(a_1 r, b_1)$ and $(a_1, r b_1)$ induce the same mapping, and also that the mapping induced by $(a_1 + a_1', b_1)$ is the sum of the mappings induced by (a_1, b_1) and (a_1', b_1). Likewise the mapping induced by $(a_1, b_1 + b_1')$ is the sum of the mappings induced by (a_1, b_1) and (a_1, b_1').

Hence, $(a_1 \otimes b_1) \cdot (a_2 \otimes b_2) := (a_1 a_2 \otimes b_1 b_2)$ is well-defined. Now, one has to verify that this gives a structure of an R-algebra. But this is a straightforward verification. □

We come to an application for group rings.

Lemma 1.7.13 *Let R be a commutative ring and let G and H be two groups. Then $R(G \times H) \simeq RG \otimes_R RH$ as R-algebras.*

Proof We define a mapping $\alpha : G \times H \longrightarrow RG \otimes_R RH$ by $\alpha(g, h) := g \otimes h$ for every $(g, h) \in G \times H$. Since $1_G \otimes 1_H$ is obviously the unit in $RG \otimes_R RH$, $\alpha(1_G, 1_H) = 1_{RG \otimes_R RH}$. Moreover,

$$\alpha((g_1, h_1) \cdot (g_2, h_2)) = \alpha(g_1 g_2, h_1 h_2) = g_1 g_2 \otimes h_1 h_2 = (g_1 \otimes h_1) \cdot (g_2 \otimes h_2)$$

$$= \alpha(g_1, h_1) \cdot \alpha(g_2, h_2)$$

and so α induces a homomorphism $R(G \times H) \longrightarrow RG \otimes_R RH$. Inversely, the elements $g \otimes h$ for $g \in G$ and $h \in H$ are a generating set for $RG \otimes_R RH$, since G is an R-basis for RG and H is an R-basis for RH. Then

$$\left(\sum_{g\in G} r_g g, \sum_{h\in H} r_h h\right) \mapsto \sum_{g\in G}\sum_{h\in H} r_g r_h (g, h)$$

is R-balanced and hence induces a mapping $\beta : RG \otimes_R RH \longrightarrow R(G \times H)$. It is clear that β is inverse to α. This proves the lemma. □

Given a commutative ring R, we have seen in Lemma 1.7.12 that for R-algebras A and B the space $A \otimes_R B$ is again an R-algebra.

Lemma 1.7.14 *Let R be a commutative ring and let A and B be R-algebras. Suppose that M is an A-module and N is a B-module. Then the space $M \otimes_R N$ is an $A \otimes_R B$-module.*

Proof Lemma 1.7.8 implies that $M \otimes_R N$ is an A-module. Now, the B-left-module structure of N is a B^{op}-right-module structure. So, $M \otimes_R N$ is an $A \otimes_R B$-module if and only if $M \otimes_R N$ is an A-B^{op}-bimodule. Lemma 1.7.8 implies that the only property to be verified is that the action of A on the left commutes with the action of B^{op} on the right, and that the actions of R inside A and inside B coincide. But since this action is defined by the equations $a \cdot (m \otimes n) = am \otimes n$ and $(m \otimes n) \cdot b = m \otimes bn$, both of the statements are clear. □

We want to look for properties of this module.

Example 1.7.15 Recall Example 1.4.11. Let Q_8 be the quaternion group of order 8, that is the group generated by a and b subject to the relations $a^4 = 1$, $a^2 = b^2 = (ab)^2$ and $a^2 b = ba^2$. This group has order 8. Its elements are $1, a, a^2, a^3, b, b^3, (ab), (ab)^3$. The elements a, b and (ab) are all of order 4, and the element a^2 is central of order 2.

Let $\mathbb{H}_{\mathbb{R}}$ be the real quaternion algebra.This is a 4-dimensional real vector space with basis 1, i, j and k. This vector space is a skew-field when one defines 1 to be the neutral element of the multiplication, $i \cdot j = k$ and $i^2 = j^2 = k^2 = -1$ (cf Example 1.4.11).

Then $\mathbb{H}_{\mathbb{R}}$ is an $\mathbb{R}Q_8$-module by identifying a with i and b with j. The relations in $\mathbb{H}_{\mathbb{R}}$ are exactly the same as the relations in Q_8. Moreover, using Schur's Lemma 1.4.9, $\mathbb{H}_{\mathbb{R}}$ is simple as an $\mathbb{R}Q_8$-module since $\mathbb{H}_{\mathbb{R}}$ is a skew-field and since Q_8 acts by multiplication on the left, $End_{\mathbb{R}Q_8}(\mathbb{H}_{\mathbb{R}}) \simeq \mathbb{H}_{\mathbb{R}}$. Therefore, since by Maschke's Theorem 1.2.8 the group algebra $\mathbb{R}Q_8$ is semisimple, and since $Q_8/Z(Q_8) \simeq C_2 \times C_2$, we get

$$\mathbb{R}Q_8 \simeq \mathbb{R} \times \mathbb{R} \times \mathbb{R} \times \mathbb{R} \times \mathbb{H}.$$

Hence

$$\begin{aligned}\mathbb{C}Q_8 &\simeq (\mathbb{C}\otimes_{\mathbb{R}}\mathbb{R})\times(\mathbb{C}\otimes_{\mathbb{R}}\mathbb{R})\times(\mathbb{C}\otimes_{\mathbb{R}}\mathbb{R})\times(\mathbb{C}\otimes_{\mathbb{R}}\mathbb{R})\times(\mathbb{C}\otimes_{\mathbb{R}}\mathbb{H}_{\mathbb{R}})\\ &\simeq \mathbb{C}\times\mathbb{C}\times\mathbb{C}\times\mathbb{C}\times(\mathbb{C}\otimes_{\mathbb{R}}\mathbb{H}_{\mathbb{R}})\simeq\mathbb{C}\times\mathbb{C}\times\mathbb{C}\times\mathbb{C}\times Mat_{2\times 2}(\mathbb{C})\end{aligned}$$

where the last isomorphism comes from the fact that $\mathbb{C}$ is algebraically closed, Corollary 1.4.18, Corollary 1.4.19 and since Q_8 is non-commutative. Hence,

$$\mathbb{C}\otimes_{\mathbb{R}}\mathbb{H}_{\mathbb{R}}\simeq Mat_{2\times 2}(\mathbb{C}).$$

But, $Mat_{2\times 2}(\mathbb{C})$ is not simple as a $\mathbb{C}Q_8$-module, but rather semisimple as a direct sum of two isomorphic simple factors.

We can give an explicit isomorphism as follows. Define

$$\alpha:\mathbb{H}_{\mathbb{R}}\otimes_{\mathbb{R}}\mathbb{C}\longrightarrow Mat_{2\times 2}(\mathbb{C})$$

$$1\otimes 1\mapsto\begin{pmatrix}1&0\\0&1\end{pmatrix}\qquad\qquad 1\otimes i\mapsto\begin{pmatrix}i&0\\0&i\end{pmatrix}$$

$$i\otimes 1\mapsto\begin{pmatrix}i&0\\0&-i\end{pmatrix}\qquad\qquad j\otimes 1\mapsto\begin{pmatrix}0&1\\-1&0\end{pmatrix}$$

and extend this multiplicatively and $\mathbb{R}$-bilinearly. It is straightforward to check that α is then a ring homomorphism. Moreover, α is injective since the 16 matrices corresponding to the image of the elements $g\otimes i$ for $g\in Q_8$ and $g\otimes 1$ are $\mathbb{R}$-linearly independent, as is easily checked.

This example shows that if A and B are K-algebras, if M is a simple A-module and if N is a simple B-module, then $M\otimes_K N$ is not necessarily a simple $A\otimes_K B$-module. Nevertheless, one can prove that it is always semisimple when M and N are finite dimensional over K and the algebras satisfy an additional hypothesis, separability. This additional hypothesis is always satisfied for algebraically closed fields K. We refer to Lemma 5.3.8 below.

Example 1.7.16 Let $G=C_3\times C_3$. Then $\mathbb{Q}G\simeq\mathbb{Q}C_3\otimes_{\mathbb{Q}}\mathbb{Q}C_3$ and $\mathbb{Q}C_3\simeq\mathbb{Q}\times\mathbb{Q}(\zeta)$ for a primitive 3-rd root of unity ζ. Hence $\mathbb{Q}(C_3\times C_3)$ is isomorphic to a direct product of 1 copy of $\mathbb{Q}$, of 2 copies of $\mathbb{Q}(\zeta)$ and of one copy of $\mathbb{Q}(\zeta)\otimes_{\mathbb{Q}}\mathbb{Q}(\zeta)$. Let c_1 be a generator of the first copy of C_3 and let c_2 be a generator of the second copy of C_3. The module $\mathbb{Q}(\zeta)$ becomes a $\mathbb{Q}G$-module in various ways. The element c_1 can act as multiplication by ζ^n and the element c_2 can act as multiplication with ζ^m. We denote the module obtained in this way by ${}_n\mathbb{Q}(\zeta)_m$. We first observe that for each $m\in\{1,2\}$ we get that ${}_0\mathbb{Q}(\zeta)_m$ is not isomorphic to ${}_m\mathbb{Q}(\zeta)_0$ and neither of these two is isomorphic to ${}_n\mathbb{Q}(\zeta)_m$ for $n,m\in\{1,2\}$. Indeed, let $\varphi:{}_0\mathbb{Q}(\zeta)_m\longrightarrow{}_k\mathbb{Q}(\zeta)_n$ be an isomorphism for $n,m\in\{0,1,2\}$ and $k\in\{1,2\}$, then $\varphi(x)=\varphi(c_1x)=c_1\cdot\varphi(x)=\zeta^k\varphi(x)$. Hence $k=0$. Likewise ${}_m\mathbb{Q}(\zeta)_0$ is not isomorphic to ${}_n\mathbb{Q}(\zeta)_k$ for any $n,m\in\{0,1,2\}$ and $k\in\{1,2\}$. We see that complex conjugation gives isomorphisms ${}_1\mathbb{Q}(\zeta)_1\simeq{}_2\mathbb{Q}(\zeta)_2$, ${}_1\mathbb{Q}(\zeta)_2\simeq{}_2\mathbb{Q}(\zeta)_1$, ${}_0\mathbb{Q}(\zeta)_1\simeq{}_0\mathbb{Q}(\zeta)_2$ and ${}_1\mathbb{Q}(\zeta)_0\simeq{}_2\mathbb{Q}(\zeta)_0$. Suppose

$\varphi : {}_1\mathbb{Q}(\zeta)_1 \longrightarrow {}_1\mathbb{Q}(\zeta)_2$ is an isomorphism. Then

$$\varphi(1)\zeta^2 = \varphi(1) \cdot c_2 = \varphi(1 \cdot c_2) = \varphi(\zeta) = \varphi(c_1 \cdot 1) = c_1\varphi(1) = \zeta\varphi(1),$$

which is a contradiction since $\varphi(1)$ is invertible and $\mathbb{Q}(\zeta)$ is commutative. Therefore

$$\mathbb{Q}(C_3 \times C_3) \simeq \mathbb{Q} \times {}_0\mathbb{Q}(\zeta)_1 \times {}_1\mathbb{Q}(\zeta)_0 \times {}_1\mathbb{Q}(\zeta)_1 \times {}_1\mathbb{Q}(\zeta)_2$$

as both sides have the same dimension over $\mathbb{Q}$, and since we know that all the modules on the right are mutually non-isomorphic. Since $\mathbb{Q}C_3 \simeq \mathbb{Q} \times \mathbb{Q}(\zeta)$ we deduce the decomposition

$$\mathbb{Q}(C_3 \times C_3) \simeq \mathbb{Q} \times {}_0\mathbb{Q}(\zeta)_1 \times {}_1\mathbb{Q}(\zeta)_0 \times {}_1\mathbb{Q}(\zeta) \otimes_{\mathbb{Q}} \mathbb{Q}(\zeta)_1$$

and therefore

$$\mathbb{Q}(\zeta) \otimes_{\mathbb{Q}} \mathbb{Q}(\zeta) \simeq {}_1\mathbb{Q}(\zeta)_1 \times {}_1\mathbb{Q}(\zeta)_2$$

decomposes as a product of two non-isomorphic simple modules.

Example 1.7.17 I learned the following example from Dugas and Martinez-Villa [9]. Let $\mathbb{F}_p := \mathbb{Z}/p\mathbb{Z}$ for some prime p, and let $k := \mathbb{F}_p(X)$ be the field of rational functions in one variable X with coefficients in $\mathbb{F}_p$. Then $K := k[T]/(T^p - X)$ is an (inseparable) extension field of k. We compute

$$\begin{aligned} K \otimes_k K &\simeq k[T_1, T_2]/(T_1^p - X, T_2^p - X) \\ &\simeq k[T_1, T_2]/(T_1^p - T_2^p, T_2^p - X) \\ &\simeq k[T_1, T_2]/((T_1 - T_2)^p, T_2^p - X) \\ &\simeq k[T_1 - T_2, T_2]/((T_1 - T_2)^p, T_2^p - X) \\ &\simeq k[T, T_2]/(T^p, T_2^p - X) \\ &\simeq K[T]/(T^p) \end{aligned}$$

which is not a semisimple algebra.

We shall study the endomorphism algebra of such a tensor product $M \otimes_K N$ as an $A \otimes_K B$-module where M is an A-module and N is a B-module.

Lemma 1.7.18 *Let K be a field and let A and B be K-algebras. Suppose that M_1, M_2 are A-modules and N_1, N_2 are B-modules. Then there is a homomorphism of vector spaces*

$$Hom_A(M_1, M_2) \otimes_K Hom_B(N_1, N_2) \to Hom_{A\otimes_K B}(M_1 \otimes_K N_1, M_2 \otimes_K N_2).$$

If $A = B = K$, then this homomorphism is an isomorphism.

Proof First we get a homomorphism of K-modules

$$\Phi : Hom_A(M_1, M_2) \otimes_K Hom_B(N_1, N_2) \to Hom_{A\otimes_K B}(M_1 \otimes_K N_1, M_2 \otimes_K N_2)$$

by Lemma 1.7.6, defining $\Phi(\alpha \otimes \beta) := \alpha \otimes \beta$.

Suppose now $A = B = K$, and therefore $A \otimes_K B = K$. Choosing K-bases B_{M_1} of M_1, B_{M_2} of M_2 and B_{N_1} of N_1, B_{N_2} of N_2, then $B_{M_1} \otimes B_{M_2} := \{b_1 \otimes b_2 | \; b_1 \in B_{M_1}; b_2 \in B_{M_2}\}$ is a basis for $M_1 \otimes_K M_2$ and likewise $B_{N_1} \otimes B_{N_2}$ is a basis for $N_1 \otimes_K N_2$. We obtain a basis for $Hom_K(M_1, M_2)$ by defining for each $b_{M_1} \in B_{M_1}$ and each $b_{M_2} \in B_{M_2}$ the mappings $f_{b_{M_1}}^{b_{M_2}}$ by

$$f_{b_{M_1}}^{b_{M_2}}(b'_{M_1}) = \begin{cases} b_{M_2} \text{ if } b_{M_1} = b'_{M_1} \\ 0 \;\text{ if } b'_{M_1} \in B_{M_1} \setminus \{b_{M_1}\}. \end{cases}$$

Similar constructions can be carried out for the vector space $Hom_K(N_1, N_2)$ and for the vector space $Hom_K(M_1 \otimes_K N_1, M_2 \otimes_K N_2)$. If

$$\Phi\left(\sum_{b_{M_1}\in B_{M_1}, b_{N_1}\in B_{N_1}, b_{M_2}\in B_{M_2}, b_{N_2}\in B_{N_2}} \lambda_{b_{M_1},b_{N_1}}^{b_{M_2},b_{N_2}} \left(f_{b_{M_1}}^{b_{M_2}} \otimes f_{b_{N_1}}^{b_{N_2}}\right)\right) = 0,$$

then $\lambda_{b_{M_1},b_{N_1}}^{b_{M_2},b_{N_2}} = 0$ for all indices and Φ is injective. Surjectivity follows from a comparison of the dimensions. □

Lemma 1.7.19 *Let K be a field, let G_1 and G_2 be two finite groups and suppose that the order of G_1 and of G_2 are invertible in K. Suppose K is a splitting field for G_1 and for G_2. Then for each simple $K(G_1 \times G_2)$-module T there is a simple KG_1-module S_1 and a simple KG_2-module S_2 such that T is isomorphic to $S_1 \otimes_K S_2$. Moreover, for each simple KG_1-module S_1 and each simple KG_2-module S_2 the $K(G_1 \times G_2)$-module $S_1 \otimes_K S_2$ is simple.*

Proof Let T be a simple $K(G_1 \times G_2)$-module. By Lemma 1.7.13 we obtain $K(G_1 \times G_2) \simeq KG_1 \otimes_K KG_2$ and we get an isomorphism

$$\begin{aligned} Hom_K(S_1 \otimes_K S_2, S_1 \otimes_K S_2) &= Hom_{K\otimes_K K}(S_1 \otimes_K S_2, S_1 \otimes_K S_2) \\ &= Hom_K(S_1, S_1) \otimes Hom_K(S_2, S_2). \end{aligned}$$

Now, $G_1 \times G_2$ acts on the spaces on the left as a group acting on $S_1 \otimes_K S_2$, and on each of the spaces $End_K(S_1)$ and $End_K(S_2)$ on the right accordingly. The morphism of Lemma 1.7.18 commutes with this action and so one may take invariants on both sides, i.e. one considers the largest sub-module with trivial $G_1 \times G_2$-action. These are the $K(G_1 \times G_2)$ linear endomorphisms on the left and the tensor product of the KG_1-linear endomorphisms with the KG_2-linear endomorphisms on the right. Hence

$$Hom_{K(G_1\times G_2)}(S_1 \otimes_K S_2, S_1 \otimes_K S_2) = Hom_{KG_1}(S_1, S_1) \otimes Hom_{KG_2}(S_2, S_2).$$

But, on the right we get $Hom_{KG_1}(S_1, S_1) = K = Hom_{KG_2}(S_2, S_2)$ and so $S_1 \otimes_K S_2$ is simple.

The number of conjugacy classes of $G_1 \times G_2$ equals the product of the number of conjugacy classes of G_1 and of G_2. Hence each simple $K(G_1 \times G_2)$-module is of the form $S_1 \otimes_K S_2$ for simple KG_i-modules S_i, $i \in \{1, 2\}$. This proves the statement.□

1.7.2 Immediate Applications for Group Rings

If E is an extension field of K, A is a K-algebra and M is an A-module, then by Lemma 1.7.12 $E \otimes_K A$ is again an algebra. Moreover, by Lemma 1.7.8 the tensor product $E \otimes_K A$ is an E-vector space. The multiplication inside $E \otimes_K A$ implies that the elements $e \otimes 1$ for $e \in E$ are all central, and hence $E \otimes_K A$ is an E-algebra. The module $E \otimes_K M$ is an $E \otimes_K A$-module by Lemma 1.7.14. This procedure is called a "change of rings", meaning change of base rings. We observe that the discussion in Sect. 1.4.3 is precisely a down-to-earth description of this abstract "change of rings".

The second occasion where tensor products are used is a generalisations of this: induction of modules. We start with the observation that given a commutative ring R, an R-algebra A and an R-subalgebra B, the ring B acts on A by right multiplication. In other words, restricting the action of A on the regular right A-module A to B gives A the structure of a right B-module. Also, A can be regarded as a regular left A-module. Both actions commute. That is, A is an A-B-bimodule if we define $a \cdot x \cdot b := axb$ for all $a, x \in A$ and $b \in B$.

Definition 1.7.20 Let G be a group, let R be a commutative ring and let H be a subgroup of G. Then for every RH-module M one defines the *induced module* $M \uparrow_H^G$ to be $RG \otimes_{RH} M$ where for all $g \in G$ and all $x \in RG$, $m \in M$ we define $g \cdot (x \otimes m) := (gx) \otimes m$.

The induced module is of extreme importance in representation theory.

Remark 1.7.21 Tensor products also appear in more sophisticated structures, so-called Hopf-algebras. For a very brief introduction see Sect. 6.2.1 or e.g. Montgomery's book [10]. Group rings are Hopf algebras. For a commutative ring K and a group G as well as two KG-modules M_1 and M_2 we get that $M_1 \otimes_K M_2$ is a also KG-module by putting $g \cdot (m_1 \otimes m_2) := (gm_1 \otimes gm_2)$ for all $g \in G$ and $m_1 \in M_1$, $m_2 \in M_2$. Indeed, for all $g \in G$ the mapping

$$\begin{aligned} g\cdot \;\; : M_1 \times M_2 &\longrightarrow M_1 \otimes_K M_2 \\ (m_1, m_2) &\mapsto gm_1 \otimes gm_2 \end{aligned}$$

is K-balanced and so it induces a mapping

$$g\cdot \;\; : M_1 \otimes_K M_2 \longrightarrow M_1 \otimes_K M_2$$

$$m_1 \otimes m_2 \mapsto gm_1 \otimes gm_2$$

which in turn induces a K-linear action of G on $M_1 \otimes_K M_2$, since $1_G \cdot = id_{M_1 \otimes_K M_2}$ and $(g_1 g_2)\cdot = (g_1 \cdot) \circ (g_2 \cdot)$ for all $g_1, g_2 \in G$ as is immediately checked.

However, a warning should be given here. Elements $g_1 + g_2$ in KG *do not* act diagonally on $M_1 \otimes_K M_2$. In fact, for $m_1 \in M_1$ and $m_2 \in M_2$ we have

$$\begin{aligned}(g_1 + g_2) \cdot (m_1 \otimes m_2) &= g_1 \cdot (m_1 \otimes m_2) + g_2 \cdot (m_1 \otimes m_2) \\ &= (g_1 m_1 \otimes g_1 m_2) + (g_2 m_1 \otimes g_2 m_2)\end{aligned}$$

whereas

$$\begin{aligned}(g_1 + g_2)m_1 \otimes (g_1 + g_2)m_2 &= (g_1 m_1 \otimes g_1 m_2) + (g_1 m_1 \otimes g_2 m_2) \\ &\quad + (g_2 m_1 \otimes g_1 m_2) + (g_2 m_1 \otimes g_2 m_2).\end{aligned}$$

We now continue to examine induced modules.

Remark 1.7.22 Let us give the action of G on $M \uparrow_H^G$ more explicitly. Let $G = \bigcup_{i \in I} g_i H$ be a disjoint union of cosets. Recall that the cardinality of I is the index of H in G. Then for all $g \in G$ and all $i \in I$ we have $gg_i \in g_{j(g,i)}H$. This means that there is an $h(g,i) \in H$ and a $j(g,i) \in I$ such that $gg_i = g_{j(g,i)}h(g,i)$ for all $i \in I, g \in G$. Now we have for all $m \in M$

$$g \cdot (g_i \otimes m) = (gg_i) \otimes m = g_{j(g,i)}h(g,i) \otimes m = g_{j(g,i)} \otimes h(g,i)m.$$

Particular cases are simpler, for example if M is a trivial KH-module K, then $g \cdot (g_i \otimes m) = g_{j(g,i)} \otimes m$ and so G acts by permuting the basis $g_i \otimes 1$ of $K \uparrow_H^G$.

Lemma 1.7.23 *If K is a field, if G is a group, if H is a subgroup of G and if M is a finite dimensional KG-module, then $M \uparrow_H^G$ is finite dimensional if and only if H is of finite index in G, that is there are only finitely many cosets $g_1H, g_2H, \ldots, g_nH$ in G/H.*

Proof If H is of finite index in G, then $G = \bigcup_{i \in I}^n g_i H$ for certain elements $g_i \in G$; $i \in I$, so that $g_i H = g_j H \Rightarrow i = j$. But then KG as a KH-right module is just isomorphic to $\bigoplus_{i \in I} KH$. Hence, by Lemma 1.7.10 and Remark 1.7.11 we get the following isomorphisms of vector spaces:

$$(\dagger) : KG \otimes_{KH} M \simeq \bigoplus_{i \in I} KH \otimes_{KH} M \simeq \bigoplus_{i \in I} M.$$

So $M \uparrow_H^G$ is finite dimensional if and only if I is finite and M is finite dimensional. Hence, $M \uparrow_H^G$ is finite dimensional if and only if M is finite dimensional and H is of finite index in G. □

Corollary 1.7.24 *If K is a field, if G is a group, if H is a subgroup of G of finite index and if M is a finite dimensional KG-module, then*

$$\dim_K(M \uparrow_H^G) = \dim_K(M) \cdot |G : H|.$$

Proof This follows immediately from the formula (†). □

We observe that in the extreme case of $H = G$ then $M \uparrow_G^G \simeq M$. This is clear since $M \uparrow_G^G = KG \otimes_{KG} M \simeq M$ by Lemma 1.7.10.

An important property of induced modules is that one can also induce morphisms.

Definition 1.7.25 Let G be a group, let H be a subgroup, let R be a commutative ring and let M and N be RH-modules. Define for every $\alpha \in Hom_{RH}(M, N)$ the morphism $id_{RG} \otimes \alpha =: \alpha \uparrow_H^G$.

We will see that $\alpha \uparrow_H^G$ is G-linear.

Lemma 1.7.26 *Let G be a group with subgroup H, let R be a commutative ring and let M and N be RH-modules. Then for every $\alpha \in Hom_{RH}(M, N)$ we have $\alpha \uparrow_H^G \in Hom_{RG}(M \uparrow_H^G, N \uparrow_H^G)$.*

Proof Let again $G = \bigcup_{i \in I}^n g_i H$ be a decomposition into cosets. We compute for all $g \in G, i \in I$ and $m \in M$

$$\begin{aligned} g \cdot \left(\alpha \uparrow_H^G (g_i H \otimes m)\right) &= g \cdot (g_i H \otimes \alpha(m)) = (g g_i H) \otimes \alpha(m) \\ &= \alpha \uparrow_H^G (g g_i H \otimes m) = \alpha \uparrow_H^G (g(g_i H \otimes m)). \end{aligned}$$

This proves the lemma. □

In characteristic 0 all KG-modules can be reached by induction. This is the statement of the next lemma.

Lemma 1.7.27 *Let G be a finite group, let H be a subgroup and let K be a field in which $|H|$ is invertible. Then for every indecomposable KG-module N which is a direct factor of the regular KG-module, there is a simple KH-module M such that N is a direct factor of $M \uparrow_H^G$.*

Proof First, if $M = M_1 \oplus M_2$ for KH-modules M, M_1 and M_2, then

$$M \uparrow_H^G = (M_1 \oplus M_2) \uparrow_H^G \simeq M_1 \uparrow_H^G \oplus M_2 \uparrow_H^G$$

by Lemma 1.7.9 and so, if N is an indecomposable KG-module and if N is a direct factor of $M \uparrow_H^G$ for some KH-module M, then there is an indecomposable KH-module M' such that N is a direct factor of $M' \uparrow_H^G$. Indecomposable modules are simple by Maschke's Theorem 1.2.8 since $|H|$ is invertible in K.

Now, $KH \uparrow_H^G = KG \otimes_{KH} KH \simeq KG$ by Lemma 1.7.10 and so the regular H-module induced to G is the regular G-module. Since N is a direct factor of the regular module, we have proved the lemma. □

Remark 1.7.28 It is important to note that this does not mean that every KG-module is isomorphic to a module of the form $M \uparrow_H^G$ for some subgroup H of G and a KH-module M. One might need to pass to quotients and submodules in order to get a specific KG-module M. This remark will be crucial in the case when the order of the group is not invertible in K.

Lemma 1.7.29 *Let R be a commutative ring, let G be a group and let H be a subgroup of G and K be a subgroup of H. Then for every two RK-modules M and N one has*

$$\left(M \uparrow_K^H\right) \uparrow_H^G \simeq M \uparrow_K^G$$

and

$$(M \oplus N) \uparrow_H^G \simeq M \uparrow_H^G \oplus N \uparrow_H^G .$$

Proof Using Lemma 1.7.9 we have

$$M \uparrow_K^H \uparrow_H^G \simeq RG \otimes_{RH} (RH \otimes_{RK} M) \simeq (RG \otimes_{RH} RH) \otimes_{RK} M \simeq RG \otimes_{RK} M.$$

The second affirmation is proved in a similar way. □

Example 1.7.30 Let $\mathfrak{S}_n$ be the symmetric group on n letters $\{1, 2, \ldots, n\}$. The group $\mathfrak{S}_n$ acts on $\{1, 2, \ldots, n\}$ and one gets that $\mathfrak{S}_{n-1}$ is a subgroup of $\mathfrak{S}_n$ by just considering those permutations of $\{1, 2, \ldots, n\}$ which fix the last point n. In other words, $Stab_{\mathfrak{S}_n}(n) \simeq \mathfrak{S}_{n-1}$, where $Stab_G(S)$ denotes the stabiliser of the subset S of Ω in the group G acting on Ω. Let K be the trivial $\mathfrak{S}_n$-module, consider it as a trivial $\mathfrak{S}_{n-1}$-module simply by restricting the action of $\mathfrak{S}_n$ to $\mathfrak{S}_{n-1}$, and construct $M := K \uparrow_{\mathfrak{S}_{n-1}}^{\mathfrak{S}_n}$. Then M is of dimension n over K. We apply Lemma 1.7.9 to this situation and obtain

$$\begin{aligned} Hom_{K\mathfrak{S}_n}(M, K) &= Hom_{K\mathfrak{S}_n}(K\mathfrak{S}_n \otimes_{K\mathfrak{S}_{n-1}} K, K) \\ &\simeq Hom_{K\mathfrak{S}_{n-1}}(K, Hom_{K\mathfrak{S}_n}(K\mathfrak{S}_n, K)) \\ &\simeq Hom_{K\mathfrak{S}_{n-1}}(K, K) \\ &= K \end{aligned}$$

where the second isomorphism is Lemma 1.7.9 and the third isomorphism comes from the following fact. For every algebra A and every A-module V one has $Hom_A(A, V) \simeq V$ as an A-module, using the A-A-bimodule structure of A by left and right multiplication. The isomorphism is given by $Hom_A(A, V) \ni f \mapsto f(1) \in V$. Hence, the trivial $K\mathfrak{S}_n$-module K is a quotient of M, and in the case when $n!$ is invertible in K, a direct factor of M. Suppose $n!$ is invertible in K. Then we have $M \simeq K \oplus N$ for another $K\mathfrak{S}_n$-module N of dimension $n - 1$. Moreover, the trivial module is not a direct factor of N. Is N simple? In order to decide this we shall use Lemma 1.7.9 again.

$$\begin{aligned} Hom_{K\mathfrak{S}_n}(M,M) &= Hom_{K\mathfrak{S}_n}(K\mathfrak{S}_n \otimes_{K\mathfrak{S}_{n-1}} K, K\mathfrak{S}_n \otimes_{K\mathfrak{S}_{n-1}} K) \\ &\simeq Hom_{K\mathfrak{S}_{n-1}}(K, Hom_{K\mathfrak{S}_n}(K\mathfrak{S}_n, K\mathfrak{S}_n \otimes_{K\mathfrak{S}_{n-1}} K)) \\ &\simeq Hom_{K\mathfrak{S}_{n-1}}(K, K\mathfrak{S}_n \otimes_{K\mathfrak{S}_{n-1}} K) \end{aligned}$$

and we know that the dimension of this space is at least 2. The dimension is 2 if and only if N is simple. Indeed,

$$\begin{aligned} Hom_{K\mathfrak{S}_n}(M,M) &= Hom_{K\mathfrak{S}_n}(K \oplus N, K \oplus N) \\ &= \begin{pmatrix} Hom_{K\mathfrak{S}_n}(K,K) & Hom_{K\mathfrak{S}_n}(N,K) \\ Hom_{K\mathfrak{S}_n}(K,N) & Hom_{K\mathfrak{S}_n}(N,N) \end{pmatrix} \\ &= \begin{pmatrix} K & 0 \\ 0 & Hom_{K\mathfrak{S}_n}(N,N) \end{pmatrix} \end{aligned}$$

and so, since $Hom_{K\mathfrak{S}_n}(N,N)$ contains at least the identity,

$$\dim_K(Hom_{K\mathfrak{S}_n}(M,M)) = 2 \Rightarrow N \text{ is simple.}$$

Recall $End_{K\mathfrak{S}_n}(M) \simeq Hom_{K\mathfrak{S}_{n-1}}(K, K\mathfrak{S}_n \otimes_{K\mathfrak{S}_{n-1}} K)$ and hence we need to examine $Hom_{K\mathfrak{S}_{n-1}}(K, K\mathfrak{S}_n \otimes_{K\mathfrak{S}_{n-1}} K)$, or what is the same, to determine how many copies of the trivial $K\mathfrak{S}_{n-1}$-module are a submodule of the induced module $K\mathfrak{S}_n \otimes_{K\mathfrak{S}_{n-1}} K$, considered as a $K\mathfrak{S}_{n-1}$ left module. Now, for every $m \in \{1, 2, \ldots, n-1\}$ call the 2-cycle $\sigma_m = (m\ n)$ the element in $\mathfrak{S}_n$ that interchanges m with n in $\{1, 2, \ldots, n\}$. Then one computes $\sigma_m \sigma_k = (k\ n\ m) \notin \mathfrak{S}_{n-1}$. Hence, we can write $\mathfrak{S}_n$ as a union of left cosets modulo $\mathfrak{S}_{n-1}$:

$$\mathfrak{S}_n = \mathfrak{S}_{n-1} \cup \bigcup_{m=1}^{n-1} \sigma_m \mathfrak{S}_{n-1}.$$

By Remark 1.7.22 one gets that $\mathfrak{S}_{n-1}$ acts as permutation of the basis $\{\sigma_m \mathfrak{S}_{n-1} \otimes 1 \mid m \in \{1, 2, \ldots, n-1\}\} \cup \{\mathfrak{S}_{n-1}\}$ by left multiplication with elements of $\mathfrak{S}_{n-1}$.

For a group G and a field K in which $|G|$ is invertible, we form $e := \frac{1}{|G|} \sum_{g \in G} g$. If M is a KG-module, then $e \cdot M$ is a KG-submodule of M on which G acts trivially. Conversely, if T is a trivial submodule of M, then $e \cdot T = T$.

Hence, in order to find the trivial submodule, one needs to determine the $K\mathfrak{S}_{n-1}$-module

$$\left(\frac{1}{|\mathfrak{S}_{n-1}|} \sum_{\sigma \in \mathfrak{S}_{n-1}} \sigma \right) \cdot K\mathfrak{S}_n \otimes_{K\mathfrak{S}_{n-1}} K.$$

Put $e_1 := \left(\frac{1}{|\mathfrak{S}_{n-1}|}\sum_{\sigma\in\mathfrak{S}_{n-1}}\sigma\right)$. Since every element σ in $\mathfrak{S}_{n-1}$ has the property that $\sigma\mathfrak{S}_{n-1} = \mathfrak{S}_{n-1}$ it is clear that

$$e_1 \cdot \left(K\mathfrak{S}_{n-1} \otimes_{K\mathfrak{S}_{n-1}} 1_K\right) = K\mathfrak{S}_{n-1} \otimes_{K\mathfrak{S}_{n-1}} 1_K.$$

Moreover, since $(k\ m)\cdot(m\ n)\cdot(k\ m)^{-1} = (k\ n)$ for all $m, k \in \{1, 2, \dots, n-1\}$ one has

$$e_1 \cdot \left(\sigma_m \otimes_{K\mathfrak{S}_{n-1}} 1_K\right) = \sum_{m=1}^{n-1} \sigma_m \otimes_{K\mathfrak{S}_{n-1}} 1_K.$$

Therefore, $\dim_K(Hom_{K\mathfrak{S}_n}(M, M)) = 2$, in particular it is generated by the two elements $1 \otimes_{K\mathfrak{S}_{n-1}} 1_K$ and $\sum_{m=1}^{n-1} \sigma_m \otimes_{K\mathfrak{S}_{n-1}} 1_K$.

We have shown that if $n!$ is invertible in K, then $K \uparrow_{\mathfrak{S}_{n-1}}^{\mathfrak{S}_n} \simeq K \oplus N$ for a simple $K\mathfrak{S}_n$-module N and the trivial module K. This completes Example 1.2.12.

The methods of the example generalise to the class of all finite groups.

Definition 1.7.31 Let G be a group and let H be a subgroup. Suppose that R is a commutative ring. Then for any RG-module M denote by $M \downarrow_H^G$ the RH-module whose underlying R-module structure is the same as M, and H acts on M as G does. In other words, the group homomorphism $H \longrightarrow Aut_R(M)$ which gives M the structure of an RH-module is the composition of the embedding $H \longrightarrow G$ and the group homomorphism $G \longrightarrow Aut_R(M)$ which gives M the structure of an RG-module. We call $M \downarrow_H^G$ the *restriction of M to H*.

Using this we get the following statement.

Lemma 1.7.32 (Frobenius reciprocity) *Let G be a group and let H be a subgroup of G. Suppose R is a commutative ring. Then for every RG-module M and every RH-module N we get $Hom_{RG}(N \uparrow_H^G, M) \simeq Hom_{RH}(N, M \downarrow_H^G)$, natural in N and M in the sense that the isomorphism commutes with maps induced from RG-homomorphisms $M \to M'$ and RH-homomorphisms $N \to N'$.*

Proof The main ingredient is Lemma 1.7.9.

$$\begin{aligned} Hom_{RG}(N \uparrow_H^G, M) &\simeq Hom_{RG}(RG \otimes_{RH} N, M) \\ &\simeq Hom_{RH}(N, Hom_{RG}(RG, M)) \\ &\simeq Hom_{RH}(N, M \downarrow_H^G). \end{aligned}$$

This proves the lemma. □

Corollary 1.7.33 *Let G be a finite group and let H be a subgroup of G. Suppose K is a field in which $|G|$ is invertible. Then for every finite dimensional KG-module M and every finite dimensional KH-module N we have $Hom_{KG}(M, N \uparrow_H^G) \simeq Hom_{KH}(M \downarrow_H^G, N)$.*

Proof This is a consequence of the fact that Wedderburn's theorem implies that for finite dimensional semisimple K-algebras A and finite dimensional A-modules M and N one has $\dim_K(Hom_A(M,N)) = \dim_K(Hom_A(N,M))$. □

Remark 1.7.34 Since H is of finite index in G the result of Corollary 1.7.33 is true without the hypothesis that KG is semisimple. In fact, define

$$Hom_{RH}(M\downarrow_H^G, N) \xrightarrow{\Phi} Hom_{RG}(M, N\uparrow_H^G)$$
$$\varphi \mapsto \left(m \mapsto \sum_{gH\in G/H} g\otimes\varphi(g^{-1}m)\right)$$

Then this formula does not depend on the representatives gH of G/H. Indeed,

$$\begin{aligned}(\Phi(\varphi))(m) &= \sum_{gH\in G/H} g\otimes\varphi(g^{-1}m) = \sum_{gH\in G/H} g\otimes\varphi(hh^{-1}g^{-1}m)\\ &= \sum_{gH\in G/H} g\otimes h\varphi(h^{-1}g^{-1}m) = \sum_{gH\in G/H} gh\otimes\varphi(h^{-1}g^{-1}m)\\ &= \sum_{ghH\in G/H} gh\otimes\varphi((gh)^{-1}m)\end{aligned}$$

for all $m\in M$ and $h\in H$.

Moreover, $\Phi(\varphi)$ is G-linear. Indeed, for all $m\in M$ and $k\in G$ one has

$$\begin{aligned}(\Phi(\varphi))(km) &= \sum_{gH\in G/H} g\otimes\varphi(g^{-1}km) = \sum_{gH\in G/H} k(k^{-1}g)\otimes\varphi((k^{-1}g)^{-1}m)\\ &= k\left(\sum_{gH\in G/H}(k^{-1}g)\otimes\varphi((k^{-1}g)^{-1}m)\right)\\ &= k\left(\sum_{(k^{-1}g)H\in G/H}(k^{-1}g)\otimes\varphi((k^{-1}g)^{-1}m)\right) = k\left(\Phi(\varphi)(m)\right).\end{aligned}$$

We need to show that Φ is bijective. For this we define a mapping in the other direction. Since $RG\otimes_{RH} N = \bigoplus_{gH\in G/H} g\otimes N$ as RH-modules, every $\varphi\in Hom_{RG}(M, RG\otimes_{RH} N)$ decomposes into $\varphi = \oplus_{gH\in G/H}\varphi_g$ with $\varphi_g : N \longrightarrow g\otimes M$. Define $\Psi(\varphi) = \varphi_1$ for every $\varphi\in Hom_{RG}(M, RG\otimes_{RH} N)$. It is clear by construction that φ_1 is RH-linear, and that Ψ is inverse to Φ.

1.7.3 Mackey's Formula

A very far-reaching concept using induction from and restriction to subgroups is Mackey's formula. It should be observed that there is an extremely extensive literature on abstract versions of Mackey's formula. For a very complete account see Bouc [11].

The question there is what happens if one induces a module from a subgroup H of a group G and restricts the result to a possibly different subgroup K. In order to find a concept well-suited to study this problem let G be a group and let H and K be two subgroups of G. Of course $H = K$ is allowed, but the subgroups may well differ.

Define a relation ${}_K\sim_H$ on G by setting

$$g_1 \,{}_K\sim_H\, g_2 \Leftrightarrow g_1 \in \{hg_2k \mid h \in H; k \in K\}$$

for all $g_1, g_2 \in G$.

Lemma 1.7.35 *Let G be a group and H and K be two subgroups. Then ${}_K\sim_H$ is an equivalence relation.*

Proof Of course,

$$\begin{aligned} g_1 \,{}_K\sim_H\, g_2 &\Rightarrow \exists h \in H, k \in K : g_1 = hg_2k \\ &\Rightarrow \exists h \in H, k \in K : h^{-1}g_1k^{-1} = g_2 \\ &\Rightarrow g_2 \,{}_K\sim_H\, g_1 \end{aligned}$$

and $g_1 \,{}_K\sim_H\, g_1$ for all $g_1 \in G$ taking $h = k = 1$.

Moreover

$$\begin{aligned} g_1 \,{}_K\sim_H\, g_2 \text{ and } g_2 \,{}_K\sim_H\, g_3 &\Rightarrow \exists h_1, h_2 \in H, k_1, k_2 \in K : g_1 = h_1g_2k_1 \\ &\qquad \text{and } g_2 = h_2g_3k_2 \\ &\Rightarrow \exists h_1, h_2 \in H, k_1, k_2 \in K : g_1 = (h_1h_2)g_3(k_2k_1). \end{aligned}$$

This proves the lemma. □

As a consequence one sees that two classes Kg_1H and Kg_2H are either identical or disjoint. Moreover, G is the disjoint union of such classes.

Definition 1.7.36 The set of equivalence classes of G under the relation ${}_K\sim_H$ is denoted by $K\backslash G/H$. The equivalence class of $g \in G$ is called the *double coset*, or *double class* of g modulo K and H and is denoted by KgH.

Example 1.7.37 If $K = 1$ we get the usual left cosets of G/H as double cosets. There each coset gH for $g \in G$ is of cardinality $|H|$. This is no longer true if $K \neq 1$. The sizes of double cosets are not constant in general.

Let $D_4 = \langle a, b | \ a^4, b^2, baba \rangle$ be the dihedral group of order 8. Let $B := \langle b \rangle$ be the cyclic group of order 2 generated by b. Then D_4/B has 4 classes of length 2

$$\begin{aligned} D_4/B &= \{1, b\} \cup \{a, ab\} \cup \{a^2, a^2b\} \cup \{a^3, a^3b\} \\ &= (1 \cdot B) \cup (a \cdot B) \cup (a^2 \cdot B) \cup (a^3 \cdot B). \end{aligned}$$

In $B \backslash D_4/B$ there are 2 classes $\{1, b\}$ and $\{a^2, a^2b\}$ of length 2 and one class $\{a, ab, a^3, a^3b\}$ of length 4.

$$\begin{aligned} B \backslash D_4/B &= \{1, b\} \cup \{a^2, a^2b\} \cup \{a, ab, a^3, a^3b\} \\ &= (B \cdot 1 \cdot B) \cup (B \cdot a \cdot B) \cup (B \cdot a^2 \cdot B). \end{aligned}$$

Lemma 1.7.38 *Let G be a group and let H and K be subgroups of G. If H or K is a normal subgroup of G, then the double cosets $K \backslash G/H$ are all of the same size.*

Proof Suppose K is normal in G. Then $KgH = gKH$ and KH is a subgroup of G. Hence the double cosets of G modulo K and H are the same as the left cosets modulo the subgroup KH. The case when H is normal is similar. □

The reason why we need to study double cosets comes from the question of understanding the group ring RG, where G a group and R is a commutative ring, as an RK–RH-bimodule, K and H being subgroups of G. Once this is understood we will be able to deduce the important Mackey formula.

We first need another rather general concept. Let R be a commutative ring, let A be an R-algebra and let α be an automorphism of A as an R-algebra. For any A-module M we form another A-module ${}^{\alpha}M$ via the following construction.

Since M is an A-module, using Lemma 1.1.14, the module structure on M is given by a homomorphism of R-algebras $A \xrightarrow{\mu} End_R(M)$. The composition

$$A \xrightarrow{\alpha} A \xrightarrow{\mu} End_R(M)$$

is a homomorphism of R-algebras again. Hence one gets a structure of an A-module on the *same set* M. This A-module is denoted by ${}^{\alpha}M$.

Definition 1.7.39 Let R be a commutative ring, let A be an R-algebra and let M be an A-module. Given an automorphism α of A, the A-module ${}^{\alpha}M$ is called the *twist* of M by α.

Let us give a more explicit description of such an ${}^{\alpha}M$. Since α is R-linear, it is clear that *as an R-module* one has that ${}^{\alpha}M$ is really equal to M. Denote by $\bullet$ the operation of A on ${}^{\alpha}M$ and by $\cdot$ the operation of A on M. Then for every $a \in A$ one has $a \bullet m := \alpha(a) \cdot m$.

Definition 1.7.40 Let A be a ring and let $u \in A^{\times}$ be an invertible element. The mapping $A \longrightarrow A$ given by $a \mapsto u \cdot a \cdot u^{-1}$ is an automorphism, called an *inner* *automorphism.* The set of inner automorphisms of A is a normal subgroup of the group of automorphisms of A and the group of inner automorphisms is denoted by $Inn(A)$. The quotient $Aut(A)/Inn(A)$ is denoted by $Out(A)$.

There are some statements in the definition that need to be proved. However, the proof is so straightforward that it can be left to the reader without any trouble.

Lemma 1.7.41 *Let A be an R-algebra and let M be an A-module. Then for all $\alpha \in Inn(A)$ one has ${}^{\alpha}M \simeq M$.*

Proof Let u be an element of A such that $\alpha(a) = u \cdot a \cdot u^{-1}$ for all $a \in A$. Then

$$\begin{aligned} M &\xrightarrow{\varphi} {}^{\alpha}M \\ m &\mapsto u \cdot m \end{aligned}$$

is an A-module homomorphism. Indeed, denoting again by $\bullet$ the operation of A on ${}^{\alpha}M$, one gets

$$\varphi(a \cdot m) = u \cdot (a \cdot m) = (u \cdot a \cdot u^{-1}) \cdot (u \cdot m) = a \bullet \varphi(m)$$

for all $a \in A$ and $m \in M$. Of course, φ is a homomorphism of R-modules since the image of R in A is in the centre of A. Moreover, since u is invertible, φ is invertible as well. The inverse application is multiplication by u^{-1}. □

If α is an inner automorphism of A and $\alpha(a) = u \cdot a \cdot u^{-1}$ for all $a \in A$, then we shall denote ${}^{\alpha}M$ by ${}^{u}M$ for every A-module M. The proof of Lemma 1.7.41 suggests to denote ${}^{u}M$ by $u \cdot M$ as well.

Example 1.7.42 The converse of Lemma 1.7.41 is not true in general. Let K be a field, let G be a group and let K be the trivial KG-module. Let α be an automorphism of G that is not inner in KG. This happens for example if G is abelian and $\alpha \in Aut(G)$ is not the identity. Then ${}^{\alpha}K \simeq K$ since every element of G acts as the identity, and so $g \in G$ acts in exactly the same way as $\alpha(g)$.

Remark 1.7.43 We shall study twisted modules in more detail in Sect. 1.10.1.

Lemma 1.7.44 *Let R be a commutative ring and let G be a group. Suppose H and K are subgroups of G. Then RG is an RK–RH-bimodule by multiplication of elements in RG. Moreover,*

$$RG \simeq \bigoplus_{(KgH \in K\backslash G/H)} RK \otimes_{(K \cap gHg^{-1})} gRH$$

as an RK–RH-bimodule.

Proof Let

$$K\backslash G/H = \bigcup_{i\in I} Kg_iH$$

with elements $g_i \in G$ for all $i \in I$, an index set, so that $Kg_iH = Kg_jH \Leftrightarrow i = j$. Then for every $g \in G$ there is a unique $i(g) \in I$ and a unique $k(g) \in K$, $h(g) \in H$ such that $g = k(g)g_{i(g)}h(g)$. Define a mapping

$$\begin{array}{rcl} RG & \xrightarrow{\alpha} & \bigoplus_{i\in I} RK \otimes_{\left(K\cap g_iHg_i^{-1}\right)} g_iRH \\ g & \mapsto & k(g) \otimes g_{i(g)}h(g) \end{array}$$

and an inverse mapping

$$\begin{array}{rcl} \bigoplus_{i\in I} RK \otimes_{\left(K\cap g_iHg_i^{-1}\right)} g_iRH & \xrightarrow{\beta} & RG \\ k \otimes gh & \mapsto & kgh \end{array}$$

First, β is well-defined. Actually gRH is an $R\left(gHg^{-1}\right)$-module. More generally, if L is a subgroup of G and if α is an automorphism of G then for each RL-module M one has that ${}^{\alpha}M$ is an $R\alpha^{-1}(L)$-module. Indeed, $\ell \bullet m = \alpha(\ell)m$ and this is defined whenever $\ell \in \alpha^{-1}(L)$. Moreover, as is immediately seen, for every $x \in K \cap gHg^{-1}$ one has

$$\beta(kx \otimes gh) = kxgh = kg(g^{-1}xgh) = \beta(k \otimes g(g^{-1}xgh))$$

and so the operation of $K \cap gHg^{-1}$ on gRH is defined exactly in the way needed to give a well-defined mapping.

It is clear by definition that α and β are mutually inverse mappings. Hence, β and α are bijective.

Moreover, β is trivially a homomorphism of RK-RH-bimodules. This proves the lemma. □

We come to the main result of Sect. 1.7.3.

Theorem 1.7.45 (Mackey) *Let R be a commutative ring, let G be a group with subgroups H and K and let M be an RH-module. Then*

$$M \uparrow_H^G \downarrow_K^G \simeq \bigoplus_{KgH\in K\backslash G/H} \left({}^g\left(M \downarrow^H_{(H\cap gKg^{-1})}\right)\right) \uparrow^K_{(gHg^{-1}\cap K)} .$$

Proof

$$M \uparrow_H^G \downarrow_K^G \simeq \left(\bigoplus_{KgH\in K\backslash G/H} RK \otimes_{K\cap gHg^{-1}} gRH \right) \otimes_{RH} M \text{ (cf Lemma 1.7.44)}$$

$$\simeq \bigoplus_{KgH\in K\backslash G/H} RK \otimes_{K\cap gHg^{-1}} (gRH \otimes_{RH} M) \text{ (cf Lemma 1.7.9)}$$
$$\simeq \bigoplus_{KgH\in K\backslash G/H} RK \otimes_{K\cap gHg^{-1}} gM \text{ (cf Lemma 1.7.10)}$$
$$\simeq \bigoplus_{KgH\in K\backslash G/H} \left({}^{g}\left(M \downarrow^{H}_{H\cap gKg^{-1}}\right)\right) \uparrow^{K}_{gHg^{-1}\cap K}.$$

This proves the theorem. □

There are plenty of applications of this formula. Its force will become evident when we discuss representations over fields of positive characteristic dividing the group order in Chap. 2. We give two applications immediately.

Corollary 1.7.46 *Let G be a finite group, H be a subgroup and let R be a commutative ring. Then for every RH-module M one has that M is a direct factor of $M \uparrow^G_H\downarrow^G_H$.*

Proof

$$M \uparrow^G_H\downarrow^G_H\simeq \bigoplus_{HgH\in H\backslash G/H} \left({}^{g}\left(M \downarrow^{H}_{(H\cap gHg^{-1})}\right)\right) \uparrow^{H}_{(gHg^{-1}\cap H)}$$

and in particular for the class $H1H = H$ one has

$$\left({}^{1}\left(M \downarrow^{H}_{(H\cap 1H1^{-1})}\right)\right) \uparrow^{H}_{(1H1^{-1}\cap H)}= M \downarrow^H_H\uparrow^H_H= M$$

is a direct factor of $M \uparrow^G_H\downarrow^G_H$. This proves the corollary. □

Definition 1.7.47 Let G be a group and let R be a commutative ring. A finitely generated RG-module M is a *permutation module* if there is an R-basis B of M such that for all $g \in G$ and all $b \in B$ one has $gb \in B$. Such a basis B is called a *permutation basis*. A *p-permutation module* is a direct summand of a permutation module. A *transitive permutation module* is a permutation module with a permutation basis which is a transitive G-set.

Lemma 1.7.48 *Given a commutative ring R and a group G, then M is an RG-permutation module if and only if M is a direct sum of modules of the form $R \uparrow^G_H$ for some subgroups H of G.*

Proof Let M be a permutation module and let B be a permutation basis. Let $B = B_1\cup B_2\cup\cdots\cup B_s$ be a disjoint composition into G-orbits. Then each B_i for $i \in \{1, \dots, s\}$ is a transitive G-set and therefore for all $i \in \{1, \dots, s\}$ one gets that $B_i \simeq G/H_i$ as G-sets for subgroups $H_1, H_2, \dots, H_s$ of G.

Moreover, for all $i \in \{1, \dots, s\}$, denoting by RB_i the R-linear combinations inside M formed by elements in B_i one gets $M \simeq RB_1 \oplus RB_2 \oplus \cdots \oplus RB_s$ as RG-modules.

Hence, without loss of generality, we may suppose $s = 1$ and we identify $B = B_1$ with G/H,

$$\begin{aligned} RG \otimes_{RH} R &\xrightarrow{\alpha} M \\ \sum_{g\in G} r_g g \otimes 1 &\mapsto \sum_{g\in G} r_g \cdot (g \cdot H) \end{aligned}$$

Then α is well defined. Indeed,

$$\begin{aligned} \alpha\left(\sum_{g\in G} r_g gh \otimes 1\right) &= \sum_{g\in G} r_g \cdot (gh \cdot H) = \sum_{g\in G} r_g \cdot (g \cdot hH) \\ &= \sum_{g\in G} r_g \cdot (g \cdot H) = \alpha(\sum_{g\in G} r_g g \otimes 1). \end{aligned}$$

Moreover α is RG-linear. We define a mapping in the other direction by

$$\begin{aligned} M &\xrightarrow{\beta} RG \otimes_{RH} R \\ \sum_{g\in G} r_g gH &\mapsto \sum_{g\in G} r_g g \otimes 1 \end{aligned}$$

Then again β is well-defined and obviously inverse to α.

To prove the converse we first mention that direct sums of permutation modules are again permutation modules, the permutation basis of the sum being the union of the permutation bases of the summands. Hence we only need to show that $R \uparrow_H^G$ is a permutation module. But this is actually done in the first step, showing that $RG \otimes_{RH} R \simeq R(G/H)$, where $R(G/H)$ is the R-free R-module $R^{|G/H|}$ as an R-module and G acts on $R^{|G/H|}$ as G acts on G/H. This proves the lemma. □

Remark 1.7.49 Some authors use the expression *trivial source module* for what we have called a p-permutation module. The reason for this alternative terminology will become clear once we have introduced the notion of a source in Definition 2.1.17 in connection with Lemma 1.7.48

Corollary 1.7.50 *Let K be a field and let G be a group. Then the G-fixed points of a finite dimensional transitive permutation module is a one-dimensional K-space.*

Proof Let $M = K(G/H)$ be a transitive G-permutation module. Then, using Remark 1.7.34,

$$\begin{aligned} M^G = Hom_{KG}(K, M) &= Hom_{KG}(K, K \uparrow_H^G) \\ &= Hom_{KH}(K \downarrow_H^G, K) = Hom_{KH}(K, K) = K. \end{aligned}$$

This proves the lemma. □

We have the following immediate consequence of Corollary 1.7.50 and Mackey's theorem 1.7.45.

Corollary 1.7.51 *Let G be a group, let H be a subgroup of finite index and let K be a field. Then* $\dim_K(End_{KG}(K(G/H))) = |H\backslash G/H|$.

Proof

$$\begin{aligned} End_{KG}(K(G/H)) &\simeq Hom_{KG}(K\uparrow_H^G, K\uparrow_H^G) \\ &\simeq Hom_{KH}(K, K\uparrow_H^G\downarrow_H^G) \\ &\simeq Hom_{KH}\left(K, \bigoplus_{HgH\in H\backslash G/H} \left({}^g\left(K\downarrow^H_{(H\cap gHg^{-1})}\right)\right)\uparrow^H_{(gHg^{-1}\cap H)}\right) \\ &\simeq \bigoplus_{HgH\in H\backslash G/H} Hom_{KH}\left(K, \left({}^g\left(K\downarrow^H_{(H\cap gHg^{-1})}\right)\right)\uparrow^H_{(gHg^{-1}\cap H)}\right) \\ &\simeq \bigoplus_{HgH\in H\backslash G/H} Hom_{KH}\left(K, K\downarrow^H_{(H\cap gHg^{-1})}\uparrow^H_{(gHg^{-1}\cap H)}\right) \end{aligned}$$

and each of the modules

$$K\downarrow^H_{(H\cap gHg^{-1})}\uparrow^H_{(gHg^{-1}\cap H)} = K\uparrow^H_{(gHg^{-1}\cap H)}$$

is a transitive KH-permutation module. By Corollary 1.7.50 its fixed point space is one-dimensional. This proves the corollary. □

1.8 Some First Steps in Homological Algebra; Extension Groups

We shall now consider the question raised in Example 1.6.23 of how many possible ways there are of obtaining a module from given composition factors. The way to do this is to generalise the notion of a module admitting a basis to that of being a direct factor of such a module, and then to consider and closely examine modules which are not of this type, but instead arise as a cokernel of a morphism between such modules.

1.8.1 Projective and Injective Modules

The reason why linear algebra can be considered as almost completely understood comes from the fact that every vector space has a basis. For semisimple algebras some of the main tools can still be applied. Nevertheless, if the algebra is not semisimple the modules having properties similar to those one would like to use in the presence of a basis are quite special.

What is a basis of a vector space? It is a subset of the vector space, such that any set theoretic mapping from this subset to another vector space can be completed in a unique way to a vector space morphism.

Definition 1.8.1 Let A be an algebra. An A-module M is *free on a subset S of M* if for every A-module N and every set theoretic mapping $f : S \longrightarrow N$ there is a unique A-module homomorphism $\varphi : M \longrightarrow N$ with $\varphi|_S = f$.

Obviously, the isomorphism class of a free module depends only on the cardinality of S. Indeed, if F_1 is free on S_1, if F_2 is free on S_2 and if there is a bijection $f : S_1 \longrightarrow S_2$, then this bijection f induces a unique A-module homomorphism $\varphi : F_1 \longrightarrow F_2$ restricting to f. Moreover, $f^{-1} : S_2 \longrightarrow S_1$ induces a unique A-module homomorphism $\psi : F_2 \longrightarrow F_1$ restricting to f^{-1}. Now, $id_{S_2} = f \circ f^{-1} : S_2 \longrightarrow S_2$ induces a unique A-module homomorphism $F_2 \longrightarrow F_2$ restricting to id_{S_2}. But, the identity on F_2 as well as $\varphi \circ \psi$ restrict to the identity on S_2 and are A-module homomorphism. Hence, $\varphi \circ \psi = id_{F_2}$. Likewise, $\psi \circ \varphi = id_{F_1}$.

We should remark that a free $\mathbb{Z}$-module is nothing else than a free abelian group, introduced in Definition 1.7.1.

Example 1.8.2 Given a set S and an algebra A we shall construct a free module on a set of the same cardinality as S. Let

$$F_S := \{f \in Map(S, A) \mid |f^{-1}(A \setminus \{0\})| < \infty\}$$

be the set theoretic maps from S to A such that only finitely many elements of S map to non-zero elements of A. We say in this case that an element of F_S has finite support. Then F_S is an A-module by defining $(a \cdot f)(s) = af(s)$ for all $s \in S$ and $a \in A$. Moreover, for every $t \in S$ define $f_t \in F_S$ by the property $f_t(s) = 0$ if $t \neq s \in S$ and $f_t(t) = 1$. Then F_S is free on $S' := \{f_t \mid t \in S\}$. Indeed, let N be an A-module and let $g : S' \longrightarrow N$ be a set theoretic mapping. Then define $\varphi : F_S \longrightarrow N$ by $\varphi(f) := \sum_{s \in S} f(s) \cdot g(f_s)$. This is well-defined since the mappings in F_S have finite support. It is an A-module homomorphism and restricts to g as is immediately seen. Moreover, it is the only possible A-module homomorphism since every $f \in F_S$ is in a unique way an A-linear combination of maps in S'. The rest follows.

In more common terms, if S is finite with $n \in \mathbb{N}$ elements, $F_S \simeq A^n$. In more abstract terms,

$$Map(S', N) \simeq Hom_A(F_S, N)$$

and this is "natural" with respect to N.

For vector spaces we have that any direct factor of a free module is again free. This is not true for modules. Take, for example, $A = K \times K$ for a field K. Then $K \times \{0\}$ is a direct factor of A, but it is not free. Many properties we use can be proved by computing in a bigger module, and then taking direct factors.

Definition 1.8.3 A *projective module* is a direct factor of a free module.

This definition is the one which is used in most applications, at least in Chap. 2. Nevertheless we frequently need one of the following equivalent properties. Before we formulate these we introduce some notation.

We shall now define a very useful convention. Given three A-modules M, N and L and A-module morphisms $M \longrightarrow N$ and $N \longrightarrow L$, we write these morphisms in the scheme

$$M \longrightarrow N \longrightarrow L$$

and say that this *sequence is exact* at N if $\ker(N \longrightarrow L) = \mathrm{im}(M \longrightarrow N)$.

Moreover, the sequence

$$0 \longrightarrow M \longrightarrow N \longrightarrow L \longrightarrow 0$$

is said to be *short exact* if it is exact at L, at M and at N.

A short exact sequence

$$0 \longrightarrow M \longrightarrow N \longrightarrow L \longrightarrow 0$$

splits if and only if there is a homomorphism $L \longrightarrow N$ such that the composition $L \longrightarrow N \longrightarrow L$ is the identity. This is equivalent to the fact that $N \simeq M \oplus L$ and the mapping $N \longrightarrow L$ is carried by the isomorphism to the projection of $M \oplus L$ onto L. Indeed, the endomorphism $N \longrightarrow L \longrightarrow N$ is idempotent, as is readily seen, and by Lemma 1.2.10 this idempotent endomorphism ϵ defines a direct sum decomposition of N into the image of ϵ and its kernel. Since the sequence is exact, the embedding $M \longrightarrow N$ is carried by the isomorphism $N \simeq M \oplus L$ to the embedding into its corresponding summand. The dual proof shows that a short exact sequence splits if and only if there is a homomorphism $N \longrightarrow M$ such that $M \longrightarrow N \longrightarrow M$ is the identity.

Proposition 1.8.4 *Let A be an algebra. Then the following properties are equivalent.*

1. *P is a projective A-module.*
2. *For any two A-modules M and N and any surjective A-module homomorphism $\varphi : M \longrightarrow N$ the mapping $Hom_A(P, M) \longrightarrow Hom_A(P, N)$ given by $\psi \mapsto \varphi \circ \psi$ is surjective as well.*
3. *For any two A-modules M and N and any surjective A-module homomorphism $\varphi : M \longrightarrow N$ and every every A-module homomorphism $\psi : P \longrightarrow N$ there is an A-module homomorphism $\chi : P \longrightarrow M$ such that $\varphi \circ \chi = \psi$. We say in this case that ψ lifts along φ to χ.*

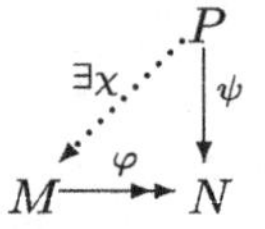

4. *Every short exact sequence*

$$0 \longrightarrow S \longrightarrow M \longrightarrow P \longrightarrow 0$$

of A-modules splits.

Proof **(1) implies (2):** Suppose P is a projective module. Then let Q be another projective module such that $P \oplus Q$ is free on S, say. Given two modules M and N with an epimorphism $\varphi : M \longrightarrow\!\!\!\!\rightarrow N$, in order to show that $Hom_A(P, M) \longrightarrow Hom_A(P, N)$ is surjective, we need to show that any homomorphism $\alpha : P \longrightarrow N$ can be "lifted" to $P \longrightarrow M$. So, extend φ to $P \oplus Q$ by putting $\varphi(q) = 0$ for all $q \in Q$. Let $n_s := \alpha(s)$ for all $s \in S$ and let $m_s \in M$ so that $\alpha(m_s) = n_s$. The existence of the elements m_s uses the fact that α is surjective. Since $P \oplus Q$ is free on S, define a mapping $\hat{\beta} : S \longrightarrow M$ by putting $\beta(s) = m_s$ for all $s \in S$ and the mapping β induces a unique homomorphism $P \oplus Q \longrightarrow M$ with $\beta(s) = \hat{\beta}(s)$. Then

$$\varphi \circ \beta(s) = \varphi(m_s) = n_s = \alpha(s).$$

Unicity shows that $\varphi \circ \beta = \alpha$. Restriction of β to P gives the desired homomorphism.

(2) is equivalent to (3): Clearly, (3) is really a reformulation of (2).

(3) implies (4): Use (3) with $N = P$ and the identity $P \overset{id}{\longrightarrow} N$ as ψ.

(4) implies (1): Any module M is a quotient of a free module F_M, and so is P. Indeed, $M = \sum_{m \in M} A \cdot m$. Hence $F_M := \oplus_{m \in M} A$ maps onto M by putting

$$\left(\sum_{m \in M} a_m\right) \mapsto \sum_{m \in M} a_m m.$$

Therefore we get a surjective homomorphism $F_P \longrightarrow P$ with kernel N, inducing a short exact sequence

$$0 \longrightarrow N \longrightarrow F_P \longrightarrow P \longrightarrow 0$$

which is split, by (4). Hence P is a direct factor of F_P. This proves the statement. □

An injective module is a module having the dual properties of a projective module.

Proposition 1.8.5 *Let A be an algebra. Then the following properties are equivalent.*

1. *For any two A-modules M and N and any injective A-module homomorphism* $\varphi : M \longrightarrow N$ *the mapping* $Hom_A(N, I) \longrightarrow Hom_A(M, I)$ *given by* $\psi \mapsto \psi \circ \varphi$ *is surjective.*
2. *For any two A-modules M and N, and any injective A-module homomorphism* $\varphi : M \longrightarrow N$ *and every A-module homomorphism* $\lambda : M \longrightarrow I$ *there is an A-module homomorphism* $\mu : N \longrightarrow I$ *such that* $\mu \circ \varphi = \lambda$.

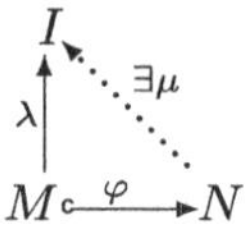

3. *Every short exact sequence*

$$0 \longrightarrow I \longrightarrow M \longrightarrow S \longrightarrow 0$$

of A-modules splits.

The proof is completely analogous to the proof of Proposition 1.8.4. □

Definition 1.8.6 A module is *injective* if it satisfies one of the equivalent conditions in Proposition 1.8.5.

Remark 1.8.7 In Proposition 1.8.4 and Proposition 1.8.5 we have seen an important construction. If A is an algebra and if L, M and N are A-modules, then any homomorphism $M \stackrel{\alpha}{\longrightarrow} N$ induces a homomorphism $Hom_A(L, M) \stackrel{Hom_A(L,\alpha)}{\longrightarrow} Hom_A(L, N)$ given by $\varphi \mapsto \alpha \circ \varphi$. Similarly, α induces a morphism $Hom_A(N, L) \stackrel{Hom_A(\alpha,L)}{\longrightarrow} Hom_A(M, L)$ given by $\varphi \mapsto \varphi \circ \alpha$.In the first case we say that we apply $Hom_A(L, -)$ to $M \stackrel{\alpha}{\longrightarrow} N$, and in the second case we say that we apply $Hom_A(-, L)$ to $M \stackrel{\alpha}{\longrightarrow} N$.

Lemma 1.8.8 *Let A be an algebra, let* $0 \to L \stackrel{\alpha}{\to} M \stackrel{\beta}{\to} N \to 0$ *be a short exact sequence of A-modules and let U be an A-module. Then*

$$0 \to Hom_A(U, L) \stackrel{Hom_A(U,\alpha)}{\longrightarrow} Hom_A(U, M) \stackrel{Hom_A(U,\beta)}{\longrightarrow} Hom_A(U, N)$$

and

$$0 \to Hom_A(N, U) \stackrel{Hom_A(\beta,U)}{\longrightarrow} Hom_A(M, U) \stackrel{Hom_A(\alpha,U)}{\longrightarrow} Hom_A(L, U)$$

are exact sequences of abelian groups.

Proof First, $Hom_A(U, \beta) \circ Hom_A(U, \alpha) = Hom_A(U, \beta \circ \alpha) = 0$, and likewise $Hom_A(\beta \circ \alpha, U) = 0$. The fact that $Hom_A(U, \alpha)$ is injective follows immediately from the fact that α is a monomorphism. Indeed, $\alpha \circ \gamma = 0$ implies $\gamma = 0$ for monomorphisms α. The fact that the sequence is exact at $Hom_A(U, M)$ is the universal property of the kernel. Similarly, $Hom_A(\beta, U)$ is injective since β is an epimorphism. Indeed, $\gamma \circ \beta = 0$ implies $\gamma = 0$ for epimorphisms β. The fact that the second sequence is exact at $Hom_A(M, U)$ is the universal property of the cokernel. □

The fact that the sequences are not short exact, i.e. $Hom_A(U, \beta)$ and $Hom_A(\alpha, U)$ are not surjective in general, is an important issue. Large parts of what follows are motivated by this fact.

While the proof of Proposition 1.8.4 implies that for any A-module M there is a projective module P_M and a surjective homomorphism $P_M \longrightarrow M$, the dual statement for injective modules is not as clear. Nevertheless, the statement is true in general, however it uses infinitely generated modules (cf e.g. [12, (3.20) Theorem]).

Example 1.8.9 An abelian group A is *divisible* if for all integers $n \neq 0$ and all $b \in A$ there is an $a \in A$ with $na = b$. In other words the endomorphism of A given by multiplication by any integer $n \neq 0$ is surjective. The abelian group $\mathbb{Q}/\mathbb{Z}$ is divisible.

An injective $\mathbb{Z}$-module I is divisible. Indeed, for any integer m there is a monomorphism of abelian groups $\iota_m : \mathbb{Z} \longrightarrow \mathbb{Z}$ given by $\iota_m(k) = km$. Let I be injective and let $s \in I$. Then there is a morphism of abelian groups $\varphi_s : \mathbb{Z} \longrightarrow I$ given by $\varphi(k) = ks$; where ks is the result of adding k times s in case k is positive, and in case $k < 0$ adding $-k$ times $-s$. Since I is injective, the mapping φ_s extends to a mapping $\psi_{m,s} : \mathbb{Z} \longrightarrow I$ so that $\varphi_s = \psi_{m,s} \circ \iota_m$. Put $t := \psi_{m,s}(1)$. Then

$$s = \varphi_s(1) = \psi_{m,s} \circ \iota_m(1) = \psi_{m,s}(m \cdot 1) = m \cdot \psi_{m,s}(1) = mt$$

and so I is divisible.

Divisible abelian groups are infinitely generated or equal to $\{1\}$. Indeed, let $D \neq \{1\}$ be a finitely generated abelian group. Then by the principal theorem on finitely generated abelian groups one gets that

$$D \simeq \mathbb{Z}^f \times \prod_{j=1}^{s} \mathbb{Z}/\delta_j\mathbb{Z}$$

for integers δ_i so that δ_i divides δ_{i+1} for all $i \in \{1, 2, \ldots, s-1\}$. None of the generators of each of the direct factors $\mathbb{Z}^f$ is divisible by 2, say, since $\frac{1}{2} \notin \mathbb{Z}$. Hence $f = 0$. Let p be a divisor of δ_1. Then the image of multiplication by p on $\mathbb{Z}/\delta_1\mathbb{Z}$ is the subgroup $p\mathbb{Z}/\delta_1\mathbb{Z}$ of index p and the equation

$$p \cdot (a + \delta_1\mathbb{Z}) = (1 + \delta_1\mathbb{Z})$$

does not have a solution a. Hence $\delta_1 = 1$, which implies that we can remove δ_1 and replace s by $s-1$. Induction on s gives that D is the trivial group.

We come to an important notion which will be used frequently in the sequel as soon as we try to understand the deeper structure of modules, in particular how modules can be glued together in a similar way as in Example 1.6.23.

Given an A-module M for a K-algebra A, then as seen above there is a projective module P_M and an epimorphism $P_M \longrightarrow M$. The kernel will play an important role.

Definition 1.8.10 Let K be a commutative ring, let A be a K-algebra, let M be an A-module, and let $P_M \longrightarrow M$ be an epimorphism of a projective module to M. Then

$$\ker(P_M \longrightarrow M) =: \Omega_M$$

is the first *syzygy* of M.

Remark 1.8.11 The word syzygy is used in astronomy to describe a straight line gravitational configuration of three celestial bodies, such as a lunar eclipse.

We see that since P_M is not unique, the syzygy will also not generally be unique. Indeed, if P is a projective A-module, and $P_M \xrightarrow{\pi_M} M$ is an epimorphism from a projective module P_M to M, then $P \oplus P_M \xrightarrow{(0,\pi_M)} M$ is an epimorphism of a projective module to M as well. Hence, if Ω_M is a first syzygy of M, then $P \oplus \Omega_M$ is a first syzygy of M as well. This shows that first syzygies are determined only "up to projective modules". Actually, we can make this observation much more precise, using the notion of a stable category. This will be done in Chap. 5 since it requires a lot more abstract knowledge than what we currently have. Nevertheless, what we can see immediately is the following Lemma.

Lemma 1.8.12 (Schanuel's Lemma) *Let A be a K-algebra and let M be an A-module. If P_M and Q_M are projective A-modules, and if $P_M \xrightarrow{\pi_M} M$ and $Q_M \xrightarrow{\kappa_M} M$ are epimorphisms, then*

$$\ker(\pi_M) \oplus Q_M \simeq \ker(\kappa_M) \oplus P_M.$$

Proof Form

$$S := \{(p,q) \in P_M \oplus Q_M \mid \pi_M(p) = \kappa_M(q)\}.$$

Then the projection onto the first component gives a mapping

$$\pi_1 : S \longrightarrow P_M$$

and the projection onto the second component gives a mapping

$$\pi_2 : S \longrightarrow Q_M.$$

We get

$$\begin{aligned}\ker(\pi_1) &= \{(p,q) \in P_M \oplus Q_M \mid \pi_M(p) = \kappa_M(q) \text{ and } p = 0\}\\ &= \{(0,q) \in P_M \oplus Q_M \mid 0 = \kappa_M(q) \text{ and } p = 0\}\\ &= \{0\} \oplus \ker(\kappa_M)\end{aligned}$$

and likewise

$$\ker(\pi_2) = \ker(\pi_M) \oplus \{0\}.$$

Moreover, π_1 and π_2 are surjective. Indeed, given $p \in P_M$, then since κ_M is surjective, there is a $q \in Q_M$ such that $\pi_M(p) = \kappa_M(q)$. Hence, $\pi_1(p,q) = p$. Likewise, π_2 is surjective.

But now, since $S \longrightarrow P_M$ is surjective and since P_M is projective,

$$S \simeq \ker(\kappa_M) \oplus P_M$$

and likewise

$$S \simeq Q_M \oplus \ker(\pi_M).$$

This proves the statement. □

Remark 1.8.13 The module S studied above is a special case of a pullback. This construction will be of quite some importance later, and in particular we shall see that the property of the kernels of the corresponding mappings being isomorphic is a general phenomenon. We refer to Definition 1.8.24 for the abstract definition, to Proposition 1.8.25 for a proof that pullbacks are really described as the module S, and Lemma 1.8.27 for a general proof of the property that kernels are preserved.

Hence, as a consequence, given two first syzygies of M, they become isomorphic if one adds appropriate projective modules as direct factors.

Moreover, $\Omega_{M\oplus N}$ and $\Omega_M \oplus \Omega_N$ are both syzygies of $M \oplus N$ since if $P_M \longrightarrow\!\!\!\!\rightarrow M$ and $P_N \longrightarrow\!\!\!\!\rightarrow N$, then $P_M \oplus P_N \longrightarrow\!\!\!\!\rightarrow M \oplus N$ via the diagonal map.

The next concept is linked to higher order syzygies. Indeed, it is obvious how to iterate the syzygy concept.

Definition 1.8.14 Let K be a commutative ring and let A be a K-algebra. Let M be an A-module. Then define $\Omega^1(M) := \Omega_M$ and define for all $n \in \mathbb{N}$ the module $\Omega^n(M) := \Omega^n_M := \Omega_{\Omega^{n-1}_M}$. The module Ω^n_M is the *n-th syzygy* of M.

As seen above the n-th syzygy is defined only up to adding suitable projective modules. By definition, an n-th syzygy comes together with a short exact sequence

$$0 \longrightarrow \Omega^n_M \longrightarrow P_{\Omega^{n-1}_M} \longrightarrow \Omega^{n-1}_M \longrightarrow 0$$

with a projective module $P_{\Omega^{n-1}_M}$, and this allows us to put these sequences together, linking

$$P_{\Omega^{n-1}_M} \longrightarrow \Omega^{n-1}_M \longrightarrow P_{\Omega^{n-2}_M}$$

for all n. This yields an exact sequence

$$\cdots \to P_{\Omega^n_M} \to P_{\Omega^{n-1}_M} \to \cdots \to P_{\Omega^2_M} \to P_{\Omega^1_M} \to P_M \to M \to 0$$

which means that the kernel of every mapping is the image of the previous one. The sequence is called "long exact".

Definition 1.8.15 The sequence

$$\cdots \to P_{\Omega^n_M} \to P_{\Omega^{n-1}_M} \to \cdots \to P_{\Omega^2_M} \to P_{\Omega^1_M} \to P_M \to M \to 0$$

defined above is a *projective resolution* of M. If all the modules $P_{\Omega^n_M}$ and P_M are free, then we say that the above is a *free resolution*.

Definition 1.9.8 below will give a dual notion for injective objects.

Example 1.8.16 Projective resolutions can be finite or infinite.

1. Let K be a field and let $A = K[X]$. Then by the elementary divisor theorem (i.e. Gauss elimination) any finitely generated indecomposable A-module is either isomorphic to $K[X]/P(X)K[X]$ where $P(X)$ is a power of an irreducible polynomial, or is free of rank one. A projective resolution of $K[X]/P(X)K[X]$ is

$$0 \longrightarrow K[X] \longrightarrow K[X] \longrightarrow K[X]/P(X)K[X] \longrightarrow 0$$

   where the first inclusion is the mapping given by multiplication by $P(X)$. A projective resolution of a free module F is the free module itself, since free modules are projective, and so $P_F = F$. We observe that any projective resolution of a finitely generated $K[X]$-module is of length 1 or 2. We say that the global dimension of $K[X]$ is 1 (one less than the length of the longest resolution). We refer to Definition 5.11.10 for a more proper definition and further material on this concept. We observe that in algebraic geometry the ring of regular functions of a line is $K[X]$. A line really should have dimension 1. This is not an accident.
2. Let K be a field again and let $A = K[X]/X^2K[X]$. Any finitely generated indecomposable A-module is either isomorphic to $K[X]/P(X)K[X]$ where $P(X)$ is a divisor of X^2, or is free. A projective resolution of a free module is again of length 1, as above, and a projective resolution of $K[X]/XK[X]$ is given by

$$\cdots \longrightarrow A \longrightarrow A \longrightarrow K[X]/XK[X] \longrightarrow 0$$

   where all morphisms between free modules are just multiplication by X. Hence the global dimension of A is infinite (again see Definition 5.11.10 for a more proper definition).

1.8.2 Ext-groups

The syzygy has an enormous impact on the structure of a module M. Actually, the syzygy Ω_M tells us which module we might "glue" to the given module to obtain a bigger module different from the direct sum. By this we mean the question of what be the structure of a submodule N of X such that $X/N \simeq M$. Of course, $X = M \oplus N$ always has this property. But, we would like to study the situation when X is not a direct sum of M and N.

Let A be a K-algebra and let M be an A-module. Then let Ω_M be the first syzygy of M, attached to the projective module P_M mapping to M. Suppose that there is a homomorphism $\varphi : \Omega_M \longrightarrow N$. Then this mapping induces a short exact sequence as follows.

Let $\Omega_M \xrightarrow{\iota} P_M$ be the natural embedding. Define

$$Q_M := (N \oplus P_M)/\{(\varphi(\omega), -\iota(\omega)) \in N \oplus P_M \mid \omega \in \Omega_M\}$$

and define a mapping $N \longrightarrow Q_M$ by mapping $n \in N$ to $(n, 0) \in Q_M$.

A first observation is that this mapping is injective. Indeed, suppose $(n, 0) = 0 \in Q_M$. Then $(n, 0) \in \{(\varphi(\omega), -\iota(\omega)) \in N \oplus P_M \mid \omega \in \Omega_M\}$, and so let $\omega_n \in \Omega_M$ so that $(\varphi(\omega_n), -\iota(\omega_n)) = (n, 0)$. Hence also $\varphi(\omega_n) = n$ and $\iota(\omega_n) = 0$. Since ι is injective, $\omega_n = 0$ which implies $n = 0$.

A second observation is that $Q_M/N \simeq M$. Indeed,

$$Q_M/(N, 0) \simeq P_M/\Omega_M \simeq M.$$

Hence

$$0 \longrightarrow N \longrightarrow Q_M \longrightarrow M \longrightarrow 0$$

is an exact sequence, that is a homomorphism $\Omega_M \longrightarrow N$ induces a module Q_M together with an injection of N into Q_M, the quotient being M. Denote by λ the injection of N into Q_M and let $\pi : P_M \longrightarrow Q_M$ be given by $p \mapsto (0, p)$ so that the diagram

$$\begin{array}{ccc} \Omega_M & \xrightarrow{\iota} & P_M \\ \downarrow \varphi & & \downarrow \pi \\ N & \xrightarrow{\lambda} & Q_M \end{array}$$

is commutative. Then λ is a split injection if and only if there is a mapping $\sigma : Q_M \longrightarrow N$ with $\sigma \circ \lambda = id_N$. If there is such a σ, then $\varphi = \sigma \circ \lambda \circ \varphi = \sigma \circ \pi \circ \iota$, and hence there is a morphism $\sigma \circ \pi =: \psi \in Hom_A(P_M, N)$ so that $\psi \circ \iota = \varphi$. If there is a morphism $\psi \in Hom_A(P_M, N)$ with $\psi \circ \iota = \varphi$ then define $\sigma : Q_M \longrightarrow N$ by $\sigma(n, p) := n + \psi(p)$. This is well-defined since if $(n, p) = (\varphi(\omega), -\iota(\omega))$, then $\sigma(\varphi(\omega), -\iota(\omega)) = \varphi(\omega) - \psi \circ \iota(\omega) = 0$. Hence λ is split injective if and only if there is a $\psi \in Hom_A(P_M, N)$ such that $\varphi = \psi \circ \iota$.

This is a more general phenomenon. We need to define what it means to say that two exact sequences with the same end and beginning terms are equivalent.

Let M and N be two A-modules. Two exact sequences

$$0 \longrightarrow N \longrightarrow L_1 \longrightarrow M \longrightarrow 0$$

and

$$0 \longrightarrow N \longrightarrow L_2 \longrightarrow M \longrightarrow 0$$

are equivalent if and only if there is a homomorphism $L_1 \longrightarrow L_2$ making the diagram

$$\begin{array}{ccccccccc} 0 & \longrightarrow & N & \longrightarrow & L_1 & \longrightarrow & M & \longrightarrow & 0 \\ & & \| & & \downarrow & & \| & & \\ 0 & \longrightarrow & N & \longrightarrow & L_2 & \longrightarrow & M & \longrightarrow & 0 \end{array}$$

commutative.

Lemma 1.8.17 *The relation on the set of short exact sequences with beginning term N and ending term M is an equivalence relation.*

Proof This is an equivalence relation since necessarily $L_1 \longrightarrow L_2$ is an isomorphism. Indeed, let a be in the kernel. Then a is in the kernel of the composition $L_1 \longrightarrow L_2 \longrightarrow M$ and, by the commutativity of the right square, in the kernel of the projection $L_1 \longrightarrow M$. Hence, $a \in N$, but the restriction of $L_1 \longrightarrow L_2$ to N is the identity. Therefore $a = 0$. Let $b \in L_2$. Then its image m_b in M is an image of $a_b \in L_1$ by the projection $L_1 \longrightarrow M$. Hence, the image b_{a_b} of a_b in L_2 under the mapping $L_1 \longrightarrow L_2$ has the property that $b - b_{a_b}$ is in the kernel of $L_2 \longrightarrow M$. Hence, $b - b_{a_b} = n \in N$. Now, $n + b_{a_b}$ maps to b under the mapping $L_1 \longrightarrow L_2$.

Reflexivity, symmetry and transitivity of the relation are now clear. □

Proposition 1.8.18 *The set of equivalence classes of short exact sequences*

$$0 \longrightarrow N \longrightarrow ? \longrightarrow M \longrightarrow 0$$

is in bijection with the space

$$Hom_A(\Omega_M, N)/(Hom_A(P_M, N) \circ \iota_M)$$

where $\iota_M : \Omega_M \hookrightarrow P_M$ is the natural embedding.

Proof Given a mapping $f : \Omega_M \longrightarrow N$, we have constructed a short exact sequence of the required type. If g factors through the embedding $\Omega_M \stackrel{\iota}{\longrightarrow} P_M$ as above, we need to show that the short exact sequence induced by $f + g$ is equivalent to the sequence induced by f.

$$\begin{array}{ccccccccc} 0 & \longrightarrow & \Omega_M & \longrightarrow & P_M & \longrightarrow & M & \longrightarrow & 0 \\ & & \downarrow f & & \downarrow & & \| & & \\ 0 & \longrightarrow & N & \longrightarrow & Q_M(f) & \longrightarrow & M & \longrightarrow & 0 \end{array}$$

$$\begin{array}{ccccccccc} 0 & \longrightarrow & N & \longrightarrow & Q_M(f+g) & \longrightarrow & M & \longrightarrow & 0 \\ & & \uparrow f+g & & \uparrow & & \| & & \\ 0 & \longrightarrow & \Omega_M & \longrightarrow & P_M & \longrightarrow & M & \longrightarrow & 0 \end{array}$$

Since g factorises through P_M, there is a mapping $\ell : P_M \longrightarrow N$ so that $g = \ell \circ \iota$, where ι is the embedding $\Omega_M \longrightarrow P_M$.

Then

$$Q_M(f+g) = N \oplus P_M / \{((f+g)(\omega), -\iota(\omega)) \mid \omega \in \Omega_M\}$$

and

$$Q_M(f) = N \oplus P_M/\{(f(\omega), -\iota(\omega)) \mid \omega \in \Omega_M\}.$$

We define

$$\begin{pmatrix} id_N & -\ell \\ 0 & id_{P_M} \end{pmatrix} : Q_M(f) \longrightarrow Q_M(f+g).$$

Observe that

$$\begin{pmatrix} id_N & -\ell \\ 0 & id_{P_M} \end{pmatrix} \cdot \begin{pmatrix} f \\ -\iota \end{pmatrix} = \begin{pmatrix} f + \ell \circ \iota \\ -\iota \end{pmatrix} = \begin{pmatrix} f+g \\ -\iota \end{pmatrix}.$$

Hence the mapping given by this matrix is well-defined. Moreover, the mapping is an isomorphism since

$$\begin{pmatrix} id_N & \ell \\ 0 & id_{P_M} \end{pmatrix} : Q_M(f+g) \longrightarrow Q_M(f)$$

is an inverse. Finally, the mapping makes the diagram

$$\begin{array}{ccccccccc} 0 & \longrightarrow & N & \longrightarrow & Q_M(f) & \longrightarrow & M & \longrightarrow & 0 \\ & & \| & & \downarrow & & \| & & \\ 0 & \longrightarrow & N & \longrightarrow & Q_M(f+g) & \longrightarrow & M & \longrightarrow & 0 \end{array}$$

commutative since

$$\begin{pmatrix} id_N & \ell \\ 0 & id_{P_M} \end{pmatrix} \cdot \begin{pmatrix} \ell \\ 0 \end{pmatrix} = \begin{pmatrix} \ell \\ 0 \end{pmatrix}$$

and denoting the natural projection by $\pi : Q_M(*) \longrightarrow M$

$$\begin{pmatrix} 0 & \pi \end{pmatrix} \cdot \begin{pmatrix} id_N & \ell \\ 0 & id_{P_M} \end{pmatrix} = \begin{pmatrix} 0 & \pi \end{pmatrix}.$$

Now, given a short exact sequence

$$0 \longrightarrow N \longrightarrow X \longrightarrow M \longrightarrow 0$$

then, using that P_M is projective and that $X \longrightarrow M$ is surjective, by Proposition 1.8.4 there is a mapping $P_M \longrightarrow X$ making the square

$$\begin{array}{ccc} P_M & \longrightarrow & M \\ \downarrow & & \| \\ X & \longrightarrow & M \end{array}$$

commutative. By the universal property of the kernel the mapping $P_M \longrightarrow M$ induces a mapping $\Omega_M \longrightarrow N$ by restriction so that the diagram

$$\begin{array}{ccccccccc} 0 & \longrightarrow & \Omega_M & \longrightarrow & P_M & \longrightarrow & M & \longrightarrow & 0 \\ & & \downarrow & & \downarrow & & \| & & \\ 0 & \longrightarrow & N & \longrightarrow & X & \longrightarrow & M & \longrightarrow & 0 \end{array}$$

is commutative.

Since P_M is unique up to projective modules only, we need to study what happens if one replaces P_M by another projective P'_M, and Ω_M by Ω'_M. We need to show that

$$Hom_A(\Omega_M, N)/Hom_A(P_M, N) \circ \iota_M \simeq Hom_A(\Omega'_M, N)/Hom_A(P'_M, N) \circ \iota'_M$$

where $\iota'_M : \Omega'_M \longrightarrow P'_M$ is the natural embedding.

Since P_M is projective, and since $P'_M \longrightarrow M$ is surjective, there is a morphism $\varphi : P_M \longrightarrow P'_M$ such that

$$P_M \longrightarrow M = P_M \xrightarrow{\varphi} P'_M \longrightarrow M.$$

The same argument gives a morphism $\psi : P'_M \longrightarrow P_M$ such that

$$P'_M \longrightarrow M = P'_M \xrightarrow{\psi} P_M \longrightarrow M.$$

But then the diagram

$$\begin{array}{ccccccccc} 0 & \longrightarrow & \Omega_M & \xrightarrow{\iota_M} & P_M & \longrightarrow & M & \longrightarrow & 0 \\ & & & & \uparrow \psi & & \| & & \\ 0 & \longrightarrow & \Omega'_M & \xrightarrow{\iota'_M} & P'_M & \longrightarrow & M & \longrightarrow & 0 \end{array}$$

is commutative, and therefore there is a unique morphism

$$\Omega'_M \xrightarrow{\Omega(\psi)} \Omega_M$$

making the diagram

$$\begin{array}{ccccccccc} 0 & \longrightarrow & \Omega_M & \xrightarrow{\iota_M} & P_M & \longrightarrow & M & \longrightarrow & 0 \\ & & \uparrow \Omega(\psi) & & \uparrow \psi & & \| & & \\ 0 & \longrightarrow & \Omega'_M & \xrightarrow{\iota'_M} & P'_M & \longrightarrow & M & \longrightarrow & 0 \end{array}$$

commutative. A morphism $\Omega_M \xrightarrow{\alpha} N$ then induces by composition a morphism $\Omega'_M \xrightarrow{\Omega(\psi)} \Omega_M \xrightarrow{\alpha} N$. If $\alpha = \beta \circ \iota_M$ for some $\beta : P_M \longrightarrow N$, then

$$\alpha \circ \Omega(\psi) = \beta \circ \iota_M \circ \Omega(\psi) = \beta \circ \psi \circ \iota'_M = \beta' \circ \iota'_M$$

for $\beta' := \beta \circ \psi$. Hence

$$\begin{array}{rcl} Hom_A(\Omega_M, N)/Hom_A(P_M, N) \circ \iota_M & \to & Hom_A(\Omega'_M, N)/Hom_A(P'_M, N) \circ \iota'_M \\ \alpha + Hom_A(P_M, N) \circ \iota_M & \mapsto & \alpha \circ \Omega(\psi) + Hom_A(P'_M, N) \circ \iota'_M \end{array}$$

is well-defined. Likewise φ induces a morphism in the opposite direction. We need to show that

$$\alpha - \alpha \circ \Omega(\psi) \circ \Omega(\varphi) \in Hom_A(P_M, N) \circ \iota_M$$

and likewise for φ and P'_M and Ω'_M in place of ψ, P_M and Ω_M. By symmetry we only show the first of the inclusions. Moreover, in order to do so it is sufficient to show

$$id_{\Omega_M} - \Omega(\psi) \circ \Omega(\varphi) \in Hom_A(P_M, \Omega_M) \circ \iota_M$$

since then we may compose with $\alpha \circ -$. Now, for this it is sufficient to show that if the identity on M lifts to an endomorphism $\chi : P \longrightarrow P$ of a projective P mapping onto M, then the map induced on the kernel differs from the identity only by a mapping factoring through P:

$$\begin{array}{ccccccccc} 0 & \longrightarrow & \Omega_M & \longrightarrow & P & \longrightarrow & M & \longrightarrow & 0 \\ & & \uparrow \Omega(\chi) & & \uparrow \chi & & \| & & \\ 0 & \longrightarrow & \Omega_M & \longrightarrow & P & \longrightarrow & M & \longrightarrow & 0 \end{array}$$

We may subtract the identity from all vertical morphisms and we need to prove that $\Omega(\nu)$ in

$$\begin{array}{ccccccccc} 0 & \longrightarrow & \Omega_M & \stackrel{\iota_M}{\longrightarrow} & P & \stackrel{\pi_M}{\longrightarrow} & M & \longrightarrow & 0 \\ & & \uparrow \Omega(\nu) & & \uparrow \nu & & \uparrow 0 & & \\ 0 & \longrightarrow & \Omega_M & \stackrel{\iota_M}{\longrightarrow} & P & \stackrel{\pi_M}{\longrightarrow} & M & \longrightarrow & 0 \end{array}$$

factors through $\Omega_M \longrightarrow P$. But since $\pi_M \circ \nu = 0$, there is a unique morphism $\mu : P \longrightarrow \Omega_M$ such that $\nu = \iota_M \circ \mu$. But

$$\iota_M \circ \mu \circ \iota_M = \nu \circ \iota_M = \iota_M \circ \Omega(\nu)$$

and since ι_M is injective,

$$\mu \circ \iota_M = \Omega(\nu).$$

This is what we claimed.

Let

$$\begin{array}{ccccccccc} 0 & \longrightarrow & N & \longrightarrow & X_1 & \longrightarrow & M & \longrightarrow & 0 \\ & & \| & & \downarrow & & \| & & \\ 0 & \longrightarrow & N & \longrightarrow & X_2 & \longrightarrow & M & \longrightarrow & 0 \end{array}$$

be two equivalent short exact sequences, then the first one gives a mapping $\Omega_M \longrightarrow N$ by the diagram

$$\begin{array}{ccccccccc} 0 & \longrightarrow & \Omega_M & \longrightarrow & P_M & \longrightarrow & M & \longrightarrow & 0 \\ & & \downarrow & & \downarrow & & \| & & \\ 0 & \longrightarrow & N & \longrightarrow & X_1 & \longrightarrow & M & \longrightarrow & 0 \\ & & \| & & \downarrow & & \| & & \\ 0 & \longrightarrow & N & \longrightarrow & X_2 & \longrightarrow & M & \longrightarrow & 0 \end{array}$$

and the second one by the analogous map $\Omega'_M \longrightarrow N$.

We need to show that given

$$\begin{array}{ccccccccc} 0 & \longrightarrow & \Omega_M & \xrightarrow{h} & P_M & \longrightarrow & M & \longrightarrow & 0 \\ & & \downarrow f & & \downarrow g & & \| & & \\ 0 & \longrightarrow & N & \longrightarrow & X & \longrightarrow & M & \longrightarrow & 0 \end{array}$$

then there is a mapping $\lambda : Q_M(f) \longrightarrow X$ making the diagram

$$(*) \qquad \begin{array}{ccccccccc} 0 & \longrightarrow & N & \longrightarrow & Q_M(f) & \longrightarrow & M & \longrightarrow & 0 \\ & & \| & & \downarrow \lambda & & \| & & \\ 0 & \longrightarrow & N & \xrightarrow{\nu} & X & \xrightarrow{\mu} & M & \longrightarrow & 0 \end{array}$$

commutative. Put

$$\begin{array}{rcl} \underline{Q_M} & \longrightarrow & X \\ \overline{(n,p)} & \mapsto & (\nu(n), g(p)). \end{array}$$

This is well-defined, since

$$(\nu, g)(f(\omega), -h(\omega)) = \nu \circ f(\omega) - g \circ h(\omega) = 0.$$

The fact that the two squares are commutative is clear.

The fact that the diagram $(*)$ is commutative implies that given an exact sequence

$$\mathcal{E} : \ 0 \longrightarrow N \longrightarrow X \longrightarrow M \longrightarrow 0$$

sending it to the corresponding $f_{\mathcal{E}} : \Omega_M \longrightarrow N$ and then sending this $f_{\mathcal{E}}$ to the corresponding exact sequence

$$0 \longrightarrow N \longrightarrow Q_M(f_{\mathcal{E}}) \longrightarrow M \longrightarrow 0$$

is the identity, up to equivalence of exact sequences.

The reverse composition gives the identity almost by definition. Indeed,

$$\begin{array}{ccccccccc} 0 & \longrightarrow & \Omega_M & \longrightarrow & P_M & \longrightarrow & M & \longrightarrow & 0 \\ & & \downarrow f & & \downarrow & & \| & & \\ 0 & \longrightarrow & N & \stackrel{\nu}{\longrightarrow} & Q_M(f) & \stackrel{\mu}{\longrightarrow} & M & \longrightarrow & 0 \end{array}$$

is commutative and so a given f is mapped to

$$0 \longrightarrow N \stackrel{\nu}{\longrightarrow} Q_M(f) \stackrel{\mu}{\longrightarrow} M \longrightarrow 0$$

and this again is mapped to f, since

$$\begin{array}{ccccccccc} 0 & \longrightarrow & \Omega_M & \longrightarrow & P_M & \longrightarrow & M & \longrightarrow & 0 \\ & & \downarrow f & & \downarrow & & \| & & \\ 0 & \longrightarrow & N & \stackrel{\nu}{\longrightarrow} & Q_M(f) & \stackrel{\mu}{\longrightarrow} & M & \longrightarrow & 0 \end{array}$$

is commutative.

This proves the proposition. □

Definition 1.8.19 Let A be a K-algebra for a commutative ring K, let M and N be A-modules and let

$$(*)\ 0 \longrightarrow \Omega_M \stackrel{\iota_M}{\longrightarrow} P_M \stackrel{\pi_M}{\longrightarrow} M \longrightarrow 0$$

be a short exact sequence where P_M is a projective A-module. Then define $Ext_A^1(M, N)$ by the exact sequence

$$Hom_A(P_M, N) \stackrel{Hom_A(\iota_M, N)}{\longrightarrow} Hom_A(\Omega_M, N) \rightarrow Ext_A^1(M, N) \rightarrow 0$$

which is induced by applying $Hom_A(-, N)$ to the sequence $(*)$ in the sense of Remark 1.8.7.

Remark 1.8.20 We have seen in Proposition 1.8.18 that $Ext_A^1(M, N)$ classifies short exact sequences

$$0 \longrightarrow N \longrightarrow X \longrightarrow M \longrightarrow 0$$

up to equivalence. Hence, $Ext_A^1(M, N)$ defined in this way does not depend on the specific module P_M and the epimorphism $P_M \rightarrow M$. Moreover, $Ext_A^1(M, N) \simeq Ext_A^1(M \oplus P, N)$ for every projective A-module P. Indeed, if $P_M \twoheadrightarrow M$, then $P_M \oplus P \twoheadrightarrow M \oplus P$ and the kernels coincide.

Remark 1.8.21 Observe that the order of the argument in $Ext_A^1(M, N)$ is inverse with respect to the order of the modules in the exact sequence

$$0 \longrightarrow N \longrightarrow ? \longrightarrow M \longrightarrow 0.$$

Remark 1.8.22 Let N be an A-module and let I_N be an injective A-module. Then for every short exact sequence $0 \to N \stackrel{\iota}{\to} I_N \stackrel{\nu}{\to} \mho_N \to 0$ we may apply $Hom_A(M,-)$ in the sense of Remark 1.8.7 and define $IExt_A^1(M,N)$ by the exact sequence

$$Hom_A(M,I_N) \stackrel{Hom_A(M,\nu)}{\longrightarrow} Hom_A(M,\mho_N) \to IExt_A^1(M,N) \to 0.$$

We claim that $Ext_A^1(M,N) \simeq IExt_A^1(M,N)$. Indeed, let P_M be a projective module and let $0 \to \Omega_M \stackrel{\mu}{\to} P_M \stackrel{\pi}{\to} M \to 0$ be exact. Since P_M is projective, and since ν is an epimorphism, for every $M \stackrel{\alpha}{\to} \mho_N$ there is a $P_M \stackrel{\hat{\alpha}}{\to} I_N$ such that $\nu \circ \hat{\alpha} = \alpha \circ \pi$. By the property of the kernel, there is a unique $\Omega_M \stackrel{\tilde{\alpha}}{\to} N$ such that $\iota \circ \tilde{\alpha} = \hat{\alpha} \circ \mu$. If α is in the image of $Hom_A(M,\nu)$, then $\alpha = \nu \circ \sigma$ for some $M \stackrel{\sigma}{\to} I_N$ and we can take $\hat{\alpha} := \sigma \circ \pi$ since $\nu \circ \hat{\alpha} = \nu \circ \sigma \circ \pi = \alpha \circ \pi$. But then $\hat{\alpha} \circ \mu = \sigma \circ \pi \circ \mu = 0$ and therefore $\tilde{\alpha} = 0$. If we take another $\hat{\alpha}'$ such that $\nu \circ \hat{\alpha} = \alpha \circ \pi$, then $\hat{\alpha} - \hat{\alpha}'$ lifts the morphism $M \stackrel{0}{\to} \mho_N$, and then for $\Omega_M \stackrel{\tilde{\alpha}'}{\to} N$ such that $\iota \circ \tilde{\alpha}' = \hat{\alpha}' \circ \mu$ there is a $P_M \stackrel{\tau}{\to} N$ such that $\tau \circ \mu = \tilde{\alpha}'$. In any case, $\alpha \mapsto \tilde{\alpha}$ induces a well-defined morphism $Ext_A^1(M,N) \to IExt_A^1(M,N)$. The dual argumentation, starting with a morphism $\Omega_M \to N$ and lifting along the short exact sequences, using the injectivity of I_N, gives a morphism $IExt_A^1(M,N) \to Ext_A^1(M,N)$, and is easy to verify that both compositions of these morphisms are the identity.

If A is a K-algebra, we see that $Ext_A^1(M,N)$ carries the structure of a K-module. This can be interpreted in terms of short exact sequences, at least in the case $K = \mathbb{Z}$. Indeed, the so-called *Baer sum* realises the abelian group structure.

Let

$$0 \longrightarrow N \stackrel{\nu_1}{\longrightarrow} X_1 \stackrel{\mu_1}{\longrightarrow} M \longrightarrow 0$$

and

$$0 \longrightarrow N \stackrel{\nu_2}{\longrightarrow} X_2 \stackrel{\mu_2}{\longrightarrow} M \longrightarrow 0$$

be two short exact sequences with the same end terms. These induce a short exact sequence

$$0 \longrightarrow N \oplus N \stackrel{\begin{pmatrix}\nu_1 & 0\\ 0 & \nu_2\end{pmatrix}}{\longrightarrow} X_1 \oplus X_2 \stackrel{\begin{pmatrix}\mu_1 & 0\\ 0 & \mu_2\end{pmatrix}}{\longrightarrow} M \oplus M \longrightarrow 0.$$

Let $\Delta_M : M \longrightarrow M \oplus M$ be the diagonal mapping $m \mapsto (m,m)$. Then we get a commutative diagram

$$\begin{array}{ccccccccc}
0 \longrightarrow & N \oplus N & \stackrel{\begin{pmatrix}\nu_1 & 0\\ 0 & \nu_2\end{pmatrix}}{\longrightarrow} & X_1 \oplus X_2 & \stackrel{\begin{pmatrix}\mu_1 & 0\\ 0 & \mu_2\end{pmatrix}}{\longrightarrow} & M \oplus M & \longrightarrow 0 \\
 & \| & & \uparrow & & \uparrow \Delta_M & \\
0 \longrightarrow & N \oplus N & \longrightarrow & Z_1 & \longrightarrow & M & \longrightarrow 0
\end{array}$$

by putting

$$Z_1 := \{(x_1, x_2, m) \in X_1 \oplus X_2 \oplus M \mid \mu_1(x_1) = m = \mu_2(x_2)\}.$$

The mapping $Z_1 \longrightarrow M$ is given by the projection onto the third component. The kernel is therefore given by the set of (x_1, x_2) such that $\mu_1(x_1) = 0 = \mu_2(x_2)$, which is $N \oplus N$ embedded by the diagonal embedding into the first two components via $\begin{pmatrix} \nu_1 & 0 \\ 0 & \nu_2 \end{pmatrix}$. By the codiagonal mapping $\nabla_N : N \oplus N \longrightarrow N$ mapping (n_1, n_2) to $n_1 + n_2$ we produce a new short exact sequence and commutative diagram

$$\begin{array}{ccccccccc} 0 & \longrightarrow & N \oplus N & \longrightarrow & Z_1 & \longrightarrow & M & \longrightarrow & 0 \\ & & \downarrow \nabla_N & & \downarrow & & \| & & \\ 0 & \longrightarrow & N & \longrightarrow & Z & \longrightarrow & M & \longrightarrow & 0 \end{array}$$

where

$$Z = (Z_1 \oplus N) / \{(\nu_1(n_1), \nu_2(n_2), 0, -(n_1 + n_2)) \mid n_1, n_2 \in N\},$$

where the middle vertical arrow is the embedding into the first component, and where the horizontal left arrow is the embedding into the second component. The sequence is exact since taking the quotient modulo the last component N yields Z_1 modulo the diagonal $N \oplus N$ embedded via (ν_1, ν_2). This is isomorphic to M.

The fact that this construction is well-defined is easily checked. The so-defined construction is called the *Baer sum*.

We shall have to prove that the resulting sequence

$$0 \longrightarrow N \longrightarrow Z \longrightarrow M \longrightarrow 0$$

corresponds to $f_1 + f_2$.

Proposition 1.8.23 *The Baer sum on the set of equivalence classes of exact sequences*

$$0 \longrightarrow N \longrightarrow ? \longrightarrow M \longrightarrow 0$$

corresponds to the abelian group structure of $Ext_A^1(M, N)$ *under the natural isomorphism.*

In order to prove this, it is convenient to introduce some concepts of homological algebra, which we have used above in a special context, namely the pullback and the pushout.

Definition 1.8.24 Let A be a K-algebra and let M, N_1 and N_2 be three A-modules.

- If moreover $\alpha_1 \in Hom_A(N_1, M)$ and $\alpha_2 \in Hom_A(N_2, M)$, then a triple (L, β_1, β_2) with L an A-module, $\beta_1 \in Hom_A(L, N_1)$, and $\beta_2 \in Hom_A(L, N_2)$ is a *pullback* if and only if the following conditions hold:

$$\alpha_1 \circ \beta_1 = \alpha_2 \circ \beta_2$$

and whenever there is a triple (T, γ_1, γ_2) with T an A-module, $\gamma_1 \in Hom_A(T, N_1)$, and $\gamma_2 \in Hom_A(T, N_2)$ and

$$\alpha_1 \circ \gamma_1 = \alpha_2 \circ \gamma_2,$$

then there is a unique $\delta \in Hom_A(T, L)$ such that

$$\beta_i \circ \delta = \gamma_i$$

for all $i \in \{1, 2\}$.

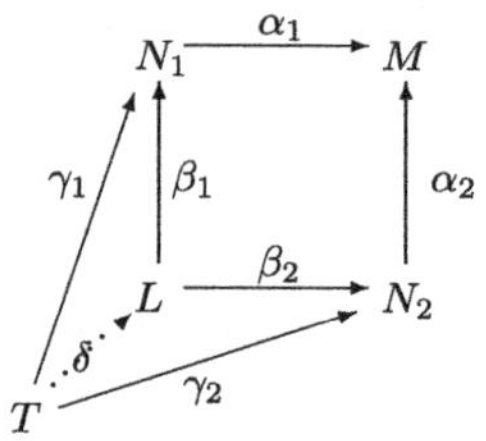

- If moreover $\alpha_1 \in Hom_A(M, N_1)$ and $\alpha_2 \in Hom_A(M, N_2)$, then a triple (L, β_1, β_2) with L an A-module, $\beta_1 \in Hom_A(N_1, L)$, and $\beta_2 \in Hom_A(N_2, L)$ is a *pushout* if and only if the following conditions hold

$$\beta_1 \circ \alpha_1 = \beta_2 \circ \alpha_2$$

and whenever there is a triple (T, γ_1, γ_2) with T an A-module, $\gamma_1 \in Hom_A(N_1, T)$, and $\gamma_2 \in Hom_A(N_2, T)$ and

$$\gamma_1 \circ \alpha_1 = \gamma_2 \circ \alpha_2,$$

then there is a unique $\delta \in Hom_A(L, T)$ such that

$$\delta \circ \beta_i = \gamma_i$$

for all $i \in \{1, 2\}$.

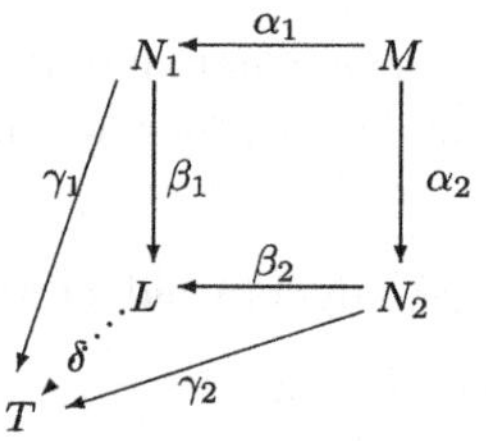

It should be noted that due to the unicity condition of δ, the pullback and the pushout are both unique up to isomorphism, if they exist.

Proposition 1.8.25 *Let A be a K-algebra.*

- *Then for any configuration $(M, N_1, N_2, \alpha_1, \alpha_2)$ of A-modules, with $\alpha_i \in Hom_A(N_i, M)$ the pullback is given as*

$$L := \{(n_1, n_2) \in N_1 \oplus N_2 \mid \alpha_1(n_1) = \alpha_2(n_2)\}.$$

In particular pullbacks exist for A-modules.

- *For any configuration $(M, N_1, N_2, \alpha_1, \alpha_2)$ of A-modules, where we have $\alpha_i \in Hom_A(M, N_i)$, the pushout is given as*

$$L' := (N_1 \oplus N_2) / \{(\alpha_1(m), -\alpha_2(m)) \mid m \in M\}.$$

In particular pushouts exist for A-modules.

Proof L is built so that the projection onto the first resp. second component is going to give mappings β_1 and β_2 as required. Moreover, given (T, γ_1, γ_2), one needs to define $\delta(t) := (\gamma_1(t), \gamma_2(t))$ in order to get the commutativity. Moreover, this definition makes the diagram commutative, and abuts in L.

The proof for the pushout is completely analogous. □

Remark 1.8.26 Pullbacks also exist in other situations.

- The pullback of groups is defined in the same way. If G, E_1 and E_2 are groups, and if $E_1 \xrightarrow{\varphi_1} G \xleftarrow{\varphi_2} E_2$ are homomorphisms of groups, then there is a unique group P and group homomorphisms $E_1 \xleftarrow{\psi_1} P \xrightarrow{\psi_2} E_2$ such that $\varphi_1 \circ \psi_1 = \varphi_2 \circ \psi_2$ and such that whenever there is another group Q with group homomorphisms $E_1 \xleftarrow{\chi_1} Q \xrightarrow{\chi_2} E_2$ satisfying $\varphi_1 \circ \chi_1 = \varphi_2 \circ \chi_2$, then there is a unique group homomorphism $Q \xrightarrow{\alpha} P$ satisfying $\chi_i = \psi_i \circ \alpha \; \forall\, i \in \{1, 2\}$. Actually

$$P := \{(e_1, e_2) \in E_1 \times E_2 \mid \varphi_1(e_1) = \varphi_2(e_2)\}$$

and the projection ψ_i onto the i-component satisfies this property. The proof is identical to that for modules.
- The pullback of mappings of sets is defined completely analogously, replacing in the above the word "group" by the word "set", and the word "group homomorphism" by the word "mapping". Also the construction of the pullback as a subset of pairs is the same as for groups or as for modules over algebras.

The fact that the cokernel on the pushout side and the kernel on the pullback side remains the same, as we have observed in our special case is a general phenomenon.

Lemma 1.8.27 *Consider the diagram of A-modules*

$$\begin{array}{cccc} & A & \stackrel{\alpha}{\longrightarrow} & B \\ (*) & \gamma\downarrow & & \beta\downarrow \\ & C & \stackrel{\delta}{\longrightarrow} & D \end{array}$$

- *The diagram is a pullback diagram if and only if γ induces an isomorphism on the kernels of α and of δ. Moreover, if the diagram is a pullback diagram, then α is surjective if δ is surjective.*
- *The diagram is a pushout diagram if and only if β induces an isomorphism on the cokernels of α and of δ. Moreover, if the diagram is a pushout diagram, then δ is injective if α is injective.*

Proof We only prove the first statement, the other being completely dual. First, let K be the kernel of δ and let L be the kernel of α. Then γ induces a mapping . $L \stackrel{\mu}{\longrightarrow} K$ since the square $(*)$ is commutative. Let $L \stackrel{\lambda}{\hookrightarrow} A$ and $K \stackrel{\kappa}{\hookrightarrow} C$ be the natural embeddings. The 0-mapping $K \longrightarrow B$ together with κ satisfy $\beta \circ 0 = \delta \circ \kappa$ since K is the kernel of δ. Hence, by the universal property of the pullback, there is a unique mapping $K \stackrel{\nu}{\longrightarrow} A$ such that $\gamma \circ \nu = \kappa$, and such that $\alpha \circ \nu = 0$. The last equation implies that there is a $K \stackrel{\omega}{\longrightarrow} L$ with $\nu = \lambda \circ \omega$. Therefore, $\kappa \circ \mu \circ \omega = \gamma \circ \lambda \circ \omega = \gamma \circ \nu = \kappa \circ id_K$. Since κ is a monomorphism, $\mu \circ \omega = id_K$. Moreover, the composition $\omega \circ \mu \circ \omega$ is a morphism with the same property as required in the pullback axioms. By unicity of this morphism, $\omega \circ \mu \circ \omega = \omega$. The fact that $\mu \circ \omega = id_K$ proves that ω is a monomorphism, and so $\omega \circ \mu = id_L$. This proves the first part of the statement.

We need to prove the converse. Suppose that γ induces an isomorphism $L \stackrel{\mu}{\longrightarrow} K$. Let P be the pullback of the diagram $C \stackrel{\delta}{\longrightarrow} D \stackrel{\beta}{\longleftarrow} B$, with morphisms $P \stackrel{\pi}{\longrightarrow} B$ and $P \stackrel{\psi}{\longrightarrow} C$. By the universal property of the pullback, there is a map $A \stackrel{\varphi}{\longrightarrow} P$ such that $\gamma = \psi \circ \varphi$ and $\alpha = \pi \circ \varphi$. Put $M := \ker(\pi)$ and denote by $M \stackrel{\varepsilon}{\hookrightarrow} P$ its natural embedding. Then φ restricts to a mapping $L \stackrel{\chi}{\longrightarrow} M$ and ψ restricts to a mapping $M \stackrel{\xi}{\longrightarrow} K$. Since $\gamma = \psi \circ \varphi$, we get $\mu = \xi \circ \chi$. Since μ is an isomorphism, ξ is a split epimorphism and χ is a split monomorphism. Hence $\ker \xi =: N$ is a direct factor of M and $\xi|_N = 0$ and we have $\pi \circ \varepsilon|_N = 0$. Therefore $N \stackrel{0}{\to} B$ and $N \stackrel{0}{\to} C$ satisfy the universal property of the pullback P, so that there is a unique mapping $N \stackrel{\rho}{\to} P$ such that $\pi \circ \rho = 0$ and $\psi \circ \rho = 0$. Necessarily $\rho = 0$, but $\varepsilon|_N$ also has these properties. Therefore $\varepsilon|_N = 0$, and since ε is a monomorphism, $N = 0$. This shows that ξ is an isomorphism, and since $\mu = \xi \circ \chi$, χ is also an isomorphism. But then φ is an isomorphism as well since we have a commutative diagram

$$\begin{array}{ccccccc} 0 \to & L & \stackrel{\lambda}{\to} & A & \stackrel{\alpha}{\to} & B & \to 0 \\ & \downarrow \chi & & \downarrow \varphi & & \| & \\ 0 \to & M & \stackrel{\varepsilon}{\to} & P & \stackrel{\pi}{\to} & B & \to 0 \end{array}$$

We still need to show that α is surjective if δ is surjective. The easiest way to prove this is to use the actual construction. Given $b \in B$, let $\overline{b}$ be its image in D. Since δ is surjective, there is a $c \in C$ such that its image under δ equals $\overline{b}$. Therefore, $(b, c) \in P$ maps to b under α. This proves the lemma. □

Remark 1.8.28 Except the statement on surjectivity/injectivity, everything has been proved using only the universal property of the pullback/pushout.

We now observe that the Baer sum construction is a pullback followed by a pushout.

Proof of Proposition 1.8.23 We come back to the proof of fact that the Baer sum actually coincides with the abelian group structure of the $Ext_A^1(M, N)$. The obvious mapping given by the diagonals

$$\begin{array}{ccccccccc}
0 & \longrightarrow & \Omega_M \oplus \Omega_M & \longrightarrow & P_M \oplus P_M & \longrightarrow & M \oplus M & \longrightarrow & 0 \\
 & & \uparrow & & \uparrow & & \uparrow & & \\
0 & \longrightarrow & \Omega_M & \longrightarrow & P_M & \longrightarrow & M & \longrightarrow & 0
\end{array}$$

yields commutative squares in the obvious sense.

By the property of a pullback, and by the property of the kernel, there are mappings

$$\begin{array}{ccccccccc}
0 & \longrightarrow & \Omega_M \oplus \Omega_M & \longrightarrow & P_M \oplus P_M & \longrightarrow & M \oplus M & \longrightarrow & 0 \\
 & & \uparrow & & \uparrow & \text{pullback} & \uparrow & & \\
0 & \longrightarrow & \Omega_M \oplus \Omega_M & \longrightarrow & Z_1 & \longrightarrow & M & \longrightarrow & 0 \\
 & & \uparrow & & \uparrow & & \| & & \\
0 & \longrightarrow & \Omega_M & \longrightarrow & P_M & \longrightarrow & M & \longrightarrow & 0
\end{array}$$

factorising the diagonal mappings. Continuing further along the pushout mappings again gives commutative diagrams

$$\begin{array}{ccccccccc}
0 & \longrightarrow & N & \longrightarrow & Z_3 & \longrightarrow & M & \longrightarrow & 0 \\
 & & \uparrow + & \text{pushout} & \uparrow & & \| & & \\
0 & \longrightarrow & N \oplus N & \longrightarrow & Z_2 & \longrightarrow & M & \longrightarrow & 0 \\
 & & \uparrow (\mathrm{f}_1, \mathrm{f}_2) & \text{pushout} & \uparrow & & \| & & \\
0 & \longrightarrow & \Omega_M \oplus \Omega_M & \longrightarrow & Z_1 & \longrightarrow & M & \longrightarrow & 0 \\
 & & \uparrow & & \uparrow & & \| & & \\
0 & \longrightarrow & \Omega_M & \longrightarrow & P_M & \longrightarrow & M & \longrightarrow & 0 \\
 & & \downarrow f_1 + f_2 & & \downarrow & & \| & & \\
0 & \longrightarrow & N & \longrightarrow & Z & \longrightarrow & M & \longrightarrow & 0
\end{array}$$

But now we observe that the identity mapping on the uppermost N and on the lowermost N makes the chain of the leftmost vertical arrows commutative.

The pushout property and the cokernel property give a commutative diagram

$$\begin{array}{ccccccccc} 0 & \longrightarrow & N & \longrightarrow & Z_3 & \longrightarrow & M & \longrightarrow & 0 \\ & & \| & & \uparrow & & \| & & \\ 0 & \longrightarrow & N & \longrightarrow & Z & \longrightarrow & M & \longrightarrow & 0 \end{array}$$

which implies that the two sequences are equivalent.

It needs still to be explained why the pullback construction on the level of the projective resolution, then pushing out along (f_1, f_2) and finally pushing out the codiagonal, gives an exact sequence equivalent to the one given by first pushing out along (f_1, f_2), then pulling back along the diagonal on M. We hence need to show that the lowermost exact sequence is equivalent to the uppermost exact sequence of the following diagram:

$$\begin{array}{ccccccc} 0 \longrightarrow & N \oplus N & \longrightarrow & Z_4 & \longrightarrow & M & \longrightarrow 0 \\ & \| & & \downarrow & \text{pullback} & \downarrow \Delta_M & \\ 0 \longrightarrow & N \oplus N & \longrightarrow & Z_3 & \longrightarrow & M \oplus M & \longrightarrow 0 \\ & (f_1, f_2) \uparrow & \text{pushout} & \uparrow & & \| & \\ 0 \longrightarrow & \Omega_M \oplus \Omega_M & \longrightarrow & P_M \oplus P_M & \longrightarrow & M \oplus M & \longrightarrow 0 \\ & \| & & \uparrow & \text{pullback} & \uparrow & \\ 0 \longrightarrow & \Omega_M \oplus \Omega_M & \longrightarrow & Z_1 & \longrightarrow & M & \longrightarrow 0 \\ & (f_1, f_2) \downarrow & \text{pushout} & \downarrow & & \| & \\ 0 \longrightarrow & N \oplus N & \longrightarrow & Z_2 & \longrightarrow & M & \longrightarrow 0 \end{array}$$

The pushout property of the lowermost right-hand square together with the identity map from the lowermost copy of $N \oplus N$ to the copy of $N \oplus N$ in the second row yields a mapping $Z_2 \longrightarrow Z_3$ making the following diagram commutative:

$$\begin{array}{ccccccc} 0 \longrightarrow & N \oplus N & \longrightarrow & Z_4 & \longrightarrow & M & \longrightarrow 0 \\ & \| & & \downarrow & \text{pullback} & \downarrow \Delta_M & \\ 0 \longrightarrow & N \oplus N & \longrightarrow & Z_3 & \longrightarrow & M \oplus M & \longrightarrow 0 \\ & \| & & \uparrow & & \uparrow \Delta_M & \\ 0 \longrightarrow & N \oplus N & \longrightarrow & Z_2 & \longrightarrow & M & \longrightarrow 0 \end{array}$$

Since now the upper right-hand square is a pullback along the same map as the lower right-hand square, the exact sequence on the upper line and on the lower line are equivalent:

$$\begin{array}{ccccccccc} 0 & \longrightarrow & N \oplus N & \longrightarrow & Z_4 & \longrightarrow & M & \longrightarrow & 0 \\ & & \| & & \uparrow & & \| & & \\ 0 & \longrightarrow & N \oplus N & \longrightarrow & Z_2 & \longrightarrow & M & \longrightarrow & 0 \end{array}$$

This proves the proposition. □

We can study "higher" Ext-groups, which are defined as follows.

Definition 1.8.29 Let A be a K-algebra and let M and N be two A-modules. Let

$$\xrightarrow{\partial_{i+1}} P_{i+1} \xrightarrow{\partial_i} P_i \xrightarrow{\partial_{i-1}} \cdots \xrightarrow{\partial_2} P_2 \xrightarrow{\partial_1} P_1 \xrightarrow{\partial_0} P_0 \xrightarrow{\epsilon} M \longrightarrow 0$$

be a projective resolution of M. Then

$$Ext^i_A(M,N) := \ker(Hom_A(\partial_i, N))/\mathrm{im}(Hom_A(\partial_{i-1}, N)).$$

for all $i \in \mathbb{N} \setminus \{0\}$.

Lemma 1.8.30 *$Ext^1_A(M,N)$ defined in Definition 1.8.29 coincides with the definition of $Ext^1_A(M,N)$ in Definition 1.8.19.*

Proof We see that

$$0 \longrightarrow \Omega_M \stackrel{\iota}{\longrightarrow} P_0 \stackrel{\epsilon}{\longrightarrow} M \longrightarrow 0$$

is exact, and following Definition 1.8.19 we obtain $Ext^1_A(M,N)$ as cokernel of

$$Hom_A(P_0, N) \stackrel{Hom_A(\iota,N)}{\longrightarrow} Hom_A(\Omega_M, N).$$

This is precisely $Hom_A(\Omega_M, N)/Hom_A(P_0, N) \circ \iota$. Definition 1.8.29 describes $Ext^1_A(M,N) = \ker(Hom_A(\partial_1, N))/\mathrm{im}(Hom_A(\partial_0, N))$. Now,

$$\ker(Hom_A(\partial_1, N)) = \{\alpha \in Hom_A(P_1, N) \mid \partial_1 \circ \alpha = 0\} = Hom_A(\Omega_M, N)$$

by the universal property of $\mathrm{coker}(\partial_1) = \Omega_M$. Further,

$$\mathrm{im}(Hom_A(\partial_0, N)) = \{\beta \circ \partial_0 \mid \beta \in Hom_A(P_0, N)\} = \mathrm{im}(Hom_A(\iota, N))$$

since the image of ∂_0 is the same as the image of ι in P_0. This proves the statement. □

We need to show that the construction from Definition 1.8.29 does not depend on the projective resolution. By Remark 1.8.20 we obtain that $Ext^1_A(M,N)$ does not depend on the projective cover of M which was chosen to evaluate $Ext^1_A(M,N)$. By definition we get

$$Ext^i_A(M,N) = Ext^1_A(\Omega^{i-1}_M, N).$$

We know that Ω_M is well-defined up to projective direct factors and isomorphism, and hence Ω^i_M is well-defined up to projective direct factors and isomorphism. We shall use the identification from Lemma 1.8.30. If Q is a projective module, then

$$Ext^1_A(\Omega^{i-1}_M \oplus Q, N) \simeq Ext^1_A(\Omega^{i-1}_M, N)$$

since if $0 \to N \to Y \to \Omega^{i-1}_M \oplus Q \to 0$ is a short exact sequence with projective Q, then $Y \simeq X \oplus Q$ and $0 \to N \to Y \to \Omega^{i-1}_M \oplus Q \to 0$ is equivalent to the short exact sequence $0 \to N \to X \oplus Q \stackrel{\pi}{\to} \Omega^{i-1}_M \oplus Q \to 0$, where $\pi = \begin{pmatrix} \pi' & 0 \\ 0 & id_Q \end{pmatrix}$. Finally Schanuel's Lemma 1.8.12 proves the statement.

We can prove certain first properties of Ext-groups. They behave in the same way as Hom-groups with respect to Frobenius reciprocity.

Proposition 1.8.31 *Let R be a commutative ring and let A and B be R-algebras. Suppose that there is an R-algebra homomorphism $B \to A$ and so A is a B-module. Then for all A-modules M and B-modules N we have*

$$Ext_A^i(A \otimes_B N, M) \simeq Ext_B^i(N, M)$$

for all $i \in \mathbb{N}$, natural in M and in N.

If A is projective as a B-module, then

$$Ext_A^i(N, A \otimes_B N) \simeq Ext_B^i(M, N)$$

for all $i \in \mathbb{N}$, natural in M and in N.

Proof For every projective B-module Q the module $A \otimes_B Q$ is projective as well. Indeed, Q is a direct factor of the free module $\bigoplus_I B$ and so $A \otimes_B Q$ is a direct factor of $A \otimes_B \bigoplus_I B = \bigoplus_I A$. Moreover, if $X \longrightarrow Y \longrightarrow Z$ is an exact sequence of projective B-modules, then

$$A \otimes_B X \longrightarrow A \otimes_B Y \longrightarrow A \otimes_B Z$$

is an exact sequence of A-modules. Indeed, this is true for free modules, and by taking direct summands also for projective modules. Finally, if

$$Q_0 \longrightarrow N \longrightarrow 0$$

is exact, then

$$A \otimes_B Q_0 \longrightarrow A \otimes_B N \longrightarrow 0$$

is exact as well. In other words, taking the tensor product preserves epimorphisms. Therefore, given a projective resolution

$$\cdots \longrightarrow Q_2 \longrightarrow Q_1 \longrightarrow Q_0 \longrightarrow N \longrightarrow 0$$

of N as a B-module,

$$\cdots \longrightarrow A \otimes_B Q_2 \longrightarrow A \otimes_B Q_1 \longrightarrow A \otimes_B Q_0 \longrightarrow A \otimes_B N \longrightarrow 0$$

is a projective resolution of $A \otimes_B N$ as an A-module, using the fact that tensor products preserve epimorphisms. Taking homomorphism spaces to M, i.e. applying $Hom_B(-, M)$ to the first sequence and $Hom_A(-, M)$ to the second sequence, gives two exact sequences which are actually isomorphic since

$$Hom_B(-, M) \simeq Hom_B(-, Hom_A(A, M)) \simeq Hom_A(A \otimes_B -, M)$$

by Frobenius reciprocity. More precisely

$$\begin{array}{ccccccc}
\cdots \leftarrow & Hom_B(Q_2, M) & \leftarrow & Hom_B(Q_1, M) & \leftarrow & Hom_B(Q_0, M) \\
& \| & & \| & & \| \\
\cdots \leftarrow & Hom_A(A \otimes_B Q_2, M) & \leftarrow & Hom_A(A \otimes_B Q_1, M) & \leftarrow & Hom_A(A \otimes_B Q_0, M)
\end{array}$$

Therefore,

$$Ext^i_A(A \otimes_B N, M) \simeq Ext^i_B(N, M)$$

as claimed.

For the other isomorphism we may apply the same arguments. Indeed, the restriction of a projective A-module to B is a projective B-module. This is a consequence of the assumption that the regular A-module is a projective B-module. Using this, the restriction of a free A-module is projective as a B-module, and therefore the restriction of a projective A-module to B is again projective as a B-module. Hence the tensor product of a projective resolution of N as a B-module gives a resolution of $A \otimes_B N$ as an A-module. The remaining argument is identical to that for the first case. □

Proposition 1.8.32 *Let R be a commutative ring, G be a group and H be a subgroup of G. Then for all RG-modules M and RH-modules N we have*

$$Ext^i_{RG}(M, N \uparrow_H^G) \simeq Ext^i_{RH}(M \downarrow_H^G, N)$$

and

$$Ext^i_{RG}(N \uparrow_H^G, M) \simeq Ext^i_{RH}(N, M \downarrow_H^G)$$

for all $i \in \mathbb{N}$, natural in M and in N.

Proof We use Proposition 1.8.31 in the case $A = RG$ and $B = RH$. Of course RG is a projective RH-module. Moreover, by definition, $RG \otimes_{RH} X = X \uparrow_H^G$. □

Lemma 1.8.33 *Let A be an R-algebra for a commutative ring R and let M_1, M_2 and N be A-modules. Then for all $i \in \mathbb{N}$ we get*

$$Ext^i_A(M_1 \oplus M_2, N) \simeq Ext^i_A(M_1, N) \oplus Ext^i_A(M_2, N)$$

and

$$Ext^i_A(N, M_1 \oplus M_2) \simeq Ext^i_A(N, M_1) \oplus Ext^i_A(N, M_2)$$

natural in M_1, M_2 and N.

Proof Consider a projective resolution

$$\cdots \longrightarrow P_2 \longrightarrow P_1 \longrightarrow P_0 \longrightarrow N \longrightarrow 0$$

of N as an A-module. Then we may apply

$$Hom_A(-, M_1 \oplus M_2) \simeq Hom_A(-, M_1) \oplus Hom_A(-, M_2)$$

to this sequence and for $M := M_1 \oplus M_2$ we obtain a commutative diagram

$$\begin{array}{ccccccc}
\cdots \leftarrow & Hom_A(P_2, M) & \leftarrow & Hom_A(P_1, M) & \leftarrow & Hom_A(P_0, M) \\
& \| & & \| & & \| \\
& Hom_A(P_2, M_1) & & Hom_A(P_1, M_1) & & Hom_A(P_0, M_1) \\
\cdots \leftarrow & \oplus & \leftarrow & \oplus & \leftarrow & \oplus \\
& Hom_A(P_2, M_2) & & Hom_A(P_1, M_2) & & Hom_A(P_0, M_2)
\end{array}$$

This gives the second isomorphism. The first isomorphism uses the fact that a projective resolution

$$\cdots \longrightarrow P_2^1 \longrightarrow P_1^1 \longrightarrow P_0^1 \longrightarrow M_1 \longrightarrow 0$$

of M_1 and a projective resolution

$$\cdots \longrightarrow P_2^2 \longrightarrow P_1^2 \longrightarrow P_0^2 \longrightarrow M_2 \longrightarrow 0$$

of M_2 induces a projective resolution

$$\cdots \longrightarrow P_2^1 \oplus P_2^2 \longrightarrow P_1^1 \oplus P_1^2 \longrightarrow P_0^1 \oplus P_0^2 \longrightarrow M_1 \oplus M_2 \longrightarrow 0$$

of $M_1 \oplus M_2$. The remaining argument is analogous to the case considered above. □

Lemma 1.8.34 *Let K be a commutative ring, let A be a K-algebra and let M and N be A-modules. Then $Ext_A^i(M, N)$ is an $End_A(N)$-left module. The module structure can be realised on $Ext_A^1(M, N)$ by a pullback-pushout construction on exact sequences.*

Proof We shall use Remark 1.8.30. Let

$$0 \longrightarrow N \longrightarrow X \longrightarrow M \longrightarrow 0$$

be a short exact sequence and let $\alpha \in End_A(N)$. Then we may form the pushout

$$\begin{array}{ccccccccc}
0 \longrightarrow & N & \longrightarrow & X & \longrightarrow & M & \longrightarrow 0 \\
& \downarrow \alpha & & \downarrow & & \| & \\
0 \longrightarrow & N & \longrightarrow & Y & \longrightarrow & M & \longrightarrow 0
\end{array}$$

which is defined to be the image of the sequence under α. Given $\alpha' \in End_A(N)$, we may compose to get

$$\begin{array}{ccccccccc} 0 & \longrightarrow & N & \longrightarrow & X & \longrightarrow & M & \longrightarrow & 0 \\ & & \downarrow \alpha & & \downarrow & & \| & & \\ 0 & \longrightarrow & N & \longrightarrow & Y & \longrightarrow & M & \longrightarrow & 0 \\ & & \downarrow \alpha' & & \downarrow & & \| & & \\ 0 & \longrightarrow & N & \longrightarrow & Z & \longrightarrow & M & \longrightarrow & 0 \end{array}$$

and the lowest sequence is the pushout along $\alpha' \circ \alpha$ since it induces the identity on M. The fact that the so-defined multiplicative action of $End_A(N)$ is distributive follows using Baer sums by a very similar argument. This proves the lemma. □

Corollary 1.8.35 *Let K be a commutative ring and let A be a K-algebra. Let M and N be two A-modules and suppose that N is a finite set. Then $|N| \cdot Ext^i_A(M, N) = 0$ for every $i \geq 1$. In particular, if G is a finite group and N is a finite KG-module, then $|N| \cdot Ext^i_{KG}(K, N) = 0$.*

Proof Indeed, multiplication by $|N|$ is the 0 morphism in $End_A(N)$, and hence the result follows by Lemma 1.8.34. □

Let K be a commutative ring and let A be a K-algebra. Then for every morphism $\alpha : M \longrightarrow N$ of A-modules and for every A-module T we get a morphism $Ext^1_A(T, M) \longrightarrow Ext^1_A(T, N)$ given by pushout

$$\begin{array}{ccccccccc} 0 & \longrightarrow & M & \longrightarrow & X & \longrightarrow & T & \longrightarrow & 0 \\ & & \downarrow & \text{P.O.} & \downarrow & & \| & & \\ 0 & \longrightarrow & N & \longrightarrow & Y & \longrightarrow & T & \longrightarrow & 0 \end{array}$$

and likewise in the first variable, contravariantly by pullback.

The following lemma gives a first version of an important phenomenon in terms of short exact sequences. Later, we shall give a more general statement, and a proof, in a much more general setup (cf Proposition 3.5.29 and 3.4.11). However, it is nice to know that the statement can be given entirely in terms of equivalence classes of short exact sequences.

Lemma 1.8.36 *Let K be a commutative ring and let A be a K-algebra. Suppose*

$$0 \longrightarrow M \longrightarrow N \longrightarrow L \longrightarrow 0$$

is an exact sequence of A-modules, and let T be an A-module. Then

$$Ext^1_A(T, M) \longrightarrow Ext^1_A(T, N) \longrightarrow Ext^1_A(T, L)$$

is exact.

Proof The composition $M \longrightarrow N \longrightarrow L$ is the 0-mapping. The composition $Ext^1_A(T, M) \longrightarrow Ext^1_A(T, N) \longrightarrow Ext^1_A(T, L)$ is therefore the pushout of the 0-mapping, which gives a split exact sequence. Hence, the composition $Ext^1_A(T, M) \longrightarrow Ext^1_A(T, N) \longrightarrow Ext^1_A(T, L)$ is 0. Now, let

$$\mathcal{E} \in \ker(Ext^1_A(T,N) \longrightarrow Ext^1_A(T,L))$$

to be represented by a short exact sequence

$$0 \longrightarrow N \longrightarrow Y \longrightarrow T \longrightarrow 0.$$

This implies that the lower short exact sequence in the following commutative diagram splits.

$$\begin{array}{ccccccccc} 0 & \longrightarrow & N & \longrightarrow & Y & \longrightarrow & T & \longrightarrow & 0 \\ & & \downarrow & \text{P.O.} & \downarrow & & \| & & \\ 0 & \longrightarrow & L & \longrightarrow & Z & \longrightarrow & T & \longrightarrow & 0 \end{array}$$

Hence, in the lower exact sequence, we may replace Z by $L \oplus T$, and the injection $L \longrightarrow Z$ by the natural identification into the first summand, and the projection $Z \longrightarrow T$ by the identification with the second summand.

$$\begin{array}{ccccccccc} 0 & \longrightarrow & N & \longrightarrow & Y & \longrightarrow & T & \longrightarrow & 0 \\ & & \downarrow & \text{P.O.} & \downarrow & & \| & & \\ 0 & \longrightarrow & L & \longrightarrow & Z & \longrightarrow & T & \longrightarrow & 0 \\ & & \| & & \downarrow & & \| & & \\ 0 & \longrightarrow & L & \longrightarrow & L \oplus T & \longrightarrow & T & \longrightarrow & 0 \end{array}$$

Moreover, $Y \longrightarrow Z$ is replaced by

$$Y \stackrel{(\sigma,\tau)}{\longrightarrow} L \oplus T.$$

We put $X := \ker(\sigma)$ and observe that then the composition

$$M \longrightarrow N \longrightarrow Y$$

factors through

$$\ker(Y \longrightarrow L \oplus T),$$

and in particular we get an induced map $M \longrightarrow X$. The map

$$M \longrightarrow \ker(Y \longrightarrow L \oplus T)$$

is injective, since if m is in the kernel of this map, its image in N is already in the kernel of $N \longrightarrow Y$, which is 0. Hence we get a commutative diagram with exact rows and columns

$$\begin{array}{ccccccccc}
0 & \longrightarrow & M & \longrightarrow & X & & & & \\
 & & \downarrow & & \downarrow & & & & \\
0 & \longrightarrow & N & \longrightarrow & Y & \longrightarrow & T & \longrightarrow & 0 \\
 & & \downarrow \text{ P.O.} & & \downarrow (\sigma, \tau) & & \| & & \\
0 & \longrightarrow & L & \longrightarrow & L \oplus T & \longrightarrow & T & \longrightarrow & 0
\end{array}$$

The cokernel of $M \longrightarrow X$ is $T = \mathrm{im}(\tau)$, so that we obtain a commutative diagram

$$\begin{array}{ccccccccc}
0 & \longrightarrow & M & \longrightarrow & X & \longrightarrow & T & \longrightarrow & 0 \\
 & & \downarrow & & \downarrow & & \| & & \\
0 & \longrightarrow & N & \longrightarrow & Y & \longrightarrow & T & \longrightarrow & 0 \\
 & & \downarrow \text{ P.O.} & & \downarrow (\sigma, \tau) & & \| & & \\
0 & \longrightarrow & L & \longrightarrow & L \oplus T & \longrightarrow & T & \longrightarrow & 0
\end{array}$$

with exact rows and columns. Since the rightmost vertical mappings are all identities, by Lemma 1.8.27, the square

$$\begin{array}{ccccc}
0 & \longrightarrow & M & \longrightarrow & X \\
 & & \downarrow & & \downarrow \\
0 & \longrightarrow & N & \longrightarrow & Y
\end{array}$$

is a pushout. Hence $\mathcal{E}$ is in the image of $Ext^1(T, M) \longrightarrow Ext^1_A(T, N)$. This proves the lemma. □

1.8.3 Low Dimensional Group Cohomology

A particularly important case for our purposes is the case of group cohomology.

Definition 1.8.37 Let G be a group and let R be a commutative ring. Then for any RG-module M, denoting by R the trivial RG-module, let $H^n(G, M) := Ext^n_{RG}(R, M)$, for $n \in \mathbb{N}$, be the *n-th group cohomology*.

What does this mean for small n? We shall discuss this question in the following paragraphs for $n \in \{0, 1, 2\}$.

Degree Zero Group Cohomology

By definition

$$H^0(G, M) = Hom_{RG}(R, M) = M^G$$

is the largest submodule of M with trivial G-action, the G-fixed points of M.

Degree One Group Cohomology

Already for $H^1(G,M)$ one needs some computation. An obvious starting point for the projective resolution of R is

$$0 \longrightarrow I(RG) \stackrel{\iota}{\longrightarrow} RG \longrightarrow R \longrightarrow 0$$

and the rightmost mapping sends all $g \in G$ to 1. The kernel of this mapping is $I(RG)$, the augmentation ideal, generated by $1-g \in RG$ for all $g \in G$, as a left ideal. Put

$$Der(G,M) = \{f : G \longrightarrow M \mid f(gh) = gf(h) + f(g)\ \forall g,h \in G\},$$

the *derivations* of G with values in M, and

$$InnDer(G,M) := \{f : G \longrightarrow M \mid \exists m \in M : f(g) = gm - m\},$$

the *inner derivations* of G with values in M. Then

$$H^1(G,M) = Der(G,M)/InnDer(G,M).$$

Indeed, if

$$\cdots \longrightarrow P_2 \stackrel{\partial_1}{\longrightarrow} P_1 \stackrel{\partial_0}{\longrightarrow} RG \longrightarrow R$$

is a projective resolution, then

$$\ker(Hom_{RG}(\partial_1, M)) = Hom_{RG}(I(RG), M)$$

since the morphisms $\varphi : P_1 \longrightarrow M$ with $\varphi \circ \partial_1 = 0$ factor through

$$\mathrm{coker}(\partial_1) = \ker(\partial_0) = I(RG).$$

Hence, since $g(h-1) = gh - g = (gh-1) - (g-1)$, we have

$$\begin{aligned}
\ker(Hom_{RG}(\partial_1, M)) &= Hom_{RG}(I(RG), M) \\
&= \{\varphi \in Hom_R(I(RG), M) \mid \\
&\qquad \forall g,h \in G :\ g\varphi(h-1) = \varphi(gh-1) - \varphi(g-1)\} \\
&= \{f : G \longrightarrow M \mid \\
&\qquad \forall g,h \in G :\ g \cdot f(h) = f(gh) - f(g)\} \\
&= Der(G,M).
\end{aligned}$$

Moreover,

$$\begin{aligned}\mathrm{im}(Hom_{RG}(\partial_0, M)) = \{f \in Hom_{RG}(I(RG), M) \mid \\ \exists \varphi \in Hom_{RG}(RG, M) : f = \varphi \circ \iota\}\end{aligned}$$

and since $Hom_{RG}(RG, M) = M$ by mapping φ to $\varphi(1) =: m$ we get for every RG-linear mapping f that $f(g-1) = (g-1)m$ and hence

$$\begin{aligned}\mathrm{im}(Hom_{RG}(\partial_0, M)) &= \{f \in Hom_{RG}(I(RG), M) \mid \\ &\qquad \exists m \in M \forall g \in G : f(g-1) = gm - m\} \\ &= InnDer(G, M).\end{aligned}$$

This proves the statement.

Degree Two Group Cohomology

In order to compute $H^2(G, M) = Ext^2_{\mathbb{Z}G}(\mathbb{Z}, M)$ we need to give a projective resolution of the trivial module $\mathbb{Z}$ in a more explicit way, and in particular in a way that works for all groups G. For algebras in general this is the so-called bar resolution, which will be given in full detail and complete generality in Definition 3.6.4. However, up to degree 2 the resolution is sufficiently simple to be given immediately, in particular for group rings.

Using the fact that

$$0 \longrightarrow I(RG) \longrightarrow RG \longrightarrow R \longrightarrow 0$$

is an exact sequence where RG is free, we need to find a projective RG-module P and an epimorphism $P \longrightarrow I(RG)$. Let $P := RG \otimes_R RG$ with the action of RG by multiplication on the left term of the tensor product. This module is clearly projective, even free. We define

$$\begin{aligned}RG \otimes_R RG &\xrightarrow{\partial_0} RG \\ g_1 \otimes g_2 &\mapsto g_1(g_2 - 1)\end{aligned}$$

for all $g_1, g_2 \in G$. It is clear that this map is well-defined and surjective since $I(RG)$ is generated by $g - 1$ for $g \in G$. Let $Q = RG \otimes_R RG \otimes_R RG$. This is a projective RG-module by multiplication on the left-most term. Let

$$\begin{aligned}RG \otimes_R RG \otimes_R RG &\xrightarrow{\partial_1} RG \otimes_R RG \\ g_1 \otimes g_2 \otimes g_3 &\mapsto g_1 g_2 \otimes g_3 - g_1 \otimes g_2 g_3 + g_1 \otimes g_2\end{aligned}$$

Then ∂_1 is RG-linear, and it is completely determined by the case $g_1 = 1$. We compute

$$\begin{aligned}\partial_0 \circ \partial_1(1 \otimes g_2 \otimes g_3) &= \partial_0(g_2 \otimes g_3 - 1 \otimes g_2g_3 + 1 \otimes g_2) \\ &= g_2(g_3 - 1) - (g_2g_3 - 1) + (g_2 - 1) \\ &= 0.\end{aligned}$$

Finally, we define a mapping

$$\begin{aligned}RG \otimes RG \otimes RG \otimes RG &\xrightarrow{\partial_2} RG \otimes_R RG \otimes_R RG \\ g_1 \otimes g_2 \otimes g_3 \otimes g_4 &\mapsto g_1g_2 \otimes g_3 \otimes g_4 - g_1 \otimes g_2g_3 \otimes g_4 \\ &\quad + g_1 \otimes g_2 \otimes g_3g_4 - g_1 \otimes g_2 \otimes g_3\end{aligned}$$

and compute similarly as in the previous computation that $\partial_1 \circ \partial_2 = 0$. The R-linear mappings

$$\begin{aligned}RG &\xrightarrow{h_0} RG \otimes RG \\ g &\mapsto 1 \otimes g \\ RG \otimes RG &\xrightarrow{h_1} RG \otimes RG \otimes RG \\ g_1 \otimes g_2 &\mapsto 1 \otimes g_1 \otimes g_2 \\ RG \otimes RG \otimes RG &\xrightarrow{h_2} RG \otimes RG \otimes RG \otimes RG \\ g_1 \otimes g_2 \otimes g_3 &\mapsto 1 \otimes g_1 \otimes g_2 \otimes g_3\end{aligned}$$

reveal that $\partial_1 \circ h_1 + h_0 \circ \partial_0 = id_{RG \otimes_R RG}$, and $\partial_2 \circ h_2 + h_1 \circ \partial_1 = id_{RG^{\otimes 3}}$. Indeed,

$$\begin{aligned}(\partial_1 \circ h_1 + h_0 \circ \partial_0)(1 \otimes g) &= \partial_1(1 \otimes 1 \otimes g) + h_0(g - 1) \\ &= 1 \otimes g - 1 \otimes g + 1 \otimes 1 + 1 \otimes g - 1 \otimes 1 \\ &= 1 \otimes g\end{aligned}$$

and

$$\begin{aligned}&(\partial_2 \circ h_2 + h_1 \circ \partial_1)(1 \otimes g \otimes h) \\ &= \partial_2(1 \otimes 1 \otimes g \otimes h) + h_1(g \otimes h - 1 \otimes gh + 1 \otimes g) \\ &= 1 \otimes g \otimes h - 1 \otimes g \otimes h + 1 \otimes 1 \otimes gh \\ &\quad - 1 \otimes 1 \otimes g + 1 \otimes g \otimes h - 1 \otimes 1 \otimes gh + 1 \otimes 1 \otimes g \\ &= 1 \otimes g \otimes h.\end{aligned}$$

A slightly more general approach is given in the proof of Proposition 3.6.1. Now, if $u \in \ker(\partial_0)$, then

$$\partial_1(h_1(u)) = -h_0(\partial_0(u)) + u = u$$

and similarly $v \in \ker \partial_1 \Rightarrow v \in \mathrm{im}(\partial_2)$. Hence the sequence

$$RG \otimes_R RG \otimes_R RG \otimes_R RG \stackrel{\partial_2}{\to} RG \otimes_R RG \otimes_R RG \stackrel{\partial_1}{\to} RG \otimes_R RG \stackrel{\partial_0}{\to} RG \to R \to 0,$$

which we denote for the moment by (†), is exact, and is therefore the beginning of a projective resolution of R.

An element in $Ext^2_{RG}(R, M)$ is given by a morphism of RG-modules $\varphi : RG \otimes_R RG \otimes_R RG \longrightarrow M$ such that $\varphi \circ \partial_2 = 0$. Two such morphisms φ, φ' give the same element in $Ext^2_{RG}(R, M)$ if and only if there is a morphism $\psi : RG \otimes_R RG \longrightarrow M$ of RG-modules such that

$$\varphi - \varphi' = \psi \circ \partial_1.$$

Since φ and ψ are both assumed to be RG-linear, their values are determined by elements of the form $1 \otimes g_2 \otimes g_3$ for φ and $1 \otimes g_2$ for ψ. The condition $\varphi \circ \partial_2 = 0$ translates into

$$\begin{aligned}
0 &= \varphi \circ \partial_2(1 \otimes g_2 \otimes g_3 \otimes g_4) \\
&= \varphi(g_2 \otimes g_3 \otimes g_4) - \varphi(1 \otimes g_2g_3 \otimes g_4) + \varphi(1 \otimes g_2 \otimes g_3g_4) - \varphi(1 \otimes g_2 \otimes g_3) \\
&= g_2\varphi(1 \otimes g_3 \otimes g_4) - \varphi(1 \otimes g_2g_3 \otimes g_4) + \varphi(1 \otimes g_2 \otimes g_3g_4) - \varphi(1 \otimes g_2 \otimes g_3)
\end{aligned}$$

and putting $f(g_2, g_3) := \varphi(1 \otimes g_2 \otimes g_3)$ we obtain that such an $f : G \times G \longrightarrow M$ gives an element in $H^2(G, M)$ if and only if for all $g_2, g_3, g_4 \in G$ we have

$$g_2 \cdot f(g_3, g_4) - f(g_2g_3, g_4) + f(g_2, g_3g_4) - f(g_2, g_3) = 0.$$

The difference of two such mappings f and f' gives the 0 element in $H^2(G, M)$ if and only if $f - f' = \psi \circ \partial_1$. Again, since ∂_1 and ψ are both RG-linear, we only need to evaluate the composition on elements of the form $1 \otimes g_2 \otimes g_3$ for all $g_2, g_3 \in G$. We therefore compute

$$\begin{aligned}
\psi \circ \partial_1(1 \otimes g_2 \otimes g_3) &= \psi(g_2 \otimes g_3 - 1 \otimes g_2g_3 + 1 \otimes g_2) \\
&= g_2 \cdot \psi(1 \otimes g_3) - \psi(1 \otimes g_2g_3) + \psi(1 \otimes g_2) \\
&= g_2\sigma(g_3) - \sigma(g_2g_3) + \sigma(g_2)
\end{aligned}$$

where we define $\sigma : G \times G \longrightarrow M$ by $\sigma(g) := \psi(1 \otimes g)$.

Definition 1.8.38 Let G be a finite group and let M be a $\mathbb{Z}G$-module. Then a mapping $f : G \times G \longrightarrow M$ is a *2-cocycle* if for all $g_2, g_3, g_4 \in G$ we have

$$g_2 \cdot f(g_3, g_4) - f(g_2g_3, g_4) + f(g_2, g_3g_4) - f(g_2, g_3) = 0.$$

A 2-cocycle $f : G \times G \longrightarrow M$ is a 2-*coboundary* if there is a mapping $\sigma : G \longrightarrow M$ such that

$$f(g_2, g_3) = g_2 \cdot \sigma(g_3) - \sigma(g_2 g_3) + \sigma(g_2).$$

Proposition 1.8.39 *Let G be a finite group and let M be a $\mathbb{Z}G$-module. Then the set $Z^2(G, M)$ of 2-cocycles is an abelian group and the set of 2-coboundaries $B^2(G, M)$ is a subgroup. Moreover*

$$H^2(G, M) \simeq Z^2(G, M)/B^2(G, M).$$

Proof The sequence (†) is a projective resolution, as is shown above. Now, $H^2(G, M) = Ext^2_{\mathbb{Z}G}(\mathbb{Z}, M)$ by definition and hence the above consideration gives an explicit computation of this group. □

Proposition 1.8.40 *Let G be a finite group and let M be a $\mathbb{Z}G$-module, then the order of each element in $H^2(G, M)$ divides $|G|$.*

Proof Let $f : G \times G \longrightarrow M$ be a 2-cocycle. Then

$$g_2 \cdot f(g_3, g_4) - f(g_2 g_3, g_4) + f(g_2, g_3 g_4) = f(g_2, g_3)$$

for all $g_2, g_3, g_4 \in G$. The right-hand side is independent of g_4. We sum this equation over all $g_4 \in G$ and obtain

$$g_2 \cdot \sum_{g_4 \in G} f(g_3, g_4) - \sum_{g_4 \in G} f(g_2 g_3, g_4) + \sum_{g_4 \in G} f(g_2, g_3 g_4) = |G| \cdot f(g_2, g_3).$$

If we put $\sigma(g_2) := \sum_{g_4 \in G} f(g_2, g_4)$, we obtain, since g_4 runs through G if and only if $g_3 g_4$ runs through G,

$$g_2 \cdot \sigma(g_3) - \sigma(g_2 g_3) + \sigma(g_2) = |G| \cdot f(g_2, g_3).$$

Hence $|G| \cdot f$ is a 2-coboundary. This proves the statement. □

As usual we denote by $gcd(n, m)$ the greatest common divisor of two integers.

Corollary 1.8.41 *Let G be a finite group and let M be a finite $\mathbb{Z}G$-module. If $gcd(|M|, |G|) = 1$, then $H^2(G, M) = 0$.*

Proof Proposition 1.8.40 shows that $|G| \cdot H^2(G, M) = 0$ and Corollary 1.8.35 shows that $|M| \cdot H^2(G, M) = 0$. Since $gcd(|M|, |G|) = 1$ we obtain $H^2(G, M) = 0$. □

Remark 1.8.42 For $n \leq 2$ we have a quite comfortable description of the group cohomology $H^n(G, M)$, or what is the same $Ext^n_{\mathbb{Z}G}(\mathbb{Z}, M)$, by means of cocycles and coboundaries. This can be done in the same way for higher degrees $n > 2$, but the formulas one obtains are not necessarily easy to use. In degree 2, however, this is useful and appears in various places.

1.8.4 Short Exact Sequences of Groups and Group Cohomology

We shall now give an interpretation of $H^2(G,M)$ in terms of group extensions. For this we observe that

$$H^2(G,M) \simeq Ext^2_{\mathbb{Z}G}(\mathbb{Z},M) \simeq Ext^1_{\mathbb{Z}G}(I(\mathbb{Z}G),M).$$

We have seen in Remark 1.8.30 that $Ext^1_{\mathbb{Z}G}(I(\mathbb{Z}G),M)$ is in bijection with equivalence classes of short exact sequences of $\mathbb{Z}G$-modules

$$0 \longrightarrow M \longrightarrow X \longrightarrow I(\mathbb{Z}G) \longrightarrow 0$$

up to some equivalence relation. Now we consider short exact sequences of groups

$$1 \longrightarrow M \xrightarrow{\iota} E \xrightarrow{\epsilon} G \longrightarrow 1$$

in the sense that ι is an injective group homomorphism and ϵ is a surjective group homomorphism such that $\ker(\epsilon) = \mathrm{im}(\iota)$, and say that two such sequences

$$1 \longrightarrow M \xrightarrow{\iota} E \xrightarrow{\epsilon} G \longrightarrow 1$$

and

$$1 \longrightarrow M \xrightarrow{\iota'} E' \xrightarrow{\epsilon'} G \longrightarrow 1$$

are equivalent if and only if there is a group homomorphism $\alpha : E \longrightarrow E'$ making the diagram

$$\begin{array}{ccccccccc} 1 & \longrightarrow & M & \xrightarrow{\iota} & E & \xrightarrow{\epsilon} & G & \longrightarrow & 1 \\ & & \| & & \downarrow\alpha & & \| & & \\ 1 & \longrightarrow & M & \xrightarrow{\iota'} & E' & \xrightarrow{\epsilon'} & G & \longrightarrow & 1 \end{array}$$

commutative. It is clear that then α is an isomorphism, since $\ker(\alpha) \subseteq \ker(\epsilon)$, using that the right-hand square is commutative, and the commutativity of the left-hand square implies that $\ker(\alpha) = 1$. If $f \in E'$, then $\epsilon'(f) = \epsilon(e)$ for some $e \in E$ and hence $\epsilon'(\alpha(e)^{-1}f) = 1$. Hence $\alpha(e)^{-1}f = \iota'(m) = \alpha(\iota(m))$ for some $m \in M$ and therefore $f = \alpha(e\iota(m))$. Therefore α is surjective. Observe that the setting implies that M is a normal abelian subgroup of E and that $E/M \simeq G$. Moreover, the conjugation of E on M is actually a conjugation action of G on M, since M is abelian.

Recall that for a group G and a commutative ring R we denote by $I(RG)$ the kernel of the augmentation map

$$RG \longrightarrow R$$

$$\sum_{g \in G} \alpha_g g \mapsto \sum_{g \in G} \alpha_g$$

This ideal is generated by the set $\{g - 1 \mid g \in G\}$ as an ideal of RG. If N is a normal subgroup of a group E, then $I(RN)E$ is the kernel of the ring homomorphism

$$\begin{aligned} RE &\longrightarrow RE/N \\ \sum_{g \in E} \alpha_g g &\mapsto \sum_{g \in E} \alpha_g gN \end{aligned}$$

This ideal is generated by the set $\{n - 1 \mid n \in N\}$ as an RE-module. Since $N \trianglelefteq G$ we have $I(RN) \cdot RE = RE \cdot I(RN)$.

Let $e \in E$ and $n \in M$. Then multiplication by e coincides with multiplication by en on $I(\mathbb{Z}M)E/(I(\mathbb{Z}E) \cdot I(\mathbb{Z}M))$. Indeed, $e - en = e(1 - n)$ and for $m \in M$ multiplication by $e - en$ on a generic element $m - 1$ produces

$$(e - en)(m - 1) = e(1 - n)(m - 1) \in I(\mathbb{Z}E) \cdot I(\mathbb{Z}M)$$

and hence multiplication by e on $I(\mathbb{Z}M)E/(I(\mathbb{Z}E) \cdot I(\mathbb{Z}M))$ coincides with multiplication by en on $I(\mathbb{Z}M)E/(I(\mathbb{Z}E) \cdot I(\mathbb{Z}M))$. For each $g \in G$ we may choose $e_g \in E$ with $\epsilon(e_g) = g$. Therefore, multiplication by e_g on $m - 1$ defines an action of G on $I(\mathbb{Z}M)E/(I(\mathbb{Z}E) \cdot I(\mathbb{Z}M))$. This multiplication action is actually the action of G on M, as will be shown in the following lemma.

Lemma 1.8.43 *Let G be a finite group and let M be an abelian normal subgroup of a group E with an isomorphism $E/M \simeq G$. Then we get an isomorphism*

$$M \simeq I(\mathbb{Z}M)E/(I(\mathbb{Z}E) \cdot I(\mathbb{Z}M))$$

of $\mathbb{Z}G$-modules.

Proof First observe that

$$I(\mathbb{Z}M) \cdot I(\mathbb{Z}M) \leq I(\mathbb{Z}E) \cdot I(\mathbb{Z}M).$$

Then we define

$$\begin{aligned} M &\xrightarrow{\alpha} I(\mathbb{Z}M)E/(I(\mathbb{Z}E) \cdot I(\mathbb{Z}M)) \\ m &\mapsto m - 1 \end{aligned}$$

This mapping is a mapping of abelian groups, since

$$mn - 1 = (m - 1) + (n - 1) + (m - 1)(n - 1) \forall m, n \in M$$

and therefore

$$\alpha(mn) = mn - 1 = (m-1) + (n-1) = \alpha(m) + \alpha(n)$$

in $I(\mathbb{Z}M)E/I(\mathbb{Z}E)I(\mathbb{Z}M)$. Moreover for all $g \in G$ choose $e_g \in E$ so that e_g maps to g under the natural epimorphism $E \longrightarrow G$.

We get

$$(e_g m - e_g)(e_g^{-1} - 1) = e_g m e_g^{-1} - 1 - e_g m + e_g = (e_g m e_g^{-1} - 1) - (e_g m - e_g)$$

for all $g \in G$ and $m \in M$. Hence

$$\alpha({}^g m) = g \cdot \alpha(m)$$

where the action of $g \in G$ on $I(\mathbb{Z}M)E/(I(\mathbb{Z}E) \cdot I(\mathbb{Z}M))$ is defined to be multiplication by e_g on $I(\mathbb{Z}M)E/(I(\mathbb{Z}E) \cdot I(\mathbb{Z}M))$. The fact that α is bijective is clear since an inverse is given immediately by

$$\begin{array}{rcl} I(\mathbb{Z}M)E/(I(\mathbb{Z}E) \cdot I(\mathbb{Z}M)) & \stackrel{\beta}{\longrightarrow} & M \\ (m-1) & \mapsto & m \end{array}$$

The above considerations show that $e \cdot (m-1)$ is actually the same as $eme^{-1} - 1$ in this quotient, and hence it is sufficient to describe the images of $m-1$ for $m \in M$. This proves the statement. □

Let

$$1 \longrightarrow M \longrightarrow E \longrightarrow G \longrightarrow 1$$

be a short exact sequence of groups. Then

$$0 \longrightarrow I(\mathbb{Z}M)E \longrightarrow \mathbb{Z}E \longrightarrow \mathbb{Z}G \longrightarrow 0$$

is an exact sequence of $\mathbb{Z}E$-modules and since $\mathbb{Z}E$ and $\mathbb{Z}G$ admit an augmentation mapping $\mathbb{Z}E \longrightarrow \mathbb{Z}$ and $\mathbb{Z}G \longrightarrow \mathbb{Z}$, respectively, by sending $g \mapsto 1$ and $e \mapsto 1$ for all $g \in G, e \in E$, we get a commutative diagram

$$\begin{array}{ccccccccc} & & & & 0 & & 0 & & \\ & & & & \downarrow & & \downarrow & & \\ 0 & \longrightarrow & I(\mathbb{Z}M)E & \longrightarrow & I(\mathbb{Z}E) & \longrightarrow & I(\mathbb{Z}G) & \longrightarrow & 0 \\ & & \| & & \downarrow & & \downarrow & & \\ 0 & \longrightarrow & I(\mathbb{Z}M)E & \longrightarrow & \mathbb{Z}E & \longrightarrow & \mathbb{Z}G & \longrightarrow & 0 \\ & & & & \downarrow & & \downarrow & & \\ & & & & \mathbb{Z} & = & \mathbb{Z} & & \end{array}$$

which then produces an exact sequence

$$0 \longrightarrow I(\mathbb{Z}M)E \longrightarrow I(\mathbb{Z}E) \longrightarrow I(\mathbb{Z}G) \longrightarrow 0$$

of $\mathbb{Z}E$-modules. Finally, taking the quotient by $I(\mathbb{Z}M)I(\mathbb{Z}E)$ gives an exact sequence

$$0 \to I(\mathbb{Z}M)E/(I(\mathbb{Z}M)I(\mathbb{Z}E)) \to I(\mathbb{Z}E)/(I(\mathbb{Z}M)I(\mathbb{Z}E)) \to I(\mathbb{Z}G) \to 0$$

of $\mathbb{Z}E$-modules. However, M acts as 1 on each of the terms as we have seen above, and hence this is a short exact sequence of RG-modules. Lemma 1.8.43 shows that actually we obtain a short exact sequence of $\mathbb{Z}G$-modules

$$0 \longrightarrow M \longrightarrow I(\mathbb{Z}E)/(I(\mathbb{Z}M)I(\mathbb{Z}E)) \longrightarrow I(\mathbb{Z}G) \longrightarrow 0.$$

If we have a commutative diagram of groups

$$\begin{array}{ccccccccc} 1 & \longrightarrow & M & \longrightarrow & E & \longrightarrow & G & \longrightarrow & 0 \\ & & \| & & \downarrow & & \| & & \\ 0 & \longrightarrow & M & \longrightarrow & E' & \longrightarrow & G & \longrightarrow & 0 \end{array}$$

then the above construction applied to both of the sequences induces a commutative diagram of $\mathbb{Z}G$-modules

$$\begin{array}{ccccccccc} 0 & \longrightarrow & M & \longrightarrow & I(\mathbb{Z}E)/(I(M)I(\mathbb{Z}E)) & \longrightarrow & I(\mathbb{Z}G) & \longrightarrow & 0 \\ & & \| & & \downarrow & & \| & & \\ 0 & \longrightarrow & M & \longrightarrow & I(\mathbb{Z}E')/\big(I(M)I(\mathbb{Z}E')\big) & \longrightarrow & I(\mathbb{Z}G) & \longrightarrow & 0 \end{array}$$

and therefore we obtain two equivalent short exact sequences of $\mathbb{Z}G$-modules. Hence we obtain a mapping Φ from the equivalence classes of short exact sequences of groups with ending terms M and G to equivalence classes of short exact sequences of $\mathbb{Z}G$-modules with ending terms M and $I(\mathbb{Z}G)$ respectively.

We shall give a mapping Ψ in the opposite direction.

Given a short exact sequence of $\mathbb{Z}G$-modules

$$\mathcal{E}:\ 0 \longrightarrow M \longrightarrow X \longrightarrow I(\mathbb{Z}G) \longrightarrow 0$$

we define a map

$$G \xrightarrow{\overline{\alpha}} I(\mathbb{Z}G)$$
$$g \mapsto g - 1$$

and we produce a short exact sequence of groups by the following pullback construction:

$$\begin{array}{ccccccccc} 0 & \longrightarrow & M & \longrightarrow & X & \xrightarrow{\pi} & I(\mathbb{Z}G) & \longrightarrow & 0 \\ & & \| & & \uparrow & & \uparrow \overline{\alpha} & & \\ & & M & \hookrightarrow & E & \twoheadrightarrow & G & & \end{array}$$

We shall define a group structure on E.

Recall that

$$E = \{(x, g) \in X \times G \mid \pi(x) = \overline{\alpha}(g) = g - 1\}.$$

Define

$$(x, g) \cdot (x', g') := (x + g \cdot x', gg')$$

for all $x, x' \in X$ and $g, g' \in G$. This makes sense, since X is a $\mathbb{Z}G$-module, and therefore $x+g\cdot x'$ is defined. We observe that this is a subset of $X \rtimes G$, the semi-direct product of X with G. Since $X \rtimes G$ is a group, the so-defined product is associative, and we only need to show that the neutral element belongs to E, each element has its inverse in E and that E is closed under the product. First we show that the product preserves E. For this we observe that $\pi(x) = g - 1 \text{and} \pi(x') = g' - 1$ implies

$$\pi(x + g \cdot x') = \pi(x) + g \cdot \pi(x') = g - 1 + g \cdot (g' - 1) = gg' - 1$$

and therefore the product is well-defined. Now

$$(x, g)^{-1} = (-g^{-1} \cdot x, g^{-1}),$$

and

$$\pi(-g^{-1} \cdot x) = -g^{-1}\pi(x) = -g^{-1}(g - 1) = g^{-1} - 1$$

shows that

$$(x, g) \in E \Rightarrow (x, g)^{-1} \in E.$$

The neutral element is (0, 1), which clearly belongs to E. Hence E is a group with normal subgroup M, embedded into E via $M \ni m \mapsto (m, 1) \in E$ and G acts on M via the module structure. Indeed, if $m \in M \subseteq X$ then

$$\begin{aligned}(x, g) \cdot (m, 1) \cdot (-g^{-1} \cdot x, g^{-1}) &= (x + gm, g) \cdot (-g^{-1} \cdot x, g^{-1}) \\ &= (x + gm - gg^{-1}x, 1) \\ &= (gm, 1)\end{aligned}$$

for each $(x, g) \in E$. This defines a mapping Ψ as announced.

We need to show that $\Phi \circ \Psi$ is the identity and that $\Psi \circ \Phi$ is the identity on the respective set of equivalence classes of short exact sequences.

Let

$$0 \longrightarrow M \longrightarrow X \longrightarrow I(\mathbb{Z}G) \longrightarrow 0$$

be a short exact sequence of $\mathbb{Z}G$-modules. Then we produce the extension of groups via the pullback construction

$$\begin{array}{ccccccc} 0 \longrightarrow & M & \longrightarrow & X & \stackrel{\pi}{\longrightarrow} & I(\mathbb{Z}G) & \longrightarrow 0 \\ & \| & & \uparrow \alpha & & \uparrow \overline{\alpha} & \\ & M & \hookrightarrow & E & \twoheadrightarrow & G & \end{array}$$

and construct from this again a short exact sequence of $\mathbb{Z}G$-modules as

$$0 \longrightarrow M \longrightarrow I(\mathbb{Z}E)/(I(M)I(\mathbb{Z}E)) \longrightarrow I(\mathbb{Z}G) \longrightarrow 0.$$

We define a mapping

$$\begin{array}{ccc} I(\mathbb{Z}E)/(I(M)I(\mathbb{Z}E)) & \stackrel{\hat{\alpha}}{\longrightarrow} & X \\ \sum_{i=1}^{n} z_i((x_i, g_i) - 1) & \mapsto & \sum_{i=1}^{n} z_i x_i \end{array}$$

where $x_i \in X$ and $g_i \in G$ so that $(x_i, g_i) \in E$ for all $i \in \mathbb{N}$. We need to show that this is well-defined. For $m \in M$ and $e = (x, g) \in E$ we get

$$((m, 1) - 1)((x, g) - 1) = ((m + x, g) - 1) - ((x, g) - 1) - ((m, 1) - 1)$$

and this is mapped by $\hat{\alpha}$ to

$$m + x - x - m = 0$$

so that the map $\hat{\alpha}$ is well-defined.

Further, the map $\hat{\alpha}$ is $\mathbb{Z}G$-linear, since the action of G on the $\mathbb{Z}G$-module $I(\mathbb{Z}E)/(I(M)I(\mathbb{Z}E))$ is given by the usual multiplication action of some (x, g) on this module $I(\mathbb{Z}E)/(I(M)I(\mathbb{Z}E))$. Now,

$$\begin{aligned} &\hat{\alpha}\left((x, g) \cdot \sum_{i=1}^{n} z_i((x_i, g_i) - 1)\right) \\ &= \hat{\alpha}\left(\sum_{i=1}^{n} z_i\,((x, g) \cdot (x_i, g_i) - (x, g))\right) \\ &= \hat{\alpha}\left(\sum_{i=1}^{n} z_i\,(((x + gx_i, gg_i) - 1) - ((x, g) - 1))\right) \\ &= \sum_{i=1}^{n} z_i(x + gx_i - x) = \sum_{i=1}^{n} z_i g x_i = g \cdot \left(\sum_{i=1}^{n} z_i x_i\right) \\ &= g \cdot \hat{\alpha}\left(\sum_{i=1}^{n} z_i((x_i, g_i) - 1)\right). \end{aligned}$$

Finally, the diagram

$$\begin{array}{ccccccccc} 0 & \longrightarrow & M & \longrightarrow & I(\mathbb{Z}E)/(I(M)I(\mathbb{Z}E)) & \longrightarrow & I(\mathbb{Z}G) & \longrightarrow & 0 \\ & & \| & & \downarrow \hat{\alpha} & & \| & & \\ 0 & \longrightarrow & M & \longrightarrow & X & \longrightarrow & I(\mathbb{Z}G) & \longrightarrow & 0 \end{array}$$

is commutative, as is immediately verified. Hence $\Phi \circ \Psi$ is the identity.

Given now a short exact sequence of groups

$$1 \longrightarrow M \longrightarrow E \longrightarrow G \longrightarrow 1$$

for an abelian normal subgroup M we form the short exact sequence

$$0 \longrightarrow M \longrightarrow I(\mathbb{Z}E)/(I(M)I(\mathbb{Z}E)) \longrightarrow I(\mathbb{Z}G) \longrightarrow 0$$

and construct the set theoretical pullback along

$$\begin{array}{ccc} G & \stackrel{\overline{\alpha}}{\longrightarrow} & I(\mathbb{Z}G) \\ g & \mapsto & g-1 \end{array}$$

We claim that

$$\begin{array}{ccccccccc} 0 & \longrightarrow & M & \longrightarrow & I(\mathbb{Z}E)/(I(M)I(\mathbb{Z}E)) & \stackrel{\pi}{\longrightarrow} & I(\mathbb{Z}G) & \longrightarrow & 0 \\ & & \| & & \uparrow \alpha & & \uparrow \overline{\alpha} & & \\ & & M & \hookrightarrow & E & \stackrel{\overline{\pi}}{\twoheadrightarrow} & G & & \end{array}$$

is a set theoretical pullback with $\alpha(e) = e - 1$. Indeed, since the kernels of the mappings coincide, using Lemma 1.8.27, the above is a pullback of groups. But, pullbacks of groups are actually the same as pullbacks of sets. This shows that $\Phi \circ \Psi$ is the identity.

Proposition 1.8.44 *Let G be a group and let M be a $\mathbb{Z}G$-module. Then $H^2(G, M)$ parameterises equivalence classes of short exact sequences*

$$1 \longrightarrow M \longrightarrow E \stackrel{\pi}{\longrightarrow} G \longrightarrow 1$$

of groups. The explicit construction is given by Φ and the exact sequences of groups for which there is a $G \stackrel{\sigma}{\longrightarrow} E$ with $\sigma \circ \pi = id_G$ correspond to the 0 element in $H^2(G, M)$.

Remark 1.8.45 Short exact sequences of groups

$$1 \longrightarrow M \longrightarrow E \stackrel{\pi}{\longrightarrow} G \longrightarrow 1$$

for which there is a $G \stackrel{\sigma}{\longrightarrow} E$ with $\sigma \circ \pi = id_G$ are called *split exact sequences of groups*. Note that in general there is no group homomorphism $E \longrightarrow M$ so that the composition with the embedding $M \longrightarrow E$ is the identity on M. Further recall that split short exact sequences of groups occur precisely if $E \simeq M \rtimes G$. These facts are very classical and can be found in e.g. Rotman [13] or Huppert [14].

Proof of Proposition 1.8.44 The correspondence was established in the discussion preceding the statement. We still need to show that if π splits, $\overline{\pi}$ also splits. But this follows from the construction of E as a subset of $I(\mathbb{Z}E)/(I(M)I(\mathbb{Z}E)) \rtimes G$. This proves the proposition. □

Remark 1.8.46 We chose the above approach in order to emphasise the link between module extensions and group extensions, and in order to avoid the explicit use of 2-cocycles. However, we could as well just look at extensions

$$1 \longrightarrow M \longrightarrow E \longrightarrow G \longrightarrow 1$$

and find for every $g \in G$ a preimage e_g in E. Then $e_g \cdot e_h \cdot e_{(gh)}^{-1} \in M$, since $g \cdot h \cdot (gh)^{-1} = 1 \in G$. Hence if one defines $f(g,h) := e_g \cdot e_h \cdot e_{(gh)}^{-1}$, then one may show that this element does not really depend on the choice of e_g for each $g \in G$, and that the multiplication in E is associative if and only if f is a 2-cocycle. Two different choices of coset representatives e_g, and an equivalent short exact sequence of groups, will differ by a 2-coboundary, and in this way we can prove, after some work of course, that the short exact sequences of groups, up to the natural equivalence of short exact sequences, are parameterised by $H^2(G,M)$. This approach is well-documented, and can be found in practically all standard books on group theory, for example in Rotman [13, Sect. 7].

As an application we prove the important Schur-Zassenhaus theorem.

Theorem 1.8.47 *Let E be a finite group and let $N \trianglelefteq E$. Suppose that $|N|$ is relatively prime to $|G|$. Then $E \simeq N \rtimes G$.*

Proof We shall prove the result by induction on $|E|$.

Let P be a Sylow p-subgroup of N. Then, for every $g \in E$, the group gPg^{-1} is again a Sylow p-subgroup of N, since N is normal in E and hence $gNg^{-1} = N$. This shows that there is $n \in N$ so that $ngPg^{-1}n^{-1} = P$ by Sylow's theorem. Hence $ng \in N_E(P)$ and therefore $E = N \cdot N_E(P)$.

But $N_E(P) \cap N = N_N(P)$ by the definition of a normaliser. Moreover, $N_N(P)$ is a normal subgroup of $N_E(P)$. Hence

$$G = E/N = N \cdot N_E(P)/N = N_E(P)/(N \cap N_E(P)) = N_E(P)/N_N(P).$$

Now, $N_N(P)$ is a subgroup of N, and therefore the order of $N_N(P)$ is relatively prime to the order of G.

If $N_E(P) \neq E$, then by the induction hypothesis $N_E(P) \simeq N_N(P) \rtimes G$. But since $N_E(P) \leq E$, this means that E contains a subgroup $\hat{G}$ isomorphic to G. Now, $N \cap \hat{G} = 1$ since they are of relatively prime order, and $|E| = |G| \cdot |N|$ implies that the group generated by N and $\hat{G}$ equals E, has N as normal subgroup and quotient isomorphic to G. This gives $E \simeq N \rtimes G$.

Therefore, we may suppose $N_E(P) = E$, or what is the same, $P \trianglelefteq E$. Since P is a p-group, $Z(P) =: Z$ is not trivial. Since the centre is a characteristic subgroup of P, and since P is normal in E, Z is normal in E. Hence $N/Z \trianglelefteq E/Z$ and since

$$(E/Z)\,/\,(N/Z) \simeq E/N = G$$

we obtain that $E/Z \simeq (N/Z) \rtimes \check{G}$ for some group $\check{G} \simeq G$. Let H be the preimage of $\check{G}$ in E. We see that $Z \trianglelefteq H$ and $H/Z \simeq G$. If $Z \neq N$, then $|H| < |E|$ and by the induction hypothesis we get $H \simeq Z \rtimes G$. Hence in this case there is a subgroup $\tilde{G}$ of H such that $\tilde{G} \simeq G$ and as before,

$$E = N \cdot \tilde{G} \simeq N \rtimes G.$$

Hence we may suppose $Z = N$ is an abelian p-group. But then Proposition 1.8.44 and Corollary 1.8.41 prove the statement. □

Remark 1.8.48 The celebrated Feit-Thompson theorem states that every finite simple group of odd order is solvable. The proof of this result is highly involved and far beyond the level appropriate for this text. The result is moreover one important part of the classification of the finite simple groups. Here, the Feit-Thompson theorem shows that if N is a normal subgroup of E and $G \simeq E/N$, such that N and G are of coprime order, then either N or G is of odd order, and hence has no simple non-abelian composition factor. Therefore either N or G is solvable. A consequence is that if G_1 and G_2 are complements to N (i.e. $N \cap G_1 = N \cap G_2 = 1$ and $N \cdot G_1 = N \cdot G_2 = E$) then G_1 is conjugate to G_2. This comes from the fact that conjugacy classes of complements are parameterised by $H^1(G, N)$ in case N is abelian, and by a proof analogous to Corollary 1.8.41 it is shown that $H^1(G, N)$ is actually trivial. The reduction to the abelian case however is much more involved. There it is necessary to reduce first N, then G by passing to the largest normal p-subgroup, then the Frattini quotient, which works of course only if one of the groups is solvable.

The following lemma is another application of the above considerations. Although it seems to be somewhat particular we will need it in Sect. 2.2.2, and it provides a nice application of the results of this section. In particular it shows how the above consideration can be useful for infinite groups as well. Lemma 2.2.11 in Sect. 2.2.2 will give a nice and instructive example of the way the group cohomology $H^2(G, M)$ arises naturally and how it can be used efficiently.

Lemma 1.8.49 *Let G be a finite p-group and let k be a field of characteristic $p > 0$. Suppose that for all $a \in k$ the polynomial $X^{|G|} - a$ has a solution in k. Consider $k^\times$ to be a G-module with trivial action. Then $H^2(G, k^\times) = 0$. In particular, let $\tilde{G}$ be a group such that there is an injective group homomorphism $\varphi : k^\times \hookrightarrow Z(\tilde{G})$ so that $\tilde{G}/\varphi(k^\times) \simeq G$. Then $\tilde{G} \simeq G \times \varphi(k^\times)$.*

Proof Let

$$\begin{aligned} k^\times &\xrightarrow{\mu} k^\times \\ x &\mapsto x^{|G|} \end{aligned}$$

be the map which sends x to its $|G|$-th power. The hypothesis on k shows that this map is surjective. The kernel of this epimorphism is the set of $|G|$-th roots of unity over k. Hence, since the characteristic of k is $p > 0$, we get $X^{|G|} - 1 = (X - 1)^{|G|}$ in the polynomial ring $k[X]$ and therefore μ is injective. By Lemma 1.8.36 we get an exact sequence

$$0 \longrightarrow H^2(G, k^\times) \stackrel{H^2(G,\mu)}{\longrightarrow} H^2(G, k^\times)$$

of abelian groups and by Proposition 1.8.40 we obtain that $H^2(G, \mu) = 0$.

This implies $H^2(G, k^\times) = 0$. Hence $\tilde{G} \simeq k^\times \rtimes G$, and since $k^\times \subseteq Z(\tilde{G})$ we have

$$\tilde{G} \simeq k^\times \times G.$$

This proves the statement. □

1.9 More on Projective Modules

1.9.1 The Module Approach

We have seen that projective modules play a very important role in the definition and computation of the group $Ext_A^n(-,-)$. It is therefore worth studying them in some more detail, in particular from the point of view of finite dimensional algebras and under the aspect of how to obtain and classify them.

Lemma 1.9.1 *Let A be an artinian algebra and let P be an indecomposable A-module. Then $P/\mathrm{rad}(P)$ is a simple A-module.*

Proof $P/\mathrm{rad}(P)$ is also an A-module. By Lemma 1.6.8 this module $P/\mathrm{rad}(P)$ is the largest semisimple quotient of P. We claim that $P/\mathrm{rad}(P)$ is simple. Indeed, since P is indecomposable, $End_A(P)$ is local by Lemma 1.4.6. Let $\mathfrak{m}$ be the unique maximal ideal of $End_A(P)$. Since A, and therefore P, is artinian, $\mathfrak{m}$ is nilpotent and $D_P := End_A(P)/\mathfrak{m}$ is a skew-field. By definition the module $P/\mathrm{rad}(P)$ is an A-$End_A(P)$-bimodule.

Now, let φ be a non-zero endomorphism of $P/\mathrm{rad}(P)$. Then the composition with the projection $\pi : P \longrightarrow P/\mathrm{rad}(P)$ gives a homomorphism $P \stackrel{\varphi\circ\pi}{\longrightarrow} P/\mathrm{rad}(P)$ as indicated in the diagram below.

$$\begin{array}{ccc} P & & P \\ \downarrow \pi & & \downarrow \pi \\ P/\mathrm{rad}(P) & \stackrel{\varphi}{\longrightarrow} & P/\mathrm{rad}(P) \end{array}$$

Since (the right-hand vertical) π is surjective, the composition $\varphi \circ \pi$ lifts to an endomorphism α of P:

$$\begin{array}{ccc} P & \stackrel{\alpha}{\longrightarrow} & P \\ \downarrow \pi & & \downarrow \pi \\ P/\mathrm{rad}(P) & \stackrel{\varphi}{\longrightarrow} & P/\mathrm{rad}(P) \end{array}$$

On the other hand, by Lemma 1.6.9 for any endomorphism α' of P one gets $\alpha'(\mathrm{rad}(P)) \subseteq \mathrm{rad}(P)$. This implies that there is a surjective ring homomorphism

$$\rho : End_A(P) \longrightarrow End_A(P/\mathrm{rad}(P)).$$

Since $P/\mathrm{rad}(P)$ is semisimple the endomorphism algebra of $P/\mathrm{rad}(P)$ is a direct product of matrix rings over skew-fields.

Suppose that there is a matrix ring of size $n \geq 2$ occurring in this direct product. Then this occurs since $P/\mathrm{rad}(P)$ has a direct factor S^n for a simple A-module S. The endomorphism φ given by mapping the first copy identically onto the second, the second identically onto the third, the k-th identically onto the $k+1$-st (for $k < n$) and the n-th copy to 0 is nilpotent of degree $n-1$. Moreover, φ is not contained in any non-trivial two-sided ideal of this matrix ring, since by Remark 1.4.29 full matrix rings are simple algebras and do not have any non-trivial two-sided ideals. We complete this to an endomorphism $\overline{\nu}$ of $P/\mathrm{rad}(P)$ by putting the identity on each other isotypic component, i.e. the identity matrix on all the other direct factors. Since ρ is surjective, $\overline{\nu} = \rho(\nu)$ for some non-invertible endomorphism ν of P. Hence, $\nu \in \mathfrak{m}$, and since ρ is surjective, $\rho(\mathfrak{m})$ is an ideal of $End_A(P/\mathrm{rad}(P))$. Since $\rho(\mathfrak{m})$ contains $\overline{\nu}$, we get $\rho(\mathfrak{m}) = End_A(P/\mathrm{rad}(P))$. Hence, the identity on $P/\mathrm{rad}(P)$ is an image of some non-invertible endomorphism ι of P. But, A is artinian, and so $\mathfrak{m}$ is nilpotent. Hence $id = \rho(\iota)$ is nilpotent, a contradiction.

Therefore $P/\mathrm{rad}(P)$ is a direct sum of pairwise non-isomorphic simple modules. Suppose $P/\mathrm{rad}(P)$ is not simple. Then $End_A(P)$ contains at least two different maximal ideals, a contradiction again.

Hence $P/\mathrm{rad}(P)$ is simple. This proves the statement. □

Lemma 1.9.2 *Let A be an artinian algebra and let S be a simple A-module. Then there is an indecomposable projective A-module P such that* $P/\mathrm{rad}(P) \simeq S$.

Proof Take any non-zero element $s \in S$, and observe that $A \cdot s \leq S$ is a non-zero submodule. S is simple, and therefore $S = A \cdot s$. Hence there is an epimorphism $\lambda_s : A \longrightarrow S$ by taking $a \mapsto as$. Since A is artinian, there is a decomposition

$$A = P_1 \oplus P_2 \oplus \cdots \oplus P_m$$

into a direct sum of indecomposable projective A-modules. Hence, there is an m_0 with $\lambda|_{P_{m_0}} \neq 0$. Now, $\lambda(P_{m_0})$ is a non-zero submodule of S, hence $\lambda(P_{m_0}) = S$. This proves the statement. □

The isomorphism classes of simple modules are in bijection to the isomorphism classes of projective indecomposable modules for artinian algebras.

Proposition 1.9.3 *Let A be an artinian algebra. Then for every simple A-module S there is a projective indecomposable A-module P_S, unique up to isomorphism with $P_S/\text{rad}(P_S) \simeq S$. For any indecomposable projective A-module P the quotient $P/\text{rad}(P)$ is simple.*

Proof Lemma 1.9.2 and 1.9.1 give mappings between isomorphism classes of simple modules to isomorphism classes of projective indecomposable modules. The only thing to show is that for two projective indecomposable modules P_1 and P_2 with $P_1/\text{rad}(P_1) \simeq P_2/\text{rad}(P_2)$ we have $P_1 \simeq P_2$.

We know that the fact that the modules are projective implies that there is a mapping $\alpha_1 : P_1 \longrightarrow P_2$ inducing an isomorphism modulo the radicals, and also likewise a mapping $\alpha_2 : P_2 \longrightarrow P_1$ inducing an isomorphism modulo the radicals. Hence, $\alpha_1(P_1) + \text{rad}(P_2) = P_2$. Nakayama's lemma 1.6.5 implies that $\alpha_1(P_1) = P_2$ and likewise $\alpha_2(P_2) = P_1$. Hence, $\alpha_1 \circ \alpha_2$ is a surjective endomorphism of P_2, whence an isomorphism by Proposition 1.3.6, and likewise $\alpha_1 \circ \alpha_2$ is an automorphism. Hence each of α_1 and α_2 are isomorphisms. □

A consequence of the above is that whenever a projective module P maps onto a simple module S, then the mapping factors through the natural projection $P_S \longrightarrow P_S/\text{rad}(P_S) \simeq S$, and moreover that P_S is a direct factor of P, and the factorising mapping is just the projection onto the corresponding factor. We can formalise this property. The corresponding notion is given in the following.

Definition 1.9.4 Let A be an algebra. Given an A-module M, a projective module P_M together with an epimorphism $\mu : P_M \longrightarrow M$ is a *projective cover* of M if for every proper submodule Q of P_M one has that $\mu|_Q$ is not an epimorphism.

Dually, an *injective hull* I_M of M is an injective A-module M together with a monomorphism $\iota : M \longrightarrow I_M$ such that for every module J and every epimorphism $\lambda : I_M \longrightarrow J$ the map $\lambda \circ \iota$ is not a monomorphism.

It is not difficult to see that $P \xrightarrow{\mu} M$ is a projective cover if and only if for every proper submodule Q of P one has $\ker \mu + Q < P$. Indeed, if μ is a projective cover and Q is a proper submodule of P, then $\mu|_Q$ is not an epimorphism, and hence $\ker \mu + Q \neq P$. Conversely, if $Q \leq P$ and $\mu|_Q$ is an epimorphism, then by the universal property of a projective module, there is a $P \xrightarrow{\nu} Q$ such that $\mu|_Q \circ \nu = \mu$. If ι is the embedding of Q in P, then $p - \iota \circ \nu(p) \in \ker \mu$ for each $p \in P$. Therefore $p = \iota \circ \nu(p) + (p - \iota \circ \nu(p))$ and hence $P = \iota \circ \nu(P) + \ker \mu$, which implies that $P = Q + \ker \mu$. Hence $Q = P$.

Some authors use the expression *injective envelope* instead of injective hull. See Lam [12] for further properties on the injective hull in general.

Proposition 1.9.5 *Let A be an artinian algebra. Suppose that an A-module M admits a projective cover (P_M, α_M). Then P_M is unique up to isomorphism. Suppose that M admits an injective hull (I_M, ι_M), then I_M is unique up to isomorphism.*

Proof Let (P_1, α_1) and (P_2, α_2) be two projective covers. The fact that P_1 is projective and that α_2 is surjective implies that there is a mapping $\beta_1 : P_1 \longrightarrow P_2$ such that $\alpha_1 = \alpha_2 \circ \beta_1$. Likewise, there is a $\beta_2 : P_2 \longrightarrow P_1$ such that $\alpha_2 = \alpha_1 \circ \beta_2$. Hence

$$\alpha_1 = \alpha_1 \circ \beta_2 \circ \beta_1.$$

Now,

$$M = \alpha_1(P_1) = \alpha_1(\beta_2 \circ \beta_1(P_1))$$

and therefore $\beta_2 \circ \beta_1(P_1)$ cannot be a proper submodule of P_1, using the definition of a projective cover. Hence β_2 is surjective, and since P_1 is projective, P_1 is a direct summand of P_2. Likewise, P_2 is a direct summand of P_1. Since A is artinian, $P_1 \simeq P_2$.

The proof for injective hulls is similar. □

Artinian algebras admit projective covers.

Proposition 1.9.6 *Let A be an artinian algebra. Then for every finitely generated A-module M there is a projective cover P_M, unique up to isomorphism.*

Proof Let $\overline{M} := M/\text{rad}(M)$. This is semi-simple, $\overline{M} = S_1 \oplus \cdots \oplus S_\ell$, and so $P := P_{S_1} \oplus \cdots \oplus P_{S_\ell}$ is a projective cover of $\overline{M}$, using Proposition 1.9.3. Hence, there is a diagram

$$\begin{array}{ccc} & & M \\ & & \downarrow \\ P & \twoheadrightarrow & \overline{M} = M/\text{rad}(M) \end{array}$$

where the right vertical morphism is an epimorphism. The universal property of projective modules implies that the mapping $P \longrightarrow \overline{M}$ lifts to M; i.e. there is a mapping $\nu : P \longrightarrow M$ such that

$$\begin{array}{ccc} P & \xrightarrow{\nu} & M \\ \| & & \downarrow \\ P & \twoheadrightarrow & \overline{M} = M/\text{rad}(M) \end{array}$$

is commutative. Now, $\nu(P) + \text{rad}(M) = M$ since the mapping $P \longrightarrow \overline{M}$ is an epimorphism. Nakayama's Lemma 1.6.5 implies that $\nu(P) = M$ and hence ν is an epimorphism. Let Q be a submodule of P, denote by ι the inclusion, and suppose that $\nu(Q) = P/\text{rad}(P)$. Then $\nu|_Q$ is surjective, and since P is projective, there is a morphism $\lambda : P \longrightarrow Q$ such that $\nu = \nu|_Q \circ \lambda$. But then $\iota \circ \lambda$ induces the identity on $P/\text{rad}(P)$ and therefore $\iota \circ \lambda(P) + \text{rad}(P) = P$. Nakayama's lemma implies

$\iota \circ \lambda(P) = P$ and therefore ι is surjective. Since ι is injective by definition we have proved the lemma.

Therefore (P, ν) is a projective cover. Proposition 1.9.5 implies unicity. □

Lemma 1.9.7 *Let K be a field and let A be a finite dimensional K-algebra. Then for every projective A-right module P, the A-left module $Hom_K(P, K)$ is injective. Moreover, every finite dimensional A-module M has an injective hull.*

Proof We first remark that if M is a left A-module, then $Hom_K(M, K)$ is an A-right module, and if N is a right A-module, then $Hom_K(N, K)$ is a left A-module. Then if P is a projective right A-module, $Hom_K(P, K)$ is an injective left A-module, and if Q is a projective left A-module, then $Hom_K(Q, K)$ is an injective right A-module. Indeed, let

$$0 \longrightarrow Hom_K(P, K) \stackrel{\iota}{\longrightarrow} X \stackrel{\pi}{\longrightarrow} Y \longrightarrow 0$$

be a short exact sequence of finite dimensional A-left modules. Now, since P is finite dimensional over K, the evaluation map gives an isomorphism

$$P \simeq Hom_K(Hom_K(P, K), K).$$

Take K-duals to obtain a short exact sequence

$$0 \to Hom_K(Y, K) \stackrel{Hom_K(\pi,K)}{\longrightarrow} Hom_K(X, K) \stackrel{Hom_K(\iota,K)}{\longrightarrow} P \to 0$$

of right A-modules. Since P is projective, $Hom_K(X, K) \stackrel{Hom_K(\iota,K)}{\longrightarrow} P$ is a split epimorphism, and therefore there is a $P \stackrel{\rho}{\longrightarrow} Hom_K(X, K)$ such that

$$Hom_K(\iota, K) \circ \rho = id_P.$$

This shows that

$$Hom_K(\rho, K) \circ \iota = id_{Hom_K(P,K)}$$

which implies that $Hom_K(P, K)$ is injective. The case for Q is similar.

Now, let M be a finite dimensional A-left module. Then let P be a projective cover of $Hom_K(M, K)$ as an A-right module. $Hom_K(P, K)$ is injective, and moreover, $M \hookrightarrow Hom_K(P, K)$ is an injective hull. Indeed, the universal property of the injective hull is an immediate consequence of the universal property of the projective cover by taking K-duals everywhere. □

Dually to a projective resolution we can construct injective coresolutions if every module is a submodule of an injective object.

Definition 1.9.8 Let A be an algebra and let M be an A-module. An injective coresolution is a sequence

$$M \stackrel{f_0}{\hookrightarrow} I_0 \stackrel{f_1}{\to} I_1 \stackrel{f_2}{\to} I_2 \stackrel{f_3}{\to} \dots$$

such that $\mathrm{im}(f_i) = \ker(f_{i+1})$ for all $i \in \mathbb{N}$, such that f_0 is injective, and such that I_i is injective for each $i \in \mathbb{N}$.

1.9.2 The Idempotent Approach

Classically projective modules are introduced with idempotent elements of A.

Proposition 1.9.9 *Let A be an artinian algebra. Then for any indecomposable projective A-module P there is an idempotent $e^2 = e \in A$ such that $P \simeq A \cdot e$. Moreover, for every idempotent $e^2 = e \in A$ the module $A \cdot e$ is projective.*

Proof Let $e^2 = e$ be an idempotent, then $1 = e + (1 - e)$. Hence,

$$A = A \cdot 1 = A \cdot e + A \cdot (1 - e)$$

where the inclusion $A \subseteq A \cdot e + A \cdot (1 - e)$ is clear as well as the inclusion $A \supseteq A \cdot e + A \cdot (1 - e)$. Let $x \in A \cdot e \cap A \cdot (1 - e)$. Then

$$x = x \cdot e = x \cdot e \cdot (1 - e) = x \cdot (e - e^2) = 0$$

and so

$$A = A \cdot 1 = A \cdot e \oplus A \cdot (1 - e).$$

As a consequence $A \cdot e$ is projective.

Suppose P is a projective indecomposable module. Then P is a direct summand of A and $A = P \oplus Q$. Let π_P be the projection $A \longrightarrow P$ and let $\iota_P : P \longrightarrow A$ be the canonical embedding. Since $End_A(A) \simeq A^{op}$ where the isomorphism is given by right multiplication, and since $\iota_P \circ \pi_P$ is an idempotent endomorphism of A, there is an idempotent element $e_P \in A$ which realises $\iota_P \circ \pi_P$ as multiplication by e_P. Therefore $P = A \cdot e_P$. □

Lemma 1.9.10 *Let A be an artinian algebra. Let $e \neq 0$ be an idempotent. Then $A \cdot e$ is decomposable if and only if there are two idempotents $e_1 \neq 0$ and $e_2 \neq 0$ in A such that $e = e_1 + e_2$ and such that $e_1 e_2 = e_2 e_1 = 0$.*

Proof If $e = e_1 + e_2$ for two idempotents e_1 and e_2, then $A \cdot e = A \cdot e_1 \oplus A \cdot e_2$ just as in the proof of Proposition 1.9.9.

Conversely, if $A \cdot e = P_1 \oplus P_2$, then there are idempotents e_1 and e_2 such that $P_1 = A \cdot e_1$ and $P_2 = A \cdot e_2$. These idempotents arise from projection and injection into the corresponding component. Since the sum is direct, it is clear that $e_1 e_2 = e_2 e_1 = 0$. Moreover, $e = e_1 + e_2$. □

Definition 1.9.11 Let A be an algebra. Two idempotents e_1 and e_2 are *orthogonal* if $e_1 e_2 = e_2 e_1 = 0$. An idempotent $e \neq 0$ in A is *primitive* if whenever $e = e_1 + e_2$ for two orthogonal idempotents e_1 and e_2, then $e_1 = 0$ or $e_2 = 0$.

Lemma 1.9.12 *Let A be an algebra. Two idempotents e_1 and e_2 are conjugate in A if and only if $A \cdot e_1 \simeq A \cdot e_2$.*

Proof Suppose that there is an invertible element $u \in A$ such that $e_1 = u \cdot e_2 \cdot u^{-1}$. Then

$$\begin{aligned} A \cdot e_1 &\xrightarrow{\gamma} A \cdot e_2 \\ a \cdot e_1 &\mapsto a \cdot u \cdot e_2 \end{aligned}$$

Now, $a \cdot e_1 = a \cdot u \cdot e_2 \cdot u^{-1}$ and so the above mapping is the restriction to Ae_1 of the right multiplication mapping by u. This is obviously a well-defined bijective mapping. The mapping is A-linear, since

$$\gamma(b \cdot ae_1) = bae_1u = b\gamma(ae_1).$$

Suppose $\varphi : Ae_1 \longrightarrow Ae_2$ is an isomorphism of A-modules. Then $\varphi(e_1) = ae_2$ for some $a \in A$. Moreover $\varphi^{-1}(e_2) = be_1$ for some $b \in e_2Ae_1$, and so $e_1 = bae_1$. Likewise $e_2 = abe_2$. Now, $e_1 = e_1^2$, and so

$$\varphi(e_1) = e_1\varphi(e_1) = e_1ae_2.$$

Therefore, $a \in e_1Ae_2$, and likewise $b \in e_2Ae_1$. Hence ba is a unit in e_1Ae_1 and ab is a unit in e_2Ae_2. We look at the decomposition

$$(e_1 + e_2)A(e_1 + e_2) = \begin{pmatrix} e_1Ae_1 & e_1Ae_2 \\ e_2Ae_1 & e_2Ae_2 \end{pmatrix}$$

and see that the element $u := \begin{pmatrix} 0 & a \\ b & 0 \end{pmatrix}$ in this matrix ring is invertible since its square $\begin{pmatrix} ab & 0 \\ 0 & ba \end{pmatrix}$ is invertible. Similarly, for $e := e_1 + e_2$ and $f := 1 - e$ we get

$$A = \begin{pmatrix} eAe & eAf \\ fAe & fAf \end{pmatrix}$$

and see that the element $\begin{pmatrix} u & 0 \\ 0 & f \end{pmatrix}$ in this matrix ring is invertible. Since e_1 and e_2 are orthogonal, and since $b \in e_2Ae_1$ and hence $be_1 = e_2b$,

$$\begin{pmatrix} 0 & a \\ b & 0 \end{pmatrix} \cdot \begin{pmatrix} e_1 & 0 \\ 0 & 0 \end{pmatrix} = \begin{pmatrix} 0 & 0 \\ 0 & e_2 \end{pmatrix} \cdot \begin{pmatrix} 0 & a \\ b & 0 \end{pmatrix}$$

which implies $ue_1u^{-1} = e_2$. □

A decomposition of 1 into a sum of mutually orthogonal primitive idempotents exists always for artinian algebras.

Proposition 1.9.13 *Let A be an artinian algebra. Then there is a set of pairwise orthogonal primitive idempotents $\{e_1, e_2, \ldots, e_n\}$ such that $1 = e_1 + e_2 + \cdots + e_n$.*

Proof If 1 is primitive, there is nothing to do. Otherwise, there is a decomposition $1 = e_1 + e_2$ into orthogonal idempotents e_1 and e_2. We consider e_1, and if e_1 is primitive, we leave it. If not, it decomposes into a sum of two orthogonal idempotents $e_1 = e_{11} + e_{12}$. We continue to decompose e_{11} and e_{12} if they are not primitive, as well as all the idempotents obtained this way, and it is clear that this procedure has to finish, since otherwise we would have an infinite decreasing sequence of ideals

$$A > Ae_1 > Ae_{11} > \cdots$$

which contradicts the condition of A being artinian. Then we continue to decompose e_2 in the same manner, and again we finally reach a finite decomposition of 1 as requested. This proves the proposition. □

We observe that the conjugate of an idempotent is again an idempotent, and the conjugate of a primitive idempotent is a primitive idempotent. A particular case occurs when A is commutative.

Proposition 1.9.14 *Let A be an artinian commutative algebra. Then there is exactly one decomposition of 1 into a sum of primitive orthogonal idempotents. Two primitive idempotents are either identical or orthogonal.*

Proof Let e and f be two primitive idempotents. Then $e = ef + e(1 - f)$, and since A is commutative, $(ef)^2 = efef = e^2 f^2 = ef$ is idempotent, as is $e(1 - f)$. Since e is primitive, either $e = ef$ or $e = e(1 - f) = e - ef$. In the second case $ef = 0$, and in the first case $f = ef + (1 - e)f$, which implies again that $(1 - e)f$ is idempotent and therefore either $f = ef$ or $ef = 0$. Since we excluded $ef = 0$ we get $e = ef = f$. Hence, two primitive idempotents are either orthogonal or identical. This proves the second statement.

Let

$$e_1 + e_2 + \cdots + e_n = 1 = f_1 + f_2 + \cdots + f_m$$

be two decompositions of 1 into primitive orthogonal idempotents, and suppose that n is minimal possible. Then $e_1 f_i = 0$ for all $i \in \{1, \ldots, m\}$ implies that

$$e_1 = e_1 \cdot 1 = e_1 f_1 + e_1 f_2 + \cdots + e_1 f_m = 0,$$

a contradiction.

Hence there is an i_0 such that $e_1 f_{i_0} \neq 0$, which implies $e_1 = f_{i_0}$. Considering the algebra $(1 - e_1)A$ and induction on n gives the result. □

Since projective indecomposable modules of a finite dimensional K-algebra, for a field K, are of the form $A \cdot e$ for a primitive idempotent e in A, and since two primitive idempotents e and f are conjugate if and only if $Ae \simeq Af$, it is useful to give a name to a set of idempotents showing all isomorphism types of projective modules.

Definition 1.9.15 Let K be a field and let A be a finite dimensional K-algebra. Then a set $E := \{e_1, e_2, \ldots, e_n\}$ of pairwise orthogonal primitive idempotents is a *complete set of orthogonal primitive idempotents* if $1 = e_1 + \cdots + e_n$.

The following amusing statement is called Rosenberg's lemma and will be useful in later sections.

Lemma 1.9.16 *Let A be an artinian algebra and let e be a primitive idempotent of A. If $\mathfrak{I}$ is a set of two-sided ideals of A and $e \in \sum_{I \in \mathfrak{I}} I$, then there is some $I \in \mathfrak{I}$ so that $e \in I$.*

Proof Since e is primitive, Ae is an indecomposable projective A-module, and therefore $eAe = End_A(Ae)$ is a local algebra. For each $I \in \mathfrak{I}$ we get that eIe is an ideal of eAe. Since $e \in \sum_{I \in \mathfrak{I}} I$ we have

$$e \in e\left(\sum_{I \in \mathfrak{I}} I\right)e = \sum_{I \in \mathfrak{I}} eIe.$$

Since e is the unit element of eAe, we obtain $\sum_{I \in \mathfrak{I}} eIe = eAe$, and since eAe is a local algebra, there is some $I \in \mathfrak{I}$ such that $eIe = eAe$. Hence $e \in eIe \subseteq I$ since I is a two-sided ideal. This implies that $e \in I$. □

A very important property of idempotents is given in the following statement, the so-called lifting of idempotents property.

Proposition 1.9.17 *Let A be an algebra and let I be a nilpotent ideal of A. Let $\overline{e}$ be an idempotent of A/I. Then there is an idempotent e of A so that e maps to $\overline{e}$ under the morphism $A \longrightarrow A/I$.*

Proof Suppose $I^n = 0$. We note that there are epimorphisms

$$A/I^m \longrightarrow A/I^{m-1}$$

for all $m \in \mathbb{N}$ and $A/I^n = A$. Hence we are done if we can construct a sequence e_m of elements such that $e_m^2 - e_m \in I^m$ and such that $e_1 = \overline{e}$. Suppose we have constructed e_{m-1} for $m \geq 2$. Then $e_m := 3e_{m-1}^2 - 2e_{m-1}^3$ satisfies

$$\begin{aligned} e_m^2 - e_m &= (3e_{m-1}^2 - 2e_{m-1}^3)^2 - (3e_{m-1}^2 - 2e_{m-1}^3) \\ &= (3e_{m-1}^2 - 2e_{m-1}^3)(3e_{m-1}^2 - 2e_{m-1}^3 - 1) \end{aligned}$$

$$= e_{m-1}^2(3 - 2e_{m-1})(3e_{m-1}^2 - 2e_{m-1}^3 - 1)$$
$$= -(3 - 2e_{m-1})(1 + 2e_{m-1})(e_{m-1}^2 - e_{m-1})^2.$$

Now,

$$(e_{m-1}^2 - e_{m-1}) \in I^{m-1}$$

implies

$$(e_{m-1}^2 - e_{m-1})^2 \in I^{2(m-1)} \subseteq I^m$$

as soon as $m \geq 1$. □

Remark 1.9.18 We shall re-examine the above proof and a very similar statement in Proposition 2.5.17.

1.10 Injectives and Projectives May Coincide: Two Classes of Such Algebras

Group rings are algebras with many quite special and particular properties. Some of these properties can be taken as separate axioms and we can study rings satisfying these axioms. Sometimes it is even possible to prove properties of group rings by considering these more general algebras. In particular group rings may be equivalent in some special sense (such as being derived equivalent, or stably equivalent (cf Chaps. 5 and 6)) to other algebras, which are not group algebras, but which are much better known. Properties of these better understood algebras then carry over to these group algebras as soon as the property is shown to be invariant under the equivalence considered.

1.10.1 Self-Injective Algebras and Frobenius Algebras

Group rings are symmetric and self-injective. We shall develop some of the properties of these algebras here. The classical reference for self-injective algebras is Nakayama [15, 16], and a very elegant presentation for symmetric algebras is given by Broué [17].

Frobenius Algebras I

Let K be a commutative ring and let A be a K-algebra. For any right A-module M we can produce the K-linear dual $Hom_K(M, K)$, which becomes an A-left module via the following law:

$$(a \cdot f)(m) = f(ma) \ \forall \ a \in A, f \in Hom_K(M, K), m \in M.$$

Example 1.10.1 Let K be a field and let A be the ring of upper triangular 2×2 matrices with coefficients in K. Then this ring has a very simple structure. If we want to present the algebra again as a quiver with relations, as we have done in Example 1.6.23, then it would be represented by

$$\bullet \longrightarrow \bullet$$

Up to isomorphism there are two projective indecomposable left modules, P_1 and P_2, corresponding to the first and to the second column of the elements of A. Then P_1 is one-dimensional, and hence simple. P_2 contains P_1 and the quotient is one-dimensional, whence it is also simple. The three modules P_1, P_2 and P_2/P_1 are the only indecomposable A-modules. This fact is a not too difficult exercise.

Similarly, looking at right modules, the first projective right A-module Q_1 corresponds to the first line and Q_2 to the second line of elements of A. Then we get a non-split monomorphism of projective right modules $Q_2 \hookrightarrow Q_1$ and an epimorphism $Hom_K(Q_1, K) \twoheadrightarrow Hom_K(Q_2, K)$ of left modules. Then $Hom_K(Q_2, K)$ is not projective since the inclusion of Q_2 into Q_1 is non-split.

Taking dual spaces preserves decomposability, and as a consequence indecomposability. Since $P_2 \twoheadrightarrow P_2/P_1$ is an epimorphism, we get a monomorphism for the duals $Hom_K(P_2/P_1, K) \hookrightarrow Hom_K(P_2, K)$. Using Lemma 1.9.7, $Hom_K(P_2/P_1, K)$ is a simple submodule of the 2-dimensional indecomposable injective module $Hom_K(P_2, K)$.

The regular module A, i.e. considered as a left A-module by ordinary multiplication, is projective. The K-linear dual $Hom_K(A, K)$ is injective by Lemma 1.9.7 and in general is not projective as shown in Example 1.10.1.

Definition 1.10.2 Let K be a commutative ring and let A be a K-algebra. The K-algebra is called *self-injective* if every finitely generated projective A-module is injective. The algebra A is called a *Frobenius algebra* if the left A-module $Hom_K(A, K)$ is isomorphic to the regular left module A.

Lemma 1.10.3 *If A is a finite dimensional Frobenius algebra over a field K, then A is self-injective.*

Proof Let P be a finitely generated projective A-module. Then P is a direct factor of A^n for some n, and then P is a direct factor of $Hom_K(A, K)^n$. We need to show that $Hom_K(A, K)$ is injective. But this is Lemma 1.9.7. □

Remark 1.10.4 The converse is false in general as will be seen in Proposition 4.5.7 below.

Example 1.10.5 The easiest examples for Frobenius algebras are matrix rings. The dual of a matrix ring is again a matrix ring, and hence is self-injective. As a consequence, split semisimple artinian algebras are self-injective. Another almost trivial example of a Frobenius algebra, hence a self-injective algebra, is $K[X]/X^n$ for some integer n. The only indecomposable modules are $K[X]/X^m$ for $m \leq n$, and hence are characterised by their dimension. The module of dimension n is the only indecomposable projective module, and is equal to its own dual.

We first see some consequences of this concept.

Suppose A is a Frobenius K-algebra for a field K. Then, by definition, there is an isomorphism

$$\varphi : A \longrightarrow Hom_K(A, K)$$

as A-modules. This produces a bilinear form

$$\begin{array}{rcl} A \times A & \longrightarrow & K \\ (a, b) & \mapsto & (\varphi(b))\,(a) \end{array}$$

Since φ is bijective, the bilinear form is non-degenerate. Since φ is A-linear,

$$(ac, b) = (\varphi(b))\,(ac) = (c\varphi(b))\,(a) = \varphi(cb)(a) = (a, cb).$$

We call such bilinear forms $A \times A \longrightarrow K$ *associative*.

Let K be a field, let A be a K-algebra and let $\beta : A \times A \longrightarrow K$ be a non-degenerate associative bilinear form. Then

$$\begin{array}{l} A \xrightarrow{\varphi} Hom_K(A, K) \\ a \mapsto (b \mapsto \beta(b, a)) \end{array}$$

is an isomorphism of A-modules. Indeed,

$$(\varphi(ab))\,(c) = \beta(c, ab) = \beta(ca, b) = (\varphi(b))\,(ca) = (a\varphi(b))\,(c)$$

and so

$$a\varphi(b) = \varphi(ab) \ \ \forall a, b \in A.$$

Since β is non-degenerate, φ is an isomorphism. We have proved the following result.

Proposition 1.10.6 *Let K be a field and let A be a K-algebra. Then A is a Frobenius algebra if and only if there is a non-degenerate bilinear form $\beta : A \times A \longrightarrow K$ with $\beta(ab, c) = \beta(a, bc)$ for all $a, b, c \in A$.*

Example 1.10.7 We shall again describe an algebra by a quiver with certain relations. The quiver is given by

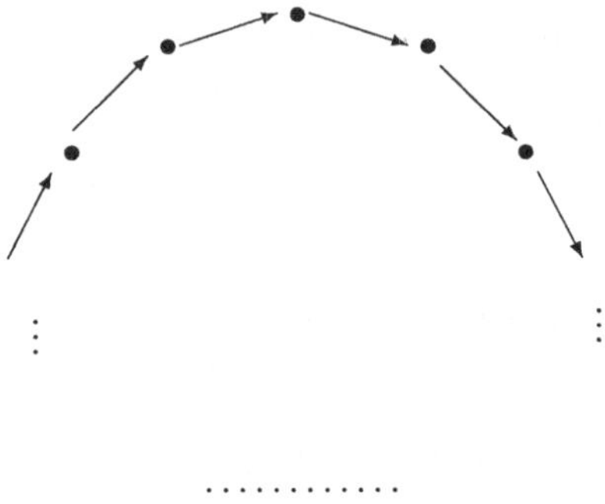

a circle formed by n arrows $\alpha_1, \alpha_2, \ldots, \alpha_n$. We form the quotient by the ideal generated by all paths of length at least $m + 1 \in \mathbb{N}$. This algebra is called a *self-injective Nakayama algebra*, denoted by N_n^m, and plays quite an important role both in the theory and as a source of examples. For a more proper definition of an algebra defined by a quiver and relation, see Definition 1.11.5.

This algebra is a Frobenius algebra. In order to simplify the notation we abbreviate $A := N_n^m$. We have n isomorphism classes of projective indecomposable modules $P_1, P_2, \ldots, P_n$. These correspond to the vertices $1, 2, \ldots, n$, and are generated by the paths $e_1, e_2, \ldots, e_n$ of length 0 in the quiver. We need to define a non-degenerate associative bilinear form on the algebra. We shall do this using a general principle, as we will see in Proposition 1.10.18. Define a linear form $\varphi : A \longrightarrow K$ by putting

$$\varphi(p) = \begin{cases} 1 \text{ if } p \in \mathrm{rad}^m(A) \setminus \mathrm{rad}^{m+1}(A) \\ 0 \text{ otherwise} \end{cases}$$

for all paths p in the quiver. Since the paths in the quiver generate the algebra, we can extend the linear form to the algebra by linearity. Now, the only relation of A on paths we have, is that paths of length at least $m + 1$ are in the relation. Hence, since the paths of length at least $m + 1$ are in the kernel of φ, the linear form φ is actually defined on A.

Now, defining $\beta(p, q) = \varphi(pq)$ for two elements p and q of A, we claim that β defines an associative non-degenerate bilinear form on A. Indeed, β is associative almost by definition: $\beta(xy, z) = \varphi(xyz) = \beta(x, yz)$ for all $x, y, z \in A$. Moreover, β is non-degenerate: Given $x \in A \setminus \{0\}$ one of the element xe_i is non-zero for some $i \in \{1, 2, \ldots, n\}$. Since a basis for Ae_i is given by the paths pe_i where p is a path of length at most m starting at some vertex and ending at the vertex i, amongst the linear summands of xe_i there is a unique one $p_m e_i$ of minimal length ℓ. In other words,

$$x = k_{p_m} p_m e_i + \text{higher order terms}$$

for some $k_{p_m} \in K \setminus \{0\}$. Then $xe_i\alpha_i\alpha_{i+1}\dots\alpha_{i+m-\ell-1}$ is a non-zero multiple of a path of length $m-1$, and hence

$$\beta(x, e_i\alpha_i\alpha_{i+1}\dots\alpha_{i+m-\ell-1}) = \varphi(xe_i\alpha_i\alpha_{i+1}\dots\alpha_{i+m-\ell-1}) = k_{p_m} \neq 0.$$

Therefore the algebra N_n^m is a Frobenius algebra.

Lemma 1.10.8 *Let A be a self-injective finite dimensional K-algebra. Then the only A-modules with a finite projective resolution are the projective modules.*

Proof Indeed, if A is self-injective and M is a non-projective indecomposable A-module, then suppose that M has a finite projective resolution

$$0 \longrightarrow P_n \longrightarrow P_{n-1} \longrightarrow \cdots \longrightarrow P_0 \longrightarrow M \longrightarrow 0$$

and choose n minimal. Since A is selfinjective, P_n is an injective module by Lemma 1.10.3. Therefore the monomorphism $P_n \longrightarrow P_{n-1}$ splits,

$$0 \longrightarrow P_{n-1}/P_n \longrightarrow \cdots \longrightarrow P_0 \longrightarrow M \longrightarrow 0$$

is a projective resolution of M as well and we get a contradiction to the minimality of n. □

On Twisted Bimodules

Before continuing we need to introduce a special class of bimodules which generalises and details what was introduced in Definition 1.7.39. Let K be a commutative ring and let A be a K-algebra. Given an automorphism α of A we define the A-A-bimodule ${}_1A_\alpha$ as follows: The bimodule is the regular module A regarded as a left module. As a right module we put

$$m \cdot a := m\alpha(a)$$

for every $m \in {}_1A_\alpha$ and $a \in A$ where the right-hand side operation of $\alpha(a)$ on m is just usual multiplication. We see that ${}_1A_\alpha$ coincides with ${}^{(id\otimes\alpha)}A$ in the notation of Definition 1.7.39 as $A \otimes_K A^{op}$ modules.

Compare the following statement with Lemma 1.7.41 and Example 1.7.42.

Lemma 1.10.9 *Let K be a commutative ring, let A be a K-algebra and let $\alpha, \beta \in Aut(A)$. Then*

$${}_1A_\alpha \simeq {}_1A_\beta \Leftrightarrow \alpha\beta^{-1} \in Inn(A).$$

Proof Let $\alpha\beta^{-1} \in Inn(A)$, then there is a unit u in A such that for all $a \in A$ we get $\alpha(a) = u\beta(a)u^{-1}$. Define

$$\begin{aligned} {}_1A_\alpha &\xrightarrow{\varphi} {}_1A_\beta \\ a &\mapsto au \end{aligned}$$

This mapping is clearly a left A-module homomorphism. It is also right A-linear since

$$\varphi(a \cdot b) = \varphi(a\alpha(b)) = a\alpha(b)u = auu^{-1}\alpha(b)u = au\beta(b) = \varphi(a) \cdot b.$$

Suppose now that

$${}_1A_\alpha \xrightarrow{\varphi} {}_1A_\beta$$

is an isomorphism of A-A-bimodules. Then

$$\begin{aligned} \alpha(a) \cdot \varphi(1) = \alpha(a) \bullet \varphi(1) &= \varphi(\alpha(a) \bullet 1) = \varphi(\alpha(a) \cdot 1) \\ &= \varphi(\alpha(a)) = \varphi(1 \bullet a) = \varphi(1) \bullet a = \varphi(1) \cdot \beta(a). \end{aligned}$$

Since φ is bijective, $\varphi(1)$ is invertible in A; otherwise the image of φ would be the principal ideal generated by $\varphi(1)$. Taking $u = \varphi(1)$ we get the statement.

This proves the lemma. □

Lemma 1.10.10 *Let K be a commutative ring, let A be a K-algebra and let α and β be automorphisms of A. Then*

$${}_1A_\alpha \otimes_A {}_1A_\beta \simeq {}_1A_{\alpha\beta}.$$

In particular

$${}_1A_\alpha \otimes_A {}_1A_{\alpha^{-1}} \simeq A$$

as bimodules.

Proof We shall first prove

$${}_1A_\beta \xrightarrow{\alpha} {}_\alpha A_{\alpha\beta}$$

with the obvious notation, i.e. on ${}_\alpha A_{\alpha\beta}$ an element $a \in A$ acts by multiplication with $\alpha(a)$ from the left and by multiplication with $\alpha\beta(a)$ from the right.

Indeed,

$$\alpha(a \bullet m) = \alpha(am) = \alpha(a)\alpha(m) = a \bullet \alpha(m),$$

and

$$\alpha(m \bullet a) = \alpha(m\beta(a)) = \alpha(m)\alpha\beta(a) = \alpha(m) \bullet a.$$

Now, let $\gamma := \alpha\beta$, and then

$${}_1A_\alpha \otimes_A {}_\alpha A_\gamma \simeq {}_{\alpha^{-1}}A_1 \otimes_A {}_1A_{\alpha^{-1}\gamma} \simeq {}_{\alpha^{-1}}A_{\alpha^{-1}\gamma} \simeq {}_1A_\gamma$$

where the isomorphism is given by multiplication of the tensors: $a \otimes b \mapsto ab$. It is straightforward to prove the fact that this is indeed a well-defined isomorphism. We have proved the lemma. □

Since bimodules of the type ${}_1A_\alpha$ will play a role in the future we shall give them a name.

Definition 1.10.11 Let K be a commutative ring, let A be a K-algebra and let $\alpha \in Aut(A)$. Then the A–A-bimodule ${}_1A_\alpha$, regular from the left, and with the action of $a \in A$ from the right by multiplication by $\alpha(a)$, is called the *bimodule twisted by* α *from the right.*

Frobenius Algebras II

Let K be a field and let A be a finite dimensional Frobenius K-algebra. A itself is an A-A-bimodule by multiplication from the left and multiplication from the right. Hence, the K-linear dual is an A-A-bimodule as well, by putting $afb(c) = f(bca)$ for all $f \in Hom_K(A, K)$, for all $a, b, c \in A$. Since A is self-injective ${}_AHom_K(A, K) \simeq {}_AA$ as an A-left module. What happens to the right module structure?

$$A^{op} = End_A({}_AA) \simeq End_A(Hom_K(A, K))$$

where the first identity comes from the fact that every endomorphism α of A as an A-module is given by

$$\alpha(a) = \alpha(a \cdot 1) = a \cdot \alpha(1)$$

for every $a \in A$. Hence, we can identify α with multiplication by $\alpha(1) \in A$ on the right, and any choice of $\alpha(1) \in A$ will give an endomorphism. The second isomorphism is just the property of being Frobenius.

Now, we claim that there is an automorphism ν of the K-algebra A such that $Hom_K(A, K) \simeq {}_1A_\nu$. Indeed, we know that $End_A(Hom_K(A, K)) \simeq A^{op}$. The isomorphism ${}_AHom_K(A, K) \simeq {}_AA$ transports the endomorphism on $Hom_K(A, K)$ given by multiplication by $a \in A$ to multiplication by $\nu(a) \in A$ on ${}_AA$. The identity endomorphism on $Hom_K(A, K)$ corresponds to the identity endomorphism of ${}_AA$, and so $\nu(1) = 1$. Given $a, b \in A$, the endomorphism given by ab on $Hom_K(A, K)$ corresponds to the composition of the two endomorphisms given by a and then by b, since $End_A(Hom_K(A, K)) \simeq A^{op}$ as rings. Therefore, the same is true for the corresponding endomorphisms on ${}_AA$, and hence $\nu(ab) = \nu(a)\nu(b)$. The endomorphism corresponding to the element $a + b$ on $Hom_K(A, K)$ corresponds to the sum of the endomorphisms given by a and by b, since the isomorphism $End_A(Hom_K(A, K)) \simeq A^{op}$ is a ring isomorphism and hence preserves sums as well. Therefore, again $\nu(a + b) = \nu(a) + \nu(b)$. The ring homomorphism ν is an automorphism. Indeed if ν had a kernel C, the endomorphism ring of $Hom_K(A, K)$ would be isomorphic to A/C, a contradiction. Hence the homomorphism ν is injective. Since A is finite dimensional and $\nu : A \longrightarrow A$ is injective and K-linear, ν is surjective as well.

Remark 1.10.12 We observe that $Hom_K({}_1A_\nu, K) \simeq {}_\nu(Hom_K(A, K)_1)$ naturally. Therefore

$$Hom_K(Hom_K(A, K), K) \simeq {}_\nu A_\nu \simeq {}_1A_1 \simeq A.$$

Example 1.10.24 below will show that ν can have any given order as an element in $Out_K(A)$.

By Lemma 1.10.9 the automorphism ν is unique up to an inner automorphism.

Definition 1.10.13 Let K be a field and let A be a finite dimensional Frobenius K-algebra. Then there is an automorphism ν of A, unique up to an inner automorphism, such that

$$Hom_K(A, K) \simeq {}_1A_\nu$$

as A-A-bimodules. The automorphism ν is called the *Nakayama automorphism* of the Frobenius algebra A.

The Nakayama automorphism permutes projective and simple modules by tensor product with the Nakayama twisted bimodule:

$$S_\nu := {}_1A_\nu \otimes_A S \text{ and } P_\nu := {}_1A_\nu \otimes_A P$$

for every simple A-module S and for every projective indecomposable A-module P.

Lemma 1.10.14 *Let K be a field and let A be a finite dimensional Frobenius K-algebra with Nakayama automorphism ν. If S is simple then S_ν is simple. If P is projective indecomposable, P_ν is projective indecomposable.*

Proof Let U be a proper submodule of S_ν. Then ${}_\nu A_1 \otimes_A U$ is a proper submodule of S. Indeed, regarded as a right module, ${}_\nu A_1$ is free of rank 1, and so tensoring over A with ${}_\nu A_1$ is just the identity on the vector space structure. The property of being a (proper) submodule is preserved. Hence, S is simple if and only if S_ν is simple.

Since

$${}_1A_\nu \otimes_A (U_1 \oplus U_2) \simeq ({}_1A_\nu \otimes_A U_1) \oplus ({}_1A_\nu \otimes_A U_2)$$

a module is indecomposable if and only if its Nakayama twist is indecomposable. For the same reason, if P is a direct summand of A^n, P_ν is also a direct summand of $({}_1A_\nu)^n$. As a left module however, ${}_1A_\nu$ is just isomorphic to the regular module. Therefore P is projective indecomposable if and only if P_ν is projective indecomposable. This proves the lemma. □

Definition 1.10.15 The permutation induced by ν on simple modules is called the *Nakayama permutation*.

Usually it is difficult to find the Nakayama automorphism explicitly. The Nakayama permutation is more easily determined.

Example 1.10.16 Recall from Example 1.10.7 the definition of a Nakayama algebra N_n^m. The Nakayama permutation sends the idempotent e_i to e_{i+m} where the index $i+m$ is to be taken modulo n.

The case of a Nakayama automorphism being the identity deserves a special name. The reason for the notation will become clear in Definition 1.10.20 and Lemma 1.10.23.

Definition 1.10.17 A Frobenius algebra A is called *weakly symmetric* if for every idempotent $e^2 = e \in A$ we get $\nu(e)$ is conjugate to e. Equivalently A is weakly symmetric if the injective hull of a simple module is isomorphic to the projective cover.

The way we obtained the non-degenerate associative bilinear form on A can be generalised to finite dimensional algebras over a field given by a quiver and relations. The case of weakly symmetric algebras was discovered by Holm and Zimmermann in [18]. For the moment a naive comprehension of an algebra defined by a quiver and relation, in the sense described briefly in Example 1.6.23, is enough for the next proposition. A proper notation is given in Definition 1.11.5 and 1.11.13 below, and the reader feeling some unease with the sloppy way we are presently dealing with quivers and relations is encouraged to refer to the more formal definition there.

Proposition 1.10.18 *Let $A = KQ/I$ be a Frobenius algebra over the field K, given by the quiver Q and ideal of relations I, and fix a K-basis $\mathcal{B}$ of A consisting of pairwise distinct non-zero paths of the quiver Q. Assume that $\mathcal{B}$ contains a basis of the socle $soc(A)$ of A. Define a K-linear mapping ψ on the basis elements by*

$$\psi(b) = \begin{cases} 1 \text{ if } b \in soc(A) \\ 0 \text{ otherwise} \end{cases}$$

for $b \in \mathcal{B}$. Then an associative non-degenerate k-bilinear form $\langle -, - \rangle$ for A is given by $\langle x, y \rangle := \psi(xy)$.

Proof By definition, since A is an associative algebra, ψ is associative on basis elements, hence is associative on all of A.

Let ν be the Nakayama automorphism of A. We observe now that

$$\psi(x\nu(e)) = \psi(ex)$$

for all $x \in A$ and all primitive idempotents $e \in A$. Indeed, since ψ is linear, we need to show this only on the elements in $\mathcal{B}$. Let $b \in \mathcal{B}$. If b is a path not in the socle of A, then $b\nu(e)$ and eb are either zero or not contained in the socle, and hence $0 = \psi(b) = \psi(b\nu(e)) = \psi(eb)$. If $b \in \mathcal{B}$ is in the socle of A, then $b = e_b b = b\nu(e_b)$ for exactly one primitive idempotent e_b and $e'b = b\nu(e') = 0$

for each primitive idempotent $e' \neq e_b$. Therefore, $\psi(e'b) = \psi(b\nu(e')) = 0$ and $\psi(e_b b) = \psi(b) = \psi(b\nu(e_b))$.

It remains to show that the map $(x, y) \mapsto \psi(xy)$ is non-degenerate. Suppose we had $x \in A \setminus \{0\}$ such that $\psi(xy) = 0$ for all $y \in A$. In particular for each primitive idempotent e_i of A we get $\psi(e_i xy) = \psi(xy\nu(e_i)) = 0$ for all $y \in A$. Hence we may suppose that $x \in e_i A$ for some primitive idempotent $e_i \in A$.

Now, xA is a right A-module. Choose a simple submodule S of xA and $s \in S\setminus\{0\}$. Then, since $s \in S \leq xA$, there is a $y \in A$ such that $s = xy$. Since $S \leq xA \leq A$, and since S is simple, $s \in soc(A) \setminus \{0\}$. Moreover, since $x \in e_i A$, also $s = e_i s$, i.e. s is in the (1-dimensional) socle of the projective indecomposable module $e_i A$. So, up to a scalar factor, s is a path contained in the basis $\mathcal{B}$ (recall that by assumption $\mathcal{B}$ contains a basis of the socle formed by paths). This implies that

$$\psi(xy) = \psi(s) = \psi(e_i s) \neq 0,$$

contradicting the choice of x, and hence proving non-degeneracy. □

Remark 1.10.19 The construction in Proposition 1.10.18 is the only possible construction. This is the subject of Proposition 3.6.14.

1.10.2 Symmetric Algebras

A special class of self-injective algebras plays an important role, in particular since group algebras are of this kind.

Definition 1.10.20 Let K be a commutative ring. A K-algebra A is *symmetric* if there is a non-degenerate, associative and symmetric bilinear form $\langle\ ,\ \rangle : A \times A \longrightarrow K$ on A.

More explicitly $\langle ab, c\rangle = \langle a, bc\rangle$, $\langle a, b\rangle = \langle b, a\rangle$ for all $a, b, c \in A$ and $\langle\ ,\ \rangle : A \times A \longrightarrow K$ is non-degenerate bilinear.

Remark 1.10.21 By Proposition 1.10.6 a symmetric algebra is always Frobenius. For symmetric algebras we ask in addition the Frobenius bilinear form to be symmetric.

Remark 1.10.22 If A is symmetric, the form $\langle -, -\rangle$ of Proposition 1.10.18 will not be symmetric in general. Examples are given in e.g. [19, Sect. 4, proof of (1)] and [20].

Proposition 1.10.23 *The identity is a Nakayama transformation of a symmetric algebra. A Frobenius algebra for which the identity is a Nakayama automorphism is symmetric, in particular an algebra is symmetric if and only if $A \simeq Hom_K(A, K)$ as A–A-bimodules.*

Proof The form $\langle\ ,\ \rangle$ is non-degenerate and bilinear, and so the mapping

$$\begin{aligned} A &\xrightarrow{\Psi} Hom_K(A, K) \\ a &\mapsto (b \mapsto \langle b, a \rangle) \end{aligned}$$

is bijective and K-linear. Moreover, as usual $Hom_K(A, K)$ is an $A \otimes_K A^{op}$-module by $(a \cdot f \cdot c)(b) = f(cba)$ for all $a, b, c \in A$ and $f \in Hom_K(A, K)$. Now, Ψ is a homomorphism of $A \otimes_K A^{op}$-modules. Indeed, for all $a, b, c, x \in A$ we have

$$\begin{aligned} \Psi(c \cdot a \cdot b)(x) = \langle x, cab \rangle &= \langle xc, ab \rangle = \langle ab, xc \rangle \\ &= \langle a, bxc \rangle = \Psi(a)(bxc) = (c \cdot \Psi(x) \cdot b)(x) \end{aligned}$$

which proves that A is isomorphic to its dual as a bimodule. By Definition 1.10.13 we get that the identity is a Nakayama automorphism.

Conversely, if the identity is a Nakayama automorphism, then

$$A \simeq Hom_K(A, K)$$

as an A–A-bimodule. Given an isomorphism

$$\Phi : A \longrightarrow Hom_K(A, K)$$

we define

$$\langle a, b \rangle := \Phi(a)(b)$$

and get

$$\langle a, cb \rangle = \Phi(a)(cb) = (\Phi(a) \cdot c)\,(b) = \Phi(ac)(b) = \langle ac, b \rangle$$

since Φ is linear as a right module. Moreover,

$$\langle a, b \rangle = \Phi(a)(b) = (a \cdot \Phi(1))\,(b) = \Phi(1)(ba) = \langle 1, ba \rangle = \langle b, a \rangle$$

using that Φ is linear as a left module, and the first step. Since Φ is bijective, the form $\langle\ ,\ \rangle$ is non-degenerate. This proves the statement. □

There are weakly symmetric algebras which are not symmetric.

Example 1.10.24 We present a slight simplification of an example which can be found in the article [21, Example 1] of König and Xi.

Let K be a field and let $\lambda \in K$. Define A to be the quotient of the free algebra on the symbols b, c by the ideal generated by

$$b^2, c^2, cb - \lambda bc.$$

This algebra is Frobenius and weakly symmetric, but the algebra is symmetric if and only if $\lambda = 1$.

Indeed, suppose the algebra is symmetric. Then there is a non-degenerate symmetric associative bilinear form $\langle\ ,\ \rangle$ on A. But then

$$\lambda\langle 1, bc\rangle = \langle 1, \lambda bc\rangle = \langle 1, cb\rangle = \langle c, b\rangle = \langle b, c\rangle = \langle bc, 1\rangle = \langle 1, bc\rangle$$

and hence $(1-\lambda)\langle 1, bc\rangle = 0$. Suppose $\lambda \neq 1$. Then $\langle 1, bc\rangle = 0$. But bc generates the socle of the algebra A. This is already a contradiction to Proposition 1.10.18 in connection with Proposition 3.6.14. But we can argue in a more elementary way. Indeed, the above shows $\langle\ , bc\rangle = 0$ and therefore bc is in the radical of the bilinear form. This contradicts the fact that $\langle\ ,\ \rangle$ is non-degenerate. If $\lambda = 1$, then the algebra is symmetric. Indeed, the bilinear form $\langle\ ,\ \rangle : A \times A \longrightarrow K$ defined by

$$\langle x, y\rangle := \varphi(xy) = \begin{cases} 1 \text{ if } xy = bc \\ 0 \text{ if } xy \neq bc \end{cases}$$

for $x, y \in \{1, b, c, bc\}$ and extended K-linearly, is symmetric (since $bc = cb$), non-degenerate and associative by Proposition 1.10.18.

A is Frobenius for all $\lambda \neq 0$. Indeed, the linear form $\varphi : A \longrightarrow K$ defined by $\varphi(bc) = 1$ and $\varphi(b) = \varphi(c) = \varphi(1) = 0$ induces a non-degenerate associative bilinear form by putting $\langle x, y\rangle = \varphi(xy)$ for all $x, y \in A$. Again taking the K-basis $\{1, b, c, bc\}$, the automorphism given by $b \mapsto \lambda b$ and $c \mapsto \lambda^{-1}c$ is a Nakayama automorphism as is readily verified. Moreover, this automorphism has the same order in $Out(A)$ as λ has in $K^\times$.

The algebra is weakly symmetric for any $\lambda \neq 0$. Indeed, the algebra is local and so there is only one isomorphism class of indecomposable projective modules and only one isomorphism class of simple modules. The injective envelope and the projective cover of this simple module is A.

We can ask if the symmetrising form is unique in some sense. The answer is given by the following lemma.

Lemma 1.10.25 *Let K be a field and let A be a symmetric K-algebra with symmetrising form $\langle\ ,\ \rangle$. Then for every $u \in Z(A)^\times$ the form $\langle\ ,\ \rangle_u$ defined by*

$$\langle a, b\rangle_u := \langle u \cdot a, b\rangle$$

for all $a, b \in A$ is symmetrising and conversely for every symmetrising form $\widetilde{\langle\ ,\ \rangle}$ there is a $u \in Z(A)^\times$ such that

$$\widetilde{\langle\ ,\ \rangle} = \langle\ ,\ \rangle_u.$$

Proof Let u be a central unit of A. Since $\langle\ ,\ \rangle$ is non-degenerate,

$$\bigcap_{x\in A} \ker(\langle u \cdot x,\ \rangle) = \bigcap_{u\cdot x\in A} \ker(\langle u \cdot x,\ \rangle) = \bigcap_{x'\in A} \ker(\langle x',\ \rangle) = 0$$

and so $\langle\ ,\ \rangle_u$ is non-degenerate.

We compute for all $x, y \in A$

$$\langle x, y\rangle_u = \langle u \cdot x, y\rangle = \langle y, u \cdot x\rangle = \langle y \cdot u, x\rangle = \langle u \cdot y, x\rangle = \langle y, x\rangle_u$$

and hence $\langle\ ,\ \rangle_u$ is symmetric.

Moreover, for all $x, y, z \in A$ we get

$$\langle x, y \cdot z\rangle_u = \langle u \cdot x, y \cdot z\rangle = \langle u \cdot x \cdot y, z\rangle = \langle x \cdot y, z\rangle_u$$

and hence $\langle\ ,\ \rangle_u$ is associative. This shows that $\langle\ ,\ \rangle_u$ is a symmetrising form on A.

Conversely, we will show that any symmetrising form is of the form $\langle\ ,\ \rangle_u$ for some $u \in Z(A)$. Indeed, $\langle\ ,\ \rangle$ induces an isomorphism of $A \otimes_K A^{op}$-modules $A \simeq Hom_K(A, K)$ by sending $a \mapsto \langle\ , a\rangle$ (cf Proposition 3.6.14). Hence any symmetrising form will give such an isomorphism, whence two different forms will produce an automorphism $A \longrightarrow A$ of $A \otimes_K A^{op}$-modules. Since $End_{A\otimes_K A^{op}}(A) = Z(A)$ we see that any symmetrising form of A is of the form $\langle\ ,\ \rangle_u$ for some $u \in A$. □

Proposition 1.10.26 *Let G be a group and let K be a commutative ring. Then KG is a symmetric algebra.*

Proof Indeed, we define a linear form $\varphi : KG \longrightarrow K$ by $\varphi(\sum_{g\in G} u_g g) := u_1$. The bilinear associative form $\langle x, y\rangle := \varphi(xy)$ is non-degenerate, since given $\sum_{g\in G} u_g g \in KG$, there is at least one coefficient $u_{g_0} \neq 0$ and hence $\langle \sum_{g\in G} u_g g, g_0^{-1}\rangle = u_{g_0} \neq 0$. The form is symmetric, since

$$\varphi\left(\left(\sum_{g\in G} u_g g\right)\cdot\left(\sum_{h\in G} \ell_h h\right)\right) = \sum_{gh=1} u_g \ell_h = \varphi\left(\left(\sum_{h\in G} \ell_h h\right)\cdot\left(\sum_{g\in G} u_g g\right)\right).$$

This proves the statement. □

Remark 1.10.27 We shall generalise this result in Lemma 2.4.7.

There are (at least) two ways of dualising objects for K-algebras A. For symmetric algebras these coincide.

Proposition 1.10.28 *Let K be a commutative ring and let A be a symmetric K-algebra. Then for any A-module M one has a natural isomorphism $\eta_M : Hom_A(M, A) \simeq Hom_K(M, K)$ in the sense that for two modules M and N and every module homomorphism $\alpha : M \longrightarrow N$ the diagram*

$$\begin{array}{ccc} Hom_A(N,A) & \stackrel{Hom_A(\alpha,A)}{\longrightarrow} & Hom_A(M,A) \\ \downarrow \eta_N & & \eta_M \downarrow \\ Hom_K(N,K) & \stackrel{Hom_K(\alpha,K)}{\longrightarrow} & Hom_K(M,K) \end{array}$$

is commutative.

Remark 1.10.29 Later we shall interpret this proposition in terms of a more abstract and elegant language, the language of category theory. There the proposition actually shows that the collection of mappings η_M forms a natural transformation, actually an isomorphism, between the functors $Hom_A(-, A)$ and $Hom_K(-, K)$. This language will be introduced and studied in Chap. 3.

Proof of Proposition 1.10.28 First we shall define η_M. Actually, this is a special case of Lemma 1.7.9.

$$\begin{aligned} Hom_A(M, A) &\simeq Hom_A(M, Hom_K(A, K)) \\ &\simeq Hom_K(A \otimes_A M, K) \\ &\simeq Hom_K(M, K) \end{aligned}$$

where the first isomorphism comes from the fact that A is symmetric and hence $A \simeq Hom_K(A, K)$ as A-A-bimodules. The compatibility with morphisms can actually be deduced from there. Since the first isomorphism only concerns the second argument, it commutes with pre-composition with a morphism $\alpha : M \longrightarrow N$. The second isomorphism is

$$f \mapsto (a \otimes m \mapsto f(m)(a))$$

for every $f \in Hom_A(M, Hom_K(A, K))$, for every $m \in M$ and $a \in A$. Now, an $f \in Hom_A(N, Hom_K(A, K))$ is mapped by $Hom_A(\alpha, A)$ to

$$Hom_A(\alpha, A)(f) \mapsto (a \otimes n \mapsto f(\alpha(n))(a)).$$

Conversely, first applying $Hom_A(\alpha, A)$ and then η_N transforms the mapping f first into $f \mapsto f \circ \alpha$ and then

$$f \circ \alpha \mapsto (a \otimes n \mapsto (f \circ \alpha)(n)(a)),$$

which is the same as the above. This proves the proposition. □

Lemma 1.10.30 *Let A be a finite dimensional symmetric K-algebra over a field K. Then there is an isomorphism*

$$\varphi_M^P : Hom_A(P, M) \longrightarrow Hom_K(Hom_A(M, P), K)$$

for all finite dimensional A-modules M and all finite-dimensional projective A-modules P. Moreover, this isomorphism commutes with A-module *homomorphisms in the sense that*

$$\begin{array}{ccc} Hom_A(P,M) & \xrightarrow{\varphi_M^P} & Hom_K(Hom_A(M,P),K) \\ \uparrow Hom_A(\pi,M) & & \uparrow Hom_K(Hom_A(M,\pi),K) \\ Hom_A(P',M) & \xrightarrow{\varphi_M^{P'}} & Hom_K(Hom_A(M,P')) \end{array}$$

and

$$\begin{array}{ccc} Hom_A(P,M) & \xrightarrow{\varphi_M^P} & Hom_K(Hom_A(M,P),K) \\ \downarrow Hom_A(P,\mu) & & \downarrow Hom_K(Hom_A(P,\mu),K) \\ Hom_A(P,M') & \xrightarrow{\varphi_{M'}^{P}} & Hom_K(Hom_A(M',P)) \end{array}$$

are commutative for all $\mu \in Hom_A(M, M')$ and $\pi \in Hom_A(P, P')$, whenever M, M' are finite dimensional A-modules and P, P' are finitely generated projective A-modules.

Proof We compute

$$\begin{aligned} Hom_A(M,P) &\simeq Hom_A(M, Hom_K(Hom_K(P,K),K)) \\ &\simeq Hom_K(Hom_K(P,K)\otimes_A M, K) \\ &\simeq Hom_K(Hom_A(P,A)\otimes_A M, K) \\ &\simeq Hom_K(Hom_A(P,M),K). \end{aligned}$$

Here the first isomorphism comes from the fact that P is finite dimensional and double duality over K is the identity by evaluation. The second isomorphism is Frobenius reciprocity (Lemma 1.7.9), and the third isomorphism is Proposition 1.10.28. Then the fourth isomorphism comes from the fact that P is finitely generated projective and for these modules $Hom_A(P, A) \otimes_A M \simeq Hom_A(P, M)$. Indeed, if P is free, this is clear, and the isomorphism commutes with taking direct summands. The fact that the isomorphisms φ_M^P commute with the homomorphisms π and μ comes from a similar statement as Proposition 1.10.28 and that the statement holds for Frobenius reciprocity, as we may verify easily by writing down the explicit mappings. □

Let $\mathcal{S}$ be the set of isomorphism classes of simple A-modules. We denote by I_S the injective hull of the A-module S. We have the following.

Lemma 1.10.31 (Nakayama [15, 16]) *Suppose K is a field and A is a finite dimensional K-algebra. If A is self-injective then for each projective indecomposable A-module P, we get that $soc(P)$ is simple. Moreover the mapping $P/\mathrm{rad}(P) \mapsto soc(P)$ defined for each projective indecomposable A-module P induces a bijection ν on $\mathcal{S}$.*

Proof Let A be self-injective. Then we see that P is also the injective hull of $\mathrm{soc}(P)$. Let S be a simple submodule of $\mathrm{soc}(P)$. Then the injective hull I_S of S is a direct factor of P. Since $P/\mathrm{rad}(P)$ is simple, P is indecomposable, and hence $I_S = P$. Therefore $S = \mathrm{soc}(P)$. Thus $\mathrm{soc}(P)$ is simple for each projective indecomposable A-module P.

If $S := \mathrm{soc}(P_1) \simeq \mathrm{soc}(P_2)$ for two projective indecomposable A-modules P_1 and P_2, then P_1 and P_2 are both the injective hull of S. The injective hull is unique, and therefore $P_1 \simeq P_2$. Hence ν is injective, and since $\mathcal{S}$ is finite, ν is also bijective. □

Remark 1.10.32 We can therefore also define a Nakayama permutation for any self-injective algebra. For every simple A-module S let $\nu(S)$ be the socle of the projective cover of S. However, the Nakayama automorphism is reserved for Frobenius algebras.

1.11 Algebras Defined by Quivers and Relations: Two Classes of Algebras

We have already seen the usefulness of defining algebras in a combinatorial manner by quivers and relations. It is time to give the naive introduction a thorough and rigorous foundation.

1.11.1 Hereditary Algebras

Semisimple artinian algebras are the most tractable algebras we shall consider. By Wedderburn's theorem their structure is completely understood, up to the knowledge of finite dimensional skew-fields over a given base field. Every ideal of a semisimple algebra A is a projective module, and every quotient of A is a projective module. The next easiest class is the class of hereditary algebras for which only the first property is required.

Definition 1.11.1 Let K be a commutative ring and let A be a K-algebra. A is called *hereditary* if every ideal of A is a projective A-module.

Example 1.11.2 Let us introduce two important families of hereditary algebras.

1. As we have seen, semisimple algebras are hereditary algebras.
2. The K-algebra A of upper triangular matrices of type $n \times n$ over a field K is hereditary. Indeed, any column of A is an indecomposable module of A. Each of these modules is projective, by definition. Hence, denoting the m-th column by P_m we get a sequence of ideals

$$0 = P_0 < P_1 < P_2 < \cdots < P_n$$

such that there are A-modules P_i' with $P_i' \simeq P_i$ for all $i \in \{1, \ldots, n\}$ and $P_1' \oplus P_2' \oplus \cdots \oplus P_n'$ is the regular module. The quotient P_i/P_{i-1} is of dimension

1 over K, whence simple. We see that $\mathrm{rad}(A)$ is given by the nilpotent matrices in A, whence the upper triangular $n \times n$ matrices with main diagonal 0. Hence

$$A/\mathrm{rad}(A) = \bigoplus_{i=1}^{n} P_i/P_{i-1}$$

is a direct sum of the n simple A-modules $S_i = P_i/P_{i-1}$ and any simple A-module is isomorphic to one of the S_i.

Let a be an element in A. We will study the module Aa. Observe that multiplying with A from the left corresponds to the row operations in Gauss' algorithm, i.e. adding multiples of any row v of a to another row of a above v. We now consider $A \subseteq Mat_{n\times n}(K)$ and ideals of A inside this bigger ring. For any invertible matrix $u \in Gl_n(K)$ we have $Aa \simeq Aau$ as A-modules, where the isomorphism is given by right multiplication by u. Column operations of Gauss' algorithm are realised by multiplication by certain matrices $u \in Gl_n(K)$ from the right. Hence, by this slightly restricted form of Gauss' algorithm the A-module Aa is isomorphic to Aa' where a' is a diagonal matrix with entries only 0 and 1. This module is isomorphic to P_i where $i = \max\{s \in \{1, \ldots, n\} \mid a'_{ss} = 1\}$. Hence, since A is Noetherian, any ideal of A is finitely generated, and therefore

$$I = Aa_1 + Aa_2 + \cdots + Aa_m$$

where $Aa_i \simeq P_{j_i}$ and this sum of principal ideals is actually isomorphic to the projective module P_i where $i = \max\{j_i | \ i \in \{1, \ldots, m\}\}$. Therefore any ideal of A is projective as an A-module.

Lemma 1.11.3 *Let K be a field and let A be a hereditary K-algebra. Then any submodule of a finitely generated projective module is projective.*

Proof Since each projective module is a direct factor of a free module, it is sufficient to show that each submodule of a free module is projective. Let A^n be a free module and let S be a submodule. We shall prove the statement by induction on n.

The case $n = 1$ is the definition of a hereditary algebra. Let $n > 1$ and suppose the statement is true for submodules of A^{n-1}. Let $\pi_n : A^n \longrightarrow A$ be the projection onto the last component: $\pi_n(a_1, a_2, \ldots, a_n) := a_n$. Then $\pi_n(S)$ is a submodule of $\pi_n(A^n) = A$. Hence $\pi_n(S)$ is an ideal of A, and since A is hereditary, $\pi_n(S)$ is projective. Therefore $\pi_n(S)$ is a direct factor of S. If $\pi_n(S) = 0$, then S is a submodule of A^{n-1} and we are done by the induction hypothesis. Otherwise the kernel of the restriction of π_n to S is $S \cap A^{n-1}$. Hence

$$S = \pi_n(S) \oplus (S \cap A^{n-1}).$$

By the induction hypothesis $S \cap A^{n-1}$ is projective as well. □

Proposition 1.11.4 *Let K be a field and let A be a hereditary K-algebra. Then $Ext_A^i(M,N)=0$ for all A-modules M and N and all $i\geq 2$.*

Proof By Lemma 1.11.3 any submodule of a projective A-module is projective. Since Ω_M is by definition a submodule of the projective cover of M, we get that Ω_M is projective. Hence $\Omega_M^2=0$, and furthermore $\Omega_M^i=0$ for all $i\geq 2$. Since $Ext_A^i(M,N)=Ext_A^1(\Omega_M^{i-1},N)$ for all $i\geq 2$, and since Ω_M is projective, we get that $Ext_A^i(M,N)=0$ as soon as $i\geq 2$. □

As promised earlier, we shall now properly construct a method to define K-algebras over a field K in a quite general way and show that this construction always yields hereditary algebras. We have already used this construction in a slightly informal way.

Definition 1.11.5

- A *quiver* Γ is a pair (Γ_v,Γ_a) where Γ_v is a set, called the set of *vertices*, and Γ_a is a set, whose elements are called *arrows*, together with two mappings $h:\Gamma_a\longrightarrow\Gamma_v$, called the *head of an arrow*, and $t:\Gamma_a\longrightarrow\Gamma_v$ called the *tail*.
- A quiver Γ is *finite* if Γ_v and Γ_a are finite sets.
- An arrow $a\in\Gamma_a$ is a *loop* if $h(a)=t(a)$. Denote by Γ_ℓ the set of loops in Γ_a.
- A *path* p in the quiver Γ is a sequence $(t(a_1),a_1,\dots,a_n)\in\Gamma_v\times\Gamma_a^n$ such that $h(a_i)=t(a_{i+1})$ for all $i\in\{1,\dots,n-1\}$. Observe that if $n=0$ a path can contain only the starting vertex $t(a_1)$. In this case we call the path a *lazy path*. We write $p=a_1a_2\dots a_n$ for $p=(t(a_1),a_1,\dots a_n)$ and say that p has length n. We say $t(a_1)$ is the *starting vertex* of the path p and $h(a_n)$ the *ending vertex* of the path p. A path is *closed* if its starting and its ending vertex coincide.
- Given two paths $p=(t(a_1),a_1,\dots,a_n)$ and $q=(t(b_1),b_1,\dots,b_m)$ with $t(b_1)=h(a_n)$ then the *concatenation* of p with q is the path $pq:=(t(a_1),a_1,\dots,a_n,b_1,\dots,b_m)$.
- Given a quiver Γ we define its *double quiver* by adding reverses of all non-loop arrows: $\overline{\Gamma}:=(\Gamma_v,\Gamma_a\coprod\Gamma_a^*)$ where $\Gamma_a^*:=\Gamma_a\setminus\Gamma_\ell$. Hence $\overline{\Gamma}_a$ is the disjoint union of the arrows of Γ and another copy of the set of those arrows of Γ, which are not loops. We define the head and tail maps on $\overline{\Gamma}$ as follows: $\overline{h}(a):=h(a)$, $\overline{h}(a^*):=t(a)$, $\overline{t}(a):=t(a)$ and $\overline{t}(a^*):=h(a)$ for all $a\in\Gamma_a$ and $a^*\in\Gamma_a^*$.
- A *walk* in Γ is a path in $\overline{\Gamma}$.
- A quiver is *connected* if for every pair of vertices $v_1,v_2\in\Gamma_v$ there is a walk in Γ which starts at v_1 and ends in v_2.

Given a field K and a quiver Γ we define an algebra $K\Gamma$, the quiver algebra.

Definition 1.11.6 Let Γ be a quiver with a finite set of vertices and let K be a field. Then the *quiver algebra* $K\Gamma$ is the K-vector space with K-basis all paths in Γ. The vector space $K\Gamma$ becomes a K-algebra if one defines $p\cdot q:=pq$ the concatenation if the ending vertex of p is equal to the starting vertex of q, $p\cdot q=p$ if q is the lazy path corresponding to the ending vertex of p, and $p\cdot q:=0$ otherwise. This product on the basis is extended K-linearly and trivially defines an associative product. The unit element in $K\Gamma$ is the sum of all lazy paths.

Remark 1.11.7 In principle one can also define a path algebra for Γ with an infinite vertex set. But in this case the unit element has to be added artificially, since the sum of all lazy paths is not defined, the sum being infinite.

Remark 1.11.8 Sometimes it is convenient to use the following equivalent definitions of paths, walks, closed or not, and length of a path.

Let Q be a quiver. We call a *path* a sequence of arrows $\alpha_1\alpha_2\dots\alpha_n$ in Q such that the ending vertex of α_i is the starting vertex of α_{i-1} for all $i \in \{2,\dots,n\}$. The starting vertex of α_1 is also the starting point of the path, and likewise the ending vertex of α_n is the ending vertex of the path. The number n is the length of the path. A path is *closed* if the starting vertex of the path equals the ending vertex of the path.

A quiver homomorphism $Q_1 \longrightarrow Q_2$ is a graph map that preserves orientations, i.e. a map that maps vertices to vertices, edges to edges, and preserves adjacency and orientation.

Let L be a linear quiver with m vertices and any orientation.

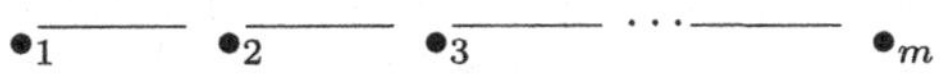

We denote by Z a quiver of the cyclic graph

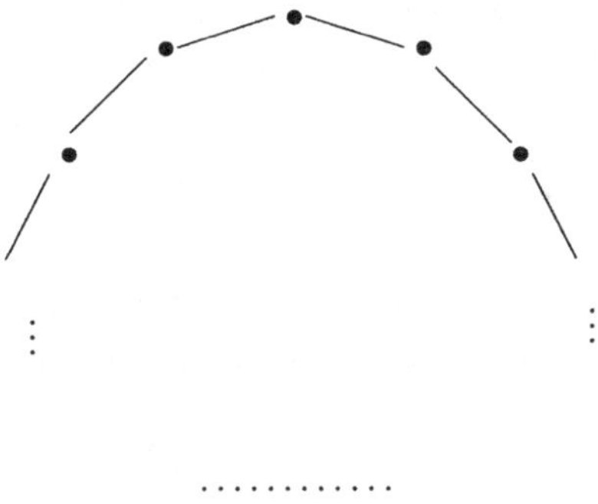

with any fixed orientation.

A *walk* in the quiver is a quiver homomorphism $w : L \longrightarrow Q$ for some orientation of L. We sometimes also denote the walk w by $w = \beta_1\beta_2\dots\beta_n$ in Q, where $\beta_i := w(\bullet_i \text{———} \bullet_{i+1})$. A *tour* is a graph homomorphism $t : Z \longrightarrow Q$ with any fixed orientation of Z. A tour is a special case of a walk, and is sometimes denoted by a sequence of arrows (as in a graph). A closed path is nothing else than a tour where Z is equi-oriented.

For each connected subquiver L' of L, the composition $L' \hookrightarrow L \longrightarrow Q$ is a subwalk of $L \longrightarrow Q$.

Given a quiver Γ with finite vertex set, K a field and $A = K\Gamma$ the path algebra, by definition of the multiplication any lazy path e_v for a vertex v is an idempotent. Hence $P_v := A \cdot e_v$ is a projective module. A K-basis of P_v is the set of all paths ending at v. Moreover,

$$Hom_A(P_v, P_w) = Hom_A(Ae_v, Ae_w) = e_v Ae_w$$

has a K-basis given by all paths starting at v and ending at w.

Lemma 1.11.9 *The algebra $K\Gamma$ is finite dimensional if and only if Γ is finite and there is no closed non-lazy path in Γ.*

Proof If Γ is not finite there are infinitely many arrows. If there is a closed path p we can form p, pp, ppp, etc. and each of these paths forms a new basis element. On the other hand, if Γ is finite and has no closed paths, there are only a finite number of paths leaving and ending at any given vertex. □

Lemma 1.11.10 *Let Γ be a quiver, let K be a field and suppose that $K\Gamma$ is finite dimensional. Then P_v is indecomposable for all $v \in \Gamma_v$ and each projective indecomposable $K\Gamma$-module is isomorphic to some P_v. Moreover* $\mathrm{rad}(K\Gamma) = \sum_{a\in\Gamma_a} K\Gamma a$.

Proof We shall prove that $\mathrm{rad}(K\Gamma) = \sum_{a\in\Gamma_a} K\Gamma a$ first. Since the sum of ideals is an ideal, it is clear that $\sum_{a\in\Gamma_a} K\Gamma a$ is an ideal of $K\Gamma$. Moreover, given a path p of length n and a path q of length m, then the length of pq is $n+m$ if $pq \neq 0$. Since the algebra $K\Gamma$ is finite dimensional there is a number N such that all paths of $K\Gamma$ have length at most N. Since all elements in $\sum_{a\in\Gamma_a} K\Gamma a$ are linear combinations of paths of length at least 1, we have $\left(\sum_{a\in\Gamma_a} K\Gamma a\right)^N = 0$. Hence $\sum_{a\in\Gamma_a} K\Gamma a$ is nilpotent and we have $\sum_{a\in\Gamma_a} K\Gamma a \subseteq \mathrm{rad}(K\Gamma)$. Moreover

$$K\Gamma / \sum_{a\in\Gamma_a} K\Gamma a \simeq \bigoplus_{v\in\Gamma_v} K\Gamma e_v/\mathrm{rad}(K\Gamma e_v)$$

and this algebra is semisimple since $K\Gamma e_v/\mathrm{rad}(K\Gamma e_v)$ is one-dimensional. We get one simple module $S_v = K\Gamma e_v/\mathrm{rad}(K\Gamma e_v)$ of dimension 1 for each vertex $v \in \Gamma_v$. Hence

$$\sum_{a\in\Gamma_a} K\Gamma a = \mathrm{rad}(K\Gamma).$$

Since each simple module S_v has a projective cover $K\Gamma e_v$, we have proved the lemma. □

Proposition 1.11.11 *Let Γ be a finite quiver and let K be a field such that $K\Gamma$ is finite dimensional. Then $K\Gamma$ is a hereditary K-algebra.*

Proof We need to show that every ideal of $A := K\Gamma$ is a projective module. In order to do so we only need to show that every submodule of every indecomposable projective module is projective. Indeed, let $P = P_1 \oplus P_2$ with P_1 indecomposable. Suppose $I \leq P_1 \oplus P_2$ and let $\pi : P_1 \oplus P_2 \longrightarrow\!\!\!\!\!\rightarrow P_1$ be the natural projection. Then $I_1 := \pi(I) \leq P_1$ and suppose we have shown that this implies I_1 is projective. Then $I = I_1 \oplus I_2$ for some submodule I_2 of I since I_1 is projective and $I \longrightarrow I_1$ is surjective. The result follows by induction on the dimension of I.

Let $P_v = Ae_v$ be an indecomposable projective A-module. We claim that it is sufficient to show that $\mathrm{rad}(P_v)$ is projective. Indeed, since the radical of A is nilpotent, the result then follows by induction on the Loewy length. But now, $\mathrm{rad}(P_v) = A\alpha_1 + A\alpha_2 + \cdots + A\alpha_n$ for $\{\alpha_1, \ldots, \alpha_n\} = h^{-1}(v)$. Moreover

$$A\alpha_1 + A\alpha_2 + \cdots + A\alpha_n = A\alpha_1 \oplus A\alpha_2 \oplus \cdots \oplus A\alpha_n$$

since two paths with a different last arrow are different and so $A\alpha_i \cap A\alpha_j = \{0\}$. So we need to show that $A\alpha_i$ is projective for all i. But $A\alpha_i = A \cdot t(\alpha_i) \cdot \alpha_i$ for all $i \in \{1, \ldots, n\}$ and we claim that

$$\begin{aligned} \mu_i : A \cdot t(\alpha_i) &\longrightarrow A\alpha_i \\ a &\mapsto a\alpha_i \end{aligned}$$

is an isomorphism. Indeed, first it is clear that μ_i is A-linear. Further μ_i is surjective since every path ending with an arrow α_i has to pass through the vertex $t(\alpha_i)$. Finally, μ_i is injective since if

$$x = \sum_j \lambda_j p_j \in A \cdot t(\alpha_i)$$

for paths p_j and $\lambda_j \in K$, then

$$x\alpha_i = \sum_j \lambda_j p_j \alpha_i \in A\alpha_i$$

and if $p_j \neq p_\ell$, then $p_j\alpha_i \neq p_\ell\alpha_i$, and all paths p_j end at $t(\alpha_i)$ by hypothesis, and so $p_j\alpha_i \neq 0$ for all j. This proves the statement which in turn proves the Proposition. □

Remark 1.11.12 If K is algebraically closed then we shall see in Sect. 4.5.1 that in some sense all finite dimensional hereditary K-algebras are equivalent to a quiver algebra.

1.11.2 Special Biserial Algebras

We shall need a very useful and illustrative way to determine indecomposable modules and morphism spaces between modules over a nice class of algebras, the so-called special biserial algebras. It will turn out that the algebras we are going to face for certain group representations are special biserial. Moreover, this will give a nice way to produce indecomposable modules in a very visual manner. The material of this section comes from Wald and Waschbüsch [8], whereas the following definition is very classical.

Recall from Definition 1.11.5 the definitions and notations used in a quiver.

Definition 1.11.13 Let K be a field and let $Q = (Q_0, Q_1)$ be a finite quiver. Let $\mathcal{F}$ be the two-sided ideal of KQ generated by all the arrows of Q. An ideal I is *admissible* if there is an integer $m \in \mathbb{N}$ such that $\mathcal{F}^m \leq I \leq \mathcal{F}^2$. An element u of KQ is an *admissible relation* if the two-sided ideal generated by u is an admissible ideal.

Remark 1.11.14 In other words, a two-sided ideal I of KQ is admissible if I is generated by a finite number of linear combinations $a_j := \sum_{i=1}^{n_j} p_i^j$ of paths p_i^j of length at least 2 and so that the starting vertex of each p_i^j and the ending vertex of each p_i^j depends only on j and not on i, and so that all paths of length at least m are in the ideal I. We shall prove in Theorem 4.5.2 that as long as we are dealing with finite dimensional algebras, and as long as we are only interested in the behaviour of modules over the algebras, then we may always use admissible ideals.

Definition 1.11.15 (Wald and Waschbüsch [8])Let K be an algebraically closed field, let Q be a quiver and let I be a set of admissible relations. Then pair (Q, I) is *special biserial* and the algebra $A = KQ/I$ is a *special biserial algebra* if

1. (a) any vertex of Q is the starting point of at most two arrows,
   (b) any vertex of Q is the ending point of at most two arrows,
2. (a) given an arrow β with ending vertex $x = h(\beta)$, then there is at most one arrow α with starting vertex $x = t(\alpha)$ such that $\alpha\beta \notin I$,
   (b) given an arrow α with starting vertex $x = t(\alpha)$, then there is at most one arrow β with ending vertex $x = h(\beta)$ such that $\alpha\beta \notin I$,
3. each infinite path contains a subpath which is in I.

Remark 1.11.16 It is clear by this definition that A is biserial in the sense that whenever P is a projective indecomposable A-module, then $\text{rad}(P) = U + V$ where U and V are uniserial A-modules and the Loewy length of $U \cap V$ is at most 1.

Indeed, starting at a vertex v, there are at the beginning at most two possible paths ending at v (axiom 1b) and then, after having chosen one arrow, there is a unique way to continue (axiom 2a). If the possibly two paths emanating from our vertex v join to the same vertex w so that the two circuits produce the same element in A, then the paths have maximal length, i.e. they represent the projective indecomposable module. Indeed, let p_1 and p_2 be the two paths. Axiom 2b shows that there is at most one arrow γ such that $p_1\gamma = p_2\gamma \neq 0$. Let $p_1 = p_1'\delta_1$ and $p_2 = p_2'\delta_2$. If there is an arrow γ with $p_1\gamma = p_2\gamma \neq 0$, then we apply 2a) to γ and obtain that $\delta_1 = \delta_2$. This shows by induction on the length of p_1 and p_2 that $p_1 = p_2$.

Remark 1.11.17 Note that the concept of being biserial depends only on the algebra. It can be verified independently of the presentation of the algebra.

The concept of a special biserial algebra is a concept of a quiver with relations. The axioms 1a and 1b are independent of the choice of the presentation, since if $A = KQ/I$, then Q is uniquely determined by A. The ideal I is not unique, and the axioms 2a and 2b depend on I, even on the elements chosen to generate I.

We shall construct all indecomposable A-modules where A is special biserial. Here we closely follow [8].

A *track* $T = (Q, I, v)$ is a quiver Q with relations I and a path v such that each arrow α of Q occurs precisely once in v, for any two arrows α and β such that the ending vertex of α is the starting vertex of β, $\alpha\beta \in I$ if and only if $\alpha\beta$ is not a subpath of v, so that v runs at most twice through any given vertex, and such that KQ/I is finite dimensional. If v is closed, we call the track *cyclic*. Otherwise v is called *linear*.

Lemma 1.11.18 *Let $Q = (Q_0, Q_1)$ be a quiver and let $T = (Q, I, v)$ be a track. Then $A = KQ/I$ is special biserial.*

If $A = KQ/I$ is special biserial, then there is a subset of arrows $\Omega \subseteq Q_1$ of Q, and for each $\alpha \in \Omega$ a track $(Q_\alpha, I \cap kQ_\alpha, v_\alpha)$, such that Q_α is a subquiver of Q and such that each arrow of Q belongs to exactly one Q_α.

Moreover, such a set of tracks is uniquely determined by the quiver and relation (Q, I) of the special biserial algebra $A = KQ/I$.

Proof A track is special biserial. Indeed, if α is an arrow and β_1 and β_2 start at the terminal vertex of α, then only one of the paths $\alpha\beta_1$ or $\alpha\beta_2$ is a subpath of v. The other one is in I by definition. Likewise we proceed for $\gamma_1\alpha$ and $\gamma_2\alpha$. This proves that axiom 2 is satisfied. Axiom 1 follows since v runs at most twice through any given vertex, and axiom 3 follows since $\dim_K(KQ/I) < \infty$.

Let (Q, I) be special biserial. As we have seen in Remark 1.11.16 each arrow α determines a unique maximal path p_α so that no subpath of p_α of length 2 is in I, but α belongs to p_α, and p_α does not contain two identical arrows. Let Q_α be the underlying quiver of p_α. By definition $T_\alpha := (Q_\alpha, I \cap KQ_\alpha, v_\alpha)$ is a track. Indeed, each arrow of Q_α occurs precisely once by the choice of p_α. Each vertex of Q_α occurs at most twice since (Q, I) is special biserial. We may define an equivalence relation on the set of arrows Q_1 by saying that two arrows α and β are equivalent if $T_\alpha = T_\beta$. An equivalence class is then the set of arrows in a track. Let Ω be a set of representatives of the classes of this equivalence relation. The construction of the tracks T_α implies that they are actually unique. Then the result follows. □

We continue with a result which may be of independent interest.

Proposition 1.11.19 [8] *Any special biserial algebra is a quotient of a symmetric special biserial algebra.*

Proof Let $\{(Q_\alpha, I_\alpha, p_\alpha) \mid \alpha \in \Omega_A\}$ be the set of tracks. Let $\Sigma_A \subset \Omega_A$ be the set of all those arrows α so that p_α is a closed path for each $\alpha \in \Sigma_A$. We shall construct recursively by induction on $\Omega_A \setminus \Sigma_A$ a special biserial algebra A' which maps onto A and for which $\Sigma_{A'} = \Omega_{A'}$.

Let p_α be such that $\alpha \notin \Sigma_A$. Let a_α be the starting vertex of p_α and let b_α be the ending vertex of α.

Case 1: Suppose there is a subset $\Delta_A \subset \Omega_A \setminus \Sigma_A$ such that $\Delta_A = \{\alpha_1, \dots, \alpha_d\}$ and such that the paths $\{p_\alpha \mid \alpha \in \Delta\}$ can be ordered so that $p_{\alpha_1} p_{\alpha_2} \cdots p_{\alpha_d}$ is a closed path.

Let β_i be the last arrow in p_{α_i} and let γ_i be the starting arrow of p_{α_i}. Then we delete from a minimal generating set of I all relations $\beta_i\gamma_{i+1}$ where $i \in \{1, \dots, d-1\}$ as well as the relation $\beta_d\gamma_1$. Actually, by the maximality of p_α, we can (and will) choose a minimal generating set of I containing the elements $\beta_i\gamma_{i+1}$ where $i \in \{1, \dots, d-1\}$. Let I' be the new ideal of relations. Then (Q, I') yields a special biserial algebra and we have an inclusion of ideals $I' \le I$. Therefore KQ/I is a quotient algebra of KQ/I', and it is clear that $|\Omega_A \setminus \Sigma_A| > |\Omega_{A'} \setminus \Sigma_{A'}|$.

Case 2: Suppose there is a subset $\Delta_A \subset \Omega_A \setminus \Sigma_A$ such that $\Delta_A = \{\alpha_1, \dots, \alpha_d\}$ and such that no ordering of the paths $\{p_\alpha \mid \alpha \in \Delta\}$ can be chosen so that $p_{\alpha_1}p_{\alpha_2}\cdots p_{\alpha_d}$ is a closed path.

Choose any ordering such that $p_{\alpha_1}p_{\alpha_2}\cdots p_{\alpha_d}$ is a non-closed path. Then we may assume that Δ_A is maximal, and hence no other path in $\Omega_A \setminus (\Sigma_A \cup \Delta_A)$ is composable with $p_{\alpha_1}p_{\alpha_2}\cdots p_{\alpha_d}$. This implies that at most one arrow of Q ends at the ending vertex v_e of p_{α_d} and at most one arrow of Q starts at the starting vertex v_s of p_{α_1}. We add a new arrow $v_e \xrightarrow{\sigma} v_s$ to Q. The new quiver is named Q'. Add the relations $\sigma\beta_1 = 0$ and $\gamma_d\sigma = 0$ to I. The ideal I' is then generated by $I \cup \{\sigma\beta_1, \gamma_d\sigma\}$ and we observe that $A' := KQ'/I'$ maps surjectively onto $A := KQ/I$. Moreover, (Q', I') is special biserial and $|\Omega_A \setminus \Sigma_A| > |\Omega_{A'} \setminus \Sigma_{A'}|$.

Case 3: No two paths p_{α_1} and p_{α_2} for $\alpha_1, \alpha_2 \in \Omega_A \setminus \Sigma_A$ are composable.

Then the axioms of a track imply that for any $\alpha \in \Omega_A \setminus \Sigma_A$ we can proceed as in case 2 and add a new arrow σ in the opposite direction of α and defining Q' to be the quiver given by the vertices of Q, the arrows of Q and in addition the arrow σ, defining I' to be the ideal generated by I and $\alpha\sigma, \sigma\alpha$ we obtain a special biserial algebra $A' = KQ'/I'$ mapping surjectively onto $A = KQ/I$ and $|\Omega_A \setminus \Sigma_A| = |\Omega_{A'} \setminus \Sigma_{A'}| + 1$.

Hence, by induction we may assume that all paths p_α in the tracks T_α of A are closed paths (where we define Ω to be a system of representing arrows for the tracks of A). Let m be the smallest integer such that $\mathrm{rad}^m(A) = 0$, and for all $\alpha \in \Omega$ we put $p_\beta := p_\alpha$ whenever β occurs in p_α (i.e. β is an arrow of Q_α). For any vertex v of Q and arrow α starting at v there is a unique closed path p_α containing α once. Denote by $\tau_\alpha := p_\alpha^m$ the unique closed path starting at v, running m times around p_α, and ending at v. Let I_s be the ideal of KQ generated by all elements

- $\beta\alpha$ when $\beta\alpha$ is not a subpath of p_α,
- all paths containing a path τ_α for some α,
- $\tau_\alpha - \tau_\beta$ when α and β start at the same vertex v.

Then (Q, I_s) defines a special biserial algebra $A_s := KQ/I_s$ and A is an epimorphic image of A_s. Proposition 1.10.18 then defines a non-degenerate bilinear form on A_s. The form is symmetric since if uv is a path in the socle, then $uv = \tau_\alpha$ for some α, and then $vu = \tau_\beta$ for some β. Therefore A_s is a symmetric special biserial algebra mapping onto A. □

Remark 1.11.20 In order to classify the indecomposable A-modules for special biserial algebras A, it is therefore sufficient to classify the indecomposable A'-modules for A' symmetric, actually the algebra constructed in the proof of Proposition 1.11.19.

Let A be a special biserial K-algebra presented as $A = KQ/I$ for a special biserial pair (Q, I). Recall the notations of Remark 1.11.8.

Definition 1.11.21 A walk $w : L \longrightarrow Q$ is called a *string* if no subpath of w is in I and if in addition there is no subwalk

$$\bullet \stackrel{\alpha}{\longleftarrow} \bullet \stackrel{\alpha}{\longrightarrow} \bullet \qquad \text{or} \qquad \bullet \stackrel{\alpha}{\longrightarrow} \bullet \stackrel{\alpha}{\longleftarrow} \bullet$$

(i.e. it is forbidden to walk an arrow forth and back, or back and forth immediately).

We shall define indecomposable A-modules by the following construction.

First we shall define a special representation of KL for the quiver L defined in Remark 1.11.8 (after having fixed an orientation). Suppose L has m vertices. Then let V_L be an m-dimensional k-vector space, with basis $\{b_v \mid v \in L_0\}$ indexed by the vertices of L. Then a primitive idempotent e_v of KL associated to the vertex v acts on V_L as $e_v \cdot b_w := \delta_{v,w} \cdot b_v$ where $\delta_{*,*}$ is the Kronecker delta, being 1 if the indices coincide, and 0 otherwise. Moreover, let $v_1 \stackrel{\alpha}{\longrightarrow} v_2$ be an arrow of L. Then $\alpha \cdot b_v := \delta_{v_1,v} \cdot b_{v_2}$. It is obvious that this produces an indecomposable KL-module, which we denote by V_L in the sequel.

A string $L \stackrel{\sigma}{\longrightarrow} Q$ is a quiver homomorphism respecting the relations. It is easy to verify that this induces an algebra homomorphism $KQ/I \stackrel{\sigma^*}{\longrightarrow} KL$ by putting

$$\sigma^*(e_v) = \sum_{w \in L_0; \sigma(w) = v} e_v$$

and if $\alpha : v \longrightarrow w$ is an arrow of Q, then put

$$\sigma^*(\alpha) = \sum_{\beta \in L_1; \sigma(\beta) = \alpha} \beta.$$

Then each KL-module M induces a KQ/I-module by maintaining the K-vector space structure, and $x \in KQ/I$ acts on $m \in M$ as multiplication by $\sigma^*(x)$ on m. Then any morphism $M \longrightarrow M'$ of KL-modules is also a morphism of KQ/I-modules. The educated reader will recognise that we defined a functor $F_\sigma : KL - mod \longrightarrow KQ/I - mod$ by defining for each KL-module M the KQ/I-module $F_\sigma(M) := M$ as a K-vector space, and $x \cdot m := \sigma(x) \cdot m$ for each $m \in M$ and $x \in KQ/I$. This concept will be explained in full detail in Sect. 3.1.2.

Definition 1.11.22 A *string module* of a special biserial algebra KQ/I is a module $F_\sigma(V_L)$ where V_L is the KL-module constructed above for a string $\sigma : L \longrightarrow Q$.

From time to time, to illustrate the situation, we denote the string module of a walk of the form

$$w(\bullet_1 \longrightarrow \bullet_2 \longleftarrow \bullet_3)$$

by

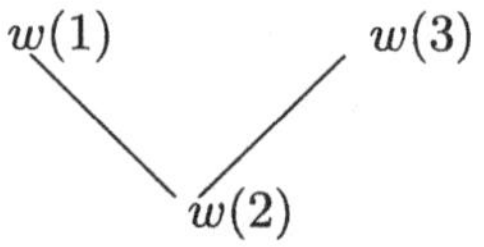

and the string module of a walk of the form

$$w(\bullet_1 \longleftarrow \bullet_2 \longrightarrow \bullet_3)$$

by

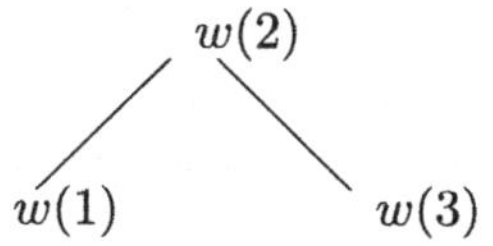

This corresponds in some sense to the Loewy structure. The highest vertex in the diagram corresponds to a quotient module, the lowest to a submodule. Similarly, the string module of a walk of the form

$$w(\bullet_1 \longleftarrow \bullet_2 \longrightarrow \bullet_3 \longrightarrow \bullet_4 \longleftarrow \bullet_5)$$

may be depicted as

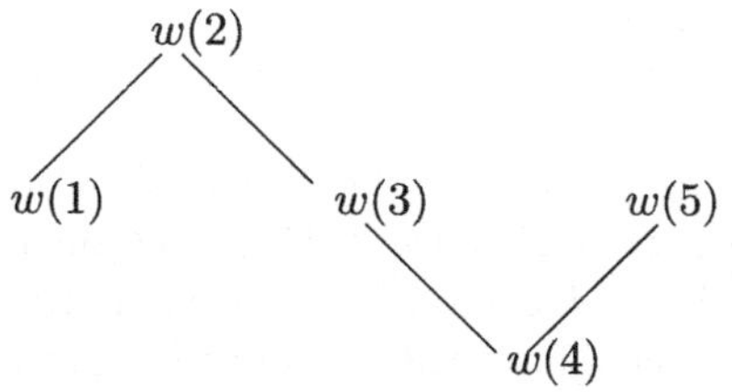

and it should be clear now how to depict string modules for more general walks.

Lemma 1.11.23 *A string module over KQ/I is indecomposable.*

Proof Let M be a string module over KQ/I for a string $L \stackrel{\sigma}{\longrightarrow} Q$. Then $End_{KQ/I}(M) = End_{KL}(M)$ since the action of KQ/I on M is the action of KL

on M via σ^*. Since M is an indecomposable KL-module, its endomorphism ring is local, and therefore M is also indecomposable as a KQ/I-module. □

There is a second kind of module which we can construct from string modules. Let $\sigma : L \longrightarrow Q$ be a string, and let $C_L^{(n)} := F_\sigma(V_L)^n$ be a direct sum of n copies of the string module associated to the string σ. Suppose now σ represents a tour, i.e. the starting vertex of σ equals the ending vertex v of σ and suppose L has one of the following two orientations

$$\bullet_v \longleftarrow \bullet \text{ —— } \bullet \text{ —— } \cdots \text{ —— } \bullet \longrightarrow \bullet_v$$

or

$$\bullet_v \longrightarrow \bullet \text{ —— } \bullet \text{ —— } \cdots \text{ —— } \bullet \longleftarrow \bullet_v$$

Let S_v be the simple KQ/I-module on which $e_{\sigma(v)}$ acts as 1, and each other primitive idempotent of KQ/I act as 0.

In the upper case, S_v maps to C_L in two ways. First, the mapping α_1 is the identity in the first copy of v, corresponding to the leftmost vertex, and 0 on all the other vertices, second the mapping α_2 is the identity in the second copy of v, corresponding to the rightmost vertex, and 0 on all the other vertices.

Now fix an integer $n \geq 1$. Then define a (diagonal) mapping $S_v^n \longrightarrow C_L^{(n)}$ by the matrix

$$\begin{pmatrix} \alpha_1 & 0 & \cdots & \cdots & 0 \\ 0 & \alpha_1 & 0 & & \vdots \\ \vdots & \ddots & \ddots & \ddots & \vdots \\ \vdots & & \ddots & \ddots & 0 \\ 0 & \cdots & \cdots & 0 & \alpha_1 \end{pmatrix} - \begin{pmatrix} \lambda\alpha_2 & \alpha_2 & 0 & \cdots & 0 \\ 0 & \lambda\alpha_2 & \alpha_2 & 0 & \vdots \\ \vdots & \ddots & \ddots & \ddots & 0 \\ \vdots & & \ddots & \ddots & \alpha_2 \\ 0 & \cdots & \cdots & 0 & \lambda\alpha_2 \end{pmatrix}$$

In other words, S_v^n is identified identically with the vector space at the leftmost vertex, and via a Jordan normal form matrix with the rightmost vertex. Let $B_L^{(n),\lambda}$ be the quotient module.

Recall Example 1.6.23 where the same construction was used to define an infinite family of 2-dimensional representations $B_{2,1,2}^{(\lambda,\mu)}$ of the Kronecker quiver algebra.

Dually we may proceed for the second type of string, but there we have mappings β_1 and β_2 to the simple S_v, and we consider the kernel of the above matrix to obtain modules $\check{B}_L^{(n),\lambda}$.

Definition 1.11.24 The modules $B_L^{(n),\lambda}$ and $\check{B}_L^{(n),\lambda}$ are called *band modules*.

Lemma 1.11.25 *Let $A = KQ/I$ be a special biserial algebra. Then a band module is indecomposable. Moreover,*

$$B_L^{(n),\lambda} \simeq B_L^{(n),\mu} \Leftrightarrow \lambda = \mu.$$

Proof We know that $End_{KQ/I}(B_L^{(n),\lambda}) = End_{KL}(B_L^{(n),\lambda})$, and so it is sufficient to prove this fact for KL-modules. But this reduces the question to the case of a quiver with a single vertex v and one loop ϵ. This algebra is isomorphic to $K[X]$. The $K[X]$-module of dimension n on which X acts via the Jordan normal form with only one block of parameter λ is indecomposable, this being the main result of Jordan's theorem in linear algebra. It is also an easy consequence of the classification of indecomposable modules over a Euclidean ring, and where we observe that $K[X]$ is a Euclidean ring, and where $B_L^{(n),\lambda}$ corresponds to $K[X]/(X-\lambda)^n$. This proves the lemma. □

Lemma 1.11.26 *Let $A = KQ/I$ be a special biserial K-algebra. Then a band module is never isomorphic to a string module.*

Proof Let V_L be a string module and let $B_{L'}^{(n),\lambda}$ be a band module. A first necessary condition is that the strings $L \longrightarrow Q$ and $L' \longrightarrow Q$ have the same image. Hence, there is an isomorphism $L \longrightarrow L'$ such that $L \longrightarrow Q = L \longrightarrow L' \longrightarrow Q$. We may therefore assume that $L = L'$. Moreover, multiplying by any idempotent corresponding to a vertex occurring in L, we see that $n = 1$. But, we see that

$$\dim_k(B_L^{(1),\lambda}) = \dim_k(V_L) - 1.$$

This proves the lemma. □

Remark 1.11.27 It was shown by Butler and Ringel [22] that all indecomposable modules over special biserial algebras are either string modules or band modules. See Erdmann's treatment [23, Sect. 2] for a more detailed introduction to this theory.

Remark 1.11.28 Crawley-Boevey gives a combinatorial way to determine morphisms between string modules in an even more general setting [24].

We shall obtain further examples of special biserial algebras later. In particular we shall see that up to some notion of equivalence (called Morita equivalence, to be explained in Chap. 4) group rings KG with an algebraically closed field K of characteristic $p > 0$ and a finite group G with cyclic Sylow p subgroup are special biserial. This will be done in Sects. 2.8, 2.12, 4.4.2 and 5.10.

References

1. Lam, T.-Y.: A First Course in Noncommutative Rings. Springer, New York (2001)
2. Hasse, H.: Vorlesungen über Zahlentheorie, Die Grundlehren der mathematischen Wissenschaften 59. Springer, Berlin (1950)
3. Curtis, C.W., Reiner, I.: Methods of Representation Theory I. Wiley, New York (1982)
4. Curtis, C.W., Reiner, I.: Methods of Representation Theory II. Wiley, New York (1986)
5. Swan, R.G.: Projective modules over group rings and maximal orders. Ann. Math. **76**, 55–61 (1962)

6. McConnell, J.C., Robson, J.C.: Noncommutative Noetherian Rings, second corrected printing. American Mathematical Society, Providence (2001)
7. Hopkins, C.: Rings with minimal condition for left ideals. Ann. Math. **40**, 712–730 (1939)
8. Wald, B., Waschbüsch, J.: Tame biserial algebras. J. Algebra. **95**, 480–500 (1985)
9. Dugas, A.S., Martinez-Villa, R.: A note on stable equivalence of morita type. J. Pure. App. Algebra. **208**, 421–433 (2007)
10. Montgomery, S.: Hopf algebras and their actions on rings. CBMS Regional Conference Series in Mathematics 82, American Mathematical Society Providence, R.I. (1993)
11. Bouc, S.: Green functors and G-sets. Lecture Notes in Mathematics, vol. 1671. Springer, Berlin (1997)
12. Lam, T.-Y.: Lectures on Modules and Rings. Springer, New York (1998)
13. Rotman, J.J.: The Theory of Groups, 2nd edn. Allyn and Bacon, Boston (1973)
14. Huppert, B.: Endliche Gruppen 1. Springer, Berlin (1967)
15. Nakayama, T.: On Frobeniusean algebras I. Ann. Math. **40**(3), 611–633 (1939)
16. Nakayama, T.: On Frobeniusean algebras II. Ann. Math. **42**(1), 1–21 (1941)
17. Broué, M.: Higman's criterion revisited. Michigan Math. J. **58**(1), 125–179 (2009)
18. Holm, T., Zimmermann, A.: Generalized Reynolds ideals and derived equivalences for algebras of dihedral and semidihedral type. J. Algebra. **320**, 3425–3437 (2008)
19. Holm, T., Skowroński, A.: Derived equivalence classification of symmetric algebras of polynomial growth. Glasg. Math. J. **53**, 277–291 (2011)
20. Holm, T., Zimmermann, A.: Deformed preprojective algebras of type L: Külshammer spaces and derived equivalences. J. Algebra. **346**, 116–146 (2011)
21. König, S., Xi, C.: A self-injective cellular algebra is weakly symmetric. J. Algebra. **228**, 51–59 (2000)
22. Butler, M., Ringel, C.: Auslander-Reiten sequences with few middle terms. Commun. Algebra. **15**, 145–179 (1987)
23. Erdmann, K.: Blocks of tame representation type and related algebras. Springer Lecture Notes in Mathematics, vol. 1428. Springer, Heidelberg (1990)
24. Crawley-Boevey, W.: Maps between representations of zero-relation algebras. J. Algebra. **126**, 259–263 (1989)

Chapter 2
Modular Representations of Finite Groups

We are now ready to apply the results from the previous chapter to group rings of finite groups. If the order of the group is invertible in the base field, Maschke's Theorem 1.2.8 tells us that the group ring is semisimple, and semisimple rings are of less interest from the homological algebra point of view. Hence, we are mainly interested in base fields in which the group order is not invertible. Representations in this situation are of great interest, and there remain many unsolved questions. They will also provide the main application terrain and test ground for the theories we are going to develop in the sequel. In this chapter, we are dealing with groups, and the variable K will be used most often for subgroups. For this reason the base ring will usually be denoted by k.

2.1 Relatively Projective Modules

2.1.1 Relatively Projective Modules for Subalgebras

Let k be a field and let A be a k-algebra, then recall that an A-module P is projective if every k-split exact sequence

$$0 \longrightarrow M \longrightarrow N \xrightarrow{\pi} P \longrightarrow 0$$

is also split over A.

From this point of view it seems to be natural to define, for a subalgebra B of A, modules over A to be relatively projective with respect to B.

Definition 2.1.1 Let k be a commutative ring and let B be a k-subalgebra of the k-algebra A. An A-module Q is *relatively B-projective* if every exact sequence of A-modules

$$0 \longrightarrow M \longrightarrow N \longrightarrow Q \longrightarrow 0$$

which is split as a sequence of B-modules, is also split as a sequence of A-modules.

A. Zimmermann, *Representation Theory*,
Algebra and Applications 19, DOI: 10.1007/978-3-319-07968-4_2,

A relatively B-projective A-module has the nicest properties when A is a projective B-module. If A is projective as a B-module, then we may define the notion of being relatively projective with respect to a subalgebra by means of $Ext_A^1(Q, M)$ and $Ext_B^1(Q, M)$. Suppose A is projective as a B-module. Then any free A-module, considered as B-module by restriction, is again projective, and hence any projective A-module, considered as B-module by restriction, is a projective B-module. Hence the restriction of A-modules to B-modules maps projective modules to projective modules. Therefore, the restriction of a projective resolution

$$\cdots \longrightarrow P_3 \longrightarrow P_2 \longrightarrow P_1 \longrightarrow P_0 \longrightarrow Q \longrightarrow 0$$

of Q as an A-module to B is a projective resolution of Q as a B-module. We may apply $Hom_A(-, N)$ or $Hom_B(-, N)$ to the same resolution. Since $Hom_A(-, N) \hookrightarrow Hom_B(-, N)$ we obtain a commutative diagram

$$\begin{array}{ccccccc} \ldots \leftarrow Hom_A(P_2, N) & \leftarrow & Hom_A(P_1, N) & \leftarrow & Hom_A(P_0, N) & \leftarrow & Hom_A(Q, N) \\ \downarrow & & \downarrow & & \downarrow & & \downarrow \\ \ldots \leftarrow Hom_A(P_2, N) & \leftarrow & Hom_B(P_1, N) & \leftarrow & Hom_B(P_0, N) & \leftarrow & Hom_B(Q, N) \end{array}$$

which induces a mapping

$$Ext_A^i(Q, N) \xrightarrow{\rho} Ext_B^i(Q, N)$$

for all $i \geq 0$. This morphism is compatible with morphisms $N \to N'$ in the sense that the diagram

$$\begin{array}{ccc} Ext_A^i(Q, N) & \xrightarrow{\rho} & Ext_B^i(Q, N) \\ \downarrow & & \downarrow \\ Ext_A^i(Q, N') & \xrightarrow{\rho} & Ext_B^i(Q, N') \end{array}$$

obtained by the construction preceding Lemma 1.8.36 is commutative. The following lemma is therefore an immediate consequence of the definition.

Lemma 2.1.2 *Let k be a commutative ring and let B be a k-subalgebra of the k-algebra A. Suppose A is projective as a B-module. Then an A-module Q is relatively B-projective if the natural mapping $Ext_A^1(Q, N) \longrightarrow Ext_B^1(Q, N)$ is injective for all N.*

The most important case for the representation theory of groups will be the case of a kG-module M which is relatively projective with respect to a subgroup H of G, i.e. relatively projective with respect to the subalgebra kH.

Definition 2.1.3 Let G be a finite group, let k be a field and let H be a subgroup of G. A kG-module M is *H-projective* if M is relatively kH-projective.

Recall that for $g \in G$ and a kH-module M we denote by ${}^gM$ the $k(gHg^{-1})$-module given by the same k-module structure as M, and a gHg^{-1}-action on ${}^gM$ denoted by $\bullet$, with $ghg^{-1} \bullet m := h \cdot m$, and where $h \cdot m$ is understood to be the usual action of H on M.

A first observation is that if Q is H-projective, then Q is $g^{-1}Hg$-projective for all $g \in G$. Indeed, given a short exact sequence

$$0 \longrightarrow M \longrightarrow N \longrightarrow Q \longrightarrow 0$$

which splits as a sequence of $k(g^{-1}Hg)$-modules, then the sequence of kG-modules

$$0 \longrightarrow {}^gM \longrightarrow {}^gN \longrightarrow {}^gQ \longrightarrow 0$$

splits as a sequence of kH-modules. For every kG-module X there is a morphism of kG-modules $X \longrightarrow {}^gX$ given by $x \mapsto g \cdot x$. Hence there is an isomorphism of sequences

$$\begin{array}{ccccccccc} 0 & \longrightarrow & M & \longrightarrow & N & \longrightarrow & Q & \longrightarrow & 0 \\ & & \downarrow & & \downarrow & & \downarrow & & \\ 0 & \longrightarrow & {}^gM & \longrightarrow & {}^gN & \longrightarrow & {}^gQ & \longrightarrow & 0 \end{array}$$

which implies that Q is $g^{-1}Hg$-projective as well.

We remark that the same proof also works in the case of algebras.

Lemma 2.1.4 *If A is a k-algebra and Q is projective relative to the subalgebra B. Then Q is projective relative to the subalgebra uBu^{-1} for all units u of A.*

Proposition 2.1.5 *Let A be a k-algebra and let B be a subalgebra of A such that A is a projective B-module. Suppose that M is an A-module and suppose that the multiplication mapping $A \otimes_B M \longrightarrow M$ is split as a morphism between A-modules. Then M is relatively B-projective.*

Proof We need to show that the map

$$res : Ext^1_A(M, -) \longrightarrow Ext^1_B(M, -)$$

induced by restriction to B is injective. By hypothesis the multiplication map $A \otimes_B M \to M$ is split and so M is a direct factor of $A \otimes_B M$. Denote by $\mu : A \otimes_B M \longrightarrow M$ the multiplication and by $\sigma : M \longrightarrow A \otimes_B M$ the A-linear splitting. By Lemma 1.8.33 these maps produce a (split injective) map $Ext^1_A(M, -) \stackrel{\mu^*}{\to} Ext^1_A(A \otimes_B M, -)$. We obtain a diagram

$$\begin{array}{ccc} Ext^1_A(M, -) & \stackrel{res}{\longrightarrow} & Ext^1_B(M, -) \\ \downarrow \mu^* & & \| \\ Ext^1_A(A \otimes_B M, -) & \stackrel{Proposition\ 1.8.31}{\simeq} & Ext^1_B(M, -) \end{array}$$

where σ^* and μ^* are the mappings induced by σ and μ on the Ext^1-groups. We need to show that this diagram is commutative. Indeed, let

$$P_1 \longrightarrow P_0 \longrightarrow M \longrightarrow 0$$

be the first terms of a projective resolution of M as an A-module. Then, by definition, $Ext^1_A(M,X)$ is a subquotient of $Hom_A(P_1,X)$ and analogously $Ext^1_A(A\otimes_B M,X)$ is a subquotient of $Hom_A(A\otimes_B P_1,X)$. Now

$$\begin{array}{rcl} Hom_A(A\otimes_B P_1,X) & \overset{nat}{\simeq} & Hom_B(P_1,X) \\ & \varphi \mapsto & (p\mapsto \varphi(1\otimes p)) \end{array}$$

and so the multiplication $A\otimes_B M \longrightarrow M$ induces the multiplication mapping $A\otimes_B P_1 \longrightarrow P_1$ which in turn induces

$$\begin{array}{rcl} Hom_A(P_1,X) & \overset{\mu^*}{\longrightarrow} & Hom_A(A\otimes_B P_1,X) \\ \varphi & \mapsto & (a\otimes p\mapsto \varphi(a\cdot p)) \end{array}$$

But this implies that $nat\circ\mu^*$ coincides with the restriction to B. Hence, *res* is injective if and only if μ^* is injective. But μ^* is split by σ^*, and therefore μ^* is injective. This proves the proposition. □

The converse of Proposition 2.1.5 is true as well, as is shown below.

Proposition 2.1.6 *Let k be a commutative ring and let A be a k-algebra containing a subalgebra B. Suppose that A is projective as a B-module. Then an A-module Q is relatively B-projective if and only if the multiplication mapping $A\otimes_B Q \longrightarrow Q$ is split as a morphism between A-modules.*

Proof In Proposition 2.1.5 we have seen that if $A\otimes_B Q \longrightarrow Q$ is split as a morphism between A-modules, then Q is relatively B-projective. In order to show the converse we observe that the epimorphism given by the multiplication mapping

$$A\otimes_B Q \overset{\mu}{\longrightarrow} Q$$

is split as a mapping of B-modules by sending $q\in Q$ to $1\otimes q\in A\otimes_B Q$. Indeed, call this mapping σ. Then

$$\sigma(bq) = 1\otimes bg = b\otimes q = b\cdot(1\otimes q)$$

for all $b\in B$ and $q\in Q$. Let $K := \ker(\mu)$. Then the exact sequence

$$0\longrightarrow K\longrightarrow A\otimes_B Q\longrightarrow Q\longrightarrow 0$$

is split when considered as a sequence of B-modules. Since Q is relatively B-projective, the sequence is split as a sequence of A-modules, and therefore the multiplication mapping μ is split as a morphism between A-modules. □

Remark 2.1.7 In Lemma 2.1.4 we have seen that if Q is relatively B-projective for some k-subalgebra B of the k-algebra A, so that A is projective as a B-module, then Q is also relatively projective for any algebra conjugate to B. Moreover, trivially any A-module is relatively A-projective. Being relatively k-projective is exactly the same as being projective. If Q is relatively B-projective, then the definition implies immediately that Q is relatively C-projective for any subalgebra C of A containing B, and so that B is projective as a C-module.

A very special case, when Proposition 2.1.6 is always true, is the subject of the following statement.

Proposition 2.1.8 *Let k be a commutative ring, let B be a k-algebra and let A be a k-subalgebra. Then the multiplication map*

$$\begin{array}{ccc} B\otimes_A B & \xrightarrow{\mu} & B \\ b_1\otimes b_2 & \mapsto & b_1 b_2 \end{array}$$

is split as a morphism of $B\otimes_k B^{op}$-modules if and only if there is an $\omega\in\mu^{-1}(1_B)$ such that $b\omega=\omega b$ for all $b\in B$.

If there is such an ω, then every B-module is A-projective and a B-module P is projective if P, considered as an A-module, is projective.

Proof Suppose that μ is split. Then there is a $\sigma: B\longrightarrow B\otimes_A B$ such that $\mu\circ\sigma=id_B$. Let $\omega:=\sigma(1_B)$. Then we get for all $b\in B$

$$b\cdot\omega=b\cdot\sigma(1)=\sigma(b)=\sigma(1)\cdot b=\omega\cdot b.$$

Suppose that ω exists. Then put $\sigma(b):=\omega\cdot b$ for all $b\in B$. Since

$$b_1\cdot\sigma(x)\cdot b_2=b_1\cdot\omega\cdot x\cdot b_2=\omega\cdot b_1\cdot x\cdot b_2=\sigma(b_1\cdot x\cdot b_2)$$

for all $x, b_1, b_2\in B$, this is a morphism of $B\otimes_k B^{op}$-modules. Moreover, for all $b\in B$ we get

$$\mu\circ\sigma(b)=\mu(\omega\cdot b)=\mu(\omega)\cdot b=1_B\cdot b=b$$

since $\omega\in\mu^{-1}(1_B)$.

Let M be a B-module. If μ is split by σ, then the multiplication map μ_M $B\otimes_A M\xrightarrow{\mu_M} M$ on M is split by $\sigma\otimes_B id_M$. Indeed, $\mu_M=\mu\otimes_B id_M$, where we identify

$$B\otimes_B M\simeq M \text{ and } B\otimes_A B\otimes_B M\simeq B\otimes_A M.$$

Hence M is a direct factor of $B\otimes_A M$. Now apply Proposition 1.8.31. This proves the statement. □

We shall now partially follow Külshammer [1].

Definition 2.1.9 Let k be a commutative ring, let B be a k-algebra and let A be a k-subalgebra. An *algebra extension* of A is a k-algebra C together with a k-algebra map $A \to C$. The extension $A \to B$ is *separable* if the multiplication

$$\begin{array}{rcl} B \otimes_A B & \xrightarrow{\mu} & B \\ b_1 \otimes b_2 & \mapsto & b_1 b_2 \end{array}$$

is split as a morphism of B-B-bimodules.

We apply the concept to a Maschke-type statement (cf Theorem 1.2.8).

Proposition 2.1.10 *Let k be a commutative ring, let G be a group and let H be a subgroup of G. Suppose that the index $|G : H|$ is finite and invertible in k. Then the extension $kH \leq kG$ is separable.*

Proof Let

$$\omega := \left(\frac{1}{|G:H|}\sum_{gH\in G/H} g\otimes g^{-1}\right) \in kG \otimes_{kH} kG.$$

It is clear that ω is well-defined, in the sense that $g \otimes_H g^{-1}$ does not depend on the representative $g \in gH$. Moreover, for all $x \in G$ we get

$$\begin{aligned} x\cdot\omega = x\cdot\left(\frac{1}{|G:H|}\sum_{gH\in G/H} g\otimes g^{-1}\right) &= \left(\frac{1}{|G:H|}\sum_{gH\in G/H} xg\otimes g^{-1}\right)\\ &= \left(\frac{1}{|G:H|}\sum_{gH\in G/H} xg\otimes (xg)^{-1}x\right)\\ &= \left(\frac{1}{|G:H|}\sum_{xgH\in G/H} xg\otimes (xg)^{-1}\right)\cdot x = \omega\cdot x. \end{aligned}$$

Finally

$$\mu(\omega) = \mu\left(\frac{1}{|G:H|}\sum_{gH\in G/H} g\otimes g^{-1}\right) = \left(\frac{1}{|G:H|}\sum_{gH\in G/H} g\cdot g^{-1}\right) = 1_{kG}.$$

By Proposition 2.1.8 we obtain the statement. □

Lemma 2.1.11 *Let k be a commutative ring, let G be a group and let H be a subgroup of finite index. If the extension $kH \hookrightarrow kG$ is separable, then $|G : H|$ is invertible in k.*

Proof Suppose that the extension $kH \leq kG$ is separable. By Proposition 2.1.8 every kG-module M is kH-projective and by Proposition 2.1.6 we get that $\mu_M : kG \otimes_{kH} M \longrightarrow M$ is split. Choose $M = k$ the trivial module and let σ_k be a splitting. Then $\sigma_k(1)$ is in the trivial submodule of $kG \otimes_{kH} k \simeq kG/H$. However, the G-trivial submodule of kG/H is k-linearly generated by $\sum_{gH \in G/H} gH$. Hence

$$\sigma_k(1) = \lambda \cdot \sum_{gH \in G/H} gH$$

for some $\lambda \in k$. Since

$$1 = \mu_M \circ \sigma_k(1) = \lambda \cdot |G : H|$$

we get that $|G : H|$ is invertible in k. □

Lemma 2.1.12 *If the index of H in G is invertible in k, then any kG-module is projective relative to H.*

Proof By Proposition 2.1.10 we know that the extension $kH \leq kG$ is separable. Proposition 2.1.8 then shows that every kG-module is relatively kH-projective. □

2.1.2 Vertex and Source

Given an indecomposable A-module Q, is there some minimal subalgebra B_Q, uniquely defined up to conjugacy, such that Q is relatively B_Q-projective? We shall concentrate on the case $A = kG$ for a field k and subalgebras kH for subgroups H of G and mainly use Mackey's formula from Sect. 1.7.3 as a new tool available in this setting. Of course, if H is a subgroup of G, then kG is a free kH-module of rank $|G : H|$. Hence the results of Sect. 2.1.1 apply.

Definition 2.1.13 Let k be a field and let G be a finite group. Let M be an indecomposable kG-module. Then a subgroup D of G is called a *vertex* of M if M is relatively kD-projective but for any proper subgroup D' of D we have that M is not relatively kD'-projective.

We have seen that a vertex always exists, since any kG-module is relatively kG-projective by Remark 2.1.7. If k is of characteristic 0, then by Maschke's Theorem 1.2.8 any module is projective, and therefore the vertex is always the trivial group in this case.

Theorem 2.1.14 *Let G be a finite group and let k be a field of characteristic $p > 0$. Suppose M is an indecomposable kG-module. Then*

1. *any vertex of M is a p-subgroup of G,*
2. *any two vertices of M are conjugate in G.*

Proof The first statement is a direct consequence of Lemma 2.1.12.

We need to prove the second statement. Suppose M has vertices D_1 and D_2 and therefore M is relatively kD_1-projective and kD_2-projective for two p-subgroups D_1 and D_2 of G. We shall now use Proposition 2.1.6. Then M is a direct factor of $M\downarrow_{D_1}^G\uparrow_{D_1}^G$ and of $M\downarrow_{D_2}^G\uparrow_{D_2}^G$. But then

$$\begin{aligned} M \Big| (M\downarrow_{D_1}^G\uparrow_{D_1}^G) &\Rightarrow (M\downarrow_{D_2}^G)\Big|(M\downarrow_{D_1}^G\uparrow_{D_1}^G\downarrow_{D_2}^G) \\ &\Rightarrow (M\downarrow_{D_2}^G\uparrow_{D_2}^G)\Big|(M\downarrow_{D_1}^G\uparrow_{D_1}^G\downarrow_{D_2}^G\uparrow_{D_2}^G) \\ &\Rightarrow M\Big|(M\downarrow_{D_1}^G\uparrow_{D_1}^G\downarrow_{D_2}^G\uparrow_{D_2}^G) \end{aligned}$$

where the last implication holds since $M \Big| (M\downarrow_{D_2}^G\uparrow_{D_2}^G)$. Hence

$$\left(M \Big| (M\downarrow_{D_1}^G\uparrow_{D_1}^G)\right) \text{ and } \left(M \Big| (M\downarrow_{D_1}^G\uparrow_{D_1}^G\downarrow_{D_2}^G\uparrow_{D_2}^G)\right).$$

But Mackey's Theorem 1.7.45 implies

$$M\downarrow_{D_1}^G\uparrow_{D_1}^G\downarrow_{D_2}^G\uparrow_{D_2}^G \simeq \bigoplus_{D_1gD_2\in D_1\backslash G/D_2} {}^gM\downarrow_{D_1\cap{}^gD_2}^G\uparrow_{D_1\cap{}^gD_2}^G$$

and since M is indecomposable, there is a $g_0\in G$ such that M is a direct factor of ${}^{g_0}M\downarrow_{D_1\cap{}^{g_0}D_2}^G\uparrow_{D_1\cap{}^{g_0}D_2}^G$. But D_1 is a vertex and any group $D_1\cap{}^gD_2$ is a subgroup of D_1. Hence $D_1\cap{}^{g_0}D_2=D_1$ which is equivalent to $D_1\subseteq{}^{g_0}D_2$. By the analogous argument interchanging D_1 and D_2 we get that D_2 is conjugate to a subgroup of D_1. This proves the second statement. □

In the case of group rings we will get a slightly more practical characterisation of relative projectivity than Proposition 2.1.6.

Proposition 2.1.15 *Let k be a field, let G be a finite group and let H be a subgroup of G. Then an indecomposable kG-module M is relatively H-projective if and only if M is a direct factor of $L\uparrow_H^G$ for some kH-module L. Here L can be taken to be $M\downarrow_H^G$.*

Proof We have already seen in Proposition 2.1.6 that M is relatively kH-projective if and only if the multiplication map $kG\otimes_{kH}M\longrightarrow M$ is split. But this shows that if M is relatively kH-projective, then M is a direct factor of $M\downarrow_H^G\uparrow_H^G$.

Clearly, if M is a direct factor of $M\downarrow_H^G\uparrow_H^G$ then M is a direct factor of $L\uparrow_H^G$ for some kH-module L.

Suppose now M is a direct factor of $L\uparrow_H^G$ and denote by $\iota: M\hookrightarrow L\uparrow_H^G$ the injection, and by $\pi: L\uparrow_H^G\longrightarrow\!\!\!\!\to M$ the projection onto this direct factor. Then there is a kH-linear endomorphism

$$\begin{array}{rcl} kG \otimes_{kH} L & \stackrel{\rho}{\longrightarrow} & kG \otimes_{kH} L \\ g \otimes x & \mapsto & \begin{cases} g \otimes x & \text{if } g \in H \\ 0 & \text{otherwise} \end{cases} \end{array}$$

and put $\theta := \pi \circ \rho \circ \iota \in End_{kH}(M)$. For each endomorphism $\tau \in End_{kH}(M)$ we define a trace map

$$Tr_H^G(\tau) := \sum_{gH \in G/H} g\tau g^{-1},$$

that is $Tr_H^G(\tau)(m) := \sum_{gH \in G/H} g\tau(g^{-1}m)$ for all $m \in M$. We observe, just as in the proof of Maschke's theorem, that η is kG-linear. It is clear by definition that we have

$$Tr_H^G(\alpha \circ \tau \circ \beta) = \alpha \circ Tr_H^G(\tau) \circ \beta$$

for all kG-linear endomorphisms α and β of M. Moreover, we see that

$$Tr_H^G(\rho) = id_{L\uparrow_H^G}$$

and hence

$$Tr_H^G(\theta) = \pi \circ Tr_H^G(\rho) \circ \iota = \pi \circ \iota = id_M.$$

Let now

$$0 \longrightarrow X \longrightarrow Y \stackrel{\gamma}{\longrightarrow} M \longrightarrow 0$$

be a short exact sequence of kG-modules. Suppose that γ has a splitting δ as a morphism of kH-modules, i.e. $\gamma \circ \delta = id_M$. Then put $\delta' := Tr_H^G(\delta \circ \theta)$ and observe

$$\gamma \circ \delta' = \gamma \circ Tr_H^G(\delta \circ \theta) = Tr_H^G(\gamma \circ \delta \circ \theta) = Tr_H^G(\theta) = id_M.$$

Since δ' is a kG-linear homomorphism, we get that γ is split as a morphism of kG-modules. This proves the proposition. □

Remark 2.1.16 We shall continue to study the properties of the map Tr_H^G in a slightly more general context and in more detail in Sect. 2.10.2.

We have seen in Proposition 2.1.15 that an indecomposable kG-module M has vertex D if and only if M is a direct factor of $M \downarrow_D^G \uparrow_D^G$ and D is minimal with respect to this property. Since M is indecomposable, M is a direct factor of $L \uparrow_D^G$ for an indecomposable direct factor L of $M \downarrow_D^G$. Indeed, if

$$M \downarrow_D^G = L_1 \oplus L_2 \oplus \cdots \oplus L_n,$$

then

$$M \downarrow_D^G \uparrow_D^G = L_1 \uparrow_D^G \oplus \cdots \oplus L_n \uparrow_D^G$$

and by the Krull-Schmidt theorem there is an $i \in \{1, 2, \ldots, n\}$ such that M is a direct factor of $L_i \uparrow_D^G$.

Definition 2.1.17 Let k be a field of characteristic $p > 0$ and let G be a finite group. Let M be an indecomposable kG-module with vertex D. An indecomposable kD-module L such that M is a direct factor of $L \uparrow_D^G$ is called a *source* of M.

A source is not unique in general. Indeed, let $N_G(D)$ be the normaliser of D in G. For an indecomposable kD-module L and $g \in N_G(D)$, since D is normal in $N_G(D)$ we get that ${}^gL$ is an indecomposable kD-module as well. Moreover,

$$\begin{aligned} {}^gL \uparrow_D^G &\xrightarrow{\psi} L \uparrow_D^G \\ x \otimes \ell &\mapsto x \cdot g^{-1} \otimes \ell \end{aligned}$$

is well-defined since for $d \in D$, $x \in kG$ and $\ell \in L$ we compute

$$\psi(xd \otimes \ell) = xdg^{-1} \otimes \ell = xg^{-1}gdg^{-1} \otimes \ell = xg^{-1} \otimes gdg^{-1}\ell = \psi(x \otimes d \cdot \ell).$$

Obviously ψ is bijective since an inverse is given by using g instead of g^{-1}. Finally ψ is kG-linear, since

$$\psi(h \cdot (x \otimes \ell)) = \psi(hx \otimes \ell) = hxg^{-1} \otimes \ell = h \cdot (xg^{-1} \otimes \ell) = h \cdot \psi(x \otimes \ell)$$

for all $x \in kG$, $\ell \in L$ and $h \in G$.

Proposition 2.1.18 *Let k be a field of characteristic $p > 0$, let G be a finite group, and let M be an indecomposable kG-module with vertex D. Then a source of M is unique up to conjugacy. More precisely, for any two sources L_1 and L_2, which are kD-modules, there is a $g \in N_G(D)$ such that ${}^gL_1 \simeq L_2$.*

Proof Any source L of M is a direct summand of $M \downarrow_D^G$. Hence L_2 is a direct factor of

$$L_1 \uparrow_D^G \downarrow_D^G \simeq \bigoplus_{DgD \in D\backslash G/D} {}^gL_1 \downarrow_{D\cap {}^gD}^D \uparrow^D$$

and since L_2 is indecomposable there is a $g \in G$ such that L_2 is a direct factor of ${}^gL_1 \downarrow_{D\cap {}^gD}^D \uparrow^D$. But since L_2 is a source of M, and hence M is a direct factor of $L_2 \uparrow_D^G$, we get that M is a direct factor of

$${}^gL_1 \downarrow_{D\cap {}^gD}^D \uparrow^D \uparrow_D^G = {}^gL_1 \downarrow_{D\cap {}^gD}^D \uparrow^G .$$

Now, D is a vertex of M, and hence M cannot be a direct factor of a module induced from a proper subgroup of D. Therefore $D \cap {}^gD = D$, which is equivalent to ${}^gD = D$, whence $g \in N_G(D)$. We obtain a $g \in N_G(D)$ such that L_2 is a direct factor of ${}^gL_1 \uparrow_{D\cap {}^gD}^D$. But $D = D \cap {}^gD$ and L_1 indecomposable implies that $L_2 \simeq {}^gL_1$. □

Lemma 2.1.19 *Let k be a field of characteristic $p > 0$ and let G be a finite group. Then every vertex of the trivial kG-module is a Sylow p-subgroup of G.*

Proof Let D be a vertex of the trivial module k. Then k is a direct factor of kG/D. The trivial module is always a submodule with multiplicity 1 of the permutation module kG/D. Indeed, Frobenius reciprocity shows that $Hom_{kG}(k, k\uparrow_D^G) = Hom_{kD}(k, k)$, which is one-dimensional, generated by φ, say. Similarly, $Hom_{kG}(k\uparrow_D^G, k)$ is one-dimensional, generated by ψ, say. However $\psi \circ \varphi = |G : D| \cdot id_k$ and so, if k is a direct factor of kG/D, then D contains a Sylow p-subgroup. □

2.1.3 Green Correspondence

Let G be a finite group and let k be a field of characteristic $p > 0$.

We shall establish a correspondence between indecomposable modules over G with vertex D and indecomposable modules over H with vertex D as soon as $N_G(D) \leq H \leq G$. This correspondence is due to Green.

Let now D be a p-subgroup of G and let $H \leq G$ with $N_G(D) \leq H$. Then define

$$\mathcal{X}_{D,H} := \{X \leq G \mid \exists g \in G \setminus H : X \leq {}^gD \cap D\}$$
$$\mathcal{Y}_{D,H} := \{Y \leq G \mid \exists g \in G \setminus H : Y \leq {}^gD \cap H\}$$

and observe that

$$D \notin \mathcal{Y}_{D,H} \text{ and } \mathcal{X}_{D,H} \subseteq \mathcal{Y}_{D,H}.$$

Indeed, if $D \leq {}^gD \cap H$ for some $g \in G \setminus H$, then $D \leq {}^gD$, and since D is of finite order, we get $D = {}^gD$. Hence $g \in N_G(D)$ which was excluded. This contradiction shows that $D \notin \mathcal{Y}_{D,H}$. Since $D \leq N_G(D) \leq H$ by hypothesis, ${}^gD \cap D \leq {}^gD \cap H$ and so $\mathcal{X}_{D,H} \subseteq \mathcal{Y}_{D,H}$.

Lemma 2.1.20 *Let T be an indecomposable D-projective kH-module. Then $T \uparrow_H^G \downarrow_H^G \simeq T \oplus T_1$ for a kH-module T_1 so that all indecomposable direct factors of T_1 have vertex in $\mathcal{Y}_{D,H}$.*

Proof Since T is D-projective, there is an indecomposable kD-module S such that $T \mid S \uparrow_D^H$. Hence there is a kH-module T'' with $S \uparrow_D^H \simeq T \oplus T''$. Mackey's formula gives

$$T \uparrow_H^G \downarrow_H^G \simeq \bigoplus_{HgH \in H\backslash G/H} {}^gT \downarrow^{{}^gH}_{{}^gH \cap H} \uparrow^H = T \oplus T'$$
$$T'' \uparrow_H^G \downarrow_H^G \simeq \bigoplus_{HgH \in H\backslash G/H} {}^gT'' \downarrow^{{}^gH}_{{}^gH \cap H} \uparrow^H = T'' \oplus T'''$$

where the direct factor T (respectively T'') occurs for the class $HgH = H1H = H$ and

$$T' := \bigoplus_{HgH \in (H\backslash G/H)\backslash H} {}^gT \downarrow^{^gH}_{^gH\cap H}\uparrow^H .$$

$$T''' := \bigoplus_{HgH \in (H\backslash G/H)\backslash H} {}^gT'' \downarrow^{^gH}_{^gH\cap H}\uparrow^H .$$

But now

$$S \uparrow_D^G \downarrow_H^G \simeq \bigoplus_{DgH \in D\backslash G/H} {}^gS \downarrow^{^gD}_{^gD\cap H}\uparrow^H \simeq S \uparrow_D^H \oplus S'$$

where again the class $D1H = H$ gives the module $S \uparrow_D^H \simeq T \oplus T''$ and where the other direct summands give $S' \simeq T' \oplus T'''$. Moreover, this description give that all indecomposable direct summands of S' have vertex in $\mathcal{Y}_{D,H}$ and therefore all indecomposable direct factors of T' have vertex in $\mathcal{Y}_{D,H}$. But since the definition of T' is

$$T \uparrow_H^G \downarrow_H^G = T \oplus T'$$

we get the desired statement if we put $T_1 = T'$. □

Theorem 2.1.21 (Green correspondence) *Let k be a field of characteristic $p > 0$, let D be a p-subgroup of G and let $H \leq G$ with $N_G(D) \leq H$. Then*

1. *For every indecomposable kG-module M with vertex D there is a unique indecomposable kH-module $f(M)$ with vertex D and which is a direct summand of $M \downarrow_H^G$.*
2. *For every indecomposable kH-module N with vertex D there is a unique indecomposable kG-module $g(N)$ with vertex D and which is a direct summand of $N \uparrow_H^G$.*
3. *Each indecomposable direct summand of $M \downarrow_H^G /f(M)$ has vertex in $\mathcal{Y}_{D,H}$.*
4. *Each indecomposable direct summand of $N \uparrow_H^G /g(N)$ has vertex in $\mathcal{X}_{D,H}$.*
5. *$fg(M) \simeq M$ and $gf(N) \simeq N$.*

Proof Let M be an indecomposable kG-module with vertex D and source L. Then M is a direct summand of $L \uparrow_D^G$. Let $L \uparrow_D^H = N_0 \oplus N_0'$ for an indecomposable kH-module N_0 so that M is a direct summand of $N_0 \uparrow_H^G$. By Lemma 2.1.20 we know that $N_0 \uparrow_H^G \downarrow_H^G = N_0 \oplus N_1$ for some kH-module N_1 so that each indecomposable direct factor of N_1 has vertex in $\mathcal{Y}_{D,H}$. But $M \downarrow_H^G$ is a direct factor of $N_0 \uparrow_H^G \downarrow_H^G = N_0 \oplus N_1$. If N_0 is not a direct summand of $M \downarrow_H^G$, then each indecomposable direct summand of $M \downarrow_H^G$ has vertex in $\mathcal{Y}_{D,H}$. But this implies that M has vertex in $\mathcal{Y}_{D,H}$ since M is a direct summand of $M \downarrow_H^G \uparrow_H^G$. This contradiction proves 1 and 3 if we put $f(M) := N_0$.

On the other hand, let N be an indecomposable kH-module with vertex D. By Lemma 2.1.20 we know that N is a direct summand of $N \uparrow_H^G \downarrow_H^G$. Let M

be an indecomposable direct summand of $N \uparrow_H^G$ so that N is a direct summand of $M \downarrow_H^G$. Denote $N \uparrow_H^G = M \oplus M'$ for a kG-module M'. By Lemma 2.1.20 each indecomposable direct summand of $N \uparrow_H^G \downarrow_H^G$ different from N has vertex in $\mathcal{Y}_{D,H}$. Since N is a direct summand of $M \downarrow_H^G$, each indecomposable direct summand of $M' \downarrow_H^G$ has vertex in $\mathcal{Y}_{D,H}$. We have already shown 1 and 3 and see that therefore each indecomposable direct summand of $M \downarrow_H^G$ different from N has vertex in $\mathcal{Y}_{D,H}$. Putting $g(N) := M$ we have shown 2in order to prove 4 we remark that if L denotes the source of N, then N is a direct factor of $L \uparrow_D^H$, and therefore $N \uparrow_H^G$ is a direct summand of $L \uparrow_D^H \uparrow_H^G = L \uparrow_D^G$. Therefore each indecomposable direct summand of $N \uparrow_H^G$ has vertex in $D \cap X$ for some $X \in \mathcal{Y}_{D,H}$. But X is a subgroup of ${}^gD \cap H$, and hence $D \cap X \leq D \cap {}^gD \cap H = D \cap {}^gD$. This describes exactly the groups in $\mathcal{X}_{D,H}$. We have proved 4.

The last point 5 is immediate by definition. □

Definition 2.1.22 Let k be a field of characteristic $p > 0$ and let G be a finite group. Let D be a p-subgroup of G and let H be a subgroup of G containing the normaliser $N_G(D)$ of D in G.

- Then for each indecomposable kG-module M with vertex D we define its *Green correspondent* to be the unique indecomposable direct summand $g(M)$ of $M \downarrow_H^G$ with vertex D.
- For each indecomposable kH-module N with vertex D we define its *Green correspondent* to be the unique indecomposable direct summand $f(N)$ of $N \uparrow_H^G$ with vertex D.

Green correspondence commutes with morphisms in a certain conceptual sense. We start with three indecomposable kG-modules M_1, M_2 and M_3 with vertices D_1, D_2 and D_3 and a subgroup H of G such that

$$N_G(D_1) \cdot N_G(D_2) \cdot N_G(D_3) \leq H.$$

Then the Green correspondents $g(M_1)$, $g(M_2)$ and $g(M_3)$ are defined; these are indecomposable kH-modules with vertex D_1, respectively D_2, respectively D_3 the unique direct summands of $M_1 \downarrow_H^G$, respectively $M_2 \downarrow_H^G$, respectively $M_3 \downarrow_H^G$ with vertex D_1, respectively D_2, respectively D_3. Let $\alpha \in Hom_{kG}(M_1, M_2)$ and $\beta \in Hom_{kG}(M_2, M_3)$. By restriction α and β induce $\alpha \in Hom_{kH}(M_1, M_2)$ and $\beta \in Hom_{kH}(M_2, M_3)$. Then by projection and inclusion of the direct summands $g(M_1)$, $g(M_2)$ and $g(M_3)$ into $M_1 \downarrow_H^G$, $M_2 \downarrow_H^G$ and $M_3 \downarrow_H^G$ we obtain $g(\alpha) \in Hom_{kH}(g(M_1), g(M_2))$ and $g(\beta) \in Hom_{kH}(g(M_2), g(M_3))$. In general we will not get $g(\beta \circ \alpha) = g(\beta) \circ g(\alpha)$. In order to see why this is, consider for all $i \in \{1, 2, 3\}$ the decomposition $M_i \downarrow_H^G = g(M_i) \oplus \check{M}_i$ for some kH-module M_i. Then

$$\alpha = \begin{pmatrix} g(\alpha) & \alpha_{1,2} \\ \alpha_{2,1} & \alpha_{2,2} \end{pmatrix} \text{ and } \beta = \begin{pmatrix} g(\beta) & \beta_{1,2} \\ \beta_{2,1} & \beta_{2,2} \end{pmatrix}$$

for

$$\alpha_{1,2} \in Hom_{kH}(\check{M}_1, g(M_2)), \qquad \beta_{1,2} \in Hom_{kH}(\check{M}_2, g(M_3)),$$
$$\alpha_{2,2} \in Hom_{kH}(\check{M}_1, \check{M}_2), \qquad \beta_{2,2} \in Hom_{kH}(\check{M}_2, \check{M}_3),$$
$$\alpha_{2,1} \in Hom_{kH}(g(M_1), \check{M}_2), \qquad \beta_{2,1} \in Hom_{kH}(g(M_2), \check{M}_3).$$

Then

$$\begin{aligned}\beta \circ \alpha &= \begin{pmatrix} g(\beta) & \beta_{1,2} \\ \beta_{2,1} & \beta_{2,2} \end{pmatrix} \circ \begin{pmatrix} g(\alpha) & \alpha_{1,2} \\ \alpha_{2,1} & \alpha_{2,2} \end{pmatrix} \\ &= \begin{pmatrix} g(\beta) \circ g(\alpha) + \beta_{1,2} \circ \alpha_{2,1} & g(\beta) \circ \alpha_{1,2} + \beta_{1,2}\alpha_{2,2} \\ \beta_{2,1} g(\alpha) + \beta_{2,2}\alpha_{2,1} & \beta_{2,1}\alpha_{1,2} + \beta_{2,2}\alpha_{2,2} \end{pmatrix}\end{aligned}$$

so that by this naive definition

$$g(\beta \circ \alpha) = g(\beta) \circ g(\alpha) + \beta_{1,2} \circ \alpha_{2,1}.$$

We will now introduce a limited version of a concept which will be of crucial importance in Chap. 5. Our argument is taken from the much more general approach of Auslander-Kleiner [2].

Suppose $D_1 = D_2 = D_3 =: D$. Then put

$$\mathcal{Y} := \{\, Y \leq G \mid Y \leq {}^{g}D \cap H \quad \text{for some } g \in G \setminus H\}$$

and for two indecomposable kG-modules M_1 and M_2 with vertices D_1 and D_2 let

$$Hom_{H,\mathcal{Y}}(g(M_1), g(M_2)) := Hom_{kH}(g(M_1), g(M_2))/Hom_{kH}^{\mathcal{Y}}(g(M_1), g(M_2))$$

and let $Hom_{kH}^{\mathcal{Y}}(g(M_1), g(M_2))$ be the set of those homomorphisms $\gamma : g(M_1) \longrightarrow g(M_2)$ such that there is a kH-module L which is a direct sum of indecomposable kH-modules L_i so that each L_i is relatively projective with respect to some $Y_i \in \mathcal{Y}$ and so that there are morphisms $\delta : g(M_1) \longrightarrow \bigoplus_i L_i$ and $\epsilon : \bigoplus_i L_i \longrightarrow g(M_2)$ with $\gamma = \epsilon \circ \delta$.

It is immediate to see that the composition of kH-linear morphisms induces a natural and well-defined composition of classes of morphisms in $Hom_{H,\mathcal{Y}}$.

In a completely analogous manner we define

$$\mathcal{X} := \{\, X \leq G \mid X \leq {}^{g}D \cap D \quad \text{for some } g \in G \setminus H\}$$

and for two indecomposable kH-modules N_1 and N_2 with vertices D_1 and D_2 let

$$Hom_{G,\mathcal{X}}(f(M_1), f(M_2)) := Hom_{kG}(f(N_1), f(N_2))/Hom_{kH}^{\mathcal{X}}(f(N_1), f(N_2))$$

and let $Hom_{kG}^{\mathcal{X}}(f(N_1), f(N_2))$ be the set of those homomorphisms $\gamma : f(N_1) \longrightarrow f(N_2)$ such that there is a kG-module L which is a direct sum of indecomposable

kG-modules L_i so that each L_i is relatively projective with respect to some $X_i \in \mathcal{X}$ and so that there are morphisms $\delta : f(N_1) \longrightarrow \bigoplus_i L_i$ and $\epsilon : \bigoplus_i L_i \longrightarrow f(N_2)$ with $\gamma = \epsilon \circ \delta$.

Again, the composition of morphisms induces a well-defined composition of classes of morphisms in $Hom_{G,\mathcal{X}}$.

Proposition 2.1.23 *Let k be a field of characteristic $p > 0$ and let G be a finite group. Let D be a p-subgroup of G and let H be a subgroup of G containing the normaliser $N_G(D)$ of D in G. Let M_1, M_2 and M_3 be indecomposable kG-modules with vertex D and let N_1, N_2 and N_3 be indecomposable kH-modules with vertex D.*

- *Then for all $\alpha \in Hom_{kG}(M_1, M_2)$ and $\beta \in Hom_{kG}(M_2, M_3)$ we have*

$$g(\beta \circ \alpha) = g(\beta) \circ g(\alpha) \in Hom_{H,\mathcal{Y}}(g(M_1), g(M_3)).$$

- *Moreover, for all $\rho \in Hom_{kH}(N_1, N_2)$ and $\sigma \in Hom_{kH}(N_2, N_3)$ we have*

$$f(\sigma \circ \rho) = f(\sigma) \circ f(\rho) \in Hom_{G,\mathcal{X}}(f(N_1), f(N_3)).$$

Proof The proof of the first part has already been done in the remarks before the proposition. The statement for f is completely analogous. □

2.2 Clifford Theory

We come to the question of whether it is possible to link the representation theory of a group G over a commutative ring R to the representation theory of some normal subgroup N of G over R. Of course in this generality the question cannot have a positive answer, since the trivial group with one element has a trivial representation theory and of course is a normal subgroup of every group. But we shall see that some information can be obtained.

2.2.1 Clifford's Main Theorem

In order to solve this problem one needs to study the concept of an inertia group. Given a group G and a normal subgroup N, let R be a commutative ring and let M be an RN-module. Then for all $g \in G$ conjugation with g gives an automorphism γ_g on N, since N is normal in G. Hence we may study the twisted KN-module ${}^{\gamma_g}M$ (cf Definition 1.7.39) and denote it again, slightly abusing the notation, by ${}^gM$. We have already used this construction in previous sections, such as for Mackey's formula and for Green correspondence.

Definition 2.2.1 Let G be a group and let N be a normal subgroup of G. Let R be a commutative ring and let M be an RN-module. Then

$$I_G(M) := \{g \in G \mid {}^gM \simeq M \text{ as } RN\text{-module}\}$$

is the *inertia group* of M in G.

Lemma 2.2.2 *For G a group, R a commutative ring, N a normal subgroup of G and M an RN-module, the inertia group $I_G(M)$ is a subgroup of G and N as well as the centraliser $C_G(N)$ of N in G is a subgroup of $I_G(M)$.*

Proof Indeed, $\gamma(gh) = \gamma(g)\gamma(h)$ and so

$${}^{\gamma(gh)}M \simeq {}^{\gamma(g)}\left({}^{\gamma(h)}M\right) \simeq {}^{\gamma(g)}M \simeq M$$

whenever $g, h \in I_G(M)$. A similar proof shows $g \in I_G(M) \Rightarrow g^{-1} \in I_G(M)$.

The fact that N is a subgroup of $I_G(M)$ is a consequence of Lemma 1.7.41. The fact that $C_G(N)$ is a subgroup of $I_G(M)$ follows from the definition of a twisted module. This proves the lemma. □

In order to illustrate Clifford's theorem we examine Mackey's formula in the case when $K = H$ is normal in G.

We get

$$\begin{aligned} M \uparrow_H^G \downarrow_H^G &\simeq \bigoplus_{HgH \in H\backslash G/H} \left({}^g\left(M \downarrow^H_{(H\cap gHg^{-1})}\right)\right) \uparrow^H_{(gHg^{-1}\cap H)} \\ &\simeq \bigoplus_{gH\in G/H} \left({}^g\left(M \downarrow_H^H\right)\right) \uparrow_H^H \simeq \bigoplus_{gH \in G/H} {}^gM. \end{aligned}$$

In particular we see that, as an RH-module, the induced module $M \uparrow_H^G$ is a direct sum of conjugates ${}^gM$ of M. Within the classes of G/H precisely the classes $I_G(M)/H$ lead to conjugates which are isomorphic to M.

Theorem 2.2.3 (Clifford's theorem) *Let G be a finite group and let N be a normal subgroup of G. Let R be a commutative ring such that the Krull-Schmidt theorem is valid for RS-modules, for all subgroups S of G. Let M be an indecomposable RN-module and suppose*

$$M \uparrow_N^{I_G(M)} \simeq M_1 \oplus M_2 \oplus \cdots \oplus M_r$$

for indecomposable $RI_G(M)$-modules $M_1, M_2, \ldots, M_r$.

Then for all $i, j \in \{1, 2, \ldots, r\}$ we have that $M_i \uparrow_{I_G(M)}^G$ is indecomposable and $M_i \uparrow_{I_G(M)}^G \simeq M_j \uparrow_{I_G(M)}^G$ implies $M_i \simeq M_j$.

Proof As in the remarks preceding the statement of the theorem we get

$$M \uparrow_N^G \downarrow_N^G \simeq \bigoplus_{gN \in G/N} {}^gM$$

and since for every $g_1, g_2 \in G$ one has ${}^{g_1}M \simeq {}^{g_2}M$ whenever $g_1 I_G(M) = g_2 I_G(M)$, we get

$$M \uparrow_N^G \downarrow_N^G \simeq \bigoplus_{gI_G(M) \in G/I_G(M)} ({}^gM)^{|I_G(M):N|}.$$

But, by Lemma 1.7.29 we have

$$M \uparrow_N^G \simeq \left(M \uparrow_N^{I_G(M)}\right) \uparrow_{I_G(M)}^G \simeq M_1 \uparrow_{I_G(M)}^G \oplus M_2 \uparrow_{I_G(M)}^G \oplus \cdots \oplus M_r \uparrow_{I_G(M)}^G .$$

Hence there are integers $n_1, n_2, \ldots, n_r$ with $\sum_{i=1}^r n_i = |I_G(M) : N|$ such that

$$M_i \uparrow_{I_G(M)}^G \downarrow_N^G \simeq \bigoplus_{gI_G(M) \in G/I_G(M)} ({}^gM)^{n_i}.$$

Since, as before, for every $g_1, g_2 \in G$ one has ${}^{g_1}M \simeq {}^{g_2}M$ if and only if $g_1 I_G(M) = g_2 I_G(M)$, we get $M_i \downarrow_N^{I_G(M)} \simeq M^{n_i}$.

Corollary 1.7.46 implies that for each $i \in \{1, 2, \ldots, r\}$ the $RI_G(M)$-module M_i is a direct factor of $M_i \uparrow_{I_G(M)}^G \downarrow_{I_G(M)}^G$. Hence for all $i \in \{1, 2, \ldots, r\}$ there is an indecomposable RG-module F_i such that F_i is a direct factor of $M_i \uparrow_{I_G(M)}^G$ and such that M_i is a direct factor of $F_i \downarrow_{I_G(M)}^G$. Since $M_i \downarrow_N^{I_G(M)} \simeq M^{n_i}$ one has that M^{n_i} is a direct factor of $(F_i \downarrow_{I_G(M)}^G) \downarrow_N^{I_G(M)} = F_i \downarrow_N^G$. But for each $i \in \{1, 2, \ldots, r\}$ one has that F_i is an RG-module, and so if M^{n_i} is a direct factor of $F_i \downarrow_N^G$, then ${}^gM^{n_i}$ is a direct factor of ${}^gF_i \downarrow_N^G \simeq F_i \downarrow_N^G$ for all $g \in G$. We use again that for every $g_1, g_2 \in G$ one has ${}^{g_1}M \simeq {}^{g_2}M$ if and only if $g_1 I_G(M) = g_2 I_G(M)$, so that whenever we sum over cosets modulo $I_G(M)$ we get different isomorphism classes of modules, and so, $\bigoplus_{gI_G(M) \in G/I_G(M)} ({}^gM)^{n_i}$ is a direct factor of $F_i \downarrow_N^G$. But

$$M_i \uparrow_{I_G(M)}^G \downarrow_N^G \simeq \bigoplus_{gI_G(M) \in G/I_G(M)} ({}^gM)^{n_i}$$

and F_i is a direct factor of $M_i \uparrow_{I_G(M)}^G$ with

$$F_i \downarrow_N^G \simeq \bigoplus_{gI_G(M) \in G/I_G(M)} ({}^gM)^{n_i}$$

as well. Therefore, $F_i \simeq M_i \uparrow_{I_G(M)}^G$ for all $i \in \{1, 2, \dots, r\}$. This shows that $M_i \uparrow_{I_G(M)}^G$ is indecomposable for all $i \in \{1, 2, \dots, r\}$.

We need to show that $M_i \uparrow_{I_G(M)}^G \simeq M_j \uparrow_{I_G(M)}^G$ implies $M_i \simeq M_j$.

Using Mackey's theorem and the fact that N is normal in G and contained in $I_G(M)$, we have

$$(\dagger): \quad M_i \uparrow_{I_G(M)}^G \downarrow_N^G \simeq \bigoplus_{gI_G(M) \in G/I_G(M)} {}^g\left(M_i \downarrow_N^{I_G(M)}\right).$$

Suppose M_j is a direct factor of $M_i \uparrow_{I_G(M)}^G \downarrow_{I_G(M)}^G$. Since M_i is a direct factor of $M_i \uparrow_{I_G(M)}^G \downarrow_{I_G(M)}^G$, and since its restriction to N corresponds to the coset $I_G(M)$ in $G/I_G(M)$ in the decomposition $(\dagger)$, the restriction of M_j to N is a direct sum of copies of ${}^gM$ for $g \in G \setminus I_G(M)$. But since ${}^gM \not\simeq M$ if and only if $g \in G \setminus I_G(M)$ we get that $M_i \uparrow_{I_G(M)}^G \simeq M_j \uparrow_{I_G(M)}^G$ implies $M_i \simeq M_j$. □

Often $I_G(M)$ is smaller than G, but not always. We will see now what happens if $I_G(M) = G$ and give a criterion on M which implies $I_G(M) \neq G$.

Corollary 2.2.4 *Let R be a commutative ring and G be a finite group such that the Krull-Schmidt property holds for all RL-modules, for all subgroups L of G. Let H be a normal subgroup of G and let S be a simple RG-module. Let T be a simple submodule of $S \downarrow_H^G$. If $I_G(T) = G$, then $S \downarrow_H^G \simeq T^n$ for some integer n.*

Proof Indeed, $\sum_{g \in G} gT$ is a KG-submodule of S. Since S is simple, $\sum_{g \in G} gT = S$. But, $gT \simeq {}^gT$ as KH-module, and so gT is a simple KH-module as well. Since $I_G(T) = G$ we get $T \simeq {}^gT \simeq gT$ for all $g \in G$ and so $S \downarrow_A^G \simeq T^n$, for some integer n. □

Example 2.2.5 Suppose G is a group and N is a subgroup of the centre of G. Then for all commutative rings R and all RG-modules M one has $I_G(M) = G$. Indeed, Lemma 2.2.2 shows that $I_G(M)$ contains $C_G(N)$ but since $N \subseteq Z(G)$ one has

$$G \supseteq I_G(M) \supseteq C_G(N) \supseteq C_G(Z(G)) = G.$$

Definition 2.2.6 Let G be a group and let R be a commutative ring. An RG-module S is *faithful* if whenever for $g \in G$ one has $g \cdot s = s$ for all $s \in S$, then $g = 1$.

Proposition 2.2.7 (Blichfeldt) *Let G be a finite group and let k be a field.*

Let S be a simple faithful kG-module and let A be an abelian normal subgroup of G not contained in the centre of G. Suppose that k is a splitting field for A. Let T be a simple submodule of $S \downarrow_A^G$. Then $I_G(T) \neq G$.

Proof Suppose $I_G(T) = G$. We know that T is a direct summand of $S \downarrow_A^G$. By Corollary 2.2.4 all simple direct summands of $S \downarrow_A^G$ are isomorphic to T. Since k is a splitting field for A, each simple kA-module is one-dimensional and therefore $S \downarrow_A^G \simeq T^{\dim_k(S)}$.

Since each simple kA-module is one-dimensional, there is a homomorphism of groups $\alpha : A \longrightarrow k^{\times}$ such that for all $t \in T$ we have $a \cdot t = \alpha(a)t$. Since $S \downarrow_A^G \simeq T^{\dim_k(S)}$, one gets for all $a \in A$ and all $s \in S$ that $a \cdot s = \alpha(a)s$. Therefore $\sigma|_A = \alpha$, denoting by $\sigma : G \longrightarrow Gl_{\dim_k(S)}(k)$ the group homomorphism which defines the kG-module structure on S. But the image A' of α in $Gl_{\dim_k(S)}(k)$ are multiples of the identity matrix and so A' is central in $Gl_{\dim_k(S)}(k)$. The kG-module S is faithful if and only if σ is injective. Hence $G \simeq \sigma(G) \subseteq Gl_{\dim_k(S)}(k)$. Since

$$A' = \sigma(A) \subseteq Z(Gl_{\dim_k(S)}(k)) \subseteq Z(\sigma(G))$$

we get that A is contained in the centre of G. This contradiction proves the proposition. $\square$

2.2.2 *Group Graded Rings and Green's Indecomposability Theorem*

Green correspondence studies the correspondence between specific indecomposable summands of modules induced from subgroups. Sometimes, these induced modules are decomposable, sometimes they are not. Green's indecomposability theorem provides a very useful criterion which tells us when the induced module is indecomposable. The main tools to obtain the result are Clifford's theorem and the notion of a grading of a ring by a group.

Definition 2.2.8 Let G be a group, let R be a commutative ring and let A be an R-algebra. We say that A is *graded by G* if for every $g \in G$ there is an R-module A_g such that

$$A \simeq \bigoplus_{g \in G} A_g$$

as an R-module, and such that

$$A_g \cdot A_h \subseteq A_{gh}$$

for all $g, h \in G$. We say that A is *strongly graded by G* if

$$A_g \cdot A_h = A_{gh}$$

for all $g, h \in G$.

If $A = \bigoplus_{g \in G} A_g$ and $B = \bigoplus_{g \in G} B_g$ are G-graded R-algebras, then we say that a *morphism* $f : A \longrightarrow B$ *is graded* if $f(A_g) \subseteq B_g$ for all $g \in G$. If $G = (\mathbb{Z}, +)$, then we say that f *is graded of degree* $n \in \mathbb{Z}$ if $f(A_m) \subseteq A_{n+m}$ for all $m \in \mathbb{Z}$.

Example 2.2.9 The following examples of graded algebras will be used later.

- Let k be a commutative ring and let G be a group with normal subgroup N. Then kG is strongly G/N-graded. Indeed,

$$kG = \bigoplus_{gN \in G/N} k(g \cdot N)$$

where $k(g \cdot N)$ is the free k-module generated by the coset $g \cdot N$. Moreover,

$$(g \cdot N) \cdot (h \cdot N) = g \cdot h \cdot (h^{-1}Nh) \cdot N = gh \cdot N$$

for all $g, h \in G$.
- We come to an example which will be examined in more detail in the sequel, and which motivates the whole subsection. Let k be a field, let G be a finite group, let N be a normal subgroup and let M be an indecomposable kN-module. Then $End_{kN}(M) =: B$ is a local algebra by Lemma 1.4.6. Consider $M \uparrow_N^G = kG \otimes_{kN} M$ and observe that

$$M \uparrow_N^G \downarrow_N^G = \bigoplus_{gN \in G/N} kgN \otimes_{kN} M = \bigoplus_{gN \in G/N} {}^gM.$$

Suppose now that the inertia group $I_G(M)$ of M is G. Then

$$\begin{aligned} End_{kG}(M \uparrow_N^G) &\simeq Hom_{kN}(M, M \uparrow_N^G \downarrow_N^G) \\ &\simeq \bigoplus_{gN \in G/N} Hom_{kN}(M, {}^gM) = \bigoplus_{gN \in G/N} B_g \end{aligned}$$

where $B_g := Hom_{kN}(M, {}^gM) = B$ since ${}^gM \simeq M$ for all $g \in G$. We see that $A := End_{kG}(M \uparrow_N^G)$ is a G/N-graded k-algebra with $A_{gN} = B_g$ for all $gN \in G/N$.

Some properties of group graded algebras are immediate.

Lemma 2.2.10 *Let G be a group, let k be a commutative ring, and let A be a G-graded k-algebra. The neutral element of G is denoted by 1.*

Then A_1 is a subring of A and for each $g \in G$ we get that A_g is an A_1-A_1-bimodule. Moreover $A^\times \cap A_1 = A_1^\times$. In particular if A is local then A_1 is local.

Proof Since $1 \cdot 1 = 1$ in G we get $A_1 \cdot A_1 \subseteq A_1$. If $g \in G \setminus \{1\}$, then

$$A_g \cdot A_g \subseteq A_{g^2} \neq A_g$$

and hence every idempotent in A is in A_1. In particular the multiplicatively neutral element 1_A of A is in A_1. This shows that A_1 is a subring of A. Since $A_1 \cdot A_g \subseteq A_g \supseteq A_g \cdot A_1$ for all $g \in G$, we obtain that A_g is an $A_1 - A_1$-bimodule.

Let $N := A \setminus A^\times$ and $N_1 := A_1 \setminus A_1^\times$ be the set of non-units of A and of A_1. If A is local, N is an A-ideal. We show first that $N_1 = N \cap A_1$. Indeed, if $n_1 \in A_1$, and

$$a = \sum_{g \in G} a_g \in A = \bigoplus_{g \in G} A_g$$

so that $n_1 \cdot a = 1_A$, then $n_1 \cdot a = \sum_{g \in G} n_1 a_g$ where $n_1 a_g \in A_g$ for each $g \in G$. Hence $n_1 \cdot a = 1$ implies $n_1 \cdot a_1 = 1$ and therefore $n_1 \in A_1^\times$. Hence

$$A_1^\times \supseteq A^\times \cap A_1.$$

The other inclusion is a consequence of the fact that A_1 is a subring.

Let $n_1, n_2 \in A_1 \setminus A_1^\times$ and let $a_1 \in A_1$. Then

$$a_1 \cdot n_1 \in A_1 \cdot N_1 \subseteq A_1 \cap (A \cdot N) \subseteq A_1 \cap N = N_1.$$

Similarly, $n_1 \cdot a_1 \in N_1$. Since $n_1, n_2 \in A_1 \setminus A_1^\times$, and since $A_1^\times = A_1 \cap A^\times$, we get that $n_1, n_2 \notin A^\times$. Since A is local, $n_1 + n_2 \notin A^\times$ and moreover $n_1 + n_2 \in A_1$ since A_1 is a subring. Hence $n_1 + n_2 \in N_1$. This shows that N_1 is a two-sided ideal of A_1 and we obtain the statement. □

Lemma 2.2.11 *Let k be an algebraically closed field of characteristic $p > 0$, and let $N \trianglelefteq G$ be a normal subgroup of the finite group G. Suppose G/N is a p-group. Let M be an indecomposable kN-module and suppose that $I_G(M) = G$. Then $J := \mathrm{rad}(End_{kN}(M)) \cdot End_{kG}(M \uparrow_N^G)$ is a two-sided ideal of $End_{kG}(M \uparrow_N^G)$, $J \leq rad(End_{kG}(M \uparrow_N^G))$ and $End_{kG}(M \uparrow_N^G)/J \simeq kG/N$.*

Proof We recall the second part of Example 2.2.9. Let k be an algebraically closed field of characteristic $p > 0$, let G be a group and $N \trianglelefteq G$ and suppose that G/N is a p-group. Let M be an indecomposable kN-module with endomorphism ring B. We assume again that $I_G(M) = G$. Then $End_{kG}(M \uparrow_N^G)$ is a G/N-graded k-algebra $A = \bigoplus_{gN \in G/N} B_g$ with $B_g = Hom_{kN}(M, {}^gM)$ for all $g \in G$. Now, B_g is a B-module and we shall consider $\mathrm{rad}(B) \cdot B_g = B_g \cdot \mathrm{rad}(B) = \mathrm{rad}_B(B_g)$.

Since M is indecomposable, B is local. Since k is algebraically closed, $B/\mathrm{rad}(B) \simeq k$ and $\mathrm{rad}_B(B_g)$ is of codimension 1 in B_g.

Now, let $g, h \in G/N$ and let $b_g \in B_g$ and $b_h \in B_h$. Then $b_g \cdot b_h \in B_{gh}$. Moreover, for each $g \in G$ fix an element $i_g \in B_g \setminus \mathrm{rad}_B(B_g)$ in B_g. Since $B \cdot i_g = i_g \cdot B = B_g$, we get that $\mathrm{rad}(B) \cdot A$ is a two-sided ideal of A. Every element of B_g is of the form $b \cdot i_g$ for some $b \in B_1$. For all $g, h \in G$ there is an $f(g, h) \in k^\times$ such that

$$i_g \cdot i_h - f(g, h) \cdot i_{gh} \in \mathrm{rad}(B_g).$$

Now, the associativity of the multiplication in A gives that for $g_1, g_2, g_3 \in G/N$ we obtain in $A/\mathrm{rad}(B) \cdot A$

$$\begin{aligned} f(g_1, g_2) f(g_1 g_2, g_3) i_{g_1 g_2 g_3} &= (i_{g_1} \cdot i_{g_2}) \cdot i_{g_3} = i_{g_1} \cdot (i_{g_2} \cdot i_{g_3}) \\ &= f(g_1, g_2 g_3) f(g_2, g_3) i_{g_1 g_2 g_3} \end{aligned}$$

and hence

$$f(g_1, g_2)f(g_1g_2, g_3) = f(g_1, g_2g_3)f(g_2, g_3)$$

for all $g_1, g_2, g_3 \in G/N$. Definition 1.8.38 shows that f is a 2-cocycle with values in $k^\times$ with trivial G/N-action and Lemma 1.8.49 shows that f is actually a 2-coboundary. Therefore there is a $\sigma : G \longrightarrow k^\times$ such that

$$f(g_1, g_2) = \sigma(g_1) \cdot \sigma(g_2) \cdot \sigma(g_1g_2)^{-1}$$

for all $g_1, g_2 \in G/N$. But then replacing i_g by $j_g := i_g \cdot \sigma(g)^{-1}$ we obtain

$$\begin{aligned} j_gj_h &= i_g \cdot \sigma(g)^{-1} \cdot i_h \cdot \sigma(h)^{-1} \\ &= i_{gh} \cdot f(g, h) \cdot \sigma(g)^{-1} \cdot \sigma(h)^{-1} \\ &= i_{gh} \cdot \sigma(g) \cdot \sigma(h) \cdot \sigma(gh)^{-1} \cdot \sigma(g)^{-1} \cdot \sigma(h)^{-1} \\ &= i_{gh} \cdot \sigma(gh)^{-1} \\ &= j_{gh} \end{aligned}$$

for all $g, h \in G/N$, as an equation in $A/\mathrm{rad}(B) \cdot A$. Therefore $A = \bigoplus_{gN \in G/N} B \cdot j_g$, where $j_g \cdot j_h - j_{gh} \in \mathrm{rad}_B(B_{gh})$. Moreover, $j_g \cdot B = B \cdot j_g = B_g$ for all $g \in G$ and hence $\mathrm{rad}(B) \cdot A$ is a two-sided ideal of A. Further, $A/\mathrm{rad}(B)A = kG/N$. This proves the lemma. □

Theorem 2.2.12 (Green's indecomposability theorem) *Let k be an algebraically closed field of characteristic $p > 0$ and let G be a finite group. If N is a normal subgroup of G such that G/N is a p-group, then for every indecomposable kN-module M the module $M \uparrow_N^G$ is indecomposable again.*

Proof Clifford's Theorem 2.2.3 implies that we just need to show that $M \uparrow_N^{I_G(M)}$ is indecomposable. Hence, we may suppose that $G = I_G(M)$. Consider Lemma 2.2.11 and put $A := End_{kG}(M \uparrow_N^G)$ and $B := End_{kN}(M)$. We therefore obtain $A/(\mathrm{rad}(B) \cdot A) \simeq kG/N$. Using Proposition 1.6.22 we get $A/\mathrm{rad}(A) = k$ since G/N is a p-group and hence A is local. □

2.3 Brauer Correspondence

A special case of the Green correspondence is the so-called Brauer correspondence which establishes a bijection between direct factors of the group ring kG as a ring and those of certain subgroups $H \leq G$ containing specific p-subgroups.

Definition 2.3.1 Let k be a field of characteristic $p > 0$ and let G be a finite group. A *block* B of kG is an indecomposable direct factor of kG as a ring. In other words, $kG = B \times B_1$ where B and B_1 are k-algebras and B is indecomposable as a k-algebra.

We remark that if M is an indecomposable A-module for a k-algebra A, and if $A = A_1 \times A_2$ for two k-algebras A_1 and A_2, then M is either an A_1-module or an A_2-module, but not both. Indeed, let $e_1^2 = e_1 \in Z(A)$ so that $e_1 A = A_1$ and $e_2 := 1 - e_1$. Then $M = e_1 M \oplus e_2 M$ as A-modules, and since M is indecomposable $e_1 M = 0$ or $e_2 M = 0$.

Definition 2.3.2 Let k be a field of characteristic $p > 0$ and let G be a finite group. Let M be an indecomposable kG-module. Let B be the unique block of kG such that M is a B-module. We say that M *belongs to the block* B. The trivial kG-module belongs to the *principal block* $B_0(kG)$ of kG.

We now link module and ring direct factors.

Lemma 2.3.3 *Let R be a commutative ring, let A be an R-algebra and let A_1 be a direct factor of A as a ring. Then A_1 is a direct factor as an $A \otimes_R A^{op}$-module of A. If A_2 is a direct factor as an $A \otimes_R A^{op}$-module of A, then A_2 is a direct factor of A as a ring.*

Proof We obtain $A_1 = e_1 A$ for some idempotent $e_1 \in Z(A)$. Then for all $a, b, c \in A$ we get $a \otimes b \in A \otimes_R A^{op}$ and compute

$$(a \otimes b) \cdot ce_1 = a(ce_1)b = (acb)e_1$$

and therefore Ae_1 is an $A \otimes_R A^{op}$-submodule of A. Moreover, by the same argument $A(1 - e_1)$ is an $A \otimes_R A^{op}$-submodule of A as well and $A = Ae_1 \oplus A(1 - e_1)$ as $A \otimes_R A^{op}$-modules.

Conversely, let $A = A_2 \oplus A_2'$ as $A \otimes_R A^{op}$-modules. Then $Z(A) = End_{A \otimes_R A^{op}}(A)$ and there is an idempotent $e_2 \in Z(A) = End_{A \otimes_R A^{op}}(A)$ so that $e_2(A) = e_2 \cdot A = A_2$. Hence A_2 is an algebra and moreover $A = Ae_2 \times A(1 - e_2)$ as algebras. □

We obtain that direct factors of an algebra A as a ring are the same as direct factors of A as $A \otimes_R A^{op}$-modules. The notion of decomposability is the same, when regarded as module or as algebras.

Now, let G be a group and let k be a commutative ring. Then

$$\Delta : G \longrightarrow G \times G$$

is defined by $\Delta(g) := (g, g)$ for all $g \in G$. Observe that $k\Delta(G)$ is a subring of $k(G \times G)$ and

$$k(G \times G) \simeq kG \otimes_k (kG)^{op}$$

where the isomorphism is given by $g \otimes h \mapsto g \otimes h^{-1}$. In this way, Δ extends k-linearly to an algebra homomorphism $kG \longrightarrow kG \otimes_k (kG)^{op}$ which sends $g \in G$ to $g \otimes g^{-1}$. Therefore $kG \otimes_k (kG)^{op}$ becomes a $k\Delta(G)$-right module, where $g \in G$ acts by right multiplication with $\Delta(g)$.

Proposition 2.3.4 *Let k be a field of characteristic $p > 0$, let G be a finite group and let B be a block of kG. Then the $kG \otimes_k kG^{op}$-module B has a vertex in the set of subgroups of $\Delta(G)$.*

Proof We first recall that kG is a $kG \otimes_k (kG)^{op}$-module by the action $(g \otimes h) \cdot x := gxh$ for $g, h \in G$ and $x \in kG$. The algebra $kG \otimes_k (kG)^{op}$ is a $k\Delta(G)$-right module as seen above.

In order to prove the proposition, we use Proposition 2.1.15. Consider k as a trivial $\Delta(G)$-module. Then define

$$\begin{aligned} \psi : kG &\longrightarrow (kG \otimes_k (kG)^{op}) \otimes_{k\Delta(G)} k \\ \sum_{g \in G} r_g g &\mapsto \sum_{g \in G} r_g((g \otimes 1) \otimes 1) \end{aligned}$$

The mapping ψ is certainly k-linear. Let $g, h_1, h_2 \in G$. Then

$$\begin{aligned} \psi((h_1 \otimes h_2) \cdot g) &= \psi(h_1 g h_2^{-1}) = (h_1 g h_2^{-1} \otimes 1) \otimes 1 = (h_1 g \otimes h_2) \otimes 1 \\ &= (h_1 \otimes h_2) \cdot ((g \otimes 1) \otimes 1) = (h_1 \otimes h_2) \cdot \psi(g) \end{aligned}$$

and so ψ is $kG \otimes_k kG^{op}$-linear. Moreover, the multiplication mapping

$$\begin{aligned} \mu : (kG \otimes_k (kG)^{op}) \otimes_{k\Delta(G)} k &\longrightarrow kG \\ (g \otimes h) \otimes x &\mapsto xgh^{-1} \end{aligned}$$

for $g, h \in G$ and $x \in k$ gives a morphism of $(kG \otimes_k (kG)^{op})$-modules. Finally, $\mu \circ \psi = id$.

Since kG is therefore $k\Delta(G)$-projective, the same holds for each direct summand. By Lemma 2.3.3 this shows that there are vertices of blocks which are subgroups of $\Delta(G)$. □

Note that the vertex is unique up to conjugacy only. Hence any subgroup conjugate in $G \times G$ to $\Delta(D)$ is a vertex as well. Such a conjugate will not be in $\Delta(G)$ in general.

Definition 2.3.5 Let k be a field of characteristic $p > 0$ and let G be a finite group. A *defect group* of a block B is a subgroup D of G such that $\Delta(D)$ is the vertex of B. The *defect of B* is the integer $\log_p(|D|)$.

Recall that a vertex, and hence also a defect group, is a p-group.

Proposition 2.3.6 *Let k be a field of characteristic $p > 0$ and let G be a finite group. Let M be an indecomposable kG-module that belongs to the block B. If B has defect group D, then there is a subgroup of D that is a vertex of M.*

Proof Let M be an indecomposable kG-module that belongs to the block B. Therefore M is a B-module, and we need to show that the multiplication mapping

$$B \otimes_{kD} M \longrightarrow M$$

is split. But we know that the multiplication mapping

$$(kG \otimes_k (kG)^{op}) \otimes_{k\Delta(D)} B \longrightarrow B$$

is split. Since all direct factors except B multiply to 0 on B we get a split morphism

$$(B \otimes_k B^{op}) \otimes_{k\Delta(D)} B \longrightarrow B.$$

Observe how the tensor product $k(G \times G) \otimes_{k\Delta(D)} B$ is formed. Indeed,

$$(g_1, g_2) \otimes db = (g_1 d, d^{-1} g_2) \otimes b$$

for all $g_1, g_2 \in G$, $d \in D$ and $b \in B$. The whole is a kG-right module by putting

$$((g_1, g_2) \otimes b) \cdot g = (g_1, g_2 g) \otimes b$$

and a kG-left module by putting

$$g \cdot ((g_1, g_2) \otimes b) = (gg_1, g_2) \otimes b.$$

We tensor this multiplication mapping with M over B, and so we get a split epimorphism

$$((B \otimes_k B^{op}) \otimes_{k\Delta(D)} B) \otimes_B M \longrightarrow B \otimes_B M.$$

Of course, $B \otimes_B M \simeq M$ by the multiplication map. Moreover, the B-right module structure of $((B \otimes_k B^{op}) \otimes_{k\Delta(D)} B)$ comes from the B-right module structure of the middle term B^{op}. Hence

$$\begin{array}{rcl} ((B \otimes_k B^{op}) \otimes_{k\Delta(D)} B) \otimes_B M & \simeq & (B \otimes_k M) \otimes_{k\Delta(D)} B \\ b_1 \otimes b_2 \otimes b_3 \otimes m & \mapsto & b_1 \otimes b_2 m \otimes b_3 \end{array}$$

and

$$\begin{array}{rcl} (B \otimes_k M) \otimes_{k\Delta(D)} B & \twoheadrightarrow & (B \otimes_k M) \\ b_1 \otimes m \otimes b_3 & \mapsto & b_1 b_3 \otimes m \end{array}$$

which induces a split epimorphism $B \otimes_D M \longrightarrow M$ as required. $\square$

Corollary 2.3.7 *The defect group of the principal block is a Sylow p-subgroup of G.*

Proof Indeed, by Lemma 2.1.19 the vertex of the trivial module is a Sylow p-subgroup of G. Hence the defect group of the principal block is at least a Sylow p-subgroup. The defect group is a p-group and therefore the defect group cannot be bigger than a Sylow p-subgroup. Proposition 2.3.6 then proves the statement. □

Corollary 2.3.8 *Let G be a p-group and k a field of characteristic p. Then the defect group of kG is G.*

Proof Indeed, kG is local by Proposition 1.6.22. □

Proposition 2.3.9 *If k is an algebraically closed field, then a block B is semisimple if and only if its defect group is the trivial group (i.e. the defect is* 0).

Proof If B is semisimple, then B^{op} is semisimple, and $B \otimes_k B^{op}$ is also semisimple. Hence the $B \otimes_k B^{op}$-module B is projective, and therefore B is a direct factor of $B \otimes_k B^{op}$. Hence the defect group of B is the trivial group.

If B has the trivial group as defect group, then for every B-module M the multiplication map $B \otimes_k M \longrightarrow M$ is split. Hence M is a direct factor of $B^{\dim_k(M)}$, and therefore M is projective. This shows that B is semisimple. □

Theorem 2.3.10 (Brauer's first main theorem) *Let k be a field of characteristic $p > 0$ and let G be a finite group. Then there is a one to one correspondence between blocks B of kG with defect group D and blocks b of $N_G(D)$ with defect group D. This bijection is obtained in such a way that b is a direct factor of $B \downarrow^{G \times G}_{N_G(D) \times N_G(D)}$, and b is the unique direct factor with this property.*

Proof Let B be a block of kG with defect group D. Then B is an indecomposable $\Delta(D)$-projective $k(G \times G)$-module and we look at $N_{G\times G}(\Delta(D))$. Now, $(g_1, g_2) \in N_{G\times G}(\Delta(D))$ if and only if there is a $d' \in D$ such that

$$(g_1, g_2)(d, d)(g_1^{-1}, g_2^{-1}) = (g_1 d g_1^{-1}, g_2 d g_2^{-1}) = (d', d').$$

Hence

$$(g_1, g_2) \in N_{G\times G}(\Delta(D)) \Leftrightarrow g_2^{-1} g_1 \in C_G(D) \text{and} g_1, g_2 \in N_G(D).$$

Therefore

$$N_{G\times G}(\Delta(D)) = \Delta(N_G(D)) \cdot (C_G(D) \times C_G(D)).$$

Now, kG is a free $kN_G(D)$-module, and therefore $kG = kN_G(D)^{|G:N_G(D)|}$ as $kN_G(D)$-left modules. Since each of these factors is a $k(N_G(D) \times N_G(D))$-module we get that kG is a direct sum of $|G : N_G(D)|$ copies of the $k(N_G(D) \times N_G(D))$-module $kN_G(D)$. Hence the restriction of B to $k(N_G(D) \times N_G(D))$ is a direct sum of blocks of $kN_G(D)$.

We observe now that $C_G(D) \leq N_G(D)$ and hence

$$\Delta(N_G(D)) \cdot (C_G(D) \times C_G(D)) \leq N_G(D) \times N_G(D).$$

The Green correspondence implies that there is a unique indecomposable $k\,(N_G(D) \times N_G(D))$-module with vertex $\Delta(D)$ within these direct factors. By Lemma 2.3.3 the direct factors of $kN_G(D)$ as $k\,(N_G(D) \times N_G(D))$-modules are precisely the blocks of $kN_G(D)$. Hence there is a unique block b of $kN_G(D)$ with defect group D and which is a direct factor of $B^{G\times G}_{N_G(D)\times N_G(D)}$. □

Definition 2.3.11 Let k be a field of characteristic $p > 0$ and let G be a finite group. Let B be a block of kG with defect group D. The unique block b of $kN_G(D)$ with defect group D which occurs as a direct summand of the restriction of B as a $k(N_G(D) \times N_G(D))$-module is called the *Brauer correspondent* of B.

2.4 Properties of Defect Groups, Brauer and Green Correspondence

Defect groups are far from arbitrary. Only few possibilities can occur and these possibilities are governed by the intersection behaviour of the Sylow p-subgroups of the group. We introduce a result due to Green.

Proposition 2.4.1 (Green [3]) *Let k be a field of characteristic $p > 0$ and let G be a finite group. Let B be a block of kG. Then every defect group of G is the intersection of two Sylow p-subgroups of G.*

Proof Let S be a Sylow p-subgroup of G. As in the proof of Proposition 2.3.4 we define a morphism μ of $kG \otimes_k (kG)^{op} \simeq k(G \times G)$-modules

$$\begin{array}{rcl} k(G \times G) \otimes_{k\Delta(G)} k & \longrightarrow & kG \\ (g, h) \otimes 1 & \mapsto & gh^{-1} \end{array}$$

for all $g, h \in G$. This is clearly well-defined and surjective. We showed there that μ is split by the mapping ψ defined as

$$\begin{array}{rcl} kG & \longrightarrow & k(G \times G) \otimes_{k\Delta(G)} k \\ g & \mapsto & (g, 1) \otimes 1 \end{array}$$

which is $k(G \times G)$ linear as well. Hence, kG is induced from a trivial $k\Delta(G)$-module. We see that the source of every block is the trivial module. In order to apply Mackey decomposition we shall need to consider representatives of double classes $\Delta(G)\backslash(G \times G)/(S \times S)$. We see first that

$$\Delta(G)\backslash(G \times G) = \bigcup_{g \in G} \Delta(G) \cdot (1, g).$$

Hence, double class representatives are elements $(1, g)$ for a set T of elements $g \in G$. Let B be a block of kG. Then by Green correspondence $B \downarrow_{S \times S}^{G \times G}$ has a unique direct summand $g(B)$ with vertex $\Delta(D)$ and source the trivial module. We compute

$$k \uparrow_{\Delta(G)}^{G \times G} \downarrow_{S \times S}^{G \times G} = \bigoplus_{h \in T} {}^{(1,h)}k \downarrow_{(1,h)\Delta(G)(1,h)^{-1} \cap S \times S}^{(1,h)\Delta(G)(1,h)^{-1}} \uparrow^{S \times S}$$

and observe that

$$(1, h)\Delta(G)(1, h)^{-1} \cap (S \times S) = \Delta(S \cap {}^hS)$$

and that the conjugate of the trivial module is still the trivial module. Hence

$$k \uparrow_{\Delta(G)}^{G \times G} \downarrow_{S \times S}^{G \times G} = \bigoplus_{h \in T} k \uparrow_{\Delta(S \cap {}^hS)}^{S \times S} .$$

Now, $S \times S$ is a p-group, its group ring is local and therefore each transitive permutation module (cf Definition 1.7.47), which is a quotient of the regular module and therefore has a simple head, is indecomposable. Hence each of the factors $k \uparrow_{\Delta(S \cap {}^hS)}^{S \times S} \simeq k(S \times S)/\Delta(S \cap {}^hS)$ is indecomposable and we get for the Green correspondent $g(B) \simeq k \uparrow_{\Delta(S \cap {}^hS)}^{S \times S}$ for some $h \in G$, which implies that the defect group of B is ${}^hS \cap S$. □

Corollary 2.4.2 *Let k be a field of characteristic $p > 0$ and let G be a finite group with a normal p-subgroup P. Then the defect group of every block of kG contains P.*

Proof Since any two Sylow p-subgroups of G are conjugate, we may choose one Sylow p-subgroup S freely. Let D be a defect group of G and let S be a Sylow p-subgroup of G containing D. Then any other Sylow p-subgroup T of G is conjugate to S, but since P is normal, P is fixed under conjugation and hence $P \leq T$. Since $D = S \cap T$, we obtain $P \leq D$. □

A certain converse to Proposition 2.3.6 is true as well.

Proposition 2.4.3 *If B is a block of kG with defect group D and if b is the Brauer correspondent of B in $kN_G(D)$, and if M is an indecomposable B-module with vertex D, then the Green correspondent of M in $kN_G(D)$ is a b-module. Conversely, if N is a b-module with vertex D, then the Green correspondent of N in kG is a B-module.*

Proof Since Green correspondence and Brauer correspondence are bijections we only need to prove one of the two statements. The other statement then follows by this bijection. Put $H := N_G(D)$.

Let N be an indecomposable kH-module with vertex D, suppose that N belongs to the block b of kH, and suppose that b has defect group D. The Green correspondent $g(N)$ of N has vertex D as well, by definition of the Green correspondence. Since $g(N)$ is indecomposable, there is a unique block $\hat{B}$ of kG such that $g(N)$ is a $\hat{B}$-module.

We will first show that $\hat{B}$ is a direct factor of $k(G \times G) \otimes_{k(H\times H)} b$. Since N is a b-module, the restriction of the standard isomorphism $kH \otimes_{kH} N \simeq N$ to the direct factor b of kH gives an isomorphism $b \otimes_{kH} N \simeq N$. Since N is a direct factor of the restriction of $g(N)$ to kH, we get that N is a direct factor of $b \otimes_{kH} g(N)$. But then

$$\begin{aligned}\left(k(G \times G) \otimes_{k(H\times H)} b\right) \otimes_{kG} g(N) &= kG \otimes_{kH} b \otimes_{kH} kG \otimes_{kG} g(N)\\ &= kG \otimes_{kH} b \otimes_{kH} g(N)\end{aligned}$$

has a direct factor $kG \otimes_{kH} N$. In particular, $g(N)$ is a module over a direct factor of $k(G \times G) \otimes_{k(H\times H)} b$ and therefore $\hat{B}$ is a direct factor of $b \uparrow_{H\times H}^{G\times G}$.

By Proposition 2.3.6 the defect group $\hat{D}$ of $\hat{B}$ contains D. Brauer correspondence was introduced as a special case of Green correspondence and this shows that the Brauer correspondent B of b is the unique indecomposable direct factor of $k(G \times G) \otimes_{k(H\times H)} b$ with vertex D. All the other indecomposable direct factors of $k(G \times G) \otimes_{k(H\times H)} b$ have vertex in

$$\mathcal{X}_{\Delta(D),H\times H} = \{X \leq G \times G \mid \exists g \in (G \times G) \setminus (H \times H) : X \leq \Delta(D) \cap {}^{g}\Delta(D)\}.$$

In particular, let $\tilde{B}$ be an indecomposable direct factor of $k(G\times G)\otimes_{k(H\times H)} b$ different from B. Then the vertex of $\tilde{B}$ is a proper subgroup of $\Delta(D)$, using that $H = N_G(D)$. Since $D \leq \hat{D}$ we get that $B = \hat{B}$. □

Corollary 2.4.4 (Brauer's third main theorem) *The Brauer correspondent of the principal block is the principal block.*

Proof We have seen by Proposition 2.4.3 that the trivial module corresponds to Brauer correspondents. Since the Green correspondent of the trivial module is the trivial module we obtain the statement. □

In a block with defect group D there is always a module with vertex D. In order to see this we first need a lemma.

Lemma 2.4.5 *Let k be a field of characteristic $p > 0$, let G be a finite group, let D be a p-subgroup of G, suppose H is a normal subgroup of G and suppose $D \leq H \trianglelefteq G$. If N is an indecomposable kH-module with vertex D, then $N \uparrow_H^G$ is a direct sum of indecomposable kG-modules $N_1, N_2, \ldots, N_s$ so that the vertex of N_i is D for all i.*

Proof $N \uparrow_H^G = N_1 \oplus \cdots \oplus N_m$ for indecomposable kG-modules N_i; for all $i \in \{1, 2, \ldots, m\}$. By definition, each N_i has a vertex included in D. Moreover

$$N_1 \downarrow_H^G \oplus \cdots \oplus N_m \downarrow_H^G = N \uparrow_H^G \downarrow_H^G = \bigoplus_{gH \in G/H} {}^gN$$

by Mackey decomposition and the decomposition above. Since H is normal in G, we see that ${}^gN$ has vertex ${}^gD$ as a kH-module since conjugation by g is an automorphism of the algebra kH. Hence $N_i \downarrow_H^G$ is isomorphic to a direct sum of modules ${}^gN$ for certain g, and each of these modules has vertex ${}^gD$. Therefore N_i has vertex ${}^gD$ for some g, which is the same as having vertex D. □

Proposition 2.4.6 *Let k be a field of characteristic $p > 0$ and let G be a finite group. Let B be a block of kG with defect group D. Then there is an indecomposable B-module with vertex D.*

Proof Let $H := N_G(D)$ and observe that $\text{rad}(kD) \cdot kH$ is a nilpotent ideal since $D \trianglelefteq H$ and since $\text{rad}(kD)$ is nilpotent. Hence $\text{rad}(kD) \cdot kH \leq \text{rad}(kH)$. The vertex of the trivial kD-module k is D. Hence $k \uparrow_D^H = T_1 \oplus \cdots \oplus T_n$ for indecomposable kH-modules $T_1, T_2, \ldots, T_n$. By Lemma 2.4.5 all modules T_i for $i \in \{1, 2, \ldots, n\}$ have vertex D. Since D is a p-group, we get $k = kD/\text{rad}(kD)$ and therefore

$$k \uparrow_D^H = kD/\text{rad}(kD) \uparrow_D^H = kH \otimes_{kD} (kD/\text{rad}(kD)) = kH/(\text{rad}(kD)kH)$$

where we use for the second isomorphism that induction transforms exact sequences into exact sequences. Since $\text{rad}(kD) \cdot kH \leq \text{rad}(kH)$ all isomorphism types of simple kH-modules appear as direct factors of the head of the modules $T_1, T_2, \ldots, T_n$. Let $\hat{e}$ be the central idempotent of kG such that $\hat{e} \cdot kG = B$ and let e be the central idempotent of kH such that $e \cdot kH$ is the Brauer correspondent b of B. Then there is an i such that $e \cdot T_i \neq 0$. Indeed, $e \cdot S = S$ holds for all simple kH-modules S which belong to b. Let S be a simple b-module which is a direct factor of $T_i/\text{rad}(T_i)$. Then

$$T_i \twoheadrightarrow T_i/\text{rad}(T_i)$$

gives that e does not act as 0 on T_i, since it does not act as 0 on the quotient $T_i/\text{rad}(T_i)$. Since T_i is indecomposable, T_i is a b-module and e acts as 1 on T_i. By Proposition 1.10.26 the Green correspondent $g(T_i)$ of T_i in kG is a B-module with vertex D. This proves the proposition. □

Let A be a symmetric k-algebra. Then there is a non-degenerate symmetric associative bilinear form

$$\langle\ ,\ \rangle : A \times A \longrightarrow k.$$

Let $e^2 = e$ be an idempotent in A. Restricting to $eAe \times eAe$ gives that $\langle\ ,\ \rangle$ induces a symmetric associative bilinear form

$$\langle\ ,\ \rangle : eAe \times eAe \longrightarrow k.$$

Moreover, since $\langle\ ,\ \rangle : A \times A \longrightarrow k$ is non-degenerate, for all $x \in eAe$ there is a $y \in A$ such that $\langle x, y\rangle \neq 0$. Then for all $x \in eAe$ and $y \in A$ we get

$$\langle x, y\rangle = \langle exe, y\rangle = \langle exe, ey\rangle = \langle ey, exe\rangle = \langle eye, exe\rangle = \langle exe, eye\rangle$$

and so we may take $y \in eAe$. Therefore for a symmetric algebra A and an idempotent $e \in A$ the algebra eAe is also symmetric. Recall from Proposition 1.10.26 that kG is a symmetric k-algebra. Hence each block of kG is a symmetric k-algebra.

We have proved the following lemma.

Lemma 2.4.7 *Let k be a field, let A be a symmetric k-algebra, let $e^2 = e \neq 0$ be an idempotent of A and let G be a finite group. Then eAe is also a symmetric k-algebra. In particular each block B of kG is a symmetric k-algebra.* □

2.5 Orders and Lattices

So far we have mainly looked at algebras over fields and their modules. It will be important to look at modules over more general local commutative rings R and to link modules over R-algebras Λ to modules over $R/\mathrm{rad}(R) \otimes_R \Lambda$. This works particularly well in a case we will describe now.

2.5.1 Discrete Valuation Rings

We recall some well-known elementary facts about commutative rings that are going to be used in the sequel. We will only recall the facts that are absolutely necessary for the sequel and keep this section to a strict minimum. The subject of commutative rings is vast and not considered here as the main focus of the book. For an opposite point of view one might consult Curtis-Reiner [4, 5]. For interested readers we recommend Cassels-Fröhlich [6] or Serre [7]. We follow the beginning of the presentation in Cassels-Fröhlich [6].

Definition 2.5.1 Let K be a field with unit group K^*. A map

$$v : K \longrightarrow \mathbb{Z} \cup \{\infty\}$$

is a discrete valuation of K if

- $v(K^*) = \mathbb{Z}$ and $v(ab) = v(a) + v(b)$ for all $a, b \in K^*$,
- $v(0) = \infty$,
- $v(x + y) \geq \min\{v(x), v(y)\}$ for all $x, y \in K$.

Here we define $\infty > a$ for all $a \in \mathbb{Z}$.

Lemma 2.5.2 *Let v be a discrete valuation of K. If $v(x) \neq v(y)$ then $v(x+y) = \min\{v(x), v(y)\}$. The set $R_v := \{x \in K \mid v(x) \geq 0\}$ is a local subring of K with field of fractions K and the principal ideal $\wp_v := R_v \setminus v^{-1}(0)$ is its unique maximal ideal.*

Proof The first axiom implies that R_v is multiplicatively and additively closed. Since $1 \in K$ is an idempotent, we see that $v(1)$ is an additive idempotent of $\mathbb{Z}$, and therefore $v(1) = 0$. We claim that $v^{-1}(0) = R_v^\times$ are the units of R_v. Indeed, given a with $v(a) = 0$, we consider $a^{-1} \in K$ and get

$$0 = v(1) = v(a \cdot a^{-1}) = v(a) + v(a^{-1}).$$

This implies $v^{-1}(0) = R_v^\times$. Then the third axiom implies that $(R_v, +)$ is a group, and hence that R_v is a subring of K. Since v is surjective, 1 is in the image of v and there is a $\pi \in R_v$ such that $v(\pi) = 1$. The first axiom implies that $v(\pi^k) = k$ for all $k \in \mathbb{Z}$. Given $a \in R_v$ with $v(a) = k$, then $v(a \cdot \pi^{-k}) = v(a) - k = 0$, and so $u_a := a \cdot \pi^{-k} \in R_v^\times$. This shows that $\wp_v$ is the unique maximal ideal of R_v. If $x, y \in K^\times$, then $x = \pi^{v(x)} \cdot u_x$ and $y = \pi^{v(y)} \cdot u_y$ for units u_x and u_y of R_v. Hence R_v is a principal ideal domain and any ideal is of the form $\pi^n R_v$. Suppose without loss of generality that $v(x) < v(y)$. Hence

$$v(x+y) = \pi^{v(x)} \cdot u_x + \pi^{v(y)} \cdot u_y = \pi^{v(x)}(u_x + \pi^{v(y)-v(x)} u_y)$$

and u_x a unit, whereas $\pi^{v(y)-v(x)} u_y \in \wp_v$. Therefore $u_x + \pi^{v(y)-v(x)} u_y \in R_v^\times$ since we have already shown that R_v is local. Hence $v(x+y) = v(x)$ as claimed. □

Definition 2.5.3 A generator π of the maximal ideal $\wp_v$ of R_v is called a *uniformiser* of R_v.

A discrete valuation gives rise to a metric $d(x, y) := 2^{-v(x-y)}$ for all $x, y \in K$. Indeed, $d(x, y) = 0 \Leftrightarrow v(x-y) = \infty \Leftrightarrow x = y$ and $d(x, y) = d(y, x)$ by the proof of Lemma 2.5.2, and we have $d(x, y) \leq d(x, z) + d(z, y)$ since $v(x+y) \geq \min(v(x), v(y))$.

As in real analysis, we can define Cauchy sequences and convergence of sequences.

Definition 2.5.4 A sequence $(a_n)_{n\in\mathbb{N}}$ of elements in K is a *Cauchy sequence* if and only if

$$\forall \epsilon > 0\ \exists N(\epsilon) \in \mathbb{N}\ \forall n, m > N(\epsilon) : \ d(a_n, a_m) < \epsilon.$$

A sequence $(a_n)_{n\in\mathbb{N}}$ *converges to* $a \in K$ if and only if

$$\forall \epsilon > 0\ \exists N(\epsilon) \in \mathbb{N}\ \forall n > N(\epsilon) : \ d(a_n, a) < \epsilon.$$

It is immediate that convergent sequences are Cauchy sequences. The converse is not always true, as we know from real analysis, for example.

Definition 2.5.5 A field K is *complete with respect to the valuation* v on K if every Cauchy sequence converges. A ring R is a *discrete valuation ring* if there is a field K with discrete valuation v such that $R = R_v$. A discrete valuation ring R is *complete* if its field of fractions is complete.

We can complete a given discrete valuation.

Proposition 2.5.6 *Given a discrete valuation v on K, there is a field $\hat{K}_v$ and a complete discrete valuation $\hat{v}$ on $\hat{K}_v$ such that K is a dense subfield of $\hat{K}_v$ and the restriction of $\hat{v}$ to K is v. Moreover if* $\mathrm{rad}(R_v) = \pi R_v$ *then* $\mathrm{rad}(\hat{R}_v) = \pi \hat{R}_v$. *We call $\hat{R}_v$ the completion of R at v.*

Proof Let K be a field with discrete valuation v, then let $\mathcal{C}_v$ be the set of all Cauchy sequences in K. Two Cauchy sequences $(a_n)_{n\in\mathbb{N}}$ and $(b_n)_{n\in\mathbb{N}}$ are called equivalent if the sequence $(a_n - b_n)_{n\in\mathbb{N}}$ converges to 0. Let $\hat{K}_v$ be the set of equivalence classes of Cauchy sequences.

This set carries a natural ring structure given by $(a_n) + (b_n) := (a_n + b_n)$ and $(a_n)\cdot(b_n) := (a_n b_n)$ for all $(a_n), (b_n) \in \mathcal{C}_v$. This gives the ring the structure of a field. Indeed, if (a_n) does not converge to 0, then there is a $\delta > 0$ such that $d(a_n, 0) \geq \delta$ for all $n \geq N_0$. In particular $a_n \neq 0$ for all $n \geq N_0$. Put $\breve{a}_n := a_n^{-1}$ for $n \geq N_0$ and $\breve{a}_n = 1$ for $n < N_0$. The sequence $(\breve{a}_n)$ is again a Cauchy sequence. Moreover, $(a_n)\cdot(\breve{a}_n)$ converges to the constant sequence 1. Hence $\hat{K}_v$ is a field.

We may embed K into $\hat{K}_v$ by sending $a \in K$ to the constant sequence $(a_n)_{n\in\mathbb{N}}$ with $a_n = a$ for all $n \in \mathbb{N}$.

The valuation v induces a valuation $\hat{v}$ on $\hat{K}_v$ by putting $\hat{v}((a_n)) := \lim_{n\to\infty} (v(a_n))$. We need to show that this is well-defined. Indeed, let (a_n) be a Cauchy sequence. Suppose (a_n) does not converge to 0. Hence there is an $N_0 \in \mathbb{N}$ such that $d(a_n, 0) > \delta$ for all $n \geq N_0$. Since $d(a_n, a_m) < \epsilon$ for $n, m > N(\epsilon)$ and since $d(a_n, a_m) = 2^{-v(a_n - a_m)} = 2^{-\min\{v(a_n), v(a_m)\}}$ we get that the sequence $(v(a_n))_{n\in\mathbb{N}}$ is ultimately constant. Hence $\hat{v}$ is again a discrete valuation and the restriction of $\hat{v}$ to constant sequences, i.e. the image of K in $\hat{K}_v$, is v. The field K is dense in $\hat{K}_v$. Indeed, if $\epsilon > 0$ and $x \in \hat{K}_v$, let x be represented by a Cauchy sequence $(x_n)_{n\in\mathbb{N}}$. Then for all $n > N_0$ we get that $2^{-\hat{v}(x - x_n)} < \varepsilon$. This shows that for all $n > N_0$ the elements $x_n \in K$ are in an ϵ-disk around x, and hence K is dense in $\hat{K}_v$. □

Example 2.5.7 The set $\mathbb{Q}$ carries a valuation v_p for every prime integer p. For any integer $m \in \mathbb{Z} \setminus \{0\}$ let $v_p(m) := s$ if $p^s \mid m$ but $p^{s+1} \nmid m$. Let $q = \frac{a}{b}$ for coprime integers $a > 0$ and b. Then put $v_p(q) := v_p(a) - v_p(b)$. This is a discrete valuation. The valuation ring with respect to v_p is

$$\mathbb{Z}_p := \left\{ \frac{a}{b} \mid b > 0;\ a \text{ and } b \text{ are coprime}, p \nmid b \right\}.$$

The completion with respect to v_p is $\hat{\mathbb{Q}}_p$, which is the field of fractions of formally infinite p-adic expansions, i.e.

$$\hat{\mathbb{Z}}_p := \left\{ \sum_{n=0}^{\infty} a_n p^n \mid \forall n \in \mathbb{N} : \ a_n \in \{0, 1, \ldots, p-1\} \right\}.$$

Another way of looking at the ring $\hat{\mathbb{Z}}_p$ is to consider the sequence of quotients

$$\longrightarrow \mathbb{Z}/p^{m+1}\mathbb{Z} \xrightarrow{\psi_m} \mathbb{Z}/p^m\mathbb{Z} \xrightarrow{\psi_{m-1}} \mathbb{Z}/p^{m-1}\mathbb{Z} \longrightarrow$$

and the unique ring R equipped with ring epimorphisms $\varphi_m : R \twoheadrightarrow \mathbb{Z}/p^m\mathbb{Z}$ such that $\varphi_m = \psi_m \circ \varphi_{m+1}$ for all m and such that whenever there is another ring S equipped with ring homomorphisms χ_m satisfying $\chi_m = \psi_m \circ \chi_{m+1}$, then there is a unique ring homomorphism $\chi : S \longrightarrow R$ with $\chi_m = \varphi_m \circ \chi$ for all m. It is an easy exercise to see that $\hat{\mathbb{Z}}_p$ has this property with respect to the sequence

$$\longrightarrow \mathbb{Z}/p^{m+1}\mathbb{Z} \xrightarrow{\psi_m} \mathbb{Z}/p^m\mathbb{Z} \xrightarrow{\psi_{m-1}} \mathbb{Z}/p^{m-1}\mathbb{Z} \longrightarrow .$$

We shall see later that this is an instance of an inverse limit of rings. We call the ring $\hat{\mathbb{Z}}_p$ the ring of p-adic integers and the field $\hat{\mathbb{Q}}_p$ the field of p-adic numbers.

Example 2.5.8 Not all complete discrete valuation rings are of characteristic 0, however we will mainly use the case of characteristic 0 in the sequel.

Let k be any field and let $k(X)$ be the field of rational functions with coefficients in k. Define for any polynomial $p(X) \in k[X]$ the number $v_X(p(X)) := s$ if $X^s \mid p(X)$ but $X^{s+1} \not| \, p(X)$. Then, for two coprime polynomials $p(X) \neq 0$ and $q(X)$, define

$$v_X\left(\frac{p(X)}{q(X)}\right) := v_X(p(X)) - v_X(q(X)).$$

We observe that v_X is a discrete valuation and the valuation ring contains $k[X]$. The completion of $k(X)$ with respect to v_X is field of Laurent power series with coefficients in k.

Proposition 2.5.9 *Let k be a perfect field of characteristic $p > 0$, then there is a complete discrete valuation ring R of characteristic 0 with $R/\mathrm{rad}(R) \simeq k$.*

A proof can be found, for example, in [7, Chap. II, Sect. 6, Thèorème 5].

Remark 2.5.10 If R is a discrete valuation ring, then we may also define Cauchy sequences in a free R-module R^n. If R is complete, then each Cauchy sequence in R^n converges.

Remark 2.5.11 Let R be a discrete valuation ring and let M be a finitely generated R-module. Recall that the modification of Gauss' algorithm to obtain the classification theorem for finitely generated abelian groups via elementary divisors of integral matrices can also be formulated for matrices over principal ideal domains. This yields a classification of modules over principal ideal domains, namely M is isomorphic to a direct product of a free module and modules of the form $R/\pi^m R$ for some $m \in \mathbb{N}$.

The observation in Example 2.5.7 can be used to define completions in a more general setting. We shall use this generalisation only in two cases, namely the classical Noether-Deuring theorem, and its generalisation to derived categories.

Lemma 2.5.12 *Let R be a commutative ring and let I be a proper ideal of R. Then we can form quotients of R modulo powers of I so that we obtain a sequence* $\ldots \twoheadrightarrow R/I^n \overset{\psi_{n-1}}{\twoheadrightarrow} R/I^{n-1} \twoheadrightarrow \ldots \twoheadrightarrow R/I$. *There is a unique ring $\hat{R}_I$ and morphisms $\hat{R}_I \overset{\varphi_n}{\longrightarrow} R/I^n$ such that $\psi_n \circ \varphi_{n+1} = \varphi_n$ for all n, and such that whenever there is a commutative ring S with mappings $S \overset{\chi_n}{\longrightarrow} R/I^n$ satisfying $\psi_n \circ \chi_{n+1} = \chi_n$ for all n, then there is a unique ring homomorphism $S \overset{\chi}{\longrightarrow} \hat{R}_I$ such that $\chi_n = \varphi_n \circ \chi$ for all n.*

Idea of Proof The construction of $\hat{R}_I$ is analogous to the explicit description of the p-adic integers in Example 2.5.7, namely $\hat{R}_I = \{(r_n)_{n\in\mathbb{N}} \in \prod_{n\in\mathbb{N}} R/I^n \mid \chi_n(r_{n+1}) = r_n \ \forall n \geq 1\}$. □

Definition 2.5.13 The ring $\hat{R}_I$ is the *I-adic completion* of R.

We shall see later that $\hat{R}_I$ is actually a special case of a projective limit.

An especially important case is given when R is commutative and semilocal, i.e. R has only finitely many maximal ideals (cf [8]).

Proposition 2.5.14 (Eisenbud [9, Corollary 7.6]) *Let R be a semilocal commutative ring and let $\mathfrak{m}_1, \ldots, \mathfrak{m}_r$ be its maximal ideals. Then for $I := \mathrm{rad}(R)$ we get for the I-adic completion $\hat{R}_I \simeq \hat{R}_{\mathfrak{m}_1} \times \cdots \times \hat{R}_{\mathfrak{m}_1}$.*

Proof (sketch): In this case $I = \bigcap_{i=1}^r \mathfrak{m}_i = \mathfrak{m}_1 \cdot \ldots \cdot \mathfrak{m}_r$. Then $I^n = \mathfrak{m}_1^n \cdot \ldots \cdot \mathfrak{m}_r^n$ and since $\mathfrak{m}_i^n + \mathfrak{m}_j^n = R$ for all $i \neq j$, we get by the Chinese Reminder Theorem that $R/I^n = R/\mathfrak{m}_1^n \times \cdots \times R/\mathfrak{m}_r^n$. Since $R/\mathfrak{m}_i^n$ is local we have $R/\mathfrak{m}_i^n = (R/\mathfrak{m}_i^n)_{\mathfrak{m}_i} = R_{\mathfrak{m}_i}/\mathfrak{m}_i^n R_{\mathfrak{m}_i}$, where we denote by $R_{\mathfrak{m}_i}$ the localisation at $\mathfrak{m}_i$. The construction in Lemma 2.5.12 then proves the statement. □

2.5.2 Classical Orders and Their Lattices

Throughout this section let R be a commutative ring without zero-divisors and field of fractions K. Then we want to formalise R-algebras which are "big subrings" of finite dimensional semisimple K-algebras. The notion for being "big" corresponds to the concept of containing a K-basis.

Definition 2.5.15 An R-algebra Λ is an *R-order* (or equivalently a classical R-order) in the semisimple K-algebra A if Λ is projective as an R-module, and if $K \otimes_R \Lambda \simeq A$ as K-algebras.

Example 2.5.16 Let us give some examples for properties of R-orders.

1. For every integer $n \in \mathbb{N}$ we have that $Mat_{n\times n}(R)$ is an R-order in $Mat_{n\times n}(K)$.
2. Let $\wp$ be a projective ideal of the ring R. Then the subring $\Lambda := \begin{pmatrix} R & R \\ \wp & R \end{pmatrix}$ of $Mat_{n\times n}(R)$ given by the set of matrices with lower left coefficients in $\wp$ is an R-order in $Mat_{n\times n}(K)$. Observe that Λ is a proper subring of $Mat_{n\times n}(R)$ of index $|R/\wp|$.
3. Fix a projective ideal $\wp$ of R and let $R - R := \{(x, y) \in R^2 \mid x - y \in \wp\}$. Then $R - R$ is an R-order in $K \times K$. Hence an R-order may be indecomposable in a decomposable K-algebra.
4. If Λ has R-torsion (the torsion submodule of Λ is the set of $\lambda \in \Lambda$ such that the kernel of $R \ni r \mapsto r \cdot \lambda \in \Lambda$ is not zero), then Λ is not an R-order since in this case Λ is not R-projective.
5. Let G be a finite group and let R be a discrete valuation ring of characteristic 0 and with field of fractions K. Then RG is an R-order in KG.

Let R be a discrete valuation ring with residue field k, let Λ be an R-order, let $A := k \otimes_R \Lambda$ and let M be a Λ-module, then $\overline{M} := k \otimes_R M$ is an A-module. If M is projective, then $\overline{M}$ is projective as an A-module. The main interest in using complete discrete valuation rings in this setting is the following proposition.

Proposition 2.5.17 *Let R be a Noetherian complete discrete valuation ring with residue field k and with field of fractions K. Let Λ be an R-order and let $A = k \otimes_R \Lambda$. Let $\pi : \Lambda \longrightarrow A$ be the residue mapping. Then for every idempotent $\overline{e}^2 = \overline{e} \in A$ there is an idempotent $e \in \Lambda$ such that $\pi(e) = \overline{e}$. In particular for every projective indecomposable A-module $\overline{P}$ there is a projective Λ-module P such that $\overline{P} \simeq k \otimes_R P$.*

Proof Recall that R is a complete discrete valuation ring and hence any projective R-module is free. Let π be a uniformiser of R. Since Λ is a finitely generated projective R-module, it is a free R-module of finite rank, we shall need to produce a Cauchy sequence of elements converging to an idempotent. If we have a sequence of elements $e_n \in \Lambda$ so that $e_n^2 - e_n \in \pi^n \Lambda$ and $e_n - e_{n-1} \in \pi^{n-1}\Lambda$, then the sequence e_n converges to an element $e \in \Lambda$ with $e^2 - e = 0$.

We proceed by induction. Let e_1 be an element such that $\pi(e_1) = \overline{e}$, let $n \geq 2$ and suppose we have constructed e_m for $m < n$ so that $e_m^2 - e_m \in \pi^m \Lambda$ and so that $e_m - e_{m-1} \in \pi^{m-1}\Lambda$. Put $e_n := 3e_{n-1}^2 - 2e_{n-1}^3$. Then

$$\begin{aligned} e_n^2 - e_n &= (3e_{n-1}^2 - 2e_{n-1}^3)^2 - (3e_{n-1}^2 - 2e_{n-1}^3) \\ &= (3e_{n-1}^2 - 2e_{n-1}^3)(3e_{n-1}^2 - 2e_{n-1}^3 - 1) \\ &= e_{n-1}^2(3 - 2e_{n-1})(3e_{n-1}^2 - 2e_{n-1}^3 - 1) \\ &= -(3 - 2e_{n-1})(1 + 2e_{n-1})(e_{n-1}^2 - e_{n-1})^2 \end{aligned}$$

where the last equation is verified by an elementary multiplication. But since $e_{n-1}^2 - e_{n-1} \in \pi^{n-1}\Lambda$, we obtain

$$e_n^2 - e_n \in \pi^{2(n-1)}\Lambda \subseteq \pi^n \Lambda.$$

If we consider e_n in $\Lambda/\pi^{n-1}\Lambda$, we get that $e_{n-1}^2 = e_{n-1}$ in $\Lambda/\pi^{n-1}\Lambda$ and so we compute in $\Lambda/\pi^{n-1}\Lambda$

$$e_n = 3e_{n-1}^2 - 2e_{n-1}^3 = 3e_{n-1} - 2e_{n-1} = e_{n-1} \in \Lambda/\pi^{n-1}\Lambda.$$

Hence

$$e_n - e_{n-1} \in \pi^{n-1}\Lambda.$$

This proves that (e_n) is a Cauchy sequence converging to $e \in \Lambda$ in the sense of Remark 2.5.10. The construction implies that

$$e^2 - e \in \bigcap_{n\in\mathbb{N}} \pi^n\Lambda = 0$$

and we have proved the result. □

Corollary 2.5.18 *Let G be a finite group and let R be a complete discrete valuation ring with residue field k. Then for every block B of RG the algebra $k\otimes_R B$ is a block of kG and for every block b of kG there is a unique block B of RG such that $b = k\otimes_R B$.*

Proof We know that the centre of RG is R-linearly generated by the conjugacy class sums of G, and the centre of kG is k-linearly generated by the conjugacy class sums of G. Hence the ring epimorphism $RG \twoheadrightarrow kG$ induces a ring epimorphism $Z(RG) \twoheadrightarrow Z(kG)$. Proposition 2.5.17 then proves the statement. □

Example 2.5.19 If R is not complete then it is not true that idempotents lift as in Proposition 2.5.17. We consider $\mathbb{Z}_pC_q$ for two different primes p and q. This is an order in $\mathbb{Q}C_q$ where we obtain

$$\mathbb{Q}C_q \simeq \mathbb{Q} \times \mathbb{Q}(\zeta_q)$$

for a primitive q-th root of unity ζ_q in $\mathbb{C}$ so that there are exactly two idempotents corresponding to the projection onto each of the direct factors. Since $\mathbb{Z}_pC_q$ is a subring of $\mathbb{Q}C_q$ there are at most two idempotents. $\mathbb{F}_pC_q$ is an $\mathbb{F}_p$-algebra. If $\mathbb{F}_p$ contains a primitive q-th root of unity, then $\mathbb{F}_pC_q = (\mathbb{F}_p)^q$ is a direct product of q direct factors $\mathbb{F}_p$. This happens if q divides $p-1$ since the multiplicative group of $\mathbb{F}_p$ is cyclic. Such pairs of primes exist, for example $p = 11$ and $q = 5$. Other occurrences of this kind of phenomenon appear for different reasons. It is easily seen that the Frobenius automorphism on $\mathbb{F}_2C_7$ has order 3. Indeed, $\mathbb{F}_2C_7 \simeq \mathbb{F}_2[X]/(X^7 - 1)$ and the Frobenius automorphism sends X to X^2, X^2 to X^4 and X^4 to X. Also the element X^3 lies in an orbit of length 3 under the Frobenius automorphism. Hence

$$\mathbb{F}_2C_7 \simeq \mathbb{F}_2 \times \mathbb{F}_8 \times \mathbb{F}_8$$

and there are three primitive idempotents in this algebra.

Let R be a discrete valuation ring with residue field $k = R/\text{rad}(R)$ and let Λ be an R-order. Of course, if M is a $k \otimes_R \Lambda$-module, then M is also a Λ-module since there is a surjective ring homomorphism $\Lambda \twoheadrightarrow k \otimes_R \Lambda$. However, dealing with orders we are really mainly interested in R-projective Λ-modules.

Definition 2.5.20 Let R be a commutative ring without zero-divisors and let Λ be an R-order. Then a finitely generated Λ-module L is called a *Λ-lattice* if L is projective as an R-module.

We have seen in Example 2.5.10 that for a discrete valuation ring R with residue field $k = R/\text{rad}(R)$ and an R-order Λ there may be $k \otimes_R \Lambda$-modules that are not of the form $k \otimes_R L$ for a Λ-lattice L.

A very important property of completions is given in the following lemma.

Lemma 2.5.21 *Let R be a discrete valuation ring and let $A \overset{\iota}{\hookrightarrow} B$ be an inclusion of finitely generated R-modules. Then $\hat{R} \otimes_R A \overset{id_{\hat{R}} \otimes \iota}{\longrightarrow} \hat{R} \otimes_R B$ is an inclusion of $\hat{R}$-modules, and $\hat{R} \otimes_R A = 0 \Rightarrow A = 0$.*

Moreover, for a uniformiser π of R we have $\hat{R}/\pi^\ell \hat{R} \simeq R/\pi^\ell R$ for all $\ell > 0$.

Proof Let π be a uniformiser of R. Since R is local any finitely generated indecomposable R-module is either isomorphic to R or to $R/\pi^n R$ for some $n > 0$. Now, trivially $\hat{R} \otimes_R R \simeq \hat{R}$.

We shall show $\hat{R}/\pi^\ell \hat{R} \simeq R/\pi^\ell R$. Let $(a_n)_{n \in \mathbb{N}}$ be a Cauchy sequence in R. Then $2^{-v(a_n - a_m)} < \epsilon$ for $n, m \geq N_\epsilon$ and so $a_n - a_m \in \pi^\ell R$ if $\epsilon < 2^{-\ell}$ and $n, m > N_\epsilon$. Hence, modulo π^ℓ, all Cauchy sequences are ultimately constant, and the Cauchy sequence gives, modulo $\pi^\ell \hat{R}$, the same element as the constant sequence with constant generic term a_{N_ϵ}. Hence

$$\hat{R}/\pi^\ell \hat{R} \simeq R/\pi^\ell R$$

as claimed. This shows at once that

$$\hat{R} \otimes_R A = 0 \Rightarrow A = 0.$$

We still need to show that $A \leq B$ implies $\hat{R} \otimes_R A \leq \hat{R} \otimes_R B$. For this we use that R is a subring of $\hat{R}$, and hence is an R-module, by considering constant Cauchy sequences. Now, by Remark 2.5.11 each finitely generated R-module is either free or isomorphic to $R/\pi^\ell R$ for some ℓ. We claim that $R/\pi^\ell R$ is not a submodule of $\hat{R}$. Indeed, let the equivalence class of the Cauchy sequence $(a_n)_{n \in \mathbb{N}}$ be a generator of this module. Then $\pi^\ell \cdot (a_n)_{n \in \mathbb{N}} = (\pi^\ell a_n)_{n \in \mathbb{N}}$ converges to 0. Hence for all $\epsilon > 0$ there is an N_ϵ such that

$$2^{-v(\pi^\ell a_n)} = 2^{-(\ell + v(a_n))} < \epsilon \ \ \forall n > N_\epsilon$$

and hence $(a_n)_{n \in \mathbb{N}}$ converges to 0. Therefore the Cauchy sequence $(a_n)_{n \in \mathbb{N}}$ represents the same element as 0. Therefore each finitely generated submodule of $\hat{R}$ is a free R-module.

Now, for each index set I and each family A_i of R-submodules of the R-module A and each R-module C we have the following equality

$$\left(\sum_{i\in I} A_i\right)\otimes_R C = \sum_{i\in I} (A_i \otimes_R C)\,.$$

Indeed, for finite sets I this is clear. If I is an infinite set, then let x be an element of the module on the left of the equation. Then there are only finitely many elements of I involved, and so we get the inclusion from left to right. Conversely let y be an element of the module on the right. Then again only a finite number of elements of I are involved and we get that y is also in the module on the left.

Now, each module is the sum of its finitely generated submodules. This proves the statement. □

Definition 2.5.22 We call an R-module S *flat* if for every monomorphism $\alpha : A \hookrightarrow B$ of R-modules, $id_S \otimes_R \alpha : S \otimes_R A \to S \otimes_R B$ is also a monomorphism. An R-module S is *faithfully flat* if S is flat and if $S \otimes_R A = 0$ implies $A = 0$.

Of course free modules are faithfully flat, but completions are also faithfully flat as we have seen in Lemma 2.5.21. Projective modules are flat but are not always faithfully flat. We shall study this property in more detail in Sect. 3.8.

Let R be a complete discrete valuation ring, let Λ be an R-order and let L be a Λ-lattice. For each $f \in End_\Lambda(L)$ define

$$\ker(f)^\infty := \{x \in L \mid \forall n \in \mathbb{N}\ \exists m \geq 0 : f^m(x) \in \mathrm{rad}^n(R)L\}$$

and

$$\mathrm{im}(f)^\infty := \bigcap_{n\in\mathbb{N}} \mathrm{im}(f^n).$$

Lemma 2.5.23 *Let R be a complete discrete valuation ring with uniformiser π. Then Fitting's lemma holds for endomorphisms of lattices L over R-orders Λ in the sense that there for every $f \in End_\Lambda(L)$ there is a decomposition $L = \ker(f)^\infty \oplus \mathrm{im}(f)^\infty$ so that f restricted to $\mathrm{im}(f)^\infty$ is an automorphism and f restricted to $\ker(f)^\infty$ is nilpotent modulo π^m for all m.*

Proof For all $n \in \mathbb{N}$ we see that $L/\pi^n L$ is a $\Lambda/\pi^n\Lambda$-module. Since $\Lambda/\pi^n\Lambda$ is artinian and Noetherian, we get by Fitting's lemma

$$L/\pi^n = \mathrm{im}(f)^\infty/\pi^n \oplus \ker(f)^\infty/\pi^n$$

for all $n \in \mathbb{N}$. Moreover $\mathrm{im}(f)^\infty/\pi^{n+1} \twoheadrightarrow \mathrm{im}(f)^\infty/\pi^n$ for all $n \in \mathbb{N}$ and likewise for $\ker(f)^\infty$. □

Corollary 2.5.24 *Let R be a complete discrete valuation ring. Then the Krull-Schmidt theorem holds for lattices over R-orders.*

Proof We review the proof of the Krull-Schmidt theorem in our setting. Fitting's lemma in the version of Lemma 2.5.23 again tells us that the endomorphism ring of an indecomposable lattice is local. Lemma 1.4.6 reads then that every endomorphism of an indecomposable lattice is either bijective or nilpotent modulo π^m for all m for a uniformiser π of R. In particular the series of endomorphisms used at the end of the proof of Lemma 1.4.6 converges in the topology induced by the valuation of R. Lemma 1.4.7 holds without any change, and the reader is invited to check that the proof of the Krull-Schmidt theorem holds word for word as well in this setting, without change. □

Lemma 2.5.25 *Let R be a discrete valuation ring and let $\hat{R}$ be its completion. Let Λ be an R-algebra and let $\hat{\Lambda} := \hat{R} \otimes_R \Lambda$. Then for all Λ-modules M and N we define $\hat{M} := \hat{R} \otimes_R M$ and $\hat{N} := \hat{R} \otimes_R N$. Suppose M is finitely generated. Then we have*

$$\hat{R} \otimes_R Ext^i_\Lambda(M, N) \simeq Ext^i_{\hat{\Lambda}}(\hat{M}, \hat{N})$$

for all $i \geq 0$.

Proof We shall prove that the canonical homomorphism

$$\begin{array}{rcl} Hom_\Lambda(M, N) & \longrightarrow & Hom_{\hat{\Lambda}}(\hat{M}, \hat{N}) \\ \varphi & \mapsto & id_{\hat{R}} \otimes_R \varphi \end{array}$$

induces an isomorphism

$$\hat{R} \otimes_R Hom_\Lambda(M, N) \simeq Hom_{\hat{\Lambda}}(\hat{M}, \hat{N}).$$

For this we need that if $A \longrightarrow B$ is a monomorphism of R-modules, then $\hat{R} \otimes_R A \longrightarrow \hat{R} \otimes_R B$ is a monomorphism of $\hat{R}$-modules by Lemma 2.5.21.

Then we see that for every $\hat{\Lambda}$-module X we have

$$Hom_{\hat{R} \otimes_R \Lambda}(\hat{R} \otimes_R M, X) = Hom_\Lambda(M, Hom_{\hat{R}}(\hat{R}, X)) = Hom_\Lambda(M, X)$$

and for every free Λ-module Λ^n we obtain

$$Hom_{\hat{\Lambda}}(\hat{\Lambda}^n, \hat{R} \otimes_R N) = \hat{R} \otimes_R N^n = \hat{R} \otimes_R Hom_\Lambda(\Lambda^n, N).$$

Hence, taking a free resolution of M as a Λ-module

$$\ldots \longrightarrow \Lambda^{n_1} \longrightarrow \Lambda^{n_0} \longrightarrow M \longrightarrow 0$$

we may tensor first by $\hat{R}$ over R, and then take $Hom_{\hat{\Lambda}}(-, \hat{N})$ or first take $Hom_\Lambda(-, N)$ and tensor with $\hat{R}$ over R afterwards. To shorten the notation we use the abbreviation

$$\hat{R} \otimes_R Hom_\Lambda(X, Y) = \widehat{Hom_\Lambda(X, Y)}.$$

The result on free modules is the same, and since tensor products with $\hat{R}$ over R turn exact sequences into exact sequences, we obtain a diagram with commutative squares:

$$\begin{array}{ccccccc}
\widehat{Hom_\Lambda(M,N)} & \hookrightarrow & \widehat{Hom_\Lambda(\Lambda^{n_0},N)} & \rightarrow & \widehat{Hom_\Lambda(\Lambda^{n_1},N)} & \rightarrow & \widehat{Hom_\Lambda(\Lambda^{n_2},N)} \rightarrow \\
\| & & \| & & \| & & \| \\
Hom_{\hat{\Lambda}}(\hat{M},\hat{N}) & \hookrightarrow & Hom_{\hat{\Lambda}}(\hat{\Lambda}^{n_0},\hat{N}) & \rightarrow & Hom_{\hat{\Lambda}}(\hat{\Lambda}^{n_1},\hat{N}) & \rightarrow & Hom_{\hat{\Lambda}}(\hat{\Lambda}^{n_2},\hat{N}) \rightarrow
\end{array}$$

For the first row we obtain $Ext^n_{\hat{\Lambda}}(\hat{M}, \hat{N})$ as the quotient of the kernel modulo the image of the mappings in the n-th term, and for the second row we obtain this way $\hat{R} \otimes_R Ext^n_{\hat{\Lambda}}(M, N)$ since $\hat{R}$ is faithfully flat as an R-module, and hence $\hat{R} \otimes_R -$ turns exact sequences into exact sequences. □

A very nice application of completions is the following generalisation of a result of Noether and Deuring, due to Roggenkamp. This is formulated in the more general context of I-adic completions, rather than discrete valuation rings. We shall briefly indicate what will be needed.

Remark 2.5.26 The statement of Lemma 2.5.25 remains true in the more general context of $\mathrm{rad}(R)$-adic completions of commutative Noetherian local rings R. Furthermore the lifting of idempotents result Proposition 2.5.17 still holds true in this more general context (cf Eisenbud [9, Corollary 7.5]).

Similar to the Hilbert basis theorem we have the following result.

Proposition 2.5.27 *Let R be a commutative Noetherian ring and let $\mathfrak{m}$ be a maximal ideal of R. Then the $\mathfrak{m}$-adic completion $\hat{R}_\mathfrak{m}$ is Noetherian as well. Moreover, $R/\mathfrak{m} = \hat{R}_\mathfrak{m}/\mathfrak{m}\hat{R}_\mathfrak{m}$ and $\hat{R}_\mathfrak{m}$ is a faithfully flat R-module.*

A proof can be found e.g. in Eisenbud [9, Theorem 7.1 and 7.2] or in Zariski-Samuel [10, Chap. VIII, Sect. 3, Corollary 5].

Theorem 2.5.28 (Noether-Deuring for fields, Roggenkamp [11]) *Let R be a local commutative Noetherian ring without zero-divisors and let Λ be a Noetherian R-algebra. Suppose S is either a commutative R-algebra which is a faithful finitely generated projective R-module or the* $\mathrm{rad}(R)$*-adic completion of R. Suppose that M and N are finitely generated Λ-modules so that $End_\Lambda(M)$ and $End_\Lambda(N)$ are finitely generated R-modules. Then M is a direct summand of N as Λ-modules if and only if $S \otimes_R M$ is a direct summand of $S \otimes_R N$ as $S \otimes_R \Lambda$-modules.*

Proof If M is a direct summand of N as Λ-module, then trivially $S \otimes_R M$ is a direct summand of $S \otimes_R N$ as $S \otimes_R \Lambda$-modules.

We need to show the converse, and so let $\hat{\sigma}$ be a split monomorphism $S \otimes_R M \overset{\hat{\sigma}}{\hookrightarrow} S \otimes_R N$. Suppose in the first step that $S = \hat{R}$ is the completion of R at the unique prime of R. We have seen from Lemma 2.5.25 (or Remark 2.5.26 for rad(R)-adic completions) that

$$\hat{R} \otimes_R Ext^n_\Lambda(M,N) = Ext^n_{\hat{R}\otimes_R\Lambda}(\hat{R} \otimes_R M, \hat{R} \otimes_R N)$$

for all $n \geq 0$. For $n = 0$ we get that $\hat{\sigma} = \sum_{i=1}^n \hat{r}_i \otimes_R \sigma_i$ for homomorphisms $\sigma_i \in Hom_\Lambda(M,N)$ and $\hat{r}_i \in \hat{R}$. Since $R/\mathrm{rad}(R) \simeq \hat{R}/\mathrm{rad}(\hat{R})$ by Proposition 2.5.6 (or Lemma 2.5.26 in the context of rad(R)-adic completions) for all $\hat{r}_i \in \hat{R}$ there is an r_i in R such that $r_i - \hat{r}_i \in \mathrm{rad}(\hat{R})$. Let

$$\sigma := \sum_{j=1}^n r_i\sigma_i \in Hom_\Lambda(M,N).$$

We claim that σ is a split monomorphism. To prove this it is sufficient to show that $id_{\hat{R}} \otimes_R \sigma$ is a split monomorphism in $Hom_{\hat{R}\otimes_R\Lambda}(\hat{R} \otimes_R M, \hat{R} \otimes_R N)$ since then the exact sequence

$$0 \longrightarrow M \stackrel{\sigma}{\longrightarrow} N \longrightarrow N/\mathrm{im}(\sigma) \longrightarrow 0$$

is mapped to the zero element in

$$\hat{R} \otimes_R Ext^1_\Lambda(N/\mathrm{im}(\sigma), M) = Ext^1_{\hat{R}\otimes_R\Lambda}(\hat{R} \otimes_R N/\mathrm{im}(\sigma), \hat{R} \otimes_R M).$$

Since $\hat{\sigma}$ is a split injection by hypothesis we get a homomorphism

$$\hat{\tau} \in Hom_{\hat{R}\otimes_R\Lambda}(\hat{R} \otimes_R N, \hat{R} \otimes_R M)$$

so that

$$\hat{\tau} \circ \hat{\sigma} = id_{\hat{R}\otimes_R M}.$$

We have

$$\begin{aligned}\hat{\tau} \circ (id_{\hat{R}} \otimes_R \sigma) - id_{\hat{R}\otimes_R M} &= \hat{\tau} \circ ((id_{\hat{R}} \otimes_R \sigma) - \hat{\sigma}) \\ &= \hat{\tau} \circ \left(\sum_{i=1}^n (r_i - \hat{r}_i) \otimes \sigma_i\right) \in \mathrm{rad}(\hat{R}) \cdot End_{\hat{\Lambda}}(\hat{M})\end{aligned}$$

since by the choice of r_i we get for all i that $r_i - \hat{r}_i \in \mathrm{rad}(\hat{R})$. Since $End_\Lambda(M)$ is finitely generated as R-module, $End_{\hat{\Lambda}}(\hat{M})$ is finitely generated as an $\hat{R}$-module. Therefore

$$\hat{\tau} \circ (id_{\hat{R}} \otimes_R \sigma) \in id_{\hat{R}\otimes_R M} + \mathrm{rad}(\hat{R}) \cdot End_{\hat{\Lambda}}(\hat{M})$$

is a unit in $End_{\hat{\Lambda}}(\hat{M})$, using Nakayama's Lemma 1.6.5. Therefore $id_{\hat{R}} \otimes_R \sigma$ is a split monomorphism and hence σ is a split monomorphism.

In order to prove the case when S is a faithful finitely generated projective R-module we observe that the completion $\hat{R} \otimes_R S$ of S is a projective and hence free $\hat{R}$-module, whence $\hat{R} \otimes_R S = \hat{R}^n$. Hence

$$S \otimes_R M \mid S \otimes_R N \Rightarrow \hat{R} \otimes_R S \otimes_R M \,\Big|\, \hat{R} \otimes_R S \otimes_R N \Rightarrow \hat{R}^n \otimes_R M \,\Big|\, \hat{R}^n \otimes_R N .$$

By the Krull-Schmidt theorem for lattices over $\hat{R}$-orders (Corollary 2.5.24) we get

$$\hat{R} \otimes_R M \,\Big|\, \hat{R} \otimes_R N$$

and this implies $M \mid N$ by the first step of the proof. □

Remark 2.5.29 We remark that Theorem 2.5.28 includes the case of a field R and a finite extension field S of K.

Under the hypothesis of Theorem 2.5.28 we get that $S \otimes_R M \simeq S \otimes_R N$ as $S \otimes_R \Lambda$-modules implies $M \simeq N$ as Λ-modules.

Example 2.5.30 We mention the original goal of Noether and Deuring. Let K be a field and let L be a Galois extension of K with $G := Gal(L/K)$. Then L is a KG-module, since G acts on L as K-linear automorphisms. We get that $L \otimes_K L \simeq \sum_{g \in G} L_g$, where $L_g = L$ for each $g \in G$, since the various K-linear embeddings of L into L describe precisely the Galois group G. Moreover, G permutes the summands L_g i.e. $h \cdot L_g = L_{hg}$ for $g, h \in G$. Hence we get the regular module

$$L \otimes_K L \simeq LG \simeq L \otimes_K KG$$

and by the Noether-Deuring theorem $L \simeq KG$ as KG-modules.

This is the easiest case of the so-called Galois module structure of a Galois extension. Most interesting is the case when K and L are algebraic number fields and we consider the algebraic integers $\mathcal{O}_L$ in L as an $\mathcal{O}_K G$-module. This is a highly complicated and difficult problem. We refer to Fröhlich's monograph [12] for further reading.

2.6 The Cartan-Brauer Triangle

Sometimes it is useful to consider only the composition factors of a module, with multiplicity, or the number of copies of a special isomorphism class of a projective module occurring in a particular projective module. The abstract tool for this is the Grothendieck group, which we shall introduce now. The Cartan mapping associates to each projective indecomposable module its composition factors. Many interesting properties of algebras are encoded in the Cartan mapping.

2.6.1 Grothendieck Groups

Definition 2.6.1 Let S be a commutative Noetherian ring and let Γ be a Noetherian S-algebra. The *Grothendieck group* $G_0(\Gamma - mod)$ of $\Gamma - mod$ is the quotient of the free abelian group on the set of isomorphism classes $\{M\}$ of finitely generated Γ-modules M modulo the subgroup generated by the relations

$$\{M\} - \{N\} - \{L\}$$

whenever

$$0 \longrightarrow L \longrightarrow M \longrightarrow N \longrightarrow 0$$

is a short exact sequence of A-modules. We denote by $[M]$ the image of an isomorphism class $\{M\}$ in $G_0(\Gamma - mod)$.

Proposition 2.6.2 *Let S be a commutative ring and let Γ be an artinian S-algebra. Then $G_0(\Gamma - mod)$ is a free abelian group with basis*

$$\{[L] \mid L \textit{ simple } \Gamma\textit{-module}\}.$$

Proof Since Γ is artinian, every finitely generated Γ-module has a composition series using Proposition 1.6.20 and Theorem 1.6.26. By Lemma 1.2.27 every indecomposable Γ-module M has a simple submodule L. Hence

$$0 \longrightarrow L \longrightarrow M \longrightarrow M/L \longrightarrow 0$$

is exact and by induction on the length of a composition series we get that

$$[M] = \sum_{L \in cf(M)} [L]$$

where $cf(M)$ is the set of composition factors of M. Hence

$$\mathcal{S}_\Gamma := \{[L] \mid L \text{ simple } \Gamma\text{-module}\}$$

is a generating set of $G_0(\Gamma - mod)$.

We need to show that $G_0(\Gamma - mod)$ is free over $\mathcal{S}_\Gamma$ as an abelian group. Let $\mathcal{SS}(\Gamma)$ be the free abelian group generated by the isomorphism classes of semisimple Γ-modules $\{T\}$ modulo the relations

$$\{T_1 \oplus T_2\} - \{T_1\} - \{T_2\}.$$

Denote by $[T]$ the image of the element $\{T\}$ in $\mathcal{SS}(\Gamma)$. It is clear by definition that $\mathcal{SS}(\Gamma)$ is a free abelian group with basis $\{[L] \mid L \text{ simple } \Gamma\text{-module}\}$.

We get a homomorphism of abelian groups

$$\mathcal{SS}(\Gamma) \xrightarrow{\sigma} G_0(\Gamma - mod)$$

via $\sigma([T]) := [T]$. Conversely define a homomorphism of abelian groups

$$G_0(\Gamma - mod) \xrightarrow{\rho} \mathcal{SS}(\Gamma)$$

by $\rho([M]) := \sum_{L \in cf(M)} [L]$ where $cf(M)$ is the set of composition factors of M. Then it is clear that this mapping is well-defined since composition series of a submodule U of M and the quotient module M/U give composition series of M. Moreover, σ is surjective since we have seen that $\mathcal{S}_\Gamma$ is in the image of σ and that this set is generating. Finally $\rho \circ \sigma = id_{\mathcal{SS}(\Gamma)}$ since the composition series of a semisimple module is readily established. Hence σ is injective as well. Therefore $G_0(\Gamma - mod)$ is free over $\mathcal{S}_\Gamma$ as an abelian group. □

Definition 2.6.3 Let S be a commutative ring and let Γ be a Noetherian S-algebra. Then $K_0(\Gamma)$ is the quotient of the free abelian group on the isomorphism classes $\{P\}$ of finitely generated projective Γ-modules P by the subgroup generated by the expressions $\{P_1 \oplus P_2\} - \{P_1\} - \{P_2\}$. Denote by $[P]$ the image of $\{P\}$ in $K_0(\Gamma)$.

2.6.2 *Cartan and Decomposition Maps*

There is an obvious homomorphism of abelian groups

$$\begin{array}{c} K_0(\Gamma) \xrightarrow{c} G_0(\Gamma - mod) \\ [P] \mapsto [P] \end{array}$$

This map is well-defined since a direct sum of two projective modules gives rise to a short exact sequence.

Since Γ is artinian, by Proposition 1.9.6 each simple Γ-module S has a projective cover P_S and this projective cover is unique up to isomorphism. We may choose the set

$$\{[L] \mid L \text{ simple } \Gamma\text{-module}\}$$

as a $\mathbb{Z}$-basis for $G_0(\Gamma - mod)$ and the set

$$\{[P_L] \mid L \text{ simple } \Gamma\text{-module}\}$$

as a $\mathbb{Z}$-basis of $K_0(\Gamma)$. With respect to this basis the group homomorphism c is given by a matrix C_Γ with coefficients in $\mathbb{N}$.

Definition 2.6.4 Let S be a commutative ring and let Γ be an artinian S-algebra. Then the natural homomorphism of abelian groups

$$\begin{array}{rcl} K_0(\Gamma) & \longrightarrow & G_0(\Gamma - mod) \\ [P] & \mapsto & [P] \end{array}$$

is the *Cartan mapping* c of Γ. The matrix C_Γ of c with respect to the basis formed by the isomorphism classes of simple modules and their projective covers respectively is called the *Cartan matrix*.

Let R be a complete discrete valuation ring, let $k = R/\mathrm{rad}(R)$ be its residue field and let $K = \mathrm{frac}(R)$ be its field of fractions. Let Λ be an R-order and let $A := k \otimes_R \Lambda$ be the finite dimensional residue algebra of Λ.

By Proposition 2.5.17 we know that every projective indecomposable A-module $\overline{P}$ is isomorphic to $k \otimes_R P$ for some projective indecomposable Λ-module P. Hence, in this case we get that $K_0(A) \simeq K_0(\Lambda)$ and the Cartan mapping is actually a mapping

$$K_0(\Lambda) \stackrel{c}{\longrightarrow} G_0(k \otimes_R \Lambda - mod).$$

However, we get another homomorphism e of abelian groups

$$\begin{array}{rcl} K_0(\Lambda) & \stackrel{e}{\longrightarrow} & K_0(K \otimes_R \Lambda) \\ [P] & \mapsto & [K \otimes_R P] \end{array}$$

which is well-defined since

$$K \otimes_R (P \oplus Q) \simeq (K \otimes_R P) \oplus (K \otimes_R Q)$$

for any Λ-modules P and Q. We will finally define a group homomorphism

$$K_0(K \otimes_R \Lambda) \stackrel{d}{\longrightarrow} G_0(k \otimes_R \Lambda - mod)$$

called the *decomposition map* in the following way:

For a finitely generated $K \otimes_R \Lambda$-module $\hat{M}$ choose an arbitrary K-basis $\{m_1, m_2, \ldots, m_t\}$ of $\hat{M}$. Then $M := \sum_{i=1}^{s} \Lambda \cdot m_i$ is a Λ-submodule of the $K \otimes_R \Lambda$-module $\hat{M}$, and therefore a Λ-lattice. Moreover, since $\{m_1, m_2, \ldots, m_t\}$ is a K-basis of $\hat{M}$, we get $K \otimes_R M = \hat{M}$. Of course $\hat{M}$ may contain many different such Λ-modules M.

Lemma 2.6.5 *Suppose R is a complete discrete valuation ring with residue field k, uniformiser π and field of fractions K. Suppose k is a splitting field for $k \otimes_R \Lambda$. Let $\hat{M}$ be a finitely generated $K \otimes_R \Lambda$-module and let M be a Λ-lattice with $K \otimes_R M = \hat{M}$. Then $[k \otimes_R M] \in G_0(k \otimes_R \Lambda - mod)$ does not depend on the choice of M in $\hat{M}$.*

Proof Put $\overline{M} := k \otimes_R M$ and $\overline{\Lambda} := k \otimes_R \Lambda$ to simplify the notation. Let $L_1, \ldots, L_n$ be representatives of the isomorphism classes of simple $\overline{\Lambda}$-modules.

We need to compute the multiplicity of a simple module L_i as a composition factor of $k \otimes_R M$. Let $\overline{P}_i$ be the projective cover of L_i as a $k \otimes_R \Lambda$-module, and let P_i be the projective cover of L_i as a Λ-module. Observe that Proposition 2.5.17 implies that the projective cover P_i exists and that $k \otimes_R P_i \simeq \overline{P}_i$.

If L_i is a composition factor of $\overline{M}$, then there is a submodule U of $\overline{M}$ such that L_i is in the socle of $\overline{M}/U$. Hence, there is a non-zero $\overline{\Lambda}$-module homomorphism $\overline{P}_i \longrightarrow \overline{M}/U$ and since $\overline{P}_i$ is projective, there is a non-zero $\overline{\Lambda}$-module homomorphism $\overline{P}_i \longrightarrow \overline{M}$. Conversely, any $\overline{\Lambda}$-linear homomorphism $\overline{P}_i \longrightarrow \overline{M}$ is induced by a composition factor of $\overline{M}$ isomorphic to L_i. The number of composition factors of $\overline{M}$ isomorphic to L_i is $\dim_k(Hom_{\overline{\Lambda}}(\overline{P}_i, \overline{M}))$ since k is a splitting field of $\overline{\Lambda}$. Hence

$$[\overline{M}] = \sum_{i=1}^{n} \left(\dim_k(Hom_{\overline{\Lambda}}(\overline{P}_i, \overline{M}))\right) \cdot [L_i].$$

But now,

$$\dim_k(Hom_{\overline{\Lambda}}(\overline{P}_i, \overline{M})) = \dim_k(Hom_{\Lambda}(P_i, \overline{M}))$$

since any homomorphism $P_i \longrightarrow \overline{M}$ factors through $P_i \longrightarrow \overline{P}_i$, the module $\overline{M}$ being annihilated by π. But, $M \longrightarrow \overline{M}$ is surjective and P_i is projective. Therefore

$$\dim_k(Hom_{\Lambda}(P_i, \overline{M})) = \text{rank}_R(Hom_{\Lambda}(P_i, M))$$

since the projectivity of P_i implies that every homomorphism $P_i \longrightarrow \overline{M}$ factors through $M \longrightarrow \overline{M}$ and since M is a lattice, $Hom_{\Lambda}(P_i, M)$ is free as an R-module. But now,

$$\text{rank}_R(Hom_{\Lambda}(P_i, M)) = \dim_K(Hom_{\Lambda}(K \otimes_R P_i, K \otimes_R M)).$$

Indeed, the tensor product $K \otimes_R -$ gives that the left-hand side is at most as big as the right-hand side, and given a morphism $\alpha \in Hom_{\Lambda}(K \otimes_R P_i, K \otimes_R M)$, using that M is finitely generated as an R-module, there is an integer $u(\alpha) \in \mathbb{Z}$ such that $\alpha(x) \in \pi^{u(\alpha)}M$ for all $x \in P_i$. Hence we get equality in the above equation. Therefore

$$[\overline{M}] = \sum_{i=1}^{n} \left(\dim_K(Hom_{K\otimes_R\Lambda}(K \otimes_R P_i, K \otimes_R M))\right) \cdot [L_i].$$

This expression is independent of the choice of M inside $K \otimes_R M$. □

Definition 2.6.6 Let R be a complete discrete valuation ring, let $k = R/\text{rad}(R)$ be its residue field, let K be its field of fractions, and let Λ be an R-order. Suppose that k is a splitting field for $k \otimes_R \Lambda$. Let $\hat{V}_1, \hat{V}_2, \ldots, \hat{V}_s$ be representatives of the isomorphism classes of the simple $K \otimes_R \Lambda$-modules and choose representatives $P_1, P_2, \ldots, P_n$ of the isomorphism classes of the indecomposable projective $k \otimes_R \Lambda$-modules. The numbers

$$d_{i,j} := \dim_K(Hom_{K\otimes_R\Lambda}(K \otimes_R P_i, \hat{V}_j))$$

are the *decomposition numbers* of Λ. The matrix $D := (d_{i,j})_{1\le i\le s;1\le j\le n}$ is the *decomposition matrix* of Λ.

2.6.3 The Cartan-Brauer Triangle

As in the previous section, R is a complete discrete valuation ring with field of fractions K and residue field k, Λ is an R-order and $A = k\otimes_R \Lambda$ is a finite dimensional k-algebra. Suppose that k is a splitting field for A. Recall that $e : K_0(\Lambda) \longrightarrow G_0(K\otimes_R \Lambda - mod)$ is given by tensoring with K over R. Now, if K is a splitting field for $\hat{\Lambda} := K \otimes_R \Lambda$, we obtain the following formula. For every $\hat{\Lambda}$-module $\hat{M}$ the multiplicity of the simple $\hat{\Lambda}$-module $\hat{V}_i$ as a direct summand of $\hat{M}$ is $\dim_K(Hom_{\hat{\Lambda}}(\hat{M}, \hat{V}_i))$. Hence

$$[\hat{M}] = \sum_{\ell=1}^{s} \left(\dim_K(Hom_{\hat{\Lambda}}(\hat{M}, \hat{V}_\ell)\right) [\hat{V}_\ell].$$

In particular, for $\hat{M} = K \otimes_R P_i$ one obtains the coefficients of the matrix associated to the $\mathbb{Z}$-linear mapping e with respect to the basis $\{[P_i] \mid i \in \{1, \dots, n\}\}$ of $K_0(\Lambda)$ and $\{[\hat{V}_j] \mid j \in \{1, \dots, s\}\}$ of $G_0(\hat{\Lambda} - mod)$. These are precisely the transposes of the coefficients of d. Therefore e is the transpose of d, i.e. $e = d^{tr}$.

Proposition 2.6.7 *Let R be a complete discrete valuation ring and let k be its residue field, and let K be its field of fractions. Let Λ be an R-order and suppose that k is a splitting field for $k \otimes_R \Lambda =: A$ and that K is a splitting field for $K \otimes_R \Lambda$. Then $C = D \cdot D^{tr}$, or more precisely the coefficients of the Cartan matrix of A, are $c_{i,j} = \sum_{\ell=1}^{s} d_{i,\ell}d_{\ell,j}$.*

Proof We shall show that c factorises as $c = d \circ e$. Consider

$$c : K_0(\Lambda) \longrightarrow G_0(\overline{\Lambda} - mod)$$

which maps $[P]$ to $[k \otimes_R P]$ inside $G_0(\overline{\Lambda} - mod)$. First, take

$$e : K_0(\Lambda) \longrightarrow G_0(\hat{\Lambda} - mod)$$

which is just given by the tensor product $[P] \mapsto [K \otimes_R P]$. Then d is given by taking a Λ-lattice L inside $K \otimes_R P$ and then considering $k \otimes_R L$ inside $G_0(\overline{\Lambda} - mod)$. Hence we can take $L = P$, since by Lemma 2.6.5 the result does not depend on this choice. This shows that the composition $d \circ e$ maps $[P]$ to $[k \otimes_R P]$ and this is precisely $c([P])$. □

Graphically, Proposition 2.6.7 says that for an R-order Λ with K the field of fractions and k the residue field of the complete discrete valuation ring R the diagram

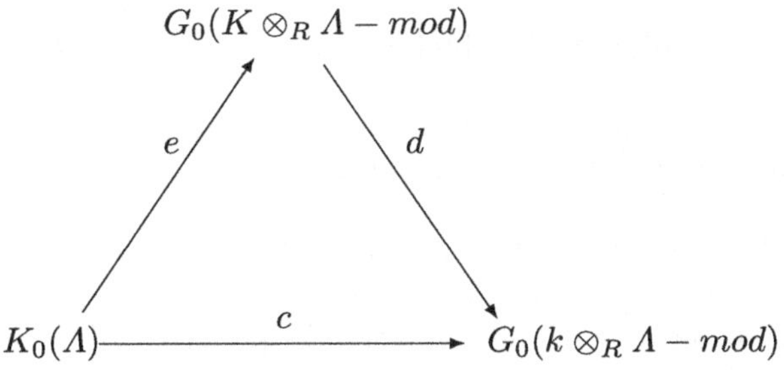

is commutative if K and k are splitting fields for $K \otimes_R \Lambda$ and for $k \otimes_R \Lambda$ respectively.

Corollary 2.6.8 *Let R be a complete discrete valuation ring with field of fractions K and let k be its residue field. Let A be a finite-dimensional k-algebra, so that k is a splitting field for A. If there is an R-order Λ with $k \otimes_R \Lambda \simeq A$, and such that K is a splitting field of $K \otimes_R \Lambda$, then the Cartan matrix of A is symmetric.*

Proof This is immediate using Proposition 2.6.7 and the fact that

$$c^{tr} = \left(d \circ d^{tr}\right) = \left(d^{tr}\right)^{tr} \circ d^{tr} = d \circ d^{tr} = c.$$ □

Example 2.6.9 Let

$$A = \begin{pmatrix} k & k \\ 0 & k \end{pmatrix}$$

be the algebra of upper triangular matrices over a field k. The Cartan matrix of A is

$$C_A = \begin{pmatrix} 1 & 1 \\ 0 & 1 \end{pmatrix},$$

whence there is no order reducing to A.

More generally, for a finite dimensional quiver algebra $k\Gamma$ the Cartan matrix can be read off from the quiver directly. The coefficient (i,j) is the number of paths from i to j. Hence the Cartan matrix of $k\Gamma$ is never symmetric when $k\Gamma$ is finite dimensional.

Remark 2.6.10 Let Γ be an artinian K-algebra over a field K. Suppose K is a splitting field for Γ then we may define a $\mathbb{Z}$-bilinear mapping

$$\langle\ ,\ \rangle : K_0(\Gamma) \times G_0(\Gamma - mod) \longrightarrow \mathbb{Z}$$

by defining

$$\langle [P], [M] \rangle := \dim_K(Hom_\Gamma(P, M))$$

for each projective Γ-module P and each Γ-module M and extending this to all of $K_0(\Gamma)$ and $G_0(\Gamma - mod)$ by putting

$$\langle [P]-[Q],[M]-[N]\rangle := \langle [P],[M]\rangle - \langle [P],[N]\rangle - \langle [Q],[M]\rangle + \langle [Q],[N]\rangle$$

for all projective Γ-modules P and Q and all Γ-modules M and N. We need to see that this is actually well-defined. Indeed, since

$$Hom_\Gamma(P_1 \oplus P_2, M) \simeq Hom_\Gamma(P_1, M) \oplus Hom_\Gamma(P_2, M)$$

the form is well-defined on the first variable. For projective Γ-modules P we have for each exact sequence

$$0 \longrightarrow L \longrightarrow M \longrightarrow N \longrightarrow 0$$

of Γ-modules an exact sequence of K-modules

$$0 \longrightarrow Hom_\Gamma(P, L) \longrightarrow Hom_\Gamma(P, M) \longrightarrow Hom_\Gamma(P, N) \longrightarrow 0.$$

Therefore the bilinear form is well-defined on the second variable as well.

Now, we have a basis for $K_0(\Gamma)$

$$\{[P_L] \mid L \text{ simple } \Gamma\text{-module}\}$$

and a basis for $G_0(\Gamma - mod)$

$$\{[L] \mid L \text{ simple } \Gamma\text{-module}\}.$$

Since $Hom_\Gamma(P_L, L) = K$ and $Hom_\Gamma(P_L, L') = 0$ if L' is a simple module not isomorphic to L, we see that the two bases are dual to each other with respect to the bilinear form $\langle\ ,\ \rangle$. The form is non-degenerate.

2.7 Defining Blocks by Non-vanishing Ext^1 Between Simple Modules

In this section we shall give a characterisation of blocks, and more generally of indecomposable direct factors of finite dimensional k-algebras over a field k. This characterisation is useful in certain technical considerations and gives another illuminating view on the indecomposable factors of an algebra.

Let A be a finite dimensional k-algebra and let

$$A = A_1 \times \cdots \times A_n$$

be a decomposition of A into indecomposable direct factors. Equivalently we decompose $1 \in Z(A)$ into primitive central idempotents

$$1 = e_1 + \cdots + e_n$$

and we know by Proposition 1.9.14 that this decomposition is unique.

Analogously to the case of a block of a group ring, given in Definition 2.3.2, we say that an indecomposable A-module belongs to the block A_i if M is an A_i-module. It is clear that there is a unique A_i such that M belongs to A_i. Indeed,

$$M = e_1M \oplus \cdots \oplus e_nM$$

as A-modules since $e_i \in Z(A)$ for all $i \in \{1, \ldots, e_n\}$ and therefore, since M is indecomposable, there is a unique e_i with $M = e_iM$. Hence $A_j \cdot M = 0$ for all $j \neq i$, where $A_i := A \cdot e_i$.

Lemma 2.7.1 *Let M and N be two indecomposable A-modules and suppose $Ext^1_A(M, N) \neq 0$. Then M and N belong to the same block A_i.*

Proof Let

$$0 \longrightarrow N \stackrel{\nu}{\longrightarrow} X \stackrel{\mu}{\longrightarrow} M \longrightarrow 0$$

be a non-split short exact sequence. Suppose M belongs to A_i and suppose N belongs to A_j where $i \neq j$. Then

$$0 \longrightarrow e_iN \longrightarrow e_iX \longrightarrow e_iM \longrightarrow 0$$

and

$$0 \longrightarrow e_jN \longrightarrow e_jX \longrightarrow e_jM \longrightarrow 0$$

are short exact sequences of A-modules. Moreover

$$\mu(e_ix) = e_i\mu(x) = \mu(x)$$

for all $x \in X$ and likewise

$$\nu(y) = \nu(e_jy) = e_j\nu(y)$$

for all $y \in N$. Hence, since $X = e_iX \oplus (1 - e_i)X$, we see that ν is an isomorphism $N \simeq (1 - e_i)X$ and μ is an isomorphism $e_iX \simeq M$. Therefore the sequence

$$0 \longrightarrow N \stackrel{\nu}{\longrightarrow} X \stackrel{\mu}{\longrightarrow} M \longrightarrow 0$$

splits, contrary to our assumption. □

Definition 2.7.2 Let k be a field and let A be a finite dimensional k-algebra. Two simple A-modules S and T are in the same *block-class* if there is a sequence of simple A-

modules $S_1, S_2, \ldots, S_n$ such that $S_1 = S$ and $S_n = T$ and such that $Ext^1_A(S_i, S_{i+1}) \neq 0$ or $Ext^1_A(S_{i+1}, S_i) \neq 0$ for all $i \in \{1, 2, \ldots, n-1\}$.

Remark 2.7.3 Curtis-Reiner [13] use the notion of a "linkage class" for this concept. Lemma 2.7.1 shows that if two simple modules S and T belong to the same block class then S and T belong to the same block.

We come to the main result of this section.

Proposition 2.7.4 *Let k be a field and let A be a finite dimensional k-algebra. Then two simple A-modules S and T belong to the same block of A if and only if they belong to the same block class of A.*

Proof Remark 2.7.3 and Proposition 2.7.1 show that two simple modules in the same block class belong to the same block.

Suppose S and T belong to the same block. We need to show that there is a sequence $S = S_1, S_2, \ldots, S_n = T$, such that $Ext^1_A(S_i, S_{i+1}) \neq 0$ or $Ext^1_A(S_{i+1}, S_i) \neq 0$ for all $i \in \{1, 2, \ldots, n-1\}$. For every simple A-module L we choose its projective cover P_L and obtain $L \simeq P_L/\text{rad}(P_L)$. From the exact sequence

$$0 \longrightarrow \text{rad}(P_L) \longrightarrow P_L \longrightarrow L \longrightarrow 0$$

we obtain an exact sequence

$$Hom_A(L, M) \hookrightarrow Hom_A(P_L, M) \rightarrow Hom_A(\text{rad}(P_L), M) \longrightarrow\!\!\!\!\rightarrow Ext^1_A(L, M)$$

using Definition 1.8.19 and Lemma 1.8.8 since P_L is projective. If M is a simple module non-isomorphic to L, then $Hom_A(P_L, M) = 0$ and we get

$$Hom_A(\text{rad}(P_L), M) \simeq Ext^1_A(L, M).$$

Since M was assumed to be simple,

$$Hom_A(\text{rad}(P_L), M) \simeq Hom_A(\text{rad}(P_L)/\text{rad}^2(P_L), M).$$

But, $\text{rad}(P_L)/\text{rad}^2(P_L)$ is semisimple and we get that $Ext^1_A(L, M) \neq 0$ if and only if M is a direct summand of $\text{rad}(P_L)/\text{rad}^2(P_L)$.

Definition 2.7.5 We associate a quiver to a finite dimensional algebra A, called the *Ext-quiver*, as follows. The vertices are the isomorphism classes $\{L\}$ of simple A-modules L. For every simple A-module L and projective cover P_L decompose

$$\text{rad}(P_L)/\text{rad}^2(P_L) = L_1 \oplus \cdots \oplus L_n$$

and draw arrows $\{L\} \longrightarrow \{L_1\}, \{L\} \longrightarrow \{L_2\}, \ldots, \{L\} \longrightarrow \{L_n\}$.

The statement of Proposition 2.7.4 is that S and T belong to the same block if and only if the vertices $\{S\}$ and $\{T\}$ belong to the same connected component of the *Ext*-quiver Γ_A associated to A. Let Γ_S be the connected component of Γ_A containing $\{S\}$ and let Γ_T be the connected component of Γ_A containing $\{T\}$.

Let

$$P_S := \bigoplus_{L\in\Gamma_S} P_L^{n_L}$$

and

$$P_{\not S} := \bigoplus_{L\notin\Gamma_S} P_L^{n_L}$$

where $P_L^{n_L}$ is a direct factor of the regular A-module, but $P_L^{n_L+1}$ is not a direct factor of the regular module. Suppose

$$\Gamma_T \subseteq \Gamma_{\not S}.$$

Then

$$P_S \oplus P_{\not S} \simeq A$$

and

$$\begin{pmatrix} End_A(P_S) & Hom_A(P_{\not S}, P_S) \\ Hom_A(P_S, P_{\not S}) & End_A(P_{\not S}) \end{pmatrix} = End_A(P_S \oplus P_{\not S}) = End_A(A) = A^{op}.$$

We shall show that the hypothesis implies

$$Hom_A(P_{\not S}, P_S) = 0 = Hom_A(P_S, P_{\not S}).$$

Suppose for a moment we have shown this. Then

$$A = End_A(P_S)^{op} \times End_A(P_{\not S})^{op}$$

is decomposable and S belongs to $End_A(P_S)$ whereas T belongs to $End_A(P_{\not S})$.

Suppose $Hom_A(P_S, P_{\not S}) \neq 0$. Then there is a projective indecomposable P_L, being a direct summand of P_S, and a projective indecomposable P_M, being a direct summand of $P_{\not S}$ such that $Hom_A(P_M, P_L) \neq 0$. This means that M is a direct factor of $\mathrm{rad}^i(P_L)/\mathrm{rad}^{i+1}(P_L)$ for some $i \geq 1$. We may choose L and M so that i is minimal possible. We claim that in this case $i = 1$. Indeed

$$\mathrm{rad}^i(P_L)/\mathrm{rad}^{i+1}(P_L) \hookrightarrow \mathrm{rad}^{i-1}(P_L)/\mathrm{rad}^{i+1}(P_L) \longrightarrow\!\!\!\!\!\rightarrow \mathrm{rad}^{i-1}(P_L)/\mathrm{rad}^i(P_L)$$

is an exact non split sequence and M is a direct factor of $\mathrm{rad}^i(P_L)/\mathrm{rad}^{i+1}(P_L)$. Hence $M \oplus U = \mathrm{rad}^i(P_L)/\mathrm{rad}^{i+1}(P_L)$. We get a short exact sequence

$$0 \to M \to \mathrm{rad}^{i-1}(P_L)/(\mathrm{rad}^{i+1}(P_L) + U) \to \mathrm{rad}^{i-1}(P_L)/\mathrm{rad}^i(P_L) \to 0$$

and claim that this sequence is non-split. If not, M would be direct factor of the quotient $\mathrm{rad}^{i-1}(P_L)/(\mathrm{rad}^{i+1}(P_L)+U)$. But then M would be a direct factor of $\mathrm{rad}^{i-1}(P_L)/\mathrm{rad}^i(P_L)$ since M is simple and since the radical is the intersection of the kernels of all homomorphisms to simple modules. This contradicts the minimality of i. Hence

$$0 \to M \to \mathrm{rad}^{i-1}(P_L)/(\mathrm{rad}^{i+1}(P_L)+U) \to \mathrm{rad}^{i-1}(P_L)/\mathrm{rad}^i(P_L) \to 0$$

is non-split. We decompose

$$\mathrm{rad}^{i-1}(P_L)/\mathrm{rad}^i(P_L) = T_1 \oplus \cdots \oplus T_m$$

into a direct sum of simple modules. Then taking pre-images of T_j inside $\mathrm{rad}^{i-1}(P_L)$ $(\mathrm{rad}^{i+1}(P_L)+U)$ gives exact sequences

$$0 \to M \to V_j \to T_j \to 0$$

for all $j \in \{1, \ldots, m\}$. If all these sequences split by morphisms

$$\sigma_j : T_j \longrightarrow V_j \subseteq \mathrm{rad}^{i-1}(P_L)/(\mathrm{rad}^{i+1}(P_L)+U),$$

then the sequence

$$0 \to M \to \mathrm{rad}^{i-1}(P_L)/(\mathrm{rad}^{i+1}(P_L)+U) \to \mathrm{rad}^{i-1}(P_L)/\mathrm{rad}^i(P_L) \to 0$$

splits by the morphism $\sum_{j=1}^m \sigma_j$. Hence there is a $j_0 \in \{1, \ldots, m\}$ such that

$$0 \to M \to V_j \to T_{j_0} \to 0$$

is non-split. However $\{T_{j_0}\}$ is in Γ_S by minimality of i and $\{M\}$ is in $\Gamma_{\mathcal{S}}$.

The case $Hom_A(P_{\mathcal{S}}, P_S) \neq 0$ gives a non-split exact sequence as above by symmetry. This completes the proof of Proposition 2.7.4. □

Example 2.7.6 In the proof of Proposition 2.7.4 we used a very important concept, the concept of an *Ext*-quiver of an algebra. We shall give some examples.

1. Let $A = kG$ where k is a field of characteristic $p > 0$ and G is a finite p-group. Then the *Ext*-quiver of A is a single vertex with n loops, where n is the p-rank of the Frattini quotient $G/\Phi(G)$ of G. The Frattini quotient of G is the largest elementary abelian quotient of the p-group G. Equivalently, the preimages of the generators of the Frattini quotient form a minimal generating set of G.
2. The *Ext*-quiver of the algebra of upper triangular $n \times n$ matrices over k is

$$\{1\} \longrightarrow \{2\} \longrightarrow \ldots \longrightarrow \{n\}.$$

We see that in the second case the path algebra given by this *Ext*-quiver is isomorphic to the algebra of upper triangular matrices. This is not an accident, as we shall see later. For a precise statement we need the concept of a Morita equivalence given in Chap. 4.

2.8 The Structure of Serial Symmetric Algebras

There are numerous applications of the results of Sect. 2.7. We shall give one application here, which will be of some importance later, but which is also interesting in its own right.

Recall from Remark 1.6.31 that an A-module is uniserial if it has only one composition series and that a ring is serial if each indecomposable projective A-module is uniserial.

We shall study the special case of a symmetric serial algebra. This case will be useful later, but it also gives an example of the power of the structure we have already obtain at this stage. We shall follow here Linckelmann's thesis [14].

Lemma 2.8.1 *Let A be a finite dimensional k-algebra. Then every projective indecomposable A-module is uniserial if and only if every indecomposable module is uniserial. Moreover, each indecomposable A-module M is uniserial if and only if for each indecomposable A-module M we have* $\mathrm{rad}^i(M)/\mathrm{rad}^{i+1}(M)$ *is either simple or* 0*, for all i.*

Proof Let M be an indecomposable A-module. We claim that each of the modules $\mathrm{rad}^i(M)/\mathrm{rad}^{i+1}(M)$ is simple or 0 for each i. Indeed, let P_M be its projective cover. If P_M is indecomposable, then we are done. Indeed, since M is a quotient of P_M, if M admits two different decomposition series, P_M also admits two different decomposition series. But, if $\mathrm{rad}^i(M)/\mathrm{rad}^{i+1}(M) = S \oplus T$ for a simple S and a non-zero semisimple module T, and i is minimal with this property, then

$$M \supseteq \mathrm{rad}(M) \supseteq \cdots \supseteq \mathrm{rad}^i(M) \supseteq \mathrm{rad}^{i+1}(M) + S \supseteq \mathrm{rad}^{i+1}(M) \supseteq \ldots$$

and

$$M \supseteq \mathrm{rad}(M) \supseteq \cdots \supseteq \mathrm{rad}^i(M) \supseteq \mathrm{rad}^{i+1}(M) + T \supseteq \mathrm{rad}^{i+1}(M) \supseteq \ldots$$

can be completed to two different composition series, refining T into a direct sum of simple modules. But this shows that the radical series of M is the only composition series of M.

So, assuming P_M is decomposable, we proceed by induction on the number of indecomposable direct summands. Let $P_M = P_0 \oplus P_1$ with P_0 indecomposable. Let

$$\begin{array}{ccc} P_M & \xrightarrow{\pi_M} & M \\ \downarrow \alpha & & \downarrow \beta \\ P_0 & \xrightarrow{\pi_0} & M_0 \end{array}$$

be a pushout diagram, where α is the natural projection. Since α and π_M are epimorphisms, β and π_0 are also epimorphisms (cf Lemma 1.8.27). Since P_0 is projective, π_0 lifts to a morphism $\sigma : P_0 \longrightarrow M$ such that $\beta \circ \sigma = \pi_0$. Since π_M is an epimorphism there is a $\tau : P_0 \longrightarrow P_M$ with $\pi_M \circ \tau = \sigma$.

But P_0 is uniserial and therefore $\ker(\pi_0) = \mathrm{rad}^{n_0}(P_0)$ for some $n_0 \in \mathbb{N}$. We may choose P_0 so that n_0 is maximal. Hence τ induces a homomorphism $M_0 \xrightarrow{\check{\gamma}} P_M/\mathrm{rad}^{n_0} P_M$. Since n_0 is maximal there is an epimorphism $P_M/\mathrm{rad}^{n_0} P_M \longrightarrow M$ so that τ induces $\gamma : M_0 \longrightarrow M$ with $\gamma \circ \pi_0 = \pi_M \circ \tau$.

But this shows

$$\beta \circ \gamma \circ \pi_0 = \beta \circ \pi_M \circ \tau = \beta \circ \sigma = \pi_0.$$

Now, π_0 is an epimorphism, and therefore $\beta \circ \gamma = id_{M_0}$. Therefore $M \simeq M_0 \oplus M'$ where $M' = \ker \beta$. We have proved the statement since by the induction hypothesis, M' is a direct sum of uniserial modules. □

Lemma 2.8.2 *Let k be a field and let A be a finite dimensional, indecomposable, self-injective k-algebra. Let $S_1, \ldots, S_n$ be representatives of the isomorphism classes of simple A-modules, and let P_i be the projective cover of S_i for each i. Then the following are equivalent.*

1. *A is serial.*
2. *Each indecomposable A-module is uniserial.*
3. *For each indecomposable A-module M the module $\mathrm{rad}^i(M)/\mathrm{rad}^{i+1}(M)$ is either simple or 0, for all i.*
4. *There is an element σ of the symmetric group $\mathfrak{S}_n$ such that*

$$\mathrm{rad}(P_i)/\mathrm{rad}^2(P_i) \simeq S_{\sigma(i)}.$$

In this case, and if A is symmetric, σ is a transitive cycle of length n; i.e. we can renumber the simple modules so that $\sigma = (1\ 2\ \ldots\ n)$. Moreover, the Loewy structure of the projective indecomposable A-modules is the Loewy structure of the Nakayama algebra $N_n^{e\cdot n+1}$ over K with n simple modules and

$$\mathrm{rad}^{e\cdot n+1}(A) = 0 \neq \mathrm{rad}^{e\cdot n}(A).$$

Proof The equivalence of the first three items has already been shown in Lemma 2.8.1. The fact that 4 implies 3 is trivial. Suppose now 3. Then $\mathrm{rad}(P_i)/\mathrm{rad}^2(P_i)$ is simple and we define σ by the property

$$\mathrm{rad}(P_i)/\mathrm{rad}^2(P_i) \simeq S_{\sigma(i)}.$$

Then $P_{\sigma(i)}$ is the projective cover of $\mathrm{rad}(P_i)/\mathrm{rad}^2(P_i)$, and therefore also of $\mathrm{rad}(P_i)$. But this implies that $\mathrm{rad}(P_{\sigma(i)})$ maps surjectively to $\mathrm{rad}(\mathrm{rad}(P_i)) = \mathrm{rad}^2(P_i)$ and therefore $P_{\sigma^2(i)}$ is the projective cover of $\mathrm{rad}^2(P_i)$. Recursively, $P_{\sigma^n(i)}$ is the projective cover of $\mathrm{rad}^n(P_i)$. Since A is self-injective, using Remark 1.10.32, there is an n such that $\mathrm{rad}^n(P_i) = \mathrm{soc}(P_i) = S_{\nu(i)}$, where ν is the Nakayama permutation. Since $\nu \in \mathfrak{S}_n$, also $\sigma \in \mathfrak{S}_n$.

Suppose that A is symmetric. Hence $\nu = id$. The composition factors of P_i are therefore formed by the orbit of i under σ. Furthermore, the composition factors of $P_{\sigma^j(i)}$ are in the same σ-orbit of i. Now, Proposition 2.7.4 implies that A is indecomposable if and only if σ is a cycle of length n in $\mathfrak{S}$.

We can now see that the Loewy structure of A and of the projective indecomposable A-modules is actually the Loewy structure of the symmetric Nakayama algebra $N_n^{e\cdot n+1}$ and its projective indecomposable modules. The only remaining possibility is that the radical length of P_i is not $e \cdot n+1$ where e is independent from i. Assuming this is the case, then suppose the radical length of P_i is $n \cdot f + 1$ and the radical length of $P_{\sigma^{-1}(i)}$ is $n \cdot e+1$ for $e > f$. Then P_i is a submodule of $P_{\sigma^{-1}(i)}/\mathrm{soc}(P_{\sigma^{-1}(i)})$. Since A is self-injective, this would imply that $P_{\sigma^{-1}(i)}$ is not indecomposable, a contradiction. □

Remark 2.8.3 We have seen that all indecomposable modules are quotients of projective indecomposable modules. Moreover, this property also holds for N_n^{en+1}. Hence, we get a bijection between the indecomposable modules of N_n^{en+1} and the indecomposable A-modules. This bijection preserves homomorphism spaces in the sense that $Hom_A(M_1, M_2) \simeq Hom_{N_n^{en+1}}(\beta M_1, \beta M_2)$ and the isomorphism is compatible with composition of morphisms. This is a particular case of a correspondence studied in Chap. 4, namely a Morita equivalence. In other words, A is Morita equivalent to N_n^{en+1} in the notation which will be introduced there.

Theorem 2.8.4 (cf e.g. Linckelmann [14, Théorème 2.1]) *Let k be a field and let A be a finite dimensional symmetric k-algebra. Then the following statements are equivalent.*

1. *There is a $t \in A$ such that* $\mathrm{rad}(A) = A \cdot t$ *or* $\mathrm{rad}(A) = t \cdot A$.
2. *The A-modules* $A/\mathrm{rad}(A)$ *and* $\mathrm{rad}(A)/\mathrm{rad}^2(A)$ *are isomorphic.*
3. *A is serial and all indecomposable projective A-modules appear with the same multiplicity as direct factors of A.*

If one of the conditions in 1 or 2 or 3 is satisfied, we obtain

$$t \cdot A = \mathrm{rad}(A) = A \cdot t.$$

Proof Let $\langle\ ,\ \rangle : A \times A \longrightarrow k$ be a symmetrising form on A, and let $M^\perp := \{a \in A \mid \langle m, a\rangle = 0\ \forall m \in M\}$ for each subset M of A. Suppose that $\mathrm{rad}(A) = A \cdot t$ and fix $a \in A$. If $t \cdot a = 0$, then

$$0 = A \cdot t \cdot a = \mathrm{rad}(A) \cdot a$$

and since $\langle m, a\rangle = \langle ma, 1\rangle$ we get $a \in \mathrm{rad}(A)^{\perp}$. Since A is symmetric, $\langle a, b\rangle = \langle b, a\rangle$ for all $b \in \mathrm{rad}(A)$. Moreover, $\mathrm{rad}(A)$ is a two-sided ideal and hence $t \cdot A \subseteq \mathrm{rad}(A) = A \cdot t$. Hence

$$\begin{aligned} \langle m, a\rangle = 0 \ \forall m \in \mathrm{rad}(A) &\Leftrightarrow 0 = \langle a, m\rangle \ \forall m \in \mathrm{rad}(A) \\ &\Rightarrow 0 = \langle a, tb\rangle = \langle at, b\rangle \ \forall b \in A, \end{aligned}$$

and this is possible only if $a \cdot t = 0$ since $\langle\ ,\ \rangle$ is non-degenerate. We obtain that the map

$$\begin{aligned} A \cdot t &\longrightarrow t \cdot A \\ a \cdot t &\mapsto t \cdot a \end{aligned}$$

is well-defined and injective. This gives $\dim_k(t \cdot A) \geq \dim_k(A \cdot t)$. Since $t \cdot A \subseteq \mathrm{rad}(A) = A \cdot t$ we get

$$t \cdot A = A \cdot t = \mathrm{rad}(A).$$

This immediately shows that

$$\mathrm{rad}^s(A) = A \cdot t^s = t^s \cdot A.$$

Hence condition 1 implies that $\mathrm{rad}(A) = A \cdot t = t \cdot A$.

We shall show that condition 1 implies that multiplication by t from the right gives an isomorphism $A/\mathrm{rad}(A) \simeq \mathrm{rad}(A)/\mathrm{rad}^2(A)$. Indeed, since $A \cdot t^2 = \mathrm{rad}^2(A)$, the mapping

$$\begin{aligned} A/\mathrm{rad}(A) &\xrightarrow{\tau} \mathrm{rad}(A)/\mathrm{rad}^2(A) \\ a + \mathrm{rad}(A) &\mapsto a \cdot t + \mathrm{rad}^2(A) \end{aligned}$$

is a well-defined A-module homomorphism. Moreover, $a \cdot t \in \mathrm{rad}^2(A) = A \cdot t^2$ implies $a \in A \cdot t = \mathrm{rad}(A)$, whence τ is injective. Since $\mathrm{rad}(A) = A \cdot t$, it is clear that τ is surjective as well. This proves condition 2.

Assume that there is an A-module isomorphism

$$A/\mathrm{rad}(A) \xrightarrow{\tau} \mathrm{rad}(A)/\mathrm{rad}^2(A).$$

Let $S_1, S_2, \ldots, S_n$ be representatives of the isomorphism classes of the simple A-modules, and let P_i be the projective cover of S_i for each i. Iterating τ shows that τ^n gives an isomorphism of A-modules

$$A/\mathrm{rad}(A) \xrightarrow{\tau^n} \mathrm{rad}^n(A)/\mathrm{rad}^{1+n}(A).$$

Composing τ with the natural projection $A \longrightarrow A/\text{rad}(A)$ we get a morphism $A \longrightarrow \text{rad}(A)/\text{rad}^2(A) \subseteq A/\text{rad}^2(A)$. Since A maps surjectively to $A/\text{rad}^2(A)$, the fact that A is projective gives the existence of a morphism $\hat{\tau} : A \longrightarrow A$ fitting into a commutative diagram

$$\begin{array}{ccc} A & \xrightarrow{\hat{\tau}} & A \\ \downarrow & & \downarrow \\ A/\text{rad}(A) & \xrightarrow{\tau} & A/\text{rad}^2(A) \end{array} .$$

But, since τ has image $\text{rad}(A)/\text{rad}^2(A)$, we have $\hat{\tau}(A) = \text{rad}(A)$. Therefore we obtain a commutative diagram

$$\begin{array}{ccc} A & \overset{\hat{\tau}}{\twoheadrightarrow} & \text{rad}(A) \\ \downarrow & & \downarrow \\ A/\text{rad}(A) & \overset{\tau}{\twoheadrightarrow} & \text{rad}(A)/\text{rad}^2(A) \end{array} .$$

Since $End_A(A) = A^{op}$, we obtain that τ is multiplication by t on the right. We restrict to an indecomposable projective direct summand $P_i = Ae_i$ of A, for some $e_i^2 = e_i \in A$. Then

$$e_i^2 \cdot t = e_i \cdot t = t \cdot f = t \cdot f^2$$

for some $f \in A$. Hence $f - f^2$ is annihilated by t. By the lifting of idempotents theorem Proposition 1.9.17 we may choose f so that $f^2 = f$. Moreover, using right modules instead, the symmetry of the situation implies that f is a primitive idempotent $e_{\sigma(i)}$. We obtain that each simple A-module S_j occurs as a direct factor of some $P_{\sigma^{-1}(j)}/\text{rad}(P_{\sigma^{-1}(j)})$ with multiplicity 1. Hence there is a permutation $\sigma \in \mathfrak{S}_n$ so that

$$P_i/\text{rad}(P_i) \simeq S_{\sigma(i)}.$$

Lemma 2.8.2 shows that A is serial. Now, let

$$A/\text{rad}(A) = \bigoplus_{i=1}^{n} S_i^{n_i}.$$

Then multiplication by t maps S_i to $S_{\sigma(i)}$ and hence

$$\text{rad}(A)/\text{rad}^2(A) = \bigoplus_{i=1}^{n} S_{\sigma(i)}^{n_i} \simeq \bigoplus_{i=1}^{n} S_i^{n_i} \simeq A/\text{rad}(A).$$

Since σ is a cycle of length n, $n_i = n_j$ for all i, j and therefore all simple A-modules occur with the same multiplicity in $A/\text{rad}(A)$. We have shown that condition 2 implies condition 3.

Now assume condition 3 Lemma 2.8.2 shows that each simple A-module occurs with the same multiplicity in $A/\mathrm{rad}(A)$ as in $\mathrm{rad}(A)/\mathrm{rad}^2(A)$. This gives an isomorphism $A/\mathrm{rad}(A) \simeq \mathrm{rad}(A)/\mathrm{rad}^2(A)$. By the proof of 2 implies 3 this shows that $\mathrm{rad}(A) = A \cdot t$ and we obtain condition 1. This proves the theorem. □

2.9 Külshammer Ideals

Let G be a finite group and let k be a field of characteristic $p > 0$. Theorem 1.5.4 is a simple formula for the number of simple KG modules when K is a splitting field for G of characteristic 0. A similar formula is desirable for the case when the field k is of characteristic $p > 0$ and is a splitting field for G.

2.9.1 Definitions of Külshammer Spaces

The route we take follows the original considerations of Brauer and the refinement proposed by Külshammer.

We start with the following simple observation. In Sect. 2.9.1 and occasionally in the subsequent sections we denote our base field by K, since we need k for our index sets. Index considerations occur frequently in this section.

Definition 2.9.1 For any field K and any K-algebra A we define the *commutator subspace* $[A, A]$ to be the K-subspace generated by the elements $\{ab - ba \mid a, b \in A\}$. We define $HH_0(A) := A/[A, A]$, the commutator quotient.

We mention that the set of commutators $\{ab - ba \mid a, b \in A\}$ is not in general a vector space. Moreover, a very important remark is that $[A, A]$ is *not* an ideal of A. Hence $A/[A, A]$ is not a ring, and is *not* an A-module.

Lemma 2.9.2 *Let K be a field.*

- *Let $A = Mat_{n\times n}(K)$ be the matrix ring over K. Then $[A, A]$ is of codimension 1 in A.*
- *For all K-algebras A_1 and A_2 we get*

$$[A_1 \times A_2, A_1 \times A_2] = [A_1, A_1] \times [A_2, A_2]$$

  and hence

$$HH_0(A_1 \times A_2) = HH_0(A_1) \times HH_0(A_2).$$

Proof For the first statement we use the usual matrix trace $trace : Mat_{n\times n}(K) \to K$ which is easily seen to be K-linear and observe that

$$\begin{aligned} trace(MN - NM) &= trace(MN) - trace(NM) \\ &= trace(MN) - trace(MN) = 0. \end{aligned}$$

Hence

$$[A, A] \subseteq \{M \in Mat_{n\times n}(K) \mid trace(M) = 0\} = \ker(trace)$$

which is of codimension 1 in A. Now, let $E_{i,j} = (e_{k,\ell})_{1\le k,\ell\le n}$ be the matrix with $e_{k,\ell} = \delta_{i,k}\delta_{j,\ell}$; where as usual we denote by $\delta_{x,y}$ the Kronecker delta which is 1 if $x = y$ and 0 otherwise. Then if $i \neq j$ we have that $E_{i,j}E_{j,i} - E_{j,i}E_{i,j}$ is the matrix with diagonal entry 1 in position (i, i), diagonal entry -1 in position (j, j) and 0 elsewhere. Moreover $E_{k,j}E_{j,i} - E_{j,i}E_{k,j}$ for $k \neq i$ is the matrix with coefficient 1 in position (k, i) and 0 elsewhere. It is obvious that these matrices span the set of matrices with trace 0. Since all these matrices are commutators we have proved the first statement.

For the second statement we just observe that elements in A_1 commute with elements in A_2. This proves the lemma. □

Recall from Definition 1.9.17 the notion of a free algebra.

Lemma 2.9.3 *Let K be a field of characteristic $p > 0$ and let A be a K-algebra. Then the $\mathbb{F}_p$-linear mapping*

$$\begin{aligned} A &\xrightarrow{\mu_p} A \\ a &\mapsto a^p \end{aligned}$$

extends to an $\mathbb{F}_p$-linear mapping

$$\begin{aligned} HH_0(A) &\xrightarrow{\mu_p} HH_0(A) \\ a + [A, A] &\mapsto a^p + [A, A]. \end{aligned}$$

Proof We need to show that if $a \in A$ and $b \in [A, A]$, then

$$(a + b)^p - a^p \in [A, A].$$

In particular we need to show that $b^p \in [A, A]$. We consider the element $(x + y)^p$ in the free K-algebra F_2 in two variables x and y. The cyclic group $C_p = \langle c\rangle$ acts on the set of products of p factors by cyclic permutation of the factors. Now, since p is prime, orbits are of length 1 or of length p. Moreover, $z - cz \in [F_2, F_2]$ since if $z = z_1z_2 \dots z_p$, then $cz = z_2 \dots z_pz_1$ and

$$z - cz = z_1z_2 \dots z_p - z_2 \dots z_pz_1 = [z_1, z_2 \dots z_p] \in [F_2, F_2].$$

Since the only orbits of length 1 are those of x^p or y^p the above implies that

$$(x+y)^p - x^p - y^p \in [F_2, F_2].$$

Therefore for all $a, b \in A$ we get

$$(ab - ba)^p - (ab)^p - (ba)^p \in [A, A]$$

and hence $\mu_p([A, A]) \subseteq [A, A]$. Moreover if $a \in A$ and $b \in [A, A]$, then

$$(a+b)^p - a^p - b^p \in [A, A]$$

and since $b^p \in [A, A]$ we obtain the statement. □

Definition 2.9.4 Let K be a field of characteristic $p > 0$ and let A be a K-algebra. Then we define the *Külshammer spaces*

$$T_n(A) := \{a \in A \mid a^{p^n} \in [A, A]\}$$

and

$$T(A) := \bigcup_{n=1}^{\infty} T_n(A).$$

By Lemma 2.9.3 we get that

$$[A, A] \subseteq T_n(A).$$

Observe that $T_n(A)$ is a K-subspace of A. Of course $T_0(A) = [A, A]$ and $T_n(A) \subseteq T_{n+1}(A)$ for all $n \in \mathbb{N}$. Moreover we recall from Proposition 1.6.18 that for artinian algebras the Jacobson radical is the largest nilpotent ideal of A and hence we get for an artinian algebra A that

$$[A, A] + \text{rad}(A) \subseteq T(A).$$

Proposition 2.9.5 *Let K be a field of characteristic $p > 0$ and let A be an artinian K-algebra. Then $T_n(A)$ is a K-subspace of A and*

$$[A, A] = T_0(A) \subseteq T_1(A) \subseteq T_2(A) \subseteq \ldots T(A) = [A, A] + \text{rad}(A)$$

is an increasing sequence of K-subspaces of A. In particular

$$T(A/\text{rad}(A)) = [A/\text{rad}(A), A/\text{rad}(A)].$$

Proof Almost all of the statements have been proved in the discussion preceding the proposition. The only missing part is the statement that $T(A) = [A, A] + \text{rad}(A)$. In order to prove this, we first observe that

$$T(A/\text{rad}(A)) = T(A)/\text{rad}(A).$$

Indeed, the ring homomorphism

$$\rho : A \longrightarrow A/\mathrm{rad}(A)$$

is compatible with the p-power mapping μ_p. Hence

$$\mu_p(a + \mathrm{rad}(A)) = \mu_p(a) + \mathrm{rad}(A).$$

Moreover ρ is a ring homomorphism and hence

$$\rho([A, A]) = [A/\mathrm{rad}(A), A/\mathrm{rad}(A)].$$

This gives

$$a^{p^n} \in [A, A] \Rightarrow (a + [A, A])^{p^n} \in [A/\mathrm{rad}(A), A/\mathrm{rad}(A)].$$

Therefore $\rho(T_n(A)) \subseteq T_n(A/\mathrm{rad}(A))$ for all $n \geq 0$, which implies

$$T(A)/\mathrm{rad}(A) \subseteq T(A/\mathrm{rad}(A))$$

since $\mathrm{rad}(A) \subseteq T(A)$. Given $\rho(a) \in T(A/\mathrm{rad}(A))$, there is an $n \in \mathbb{N}$ such that

$$\rho(a^{p^n}) = \rho(a)^{p^n} \in \rho([A, A]) = [A/\mathrm{rad}(A), A/\mathrm{rad}(A)]$$

and therefore

$$a^{p^n} \in [A, A] + \ker(\rho) = [A, A] + \mathrm{rad}(A) \subseteq T(A).$$

Since $\mathrm{rad}(A) \subseteq T(A)$, we have $a^{p^n} \in T(A)$, which implies $a \in T(A)$. □

2.9.2 The Number of Simple Modules

Proposition 2.9.6 *Let K be a field and let A be an artinian K-algebra such that K is a splitting field for A. Then $\dim_K(A/T(A))$ is the number of isomorphism classes of simple A-modules.*

Proof Wedderburn's theorem shows that $A/\mathrm{rad}(A)$ is isomorphic to a direct product of s matrix algebras over K. The number of simple A-modules equals s. Proposition 2.9.5 and Lemma 2.9.2 show that

$$\dim_K((A/\mathrm{rad}(A))/(T(A/\mathrm{rad}(A))) = s.$$

Moreover,

$$A/T(A) = A/([A, A] + \mathrm{rad}(A)) = (A/\mathrm{rad}(A))/(T(A/\mathrm{rad}(A))).$$

This proves the statement. □

Theorem 2.9.7 (Brauer) *Let k be a splitting field for the group G and suppose that the characteristic of k is $p > 0$. Then the number of isomorphism classes of simple kG-modules is equal to the number of conjugacy classes of G of elements $g \in G$ such that the order of g is not divisible by p.*

Proof We need to show that $kG/T(kG)$ has a k-basis given by representatives of conjugacy classes of elements of G which have order relatively prime to p.

Of course, G is a generating set of $kG/T(kG)$ as a vector space since it generates kG. If $g, h \in G$ then $hgh^{-1} - g = [h, gh^{-1}]$ so that two conjugate elements in G give the same element in $kG/T(kG)$. Moreover, let $g = g_p \cdot g_q \in G$ for g_p of prime power order p^m and g_q relatively prime to p. Further g_p and g_q commute. Indeed, the subgroup $\langle g\rangle$ of G generated by g is cyclic, hence abelian, and therefore the classification of finitely generated abelian groups gives the statement. By Lemma 2.9.3

$$\begin{aligned}(g - g_q)^{p^m} - g^{p^m} + g_q^{p^m} &= (g - g_q)^{p^m} - g_p^{p^m} g_q^{p^m} + g_q^{p^m}\\ &= (g - g_q)^{p^m} - g_q^{p^m} + g_q^{p^m}\\ &= (g - g_q)^{p^m} \in [kG, kG]\end{aligned}$$

and so

$$g \equiv g_q \mod T(kG).$$

Therefore, we get that a set of representatives of conjugacy classes of elements of G which have order relatively prime to p is a generating set of $kG/T(kG)$.

We need to prove that this set is linearly independent. Let C be the set of conjugacy classes g^G of G and let

$$Q := \{g^G \in C \mid \text{the order of } g \text{ is not divisible by } p\}$$

and let Q_e be a set of representatives of elements $g \in G$ with $g^G \in Q$, i.e. for all $g_1, g_2 \in Q_e$ we have $g_1^G = g_2^G \Rightarrow g_1 = g_2$ and $\{g^G \mid g \in Q_e\} = Q$.

Let $\alpha = \sum_{g\in Q_e} \alpha_g g \in T(kG)$ for coefficients $\alpha_g \in k$. Since the order of each $g \in Q_e$ is not divisible by p, there is an $m \in \mathbb{N}$ such that $g^{p^m} = g$ for all $g \in Q_e$. We may even choose m so big that $\alpha^{p^m} \in [kG, kG]$. But

$$\alpha^{p^m} = \left(\sum_{g\in Q_e} \alpha_g g\right)^{p^m} \equiv \left(\sum_{g\in Q_e} (\alpha_g)^{p^m} g\right) \mod [kG, kG].$$

Moreover $[kG, kG]$ is contained in the set of elements $\sum_{g\in G} \beta_g g$ such that $\beta_g = \beta_h$ whenever g is conjugate to h in G. Since we were picking one conjugate in each conjugacy class in Q_e we get that $\alpha_g = 0$ for all $g \in Q_e$. This proves the statement. □

Corollary 2.9.8 *Let k be a field of characteristic $p > 0$ and let G be a finite p-group. Then the trivial module is the only simple kG-module, kG is a local algebra and*

$$I(kG) := \ker(kG \longrightarrow k) = \langle g - 1 \mid g \in G \rangle_{k\text{- vector space}} = \mathrm{rad}(kG).$$

Proof Let $\overline{k}$ be an algebraic closure of k. Then by Theorem 2.9.7 we know that the trivial module is the only simple $\overline{k}G$-module, using that all elements of G except the neutral element have order divisible by p. But then k is actually already a splitting field since the endomorphism ring of the trivial module is k. Hence there is only one simple kG-module, the trivial module, which is of multiplicity 1 since its dimension is 1. Since the k-vector space generated by the elements $g - 1$, $g \in G$, is of codimension 1, and since these elements are obviously in the kernel of $kG \longrightarrow k$, we get the generating set of $I(kG)$ as claimed and so we obtain the statement. □

Remark 2.9.9 We have seen this fact already by more elementary methods in Proposition 1.6.22.

2.9.3 Külshammer Ideals of Symmetric Algebras

In this subsection we again follow Külshammer [15–18].

Let K be a field of characteristic $p > 0$ and let A be a symmetric K-algebra. Then there is a non-degenerate symmetric associative bilinear form

$$\langle\ ,\ \rangle : A \times A \longrightarrow K.$$

We shall consider for all subsets U of A orthogonal spaces $U^\perp$ with respect to this symmetrising form, i.e.

$$U^\perp := \{a \in A \mid \langle a, u \rangle = 0\ \forall u \in U\}.$$

Lemma 2.9.10 *Let K be a field of characteristic $p > 0$ and let A be a finite dimensional symmetric K-algebra. Denote by $Z(A)$ the centre of A and by $[A, A]$ the commutator space of A. Then*

$$[A, A]^\perp = Z(A) \text{ and } \mathrm{rad}(A)^\perp = \mathrm{soc}(A).$$

Therefore the symmetrising form $\langle\ ,\ \rangle : A \times A \longrightarrow K$ induces a non-degenerate bilinear form

$$\langle\ ,\ \rangle : Z(A) \times A/[A, A] \longrightarrow K.$$

Proof Let $a, b, c \in A$. Then

$$\begin{aligned}\langle a, bc - cb\rangle &= \langle a, bc\rangle - \langle a, cb\rangle = \langle ab, c\rangle - \langle cb, a\rangle \\ &= \langle c, ab\rangle - \langle c, ba\rangle = \langle c, ab - ba\rangle\end{aligned}$$

and therefore

$$\begin{aligned}a \in [A, A]^{\perp} &\Leftrightarrow \forall b, c \in A : \langle a, bc - cb\rangle = 0 \Leftrightarrow \forall b \in A : (ba - ab) \in A^{\perp} = 0 \\ &\Leftrightarrow a \in Z(A).\end{aligned}$$

Let I be a left ideal of A. We claim that $I^{\perp}$ is a right ideal. Let $a \in A$ and $b \in I^{\perp}$. Then, since I is an ideal,

$$\langle ba, I\rangle = \langle b, aI\rangle \subseteq \langle b, I\rangle = 0$$

and $ba \in I^{\perp}$. Likewise, since $\langle\ ,\ \rangle$ is symmetric, if I is a right ideal, then $I^{\perp}$ is a left ideal. Let J be a subset of A. Then

$$\langle bJ, \mathrm{rad}(A)\rangle = \langle b, J \cdot \mathrm{rad}(A)\rangle$$

for all $b \in A$ and so $\mathrm{rad}(A)^{\perp} \cdot \mathrm{rad}(A) \subseteq A^{\perp} = 0$ since $\mathrm{rad}(A)^{\perp}$ is a two-sided ideal of A. This shows that $\mathrm{rad}(A)^{\perp}$ consists of the elements in A which annihilate $\mathrm{rad}(A)$. Now, the elements of A which annihilate $\mathrm{rad}(A)$ are precisely the elements in the socle of A. This proves the statement. □

Proposition 2.9.11 *Let K be a field of characteristic $p > 0$ and let A be a finite dimensional symmetric K-algebra. Then $T_n(A)^{\perp}$ is an ideal of the centre $Z(A)$ of A and $T(A)^{\perp} = \mathrm{soc}(A) \cap Z(A)$.*

Proof Since by Proposition 2.9.5 we have $T(A) = [A, A] + \mathrm{rad}(A)$, we get

$$T(A)^{\perp} = ([A, A] + \mathrm{rad}(A))^{\perp} = [A, A]^{\perp} \cap \mathrm{rad}(A)^{\perp} = Z(A) \cap \mathrm{soc}(A).$$

For the next statement we observe that $T_n(A)$ is a $Z(A)$-module. Indeed, for all $z \in Z(A)$ and for all $a, b \in A$ we compute

$$z \cdot (ab - ba) = zab - zba = (za)b - b(za) \in [A, A]$$

and hence $[A, A]$ is a $Z(A)$-module. Moreover, using Lemma 2.9.10, we get

$$x \in T_n(A) \Rightarrow x^{p^n} \in [A, A] \Rightarrow (zx)^{p^n} = z^{p^n} x^{p^n} \in z^{p^n}[A, A] \subseteq [A, A]$$

and hence $T_n(A)$ is a $Z(A)$-module. Moreover, if $z \in Z(A)$ and $b \in T_n(A)^{\perp}$, we compute

$$\langle bz, T_n(A)\rangle = \langle b, zT_n(A)\rangle = 0 \text{ since } \langle b, T_n(A)\rangle = 0$$

since $T_n(A)$ is a $Z(A)$-module. Hence $T_n(A)^{\perp}$ is an ideal in $Z(A)$. □

Reynolds defined an ideal of group algebras via a property similar to the one used in Brauer's Theorem 1.5.4. He obtained that the so-defined ideal equals $\mathrm{soc}(A) \cap Z(A)$ where $A = kG$ is a group algebra of a finite group G over an algebraically closed field k of characteristic $p > 0$.

Definition 2.9.12 Let A be a finite dimensional K-algebra over a field K. Then $R(A) := Z(A) \cap \mathrm{soc}(A)$ is the *Reynolds ideal* of A. If K has characteristic $p > 0$ then $T_n(A)^\perp$ is the n-th *Külshammer ideal* of A.

Recall that the statement of Proposition 2.9.11 means that for finite dimensional symmetric algebras A over a perfect field K of characteristic $p > 0$ we get a sequence of ideals

$$R(A) = T(A)^\perp = \bigcap_{n=0}^{\infty} T_n(A)^\perp \subseteq \cdots \subseteq T_2(A)^\perp \subseteq T_1(A)^\perp \subseteq T_0(A)^\perp = Z(A)$$

of the centre of A. The centre hence carries a very nice additional ideal structure coming from the fact that $Z(A)$ is the centre of an algebra A. Observe that the same commutative algebra may be the centre of various different algebras.

Example 2.9.13 Let K be a field of characteristic 2, fix $c \in K$ and let $A(c)$ be the algebra given by the quiver

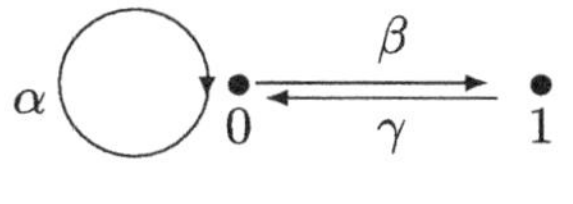

modulo the relations

$$\gamma\beta = 0, \alpha\beta\gamma = \beta\gamma\alpha, \alpha^2 = c \cdot \alpha\beta\gamma.$$

This algebra occurs in Erdmann's classification of blocks with dihedral defect groups and is called $\mathcal{D}(2A)^1(c)$ in the notation used there. The algebra actually occurs as a principal block for a certain group ring.

If $c \neq 0$, we may replace β by $c \cdot \beta$ and observe that $A(c) \simeq A(1)$ in this case. Hence we may consider the two cases $c \in \{0, 1\}$. The two projective indecomposable modules have radical length 4 independently of the value of c. We get

$$\begin{aligned} P_0/\mathrm{rad}(P_0) &= S_0; & P_1/\mathrm{rad}(P_1) &= S_1;\\ \mathrm{rad}(P_0)/\mathrm{rad}^2(P_0) &= S_0 \oplus S_1; & \mathrm{rad}(P_1)/\mathrm{rad}^2(P_1) &= S_0;\\ \mathrm{rad}^2(P_0)/\mathrm{rad}^3(P_0) &= S_0 \oplus S_1; & \mathrm{rad}^2(P_1)/\mathrm{rad}^3(P_1) &= S_0;\\ \mathrm{rad}^3(P_0) &= S_0; & \mathrm{rad}^3(P_1) &= S_1. \end{aligned}$$

The algebra has dimension 10. Every element in the centre is a linear combination of closed paths since we may multiply by the idempotents e_0 and e_1 from the left and the right. More precisely if z is in the centre $e_i z = z e_i$ and this implies that z is a linear combination of paths, and the paths starting at i also end at i. The closed paths of A are

$$e_0, \alpha, \beta\gamma, \alpha\beta\gamma, e_1, \gamma\alpha\beta.$$

Every element in the socle is in the centre since multiplying with any arrow gives 0, and multiplying by idempotents from the left and from the right gives the same result since the socle has a basis given by the closed paths $\alpha\beta\gamma$ and $\gamma\alpha\beta$. Hence we need to consider central elements as a linear combination of

$$e_0, \alpha, \beta\gamma, e_1.$$

Since a central element must commute with β we get that the coefficient of e_0 and of e_1 has to be equal. On the other hand, $e_0+e_1 = 1$ is central. Can we find coefficients $d_0, d_1 \in K$ so that

$$d_0 \cdot \alpha + d_1 \cdot \beta\gamma \in Z(A(c))?$$

We have that

$$\gamma \cdot (d_0 \cdot \alpha + d_1 \cdot \beta\gamma) = d_0\gamma\alpha \stackrel{!}{=} (d_0 \cdot \alpha + d_1 \cdot \beta\gamma) \cdot \gamma = 0$$

implies that $d_0 = 0$. However, $\beta\gamma \in Z(A(c))$ since it commutes with α, with e_0 and multiplication with any other arrow gives 0. We have proved

$$\begin{aligned} Z(A(c)) &= K \cdot 1 + K \cdot \alpha\beta\gamma + K \cdot \gamma\alpha\beta + K \cdot \beta\gamma \\ &\simeq K[X, Y, Z]/(XY, XZ, YZ, X^2, Y^2, Z^2) \end{aligned}$$

independently of c. The algebra $A(c)$ is symmetric. Indeed, choose the basis

$$B := \{e_0, e_1, \alpha, \beta, \gamma, \alpha\beta, \beta\gamma, \gamma\alpha, \alpha\beta\gamma, \gamma\alpha\beta\}$$

and define a linear form $\psi : A(c) \longrightarrow K$ by $\psi(\alpha\beta\gamma) = 1 = \psi(\gamma\alpha\beta)$ and $\psi(b) = 0$ for all $b \in B \setminus \{\alpha\beta\gamma, \gamma\alpha\beta\}$. Proposition 1.10.18 shows that in this way we get a non-degenerate associative bilinear form $\langle x, y\rangle := \psi(xy)$ which is symmetric in this specific case, as is readily verified.

Since $[A(c), A(c)]^\perp = Z(A(c))$ by Lemma 2.9.10 the commutator space $[A(c), A(c)]$ is of dimension $10 - 4 = 6$. Of course, all non-closed paths are commutators with an idempotent, and so

$$\beta, \gamma, \alpha\beta, \gamma\alpha \in [A(c), A(c)].$$

Moreover,

$$\alpha\beta\gamma - \gamma\alpha\beta = [\alpha\beta, \gamma] \in [A(c), A(c)].$$

Finally

$$[\beta, \gamma] = \beta\gamma - \gamma\beta = \beta\gamma \in [A(c), A(c)].$$

We shall compute $T_1(A(c))$. As we know

$$[A(c), A(c)] \subseteq T_1(A(c)) \subseteq \mathrm{rad}(A(c))$$

we need only decide whether a linear combination

$$s_1\alpha + s_2\alpha\beta\gamma$$

is in $T_1(A(c))$. Since $(\alpha\beta\gamma)^2 = 0$ we get that $\alpha\beta\gamma \in T_1(A(c))$ for all c. But $\alpha^2 = c \cdot \alpha\beta\gamma$ and so

$$\alpha \in T_1(A(0))$$

whereas

$$\alpha \notin T_1(A(1)).$$

Hence

$$T_1(A(c)) = \begin{cases} [A(c), A(c)] + K \cdot \alpha\beta\gamma & \text{if } c = 1, \\ [A(c), A(c)] + K \cdot \alpha\beta\gamma + K \cdot \alpha & \text{if } c = 0. \end{cases}$$

Of course $T_2(A(c)) = \mathrm{rad}(A(c))$ since the radical of $A(c)$ has nilpotence degree 4. Since $\beta\gamma$ is not orthogonal to α we get

$$T_1(A(c))^\perp = \begin{cases} \mathrm{soc}(A(c)) & \text{if } c = 1, \\ \mathrm{rad}(Z(A(c))) & \text{if } c = 0. \end{cases}$$

Under the identification $Z(A(c)) \simeq K[X, Y, Z]/(XY, XZ, YZ, X^2, Y^2, Z^2)$ we get that

$$T_1(A(1))^\perp = < X, Y > \text{ and } T_1(A(0))^\perp = < X, Y, Z > .$$

As a whole we see that the commutative ring $Z = Z(A(c))$, which is independent of c, gets an additional ideal structure from the fact that it is the centre of $A(c)$ and the ideal structure of Külshammer ideals depends on c.

We note that Holm and Zimmermann give another closely related example in [20] where the centres of two algebras are isomorphic and where the first Külshammer ideal is of the same codimension for both algebras, but where the quotients of the centre modulo the first Külshammer ideal gives two non-isomorphic rings.

Let A be a symmetric K-algebra for a field K of characteristic $p > 0$. Let $\langle\ ,\ \rangle$ be a symmetrising form for A. By Lemma 2.9.10 this symmetrising form induces a non-degenerate bilinear form

$$\langle\ ,\ \rangle : Z(A) \times A/[A, A] \longrightarrow K.$$

We have seen that the p-power map $\mu_p : a \longrightarrow a^p$ is a semilinear map

$$A/[A,A] \longrightarrow A/[A,A],$$

i.e. $\mu_p(a+b+[A,A]) = \mu_p(a)+\mu_p(b)+[A,A]$ and $\mu_p(\lambda \cdot a+[A,A]) = \lambda^p \cdot \mu_p(a)+[A,A]$ for all $a,b \in A$ and $\lambda \in K$.

Hence there is a left adjoint $\zeta_p : Z(A) \longrightarrow Z(A)$ to μ_p with respect to $\langle\ ,\ \rangle$. In other words there is a semilinear map $\zeta_p : Z(A) \longrightarrow Z(A)$ so that for all $a \in A$ and $z \in Z(A)$ we get

Lemma 2.9.14 $\langle z, \mu_p(a+[A,A])\rangle = \langle \zeta_p(z), a+[A,A]\rangle^p$.

Of course, an iteration of this relation gives

$$\langle z, \mu_p^n(a+[A,A])\rangle = \langle \zeta_p^n(z), a+[A,A]\rangle^{p^n}.$$

Lemma 2.9.15 *Let A be a symmetric K-algebra for a field K of characteristic $p>0$ and let ζ_p be the left adjoint to the p-power map $\mu_p : A/[A,A] \longrightarrow A/[A,A]$ with respect to a fixed symmetrising form. Then $T_n(A)^\perp = \mathrm{im}(\zeta_p^n)$ for all n, where the orthogonal space is taken with respect to this same fixed symmetrising form.*

Proof Since $T_n(A) = \ker(\mu_p^n)$, dualising this equality using the symmetrising form we get $T_n(A)^\perp = \mathrm{im}(\zeta_p^n)$. □

Remark 2.9.16 We should alert the reader that ζ_p depends heavily on the symmetrising form. Different symmetrising forms lead to different mappings ζ_p. However the image does not depend on the choice of the form.

2.9.4 Further Properties of Külshammer Ideals of Group Algebras

In this section we follow Külshammer [19].

Let k be a field of characteristic $p>0$ and let G be a finite group. We can describe $T_n(kG)$ in terms of the basis G of kG. Indeed, recall first that $[kG,kG] \le T_n(kG)$. Then for all $g,h \in G$ we get

$$hgh^{-1} - g = [h, gh^{-1}] \in [kG,kG] \le T_n(kG).$$

Hence, denoting by $C_g := \{hgh^{-1} \mid h \in G\}$ the conjugacy class of $g \in G$ and by $Cl(G) := \{C_g \mid g \in G\}$ the conjugacy classes of G, we obtain that $\sum_{g\in G} \alpha_g g \in T_n(kG)$ if and only if

$$
\begin{aligned}
0 \equiv \left(\sum_{g\in G} \alpha_g g\right)^{p^n} &\equiv \sum_{g\in G} (\alpha_g g)^{p^n} \\
&\equiv \sum_{C_g\in Cl(G)} \left(\sum_{h\in C_g} \alpha_h^{p^n}\right) g^{p^n} \mod [kG, kG]).
\end{aligned}
$$

We have proved the following lemma.

Lemma 2.9.17 *Let k be a field of characteristic $p > 0$, G be a finite group and, for all $C \in Cl(G)$, let $C^{p^{-n}} := \{g \in G \mid g^{p^n} \in C\}$. Then*

$$
T_n(kG) = \left\{ \sum_{g\in G} \alpha_g g \mid \forall C \in Cl(G) : \sum_{g\in C^{p^{-n}}} \alpha_g = 0 \right\}.
$$

For a finite group H we denote its exponent by $\exp(H)$. The exponent of a finite group is the least common multiple of the orders of its elements.

Lemma 2.9.18 *Let k be a field of characteristic $p > 0$, G be a finite group and let S_p be a Sylow p-subgroup of G. Then*

$$
\exp(S_p) = \min\{p^n \mid T_n(kG) = T(kG)\}.
$$

Proof Uniquely decomposing each $g \in G$ into an element g_p of order p^m for some p and g'_p of order relatively prime to p, we get $g^{p^n} = g_{p'}^{p^n} \cdot g_p^{p^n}$. Then by the proof of Theorem 2.9.7 we see that $T_n(kG) = T(kG)$ if and only if the order of g^{p^n} is not divisible by p for all $g \in G$. This happens if and only if $p^n \geq \exp(S_p)$ since $g_p \in hS_ph^{-1}$ for some $h \in G$ by Sylow's theorem. □

For an integer m we define $m_p := \max\{p^n \mid p^n \text{ divides } m\}$, the biggest power of p that divides m. We obtain immediately from Lemma 2.9.18 the following Corollary.

Corollary 2.9.19 *Let k be a field of characteristic $p > 0$ and let G be a finite group. Then G has a cyclic Sylow p-subgroup if and only if*

$$
\min\{p^n \mid T_n(kG) = T(kG)\} = |G|_p.
$$

We shall examine the consequences of Lemma 2.9.17 and obtain a k-basis for $T_n(kG)^\perp$ as a corollary.

Lemma 2.9.20 *Let k be a field of characteristic $p > 0$ and let G be a finite group. Then $T_n(kG)^\perp$ has a k-basis*

$$\left\{ \sum_{g \in C^{p^{-n}}} g | \ C \in Cl(G) \text{ and } C^{p^{-n}} \neq \emptyset \right\}$$

where $C^{p^{-n}} := \{h \in G | \ h^{p^n} \in C\}$.

Proof It is clear that the elements in $\{\sum_{g \in C^{p^{-n}}} g | \ C \in Cl(G)\}$ are linearly independent since G is linearly independent in kG. Moreover,

$$a \in T_n(kG)^{\perp} \Leftrightarrow \forall t \in T_n(kG) : \ \langle a, t \rangle = 0.$$

Lemma 2.9.17 shows that

$$T_n(kG) = \left\{ \sum_{g \in G} \alpha_g g \mid \forall C \in Cl(G) : \ \sum_{g \in C^{p^{-n}}} \alpha_g = 0 \right\}.$$

But recall the definition of $\langle \ , \ \rangle$ for kG from Proposition 1.10.26:

$$\langle g, h \rangle = \begin{cases} 1 & \text{if } gh = 1 \\ 0 & \text{if } gh \neq 1 \end{cases}$$

and then extend bilinearly to a bilinear form $kG \times kG \longrightarrow k$. The following will be used later:

Corollary 2.9.21 *For any subset* $S \subseteq G$ *we have*

$$\left\langle \sum_{s \in S} s, \sum_{g \in G} \alpha_g g \right\rangle = \sum_{s \in S} \left\langle s, \sum_{g \in G} \alpha_g g \right\rangle = \sum_{s \in S} \alpha_{s^{-1}}.$$

Proof Indeed, this is just a corollary of the definition of the standard symmetrising form of kG. □

We continue with the proof of Lemma 2.9.20. Therefore

$$\begin{aligned} T_n(kG) &= \left\{ x \in kG \mid \forall C \in Cl(G) : \ \langle x, \sum_{g \in C^{p^{-n}}} g \rangle = 0 \right\} \\ &= \bigcap_{C \in Cl(G)} \left(\sum_{g \in C^{p^{-n}}} g \right)^{\perp} \end{aligned}$$

and by consequence

$$T_n(kG)^{\perp} = \sum_{C \in Cl(G)} k \cdot \left(\sum_{g \in C^{p^{-n}}} g \right).$$

In other words, $T_n(kG)^{\perp}$ has a basis consisting of the elements $\sum_{g \in C^{p^{-n}}} g$ for all $C \in Cl(G)$. This finishes the proof of Lemma 2.9.20. □

Recall again that for every $g \in G$ one has a unique decomposition $g = g_p \cdot g_{p'} = g_{p'} \cdot g_p$ where g_p is an element of p-power order and $g_{p'}$ has order not divisible by p. This follows from the classification of finitely generated abelian groups applied to the cyclic group generated by g.

Corollary 2.9.22 (Reynolds) *Let k be a field of characteristic $p > 0$ and let G be a finite group. Then for all $g \in G$ let S_g be the set of those elements h of G such that $h_{p'}$ is conjugate to $g_{p'}$. Then the k-vector space generated by $\sum_{h \in S_g} h$ for all $g \in G$ is equal to* $\mathrm{soc}(kG) \cap Z(kG)$.

Proof This follows from the above together with the fact that $g_{p'}$ is conjugate to $h_{p'}$ if and only if there is an integer n such that g^{p^n} is conjugate to h^{p^n}. Then for $n \geq \exp(S_p)$ for some Sylow p subgroup S_p of G we obtain the statement from the fact that

$$T_n(kG)^{\perp} = T(kG)^{\perp} = \mathrm{soc}(kG) \cap Z(kG)$$

which was shown above. □

Recall the definition of ζ_p from Lemma 2.9.14. It is easy to compute the element $\zeta_p^n(\sum_{g \in C} g)$ for each $C \in Cl(G)$.

Lemma 2.9.23 *Let k be a field of characteristic $p > 0$ and let G be a finite group. Then for all $C \in Cl(G)$ we have $\zeta_p^n(\sum_{g \in C} g) = \sum_{g \in G;\, g^{p^n} \in C} g$.*

Proof We use the symmetrising form

$$\langle\ ,\ \rangle : Z(kG) \times kG/[kG, kG] \longrightarrow k,$$

Lemma 2.9.14 and Corollary 2.9.21 to get

$$\begin{aligned}
\left\langle \zeta_p^n \left(\sum_{g \in C} g \right), \sum_{h \in G} \alpha_h h \right\rangle^{p^n} &= \left\langle \sum_{g \in C} g, \left(\sum_{h \in G} \alpha_h h \right)^{p^n} \right\rangle = \left\langle \sum_{g \in C} g, \sum_{h \in G} \alpha_h^{p^n} h^{p^n} \right\rangle \\
&= \sum_{h \in G;\, h^{p^n} \in C} \alpha_{h^{-1}}^{p^n} = \left(\sum_{h \in G;\, h^{p^n} \in C} \alpha_{h^{-1}} \right)^{p^n} \\
&= \left\langle \sum_{g \in C^{p^{-n}}} g, \sum_{h \in G} \alpha_h h \right\rangle^{p^n}
\end{aligned}$$

and this shows

$$\sum_{g \in C^{p^{-n}}} g = \zeta_p^n (\sum_{g \in C} g)$$

as claimed. □

2.10 Brauer Constructions and *p*-Subgroups

We shall introduce a very useful construction, due to Brauer, which links the representation theory of a group G over a field of characteristic p to the representation theory of centralisers of p-subgroups. This construction is more technical compared to what we did before. However, it is crucial for more subtle properties of Brauer and Green correspondences.

2.10.1 The Brauer Homomorphism and Osima's Theorem

Recall that $C_G(Q)$ is the centraliser of Q in G.

Definition 2.10.1 Given a field k of characteristic $p > 0$ and a finite group G we define for any p-subgroup Q of G the *Brauer homomorphism* by

$$\beta_Q \left(\sum_{g \in G} \alpha_g g \right) := \sum_{g \in C_G(Q)} \alpha_g g$$

and obtain a mapping $\beta_Q : kG \longrightarrow kC_G(Q)$.

Lemma 2.10.2 *Let k be a field k of characteristic $p > 0$ and let G be a finite group with p-subgroup Q. Then the restriction of the Brauer homomorphism β_Q to $Z(kG)$ induces an algebra homomorphism $Z(kG) \longrightarrow Z(kC_G(Q))$.*

Proof It is clear that β_Q is k-linear. Moreover, given a conjugacy class $C \in Cl(G)$, if $C \cap C_G(Q) \neq \emptyset$, then $C \cap C_G(Q)$ is a union of conjugacy classes of $C_G(Q)$. Hence $\beta_Q(C) \in Z(kC_G(Q))$.

We need to show that β_Q is multiplicative. Let $C_1, C_2 \in Cl(G)$ be two conjugacy classes of G and put $x_1 := \sum_{g \in C_1} g$ and $x_2 := \sum_{g \in C_2} g$. Since by Lemma 1.5.3 the conjugacy class sums form a basis for the centre of kG, we only need to show $\beta_Q(x_1)\beta_Q(x_2) = \beta_Q(x_1 x_2)$. But

$$x_1 x_2 = \sum_{g \in G} \gamma_g g \text{ and } \beta_Q(x_1)\beta_Q(x_2) = \sum_{h \in C_G(Q)} \delta_h h$$

for

$$\gamma_g = |\{(u, v) \in C_1 \times C_2 \mid uv = g\}|$$

and

$$\delta_h = |\{(u, v) \in (C_1 \times C_2) \cap (C_G(Q) \times C_G(Q)) \mid uv = h\}|.$$

We observe that Q acts on C_1 and on C_2 by conjugation and $C_1 \cap C_G(Q)$ respectively $C_2 \cap C_G(Q)$ are the fixpoints of this action. Hence Q acts on $C_1 \times C_2$ by conjugation diagonally on each factor and the fixpoints are $(C_1 \times C_2) \cap (C_G(Q) \times C_G(Q))$. Now, if $(u, v) \in C_1 \times C_2$ and $q \in Q$, then

$$(quq^{-1})(qvq^{-1}) = quvq^{-1} = qgq^{-1}$$

if $uv = g$. Therefore the action of Q on $C_1 \times C_2$ restricts to an action on

$$\{(u, v) \in C_1 \times C_2 \mid uv = g\}$$

whenever $g \in C_G(Q)$. The set

$$\{(u, v) \in (C_1 \times C_2) \cap (C_G(Q) \times C_G(Q)) \mid uv = g\}$$

corresponds to the fixpoints, i.e. orbits of length 1, of this action. All other orbits are of length a non-trivial power of p, since Q is a p-group. Hence

$$\forall g \in C_G(Q) : \quad \gamma_g = \delta_g \in k$$

which proves the lemma. □

Corollary 2.10.3 *Given a field k of characteristic $p > 0$ and a finite group G with p-subgroup Q, then the kernel of $\beta_Q : Z(kG) \longrightarrow Z(C_G(Q))$ has a k-basis given by*

$$\left\{ \sum_{g \in C} g \mid C \in Cl(G) \text{ and } C \cap C_G(Q) = \emptyset \right\}.$$

Proof Indeed, this follows immediately from the definition of β_Q. □

Remark 2.10.4 Now, consider the relation $C \cap C_G(Q) = \emptyset$ for some $C \in Cl(G)$. This means that no element in C centralises all of Q. This is equivalent to the statement that for all $g \in C$ there is a $q \in Q$ such that $qg \neq gq$. This in turn is equivalent to the fact that for all $g \in C$ the group Q is not totally contained in $C_G(g)$. Since Q is a p-group, using Sylow's theorems this last statement is equivalent to the statement that the group Q is not contained in any Sylow p-subgroup of $C_G(g)$ of any $g \in C$. Denote by $Syl_p(H)$ the set of Sylow p-subgroups of H. Then we get

$$[C \cap C_G(Q) = \emptyset] \Leftrightarrow \left[\forall\, S \in \bigcup_{g \in C} Syl_p(C_G(g)) :\ Q \not\leq S\right].$$

The following consequence of Lemma 2.10.2 is a theorem of Osima with a proof due to Donald Passman.

Proposition 2.10.5 (Osima) *Let k be a field of characteristic $p > 0$, let G be a finite group and let $e^2 = e \in Z(kG)$ be a central idempotent of kG. If $e = \sum_{g \in G} \epsilon_g g$, then $\epsilon_g = 0$ if p divides the order of g.*

Proof Let $g \in G$, suppose $\epsilon_g \neq 0$ and let $g = g_p \cdot g_{p'}$ be the decomposition of g into a product of an element g_p of a power of p and an element $g_{p'}$ of order r not divisible by p. Let Q be the group generated by g_p and suppose that $Q \neq 1$. Since $g_{p'}$ commutes with g_p we get that $g \in C_G(Q)$. Since $e^2 = e$, using Lemma 2.10.2, $\beta_Q(e)$ is a central idempotent of $kC_G(Q)$. Hence we may assume that $G = C_G(Q)$ and $g_p \in Z(G)$. Now, p is a unit in $\mathbb{Z}/r\mathbb{Z}$ and so it is of finite order m' in $(\mathbb{Z}/r\mathbb{Z})^\times$. Let $|G| = p^s t$ for some integer $s \in \mathbb{N}$ and some integer $t \in \mathbb{N}$ such that p does not divide t. Increasing m' if necessary, there is an integer $m \geq s$ such that r divides $p^m - 1$ and therefore

$$\left(g_{p'}^{-1} e\right)^{p^m} = \left(g_{p'}^{-1}\right)^{p^m} e^{p^m} = g_{p'}^{-1} e.$$

Moreover, if $g_{p'}^{-1} e = \sum_{h \in G} \phi_h h$, then $\phi_{g_p} \neq 0$ since we assumed that $\epsilon_g \neq 0$. Now

$$\left(g_{p'}^{-1} e\right)^{p^m} = \left(\sum_{h \in G} \phi_h h\right)^{p^m} \equiv \sum_{h \in G} \phi_h^{p^m} h^{p^m} \quad \mod [kG, kG]$$

and since we assumed $m \geq s$, the order of h^{p^m} is not divisible by p for all $h \in G$. Since $\phi_{g_p} \neq 0$, and since h^{p^m} is of order relatively prime to p, we obtain for the difference

$$c := g_{p'}^{-1} e - \sum_{h \in G} \phi_h^{p^m} h^{p^m} = \left(g_{p'}^{-1} e\right)^{p^m} - \left(\sum_{h \in G} \phi_h h\right)^{p^m} \in [kG, kG].$$

We define the coefficients $\gamma_h \in k$ by

$$c = \sum_{h \in G} \gamma_h h$$

and we get that $\gamma_{g_p} \neq 0$. Since $[kG, kG]$ is k-linearly generated by elements $xy - yx$ for $x, y \in G$, there are $x, y \in G$ such that $xy \neq yx$ and such that $g_p = xy$. Since $g_p \in Z(G)$ we obtain

$$yx = x^{-1}(xy)x = x^{-1}g_p x = g_p = xy$$

which contradicts the above established fact that $xy - yx \neq 0$. □

2.10.2 Higman's Criterion, Traces and the Brauer Quotient

We come to a classical object of study which we have already used in the proof of Proposition 2.1.15. For a group Γ and a $k\Gamma$-module M denote by

$$M^{\Gamma} := \{m \in M \mid \gamma m = m\ \forall \gamma \in \Gamma\}$$

the space of Γ-fixed points.

Lemma 2.10.6 *Let k be a field and let G be a finite group. For every subgroup H of G and every kG-module M we get that for all $m \in M^H$ we have $\sum_{gH \in G/H} gm \subseteq M^G$.*

Proof If $m \in M^H$ consider $\sum_{gH \in G/H} gm$. Now multiplication by $g' \in G$ permutes the classes $gH \in G/H$ and the image is in M^G. □

Definition 2.10.7 Let k be a field of characteristic $p > 0$ and let G be a finite group. For every subgroup H of G and every kG-module M define the *trace from H to G* as

$$\begin{aligned} Tr_H^G : M^H &\longrightarrow M^G \\ m &\mapsto \sum_{gH \in G/H} gm \end{aligned}$$

We come to a useful observation linking traces and vertices, called Higman's criterion. Recall that if M and N are kG-modules, then $Hom_k(M, N)$ is a kG-module by $(gf)(m) := gf(g^{-1}m)$ for all $g \in G, f \in Hom_k(M, N)$ and $m \in M$. Now by Lemma 2.10.6 we get

$$Tr_H^G : Hom_{kH}(M \downarrow_H^G, N \downarrow_H^G) \longrightarrow Hom_{kG}(M, N).$$

Observe that $Hom_{kH}(M \downarrow_H^G, N \downarrow_H^G)$ is an $End_{kG}(N)$-$End_{kG}(M)$ bi-module by composition of mappings. Indeed, the composition $f \circ h$ of two maps $f \in Hom_{kH}(M \downarrow_H^G, N \downarrow_H^G)$ and $h \in Hom_{kG}(M, M)$ is in $Hom_{kH}(M \downarrow_H^G, N \downarrow_H^G)$, and likewise for endomorphisms of N.

Lemma 2.10.8 *Let G be a finite group, let k be a field and let H be a subgroup of G. Then for every kG-module M we get that $Tr_H^G : End_{kH}(M \downarrow_H^G) \longrightarrow End_{kG}(M)$ is a morphism of $End_{kG}(M)$-$End_{kG}(M)$ bimodules.*

Proof We need to show

$$Tr_H^G(h_1 \circ f \circ h_2) = h_1 \circ Tr_H^G(f) \circ h_2$$

for all $f \in Hom_{kH}(M \downarrow_H^G, M \downarrow_H^G)$ and $h_1, h_2 \in Hom_{kG}(M, M)$. To prove this we compute

$$\begin{aligned}\left(h_1 \circ Tr_H^G(f) \circ h_2\right)(m) &= \sum_{gH \in G/H} h_1((gf)(h_2 m)) \\ &= \sum_{gH \in G/H} h_1(g(f(g^{-1}(h_2 m)))) \\ &= \sum_{gH \in G/H} g(h_1(f(h_2(g^{-1}m)))) \\ &= \sum_{gH \in G/H} g\left((h_1 \circ f \circ h_2)(g^{-1}m)\right) \\ &= \left(Tr_H^G(h_1 \circ f \circ h_2)\right)(m)\end{aligned}$$

and obtain the result. □

Proposition 2.10.9 (D. Higman) *Let k be a field of characteristic $p > 0$, let G be a finite group and let $H \leq G$. Then a kG-module M is relatively H-projective if and only if $id_M \in Tr_H^G(Hom_{kH}(M \downarrow_H^G, M \downarrow_H^G))$.*

Proof Suppose $id_M \in Tr_H^G(Hom_{kH}(M \downarrow_H^G, M \downarrow_H^G))$. By Lemma 2.10.8 we obtain that $Tr_H^G(Hom_{kH}(M \downarrow_H^G, M \downarrow_H^G))$ contains $id_M \circ End_{kG}(M) = End_{kG}(M)$. Hence Tr_H^G is surjective. By definition the natural epimorphism of kG-modules

$$kG \otimes_{kH} M \longrightarrow kG \otimes_{kG} M \simeq M$$

is split if and only if the identity id_M is in the image of the induced morphism

$$Hom_{kG}(M, M \downarrow_H^G \uparrow_H^G) \xrightarrow{\mu} Hom_{kG}(M, M).$$

Recall the Frobenius reciprocity isomorphism

$$\begin{aligned}Hom_{kH}(M \downarrow_H^G, M \downarrow_H^G) &\xrightarrow{\phi} Hom_{kG}(M, M \downarrow_H^G \uparrow_H^G) \\ \varphi &\mapsto \left(m \mapsto \sum_{gH \in G/H} g \otimes \varphi(g^{-1}m)\right)\end{aligned}$$

from Remark 1.7.34. Hence $Tr_H^G = \mu \circ \phi$. Since id_M is in the image of Tr_H^G, the identity id_M is also in the image of μ.

Suppose conversely M is relatively H-projective. Proposition 2.1.6 shows that $M\downarrow_H^G\uparrow_H^G\longrightarrow M$ is split and hence the identity is in the image of μ and therefore in the image of Tr_H^G. □

Higman's criterion gives an interpretation of the defect group of a block in term of traces.

Lemma 2.10.10 *Let k be a field of characteristic $p>0$ and let G be a finite group. Let $B=kG\cdot e$ be a block of kG with defect group D. Then $e\in Tr_{\Delta(H)}^{\Delta(G)}(kG^{\Delta(H)})$ if and only if H contains a conjugate of D.*

Proof As we have seen, B is projective relative to $\Delta(G)$. Hence H contains a defect group if and only if there is an $\alpha\in End_{k\Delta(H)}(kG)$ such that $id_B=Tr_{\Delta(H)}^{\Delta(G)}(\alpha)$. But then

$$\begin{aligned} e=Tr_{\Delta(H)}^{\Delta(G)}(\alpha)(e)&=\sum_{(g,g)\Delta(H)\in\Delta(G)/\Delta(H)}(g,g)\cdot\alpha((g^{-1},g^{-1})\cdot e)\\ &=\sum_{gH\in G/H}g\alpha(e)g^{-1}=Tr_H^G(\alpha(e))\end{aligned}$$

which shows that $e\in Tr_{\Delta(H)}^{\Delta(G)}((kG)^{\Delta(H)})$. If conversely $e=Tr_{\Delta(H)}^{\Delta(G)}(a)$ for some $a\in(kG)^{\Delta(H)}$, then $\alpha(x):=x\cdot a$ for all $x\in kG^{\Delta(H)}$ defines a $k\Delta(H)$-linear endomorphism of kG. Then for all $x\in B$

$$\begin{aligned} Tr_{\Delta(H)}^{\Delta(G)}(\alpha)(x)&=\sum_{(g,g)\Delta(H)\in\Delta(G)/\Delta(H)}(g,g)\cdot\alpha((g^{-1},g^{-1})\cdot x)\\ &=\sum_{g\in G/H}g\cdot((g^{-1}\cdot x\cdot g)\cdot a)\cdot g^{-1}=x\cdot Tr_{\Delta(H)}^{\Delta(G)}(a)=x\cdot e=x\end{aligned}$$

and so $Tr_{\Delta(H)}^{\Delta(G)}(\alpha)=id_B$, which implies that B is $\Delta(H)$-projective by Higman's criterion. □

The following construction is useful mainly for p-subgroups. We shall explain below the reason for this restriction. Given a subgroup P of G, we may consider the submodule M^P of fixed-points of P. The group $N_G(P)$ acts on M^P since if $g\in N_G(P)$ and $m\in M^P$, then for all $h\in P$ we obtain

$$h\cdot(gm)=(hg)m=(g(g^{-1}hg))m=g((g^{-1}hg)m)=gm$$

since $g\in N_G(P)$ and therefore $(g^{-1}hg)\in P$ and hence $gm\in M^P$. Moreover, for all $Q\le P$ we obtain trace functions

$$Tr_Q^P:M^Q\longrightarrow M^P$$

so that we may form $M(P) := M^P / \sum_{Q<P} Tr_Q^P(M^Q)$. Now, if P is not a p-group, then there exists a proper subgroup Q of P such that p does not divide $|P : Q|$. Maschke's argument then shows that $Tr_Q^P(M^Q) = M^P$. Hence the quotient construction $M(P)$ is reasonable only for p-groups P. Indeed, if $Q < P$, then $M^Q \geq M^P$ and a k-basis of M^P may be completed to a k-basis of M^Q. Let $\iota : M^P \longrightarrow M^Q$ be the natural embedding and $\pi : M^Q \longrightarrow M^P$ be the projection on M^P. There is no reason why π should be P-linear. However

$$\hat{\pi}(m) := \frac{1}{|P : Q|} \sum_{gP \in P/Q} g\pi(g^{-1}m)$$

defines a kP-linear projection which splits ι.

Definition 2.10.11 Let G be a finite group and let P be a p-subgroup of G. For every kG-module M we define the *Brauer quotient*

$$M(P) := M^P / \sum_{Q<P; Q \neq P} Tr_Q^P(M^Q)$$

which is naturally a $k(N_G(P)/P)$-module.

2.10.3 Brauer Pairs

We shall discuss an application of the Brauer quotient from Definition 2.10.11 due to Alperin, Broué and Puig. We shall follow the most elegant approach given by Külshammer [1].

Let k be a field of characteristic $p > 0$ and let G be a finite group.

Definition 2.10.12 A *Brauer pair* is a pair (P, e) where $P \leq G$ is a p-subgroup of G and e is a primitive central idempotent of $Z(kC_G(P))$.

In this section we shall prove Sylow-like theorems for Brauer pairs. Surprisingly the main tool to achieve this is the Brauer quotient.

Recall from Definition 2.10.1 the Brauer map

$$\beta_Q : kG \longrightarrow kC_G(Q)$$

for every p-subgroup Q of G.

Lemma 2.10.13 *The restriction of the Brauer map* β_Q *to*

$$\beta_Q : (kG)^Q \longrightarrow kC_G(Q)$$

is a surjective homomorphism of k*-algebras.*

Proof Indeed, a basis for the Q fixed points in kG is obtained by taking the sums of the orbits of the elements of G under the conjugation action. The orbit of an element $x \in G$ is isomorphic to $Q/C_Q(x)$ as a set with Q-action. If the orbit of the element x is of cardinality different from 1, then the sum of the elements of the orbit is $Tr_{C_Q(x)}^Q(x)$, and therefore belongs to $\sum_{S<Q} Tr_S^Q(kS)$. If the orbit of x is of length 1, then Q centralises x, and hence $x \in kC_G(Q)$. Therefore

$$(kG)^Q = kC_G(Q) \oplus \sum_{S<Q} Tr_S^Q(kS)$$

as k-vector spaces and the mapping β_Q is just the projection onto the first component. Moreover, $\sum_{S<Q} Tr_S^Q(kS)$ is a two-sided ideal of $(kG)^Q$. □

Corollary 2.10.14 *The kernel of the restriction of the map β_Q to $(kG)^Q$ equals $\sum_{S<Q} Tr_S^Q(kS)$.* □

We shall use the following observation in the proof of the main Theorem 2.10.16 below.

Lemma 2.10.15 *Let A and B be artinian k-algebras and let $\varphi : A \longrightarrow B$ be a homomorphism of algebras. Let $e \neq 0$ be an idempotent in B. Then there is a primitive idempotent f of A such that $\varphi(f) \cdot e \neq 0$. If e is primitive and B is commutative, then $\varphi(f) \cdot e = e$ and f is unique with this property.*

Proof Since A is artinian, we can find primitive idempotents $f_1, \ldots, f_n$ of A such that $1 = \sum_{i=1}^n f_i$. Then

$$e = 1 \cdot e = \varphi(1) \cdot e = \varphi\left(\sum_{i=1}^n f_i\right) \cdot e = \sum_{i=1}^n \varphi(f_i) \cdot e.$$

This proves the existence of f. If B is commutative and e is primitive, then $\varphi(f) \cdot e$ and $(1 - \varphi(f)) \cdot e$ are both idempotents of B. Moreover,

$$e = e \cdot \varphi(f) + e \cdot (1 - \varphi(f))$$

and since e is primitive, and since $e \cdot \varphi(f) \neq 0$, we get $e = e \cdot \varphi(f)$. This also proves that $e \cdot \varphi(1 - f) = 0$, and hence unicity. □

Theorem 2.10.16 (Broué-Puig) *Let G be a finite group and let k be a field of characteristic $p > 0$. Let (P, e) be a Brauer pair of kG and let $Q \leq P$. Then there exists a unique primitive idempotent $f \in Z(kC_G(Q))$ such that for each primitive idempotent $e_P \in (kG)^P$ with $\beta_P(e_P)e \neq 0$ one has*

$$\beta_Q(e_P) \cdot f = \beta_Q(e_P).$$

We write in this case $(Q, f) \leq (P, e)$.

Proof (Külshammer [1]) We first prove unicity.

Let f_1 and f_2 be two primitive idempotents in $Z(kC_G(Q))$ such that for each primitive idempotent $e_P \in (kG)^P$ with $\beta_P(e_P) \neq 0$ one has

$$\beta_Q(e_P) \cdot f_1 = \beta_Q(e_P) = \beta_Q(e_P) \cdot f_2.$$

We have seen in Lemma 2.10.13 that β_Q is a ring homomorphism. By Lemma 2.10.15 there is a primitive idempotent e'_{i_0} of $(kG)^P$ with $\beta_P(e'_{i_0}) \cdot e \neq 0$. But since $C_G(Q) \geq C_G(P)$ we get $0 \neq \beta_Q(e'_{i_0})$ and therefore by the hypothesis

$$0 \neq \beta_Q(e'_{i_0}) = \beta_Q(e'_{i_0}) f_2 = \beta_Q(e'_{i_0}) f_1 f_2.$$

This shows that $f_1 f_2 \neq 0$ and since f_1 and f_2 are primitive central idempotents, $f_1 = f_2$.

We shall now prove the existence by induction on $|P : Q|$.

If $P = Q$, then we obtain that $\beta_P(e'_{i_0}) \cdot e \neq 0$. But $\beta_P(e'_{i_0})$ is a primitive idempotent in $kC_G(P)$. Indeed, this follows from the fact that

$$(kG)^P = kC_G(P) \oplus \sum_{S<P} Tr_S^P(kS)$$

and that $\sum_{S<P} Tr_S^P((kG)^S)$ is an ideal of $(kG)^P$. Therefore,

$$\beta_P(e'_{i_0}) \cdot e = \beta_P(e'_{i_0})$$

as claimed.

Suppose that $P > Q$. Then, since P is a p-group, $Q \neq N_P(Q) =: R$ (cf e.g. [21, III Hauptsatz 2.3]), and $Q < N_P(Q)$. Hence, by the induction hypothesis, there is a primitive idempotent $g \in Z(C_G(R))$ with $\beta_R(e_P) \cdot g = \beta_R(e_P)$ for every primitive idempotent $e_P \in (kG)^P$ with $\beta_P(e_P) e \neq 0$.

Now, Lemma 2.10.13 shows that β_R is a surjective ring homomorphism $(kG)^R \longrightarrow kC_G(R)$ and the restriction of β_R to $Z((kC_G(Q))^R)$ is a surjective ring homomorphism with image in $Z(kC_G(R))$. Indeed, any surjective ring homomorphism induces a (not necessarily surjective) homomorphism between the centres of the rings.

As in the first step there is a decomposition of $1 \in Z(kC_G(Q))^R$ into primitive idempotents. Since $\beta_Q : Z(kC_G(Q)^R) \to Z(kC_G(Q))$ is a ring homomorphism, by Lemma 2.10.15 there is a primitive idempotent $f \in Z((kC_G(Q))^R)$ such that $\beta_R(f) \cdot g = g$. Let f' be a primitive idempotent of $Z(kC_G(Q))$ so that $f \cdot f' = f'$, and let

$$I_{f'} := \{h \in N_G(Q) \mid h \cdot f' \cdot h^{-1} = f'\}.$$

It is clear that then $Tr_{I_{f'}}^R(f') = f$. Since $\ker \beta_R = \sum_{S<R} \mathrm{im}(Tr_S^R)$, the hypothesis $R \neq I_{f'}$ implies

$$\beta_R(f) = \beta_R(Tr_{I_{f'}}^R(f')) = 0,$$

a contradiction. Hence $R = I_{f'}$, and $f = f'$, which implies that f is actually a primitive idempotent in $Z(kC_G(Q))$.

Let e_P be a primitive idempotent of $(kG)^P$. We shall show that $\beta_Q(e_P)\cdot f = \beta_Q(e_P)$. Let S be a subgroup of P such that

$$Q < S \le N_P(Q) = R.$$

By Lemma 2.10.15 there is a primitive idempotent e_R of $(kG)^R$ such that

$$e_R \cdot e_P = e_P \cdot e_R = e_R$$

and $\beta_R(e_R) \neq 0$. Then

$$\begin{aligned}\beta_R(e_R)g &= \beta_R(e_R \cdot e_P)g = \beta_R(e_R) \cdot \beta_R(e_P) \cdot g \\ &= \beta_R(e_R) \cdot \beta_R(e_P) = \beta_R(e_R \cdot e_P) = \beta_R(e_R) \neq 0.\end{aligned}$$

The induction hypothesis shows that there is a primitive idempotent h of $Z(kC_G(S))$ such that

$$\beta_S(e_R)h = \beta_S(e_R) \neq 0.$$

Hence

$$0 \neq \beta_S(e_R \cdot e_P)h = \beta_S(e_R) \cdot \beta_S(e_P)h$$

and therefore $\beta_S(e_P)h \neq 0$.

Again applying the induction hypothesis there is a primitive idempotent $h' \in Z(kC_G(S))$ with $\beta_S(e_P)h' = \beta_S(e_P) \neq 0$. Hence $\beta_S(e_R \cdot e_P)h' \cdot h \neq 0$ which implies $h'h \neq 0$. Since h' and h are both primitive and central, $h = h'$ and this implies $\beta_S(e_P)h = \beta_S(e_P)$.

We obtain

$$0 \neq \beta_R(e_R)g = \beta_R(e_R)\beta_R(f)g = \beta_R(\beta_Q(e_R)f)g.$$

Since $\beta_Q(e_R)$ is a primitive idempotent in $\beta_Q((kG)^R) = kC_G(Q)^R$, we get

$$\beta_Q(e_R)f = \beta_Q(e_R).$$

This shows

$$\beta_S(e_R)\beta_S(f)h = \beta_S(e_R)h \neq 0$$

and therefore

$$\beta_S(f)h \neq 0.$$

Since $\beta_S(f) \in Z(kC_G(S))$, we get $\beta_S(f)h = h$ and hence

$$\beta_S(\beta_Q(e_P)(1-f)) = \beta_S(e_P)(1 - \beta_S(f)) = \beta_S(e_P) \cdot h \cdot (1 - \beta_S(f)) = 0.$$

But this shows that

$$\beta_Q(e_P)(1-f) \in \bigcap_{Q<S\leq R} \ker(\beta_S|_{kC_G(Q)^S}) = Tr_Q^R(kC_G(Q)) = \beta_Q(\mathrm{im}(Tr_Q^P)).$$

Therefore

$$\beta_Q(e_P)(1-f) = (1-f)\beta_Q(e_P) \in \beta_Q(e_P \cdot \mathrm{im}(Tr_Q^P) \cdot e_P).$$

Since e_P is primitive, $e_P \cdot (kG)^P \cdot e_P$ is a local algebra, and since $e_P \cdot \mathrm{im}(Tr_Q^P) \cdot e_P$ is a proper ideal of this local algebra, it is a nilpotent ideal. Since $\beta_Q(e_P)(1-f)$ is an idempotent in this nilpotent ideal, the idempotent $\beta_Q(e_P)(1-f)$ has to be 0. Hence $\beta_Q(e_P) = \beta_Q(e_P) \cdot f$. This proves the statement. □

Corollary 2.10.17 *Let (P, e), (Q, f) and (R, g) be Brauer pairs of kG. Then*

1. *$(Q, f) \leq (P, e)$ if and only if $Q \leq P$ and there is a primitive idempotent e_P in $(kG)^P$ with $\beta_Q(e_P)e \neq 0 \neq \beta_Q(e_P)f$.*
2. *$(Q, f) \leq (R, g)$ and $(R, g) \leq (P, e)$ implies $(Q, f) \leq (P, e)$.*

Indeed, the first statement is a direct consequence of Theorem 2.10.16 and the second statement uses the first statement. □

Definition 2.10.18 Suppose (P, e) is a Brauer pair of kG. Then there is a unique block $B = kG \cdot b$ of kG such that $(1, b) \leq (P, e)$. We say that (P, e) *belongs to the block* $B = kG \cdot b$ if $(1, b) \leq (P, e)$.

Note that (P, e) belongs to the block kGb if and only if $\beta_P(b)e \neq 0$. If $(Q, f) \leq (P, e)$ and if (P, e) belongs to the block b, then Corollary 2.1017 shows that (Q, f) also belongs to B. We get a version of Sylow's theorem for Brauer pairs.

Proposition 2.10.19 *Let G be a finite group and let k be a field of characteristic $p > 0$. Let B be a block of kG with defect group P. Then there exists a Brauer pair (P, e) that belongs to B. For every Brauer pair (Q, f) that belongs to B there exists an $x \in G$ such that $(Q, f) \leq (xPx^{-1}, xex^{-1})$.*

Proof Let b be the central idempotent of kG so that $B = kGb$. By Lemma 2.10.10 there is an $a \in (kG)^P$ with $Tr_P^G(a) = b$. Then $b \in Z(kG) \subseteq (kG)^P$, and we may decompose $b = \sum_{i=1}^n e_i$ where $e_1, \ldots, e_n$ are primitive idempotents of $(kG)^P$. Now

$$\begin{aligned} b = b^2 = b \cdot Tr_P^G(a) &= \sum_{i=1}^n e_i Tr_P^G(a) \\ &= \sum_{i=1}^n Tr_P^G(e_i a) \in \sum_{i=1}^n Tr_P^G((kG)^P e_i (kG)^P). \end{aligned}$$

Since $Tr_P^G((kG)^P e_i (kG)^P)$ is an ideal of $Z(kG)$, Rosenberg's Lemma 1.9.16 shows that there is an $i_0 \in \{1, \ldots, n\}$ with $b \in Tr_P^G((kG)^P e_{i_0}(kG)^P)$. If $e_{i_0} \in Tr_Q^P((kG)^Q)$ for some $Q < P$, then $b \in Tr_Q^P((kG)^Q)$ and hence $\beta_P(b) = 0$. But $\beta_P(b)e \neq 0$ since $(1, b) \leq (P, e)$. This contradiction shows that $e_{i_0} \notin Tr_Q^P((kG)^Q)$ for all $Q < P$. Again Rosenberg's Lemma 2.10.16 implies

$$e_{i_0} \notin \sum_{Q<P} Tr_Q^P((kG)^Q) = \ker \beta_P.$$

By Lemma 2.10.15 there is a primitive idempotent e of $Z(kC_G(P))$ such that $\beta_P(e_{i_0})e = \beta_P(e_{i_0}) \neq 0$. By construction (P, e) is a Brauer pair that belongs to B.

Let (Q,f) be a Brauer pair that belongs to $B = kGb$. By Corollary 2.10.17 this shows that $\beta_Q(b)f \neq 0$. We may apply Mackey's formula to get

$$b \in Tr_P^G((kG)^P e_{i_0}(kG)^P) \subseteq \sum_{QxP \in Q\backslash G/P} Tr_{Q\cap P^x}^Q((kG)^{Q\cap P^x} e_{i_0}^x (kG)^{Q\cap P^x})$$

and again by Rosenberg's Lemma 1.9.16 we get that there is an $x \in G$ such that $Q \leq P^x$ and $\beta_Q((kG)^Q e_{i_0}^x (kG)^Q \cdot f) \neq 0$. This implies $\beta_Q(e_{i_0}^x) \cdot f \neq 0$ and hence $(Q,f) \leq (xPx^{-1}, xex^{-1})$. □

Corollary 2.10.20 *Let k be an algebraically closed field of characteristic $p > 0$, let G be a finite group and let B be a block of kG with defect group D and central primitive idempotent b. Then there is a primitive idempotent f of $A := (kG)^D$ and a Brauer pair (D, e) that belongs to B such that*

$$b \in Tr_P^G(AfA) \subseteq AfA$$

and $\beta_D(f)e \neq 0$.

Proof This is precisely the construction of e_{i_0} in the proof of Proposition 2.10.19, and it is sufficient to take $f := e_{i_0}$. □

Remark 2.10.21 We shall use the results of this section, and in particular Theorem 2.10.16 will be used in Sect. 4.4.2.

There is another application of the preceding results concerning the relation between Green and Brauer correspondence.

Proposition 2.10.22 *Let k be a field of characteristic $p > 0$ and let G be a finite group. Let B be a block of kG with central primitive idempotent e and let D be a p-subgroup of G. Suppose H is a subgroup of G so that $C_G(D) \leq H \leq N_G(D)$. Then for every B-module M we get $M\downarrow_H^G = (\beta_D(e) \cdot M\downarrow_H^G) \oplus N$ and each indecomposable direct summand of N is relatively Q-projective for some $Q \leq H$ and $D \not\leq Q$.*

Proof $M = \beta_D(e) \cdot M \oplus (1 - \beta_D(e)) \cdot M$ as kH-modules. We define $N := (1 - \beta_D(e)) \cdot M$ and need to show that the vertex K of each indecomposable direct factor of N satisfies $D \not\leq K \leq H$. Since $eM = M$ we get that multiplication by $e - \beta_D(e)$ acts as the identity on N. By Lemma 2.10.13 and its proof we get that $kG^D = kC_G(D) \oplus \sum_{Q<D} Tr_Q^D((kG)^Q)$ and taking H-fixed points in this equation, we divide the orbits into those with trivial D-action, and those with non-trivial D-action. Therefore we obtain

$$kG^H = kC_G(D)^H \oplus \sum_{Q \leq H; D \not\leq Q} Tr_Q^H((kG)^Q).$$

Hence, $e - \beta_D(e) \in \sum_{Q \leq H; D \not\leq Q} Tr_Q^H((kG)^Q)$ and therefore

$$id_N \in \sum_{Q \leq H, D \not\leq Q} Tr_Q^H(End_{kQ}(N \downarrow_Q^H)).$$

Rosenberg's lemma and Higman's criterion show that for each indecomposable direct summand N' of N there is a $Q \leq H$ with $D \not\leq Q$ so that N' is projective relative Q. □

The following corollary generalises Proposition 2.4.3.

Corollary 2.10.23 *Let k be a field of characteristic $p > 0$ and let G be a finite group. Let D be a p-subgroup of G and let e be a central primitive idempotent of kG. Let M be an indecomposable kG-module and let $f(M)$ be its Green correspondent in $kN_G(D)$. Then $e \cdot M = M \Leftrightarrow \beta_D(e) \cdot f(M) = f(M)$.*

Proof This follows from Proposition 2.10.22 and the description of the Green correspondent. □

2.11 Blocks, Subgroups and Clifford Theory of Blocks

In this section we shall study normal defect groups and how blocks relate to this property.

2.11.1 Brauer Correspondence for Bigger Groups

Lemma 2.10.13 and its corollary has an interesting consequence for idempotents.

Proposition 2.11.1 *Let k be a field of characteristic $p > 0$ and let G be a finite group. If D is a normal p-subgroup of G, then every central idempotent of kG belongs to $kC_G(D)$.*

Proof Let e be a central idempotent of kG. Then by Lemma 2.10.13 we get

$$e \in Z(kG) = (kG)^G \subseteq (kG)^D = kC_G(D) \oplus J,$$

where we denote by $(kG)^G$ the G-fixed points of the conjugation action of G on kG. Hence $e = e_0 + r$ where $e_0 \in kC_G(D)$ and $r \in J$. Now, $J \subseteq \text{rad}(kG)$ and hence J is a nilpotent ideal of $(kG)^D$. Then

$$e_0 + r = e = e^2 = e_0^2 + e_0 r + r e_0 + r^2$$

and therefore $e_0^2 = e_0$ since $e_0 r + r e_0 + r^2 \in J$, J being a two-sided ideal. Moreover,

$$e_0 + e_0 r = e_0 e = e e_0 = e_0 + r e_0$$

implies that e_0 commutes with r. Therefore, using that J is nilpotent and that $\binom{p^s}{\ell}$ is divisible by p as soon as $\ell \notin \{0, p^s\}$,

$$e_0 + r = e = e^{p^s} = e_0 + \sum_{\ell=1}^{p^s} \binom{p^s}{\ell} e_0 r^\ell = e_0$$

if $J^{p^s} = 0$. This proves the lemma. □

Remark 2.11.2 Let H be a subgroup of G such that $N_G(D) \leq H \leq G$. Then by Brauer's first main theorem for a block B of kG with defect group D, there is a unique block b of $kN_G(D)$ with defect group D so that b is a direct factor of $B \downarrow^{G\times G}_{N_G(D)\times N_G(D)}$. Observe that

$$B \downarrow^{G\times G}_{N_G(D)\times N_G(D)} = \left(B \downarrow^{G\times G}_{H\times H}\right) \downarrow^{H\times H}_{N_G(D)\times N_G(D)}$$

Hence if we decompose

$$B \downarrow^{G\times G}_{H\times H} = \beta_1 \oplus \cdots \oplus \beta_s$$

into blocks of kH, then there is a unique $i \in \{1, \ldots, s\}$ such that b is a direct factor of a $(\beta_i) \downarrow^{H\times H}_{N_G(D)\times N_G(D)}$. Hence there is also a unique block β of kH with defect group D such that β is a direct factor of $B \downarrow^{G\times G}_{H\times H}$. We may therefore extend Brauer correspondence to subgroups H containing $N_G(D)$.

Remark 2.11.3 Let k be a field of characteristic $p > 0$ and let G be a finite group. Let D be a p-subgroup of G and let H be a subgroup of G with $DC_G(D) \leq H \leq N_G(D)$. Let $e \neq 0$ be a central primitive idempotent of kH, and let $b = kH \cdot e$. Then by Proposition 2.11.1 we get that $e \in kC_G(D)$. Recall the Brauer homomorphism β_D from Definition 2.10.1. By Lemma 2.10.15 there is a unique primitive idempotent e_{i_0} of $Z(kG)$ with $e = e\beta_D(e_{i_0}) \neq 0$. For all the other primitive central idempotents $e_i \neq e_{i_0}$ in $Z(kG)$ we get $e \cdot \beta_D(e_i) = 0$. We say that B_{i_0} is the *Brauer correspondent* to b and write $B_{i_0} = b^G$. In general the Brauer correspondence for such general subgroups H is not bijective. In the case $H = N_G(D)$ this newly defined Brauer correspondence is exactly the bijective Brauer correspondence Theorem 2.3.10 we have seen earlier as a special case of Green correspondence. This is precisely the statement of Proposition 2.11.4 below.

The treatment of the next statement follows Benson [22, Sect. 6.2].

Proposition 2.11.4 *Let k be a field of characteristic $p > 0$ and let D be a p-subgroup of G. Let B be a block of kG with defect group D and let b be a block of $kN_G(D)$ with defect group D. Then b is the Brauer correspondent of B if and only if $B = b^G$.*

Proof Put $H := N_G(D)$ and let e be the central primitive idempotent such that $B = kG \cdot e$. Lemma 2.10.10 implies that if $B = kGe$ has defect group D, then H contains a conjugate of D if and only if $e \in Tr_{\Delta(H)}^{\Delta(G)}((kG)^{\Delta(H)})$. Rosenberg's Lemma 1.9.16 implies that B has defect group D if and only if $e \notin \sum_{Q<D} Tr_{\Delta(Q)}^{\Delta(G)}((kG)^{\Delta(Q)})$ but $e \in Tr_{\Delta(D)}^{\Delta(G)}((kG)^{\Delta(D)})$. We have seen in Lemma 2.10.13 and its corollary that the Brauer map $\beta_D : (kG)^D \longrightarrow kC_G(D)$ has a kernel which is generated by $\sum_{Q<D} Tr_Q^D(kG)$.

We obtained as an intermediate step that the block generated by e has defect group D if and only if $\beta_D(e) \neq 0$ and $e \in Tr_{\Delta(D)}^{\Delta(G)}((kG)^{\Delta(D)})$.

Next we get for each $x \in (kG)^D$, using Mackey's formula, that

$$\beta_D(Tr_D^G(x)) = \beta_D\left(\sum_{N_G(D)gD \in N_G(D)\backslash G/D} Tr_{N_G(D)\cap {}^gD}^{N_G(D)}(gx)\right) = \beta_D(Tr_D^{N_G(D)}(x))$$

since in the sum the terms for $N_G(D)gD \neq N_G(D)$ all disappear under the Brauer homomorphism. Now, $\beta_D(Tr_D^{N_G(D)}(x)) = Tr_D^{N_G(D)}(\beta_D(x))$. Indeed, the space $(kG)^{N_G(D)}$ is spanned by $N_G(D)$-orbits. Those orbits on which D acts trivially span precisely $(kC_G(D))^{N_G(D)}$ and the other orbits span $\sum_{D<Q\leq N_G(D)} Tr_Q^{N_G(D)}(kG)$. Now, β_D projects onto $(kC_G(D))^{N_G(D)}$. This argument also shows that β_D induces a surjective map $Tr_D^G((kG)^D) \xrightarrow{\beta_D} Tr_D^{N_G(D)}((kC_G(D))^D)$. By Proposition 2.11.1 every central primitive idempotent of $kN_G(D)$ is an element in $kC_{N_G(D)}(D)$. If such a block has defect group D then the primitive central idempotent of this block belongs to $Tr_{\Delta(D)}^{\Delta(N_G(D))}((kC_G(D))^{\Delta(D)})$. Hence for a given block $B = kGe$ with defect group D there is a unique block b of $kN_G(D)$ with defect group D so that $B = b^G$. The Brauer correspondent of $B = kGe$ is the unique direct factor b of $B \downarrow_{N_G(D)\times N_G(D)}^{G\times G}$ with defect group D. Hence $\beta_D(e) \cdot b \neq 0$, and therefore b is the Brauer correspondent of B. □

Remark 2.11.5 Let G be a finite group, let k be an algebraically closed field of characteristic $p > 0$, and let D be a p-subgroup of G. Suppose $DC_G(D) \leq H \leq G$ and let b be a block of kH with defect group D. Then by the Brauer correspondence and Proposition 2.11.4 there is a unique block b' of $kN_H(D)$ with defect group D so that $b = (b')^H$. But $N_H(D) \leq N_G(D)$ and so $DC_G(D) \leq N_H(D) \leq N_G(D) \leq G$. Hence we are in the same setting as Remark 2.11.3 and we know that there is a unique block B of kG such that $B = (b')^G$.

Lemma 2.11.6 (cf Benson [22, Lemma 6.2.7]) *Let k be a field of characteristic $p > 0$ and let G be a finite group. Let D be a p-group and suppose $D \cdot C_G(D) \leq H \leq G$ for some subgroup H of G. Let b be a block of kH with defect group D. Then b^G is the unique block B of kG such that b is a direct factor of $B \downarrow_{H \times H}$.*

Proof Since by the proof of Proposition 2.3.4

$$kG = k((G \times G)/\Delta(G)) = k(G \times G) \otimes_{k\Delta(G)} k = k \uparrow_{\Delta(G)}^{G \times G}$$

we may apply Mackey's theorem and obtain

$$\begin{aligned} kG \downarrow_{H \times H}^{G \times G} &= k \uparrow_{\Delta(G)}^{G \times G} \downarrow_{H \times H} \\ &= \bigoplus_{(H \times H)(1,g)\Delta(G) \in (H \times H)\backslash(G \times G)/\Delta(G)} k \uparrow_{(H \times H) \cap {}^{(1,g)}\Delta(G)}^{H \times H} \\ &= \bigoplus_{(H \times H)(1,g)\Delta(G) \in (H \times H)\backslash(G \times G)/\Delta(G)} k \uparrow_{{}^{(1,g)}\Delta(H \cap {}^{g^{-1}}H)}^{H \times H}. \end{aligned}$$

If

$${}^{(1,g)}\Delta(H \cap {}^{g^{-1}}H) = \{(h, ghg^{-1}) \mid h \in H \cap {}^{g^{-1}}H\}$$

contains a subgroup of the form

$${}^{(h_1,h_2)}\Delta(D) = \{(h_1 d h_1^{-1}, h_2 d h_2^{-1}) \mid d \in D\}$$

for some $(h_1, h_2) \in H \times H$ then

$$h = h_1 d h_1^{-1} \text{ and } ghg^{-1} = h_2 d h_2^{-1}$$

which implies

$$g h_1 d h_1^{-1} g^{-1} = h_2 d h_2^{-1}$$

and hence

$$h_2^{-1} g h_1 \in C_G(D) \leq H.$$

This shows $g \in H$. Therefore the only summands in

$$kG \downarrow_{H \times H}^{G \times G} = \bigoplus_{(H \times H)(1,g)\Delta(G) \in (H \times H)\backslash(G \times G)/\Delta(G)} k \uparrow_{{}^{(1,g)}\Delta(H \cap {}^{g^{-1}}H)}^{H \times H}$$

which can have defect group D are those with $g \in H$, or what is the same, the class given by $(1, 1)$. This term is obtained by putting $g = 1$ in the above direct sum, and is hence

$$k \uparrow_{\Delta(H)}^{H \times H} = kH.$$

This shows that $kG \downarrow_{H \times H}^{G \times G}$ has only one direct factor with vertex $\Delta(D)$, and therefore only one direct factor isomorphic to b. This implies that there is only one block $\check{B}$ of kG such that $\check{B} \downarrow_{H \times H}$ has a direct factor isomorphic to b.

We need to show that $\check{B} = b^G$. Suppose first that $H \leq N_G(D)$. Then let B be a block of kG such that b is not isomorphic to a direct factor of $B \downarrow_{H \times H}$. Let e and e' be the central primitive idempotents such that $b = e \cdot kH$ and $B = e' \cdot kG$. Consider kH as a direct factor of kG, and observe that the projection of $e \cdot e'$ onto this direct factor is 0. Hence $e\beta_D(e') = \beta_D(ee') = 0$ and therefore $B \neq b^G$.

If H is not contained in $N_G(D)$, then we may first pass to $N_H(D) \leq N_G(D)$, and then to H as is indicated in Remark 2.11.5. We have proved the lemma. □

2.11.2 Blocks and Normal Subgroups

This section closely follows the treatment in Benson [22].

The case of a normal p-subgroup D of G is particularly important. In this connection we prove the following general statement.

Lemma 2.11.7 *Let k be a field of characteristic $p > 0$ and let G be a finite group. Let $D \trianglelefteq G$ and D be a p-subgroup. Then the natural projection $G \longrightarrow G/D$ induces a ring epimorphism $kG \longrightarrow kG/D$ with nilpotent kernel $I(kD)G$.*

Proof The fact that $G \longrightarrow G/D$ induces a ring epimorphism $\pi_D : kG \longrightarrow kG/D$ follows from Lemma 1.2.3. The kernel $I(kD)G$ of this ring epimorphism is a two-sided ideal generated by the elements $(d-1)$ for $d \in D$. The ideal $I(kD)$ is a nilpotent ideal of kD by Proposition 1.6.22. Now $I(kD)G$ is generated as a k-vector space by the elements $(d-1)g$ for all $d \in D$ and $g \in G$, and hence for $d, d' \in D$ and $g, g' \in G$ we get

$$(d-1)g \cdot (d'-1)g' = (d-1)((gd'g^{-1}) - 1)gg'$$

where we observe that $gd'g^{-1} \in D$ since D is normal in G. This shows that $I(kD)G$ is nilpotent, and hence is in the radical of kG. □

Remark 2.11.8 The combination of Lemma 2.11.7 and Lemma 1.9.17 shows that if D is a normal p-subgroup of G, then we may lift idempotents of kG/D to idempotents of kG. However, a central idempotent of kG/D need not be lifted to a central idempotent of kG. Nevertheless, Proposition 2.11.1 gives the relevant statement for central idempotents. Lifting is not the correct concept there.

Definition 2.11.9 Let k be a field and let G be a finite group. Let e be an idempotent of kG. Denote by

$$I_G(b) := I_G(e) := \{g \in G \mid geg^{-1} = e\}$$

the *inertia group* of e, respectively of b.

It is clear that if e is an idempotent of kG, then geg^{-1} is again an idempotent of kG and that $I_G(e)$ is a subgroup of G. A particularly interesting use of this fact is the following. Let N be a normal subgroup of G and let b be a block of kN. Given $g \in G$, then $g \cdot b \cdot g^{-1}$ is a direct factor of

$$g \cdot kN \cdot g^{-1} = k(gNg^{-1}) = kN,$$

and hence is again a block of kN. If $b = e \cdot kN$, then

$$g \cdot b \cdot g^{-1} = b \Leftrightarrow g \cdot e \cdot g^{-1} = e \Leftrightarrow g \in I_G(e)$$

and moreover $g \cdot e \cdot g^{-1}$ is again a primitive central idempotent of kN. Hence,

$$(*) \quad b \cap (g \cdot b \cdot g^{-1}) = 0 \text{ if } g \in G \setminus I_G(e).$$

Therefore we may consider

$$\begin{aligned} \hat{b} := \bigoplus_{gI_G(b) \in G/I_G(b)} g \cdot b \cdot g^{-1} &= \bigoplus_{gI_G(b) \in G/I_G(b)} g \cdot kNe \cdot g^{-1} \\ &= \bigoplus_{gI_G(b) \in G/I_G(b)} kN \cdot (g \cdot e \cdot g^{-1}), \end{aligned}$$

where the last equality follows since $N \trianglelefteq G$. Let

$$\sum_{gI_G(b) \in G/I_G(b)} g \cdot e \cdot g^{-1} =: \hat{e}.$$

Then

$$kN \cdot \hat{e} \subseteq \hat{b} \subseteq kG \cdot \hat{e}$$

by definition. Moreover,

$$B := kG \cdot \hat{e}$$

is a direct sum of blocks of kG. Indeed, $\hat{e}$ is an idempotent of kG since

$$(g \cdot e \cdot g^{-1}) \cdot (h \cdot e \cdot h^{-1}) = g \cdot \Big(e \cdot \Big((g^{-1}h) \cdot e \cdot (g^{-1}h)^{-1}\Big)\Big) \cdot g^{-1} = 0$$

if $g^{-1}h \notin I_G(b)$ by $(*)$. Then $\hat{e}$ is central since e is central in kN and conjugation by any $g \in G$ fixes $\hat{e}$ by construction.

However, in general, $\hat{e}$ is not primitive. If f is a central primitive idempotent of kG, then either $f\hat{e} = 0$, or $f\hat{e} = f$. We obtain $f\hat{e} = f$ if and only if $fe = e$.

Hence, for every block b of kN there are (possibly many) blocks $B = kGf$ of kG such that $B \cdot b = b$. In this case we say that B *covers* b and write $B = b^G$. We have proved the following.

Proposition 2.11.10 *Let k be a field of characteristic $p > 0$, let G be a finite group and let N be a normal subgroup of G. Let $b = kN \cdot e$ be a block of kN with block idempotent $e^2 = e \in Z(kN)$.*

- *Then for $\sum_{gI_G(b) \in G/I_G(b)} g \cdot e \cdot g^{-1} =: \hat{e}$ we get that $\hat{e}^2 = \hat{e} \in Z(kG)$ and $kG \cdot \hat{e}$ is a direct sum of blocks $B = kGf$ of kG. These are characterised by the equation $B \cdot \hat{e} = B$ or $f \cdot e = e$. We say that B covers b and write $B = b^G$.*
- *If b has defect group D and if $C_G(D) \leq N$, then there is a unique block B covering b.*
- *Moreover, precisely the $|G/I_G(b)|$ blocks $g \cdot b \cdot g^{-1}$ for $gI_G(b) \in G/I_G(b)$ are covered by B.*

Proof The first part was shown above. The third part is part of the definition of the idempotent $\hat{e}$. The only statement left to prove is that B is the unique block covering b if $C_G(D) \leq N$. But this statement is actually a consequence of Lemma 2.11.6. □

We even get an explicit description of $kG \cdot \hat{e}$ in terms of $kI_G(e) \cdot e$.

Lemma 2.11.11 *Let k be a field of characteristic $p > 0$ and let G be a finite group. Let $N \trianglelefteq G$ and let $b = kN \cdot e$ be a block of kN with $e^2 = e \in Z(kN)$. Suppose B is a block of kG covering b. Put $\hat{e} := \sum_{gI_G(b) \in G/I_G(b)} geg^{-1}$. Then there is a defect group D of B contained in $I_G(b)$. Moreover*

$$kG\hat{e} \simeq Mat_{|G/I_G(b)|}(kI_G(b) \cdot e).$$

Proof We need to restrict $kG \cdot \hat{e}$ to $N \times N$. First,

$$kG \simeq \bigoplus_{gN \in G/N} g \cdot kN$$

as $k(N \times N)$-modules, using the fact that N is normal in G. Further, the definition of $I_G(b)$ implies that b is actually a $\Delta(I_G(b)) \cdot (N \times N)$-module. Then we get the following isomorphisms as $k(N \times N)$-modules

$$\begin{aligned}
\hat{e} \cdot kG = \hat{e} \cdot \left(\bigoplus_{gN \in G/N} g \cdot kN \right)
&= \bigoplus_{gN \in G/N} \hat{e} \cdot (g \cdot kN) \\
&= \bigoplus_{gN \in G/N} \bigoplus_{hI_G(b) \in G/I_G(b)} \left(h \cdot e \cdot h^{-1} \cdot g \cdot kN \right) \\
&= \bigoplus_{gN \in G/N} \bigoplus_{hI_G(b) \in G/I_G(b)} \left(h \cdot e \cdot (h^{-1} \cdot g) \cdot kN \cdot (g^{-1} \cdot h) \cdot (h^{-1} \cdot g) \right) \\
&= \bigoplus_{gN \in G/N} \bigoplus_{hI_G(b) \in G/I_G(b)} \left(h \cdot e \cdot kN \cdot (h^{-1} \cdot g) \right) \\
&= \bigoplus_{(h_1, h_2) \in (G \times G)/(\Delta(I_G(b)) \cdot (N \times N))} \left(h_1 \cdot e \cdot kN \cdot h_2^{-1} \right) \\
&= k(G \times G) \otimes_{k(\Delta(I_G(b)) \cdot (N \times N))} b
\end{aligned}$$

where we put $h_1 = h$ and $h_2 = g^{-1}h$ in the second last isomorphism. Now, the defect group is the smallest group D such that B is $\Delta(D)$-projective. Hence

$$\Delta(D) \leq \Delta(I_G(b)) \cdot (N \times N).$$

Since $N \leq I_G(b)$, since a defect group D of B is a subgroup of $\Delta(G)$ by Proposition 2.3.4, and since

$$\Delta(G) \cap (\Delta(I_G(b)) \cdot (N \times N)) = \Delta(I_G(b))$$

we get that $D \leq I_G(b)$.

We can consider $e \cdot kG$ as a $kI_G(b)$-left module and observe that

$$e \cdot kG = \bigoplus_{gI_G(b) \in G/I_G(b)} (e \cdot kI_G(b)) \cdot g^{-1}$$

so that $e \cdot kG$ is free of rank $|G/I_G(b)|$ as an $e \cdot kI_G(b)$-left module. Hence there is a ring isomorphism

$$End_{e \cdot kI_G(b)}(e \cdot kG)^{op} \simeq Mat_{|G/I_G(b)|}(e \cdot kI_G(b)).$$

Now, we observe $\hat{e}$ is central in kG and hence we may define a mapping

$$\begin{aligned}
\hat{e} \cdot kG &\longrightarrow End_{e \cdot kI_G(b)}(e \cdot kG)^{op} \\
\hat{e} \cdot x &\mapsto (ey \mapsto ey\hat{e}x = e\hat{e}yx = eyx)
\end{aligned}$$

We see immediately that this map is a ring homomorphism. Consider the kernel of this ring homomorphism:

If $ey \mapsto eyx = 0$ for all $y \in kG$, then

$$\begin{aligned}\hat{e}x = (\hat{e}\hat{e})x = \hat{e}(\hat{e}x) &= \sum_{gI_G(b)\in G/I_G(b)} geg^{-1}(\hat{e}x)\\ &= \sum_{gI_G(b)\in G/I_G(b)} ge((g^{-1}\hat{e})x) = \sum_{gI_G(b)\in G/I_G(b)} g\cdot 0 = 0\end{aligned}$$

since $(g^{-1}\hat{e})$ is a possible y. Hence the ring homomorphism is injective.

We claim that the ring homomorphism is surjective as well. For this have

$$\begin{aligned}\hat{e}\cdot kG &= \bigoplus_{(h_1,h_2)\in(G\times G)/(\Delta(I_G(b))\cdot(N\times N))} \left(h_1\cdot e\cdot kN\cdot h_2^{-1}\right)\\ &= \bigoplus_{(h_1,h_2)\in(G\times G)/(I_G(b)\times I_G(b))} \left(h_1\cdot e\cdot kI_G(b)\cdot h_2^{-1}\right)\end{aligned}$$

as $k(N\times N)$-modules. Moreover, the lines and columns of $Mat_{|G/I_G(b)|}(e\cdot kI_G(b))$ are parameterised by classes $gI_G(b)$. The element $x\in ekI_G(b)$ in row $g_1I_G(b)$ and column $g_2I_G(b)$, and 0 elsewhere, is realised by right multiplication by $g_1xg_2^{-1}$. This proves the statement. □

Proposition 2.11.12 *Let k be a field of characteristic $p>0$ and let G be a finite group with normal subgroup N. If b is a block of kN generated by the central idempotent e, let B be a block of kG covering b. Then there is a defect group D of B such that $D\cap N$ is a defect group of b.*

Proof Let D be a defect group of B. Since by the proof of Proposition 2.3.4

$$kG = k(G\times G)\otimes_{\Delta(G)} k,$$

the $k(G\times G)$-module B is a direct factor of $k\uparrow_{\Delta(D)}^{G\times G}$ and $B\downarrow_{\Delta(D)}^{G\times G}$ has the trivial module k as a direct factor. This shows that $B\downarrow_{(\Delta(D)\cap(N\times N)}^{G\times G}$ also has the trivial module as a direct factor. Hence, some direct factor of the restriction $B\downarrow_{N\times N}^{G\times G}$ has vertex in $\Delta(D)\cap(N\times N)=\Delta(D\cap N)$.

We shall consider now $B\downarrow_{N\times N}^{G\times G}$. As in the proof of Lemma 2.11.11, we get that $B\downarrow_{N\times N}^{G\times G}$ is a direct sum of $k(N\times N)$-modules of the form $g_1bg_2^{-1}$ for $g_1,g_2\in G$. Denote the defect group of b by D_b and recall that then $\Delta(D_b)$ is the vertex of b as a $k(N\times N)$-module. Then the vertex of each of the $k(N\times N)$-modules $g_1bg_2^{-1}$ are $G\times G$-conjugates of $\Delta(D_b)$ and hence some G-conjugate of $D\cap N$ is in D_b.

Conversely, since b is a direct factor of $B\downarrow_{N\times N}^{G\times G}$, b is relatively projective to some $\Delta(G)$-conjugate of $\Delta(D\cap N)$. Hence we get equality and the result is proven. □

Although in the previous results of this subsection we did not need the base field to be algebraically closed, in the next lemma we do need this hypothesis.

Lemma 2.11.13 *Let k be an algebraically closed field of characteristic $p > 0$ and let G be a finite group. Let N be a normal subgroup of G and let $b = kN \cdot e$ be a block of kN where $e^2 = e \in Z(kN)$, and suppose that b is covered by the block B of kG with defect group D chosen so that $D \cap N$ is a defect group of b. If $C_G(D \cap N) \leq N$ then $|I_G(b) : D \cdot N|$ is not divisible by p.*

Proof By Lemma 2.11.6 B is the only block of kG such that b is a direct factor of $B^{G\times G}_{N\times N}$, and hence B is the unique block covering b. As in the proof of Lemma 2.11.11, by Definition of $I_G(b)$ we see that b is a $\Delta(I_G(b)) \cdot (N \times N)$-module. Since b is a block of kN, the restriction of b as a $\Delta(I_G(b)) \cdot (N \times N)$-module to $N \times N$ is still indecomposable.

Let $S \in Syl_p(I_G(b))$. Then the restriction of b to $\Delta(S) \cdot (N \times N)$ is still indecomposable. Since by hypothesis $D \cap N$ is a defect group of b, the $\Delta(S) \cdot (N \times N)$-module b is projective relative to $\Delta(D) \cdot (N \times N)$. Green's indecomposability Theorem 2.2.12 shows that if L is a $\Delta(D) \cdot (N \times N)$-module such that b is a direct factor of $L \uparrow^{\Delta(S)\cdot(N\times N)}_{\Delta(D)\cdot(N\times N)}$, then this induced module is indecomposable. Hence

$$b \simeq L \uparrow^{\Delta(S)\cdot(N\times N)}_{\Delta(D)\cdot(N\times N)} .$$

But $b \downarrow^{\Delta(S)\cdot(N\times N)}_{\Delta(D)\cdot(N\times N)}$ is indecomposable, as we have just seen. Therefore, by Mackey's formula,

$$\begin{aligned} b \downarrow^{\Delta(S)\cdot(N\times N)}_{\Delta(D)\cdot(N\times N)} &= L \uparrow^{\Delta(S)\cdot(N\times N)}_{\Delta(D)\cdot(N\times N)} \downarrow^{\Delta(S)\cdot(N\times N)}_{\Delta(D)\cdot(N\times N)} \\ &= \bigoplus_{DNsDN \in D\backslash S/D} {}^{s}L \downarrow_{\Delta(DN \cap {}^{s}DN)\cdot(N\times N)} \uparrow^{\Delta({}^{s}D)\cdot(N\times N)} . \end{aligned}$$

Since the left-hand side is indecomposable, the right-hand side is indecomposable as well and there is only one double-class. Therefore $S \cdot N = D \cdot N$. Hence $|I_G(b) : DN|$ is not divisible by p. □

Proposition 2.11.14 *Let k be a field of characteristic $p > 0$ and let G be a finite group. Let Q be a p-group of G and suppose $G = Q \cdot C_G(Q)$. Then the natural map $G \longrightarrow G/Q$ induces a bijective correspondence between blocks of kG with defect group D and blocks of kG/Q with defect group D/Q.*

Proof Since $Q \trianglelefteq (Q \cdot C_G(Q)) = G$ we may apply Lemma 2.4.2 to get $Q \leq D$. Lemma 2.11.1 shows that each primitive central idempotent of kG belongs to $kC_G(Q)$. But $G = Q \cdot C_G(Q)$ implies

$$Z(G) = Z(Q \cdot C_G(Q)) \geq Z(C_G(Q))$$

and so the primitive central idempotents of kG are precisely the central primitive idempotents of $kC_G(Q)$. Finally, the kernel of $kG \longrightarrow kG/Q$ is generated as an ideal by the elements $q - 1$, for $q \in Q$, and is therefore nilpotent. Moreover,

$$kG/Q = k((Q \cdot C_G(Q))/Q) = k(C_G(Q)/(Q \cap C_G(Q))) = k(C_G(Q)/Z(Q))$$

and therefore the commutative diagram

$$\begin{array}{ccc} Z(kC_G(Q)) & \hookrightarrow & kG \\ \downarrow & & \downarrow \\ Z(k(G/Q)) & \hookrightarrow & kG/Q \end{array}$$

has surjective vertical morphisms. Hence every central primitive idempotent of kG/Q lifts to a central primitive idempotent of kG, and of course every central primitive idempotent of kG maps to a central primitive idempotent of kG/Q. This proves the statement. □

We summarise what we proved so far.

Proposition 2.11.15 (cf Benson [22, Theorem 6.4.3]) *Let k be an algebraically closed field of characteristic $p > 0$, let G be a finite group. Then there is a bijection between each of the following sets.*

1. *Blocks of kG with defect group D.*
2. *Blocks of $kN_G(D)$ with defect group D.*
3. *$N_G(D)$-conjugacy classes of blocks b of $kDC_G(D)$ with defect group D such that p does not divide $|I_{N_G(D)}(b) : DC_G(D)|$.*
4. *$N_G(D)$-conjugacy classes of blocks b of $kDC_G(D)/D$ of defect 0 such that p does not divide $|I_{N_G(D)}(b) : DC_G(D)|$.*

Proof The bijection between (1) and (2) is the Brauer correspondence. The bijection between (3) and (4) is Proposition 2.11.14. The bijection between (2) and (3) is obtained as follows. Since $C_G(D) = C_{N_G(D)}(D)$, we may assume that $G = N_G(D)$ and we may therefore assume that $D \trianglelefteq G$. Let $N := DC_G(D)$, let B be a block of kG with defect group D and let b be a block of kN that is covered by B. Lemma 2.11.13 shows that p does not divide $|I(b) : DC_G(D)|$ and by Proposition 2.11.12 b also has defect group D. Proposition 2.11.10 shows that the blocks covered by B form a conjugacy class under the $N_G(D)$-action.

Conversely if b is a block of kN, then by Proposition 2.11.10 there is a unique block B of kG covering b. By Lemma 2.11.11 the defect group D of B can be chosen so that $D \leq I(b)$. But p does not divide $|I_{N_G(D)}(b) : DC_G(D)|$, so that $I_{N_G(D)}(b) \cap N = D$. □

2.12 Representation Type of Blocks, Cyclic Defect

Since $D \trianglelefteq N_G(D)$ we see that it is important in view of the Brauer and Green correspondence to study the blocks of a group ring with normal defect group. The case of a normal cyclic defect group is particularly simple.

2.12.1 The Structure of Blocks with Normal Cyclic Defect Group

The results of this section are partially a special case of the results of Sect. 2.8. However, the abstract case with which we dealt there is made explicit here for group rings and the structure constants which we obtained there get a group theoretical interpretation here.

Proposition 2.12.1 *Let k be a field of characteristic $p > 0$ and let G be a finite group. Let B be a block of G with normal cyclic defect group $D \trianglelefteq G$. Then* $\mathrm{rad}(B) = (d-1)B$ *where d is a generator of D. Moreover each indecomposable projective B-module is uniserial of Loewy length $|D|$. If m is the number of isomorphism classes of simple B-modules, then there are exactly $m \cdot |D|$ isomorphism classes of indecomposable B-modules.*

Proof Let $kG \cdot e = B$ for an idempotent $e \in Z(kG)$. Since D is normal, $\mathrm{rad}(kD) \cdot B = \mathrm{rad}(kD) \cdot kG \cdot e$ is an ideal in $\mathrm{rad}(B)$ by Lemma 2.11.7, i.e. $\mathrm{rad}(kD) \cdot B \subseteq \mathrm{rad}(B)$. Moreover, since D is cyclic, $\mathrm{rad}(kD) = (d-1) \cdot kD$ and since D is normal in G, we get

$$(d-1) \cdot kG = kG \cdot (d-1)$$

and therefore

$$(d-1) \cdot B = B \cdot (d-1) \subseteq \mathrm{rad}(B).$$

We shall show that $B/(d-1)B$ is semisimple. Indeed, since B is a direct factor of kG, we get that $B/(d-1)B$ is a direct factor of $kG/(d-1)kG = k(G/D)$. Let

$$0 \longrightarrow S \longrightarrow U \longrightarrow T \longrightarrow 0$$

be an exact sequence of $B/(d-1)B$-modules. Hence D acts trivially on each of these modules. Since k is a field, and all short exact sequences of k-modules are split as a sequence of k-modules, this shows that the sequence is actually split as a sequence of kD-modules. Since S, T and U are B-modules, they have their vertex in D, hence are relatively D-projective by Proposition 2.3.6. This shows that the sequence

$$0 \longrightarrow S \longrightarrow U \longrightarrow T \longrightarrow 0$$

is actually split as a sequence of kG-modules. Hence we obtain the result. Theorem 2.8.4 shows that B is serial. Lemma 2.4.7 shows that B is actually a symmetric algebra. Lemma 2.8.2 shows that each indecomposable B-module is uniserial. All projective indecomposable B-modules have the same Loewy length, and this Loewy length is obviously the nilpotency degree of $\mathrm{rad}(kD)$, whence $|D|$. Moreover, each indecomposable B-module is uniserial, hence isomorphic to some $P/\mathrm{rad}^n P$ for some integer n and a projective indecomposable B-module P. There are precisely $|D| \cdot m$ such modules, if there are m isomorphism classes of simple B-modules. □

As a corollary we mention a result due to Michler.

Corollary 2.12.2 (Michler) *Let B be a block with normal cyclic defect group $D = \langle d | d^{p^m} \rangle$, and let M be an indecomposable B-module. Then there is an integer s and a primitive idempotent e of B such that $M \simeq B \cdot (d-1)^s \cdot e$. In particular there are only finitely many isomorphism classes of indecomposable B-modules.* □

Corollary 2.12.3 *Let B be a block with normal cyclic defect group, then each projective indecomposable B-module is uniserial.* □

Proposition 2.12.4 *Let k be an algebraically closed field and let G be a finite group. Let B be a block of kG with normal cyclic defect group D. Then there are e isomorphism classes of simple B-modules represented by $S_1, S_2, \dots, S_n$. The simple modules can be numbered in such a way that for the projective covers P_i of S_i one gets $\mathrm{rad}^i P_j/\mathrm{rad}^{i+1} P_j \simeq S_{i+j}$ for all $j \in \{1, \dots, n\}$ and for all $i \in \{0, 1, \dots, |D|\}$. Moreover, n divides $|D| - 1$.*

Proof The only new statement is the fact that n divides $|D|-1$. But this follows from the fact that B is symmetric, and that the Loewy structure of B is the same as for the Nakayama algebra N_n^{en+1}. Since $en + 1 = |D|$, we see that n divides $|D| - 1$. The rest is just a reformulation of Theorem 2.8.4, Lemma 2.4.7 and Lemma 2.8.2. □

Remark 2.12.5 We will come back to the theory of blocks with cyclic defect group D in Sect. 5.10. Blocks with cyclic defect group are actually very well understood and the structure described in Proposition 2.12.4 is its simplest model. Indeed, the attentive reader will observe that the composition series of the blocks with cyclic normal defect group are the same as for the Nakayama algebra $N_n^{|D|+1}$. In Example 4.4.3 we will see that this is actually a very close correspondence, a Morita equivalence, which will be introduced in Chap. 4.

2.12.2 Blocks of Finite Representation Type

We are ready to deal with the number of isomorphism classes of indecomposable modules. Cyclic defect groups play a prominent role here.

Definition 2.12.6 Let k be a commutative ring and let A be a k-algebra satisfying the Krull-Schmidt theorem. If there are only a finite number of isomorphism classes of finitely generated indecomposable A-modules, then we say that A is of *finite representation type.*

Algebras satisfying the Krull-Schmidt theorem and which are not of finite representation type are of *infinite representation type.*

Remark 2.12.7 We shall develop this theme further in Sect. 5.10. The notion of representation type is crucial for the representation theory of finite dimensional algebras.

We have seen in Corollary 2.12.2 that blocks with cyclic defect groups are of finite representation type. As a first result in this section we shall prove the converse: blocks of finite representation type have cyclic defect group.

We shall start with an example.

Example 2.12.8 Let p be a prime number, let k be an algebraically closed field of characteristic $p > 0$ and let

$$G = C_p \times C_p = \langle x, y \mid x^p, y^p, xyx^{-1}y^{-1} \rangle$$

be the direct product of two cyclic groups of order p. We claim that there are infinitely many isomorphism classes of indecomposable kG-modules.

We see that

$$\begin{aligned} kG &\simeq k[X,Y]/(X^p - 1, Y^p - 1) \simeq k[X,Y]/((X-1)^p, (Y-1)^p) \\ &\simeq k[X,Y]/(X^p, Y^p) \end{aligned}$$

just as we did for the group ring of a cyclic group of prime power order. We may write this algebra as a quiver algebra with one vertex, two loops ξ and η and relations $\xi^p, \eta^p, \xi\eta - \eta\xi$.

A representation of dimension d of this algebra is therefore a vector space V of dimension d together with two commuting endomorphisms x and y of V which are nilpotent of degree at most p. Isomorphism classes are realised by simultaneous conjugation of these endomorphisms. Hence we may suppose that one of these endomorphisms, say x, is in Jordan normal form.

Let $d = p$. Suppose moreover now that x has only one Jordan block, that is x is nilpotent of degree p, and not of lower degree. Hence we can choose a basis of V so that x is represented by the matrix

$$N := \begin{pmatrix} 0 & 1 & 0 & \dots & 0 \\ \vdots & \ddots & \ddots & \ddots & \vdots \\ \vdots & & \ddots & \ddots & 0 \\ \vdots & & & \ddots & 1 \\ 0 & \dots & \dots & \dots & 0 \end{pmatrix}.$$

Then an elementary matrix multiplication shows that y commutes with x if and only if

$$y = \sum_{j=0}^{p-1} \alpha_j N^j$$

for $\alpha_j \in k$. We claim that if $\alpha_j = 0$ for all $j \neq 1$ we obtain an indecomposable representation $R(\alpha_1)$, and two of these representations $R(\alpha_1)$ and $R(\beta_1)$ are isomorphic if and only if $\alpha_1 = \beta_1$.

First, $R(\alpha_1)$ is indecomposable since its restriction to the action of x is the regular $k[X]/X^p$-module, and hence is already indecomposable.

Second, if $R(\alpha_1) \simeq R(\beta_1)$, then there is an invertible endomorphism σ of V commuting with x and conjugating $\alpha_1 x$ to $\beta_1 x$. But since σ commutes with x, it commutes with $\alpha_1 x$ as well.

Since k is algebraically closed, and hence infinite, we have proved the statement.

What is the vertex of the module $R(\alpha)$? Since $R(\alpha)$ is of dimension p and since kG is a local algebra of dimension p^2, the module $R(\alpha)$ is not a projective module. Hence, the vertex of $R(\alpha)$ is at least isomorphic to C_p. There are p different subgroups isomorphic to C_p in C_p^2. Since there are only finitely many isomorphism classes of indecomposable modules of kC_p, only finitely many isomorphism classes of modules can have vertex C_p for each of the finitely many copies of $C_p \leq C_p^2$. Since there are infinitely many indecomposable modules of the form $R(\alpha)$ we know that infinitely many of these have vertex $G = C_p^2$.

We can now formulate the converse to Corollary 2.12.2.

Proposition 2.12.9 *Let k be a field of characteristic $p > 0$ and let G be a finite group. Let B be a block with defect group D. If B is of finite representation type, then D is a cyclic group. In particular a block of a finite group over a field k of characteristic $p > 0$ is of finite representation type if and only if its defect group is cyclic.*

Proof Suppose to the contrary that D is not cyclic. Then, denoting by D' the commutator subgroup, D/D' is abelian not cyclic, and hence there is a quotient $D \twoheadrightarrow C_p \times C_p$. Since by Example 2.12.8 there are infinitely many indecomposable $k(C_p \times C_p)$-modules with vertex $C_p \times C_p$, using the morphism $D \twoheadrightarrow C_p \times C_p$ there are infinitely many indecomposable kD-modules. If M is a $k(C_p \times C_p)$-module with vertex $C_p \times C_p$, then M, regarded as a kD-module, has vertex D. Indeed, if M is a direct factor of $M \downarrow_S \uparrow^D$ for a proper subgroup S of D, since the kernel of the projection $D \twoheadrightarrow C_p \times C_p$ acts trivially on M, the group S is actually the preimage in D of a subgroup $\overline{S}$ of $C_p \times C_p$. Since the vertex of M is $C_p \times C_p$, we get that $\overline{S} = C_p \times C_p$ and hence $S = D$.

We form $M \uparrow_D^{N_G(D)}$. Since D is a p-group, D acts trivially on any simple $kN_G(D)$-module, and hence any simple $kN_G(D)$-module is actually a $kN_G(D)/D$-module. However, the socle of M is a trivial D-module k^s. Hence $kN_G(D) \otimes_{kD} k^s \simeq (kN_G(D)/D)^s$ is a submodule of $M \uparrow_D^{N_G(D)}$. But the socle of $kN_G(D)$ is the same as the socle of $kN_G(D)/D$ and hence any simple $kN_G(D)$-module is a submodule of $M \uparrow_D^{N_G(D)}$. Therefore, for every block b' of $kN_G(D)$ we get $b' \cdot M \uparrow_D^{N_G(D)} \neq 0$.

Let b be the Brauer correspondent of B in $kN_G(D)$. Then $b \cdot M \uparrow_D^{N_G(D)} \neq 0$ by the above. Let M' be an indecomposable direct factor of $b \cdot M \uparrow_D^{N_G(D)}$. Now $M \uparrow_D^{N_G(D)} \downarrow_D$ is a direct sum of $N_G(D)$-conjugates of M, the module M is a source of M'. This implies that two non-isomorphic, non-$N_G(D)$-conjugate $k(C_p \times C_p)$-modules M_1 and M_2 as constructed above yield two non-isomorphic b-modules M_1' and M_2'. Since for a given M there are only finitely many $N_G(D)$-conjugates, we obtain the statement. □

Example 2.12.10 The above construction is actually an instance of what we have seen already. We return to Example 1.6.23. Recall that the Kronecker quiver Q is given by

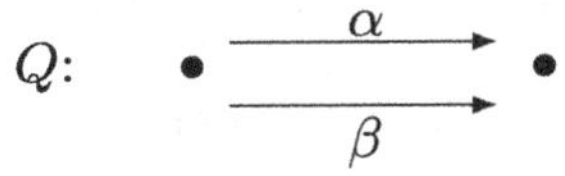

and we have seen in Example 1.6.23 that for every field K there are infinitely many isomorphism classes of indecomposable KQ-modules of dimension 2, and that these modules are parameterised by a projective line. Now, in identifying the two idempotents of KQ we observe that the algebra KQ has a quotient $A := K\overline{Q}/I$ where $\overline{Q}$ is the quiver

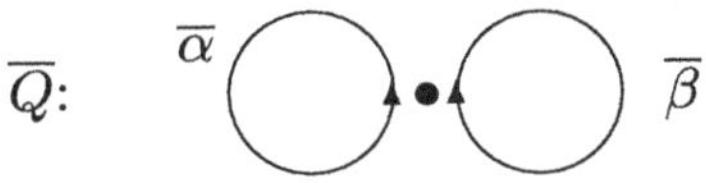

and where I is the ideal

$$I =< (\overline{\alpha})^2, (\overline{\beta})^2, \overline{\alpha}\overline{\beta}, \overline{\beta}\overline{\alpha} > .$$

Graphically this is most evident since by identifying the two vertices, the two arrows become loops. The relations arise since we cannot compose any arrows in Q, and so any composition in $\overline{Q}$ has to be in the ideal. The two-dimensional indecomposable KQ-modules from Example 1.6.23 can again be found in A. What are the two-dimensional A-modules? A has to be a nilpotent 2×2 matrix. Hence, by conjugating properly, $\overline{\alpha}$ may be represented by the matrix $\begin{pmatrix} 0 & \lambda \\ 0 & 0 \end{pmatrix}$ for $\lambda \in \{0, 1\}$. If $\lambda = 1$, then $\overline{\beta}$ also needs to be represented by such a matrix, annihilating the representing matrix of $\overline{\alpha}$ from the left and from the right. This shows that $\overline{\beta}$ is also represented by a matrix $\begin{pmatrix} 0 & \mu \\ 0 & 0 \end{pmatrix}$ for $\mu \in K$. If $\lambda = 1$ and $\mu \neq 0$, then each of these representations B^μ is indecomposable and $B^\mu \simeq B^{\mu'}$ if and only if $\mu = \mu'$. Indecomposability is verified by looking at the endomorphism ring. An endomorphism is a 2×2-matrix, commuting with the action of $\overline{\alpha}$ and $\overline{\beta}$. Only scalar multiples of the identity matrix satisfy this condition. An isomorphism is given by a regular 2×2-matrix, commuting with $\begin{pmatrix} 0 & 1 \\ 0 & 0 \end{pmatrix}$ and conjugating $\begin{pmatrix} 0 & \mu \\ 0 & 0 \end{pmatrix}$ to $\begin{pmatrix} 0 & \mu' \\ 0 & 0 \end{pmatrix}$. There is no such matrix if $\mu \cdot \mu' \neq 0$ and $\mu \neq \mu'$.

This shows that A is representation infinite. Trivially, if B is another K-algebra which admits a quotient A, then B is representation infinite as well. Indeed, any representation of A becomes a representation of B via the ring homomorphism $B \longrightarrow\!\!\!\!\rightarrow A$, indecomposability is preserved by the surjectivity of $B \longrightarrow\!\!\!\!\rightarrow A$ and by the same argument two of these B-modules are isomorphic if they are over A.

An example is $K(C_p \times C_p)$ where K is an algebraically closed field of characteristic $p > 0$. Since

$$K(C_p \times C_p) \simeq K[X, Y]/(X^p, Y^p) \longrightarrow\!\!\!\!\rightarrow K[X, Y]/(X^2, Y^2, XY) \simeq A$$

we obtain that $K(C_p \times C_p)$ is representation infinite. This is another way to approach parts of Proposition 2.12.9.

Remark 2.12.11 The complete structure of blocks with cyclic defect group will be determined in Sect. 5.10, namely Theorem 5.10.37. This quite involved theory can be shown using more elaborate methods, namely equivalences between stable categories, which we are going to study in Chap. 5, and the theory of nilpotent blocks, which will be developed in Sect. 4.4.2.

References

1. Külshammer, B.: Nilpotent Blocks Revisited in "Groups, Rings and Group Rings", pp. 263–274. Chapman and Hall, Boca Raton (2006)
2. Auslander, M., Kleiner, M.: Adjoint functors and an extension of the Green correspondence for group representations. Adv. Math. **104**, 297–314 (1994)
3. Green, J.A.: Some remarks on defect groups. Mathematische Zeitschrift **107**, 133–150 (1968)
4. Curtis, C.W., Reiner, I.: Methods of Representation Theory I. Wiley, Berlin (1982)
5. Curtis, C.W., Reiner, I.: Methods of Representation Theory II. Wiley, Berlin (1986)
6. Cassels, J.W.S., Fröhlich, A.: Algebraic Number Theory. Academic Press, London (1967)
7. Serre, J.-P.: Corps Locaux. Hermann, Paris (1968)
8. Lam, T.-Y.: A First Course in Noncommutative Rings. Springer, New York (2001)
9. Eisenbud, D.: Commutative Algebra with a View Toward Algebraic Geometry. Springer, Berlin (1996)
10. Zariski, O., Samuel, P.: Commutative Algebra, vol. II. Van Nostrand Compagny, New Jersey (1960)
11. Roggenkamp, K.W.: An extension of the Noether-Deuring theorem. Proc. Am. Math. Soc. **31**, 423–426 (1972)
12. Fröhlich, A.: Galois Module Structure of Algebraic Integers. Ergebnisse der Mathematik und ihrer Grenzgebiete (3). Springer, Berlin (1983)
13. Curtis, C.W., Reiner, I.: Representation theory of finite groups and associative algebras. AMS Chelsea, Providence (1962)
14. Linckelmann, M.: Variations sur les blocs à groupes de défaut cyclique, in Séminaire sur les groupes finis (Séminaire Claude Chevalley) Tôme IV, pp 1–83, Publications mathématiques de l'université Paris VII (1989)

15. Külshammer, B.: Bemerungen über die Gruppenalgebra als symmetrische Algebra I; J. Algebra. **71**, 1–7 (1981)
16. Külshammer, B.: Bemerungen über die Gruppenalgebra als symmetrische Algebra II; J. Algebra. **72**, 59–69 (1982)
17. Külshammer, B.: Bemerungen über die Gruppenalgebra als symmetrische Algebra III; J. Algebra. **88**, 279–291 (1984)
18. Külshammer, B.: Bemerungen über die Gruppenalgebra als symmetrische Algebra IV; J. Algebra. **93**, 310–323 (1985)
19. Külshammer, B.: Group-theoretical descriptions of ring-theoretical invariants of group algebras. Prog. Math. **95**, 425–442 (1991) (Birkhäuser Verlag Basel)
20. Holm, T., Zimmermann, A.: Generalized Reynolds ideals and derived equivalences for algebras of dihedral and semidihedral type. J. Algebra. **320**, 3425–3437 (2008)
21. Huppert, B.: Endliche Gruppen 1. Springer, Berlin (1967)
22. Benson, D.: Representations and Cohomology, vol. I. Cambridge University Press, Cambridge (1991)

Chapter 3
Abelian and Triangulated Categories

We have already observed that many of the properties we have considered are "formal". Category theory gives a mathematical precision to this feeling of being "formal", and of course this has now developed into a distinguished mathematical theory linked to many independent disciplines such as logic, set theory, and even differential geometry and mathematical physics. Many books on categories contain parts of the material presented here. The theory was most intensively used in algebraic topology and there as well as in algebraic geometry the concept of derived categories emerged with tremendous consequences for representation theory. Large parts of the representation theory of groups and algebras now deal with topics in derived and related categories. The foundations of this, as far as necessary for our purpose, will be explained in this chapter.

Much of the material displayed in this chapter can be found in some way, perhaps in partially disguised form, in the books of Mac Lane [1], Verdier [2] and Weibel [3].

3.1 Definitions

3.1.1 Categories

We start with the basic definition.

Definition 3.1.1 A category $\mathcal{C}$ consist of a class of *"objects"* and for every two objects X and Y of $\mathcal{C}$ a set $Mor_{\mathcal{C}}(X, Y)$ of *"morphisms"*, and for any three objects X, Y and Z of $\mathcal{C}$ a mapping

$$\circ : Mor_{\mathcal{C}}(Y, Z) \times Mor_{\mathcal{C}}(X, Y) \to Mor_{\mathcal{C}}(X, Z)$$

denoted $(f, g) \mapsto f \circ g$ satisfying the following axioms:

- For any quadruple X, Y, Z, W of objects in $\mathcal{C}$ the mapping $\circ$ is associative, i.e.

A. Zimmermann, *Representation Theory*,
Algebra and Applications 19, DOI: 10.1007/978-3-319-07968-4_3,

$$\begin{array}{ccc} Mor_{\mathcal{C}}(Z,W)\times Mor_{\mathcal{C}}(Y,Z)\times Mor_{\mathcal{C}}(X,Y) & \stackrel{id\times\circ}{\to} & Mor_{\mathcal{C}}(Z,W)\times Mor_{\mathcal{C}}(X,Z)\\ \downarrow \circ\times id & & \downarrow \circ\\ Mor_{\mathcal{C}}(Y,W)\times Mor_{\mathcal{C}}(X,Y) & \stackrel{\circ}{\to} & Mor_{\mathcal{C}}(X,W)\end{array}$$

is commutative.

- For each object X in $\mathcal{C}$ there is an element $1_X \in Mor_{\mathcal{C}}(X, X)$ such that for every $f \in Mor_{\mathcal{C}}(Z, Y)$ one has $1_Z \circ f = f = f \circ 1_Y$.

Observe that we deal with a class of objects whereas for every two objects we get a set of morphisms. With the class of objects only limited constructions are available, whereas for morphisms the situation is much better since more constructions are possible when dealing with sets than rather than with classes.

Example 3.1.2 The following gives a first small list of examples of categories.

1. We let $\mathcal{E}ns$ be the category of sets, morphisms being mappings and $\circ$ being the usual composition of mappings.
2. Let $\mathcal{C}$ be a category. Then define $\mathcal{C}^{op}$ to have the same objects as $\mathcal{C}$ and $Mor_{\mathcal{C}^{op}}(X, Y) := Mor_{\mathcal{C}}(Y, X)$ for all objects X, Y. Moreover compose morphisms in $\mathcal{C}^{op}$ in the opposite order as in $\mathcal{C}$, i.e. $f \circ^{op} g := g \circ f$. This gives a category, called *the opposite category* to $\mathcal{C}$.
3. Let A be a K-algebra, then A-mod denotes the category with objects being finitely generated A-modules, morphisms being homomorphisms of A-modules and composition being the composition of mappings. This example, of course, is of enormous importance for our purpose.
4. Slightly more generally, let A-Mod be the category of A-modules, finitely generated or not, with morphisms being the A-module homomorphisms.
5. The special case $A = \mathbb{Z} = K$ is of particular importance: Put $\mathbb{Z}$-$Mod =: \mathcal{A}b$, the category of abelian groups.
6. Consider a category with just one object $*$. Then $Mor(*, *)$ is a set with an associative composition law, and a unit element. This structure is usually called a *monoid.* A special case is a group, which is a monoid where every morphism is invertible in the obvious sense.
7. Let $\mathcal{C}$ be a category and fix an object B in $\mathcal{C}$. We shall define the *comma category* $(B, \mathcal{C})$ under B, which has as objects all elements in $Mor_{\mathcal{C}}(C, B)$ for all objects $C \in \mathcal{C}$, and for morphisms, for each $f \in Mor_{\mathcal{C}}(C, B)$ and $g \in Mor_{\mathcal{C}}(D, B)$ we put $Mor_{(B,\mathcal{C})}(f, g) := \{h \in Mor_{\mathcal{C}}(C, D) \mid g \circ h = f\}$.
8. Let A be a K-algebra. Then a category can be defined by taking as objects the isomorphism classes of A-modules and as morphisms the isomorphism classes of A-A-bimodules. Composition is given by $- \otimes_A -$. The identity element is the isomorphism class of the A-A-bimodule A.
9. Topology is abundant with categories: The category of topological spaces with continuous maps, the category of Hausdorff spaces with homeomorphisms, etc.

We present some nice properties formulated purely in category theory terms. As an example we define the concepts of epimorphism, monomorphism and isomorphism.

Definition 3.1.3 Let $\mathcal{C}$ be a category, let X and Y be objects of $\mathcal{C}$, and let $\alpha \in Mor_{\mathcal{C}}(X, Y)$. Then α is:

- an *isomorphism* if there is a $\beta \in Mor_{\mathcal{C}}(Y, X)$ such that $\alpha \circ \beta = id_Y$ and $\beta \circ \alpha = id_X$;
- an *epimorphism* if for any object Z of $\mathcal{C}$ and any two elements β, γ in $Mor_{\mathcal{C}}(Y, Z)$ one has $\beta \circ \alpha = \gamma \circ \alpha \Rightarrow \beta = \gamma$;
- a *monomorphism* if for any object Z of $\mathcal{C}$ and any two elements β, γ in $Mor_{\mathcal{C}}(Z, X)$ one has $\alpha \circ \beta = \alpha \circ \gamma \Rightarrow \beta = \gamma$;
- a *split monomorphism* if there is a $\beta \in Mor_{\mathcal{C}}(Y, X)$ such that $\beta \circ \alpha = id_X$;
- a *split epimorphism* if there is a $\beta \in Mor_{\mathcal{C}}(Y, X)$ such that $\alpha \circ \beta = id_Y$.

Remark 3.1.4 We observe the following immediate properties.

1. The reader can verify immediately that monomorphisms in the category of sets are injective maps, and that epimorphisms in the category of sets are surjective maps. The same holds for the category of modules over a ring. However, in the category of rings an epimorphism does not need to be surjective, as is seen by the unique ring homomorphism $\mathbb{Z} \longrightarrow \mathbb{Q}$, which is a non-surjective epimorphism.
2. Split monomorphisms are monomorphisms, as is readily seen by multiplying by the left inverse.
3. Split epimorphisms are epimorphisms, multiplying by the right inverse.
4. An isomorphism is a morphism which is a split monomorphism and a split epimorphism.

Another very important concept is the concept of a product and a coproduct.

Definition 3.1.5 Let $\mathcal{C}$ be a category, let I be a set and let $(X_i)_{i \in I}$ be a family of objects in $\mathcal{C}$, indexed by the set I.

- An object P together with a family of morphisms $(\varphi_i \in Mor_{\mathcal{C}}(P, X_i))_{i \in I}$ is a *product* if for every object X of $\mathcal{C}$ and every family of morphisms $(\psi_i \in Mor_{\mathcal{C}}(X, X_i))_{i \in I}$ there is a unique morphism $\psi \in Mor_{\mathcal{C}}(X, P)$ such that $\varphi_i \circ \psi = \psi_i$ for all $i \in I$.

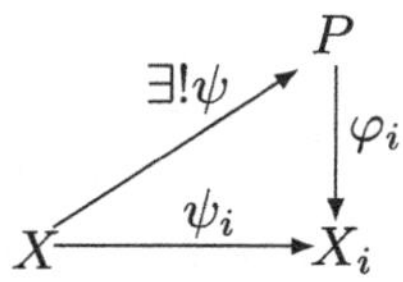

  If (P, φ_i) is a product we denote it by $P = \prod_{i \in I} X_i$, mostly without making explicit the morphisms φ_i.
- An object C together with a family of morphisms $(\iota_i \in Mor_{\mathcal{C}}(X_i, C))_{i \in I}$ is a *coproduct* if for every object X of $\mathcal{C}$ and every family of morphisms $(\chi_i \in Mor_{\mathcal{C}}(X_i, X))_{i \in I}$ there is a unique morphism $\iota \in Mor_{\mathcal{C}}(C, X)$ such that $\iota \circ \iota_i = \chi_i$ for all $i \in I$.

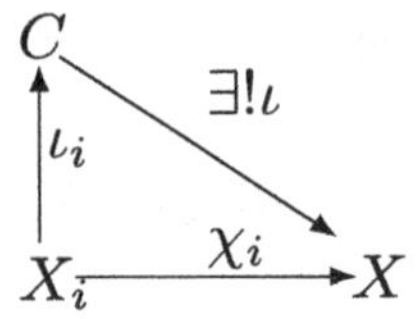

If (C, ι_i) is a coproduct we denote it by $C = \coprod_{i \in I} X_i$, mostly without making explicit the morphisms χ_i.

- If (P, φ) is a product and (C, ι_i) is a coproduct, and if $P \simeq C$ in the category, we say $P \simeq C$ is a *biproduct.*

Remark 3.1.6 The very definition of a product and of a coproduct gives that

$$Mor_{\mathcal{C}}\left(X, \prod_{i \in I} Y_i\right) \simeq \prod_{i \in I} Mor_{\mathcal{C}}(X, Y_i)$$

and

$$Mor_{\mathcal{C}}\left(\coprod_{i \in I} Y_i, X\right) \simeq \prod_{i \in I} Mor_{\mathcal{C}}(Y_i, X)$$

for all objects X, Y_i of $\mathcal{C}$.

Products and coproducts do not always exist. Sometimes they exist when one restricts to certain classes of sets I, such as finite sets I.

Products in the category of sets are classical cartesian products while coproducts in the category of sets are disjoint unions of sets. In the category of modules over a ring the coproduct is the direct sum and the products are again cartesian products. Hence, depending on the index set I, one needs to consider infinitely generated modules. The product in the category of algebras is the tensor product.

There is a concept of an initial object and a terminal object.

Definition 3.1.7 Let $\mathcal{C}$ be a category.

- An *initial* object I in $\mathcal{C}$ is an object such that for all objects X of $\mathcal{C}$ there is a unique morphism in $Mor_{\mathcal{C}}(I, X)$.
- A *terminal* object T in $\mathcal{C}$ is an object such that for all objects X of $\mathcal{C}$ there is a unique morphism in $Mor_{\mathcal{C}}(X, T)$.
- A *zero* object is an object which is both initial and terminal.

Remark 3.1.8 If $\mathcal{C}$ has two initial objects I_1 and I_2, then I_1 is isomorphic to I_2. Indeed, by the definition of an initial object we have a unique morphism $\alpha \in Mor_{\mathcal{C}}(I_1, I_2)$, a unique morphism $\beta \in Mor_{\mathcal{C}}(I_2, I_1)$, and unique endomorphisms id_{I_1} of I_1 and id_2 of I_2. Hence $\alpha \circ \beta = id_{I_1}$ and $\beta \circ \alpha = id_{I_2}$. This shows that α is an isomorphism. Likewise, terminal objects are unique up to unique isomorphism and zero objects are unique up to unique isomorphism.

Example 3.1.9 The example we have in mind is of course the module 0 in $A\text{-}Mod$ for a ring A. This is clearly a zero object. An example of a category with an initial object and without a zero object is the category $\mathcal{I}$ which has objects the non-zero ideals of $\mathbb{Z}$ and a morphism from one ideal to another is given by inclusion. Then $\mathbb{Z}$ is terminal and an initial object would be an ideal contained in each ideal. The only ideal which has this property is the ideal 0, but this ideal was excluded. If we admit the 0 ideal we get a category with an initial and terminal object but without a zero object.

Observe that if a category $\mathcal{C}$ has a zero object then $Mor_{\mathcal{C}}(X, Y)$ is never empty since the composition of the unique morphism from and to the zero object gives an element. We usually denote this morphism by 0.

Definition 3.1.10 Let $\mathcal{C}$ be a category and let X and Y be objects. Let $(\alpha_i)_{i\in I}$ be a family of morphisms in $Mor_{\mathcal{C}}(X, Y)$.

- The *equaliser* of the family $(\alpha_i)_{i\in I}$ is an object Z of $\mathcal{C}$ together with a morphism $\beta \in Mor_{\mathcal{C}}(Z, X)$ such that $\alpha_i \circ \beta = \alpha_j \circ \beta$ for all $i, j \in I$ and such that whenever there is a morphism $\gamma \in Mor_{\mathcal{C}}(U, X)$ with $\alpha_i \circ \gamma = \alpha_j \circ \gamma$ for all $i, j \in I$, there is a unique $\delta \in Mor_{\mathcal{C}}(U, Z)$ with $\beta \circ \delta = \gamma$.
- The *coequaliser* of the family $(\alpha_i)_{i\in I}$ is an object W of $\mathcal{C}$ together with a morphism $\beta' \in Mor_{\mathcal{C}}(Y, W)$ such that $\beta' \circ \alpha_i = \beta' \circ \alpha_j$ for all $i, j \in I$ and such that whenever there is a morphism $\gamma' \in Mor_{\mathcal{C}}(Y, U)$ with $\gamma' \circ \alpha_i = \gamma' \circ \alpha_j$ for all $i, j \in I$, there is a unique $\delta' \in Mor_{\mathcal{C}}(W, U)$ with $\delta' \circ \beta' = \gamma'$.

In the case of categories with a zero object we may progress further.

Definition 3.1.11 Let $\mathcal{C}$ be a category with a zero object. The *kernel* of a morphism α is the equaliser of α and 0. The *cokernel* of a morphism α is the coequaliser of α and 0.

It is clear that this notion reduces in module categories to the familiar notions of kernel and cokernel.

The notion of a pullback and of a pushout is easily seen to be a concept that can be formulated in a category. We leave the details to the reader.

3.1.2 Functors

How to compare categories? The required notion is that of a functor.

Definition 3.1.12 Let $\mathcal{C}$ and $\mathcal{D}$ be categories. Then a *(covariant) functor* $F : \mathcal{C} \longrightarrow \mathcal{D}$ is given by associating to every object X in $\mathcal{C}$ an object $F(X)$ in $\mathcal{D}$ and for every pair of objects X and X' of $\mathcal{C}$ a mapping $F : Mor_{\mathcal{C}}(X, X') \longrightarrow Mor_{\mathcal{D}}(F(X), F(X'))$ such that $F(1_X) = 1_{F(X)}$ for all objects X of $\mathcal{C}$, and such that $F(f \circ g) = F(f) \circ F(g)$ for all triples X, Y, Z of objects in $\mathcal{C}$ and all $f \in Mor_{\mathcal{C}}(Y, Z)$ and $g \in Mor_{\mathcal{C}}(Y, X)$. A *contravariant functor* is a functor (defined above) $F : \mathcal{C}^{op} \longrightarrow \mathcal{D}$.

One should mention that set theory is not used properly here. Associating to every object another object is not really allowed on classes. One has to carefully adapt the set theory one uses in order to be able to do what is needed. However, this can be done; with some considerable effort, but it can be done. It would certainly enlarge this text unreasonably if we were to do this here, so we shall refrain from giving the details. Perhaps the most comfortable solution uses Grothendieck universes [4].

Remark 3.1.13 Let $\mathcal{C}, \mathcal{D}$ and $\mathcal{E}$ be three categories and let $F : \mathcal{C} \longrightarrow \mathcal{D}$ and $G : \mathcal{D} \longrightarrow \mathcal{E}$ be two functors. Then $G \circ F$ (in the natural sense, i.e. an object X of $\mathcal{C}$ is associated to the object $G(F(X))$ and the mapping induced on the morphisms is the composition of the mappings induced by F and by G) is again a functor.

Example 3.1.14 Functors are everywhere.

- The identity functor on a category $\mathcal{C}$ associates to any object and any morphism just itself.
- The opposite functor is a contravariant functor ${}^{op} : \mathcal{C} \longrightarrow \mathcal{C}^{op}$ associating to each object itself, and to any morphism itself.
- Given a field K and a K-algebra A then any A-module M is also a K-vector space. Moreover, A-linear maps are, in particular, K-linear. So we get a functor A-*mod* $\longrightarrow K$-Mod which is the identity on morphisms and on objects. Functors of this kind are called "forgetful functors", since they preserve all data, but forget some structure.
- Given any category $\mathcal{C}$ and an object X of $\mathcal{C}$, then we may define a functor
$$Mor_{\mathcal{C}}(X, -) : \mathcal{C} \longrightarrow \mathcal{E}ns$$
by sending an object Y to $Mor_{\mathcal{C}}(X, Y)$ and a morphism $f : Y \longrightarrow Z$ to the morphism $Mor_{\mathcal{C}}(X, f) : Mor_{\mathcal{C}}(X, Y) \longrightarrow Mor_{\mathcal{C}}(X, Z)$ defined by $Mor_{\mathcal{C}}(X, f)(g) := f \circ g$. This functor is said to be *represented by* X. A functor is *representable* if there is an X such that it is "isomorphic" to $Mor_{\mathcal{C}}(X, -)$ in a sense described in Definition 3.1.20 below.
- Analogous to the previous example we define a contravariant functor
$$Mor_{\mathcal{C}}(-, X) : \mathcal{C} \longrightarrow \mathcal{E}ns$$
by sending an object Y to $Mor_{\mathcal{C}}(Y, X)$ and a morphism $f : Y \longrightarrow Z$ to $g \mapsto g \circ f$.
- Let A and B be two algebras and let M be an A-B-bimodule. Then $M \otimes_B - :$ B-*mod* $\longrightarrow$ A-*mod* is a functor. This functor is particularly well-behaved and has many properties which we shall elaborate on and use later.

Just as for modules we may define injective and projective objects in a category. The key to the definition is the notion of representable functors.

Definition 3.1.15 Let $\mathcal{C}$ be a category.

An object P of $\mathcal{C}$ is *projective* if for all epimorphisms $X \stackrel{f}{\longrightarrow} Y$ we have that $Mor_{\mathcal{C}}(P, f) : Mor_{\mathcal{C}}(P, X) \longrightarrow Mor_{\mathcal{C}}(P, Y)$ is surjective.

An object I is *injective* in $\mathcal{C}$ if for every monomorphism $X \xrightarrow{f} Y$ the map $Mor_{\mathcal{C}}(f, I) : Mor_{\mathcal{C}}(Y, I) \longrightarrow Mor_{\mathcal{C}}(X, I)$ is surjective.

Later, wee will need the notion of a limit and a colimit in a category. This notion makes a (weak) use of a functor in a category.

A partially ordered set $(I, \leq)$ is called a *codirect system* if for every $i, j \in I$ there is a $k \in I$ such that $i \leq k$ and $j \leq k$. The system is called *direct* if for every $i, j \in I$ there is a $k \in I$ such that $k \leq i$ and $k \leq j$. Observe that every partially ordered set $(I, \leq)$ can be considered as a category with objects being I and morphisms $Mor_I(i, j)$ being empty if $i \not\leq j$ and being of cardinality 1 otherwise.

Definition 3.1.16 Let $\mathcal{C}$ be a category and

- let $(I, \leq)$ be a codirect system. Let $F : (I, \leq) \longrightarrow \mathcal{C}$ be a functor. Then an object L in $\mathcal{C}$ together with morphisms $\alpha_i \in Mor_{\mathcal{C}}(F(i), L)$ is called a *colimit* and is denoted by $\text{colim}_{i\in I} F(i)$ if
  - whenever $i \leq j$ then $\alpha_j \circ F(i \leq j) = \alpha_i$ and
  - whenever there is an object K of $\mathcal{C}$ and morphisms $\beta_i \in Mor_{\mathcal{C}}(F(i), K)$ such that
  $$i \leq j \Rightarrow \beta_j \circ F(i \leq j) = \beta_i,$$
  then there is a unique $\gamma \in Mor_{\mathcal{C}}(L, K)$ such that $\gamma \circ \alpha_i = \beta_i$ for all $i \in I$.
- let $(I, \leq)$ be a direct system. Let $F : (I, \leq) \longrightarrow \mathcal{C}$ be a functor. Then an object P in $\mathcal{C}$ together with morphisms $\alpha_i \in Mor_{\mathcal{C}}(P, F(i))$ is called a *limit* and is denoted by $\lim_{i\in I} F(i)$ if
  - whenever $i \leq j$ then $F(i \leq j) \circ \alpha_i = \alpha_j$ and
  - whenever there is an object K of $\mathcal{C}$ and morphisms $\beta_i \in Mor_{\mathcal{C}}(K, F(i))$ such that
  $$i \leq j \Rightarrow F(i \leq j) \circ \beta_i = \beta_j,$$
  then there is a unique $\gamma \in Mor_{\mathcal{C}}(K, P)$ such that $\alpha_i \circ \gamma = \beta_i$ for all $i \in I$.

Sometimes we use the terminology *projective limits* for limits and *inductive limits* for colimits. Products are limits whereas coproducts are colimits as is easily seen. The set of p-adic integers $\hat{\mathbb{Z}}_p$ is the projective limit over the direct system of abelian cyclic p-groups and canonical morphisms $\mathbb{Z}/p^i\mathbb{Z} \longrightarrow \mathbb{Z}/p^j\mathbb{Z}$ whenever $i \geq j$.

Remark 3.1.17 Limits (resp colimits) are unique up to unique isomorphism, if they exist. The proof of this fact is strictly analogous to the proof that initial, terminal, and zero objects are unique up to unique isomorphism. The details of the proof are left to the reader as an easy exercise.

We can construct colimits as cokernels between direct sums.

Proposition 3.1.18 *Let A be an algebra and let $\mathcal{C} = A\text{-}Mod$.*

Further, let $\big((M_i)_{i\in I}, \iota_{(i,j)} \in Hom_{\mathcal{C}}(M_i, M_j)\big)$ be a codirected system. If we define φ by

$$\begin{array}{ccc} \coprod_{i\in\mathbb{N}} M_i & \longrightarrow & \coprod_{i\in\mathbb{N}} M_i \\ (m_i)_{i\in\mathbb{N}} & \mapsto & (\iota_{(i,j)}(m_i) - m_j)_{j\in\mathbb{N}} \end{array}$$

we get that φ is a monomorphism with cokernel the colimit $\mathrm{colim}_{i\in\mathbb{N}} M_i$.

Proof It is clear that φ is a monomorphism. The sequence

$$0 \longrightarrow M_i \stackrel{(\iota_{(i,j)}, -id)}{\longrightarrow} M_j \oplus M_i \stackrel{\binom{id}{\iota_{(i,j)}}}{\longrightarrow} M_j \longrightarrow 0$$

is exact. Hence, composing to the right $M_j \longrightarrow \mathrm{colim}_{n\in\mathbb{N}} M_n$ we obtain an exact sequence

$$0 \longrightarrow \coprod_{i\in\mathbb{N}} M_i \stackrel{\varphi}{\longrightarrow} \coprod_{i\in\mathbb{N}} M_i \longrightarrow \mathrm{colim}_{n\in\mathbb{N}} M_n.$$

The explicit construction of the cokernel by explicit quotients gives that $\mathrm{colim}_{n\in\mathbb{N}} M_n$ is indeed the cokernel. □

Remark 3.1.19 If $\mathcal{C}$ is a so-called abelian category (cf Definition 3.3.4 below) which allows direct products, then the statement and the proof of Proposition 3.1.18 still hold true.

3.1.3 Natural Transformations

Functors are used to compare categories. But how do we compare functors?

Definition 3.1.20 Consider two categories $\mathcal{C}$ and $\mathcal{D}$ and functors $F : \mathcal{C} \longrightarrow \mathcal{D}$ and $G : \mathcal{C} \longrightarrow \mathcal{D}$. A *natural transformation* $\eta : F \longrightarrow G$ is a collection of morphisms $\eta_X \in Mor_{\mathcal{D}}(F(X), G(X))$, one for each object X of $\mathcal{C}$, so that for all objects X and Y of $\mathcal{C}$ the diagram

$$\begin{array}{ccc} Mor_{\mathcal{C}}(X,Y) & \stackrel{F}{\longrightarrow} & Mor_{\mathcal{D}}(F(X),F(Y)) \\ \downarrow G & & \downarrow Mor_{\mathcal{D}}(F(X),\eta_Y) \\ Mor_{\mathcal{D}}(G(X),G(Y)) & \stackrel{Mor_{\mathcal{D}}(\eta_X, G(Y))}{\longrightarrow} & Mor_{\mathcal{D}}(F(X),G(Y)) \end{array}$$

is commutative (see Example 3.1.14 for the definition of the morphisms in the diagram).

A *natural isomorphism* is a natural transformation η such that η_C is an isomorphism for every object C of $\mathcal{C}$. The two corresponding functors are then called isomorphic.

Lemma 3.1.21 *Let $\mathcal{C}$ and $\mathcal{D}$ be categories and let $F : \mathcal{C} \longrightarrow \mathcal{D}$ and $G : \mathcal{C} \longrightarrow \mathcal{D}$ be functors. Then a collection $\eta_X \in Mor_{\mathcal{D}}(FX, GX)$ is a natural transformation $\eta : F \longrightarrow G$ if and only if for each pair of objects X and Y of $\mathcal{C}$ and each $f \in Mor_{\mathcal{C}}(X, Y)$ the diagram*

$$\begin{array}{ccc} FX & \xrightarrow{Ff} & FY \\ \downarrow \eta_X & & \downarrow \eta_Y \\ GX & \xrightarrow{Gf} & GY \end{array}$$

is commutative.

Proof Assume the diagram in the lemma is commutative. Then

$$Mor_{\mathcal{D}}(FX, \eta_Y)(f) = \eta_Y \circ F(f) = G(f) \circ \eta_X = Mor_{\mathcal{D}}(\eta_X, GY)(f)$$

shows that η is a natural transformation. The inverse implication is shown by the same equation, read in the opposite way. □

Remark 3.1.22 1. This definition again rises some issues from the naive set theoretic point of view. We recall that the objects of $\mathcal{C}$ form a class and not a set. Hence, it is not clear what is meant by a "collection η_X". This problem was solved by set theory. Again we shall refrain from explaining how.

2. Natural transformations compose naturally. Let F, G and H be functors $\mathcal{C} \longrightarrow \mathcal{D}$ and let $\eta : F \longrightarrow G$ and $\zeta : G \longrightarrow H$ be two natural transformations. Then $\zeta \circ \eta : F \longrightarrow H$ is a natural transformation defined by $(\zeta \circ \eta)_X := \zeta_X \circ \eta_X$. This composition is associative since the composition in $\mathcal{D}$ is associative.

   Moreover $id : F \longrightarrow F$ given by $id_X := id_{F(X)}$ is a natural transformation being neutral with respect to composition of natural transformations.

Example 3.1.23 Let $\mathcal{C}$ and $\mathcal{D}$ be two categories . Then $Fun(\mathcal{C}, \mathcal{D})$ is the category with objects being functors $F : \mathcal{C} \longrightarrow \mathcal{D}$ and morphisms being natural transformations. By Remark 3.1.22.2 we get that $Fun(\mathcal{C}, \mathcal{D})$ is a category, the *functor category*. Functor categories will be studied in more detail later in Sect. 5.11.1. Again, naive set theory is not sufficient to perform this construction properly.

Now we are ready to define equivalences of categories. Naively one might be tempted to define two categories $\mathcal{C}$ and $\mathcal{D}$ to be isomorphic if there are functors $F : \mathcal{C} \longrightarrow \mathcal{D}$ and $G : \mathcal{D} \longrightarrow \mathcal{C}$ so that $F \circ G = id_{\mathcal{D}}$ and $G \circ F = id_{\mathcal{C}}$. This would produce a very strict concept and many examples which we would like to be "equivalences" would not be. The reason for this phenomenon is that intuitively one likes to identify certain isomorphisms with the identity if they are particularly canonical, such as $A \otimes_A M \simeq M$. However, formally these objects are not equal, they are just isomorphic.

The more appropriate concept is that of equivalence of categories.

Definition 3.1.24 The categories $\mathcal{C}$ and $\mathcal{D}$ are *equivalent* if there is a functor $F : \mathcal{C} \longrightarrow \mathcal{D}$ and a functor $G : \mathcal{D} \longrightarrow \mathcal{C}$ as well as a natural isomorphism $\eta : id_{\mathcal{D}} \longrightarrow F \circ G$ and a natural isomorphism $\zeta : id_{\mathcal{C}} \longrightarrow G \circ F$. The functor F is called a quasi-inverse of G, and G is a quasi-inverse of F.

Example 3.1.25 We illustrate this by an example.

1. Let K be a field and let K-mod be the category of finite dimensional vector spaces with morphisms being K-linear maps. Then $Hom_K(-,K) : K\text{-}mod \longrightarrow (K\text{-}mod)^{op}$ is a functor. Actually $Hom_K : (K\text{-}mod)^{op} \longrightarrow K\text{-}mod$ is a functor as well and we get for every vector space V an isomorphism $eval_V : V \longrightarrow Hom_K(Hom_K(V,K),K)$ by evaluation $v \mapsto (\varphi \mapsto \varphi(v))$. We use here that V is finite dimensional.

   Now, $eval : Id_{K\text{-}mod} \longrightarrow Hom_K(Hom_K(-,K),K)$ is a natural isomorphism, as is $eval : Id_{(K\text{-}mod)^{op}} \longrightarrow Hom_K(Hom_K(-,K),K)$. Therefore K-mod is equivalent to its opposite category $(K\text{-}mod)^{op}$.

   Note that $eval$ is just a natural transformation and not a natural isomorphism on K-Mod, the category of all (possibly infinite dimensional) vector spaces, since an infinite dimensional vector space properly embeds into its double dual by the evaluation mapping (cf any textbook in elementary functional analysis).

   We should note that the above result is also valid for vector spaces over skew-fields D.
2. Let D be a (skew-)field, let $V := D^n$ and let $A := Mat_{n \times n}(D) \simeq End_D(V)$ be the algebra of n by n matrices over D. To ensure that the notation and the matrix multiplication are coherent, we use line vectors. We claim that $End_D(V)$-mod, the category of $End_D(V)$-modules, and D-mod are equivalent categories.

   We define two functors, one in each direction. Observe that V is a D-$End_D(V)$-bimodule by setting, for $\lambda \in D$ and $\varphi \in End_D(V)$, for each $v \in V$

   $$\lambda \cdot v \cdot \varphi := \varphi(\lambda v).$$

   It is straightforward to check that this does indeed give a bimodule structure. Since V is a D-$End_D(V)$-bimodule, its dual $Hom_D(V,D)$ is an $End_D(V)$-D-bimodule.

   $$\begin{aligned} V \otimes_{End_D(V)} - &: End_D(V)\text{-}mod \longrightarrow D\text{-}mod \\ Hom_D(V,D) \otimes_D - &: D\text{-}mod \longrightarrow End_D(V)\text{-}mod \end{aligned}$$

are obviously functors, where for all D-modules W the space $Hom_D(V, W)$ becomes an $End_D(V)$-left module by composition of mappings. In particular $Hom_D(V, D)$ is an $End_D(V)$-left module. Now,

$$V \otimes_{End_D(V)} Hom_D(V, D) \simeq D$$

and

$$Hom_D(V, D) \otimes_D V \simeq End_D(V)$$

as might be most easily seen by identifying V with line vectors and its dual with column vectors. Therefore we get for the composition

$$\begin{aligned}(V \otimes_{End_D(V)} -) \circ (Hom_D(V, D) \otimes -) &\simeq V \otimes_{End_D(V)} Hom_D(V, D) \\ &\simeq D \otimes_D - \\ &\simeq id_{D\text{-}mod}\end{aligned}$$

and

$$\begin{aligned}&(Hom_D(V, D) \otimes_D -) \circ (V \otimes_{End_D(V)} -) \\ &\qquad \simeq (Hom_D(V, D) \otimes_D V) \otimes_{End_D(V)} - \\ &\qquad \simeq End_D(V) \otimes_{End_D(V)} - \\ &\qquad \simeq id_{End_D(V)\text{-}mod}\end{aligned}$$

The astonishing fact is that this example is, from a certain point of view, already the most general case of an equivalence between module categories. This is the subject of Chap. 4.

A very important fact on natural transformations is Yoneda's lemma.

Theorem 3.1.26 (Yoneda's lemma) *Let $\mathcal{C}$ be a category, let X be an object in $\mathcal{C}$ and let $F : \mathcal{C} \longrightarrow \mathcal{E}ns$ be a functor. Then*

$$\begin{aligned}Nattrans(Mor_{\mathcal{C}}(X, -), F) &\longrightarrow F(X) \\ \eta &\mapsto \eta_X(id_X)\end{aligned}$$

is a bijection where $Nattrans$ denotes the natural transformations between functors.

Proof We see immediately that η is well-defined, i.e. $\eta_X(id_X)$ is an object of $F(X)$.

We shall define the inverse mapping

$$\begin{aligned}F(X) &\longrightarrow Nattrans(Mor_{\mathcal{C}}(X, -), F) \\ x &\mapsto h^x\end{aligned}$$

in the following way. For any object Y, any $x \in F(X)$, and any $f \in Mor_{\mathcal{C}}(X, Y)$ we put

$$h^x_Y(f) := F(f)(x).$$

We need to verify that h^x actually defines a natural transformation.

Indeed, we can proceed directly using the definition by considering the diagram

$$\begin{array}{ccc} Mor_{\mathcal{C}}(Y,Z) & \xrightarrow{F} & Mor_{\mathcal{E}ns}(FY,FZ) \\ \downarrow Mor_{\mathcal{C}}(X,-) & & \downarrow Mor_{\mathcal{E}ns}(h^x_Y,FZ) \\ Mor_{\mathcal{E}ns}(Mor_{\mathcal{C}}(X,Y),Mor_{\mathcal{C}}(X,Z)) & \xrightarrow{(h^x_Z)_*} & Mor_{\mathcal{E}ns}(Mor_{\mathcal{C}}(X,Y),FZ) \end{array}$$

where $(h^x_Z)_* = Mor_{\mathcal{E}ns}(Mor_{\mathcal{C}}(X,Y),h^x_Z)$. Elements are mapped as indicated in the following diagram.

$$\begin{array}{ccc} f & \xrightarrow{F} & F(f) \\ & & \downarrow Mor_{\mathcal{E}ns}(h^x_Y,FZ) \\ \downarrow Mor_{\mathcal{C}}(X,-) & & (\psi \mapsto (F(f)\circ F(\psi))(x)) \\ & & \| \\ (\varphi \mapsto f\circ\varphi) & \xrightarrow{(h^x_Z)_*} & (\psi \mapsto F(f\circ\psi)(x)) \end{array}$$

The equality on the lower right corner proves the property of being a natural transformation.

An alternative proof can be given using Lemma 3.1.21, and this alternative proof is generally given in the literature.

We get $h^x(id_X) = x$, which implies that the composition

$$F(X) \longrightarrow Nattrans(Mor_{\mathcal{C}}(X,-),F) \longrightarrow F(X)$$

is the identity. In particular $F(X) \longrightarrow Nattrans(Mor_{\mathcal{C}}(X,-),F)$ is injective.

We shall show that this map is surjective. Given a natural transformation $\varphi \in Nattrans(Mor_{\mathcal{C}}(X,-),F)$, put $x := \varphi_X(id_X)$. Then

$$h^x_Y(f) = F(f)(x) = F(f)(\varphi_X(id_X)) = \varphi_Y Mor_{\mathcal{C}}(X,f)(id_X) = \varphi_Y(f).$$

Hence $h^x_Y = \varphi_Y$ and we have shown that $F(X) \longrightarrow Nattrans(Mor_{\mathcal{C}}(X,-),F)$ is surjective. □

Yoneda's lemma has many applications as we shall see in the sequel. In particular for functor categories it is the essential ingredient. As an illustration we prove the following statement.

Lemma 3.1.27 *Let $\mathcal{C}$ be a category. Consider the category $Fun(\mathcal{C},\mathcal{E}ns)$ of functors from $\mathcal{C}$ to the category of sets. Then for each object X of $\mathcal{C}$ the object $Mor_{\mathcal{C}}(X,-)$ is projective in $Fun(\mathcal{C},\mathcal{E}ns)$.*

Proof Suppose $\eta : F \longrightarrow G$ is an epimorphism in $Fun(\mathcal{C}, \mathcal{E}ns)$. This means that $F(Y) \xrightarrow{\eta_Y} G(Y)$ is surjective for each object Y of $\mathcal{C}$. We apply $Nattrans(Mor_{\mathcal{C}}(X, -), ?)$ to this morphism and get a commutative diagram

$$\begin{array}{rcl} Nattrans(Mor_{\mathcal{C}}(X, -), F) & \simeq & F(X) \\ Nattrans(Mor_{\mathcal{C}}(X, -), \eta) \downarrow & & \downarrow \eta_X \\ Nattrans(Mor_{\mathcal{C}}(X, -), G) & \simeq & G(X) \end{array}$$

for which the horizontal morphisms are obtained by Yoneda's lemma. Since η_X is surjective, so is $Nattrans(Mor_{\mathcal{C}}(X, -), \eta)$. This proves the lemma. □

We need a practical criterion to decide when two categories are equivalent, or better to say when a functor is an equivalence.

Proposition 3.1.28 *Let $\mathcal{C}$ and $\mathcal{D}$ be two categories and suppose $F : \mathcal{C} \longrightarrow \mathcal{D}$ is a functor. Then F is an equivalence if F induces bijections on the homomorphism sets and if moreover for every object D in $\mathcal{D}$ there is an object C in $\mathcal{C}$ such that $F(C) \simeq D$.*

Proof By hypothesis, for each object D in $\mathcal{D}$ there is an object C_D in $\mathcal{C}$ so that

$$\alpha_D : F(C_D) \xrightarrow{\simeq} D.$$

Define

$$G(D) := C_D.$$

For two objects D_1 and D_2 in $\mathcal{D}$ we get an isomorphism

$$\begin{array}{rcl} Mor_{\mathcal{D}}(D_1, D_2) & \longrightarrow & Mor_{\mathcal{D}}(FG(D_1), FG(D_2)) \\ \varphi & \mapsto & \alpha_{D_2}^{-1} \circ \varphi \circ \alpha_{D_1} \end{array}$$

and an isomorphism,

$$\begin{array}{rcl} Mor_{\mathcal{D}}(FG(D_1), FG(D_2)) & \longrightarrow & Mor_{\mathcal{C}}(G(D_1), G(D_2)) \\ \alpha_{D_2}^{-1} \circ \varphi \circ \alpha_{D_1} & \mapsto & \overline{\varphi} := F^{-1}(\alpha_{D_2}^{-1} \circ \varphi \circ \alpha_{D_1}) \end{array}$$

using that F induces an isomorphism on the morphisms. Put $G(\varphi) := \overline{\varphi}$.

Now, for $\varphi_1 \in Mor_{\mathcal{D}}(D_2, D_3)$ and $\varphi_2 \in Mor_{\mathcal{D}}(D_1, D_2)$

$$\begin{aligned} G(\varphi_1 \circ \varphi_2) &= \overline{\varphi_1 \circ \varphi_2} \\ &= F^{-1}(\alpha_{D_3}^{-1} \circ \varphi_1 \circ \varphi_2 \circ \alpha_{D_1}) \\ &= F^{-1}(\alpha_{D_3}^{-1} \circ \varphi_1 \circ \alpha_{D_2} \circ \alpha_{D_2}^{-1} \circ \varphi_2 \circ \alpha_{D_1}) \\ &= F^{-1}(\alpha_{D_1} \circ \varphi_1 \circ \alpha_{D_2}^{-1}) \circ F^{-1}(\alpha_{D_2} \circ \varphi_2 \circ \alpha_{D_3}^{-1}) \\ &= G(\varphi_1) \circ G(\varphi_2) \end{aligned}$$

where F^{-1} exists on morphism sets by hypothesis, and where F^{-1} is compatible with composition since F is a functor, and hence is compatible with composition.

We need to find a natural transformation $\eta : FG \longrightarrow id_{\mathcal{D}}$. This is equivalent to defining maps $\eta_1 \in Mor_{\mathcal{D}}(FG(D_1), D_1)$ and $\eta_2 \in Mor_{\mathcal{D}}(FG(D_2), D_2)$ so that for all $\sigma \in Mor_{\mathcal{D}}(D_1, D_2)$,

$$\begin{array}{ccc} FG(D_1) & \stackrel{FG(\sigma)}{\longrightarrow} & FG(D_2) \\ \downarrow \eta_1 & & \eta_2 \downarrow \\ D_1 & \stackrel{\sigma}{\longrightarrow} & D_2 \end{array}$$

is commutative. But, by construction of G, setting $\eta_D := \alpha_D$ for all objects D in $\mathcal{D}$ will suffice.

We need to find a natural transformation $\zeta : GF \longrightarrow id_{\mathcal{C}}$. We have an isomorphism

$$\alpha_{F(C)} : FGF(C) \stackrel{\simeq}{\longrightarrow} F(C)$$

for all objects C of $\mathcal{C}$. Since F is bijective on morphism sets there is a unique

$$\beta_C \in Mor_{\mathcal{D}}(GF(C), C)$$

with $F(\beta_C) = \alpha_{F(C)}$.

$$\begin{array}{ccc} GF(C_1) & \stackrel{GF(\sigma)}{\longrightarrow} & GF(C_2) \\ \beta_{C_1} \downarrow & & \downarrow \beta_{C_2} \\ C_1 & \stackrel{\sigma}{\longrightarrow} & C_2 \end{array}$$

is commutative for all $\sigma \in Mor_{\mathcal{C}}(C_1, C_2)$ since

$$F(\beta_{C_2}) \circ FGF(\sigma) = \alpha_{F(C_2)} \circ FGF(\sigma) = F(\sigma) \circ \alpha_{F(C_1)} = F(\sigma \circ \beta_{C_1}).$$

Here the first equation is the definition of β, the second equation is the first step above and the third equation is again the definition of β and the fact that F is a functor. This finishes the proof. □

Remark 3.1.29 Again we should remark here that in a certain sense we have applied a version of the axiom of choice on all objects of a category: "Put $G(D) := C_D$" means that we choose for every object in a category another object in another category. This goes far beyond standard set theory. In weaker set theories the equivalence does not hold. A functor inducing isomorphisms on morphism sets such that each object in the target category is isomorphic to the image of an object in the source category is then called a weak equivalence. I thank Serge Bouc for communicating to me material due to René Guitart on this subject.

This proposition motivates the following definition.

Definition 3.1.30 Let $\mathcal{C}$ and $\mathcal{D}$ be categories.

- A functor $F : \mathcal{C} \longrightarrow \mathcal{D}$ is called *essentially surjective* or *dense* if for all objects D in $\mathcal{D}$ there is an object C in $\mathcal{C}$ so that $F(C) \simeq D$.
- A functor $F : \mathcal{C} \longrightarrow \mathcal{D}$ is called *faithful* if $Mor_{\mathcal{C}}(X, Y) \xrightarrow{F} Mor_{\mathcal{D}}(FX, FY)$ is injective for each pair of objects X and Y in $\mathcal{C}$.
- A functor $F : \mathcal{C} \longrightarrow \mathcal{D}$ is called *full* if $Mor_{\mathcal{C}}(X, Y) \xrightarrow{F} Mor_{\mathcal{D}}(FX, FY)$ is surjective for each pair of objects X and Y in $\mathcal{C}$.

Remark 3.1.31 Proposition 3.1.28 can hence be reformulated as follows: A functor F is an equivalence if and only if F is full, faithful and dense. Of course this is just a reformulation.

Sometimes categories have more structure.

Definition 3.1.32 Let R be a commutative ring and let $\mathcal{C}$ be a category. The category $\mathcal{C}$ *is R-linear* if for every pair X, Y of objects of $\mathcal{C}$ the set $Mor_{\mathcal{C}}(X, Y)$ is an R-module and composition is R-bilinear.

A functor between two R-linear categories is *R-linear* if the map which it induces between the morphism spaces is R-linear.

Finally, we shall need the following useful concept.

Definition 3.1.33 Let $\mathcal{C}$ be a category. Then a subclass $\mathcal{D}_0$ of objects of $\mathcal{C}$ defines a *full subcategory* $\mathcal{D}$ of $\mathcal{C}$ if one takes as objects the subclass of objects $\mathcal{D}_0$ and as morphisms

$$Mor_{\mathcal{D}}(X, Y) := Mor_{\mathcal{C}}(X, Y)$$

for every pair of objects X, Y in $\mathcal{D}_0$. Composition of morphisms in $\mathcal{D}$ is the same as composition of morphisms in $\mathcal{C}$.

It is clear that this actually defines a category. Examples are everywhere:

If $\mathcal{T}$ is the category of topological spaces with morphisms continuous maps, then the three-dimensional manifolds with continuous maps form a full subcategory.

If A is a Noetherian ring, then A-Mod is the category of A-modules with morphisms being the module homomorphisms. The subclass of finitely generated A-modules A-mod with morphisms A-module homomorphisms form a full subcategory.

3.2 Adjoint Functors

A very useful concept is the concept of adjoint functors. We have already seen the most prominent example, the correspondence from Lemma 1.7.9, simplified here slightly

$$Hom_A({}_AM_B \otimes_B {}_BN,\ {}_AL) \simeq Hom_B(N, Hom_A(M, L))$$

for two rings A and B and modules ${}_AM_B$, ${}_BN$ and ${}_AL$. This correspondence has already been used and exploited in various places and in various different guises.

Definition 3.2.1 Let $F : \mathcal{C} \longrightarrow \mathcal{D}$ and $G : \mathcal{D} \longrightarrow \mathcal{C}$ be two functors between the categories $\mathcal{C}$ and $\mathcal{D}$. We say that F is *left adjoint* to G, and G is *right adjoint* to F, if we have an isomorphism between the following functors in two variables $(-, ?)$

$$Mor_{\mathcal{D}}(F-, ?) \simeq Mor_{\mathcal{C}}(-, G?).$$

We say that the functors (F, G) form an *adjoint pair*.

We remark that this means that for every fixed object '?', the functors in one variable '$-$' are equivalent, and that for fixed variable '$-$' the functors in the variable '?' are equivalent.

Example 3.2.2 Lemma 1.7.9 ensures that for any pair of rings A and B the pair of functors $(M \otimes_B -, Hom_A(M, -))$ for an A-B-bimodule M is a pair of adjoint functors. The adjoint functors we shall use are very often variants of this example. However, in general adjoint functors can be very different from this example and it is customary and sometimes helpful to formulate the results in the language of abstract pairs of adjoint functors.

We immediately obtain the following important consequence.

In the equivalence

$$Mor_{\mathcal{D}}(F-, ?) \simeq Mor_{\mathcal{C}}(-, G?)$$

put $? = F-$ to obtain

$$Mor_{\mathcal{D}}(F-, F-) \simeq Mor_{\mathcal{C}}(-, GF-).$$

The identity in $F-$ is mapped by the above equivalence of functors to some mapping $\epsilon_- \in Mor_{\mathcal{C}}(-, GF-)$. Analogously

$$Mor_{\mathcal{D}}(F-, ?) \simeq Mor_{\mathcal{C}}(-, G?)$$

defines, putting $- = G?$, a mapping $\eta_? \in Mor_{\mathcal{D}}(FG?, ?)$. We call the collection ϵ_- of morphisms the *unit ϵ of the adjunction* and the collection $\eta_?$, one for each object '?', is called the *counit of the adjunction*.

Lemma 3.2.3 *The unit ϵ of an adjunction (F, G) is a natural transformation*

$$\epsilon : id_{\mathcal{C}} \longrightarrow GF$$

and the counit η is a natural transformation

$$\eta : FG \longrightarrow id_{\mathcal{D}}.$$

Proof We need to show that the diagram

$$\begin{array}{ccc} Mor_{\mathcal{C}}(X,Y) & \stackrel{GF}{\longrightarrow} & Mor_{\mathcal{C}}(GFX,GFY) \\ \downarrow id_{\mathcal{C}} & & \downarrow Mor(\epsilon_X, GFY) \\ Mor_{\mathcal{C}}(X,Y) & \stackrel{Mor(X,\epsilon_Y)}{\longrightarrow} & Mor_{\mathcal{C}}(X,GFY) \end{array}$$

is commutative. Let

$$\varphi_{-,?} : Mor_{\mathcal{D}}(F-,?) \longrightarrow Mor_{\mathcal{C}}(-,G?)$$

be the isomorphism given by the adjunction and let $\alpha \in Mor_{\mathcal{C}}(X,Y)$. Then

$$\begin{aligned} Mor(\epsilon_X, GFY)(GF\alpha) &= (GF\alpha) \circ \epsilon_X \\ &= (GF\alpha) \circ \varphi_{X,FX}(id_{FX}) \\ &= \varphi_{X,FY}(F(\alpha) \circ id_{FX}) \\ &= \varphi_{X,FY}(id_{FY} \circ F(\alpha)) \\ &= \varphi_{Y,FY}(id_{FY}) \circ \alpha \\ &= \epsilon_Y \circ \alpha \\ &= Mor(X,\epsilon_Y)(\alpha) \end{aligned}$$

since φ is "natural" with respect to composition of mappings. Some readers may prefer to see these equalities in the commutativity of the following diagram.

$$\begin{array}{ccc} Mor_{\mathcal{D}}(FX,FY) & \stackrel{\varphi_{X,FX}}{\longrightarrow} & Mor_{\mathcal{C}}(X,GFX) \\ \downarrow Mor_{\mathcal{D}}(FX,F\alpha) & & \downarrow Mor_{\mathcal{C}}(X,GF\alpha) \\ Mor_{\mathcal{D}}(FX,FY) & \stackrel{\varphi_{X,FY}}{\longrightarrow} & Mor_{\mathcal{C}}(X,GFY) \\ \uparrow Mor_{\mathcal{D}}(\alpha,FY) & & \uparrow Mor_{\mathcal{C}}(\alpha,GFY) \\ Mor_{\mathcal{D}}(FY,FY) & \stackrel{\varphi_{Y,FY}}{\longrightarrow} & Mor_{\mathcal{C}}(Y,GFY) \end{array}$$

The case of η is similar and is left as an exercise. □

Remark 3.2.4 Of course, ϵ and η depend on the choice of the isomorphism

$$Mor_{\mathcal{D}}(F-,?) \simeq Mor_{\mathcal{C}}(-,G?)$$

and in general many choices are possible.

Proposition 3.2.5 *Suppose $\mathcal{C}$ and $\mathcal{D}$ are two categories and let $F : \mathcal{C} \longrightarrow \mathcal{D}$ and $G : \mathcal{D} \longrightarrow \mathcal{C}$ be two functors.*

- *Suppose that* $\epsilon : id_{\mathcal{C}} \longrightarrow GF$ *is a natural transformation so that for each* $\alpha : X \longrightarrow GY$ *there is exactly one* $\beta \in Mor_{\mathcal{D}}(FX, Y)$ *such that* $G(\beta) \circ \epsilon_X = \alpha$. *Then* (F, G) *is an adjoint pair and* ϵ *is a unit of the adjunction.*
- *Suppose that* $\eta : FG \longrightarrow id_{\mathcal{D}}$ *is a natural transformation so that for each* $\alpha : FY \longrightarrow X$ *there is exactly one* $\beta \in Mor_{\mathcal{C}}(Y, GX)$ *such that* $\eta_X \circ F(\beta) = \alpha$. *Then* (F, G) *is an adjoint pair and* η *is a counit of the adjunction.*

Proof We only prove the first statement, the second is dual and left to the reader as an exercise.

Let $\epsilon : id_{\mathcal{C}} \longrightarrow GF$ be the natural transformation of the hypothesis. Then put $\varphi(\alpha) = \beta$ from the hypothesis. This gives a well-defined mapping

$$\varphi : Mor_{\mathcal{C}}(X, GY) \longrightarrow Mor_{\mathcal{D}}(FX, Y).$$

We get $\beta : FX \longrightarrow Y$, hence $G\beta : GFX \longrightarrow GY$ and

$$\alpha = G\beta \circ \epsilon_X : X \longrightarrow GY.$$

We need to show that this is functorial in X and in Y, and that φ is bijective.

The fact that for all α there is a unique β satisfying $G(\beta) \circ \epsilon_X = \alpha$ is precisely the statement that φ is bijective. Since ϵ is a natural transformation, the mapping φ is natural in X. The fact that G is a functor implies that φ is natural in Y as well. This proves the statement. □

The next lemma gives a very practical criterion for obtaining adjoint functors. It will be used in Proposition 6.10.12.

Lemma 3.2.6 *Let* $\mathcal{C}$ *and* $\mathcal{D}$ *be categories and let* $F : \mathcal{C} \longrightarrow \mathcal{D}$ *and* $G : \mathcal{D} \longrightarrow \mathcal{C}$ *be functors. Suppose that there are natural transformations* $id_{\mathcal{C}} \stackrel{\epsilon}{\longrightarrow} GF$ *and* $FG \stackrel{\eta}{\longrightarrow} id_{\mathcal{D}}$ *so that the compositions* $F \stackrel{F(\epsilon_-)}{\longrightarrow} FGF \stackrel{\eta_{F-}}{\longrightarrow} F$ *and* $G \stackrel{\epsilon_{G-}}{\longrightarrow} GFG \stackrel{G(\eta_-)}{\longrightarrow} G$ *are the identity. Then* (F, G) *is an adjoint pair,* ϵ *is its unit and* η *is its counit.*

Conversely, if (F, G) *is an adjoint pair with unit* ϵ *and counit* η, *then the compositions* $F \stackrel{F(\epsilon_-)}{\longrightarrow} FGF \stackrel{\eta_{F-}}{\longrightarrow} F$ *and* $G \stackrel{\epsilon_{G-}}{\longrightarrow} GFG \stackrel{G(\eta_-)}{\longrightarrow} G$ *are the identity.*

Proof We define a mapping

$$\begin{aligned} Mor_{\mathcal{D}}(FX, Y) &\longrightarrow Mor_{\mathcal{C}}(X, GY) \\ f &\mapsto Gf \circ \epsilon_X \end{aligned}$$

and a mapping

$$\begin{aligned} Mor_{\mathcal{C}}(X, GY) &\longrightarrow Mor_{\mathcal{D}}(FX, Y) \\ g &\mapsto \eta_Y \circ Fg \end{aligned}$$

We shall show that these mappings are inverses of each other. Indeed, since η is a natural transformation we get

$$f \circ \eta_{FX} = \eta_Y \circ FGf$$

and hence

$$\eta_Y \circ F(Gf \circ \epsilon_X) = \eta_Y \circ FGf \circ F\epsilon_X = f \circ \eta_{FX} \circ F\epsilon_X = f$$

and likewise for the other composition. We need to show that ϵ is the unit of this adjunction. The value of the unit on X is the image of the identity under

$$\begin{array}{rcl} Mor_{\mathcal{D}}(FX, Y) & \longrightarrow & Mor_{\mathcal{C}}(X, GY) \\ f & \mapsto & Gf \circ \epsilon_X \end{array}$$

for $Y = FX$. This image is

$$G(id_{FX}) \circ \epsilon_X = id_{GFX} \circ \epsilon_X = \epsilon_X.$$

The case of the counit is similar.

Let (F, G) be an adjoint pair, let ϵ be its unit and let η be its counit. Let

$$\varphi_{-,?} : Mor_{\mathcal{D}}(F-, ?) \longrightarrow Mor_{\mathcal{C}}(-, G?)$$

be the adjunction morphism. We need to evaluate $\eta_{F-} \circ F(\epsilon_-)$ and $G(\epsilon_-) \circ \eta_{G-}$.

The maps φ are natural in each variable, and so

$$\begin{array}{ccc} Mor_{\mathcal{D}}(FX, Y') & \stackrel{\varphi_{X,Y'}}{\longrightarrow} & Mor_{\mathcal{C}}(X, GY') \\ \downarrow {\scriptstyle Mor(FX,\alpha)} & & \downarrow {\scriptstyle Mor(X,G\alpha)} \\ Mor_{\mathcal{D}}(FX, Y) & \stackrel{\varphi_{X,Y}}{\longrightarrow} & Mor_{\mathcal{C}}(X, GY) \end{array}$$

is commutative for all $\alpha \in Mor_{\mathcal{D}}(Y', Y)$. Hence, in particular for $Y' = FX$ and $\alpha = f \in Mor_{\mathcal{D}}(FX, Y)$, we get

$$\begin{aligned} \varphi_{X,Y}(f) &= \varphi_{X,Y}(f \circ id_{FX}) = \varphi_{X,Y}(Mor_{\mathcal{D}}(FX, f)(id_{FX})) \\ &= Mor_{\mathcal{C}}(X, Gf)(\varphi_{X,FX}(id_{FX})) = G(f) \circ \varphi_{X,FX}(id_{FX}) \\ &= G(f) \circ \eta_X. \end{aligned}$$

Similarly, for each $g \in Mor_{\mathcal{C}}(X, GY)$ we get $\varphi^{-1}_{X,Y}(g) = \epsilon_Y \circ F(g)$. In particular for $X = GY$ we get for $g = id_{GY}$ the relation $\epsilon_Y = \varphi^{-1}_{GY,Y}(id_{GY})$ and therefore

$$id_{GY} = \varphi_{GY,Y}(\epsilon_Y) = G(\epsilon_Y) \circ \eta_{GY}.$$

Likewise,

$$id_{FX} = \eta_{FX} \circ F(\epsilon_X).$$

This proves the lemma. □

Adjoints are unique as is shown in the next proposition.

Proposition 3.2.7 *Let* $F : \mathcal{C} \longrightarrow \mathcal{D}$ *be a functor and* $G_1 : \mathcal{D} \longrightarrow \mathcal{C}$ *and* $G_2 : \mathcal{D} \longrightarrow \mathcal{C}$ *be right adjoint functors to* F. *Then* G_1 *and* G_2 *are isomorphic functors.*

If $G_1 : \mathcal{D} \longrightarrow \mathcal{C}$ *and* $G_2 : \mathcal{D} \longrightarrow \mathcal{C}$ *are left adjoint functors to* F, *then* G_1 *and* G_2 *are isomorphic functors.*

Proof We only deal with the right adjoint case, the left adjoint case is analogous. We have

$$Mor_{\mathcal{C}}(-, G_2?) \simeq Mor_{\mathcal{D}}(F-, ?) \simeq Mor_{\mathcal{C}}(-, G_1?)$$

and by Yoneda's lemma we get $G_1 \simeq G_2$. This finishes the proof. □

We shall need the following statement.

Proposition 3.2.8 *Suppose* $\mathcal{C}$ *and* $\mathcal{D}$ *are two categories and let* $F : \mathcal{C} \longrightarrow \mathcal{D}$ *and* $G : \mathcal{D} \longrightarrow \mathcal{C}$ *be two functors. Suppose* (F, G) *is an adjoint pair. Then* F *is an equivalence if and only if* G *is an equivalence.*

Proof Suppose that $F : \mathcal{C} \longrightarrow \mathcal{D}$ is an equivalence. Then there is a natural isomorphism $\epsilon : id_{\mathcal{C}} \longrightarrow G'F$ for some functor $G' : \mathcal{D} \longrightarrow \mathcal{C}$. Proposition 3.2.5 shows that (F, G') is an adjoint pair with unit ϵ under some universality condition. Since ϵ_X is invertible for each object X, the universality condition is trivial. The functors G and G' are both right adjoint to F, and by Proposition 3.2.7 they are isomorphic. Since G is the quasi-inverse of F, G is invertible, and hence so is G'. The case for G instead of F is similar. □

The proof actually tells us more.

Corollary 3.2.9 *Let* $F : \mathcal{C} \longrightarrow \mathcal{D}$ *be an equivalence of categories with quasi-inverse* G. *Then* (F, G) *and* (G, F) *are both pairs of adjoint functors and the unit and the counit of the adjunction are isomorphisms.*

Indeed, the natural isomorphisms $\epsilon : id_{\mathcal{D}} \longrightarrow FG$ and $\epsilon' : id_{\mathcal{C}} \longrightarrow GF$ serve as units of the corresponding adjunctions.

Another property will be of some importance later. The statement actually holds more generally. It can be shown that a functor commutes with colimits if it has a left adjoint. We shall prove a special case.

Lemma 3.2.10 *Let* $(I, \leq)$ *be a codirect system and let* $F : (I, \leq) \to \mathcal{C}$ *be a functor. Let* $L : \mathcal{C} \to \mathcal{D}$ *be an equivalence. Then there is a natural isomorphism* $L(\mathrm{colim}_{i \in I} F(i)) \simeq \mathrm{colim}_{i \in I} LF(i)$.

Proof By the universal property of the colimit there is a natural morphism $\mathrm{colim}_{i \in I} LF(i) \to L(\mathrm{colim}_{i \in I} F(i))$. Since L induces a natural bijection on morphism sets, and since L is dense, $L(\mathrm{colim}_{i \in I} F(i))$ and $\mathrm{colim}_{i \in I} LF(i)$ both satisfy the universal property of the colimit. Hence the natural morphism is an isomorphism. □

3.3 Additive and Abelian Categories

Let $\mathcal{C}$ be a category. Given any two objects X and Y of $\mathcal{C}$ there is often a nice structure on $Mor_{\mathcal{C}}(X, Y)$, namely the structure of an abelian group. This can be exploited successfully if in addition the composition of morphisms is compatible with this structure, that is, the composition of morphism ought to be $\mathbb{Z}$-bilinear.

3.3.1 The Basic Definition

Definition 3.3.1 A category $\mathcal{C}$ is called *additive* if

- $\mathcal{C}$ is $\mathbb{Z}$-linear, and
- for every finite index set I products and coproducts exist and coincide (i.e. finite biproducts exist).

An *additive functor* between additive categories is a $\mathbb{Z}$-linear functor.

Example 3.3.2 Additive categories occur frequently. We shall mention the case of the full subcategory of the category of modules of an algebra generated by projective modules over an algebra. Obviously the axioms are verified.

Remark 3.3.3 Every additive functor F between additive categories preserves finite direct sums. Indeed, let $M = M_1 \oplus M_2$ be an object, and denote by $\pi_i : M \to M_i$ the projection and $\iota_i : M_i \to M$ the embedding of M_i in M. Then $id_M = \iota_1 \circ \pi_1 + \iota_2 \circ \pi_2$ and hence $id_{FM} = F(\iota_1) \circ F(\pi_1) + F(\iota_2) \circ F(\pi_2)$. Again $F(\pi_i)$ is a split epimorphism and $F(\iota_i)$ is a split monomorphism and hence $F(M) \simeq F(M_1) \oplus F(M_2)$.

Definition 3.3.4 An additive category is *abelian* if the following three conditions hold

- every morphism has a kernel and a cokernel
- every monomorphism is the kernel of a morphism
- every epimorphism is the cokernel of a morphism.

Remark 3.3.5 Observe that the second and the third axiom of an abelian category imply that a monomorphism $\varphi : N \longrightarrow M$ is the kernel of the cokernel mapping $M \longrightarrow \text{coker}(\varphi)$ and dually the epimorphism $\psi : M \longrightarrow L$ is the cokernel of the monomorphism $\ker(\psi) \longrightarrow M$. Indeed, let $\varphi : N \longrightarrow M$ be a monomorphism which is the kernel of $\gamma : M \longrightarrow S$. Then γ factors through $\alpha : M \longrightarrow \text{coker}(\varphi)$ by definition of the cokernel: $\gamma = \beta \circ \alpha$ for a morphism $\beta : \text{coker}(\varphi) \longrightarrow S$. Now, there is a morphism $\delta : \ker(\alpha) \longrightarrow N$ such that $\varphi \circ \delta = \epsilon$ where $\epsilon : \ker(\alpha) \longrightarrow M$. Moreover, $\alpha \circ \varphi = 0$ since α is the cokernel mapping of φ. Therefore there is a morphism $\lambda : N \longrightarrow \ker(\alpha)$ with $\epsilon \circ \lambda = \varphi$. Now, $\delta \circ \lambda = id_N$ by unicity of the mappings defined by the universal properties of kernels and likewise $\lambda \circ \delta = id_{\ker(\alpha)}$. This proves the statement.

Abelian categories are of central interest to us. The main example is the category A-*mod* for a Noetherian algebra A or the category A-Mod for an algebra A. Actually a sort of inverse is true as well.

Theorem 3.3.6 *(Mitchell) Every abelian category is a full subcategory of a category A-Mod for some algebra A.*

We shall use Mitchell's result from time to time in order to perform certain constructions in abelian categories in general. In the majority of cases restricting attention to module categories will be completely satisfactory and we can avoid using Mitchell's result. Theorem 3.3.6 is proved in many monographs on category theory or homological algebra, e.g. [3].

Example 3.3.7 Certain frequently used subcategories of module categories are not abelian, and not all all abelian categories are module categories.

- For an algebra A the category A-$Proj$, which is the full subcategory of A-Mod formed by projective A-modules, is additive, but is in general not abelian. Likewise A-$proj$, the full subcategory of A-*mod* formed by finitely generated projective A-modules, is additive but not abelian in genral.
- There are more abelian categories. The category of coherent sheaves over a projective variety is abelian. We will not need this theory further but it might be useful to remember that there are examples beyond our representation theoretic universe.

Definition 3.3.8 (Keller [5, Appendix]) Let $\mathcal{A}$ be an additive category. A pair (d, i) of morphisms $d \in Mor_{\mathcal{A}}(Y, Z)$ and $i \in Mor_{\mathcal{A}}(X, Y)$ is called *exact* if i is a kernel of d and d is a cokernel of i. Let $\mathcal{E}$ be the class of exact pairs in $\mathcal{A}$ and call the elements of $\mathcal{E}$ the *conflations*. A *deflation* is a morphism which occurs as the first components of a conflation and an *inflation* is a morphism which occurs as the first components of a conflation. $\mathcal{A}$ is called an *exact category* if

- id_0 is a deflation,
- the composition of two deflations is a deflation,
- for each $f \in Mor_{\mathcal{A}}(Z', Z)$ and each deflation $d \in Mor_{\mathcal{A}}(Y, Z)$ there is a pullback diagram

$$\begin{array}{ccc} Y' & \stackrel{d'}{\longrightarrow} & Z' \\ \downarrow f' & & \downarrow f \\ Y & \stackrel{d}{\longrightarrow} & Z \end{array}$$

  where d' is a deflation,
- for each $f \in Mor_{\mathcal{A}}(X, X')$ and each inflation $i \in Mor_{\mathcal{A}}(X, Y)$ there is a pushout diagram

$$\begin{array}{ccc} X & \stackrel{i}{\longrightarrow} & Y \\ \downarrow f & & \downarrow f' \\ X' & \stackrel{i'}{\longrightarrow} & Y' \end{array}$$

  where i' is an inflation.

A functor $\mathcal{C} \xrightarrow{F} \mathcal{D}$ between two exact categories is *exact* if it maps each sequence in $\mathcal{C}$ to an exact sequence in $\mathcal{D}$.

3.3.2 Useful Tools in Abelian Categories

We start to introduce some classical tools which are going to be used in various places in the sequel.

The Snake Lemma

The very useful snake lemma can be formulated in abelian categories. Observe that in an abelian category the concept of an exact sequence makes sense and we write a short exact sequence as usual as

$$0 \longrightarrow A \longrightarrow B \longrightarrow C \longrightarrow 0.$$

Lemma 3.3.9 (Snake lemma) *Let $\mathcal{A}$ be an abelian category. Let*

$$A \xrightarrow{\alpha} B \xrightarrow{\beta} C \longrightarrow 0$$

and

$$0 \longrightarrow A' \xrightarrow{\alpha'} B' \xrightarrow{\beta'} C'$$

be exact sequences and let $A \xrightarrow{\rho} A'$, $B \xrightarrow{\sigma} B'$ and $C \xrightarrow{\tau} C'$ be morphisms such that the diagram

$$\begin{array}{ccccccc} & & A & \xrightarrow{\alpha} & B & \xrightarrow{\beta} & C & \longrightarrow 0 \\ & & \downarrow \rho & & \downarrow \sigma & & \downarrow \tau & \\ 0 \longrightarrow & & A' & \xrightarrow{\alpha'} & B' & \xrightarrow{\beta'} & C' & \end{array}$$

is commutative. Then we obtain an exact sequence

$$\ker(\rho) \to \ker(\sigma) \to \ker(\tau) \xrightarrow{\delta} \operatorname{coker}(\rho) \to \operatorname{coker}(\sigma) \to \operatorname{coker}(\tau).$$

Moreover $\ker(\rho) \to \ker(\sigma)$ *is a monomorphism if α is a monomorphism and* $\operatorname{coker}(\sigma) \to \operatorname{coker}(\tau)$ *is an epimorphism if β' is an epimorphism.*

Definition 3.3.10 The morphism δ is called the *connecting morphism.*

Proof By Mitchell's Theorem 3.3.6 we may assume that the abelian category is in fact Λ-Mod for an algebra Λ. In order to simplify the proof we shall make this assumption. However we should mention that it is possible to formulate the proof in a purely abstract fashion using only concepts in abelian categories. Without loss of generality we may assume that the kernels are physically submodules and the cokernels are physically quotients.

Restriction to the kernel gives an exact sequence

$$\ker(\rho) \to \ker(\sigma) \to \ker(\tau)$$

with the left-most morphism being mono if $A \longrightarrow B$ is mono. Analogously the dual argument gives that taking cokernels produces an exact sequence $\text{coker}(\rho) \to \text{coker}(\sigma) \to \text{coker}(\tau)$ with the right-most morphism being an epimorphism if $B' \longrightarrow C'$ is an epimorphism.

How to construct δ. Given $x \in \ker(\tau) \subseteq C$. The morphism β is surjective, and therefore there is a $b \in B$ such that $\beta(b) = x$. Since

$$\beta'(\sigma(b)) = \tau(\beta(b)) = \tau(x) = 0$$

by the choice of x, we get that there is a unique $a' \in A'$ so that $\alpha(a') = \sigma(b)$. Let

$$\delta(x) := \pi_\rho(a'),$$

where $\pi_\rho : A' \longrightarrow \text{coker}(\rho)$ is the natural mapping.

We first show that δ is independent of the choices made. Indeed, let $\hat{b}$ be in B so that $\beta(\hat{b}) = x$ as well. Then $b - \hat{b} \in \ker(\beta)$ and hence $\sigma(b) - \sigma(\hat{b}) = \alpha'(u')$ for some $u' = \rho(u)$ for $u \in A$. Therefore

$$\pi_\rho(\sigma(b) - \sigma(\hat{b})) = \pi_\rho(\rho(u)) = 0$$

which implies that δ does not depend on the choice of the lift of x to $b \in B$.

Now let $\lambda \in \Lambda$. Then the element λx can be lifted to λb and then $\delta(\lambda x) = \lambda\delta(x)$ since all the other morphisms are Λ-linear. Similarly, $x + x'$ can be lifted to a sum of the lifts of x and x', which proves that δ is additive as well.

We need to show that $\ker(\delta) = \text{coker}(\beta|_{\ker(\sigma)})$. Indeed, going through the construction of δ, one sees that $\delta(x) = 0$ if and only if $\pi_\rho(a') = 0$, which is equivalent to $a' = \rho(a)$ for some $a \in A$ and this in turn is equivalent to $b = \alpha(a)$, whence $\ker(\delta) = \text{coker}(\beta|_{\ker(\sigma)})$.

Finally we need to show

$$\ker(\text{coker}(\rho) \longrightarrow \text{coker}(\sigma)) = \text{coker}(\delta).$$

Now $a' + A' \in \ker(\mathrm{coker}(\rho) \longrightarrow \mathrm{coker}(\sigma))$ if and only if $\alpha'(a') \in \sigma(B)$. This is equivalent to $\alpha'(a') = \sigma(b)$ for some b. But then this is equivalent to $a' + A' = \delta(\beta(b))$. □

Remark 3.3.11 The connecting morphism will prove to be most useful in homology.

The reader might like to draw the morphism δ in the diagram

$$\begin{array}{ccccccc}
 & & 0 & & 0 & & 0 \\
 & & \downarrow & & \downarrow & & \downarrow \\
 & & \ker(\rho) & \longrightarrow & \ker(\sigma) & \longrightarrow & \ker(\tau) \\
 & & \downarrow & & \downarrow & & \downarrow \\
 & & A & \stackrel{\alpha}{\longrightarrow} & B & \stackrel{\beta}{\longrightarrow} & C & \longrightarrow 0 \\
 & & \downarrow \rho & & \downarrow \sigma & & \downarrow \tau \\
0 & \longrightarrow & A' & \stackrel{\alpha'}{\longrightarrow} & B' & \stackrel{\beta'}{\longrightarrow} & C' \\
 & & \downarrow & & \downarrow & & \downarrow \\
 & & \mathrm{coker}(\rho) & \longrightarrow & \mathrm{coker}(\sigma) & \longrightarrow & \mathrm{coker}(\tau) \\
 & & \downarrow & & \downarrow & & \downarrow \\
 & & 0 & & 0 & & 0
\end{array}$$

and then observes that δ moves like a snake from the right upper corner of the diagram to the lower left corner.

Watts' Theorem

We shall classify particularly nice functors between module categories. This is Watts' theorem, discovered independently by Watts [6] and Eilenberg [7].

We first give a useful characterisation of finitely presented modules.

Definition 3.3.12 Let A be an algebra over a commutative ring K. A module M is *finitely presented* if there are integers $n, m \in \mathbb{N}$ and $\alpha \in Hom_A(A^n, A^m)$ such that $M \simeq \mathrm{coker}(\alpha)$.

Observe that being finitely presented is slightly stronger than being finitely generated. For Noetherian algebras A however, Lemma 1.3.3 shows that the concept of being finitely generated and of being finitely presented coincide. We can characterise finitely presented modules in purely category theoretic terms.

Lemma 3.3.13 *Let K be a commutative ring and let A be a K-algebra. Let M be an A-module. Then the functor $Hom_A(M, -)$ commutes with arbitrary direct sums if and only if M is finitely presented.*

Definition 3.3.14 Let $\mathcal{C}$ be an additive category admitting arbitrary coproducts. An object X of $\mathcal{C}$ is *compact* if $Mor_{\mathcal{C}}(X, -) : \mathcal{C} \longrightarrow \mathbb{Z}\text{-}Mod$ commutes with arbitrary coproducts.

Proof of Lemma 3.3.13 If M is finitely presented, then we get an exact sequence

$$A^n \xrightarrow{\alpha} A^m \longrightarrow M \longrightarrow 0$$

where α can be represented by a matrix with coefficients in A. We apply $Hom_A(-, \coprod_{i\in I} N_i)$ for an index set I and A-modules N_i. Defining $N := \coprod_{i\in I} N_i$, we get a commutative diagram

$$\begin{array}{ccccc}
Hom_A(M,N) \hookrightarrow & Hom_A(A^m,N) & \xrightarrow{Hom_A(\alpha,N)} & Hom_A(A^m,N) \\
\| & \downarrow\simeq & & \downarrow\simeq \\
Hom_A(M,N) \hookrightarrow & \coprod_{i\in I} N_i^m & \xrightarrow{Hom_A(\alpha,N)} & \coprod_{i\in I} N_i^n \\
\| & \downarrow\simeq & & \downarrow\simeq \\
Hom_A(M,N) \hookrightarrow & \coprod_{i\in I} Hom_A(A^m,N_i) & \xrightarrow{\coprod Hom_A(\alpha,N_i)} & \coprod_{i\in I} Hom_A(A^m,N_i)
\end{array}$$

with exact rows. Therefore,

$$Hom_A(M,N) = Hom_A(M, \coprod_{i\in I} N_i) \simeq \coprod_{i\in I} Hom_A(M,N_i).$$

Suppose now that $Hom_A(M,-)$ commutes with arbitrary coproducts. Using Lemma 3.1.18 we see that $Hom_A(M,-)$ commutes with arbitrary colimits. Now, M is the colimit of its finitely presented submodules (cf [8, Chapitre 1, § 2 Exercice 10]). Indeed, let M be an A-module and suppose that $M = (\coprod_{i\in I} A)/R$ for some index set I and some submodule R of $\coprod_{i\in I} A$. Then the system of pairs (S, J) for J a finite subset of I and S a finitely generated submodule of $R \cap \coprod_{j\in J} A$ is partially ordered by $(S,J) \leq (S',J')$ if and only if $S \leq S'$ and $J \subseteq J'$. This partially ordered set is an inductive system I_M and $M \simeq \text{colim}_{(S,J)\in I_M}(\coprod_{j\in J} A)/S$. Let I_N be the system of the finitely presented submodules N_i, $i \in I$, of N. This implies that $N = \text{colim}_{i\in I_N} N_i$. Hence we get a natural transformation $\text{colim}_{i\in I_N} Hom_A(M,N_i) \xrightarrow{\eta_M} Hom_A(M,N)$, which is an isomorphism. In case $N = M$ we hence get a natural isomorphism $\text{colim}_{i\in I_M} Hom_A(M,M_i) \xrightarrow{\eta_M} Hom_A(M,M)$ and id_M is in the image of some $Hom_A(M,M_i)$ for large $i \in I_M$. This shows that id_M can be factored through the embedding $M_{i_0} \hookrightarrow M$ for some large i_0. This shows that M is a direct factor of some finitely presented M_{i_0}, but this implies $M = M_{i_0}$ is finitely presented. □

Definition 3.3.15 Let $\mathcal{C}$ and $\mathcal{D}$ be abelian categories. A functor $F : \mathcal{C} \longrightarrow \mathcal{D}$ is *left exact* (resp. *right exact*) if for any short exact sequence $0 \to X \to Y \to Z \to 0$ in $\mathcal{C}$ the sequence $0 \to F(X) \to F(Y) \to F(Z)$ (resp. $F(X) \to F(Y) \to F(Z) \to 0$ is exact. A functor $F : \mathcal{C} \longrightarrow \mathcal{D}$ is *exact* if F is left and right exact.

The concept of an exact functor between exact categories has already been defined in Definition 3.3.8 and it is immediate that the two concepts coincide in this special case. Right exact functors commuting with direct sums are particularly nice and we shall see further applications later on.

Theorem 3.3.16 (Watts [6], Eilenberg [7]) *Let K be a commutative ring and let A and B be two K-algebras. Let $F : A\text{-}Mod \longrightarrow B\text{-}Mod$ be an additive right exact functor that commutes with direct sums. Then there is a B-A-bimodule M such that $F \simeq M \otimes_A -$. If A is Noetherian, and if $F : A\text{-}mod \longrightarrow B\text{-}mod$ is an additive right exact functor, then there is a $B-A$-bimodule M such that $F \simeq M \otimes_A -$.*

Proof Put $M := F(R_A)$, where we denote by R_A the regular A-module. Of course, $R_A = A$ as a K-module, but in order to be able to distinguish the action from the structure on which we are acting, we use a different notation for the module and the ring. Then we get an isomorphism

$$A \simeq End_A(R_A) \xrightarrow{F} End_B(F(R_A)) = End_B(M)$$

and where A acts on the regular module R_A by right multiplication:

$$\mu_a(x) := x \cdot a$$

for all $a \in A$ and $x \in R_A$. Now, $F(\mu_a) =: \nu_a$ for all $a \in A$. We claim that the action of A on M by ν_a gives a structure of a B-A-bimodule on M. Indeed, as we are dealing with right module structures,

$$\mu_{a_1 a_2} = \mu_{a_2} \circ \mu_{a_1} \quad \forall a_1, a_2 \in A$$

implies

$$\nu_{a_1 a_2} = F(\mu_{a_1 a_2}) = F(\mu_{a_2} \circ \mu_{a_1}) = F(\mu_{a_2}) \circ F(\mu_{a_1}) = \nu_{a_2} \circ \nu_{a_1}$$

and so

$$(m \cdot a_1) \cdot a_2 = m \cdot (a_1 \cdot a_2)$$

for all $a_1, a_2 \in A$. Since F is additive, $F(\alpha + \beta) = F(\alpha) + F(\beta)$ for any two homomorphisms $\alpha, \beta \in Hom_A(X, Y)$ for any wo A-modules X, Y. This shows that A acts on M from the right. The action of A on M commutes with the action of B on the left. Indeed, $\nu_a \in Hom_B(M, M)$ for all $a \in A$ and so

$$(b \cdot m) \cdot a = \nu_a(b \cdot m) = b \cdot \nu_a(m) = b \cdot (m \cdot a)$$

for all $b \in B$ and $m \in M$.

Therefore we can define $M \otimes_A - : A\text{-}Mod \longrightarrow B\text{-}Mod$. Now, we observe that $M \otimes_A R_A = M = F(R_A)$ and since $M \otimes_A -$ and F both commute with direct sums (resp. finite direct sums), we get that

$$M \otimes_A L = F(L)$$

for all free (resp. finitely generated free) A-modules L.

Similar to the construction of a projective resolution (cf Definition 1.8.15) we construct a free resolution (i.e. a projective resolution which in addition is formed by free modules only) as follows.

Let now X be an A-module and let $(x_i)_{i \in I}$ be a generating set of X. Then there is an A-module epimorphism

$$L_2 := \bigoplus_{i \in I} A \xrightarrow{\epsilon} X$$

given by $\epsilon((a_i)_{i \in I}) := \sum_{i \in I} a_i x_i$. The sum is finite since we are dealing with direct sums. $\ker(\epsilon)$ is again an A-module with a generating set $(y_j)_{j \in J}$ and we define an epimorphism

$$L_1 := \bigoplus_{j \in J} A \xrightarrow{\delta} \ker \epsilon$$

in the same way. Composing with the embedding $\ker \epsilon \xhookrightarrow{\iota} L_2$ we obtain an exact sequence

$$L_1 \xrightarrow{\iota \circ \delta} L_2 \xrightarrow{\epsilon} X \longrightarrow 0,$$

a so-called free resolution of X. Observe that if A is Noetherian and if X is finitely generated, then L_1 and L_2 can be chosen to be finitely generated as well. By Remark 3.3.3 additive functors commute with finite direct sums. In any case we get the following commutative diagram

$$\begin{array}{ccccccc} F(L_1) & \longrightarrow & F(L_2) & \longrightarrow & F(X) & \longrightarrow 0 \\ \| & & \| & & \downarrow \varphi_X & \\ M \otimes_A L_1 & \longrightarrow & M \otimes_A L_2 & \longrightarrow & M \otimes_A X & \longrightarrow 0 \end{array}$$

in which the rows are exact since F and $M \otimes_A -$ are right exact. Hence the cokernel property proves the existence and uniqueness of φ_X. We need to show that $\varphi : F \longrightarrow M \otimes_A -$ is an isomorphism of functor. By construction φ_X is an isomorphism for each X. Let $\alpha : X \longrightarrow Y$ be a homomorphism of A-modules. Then we construct resolutions

$$L_1 \longrightarrow L_2 \longrightarrow X \longrightarrow 0 \text{ and } L_1' \longrightarrow L_2' \longrightarrow Y \longrightarrow 0$$

of free A-modules. The morphism $\alpha : X \longrightarrow Y$ lifts to a morphism of resolutions

$$\begin{array}{ccccccc} L_1 & \longrightarrow & L_2 & \longrightarrow & X & \longrightarrow 0 \\ \downarrow \alpha_1 & & \downarrow \alpha_2 & & \downarrow \alpha & \\ L_1' & \longrightarrow & L_2' & \longrightarrow & Y & \longrightarrow 0. \end{array}$$

Indeed, α_2 exists since L_2 is projective and since $L_2' \longrightarrow Y$ is surjective. Since we get for the compositions

$$(L_1 \longrightarrow L_2 \longrightarrow L_2' \longrightarrow Y) = (\underbrace{L_1 \longrightarrow L_2 \longrightarrow X}_{=0} \longrightarrow Y) = 0$$

the map

$$L_1 \longrightarrow L_2 \longrightarrow L_2'$$

has image in $\ker(L_2' \longrightarrow Y)$. But then we may use that L_1 is projective and that $L_1 \longrightarrow \ker(L_2' \longrightarrow Y)$ is surjective to get α_1 by the universal property of projective modules.

We may apply F or $M \otimes_A -$ to this commutative diagram, to obtain two commutative diagrams

$$\begin{array}{ccccccc} F(L_1) & \longrightarrow & F(L_2) & \longrightarrow & F(X) & \longrightarrow 0 \\ \downarrow F(\alpha_1) & & \downarrow F(\alpha_2) & & \downarrow F(\alpha) & \\ F(L_1') & \longrightarrow & F(L_2') & \longrightarrow & F(Y) & \longrightarrow 0 \end{array}$$

and

$$\begin{array}{ccccccc} M \otimes_A L_1 & \to & M \otimes_A L_2 & \to & M \otimes_A X & \to 0 \\ \downarrow M \otimes_A \alpha_1 & & \downarrow M \otimes_A \alpha_2 & & \downarrow M \otimes_A \alpha & \\ M \otimes_A L_1' & \to & M \otimes_A L_2' & \to & M \otimes_A Y & \to 0. \end{array}$$

On the images of the free modules L_1, L_2, L_1' and L_2' we get isomorphisms.

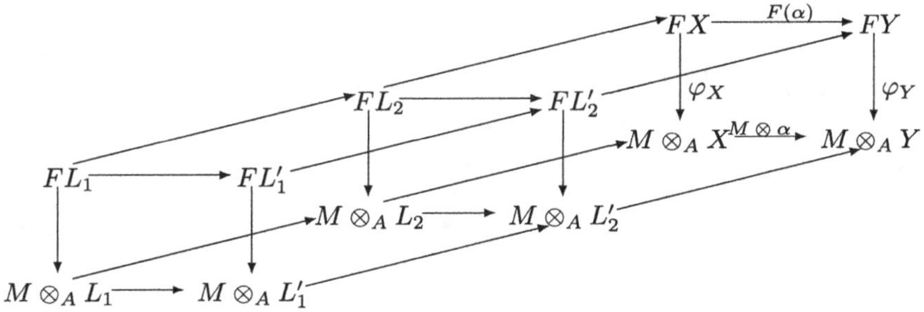

Since the lower left cube has commutative faces, and since the upper right cube is constructed by the cokernel universal property, making the diagonal sequences exact, the right upper most face of this upper right cube is also commutative. This shows that we have a natural transformation. □

3.4 Triangulated Categories

In modern representation theory the notion of triangulated categories is crucial. They provide an abstract framework for derived and for stable module categories in which most of the modern theories and correspondences are built. Moreover, they are even sufficiently flexible to allow equivalences of intermediate degree.

Historically triangulated categories were introduced by Verdier in order to obtain a nice framework for the various derived functors occurring in algebraic geometry. After Verdier's pioneering work triangulated categories became very useful in algebraic geometry of course, but also in complex analysis, in differential geometry as well as in algebraic topology.

Let $\mathcal{C}$ be an additive category and let T be a self-equivalence of $\mathcal{C}$, i.e. $T : \mathcal{C} \longrightarrow \mathcal{C}$ is an equivalence of additive categories. A *triangle* is given by three objects X, Y, Z of $\mathcal{C}$ and three morphisms $\alpha \in Mor_{\mathcal{C}}(X, Y)$, $\beta \in Mor_{\mathcal{C}}(Y, Z)$ and $\gamma \in Mor_{\mathcal{C}}(Z, TX)$. A *morphism from a triangle* $(X, Y, Z, \alpha, \beta, \gamma)$ to a triangle $(X', Y', Z', \alpha', \beta', \gamma')$ is a triple $\xi \in Mor_{\mathcal{C}}(X, X')$, $\eta \in Mor_{\mathcal{C}}(Y, Y')$ and $\zeta \in Mor_{\mathcal{C}}(Z, Z')$ so that the squares in the diagram

$$\begin{array}{ccccccc} X & \stackrel{\alpha}{\longrightarrow} & Y & \stackrel{\beta}{\longrightarrow} & Z & \stackrel{\gamma}{\longrightarrow} & TX \\ \downarrow \xi & & \downarrow \eta & & \downarrow \zeta & & \downarrow T\xi \\ X' & \stackrel{\alpha'}{\longrightarrow} & Y' & \stackrel{\beta'}{\longrightarrow} & Z' & \stackrel{\gamma'}{\longrightarrow} & TX' \end{array}$$

are commutative. Two triangles are isomorphic if there is a morphism of triangles which is formed by a triple which consists of three isomorphisms in $\mathcal{C}$.

Intuitively a triangle provides an infinite sequence applying T and T^{-1} appropriately:

$$\ldots \stackrel{T^{-2}\gamma}{\to} T^{-1}X \stackrel{T^{-1}\alpha}{\to} T^{-1}Y \stackrel{T^{-1}\beta}{\to} T^{-1}Z \stackrel{T^{-1}\gamma}{\to} X \stackrel{\alpha}{\to} Y \stackrel{\beta}{\to} Z \stackrel{\gamma}{\to} TX \stackrel{T\alpha}{\to} \ldots$$

but can of course be depicted as a triangle.

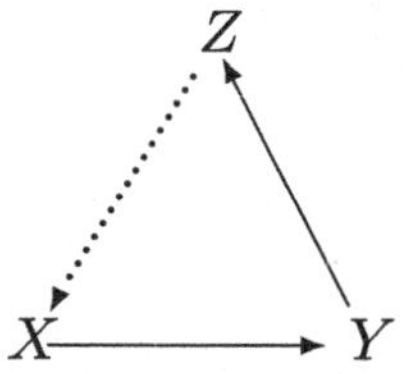

where the dotted arrow indicates that it represents a morphism of degree 1.

Writing a triangle in the form of a sequence reminds us of the long exact sequence in homology, and this resemblance is not an accident, as we shall explain below. First, however, we shall make precise what is meant by a triangulated category.

Definition 3.4.1 (*Verdier*) An additive category $\mathcal{T}$ furnished with a self-equivalence $T : \mathcal{T} \longrightarrow \mathcal{T}$ (called shift functor or suspension functor) and a class of "distinguished triangles", is a *triangulated category* if it satisfies the following axioms.

- **TR1:** A triangle which is isomorphic to a distinguished triangle is itself distinguished.
  The triangle
  $$X \xrightarrow{id} X \longrightarrow 0 \longrightarrow TX$$
  is distinguished.
  Every morphism $X \xrightarrow{\alpha} Y$ can be completed into a distinguished triangle
  $$X \xrightarrow{\alpha} Y \xrightarrow{\beta} Z \xrightarrow{\gamma} TX,$$
  called *the triangle above* α.
- **TR2:** If
  $$X \xrightarrow{\alpha} Y \xrightarrow{\beta} Z \xrightarrow{\gamma} TX$$
  is a distinguished triangle, the
  $$Y \xrightarrow{\beta} Z \xrightarrow{\gamma} TX \xrightarrow{-T\alpha} TY \quad \text{and} \quad T^{-1}Z \xrightarrow{-T^{-1}\gamma} X \xrightarrow{\alpha} Y \xrightarrow{\beta} Z$$
  are distinguished triangles.
- **TR3** If
  $$X \xrightarrow{\alpha} Y \xrightarrow{\beta} Z \xrightarrow{\gamma} TX$$
  is a distinguished triangle, and if
  $$X' \xrightarrow{\alpha'} Y' \xrightarrow{\beta'} Z' \xrightarrow{\gamma'} TX'$$
  is a distinguished triangle, then for any pair $\xi : X \longrightarrow X'$ and $\eta : Y \longrightarrow Y'$ with $\alpha' \circ \xi = \eta \circ \alpha$ there is a morphism $\zeta : Z \longrightarrow Z'$ such that (ξ, η, ζ) is a morphism of triangles:
  $$\begin{array}{ccccccc} X & \xrightarrow{\alpha} & Y & \xrightarrow{\beta} & Z & \xrightarrow{\gamma} & TX \\ \downarrow \xi & & \downarrow \eta & & \Downarrow \zeta & & \downarrow T\xi \\ X' & \xrightarrow{\alpha'} & Y' & \xrightarrow{\beta'} & Z' & \xrightarrow{\gamma'} & TX' \end{array}$$
- **TR4** *(the octahedral axiom)* Given three objects X_1, X_2, X_3 and suppose $\alpha_2 : X_1 \longrightarrow X_3$ factorises as $\alpha_2 = \alpha_1 \circ \alpha_3$ for $\alpha_3 \in Mor_{\mathcal{T}}(X_1, X_2)$ and $\alpha_1 \in Mor_{\mathcal{T}}(X_2, X_3)$. Then forming the triangles
  $$X_2 \xrightarrow{\alpha_1} X_3 \xrightarrow{\beta_1} Z_1 \xrightarrow{\gamma_1} TX_2$$

$$X_1 \xrightarrow{\alpha_3} X_2 \xrightarrow{\beta_3} Z_3 \xrightarrow{\gamma_3} TX_1$$

$$X_1 \xrightarrow{\alpha_2} X_3 \xrightarrow{\beta_2} Z_2 \xrightarrow{\gamma_2} TX_1$$

above α_1, α_2 and α_3, there are morphisms $\delta_1 : Z_3 \to Z_2$, $\delta_3 : Z_2 \to Z_1$ and $\delta_2 : Z_1 \to TZ_3$ such that

$$Z_3 \xrightarrow{\delta_1} Z_2 \xrightarrow{\delta_3} Z_1 \xrightarrow{\delta_2} TZ_3$$

is a distinguished triangle, and such that

$$\gamma_2 \circ \delta_1 = \gamma_3 \ ; \ \delta_3 \circ \beta_2 = \beta_1 \ ; \ \delta_2 = T(\beta_3) \circ \gamma_1 \ ;$$

$$\beta_2 \circ \alpha_1 = \delta_1 \circ \beta_3 \text{ and } \gamma_1 \circ \delta_3 = T(\alpha_3) \circ \gamma_2$$

as illustrated in the following picture.

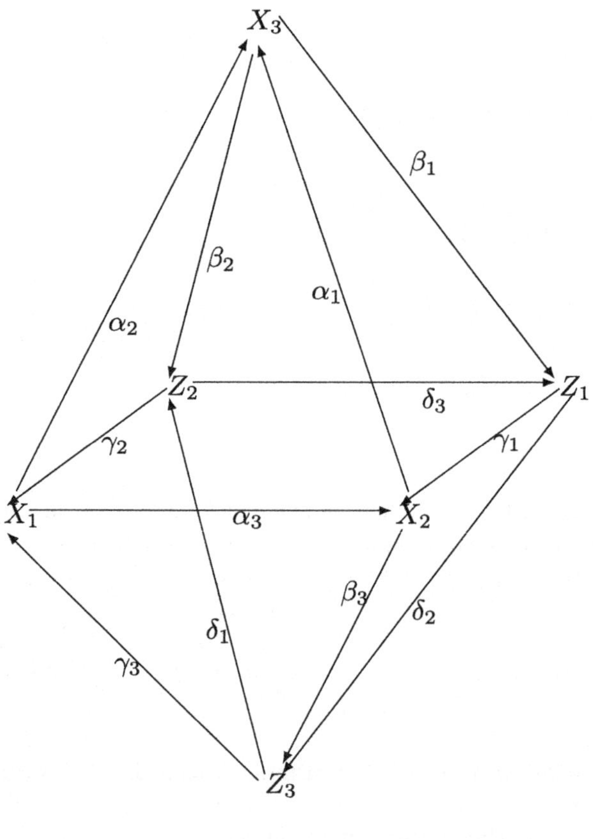

We remind the reader that the morphisms denoted by γ are all of degree 1, and that δ_2 is also of degree 1.

Remark 3.4.2 Technically, and graphically much easier to memorise, we can depict this using the following flat diagram, which displays all commutativity relations, except the commutativity of the square.

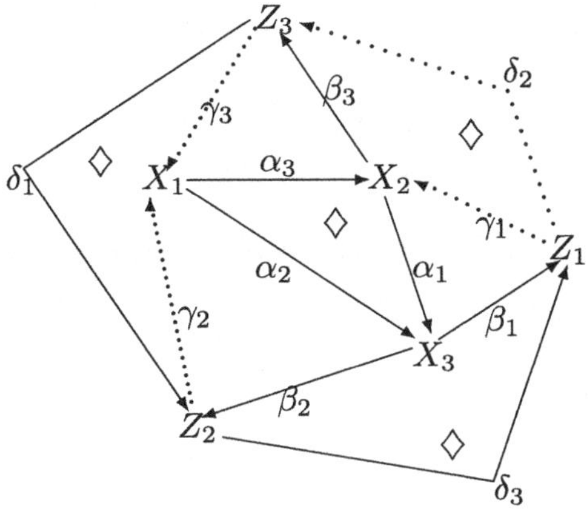

Here triangles containing a $\diamond$ are meant to be commutative and the other triangles are meant to be distinguished. The fourth and the fifth equation are not clearly visible in this diagram.

Remark 3.4.3 In the octahedron we can view three diagonal slices involving four arrows. The horizontal slice (in the picture in the definition) gives the fifth commutativity relation. The slice involving X_2, X_3, Z_2 and Z_3 involves the fourth commutativity relation. The slice containing X_1, X_3, Z_1 and Z_3 does not yield a commutativity relation.

Remark 3.4.4 We do not require uniqueness for ζ in **TR3**. This causes serious problems. The non-uniqueness is however necessary in order to include the examples one would like to. Happel showed that in **TR2** it is sufficient to ask for shifts in only one direction.

The following statement is a so-called five-lemma for derived categories. The classical five-lemma is formulated for abelian categories and is a variant of the snake lemma (Lemma 3.3.9). The proof is also a nice illustration of how to use the axioms.

Lemma 3.4.5 (May [9]) *The axiom* **TR3** *is a consequence of the axioms* **TR1**, **TR2** *and* **TR4**.

Moreover, there is an object C_3 *and morphisms* $C_2 \xrightarrow{\tau_2} C_3$, $C_3 \xrightarrow{\tau_3} TC_1$, $B_3 \xrightarrow{\beta_3} C_3$ *and* $C_3 \xrightarrow{\gamma_3} TA_3$ *such that*

$$\begin{array}{ccccccc}
A_1 & \stackrel{\alpha_1}{\longrightarrow} & B_1 & \stackrel{\beta_1}{\longrightarrow} & C_1 & \stackrel{\gamma_1}{\longrightarrow} & TA_1 \\
\downarrow \rho_1 & & \downarrow \sigma_1 & & \downarrow \tau_1 & & \downarrow T\rho_1 \\
A_2 & \stackrel{\alpha_2}{\longrightarrow} & B_2 & \stackrel{\beta_2}{\longrightarrow} & C_2 & \stackrel{\gamma_2}{\longrightarrow} & TA_2 \\
\downarrow \rho_2 & & \downarrow \sigma_2 & & \downarrow \tau_2 & & \downarrow T\rho_2 \\
A_3 & \stackrel{\alpha_3}{\longrightarrow} & B_3 & \stackrel{\beta_3}{\longrightarrow} & C_3 & \stackrel{\gamma_3}{\longrightarrow} & TA_3 \\
\downarrow \rho_3 & & \downarrow \sigma_3 & & \downarrow \tau_3 & & \\
TA_1 & \stackrel{T\alpha_1}{\longrightarrow} & TB_1 & \stackrel{T\beta_1}{\longrightarrow} & TC_1 & &
\end{array}$$

is commutative and so that all rows and columns are distinguished triangles.

Proof Let

$$\begin{array}{ccc}
A_1 & \stackrel{\alpha_1}{\longrightarrow} & B_1 \\
\downarrow \rho_1 & & \downarrow \sigma_1 \\
A_2 & \stackrel{\alpha_2}{\longrightarrow} & B_2
\end{array}$$

be a commutative diagram in the triangulated category $\mathcal{T}$. Then we may use TR1 to get horizontal and vertical distinguished triangles as follows:

$$\begin{array}{ccccccc}
A_1 & \stackrel{\alpha_1}{\longrightarrow} & B_1 & \stackrel{\beta_1}{\longrightarrow} & C_1 & \stackrel{\gamma_1}{\longrightarrow} & TA_1 \\
\downarrow \rho_1 & & \downarrow \sigma_1 & & & & \\
A_2 & \stackrel{\alpha_2}{\longrightarrow} & B_2 & \stackrel{\beta_2}{\longrightarrow} & C_2 & \stackrel{\gamma_2}{\longrightarrow} & TA_2 \\
\downarrow \rho_2 & & \downarrow \sigma_2 & & & & \\
A_3 & & B_3 & & & & \\
\downarrow \rho_3 & & \downarrow \sigma_3 & & & & \\
TA_1 & & TB_1 & & & &
\end{array}$$

We consider the morphism $\sigma_1 \circ \alpha_1 = \alpha_2 \circ \rho_1 =: \delta$ and we may apply TR4 to both of these factorisations of δ. Let then

$$A_1 \stackrel{\delta}{\longrightarrow} B_2 \stackrel{\epsilon}{\longrightarrow} D \stackrel{\varphi}{\longrightarrow} TA_1$$

be a distinguished triangle, and apply TR4 to $\sigma_1 \circ \alpha_1 = \delta$. This gives morphisms $\zeta_1 : C_1 \longrightarrow D$, $\eta_1 : D \longrightarrow B_3$ and $\chi_1 : B_3 \longrightarrow TC_1$ such that

$$C_1 \stackrel{\zeta_1}{\longrightarrow} D \stackrel{\eta_1}{\longrightarrow} B_3 \stackrel{\chi_1}{\longrightarrow} TC_1$$

is a distinguished triangle, and such that

$$\begin{array}{rcl}
\epsilon \circ \sigma_1 = \zeta_1 \circ \beta_1 & \qquad & \sigma_2 = \eta_1 \circ \epsilon \\
\gamma_1 = \varphi \circ \zeta_1 & \qquad & T\alpha_1 \circ \varphi = \sigma_3 \circ \eta_1 \\
\chi_1 = T\beta_1 \circ \sigma_3. & &
\end{array}$$

We apply TR4 to the factorisation $\alpha_2 \circ \rho_1 = \delta$ to obtain morphisms $\eta_2 : A_3 \longrightarrow D$, $\zeta_2 : D \longrightarrow C_2$ and $\chi_2 : C_2 \longrightarrow TA_3$ such that

$$A_3 \xrightarrow{\eta_2} D \xrightarrow{\zeta_2} C_2 \xrightarrow{\chi_2} TA_3$$

is a distinguished triangle and such that

$$\begin{array}{rlrl} \epsilon \circ \alpha_2 &= \eta_2 \circ \rho_2 & \zeta_2 \circ \epsilon &= \beta_2 \\ \rho_3 &= \varphi \circ \eta_2 & T\rho_1 \circ \varphi &= \gamma_2 \circ \zeta_2 \\ T\rho_2 \circ \gamma_2 &= \chi_2. & & \end{array}$$

Then we claim that

$$\tau_1' := \zeta_2 \circ \zeta_1$$

and

$$\alpha_3 := \eta_1 \circ \eta_2$$

make the diagram

$$\begin{array}{ccccccc} A_1 & \xrightarrow{\alpha_1} & B_1 & \xrightarrow{\beta_1} & C_1 & \xrightarrow{\gamma_1} & TA_1 \\ \downarrow \rho_1 & & \downarrow \sigma_1 & & \downarrow \tau_1' & & \downarrow T\rho_1 \\ A_2 & \xrightarrow{\alpha_2} & B_2 & \xrightarrow{\beta_2} & C_2 & \xrightarrow{\gamma_2} & TA_2 \\ \downarrow \rho_2 & & \downarrow \sigma_2 & & & & \\ A_3 & \xrightarrow{\alpha_3} & B_3 & & & & \\ \downarrow \rho_3 & & \downarrow \sigma_3 & & & & \\ TA_1 & \xrightarrow{T\alpha_1} & TB_1 & & & & \end{array}$$

commutative. Indeed,

$$\tau_1 \circ \beta_1 = \zeta_2 \circ \zeta_1 \circ \beta_1 = \zeta_2 \circ \epsilon \circ \sigma_1 = \beta_2 \circ \sigma_1$$

$$\gamma_2 \circ \tau_1 = \gamma_2 \circ \zeta_2 \circ \zeta_1 = T\rho_1 \circ \varphi \circ \zeta_1 = T\rho_1 \circ \gamma_1$$

$$\sigma_3 \circ \alpha_3 = \sigma_3 \circ \eta_1 \circ \eta_2 = T\alpha_1 \circ \varphi \circ \eta_2 = T\alpha_1 \circ \rho_3$$

and

$$\alpha_3 \circ \rho_2 = \eta_1 \circ \eta_2 \circ \rho_2 = \eta_1 \circ \epsilon \circ \alpha_2 = \sigma_2 \circ \alpha_2.$$

This proves the first statement.

For the second statement apply the axiom TR4 to the factorisation $\alpha_3 = \eta_1 \circ \eta_2$. Denoting by

$$A_3 \xrightarrow{\alpha_3} B_3 \xrightarrow{\beta_3} C_3 \xrightarrow{\gamma_3} TA_3$$

the distinguished triangle starting with α_3, the axiom TR4 gives morphisms $\tau_2 : C_2 \longrightarrow C_3$, $\tau_3 : C_3 \longrightarrow TC_1$ and $\mu : TC_1 \longrightarrow TC_2$ so that

$$C_2 \xrightarrow{\tau_2} C_3 \xrightarrow{\tau_3} TC_1 \xrightarrow{\mu} TC_2$$

is a distinguished triangle. Moreover, the commutativity relations associated to TR4 give that we may put $T^{-1}\mu =: \tau_1$ and still have commutativity of the diagram. □

Remark 3.4.6 Observe that we did not show that τ_1 and $\zeta_2 \circ \zeta_1$ coincide.

Given a distinguished triangle in a triangulated category, we did not explicitly ask that the composition of two consecutive maps in the triangle is 0. This is true and follows from the axioms TR1, TR2 and TR3.

Lemma 3.4.7 *Let $\mathcal{T}$ be a triangulated category. Then for every distinguished triangle*

$$X \xrightarrow{f} Y \xrightarrow{g} Z \xrightarrow{h} TX$$

we have

$$g \circ f = 0 = h \circ g.$$

Proof We know that for every object U

$$U \xrightarrow{id} U \longrightarrow 0 \longrightarrow TU$$

is a distinguished triangle by TR1. Now the diagram

$$\begin{array}{ccccccc} X & \xrightarrow{id} & X & \longrightarrow 0 \longrightarrow & & & TX \\ \downarrow id & & \downarrow f & & & & \downarrow T\,id \\ X & \xrightarrow{f} & Y & \xrightarrow{g} Z \xrightarrow{h} & & & TX \end{array}$$

is trivially commutative. The axiom TR3 then allows us to complete this by a morphism $0 \longrightarrow Z$

$$\begin{array}{ccccccc} X & \xrightarrow{id} & X & \longrightarrow & 0 & \longrightarrow & TU \\ \downarrow id & & \downarrow f & & \downarrow & & \downarrow T\,id \\ X & \xrightarrow{f} & Y & \xrightarrow{g} & Z & \xrightarrow{h} & TX \end{array}$$

making all the squares commutative. But this shows that $g \circ f = 0$. The fact that $h \circ g = 0$ follows by TR2. □

Remark 3.4.8 Observe that we did not use the octahedral axiom TR4 in the proof of Lemma 3.4.7. Hence the statement is true for slightly weaker structures.

Lemma 3.4.9 *Let $\mathcal{T}$ be a triangulated category and let $X \xrightarrow{\alpha} Y$ be a morphism in $\mathcal{T}$. Then*

- *α is an isomorphism if and only if*

$$X \xrightarrow{\alpha} Y \longrightarrow 0 \longrightarrow TX$$

is a distinguished triangle.
- *α is a split monomorphism if and only if*

$$X \xrightarrow{\alpha} Y \longrightarrow Z \xrightarrow{0} TX$$

is a distinguished triangle.
- *α is a split epimorphism if and only if*

$$X \xrightarrow{\alpha} Y \xrightarrow{0} Z \longrightarrow TX$$

is a distinguished triangle.

Proof **First step:** We need to prove the second and the third item only. Indeed, suppose that we have shown the second and third statements. If $\beta : U \longrightarrow V$ is an isomorphism, then

$$\begin{array}{ccccccc} U & \xrightarrow{\beta} & V & \longrightarrow & 0 & \longrightarrow & TU \\ \| & & \downarrow \beta^{-1} & & \| & & \| \\ U & \xrightarrow{id} & U & \longrightarrow & 0 & \longrightarrow & TU \end{array}$$

is commutative, all vertical morphisms are isomorphisms, and the lower triangle is distinguished. Hence the upper triangle is also distinguished and we obtain that if $\beta : U \longrightarrow V$ is an isomorphism, then

$$U \xrightarrow{\beta} V \longrightarrow 0 \longrightarrow TU$$

is a distinguished triangle.

On the other hand, if

$$U \xrightarrow{\beta} V \longrightarrow 0 \longrightarrow TU$$

is a distinguished triangle, by the second and the third point, β admits a left and a right inverse. This implies that β is an isomorphism.

Second step: Suppose

$$X \xrightarrow{\alpha} Y \longrightarrow Z \xrightarrow{0} TX$$

is a distinguished triangle. Then

$$\begin{array}{ccccccc} X & \stackrel{\alpha}{\longrightarrow} & Y \longrightarrow & Z & \stackrel{0}{\longrightarrow} & TX \\ \downarrow id & & & \downarrow 0 & & \downarrow id \\ X & \stackrel{id}{\longrightarrow} & X \longrightarrow & 0 & \longrightarrow & TX \end{array}$$

is commutative, and by **TR2** and **TR3** there is a morphism $\beta : Y \longrightarrow X$ which makes the diagram

$$\begin{array}{ccccccc} X & \stackrel{\alpha}{\longrightarrow} & Y & \longrightarrow & Z & \longrightarrow & TX \\ \| & & \downarrow \beta & & \downarrow & & \| \\ X & \stackrel{id}{\longrightarrow} & X & \longrightarrow & 0 & \longrightarrow & TX \end{array}$$

a morphism of distinguished triangles. Hence $\beta \circ \alpha = id_X$.

Third step: Suppose α admits a left inverse $\beta : Y \longrightarrow X$. Then we get a commutative diagram

$$\begin{array}{ccccccc} X & \stackrel{\alpha}{\longrightarrow} & Y & \longrightarrow & Z & \longrightarrow & TX \\ \| & & \downarrow \beta & & & & \| \\ X & \stackrel{id}{\longrightarrow} & X & \longrightarrow & 0 & \longrightarrow & TX \end{array}$$

where both horizontal lines are distinguished triangles. Axiom **TR3** gives that there is a mapping $Z \longrightarrow 0$ that makes

$$\begin{array}{ccccccc} X & \stackrel{\alpha}{\longrightarrow} & Y & \longrightarrow & Z & \longrightarrow & TX \\ \| & & \downarrow \beta & & \downarrow & & \| \\ X & \stackrel{id}{\longrightarrow} & X & \longrightarrow & 0 & \longrightarrow & TX \end{array}$$

a morphism of triangles. This shows that the mapping $Z \longrightarrow TX$ is 0 and hence the distinguished triangle is

$$X \stackrel{\alpha}{\longrightarrow} Y \longrightarrow Z \stackrel{0}{\longrightarrow} TX.$$

Fourth step: Suppose

$$X \stackrel{\alpha}{\longrightarrow} Y \stackrel{0}{\longrightarrow} Z \longrightarrow TX$$

is a triangle. Then

$$T^{-1}Z \longrightarrow X \stackrel{\alpha}{\longrightarrow} Y \stackrel{0}{\longrightarrow} Z$$

is a triangle and hence we get a commutative diagram

$$\begin{array}{ccccccc} T^{-1}Z & \longrightarrow & X \stackrel{\alpha}{\longrightarrow} & Y & \stackrel{0}{\longrightarrow} & Z \\ \uparrow 0 & & & \uparrow id & & \uparrow 0 \\ 0 & \longrightarrow & Y \stackrel{id}{\longrightarrow} & Y & \longrightarrow & 0 \end{array}$$

Therefore there is a morphism $\gamma : Y \longrightarrow X$ making the diagram a morphism of triangles

$$\begin{array}{ccccccc} T^{-1}Z & \longrightarrow & X & \stackrel{\alpha}{\longrightarrow} & Y & \stackrel{0}{\longrightarrow} & Z \\ \uparrow 0 & & \uparrow \gamma & & \uparrow id & & \uparrow 0 \\ 0 & \longrightarrow & Y & \stackrel{id}{\longrightarrow} & Y & \longrightarrow & 0 \end{array}$$

and hence $\alpha \circ \gamma = id_Y$.

Fifth step: Suppose

$$X \stackrel{\alpha}{\longrightarrow} Y \longrightarrow Z \longrightarrow TX$$

is a triangle and suppose that there is a $\gamma : Y \longrightarrow X$ such that $\alpha \circ \gamma = id_Y$. Then the left-hand square below can be completed to a distinguished triangle

$$\begin{array}{ccccccc} X & \stackrel{\alpha}{\longrightarrow} & Y & \longrightarrow & Z & \longrightarrow & TX \\ \uparrow \gamma & & \uparrow id & & \uparrow 0 & & \uparrow T\gamma \\ Y & \stackrel{id}{\longrightarrow} & Y & \longrightarrow & 0 & \longrightarrow & TY \end{array}$$

and hence

$$X \stackrel{\alpha}{\longrightarrow} Y \stackrel{0}{\longrightarrow} Z \longrightarrow TX$$

is a triangle.

This proves the lemma. □

Lemma 3.4.10 *Let $\mathcal{T}$ be a triangulated category and let*

$$\begin{array}{ccccccc} A_1 & \longrightarrow & A_2 & \longrightarrow & A_3 & \longrightarrow & TA_1 \\ \downarrow & & \downarrow & & \downarrow & & \downarrow \\ B_1 & \longrightarrow & B_2 & \longrightarrow & B_3 & \longrightarrow & TB_1 \end{array}$$

be a morphism between distinguished triangles. If two of the vertical morphisms are isomorphisms, then the third one is an isomorphism as well.

In particular, if

$$A_1 \stackrel{\alpha}{\longrightarrow} A_2 \longrightarrow A_3 \longrightarrow TA_1$$

and

$$A_1 \stackrel{\alpha}{\longrightarrow} A_2 \longrightarrow A_3' \longrightarrow TA_1$$

are distinguished triangles, then $A_3 \simeq A_3'$.

Proof The second statement follows from the first using TR3. Suppose $A_1 \longrightarrow B_1$ and $A_3 \longrightarrow B_3$ are isomorphisms, and let

$$A_2 \longrightarrow B_2 \longrightarrow C_2 \longrightarrow TA_2$$

be a distinguished triangle. Lemma 3.4.5 and Lemma 3.4.9 show that

$$0 \longrightarrow C_2 \longrightarrow 0 \longrightarrow T0$$

is a distinguished triangle. Again using Lemma 3.4.9 we get that $C_2 \longrightarrow 0$ is an isomorphism. □

The concept of triangulated categories suggests that they should encode what is called the long exact sequence in homology. The details are as follows.

Proposition 3.4.11 *Let $\mathcal{T}$ be a triangulated category and let*

$$X \xrightarrow{\alpha} Y \xrightarrow{\beta} Z \xrightarrow{\gamma} TX$$

be a distinguished triangle. Then applying $Hom_{\mathcal{T}}(U,-)$ to the triangle gives an exact sequence

$$\begin{array}{ccccc} \ldots \to Hom_{\mathcal{T}}(U,T^{-2}Z) & \to & Hom_{\mathcal{T}}(U,T^{-1}X) & \to & Hom_{\mathcal{T}}(U,T^{-1}Y) \\ & & & & \downarrow \\ Hom_{\mathcal{T}}(U,Y) & \leftarrow & Hom_{\mathcal{T}}(U,X) & \leftarrow & Hom_{\mathcal{T}}(U,T^{-1}Z) \\ \downarrow & & & & \\ Hom_{\mathcal{T}}(U,Z) & \to & Hom_{\mathcal{T}}(U,TX) & \to & Hom_{\mathcal{T}}(U,TY) \to \ldots \end{array}$$

and applying $Hom_{\mathcal{T}}(-,U)$ to the triangle gives an exact sequence

$$\begin{array}{ccccc} \cdots \leftarrow Hom_{\mathcal{T}}(T^{-2}Z,U) & \leftarrow & Hom_{\mathcal{T}}(T^{-1}X,U) & \leftarrow & Hom_{\mathcal{T}}(T^{-1}Y,U) \\ & & & & \uparrow \\ Hom_{\mathcal{T}}(Y,U) & \to & Hom_{\mathcal{T}}(X,U) & \to & Hom_{\mathcal{T}}(T^{-1}Z,U) \\ \uparrow & & & & \\ Hom_{\mathcal{T}}(Z,U) & \leftarrow & Hom_{\mathcal{T}}(TX,U) & \leftarrow & Hom_{\mathcal{T}}(TY,U) \leftarrow \ldots \end{array}$$

Proof By Lemma 3.4.7 the composition of two subsequent morphisms is 0. We need to show that the sequence is exact. Let now $\varphi \in Hom_{\mathcal{T}}(U,Z)$ be a morphism with

$$Hom_{\mathcal{T}}(U,\gamma)(\varphi) = \gamma \circ \varphi = 0.$$

Then we obtain a commutative diagram

$$\begin{array}{ccccccc} X & \xrightarrow{\alpha} & Y & \xrightarrow{\beta} & Z & \xrightarrow{\gamma} & TX \\ \uparrow & & & & \uparrow \varphi & & \uparrow \\ 0 & \longrightarrow & U & \xrightarrow{id} & U & \longrightarrow & 0 \end{array}$$

TR3 shows that there exists a $\psi \in Hom_{\mathcal{T}}(U,Y)$ so that the diagram

$$\begin{array}{ccccccc} X & \stackrel{\alpha}{\longrightarrow} & Y & \stackrel{\beta}{\longrightarrow} & Z & \stackrel{\gamma}{\longrightarrow} & TX \\ \uparrow & & \uparrow \psi & & \uparrow \varphi & & \uparrow \\ 0 & \longrightarrow & U & \stackrel{id}{\longrightarrow} & U & \longrightarrow & 0 \end{array}$$

is commutative. But this is exactly what we need to show that the complex is exact at $Hom_{\mathcal{T}}(U, Z)$. By **TR2** we obtain that the complex is exact everywhere.

The statement for the contravariant functor is dual. Again, by Lemma 3.4.7 composition of two subsequent morphisms is 0. Let

$$\varphi \in \ker(Hom_{\mathcal{T}}(Y, U) \longrightarrow Hom_{\mathcal{T}}(X, U)).$$

Then we obtain a commutative diagram

$$\begin{array}{ccccccc} X & \stackrel{\alpha}{\longrightarrow} & Y & \stackrel{\beta}{\longrightarrow} & Z & \stackrel{\gamma}{\longrightarrow} & TX \\ \downarrow & & \downarrow \varphi & & & & \downarrow \\ 0 & \longrightarrow & U & \stackrel{id}{\longrightarrow} & U & \longrightarrow & 0 \end{array}$$

which can be completed to a morphism of distinguished triangles, using **TR3**, with a morphism ψ:

$$\begin{array}{ccccccc} X & \stackrel{\alpha}{\longrightarrow} & Y & \stackrel{\beta}{\longrightarrow} & Z & \stackrel{\gamma}{\longrightarrow} & TX \\ \downarrow & & \downarrow \varphi & & \downarrow \psi & & \downarrow \\ 0 & \longrightarrow & U & \stackrel{id}{\longrightarrow} & U & \longrightarrow & 0 \end{array}$$

This proves the statement. □

Definition 3.4.12 A covariant (resp. contravariant) functor $F : \mathcal{T} \longrightarrow \mathcal{A}$ from a triangulated category $\mathcal{T}$ to an abelian category $\mathcal{A}$ is called *homological* (resp. *cohomological*) if it transforms distinguished triangles to long exact sequences.

Lemma 3.4.11 shows that representable functors are (co-)homological.

Definition 3.4.13 Let $\mathcal{T}_1$ and $\mathcal{T}_2$ be triangulated categories. Then a functor $F : \mathcal{T}_1 \longrightarrow \mathcal{T}_2$ is a *functor of triangulated categories* if F sends distinguished triangles to distinguished triangles and $T_2 \circ F = F \circ T_1$ where T_i is the shift functor in the triangulated category $\mathcal{T}_i$.

3.5 Complexes

We shall intensively use a specific triangulated category which is derived from complexes of modules. This will be the subject of this section.

3.5.1 The Category of Complexes

Definition 3.5.1 Let K be a commutative ring and let A be a K-algebra. An A-module M is *graded* if $M = \bigoplus_{i\in\mathbb{Z}} M_i$ where each M_i is an A-submodule of M. We call M_i the *degree i homogeneous component* of the graded module. Let M and N be graded A-modules. A *morphism of degree d of graded A-modules* is a homomorphism of A-modules $\varphi : M \longrightarrow N$ such that

$$\operatorname{im}(\varphi|_{M_i}) \subseteq N_{i+d}.$$

Definition 3.5.2 A *complex* $M^\bullet := (M, d)$ is a graded module $M = \bigoplus_{i\in\mathbb{Z}} M_i$ equipped with an endomorphism d of graded modules of degree 1 and square 0. Recall from Definition 2.2.8 the definition of a graded module and that for simplicity if M is $\mathbb{Z}$-graded, then we simply say that M is graded. The endomorphism d is the *differential* of the complex.

For two complexes (M, d_M) and (N, d_N) a morphism of complexes is a homomorphism of graded A-modules $\varphi : M \longrightarrow N$ of degree 0 such that $d_N \circ \varphi = \varphi \circ d_M$.

In case no confusion may arise a complex (M, d) is sometimes denoted by M only.

We visualise a complex as

$$\dots \xrightarrow{d_{i-1}} M_i \xrightarrow{d_i} M_{i+1} \xrightarrow{d_{i+1}} M_{i+2} \xrightarrow{d_{i+2}} M_{i+3} \xrightarrow{d_{i+3}} \dots$$

where $d_i \circ d_{i-1} = 0$ for all $i \in \mathbb{Z}$.

Remark 3.5.3 We observe that we may as well define a complex to be a $\mathbb{Z}$-graded object and a morphism of degree -1 with square 0. Whether d is of degree 1 or of degree -1 is not important. Indeed, one can transfer one to the other replacing M_i by M_{-i} and leaving d unchanged. We shall not be very consistent if a complex has differentials of degree 1 or differentials of degree -1. The context will sometimes justify one case and sometimes the other.

In the literature the concepts of complexes with degree 1 differential and with degree -1 differential are distinguished by saying that one is a chain complex and the other a cochain complex. This terminology reveals the topological context in which the theory was established.

Remark 3.5.4 It is clear that the complexes of A-modules together with morphisms of complexes form a category, which we shall denote by $C(A\text{-}Mod)$. It is clear that we may form the category of complexes $C(\mathcal{A})$ of any additive subcategory $\mathcal{A}$ of A-Mod. In this case, an object in $C(\mathcal{A})$ is given by a $\mathbb{Z}$-graded A-module $M = \bigoplus_{i\in\mathbb{Z}} M_i$ where each M_i is an object of $\mathcal{A}$, and an A-linear differential $d : M \longrightarrow M$ of degree -1 such that each degree i homogeneous component $d_i : M_i \longrightarrow M_{i-1}$ is a morphism in $\mathcal{A}$.

Let (M, d) be a complex. Since $d \circ d = 0$ we get that $\text{im}(d) \subseteq \ker(d)$. Hence $H(M) := \ker(d)/\text{im}(d)$ is an A-module. Since d is of degree 1, $H(M)$ is also graded:

$$H(M) = \bigoplus_{i \in \mathbb{Z}} H_i(M).$$

Nevertheless, d induces only a trivial endomorphism on $H(M)$, the endomorphism 0. Indeed, the image of $\text{im}(d)$ in $H(M)$ is 0 by definition. Nevertheless, the homology of a complex can be considered as a complex again with differential 0.

Definition 3.5.5 The homology of a complex (M, d) of A-modules is

$$H(M) = \ker(d)/\text{im}(d).$$

The A-submodule $\ker(d)$ is called the cycles of the complex and the module $\text{im}(d)$ is called the boundaries of the complex.

- If the differential is of degree $+1$, then the complex is bounded above if there is an i_0 such that $M_i = 0$ for all $i > i_0$ and the complex is bounded below if there is an i_1 such that $M_i = 0$ for all $i < i_0$.
- If the differential is of degree -1, then the complex is bounded above if there is an i_0 such that $M_i = 0$ for all $i < i_0$ and the complex is bounded below if there is an i_1 such that $M_i = 0$ for all $i > i_0$.
- In any case, a complex is bounded if it is bounded below and bounded above.

The complex (M, d) has right (resp. left) bounded homology if the complex $(H(M), 0)$ is bounded above (resp. below). The complex has bounded homology if it has left and right bounded homology.

Remark 3.5.6 The above terminology comes from algebraic topology. For details we refer to any textbook on algebraic topology, such as Spanier [10]. Though interesting we shall not develop the link here.

Remark 3.5.7 A morphism of complexes $\varphi : (M, d_M) \longrightarrow (N, d_N)$ induces a homomorphism of graded modules $H(\varphi) : H(M) \longrightarrow H(N)$ of degree 0 by $H(\varphi) : m + d_M(M) := \varphi(m) + d_N(N)$, which is well-defined since

$$\varphi(d_M(M)) = d_N(\varphi(M)) \subseteq d_N(N)$$

and since

$$m \in \ker(d_M) \Rightarrow d_N(\varphi(m)) = \varphi(d_M(m)) = \varphi(0) = 0 \Rightarrow d_M(m) \in \ker(d_M).$$

Moreover, $H(id_M) = id_{H(M)}$ and $H(\varphi \circ \psi) = H(\varphi) \circ H(\psi)$ where $\varphi : M \longrightarrow N$ and $\psi : L \longrightarrow M$ are two morphisms of complexes.

Lemma 3.5.8 *Let K be a commutative ring and let A be a K-algebra. Then the homology is a functor from the category of complexes of A-modules to the category of graded A-modules.* □

Recall from Lemma 3.5.4 the notion of a complex in an additive subcategory of $A\text{-}Mod$.

Definition 3.5.9 Let K be a commutative ring and let A be a K-algebra. Denote by $\mathcal{A}$ an additive subcategory of $A\text{-}Mod$. A complex in $\mathcal{A}$ is a complex such that each homogeneous component is in $\mathcal{A}$ and such that the differential in each homogeneous component is a morphism in $\mathcal{A}$.

- $C(\mathcal{A})$ is the category of complexes in $\mathcal{A}$.
- $C^+(\mathcal{A})$ is the category of left bounded complexes in $\mathcal{A}$.
- $C^-(\mathcal{A})$ is the category of right bounded complexes in $\mathcal{A}$.
- $C^b(\mathcal{A})$ is the category of bounded complexes in $\mathcal{A}$.
- $C^{\emptyset,+}(\mathcal{A})$ is the category of complexes in $\mathcal{A}$ with left bounded homology.
- $C^{\emptyset,-}(\mathcal{A})$ is the category of complexes in $\mathcal{A}$ with right bounded homology.
- $C^{\emptyset,b}(\mathcal{A})$ is the category of complexes in $\mathcal{A}$ with bounded homology.
- $C^{-,b}(\mathcal{A})$ is the category of right bounded complexes in $\mathcal{A}$ with bounded homology.
- $C^{+,b}(\mathcal{A})$ is the category of left bounded complexes in $\mathcal{A}$ with bounded homology.

Proposition 3.5.10 *Let K be a commutative ring and let A be a K-algebra. Denote by $\mathcal{A}$ an additive subcategory of $A\text{-}Mod$. If $\mathcal{A}$ is abelian, then $C^{x,y}(\mathcal{A})$ is an abelian category for $x, y \in \{\emptyset, b, +, -\}$.*

Proof We just deal with $C^{-,b}(\mathcal{A})$, the other cases are similar and the proof is left to the reader.

The set of morphisms of complexes $Hom_{C^{-,b}(\mathcal{A})}(M, N)$ is an abelian group since $\mathcal{A}$ is additive, whence $Mor_{\mathcal{A}}(M_i, N_i)$ is an abelian group for all $i \in \mathbb{Z}$. Hence $\prod_{i\in\mathbb{Z}} Mor_{\mathcal{A}}(M_i, N_i)$ is an abelian group and

$$Hom_{C^{-,b}(\mathcal{A})}(M, N) = \left\{ (\varphi_i)_{i\in\mathbb{Z}} \in \prod_{i\in\mathbb{Z}} Mor_{\mathcal{A}}(M_i, N_i) \,\middle|\, \varphi_i \circ (d_M)^{(i-1)} = (d_N)^{(i)} \circ \varphi_{i-1} \forall i \in \mathbb{Z} \right\}$$

is an abelian group since the composition in $\mathcal{A}$ is bilinear with respect to the abelian group structure.

The composition of morphisms of complexes is bilinear since the composition of morphisms in $\mathcal{A}$ is bilinear and sums are taken degree by degree. The product of complexes is the complex formed by the product componentwise, with diagonal differential. The universal property is readily verified. The coproduct of complexes is the complex formed by the coproduct of the homogeneous components and diagonal morphisms. Again the universal property is easy to verify. Hence, finite biproducts exist in $C^{-,b}(\mathcal{A})$. Observe that the homology of a finite coproduct is still bounded since the index set is finite. Hence $C^{-,b}(\mathcal{A})$ is additive.

Suppose now $\mathcal{A}$ is abelian. A morphism of complexes $\varphi : M \longrightarrow N$ induces a complex of the kernels $(\ker(\varphi), (d_M)|_{\ker(\varphi)})$ since the diagram

$$\begin{array}{ccc} M_i & \overset{(d_M)^{(i)}}{\longrightarrow} & M_{i+1} \\ \downarrow \varphi_i & & \downarrow \varphi_{i+1} \\ N_i & \overset{(d_N)^{(i)}}{\longrightarrow} & N_{i+1} \end{array}$$

is commutative and so, by the universal property of the kernel $(d_M)^{(i)}$, induces a morphism $(d_{\ker\varphi})^{(i)} : \ker\varphi_i \longrightarrow \ker\varphi_{i+1}$. Moreover $d_{\ker\varphi} \circ d_{\ker\varphi} = 0$, since this is true by composing at the end with the embedding $\ker(\varphi) \longrightarrow M$, and then using that the kernel is a monomorphism.

The fact that this complex has the universal property is clear, using the universal property on each homogeneous component; in order to show that the locally constructed morphism is indeed a morphism of complexes one uses an argument similar to the one we just used to show that $d_{\ker\varphi}$ is a differential.

The case of cokernels is shown dually.

We need to show that each monomorphism is the kernel of a morphism and each epimorphism is the cokernel of a morphism. We just deal with the kernel case. The cokernel is then analogous. Let $\varphi : K \longrightarrow M$ be a monomorphism in $C^{-,b}(\mathcal{A})$. We have seen that morphisms of complexes admit a cokernel. Let $\gamma : M \longrightarrow C(\varphi)$ be the cokernel of φ. We claim that $\varphi = \ker(\gamma)$. Indeed, this is true at each degree since $\mathcal{A}$ is abelian using Remark 3.3.5. Given a complex D and a morphism of complexes $\delta : D \longrightarrow M$ such that $\gamma \circ \delta = 0$, there is a unique morphism $\beta : D \longrightarrow K$ in the category of graded objects of $\mathcal{A}$ such that $\delta = \varphi \circ \beta$. Now, δ which is a morphism of complexes and φ is a monomorphism and is a morphism of complexes. Denoting by d^K the differential on K, by d^M the differential on M and by d^D the differential on D, we have

$$\varphi \circ d^K \circ \beta = d^M \circ \varphi \circ \beta = d^M \circ \delta = \delta \circ d^D = \varphi \circ \beta \circ d^D$$

which implies $d^K \circ \beta = \beta \circ d^D$. Therefore also β is a morphism of complexes. □

For later use we shall prove the Krull-Schmidt theorem for the category of complexes of modules over a finite dimensional algebra. The proof provides an interesting interpretation of a complex as module over a specific algebra. This point of view is useful for various other applications as well.

Proposition 3.5.11 *Let K be a field and let A be a finite dimensional K-algebra. Then $C^b(A\text{-}mod)$ satisfies the Krull-Schmidt theorem.*

Proof Let $\vec{A}_n$ be the quiver given by n vertices $1, 2, \ldots, n$ and an arrow a_i from i to $i+1$ for all $i \in \{1, 2, \ldots, n-1\}$:

$$\vec{A}_n : \bullet_1 \overset{a_1}{\longrightarrow} \bullet_2 \overset{a_2}{\longrightarrow} \cdots \overset{a_{n-2}}{\longrightarrow} \bullet_{n-1} \overset{a_{n-1}}{\longrightarrow} \bullet_n$$

As usual the quiver algebra $K\overrightarrow{A}_n$ has a K-basis given by all paths in this quiver, including the paths of length 0, and multiplication given by concatenation of paths, if concatenation is possible, and 0 otherwise. In this case the algebra is isomorphic to the algebra of upper triangular n by n matrices. We consider the algebra $B_n := K\overrightarrow{A}_n/\text{rad}^2(K\overrightarrow{A}_n)$ which annihilates all paths of length at least 2 (where the length of a path is given by the number of arrows which compose it). Observe that a B_n right module is precisely a complex of vector spaces of length at most n. Indeed, let e_i be the idempotent of B_n corresponding to the vertex i and let M be a B_n-module. Then Me_i is a K-vector space and multiplication by a_i from the right is a linear map $Me_i \longrightarrow Me_{i+1}$ for all i so that the composition $Me_i \longrightarrow Me_{i+1} \longrightarrow Me_{i+2}$ is 0, since $a_ia_{i+1} = 0$ in B_n.

Therefore a complex of A-modules of length n is the same as an A-B_n-bimodule. If A is finite dimensional, $A \otimes_K B_n^{op}$ is a finite dimensional K-algebra as well. Therefore the Krull Schmidt theorem is valid for the category of A-B_n-bimodules. Hence, in order to study the existence and unicity of a direct sum decomposition into indecomposable objects, the case of complexes in $C^b(A\text{-}mod)$ is reduced to the existence and unicity of a direct sum composition of A-B_n-bimodules for a certain n, namely the n for which the particular complex one would like to decompose belongs to $A \otimes_K B_n^{op}\text{-}mod$. This proves the statement. □

3.5.2 Homotopy Category of Complexes

The category of complexes is already of great interest. For our purposes we need to modify this category further.

Definition 3.5.12 Let $M^\bullet = (M, d_M)$ and $N^\bullet = (N, d_N)$ be two complexes in an additive category $\mathcal{A}$. Then a morphism $\alpha \in Hom_{C(\mathcal{A})}(M^\bullet, N^\bullet)$ is *homotopic to zero* if and only if there is a graded morphism $h : M \longrightarrow N$ of degree -1 so that

$$\alpha = h \circ d_M + d_N \circ h.$$

The morphism h is called a *homotopy*. Two morphisms $\alpha, \beta \in Hom_{C(\mathcal{A})}(M^\bullet, N^\bullet)$ are called *homotopy equivalent* if and only if $\alpha - \beta$ is homotopic to zero.

Clearly the set of morphisms of complexes which are homotopic to zero is an abelian subgroup of the group of morphisms. Therefore the homotopy equivalence classes do make sense.

Again this terminology comes from algebraic topology and the origin of homology in the category of certain topological spaces.

Definition 3.5.13 Let $\mathcal{A}$ be an additive category. Then for all $x, y \in \{\emptyset, b, +, -\}$ let $K^{x,y}(\mathcal{A})$ be the category with objects the same objects as $C^{x,y}(\mathcal{A})$ and with morphisms homotopy equivalence classes of morphisms of complexes. $K^{x,y}(\mathcal{A})$ is called the *homotopy category of complexes* of objects in $\mathcal{A}$.

We encounter here the first triangulated category we are going to deal with. The fact that the category is triangulated will be shown in Proposition 3.5.25. A particularly important case is when $\mathcal{A} = A\text{-}Proj$ is the full subcategory of $A\text{-}Mod$ generated by projective modules, or when $\mathcal{A} = A\text{-}proj$ is the full subcategory of $A\text{-}mod$ generated by finitely generated projective modules.

Let A be a K-algebra and let M be an A-module. Recall from Definition 1.8.15 the definition of a projective resolution $P^\bullet$ of a module M, i.e. a complex $P^\bullet$ with homology M concentrated in degree 0.

Moreover, the isomorphism class of $P^\bullet$ in the homotopy category of complexes of projective A-modules only depends on the isomorphism class of M in $A\text{-}Mod$ and not on the various choices made in the construction of $P^\bullet$. Observe that $P^\bullet$ is a right bounded complex of projective modules, whence is an object in $K^-(A\text{-}Proj)$.

Proposition 3.5.14 *Let K be a commutative ring and let A be a K-algebra. Let M be an A-module. Then two projective resolutions of M are isomorphic in $K^-(A\text{-}Proj)$.*

The proof uses a very useful lemma.

Lemma 3.5.15 *Let $\mathcal{A}$ be an abelian category, let*

$$X^\bullet : \quad \dots \longrightarrow X_4 \xrightarrow{f_4} X_3 \xrightarrow{f_3} X_2 \xrightarrow{f_2} X_1 \xrightarrow{f_1} X_0 \longrightarrow M \longrightarrow 0$$

be an exact complex (i.e. $\ker(f_i) = \operatorname{im}(f_{i+1})$ *for all i) and let*

$$P^\bullet : \quad \dots \longrightarrow P_4 \xrightarrow{d_4} P_3 \xrightarrow{d_3} P_2 \xrightarrow{d_2} P_1 \xrightarrow{d_1} P_0 \longrightarrow N \longrightarrow 0$$

be a complex so that the objects P_i are all projective. Then for all morphisms $\alpha : N \longrightarrow M$ there is a morphism of complexes $\alpha^\bullet : P^\bullet \longrightarrow X^\bullet$ lifting α, i.e. the diagram

$$\begin{array}{ccccccccccccc}
\dots\longrightarrow & P_4 & \xrightarrow{d_4} & P_3 & \xrightarrow{d_3} & P_2 & \xrightarrow{d_2} & P_1 & \xrightarrow{d_1} & P_0 & \longrightarrow & N & \longrightarrow 0 \\
 & \downarrow \alpha_4 & & \downarrow \alpha_3 & & \downarrow \alpha_2 & & \downarrow \alpha_1 & & \downarrow \alpha_0 & & \downarrow \alpha & \\
\dots\longrightarrow & X_4 & \xrightarrow{f_4} & X_3 & \xrightarrow{f_3} & X_2 & \xrightarrow{f_2} & X_1 & \xrightarrow{f_1} & X_0 & \xrightarrow{f_0} & M & \longrightarrow 0
\end{array}$$

is commutative.

Proof of Lemma 3.5.15 The composition of α with $P_0 \longrightarrow N$ defines a mapping $P_0 \longrightarrow M$. Since P_0 is projective and since $X_0 \longrightarrow M$ is an epimorphism, there is an $\alpha_0 : P_0 \longrightarrow X_0$ making the right-hand diagram commutative. Let

$$C_i := \operatorname{im}(f_i) := \ker(f_{i-1})$$

for all $i \in \mathbb{N} \setminus \{0\}$, observing that the homology of $X^\bullet$ is 0. Then $\alpha_0 \circ d_1$ actually has image in C_1 since the rightmost square is commutative and since therefore $f_0 \circ \alpha_0 \circ$

$d_1 = 0$. Since $f_1 : X_1 \longrightarrow C_1$ is an epimorphism and since P_1 is projective there is a morphism α_1 such that

$$f_1 \circ \alpha_1 = \alpha_0 \circ d_1.$$

Suppose α_j is constructed for $j \leq i$. Then

$$\alpha_i \circ d_{i+1} : P_{i+1} \longrightarrow X_i$$

actually has image in C_{i+1} since

$$f_i \circ \alpha_i \circ d_{i+1} = \alpha_{i-1} \circ d_i \circ d_{i+1} = 0$$

by the induction hypothesis. Again the fact that $f_{i+1} : X_{i+1} \longrightarrow C_{i+1}$ is an epimorphism and that P_{i+1} is projective establishes the existence of α_{i+1}. This proves the lemma. □

Proof of Proposition 3.5.14 Given two projective resolutions of the same module M

$$P^\bullet : \quad \dots P_4 \xrightarrow{d_4} P_3 \xrightarrow{d_3} P_2 \xrightarrow{d_2} P_1 \xrightarrow{d_1} P_0 \longrightarrow M \longrightarrow 0$$

and

$$Q^\bullet : \quad \dots Q_4 \xrightarrow{e_4} Q_3 \xrightarrow{e_3} Q_2 \xrightarrow{e_2} Q_1 \xrightarrow{e_1} Q_0 \longrightarrow M \longrightarrow 0$$

using Lemma 3.5.15 we may lift the identity morphism on M to a morphism of complexes $\alpha : P^\bullet \longrightarrow Q^\bullet$ and to a morphism of complexes $\beta : Q^\bullet \longrightarrow P^\bullet$. Hence the difference $\beta \circ \alpha - id_{P^\bullet}$ lifts the zero endomorphism on M to an endomorphism of $P^\bullet$. Similarly $\alpha \circ \beta - id_{Q^\bullet}$ lifts the zero endomorphism on M to an endomorphism of $Q^\bullet$. We shall prove the following lemma which implies the statement immediately.

Lemma 3.5.16 *The zero morphism lifts to a zero homotopic complex morphism. Moreover, any lift of the zero morphism is zero homotopic.*

Proof We have the following situation of two projective resolutions.

$$\begin{array}{ccccccccccccc}
\dots \longrightarrow & P_4 & \xrightarrow{d_4} & P_3 & \xrightarrow{d_3} & P_2 & \xrightarrow{d_2} & P_1 & \xrightarrow{d_1} & P_0 & \xrightarrow{d_0} & M & \longrightarrow 0 \\
& \downarrow \alpha_4 & & \downarrow \alpha_3 & & \downarrow \alpha_2 & & \downarrow \alpha_1 & & \downarrow \alpha_0 & & \downarrow 0 & \\
\dots \longrightarrow & Q_4 & \xrightarrow{e_4} & Q_3 & \xrightarrow{e_3} & Q_2 & \xrightarrow{e_2} & Q_1 & \xrightarrow{e_1} & Q_0 & \xrightarrow{e_0} & N & \longrightarrow 0
\end{array}$$

Put $C_{i+1} := \ker(e_i)$ for all i. Since $e_0 \circ \alpha_0 = 0$, α_0 actually factors through $C_1 \longrightarrow Q_0$. Since $e_1 : Q_1 \longrightarrow C_1$ is an epimorphism we get $h_0 : P_0 \longrightarrow Q_1$ so that $\alpha_0 = e_1 \circ h_0$. Consider $\gamma_1 := \alpha_1 - h_0 \circ d_1$. We obtain

$$e_1 \circ \gamma_1 = e_1 \circ \alpha_1 - e_1 \circ h_0 \circ d_1 = \alpha_0 \circ d_1 - \alpha_0 \circ d_1 = 0.$$

Hence, again by the fact that Q_2 is projective and that $e_2 : Q_2 \longrightarrow C_2$ is an epimorphism, there is a morphism $h_1 : P_1 \longrightarrow Q_2$ so that $\alpha_1 = h_0 \circ d_1 + e_2 \circ h_1$. An obvious induction finishes the proof. □

As we have seen, this completes the proof of Proposition 3.5.14. □

We shall frequently use, without explicit mention, the following fundamental lemma which shows that homology is a notion which belongs to the homotopy category of complexes, rather than the category of complexes.

Lemma 3.5.17 *Zero homotopic morphisms induce the morphism* 0 *on homology.*

Proof Indeed, let $f = d_C \circ h + h \circ d_D$ be a zero homotopic morphism from D to C. Let $x \in \ker(d_D)$. Then

$$f(x) = (d_C \circ h + h \circ d_D)(x) = d_C(h(x)) \in \text{im}(d_C)$$

and since $H(C) = \ker(d_C)/\text{im}(d_C)$, we get the statement. □

Proposition 3.5.18 *Let K be a commutative ring and let A be a K-algebra. Then A-Mod is a full subcategory of $K^-(A\text{-}Proj)$ and if A is Noetherian A-mod is a full subcategory of $K^-(A\text{-}proj)$. The functor is given by associating a projective resolution to each A-module.*

Proof Bearing in mind the usual set theoretical problem concerning the choice of a fixed resolution for each module, we associate to each module homomorphism a morphism of complexes using Lemma 3.5.15. We need to show that this is well defined. If we lift a fixed morphism in two different ways, the difference of the two lifts lift the zero homomorphism. Lemma 3.5.16 shows that this difference is zero-homotopic, whence the two lifts are equal in the homotopy category. Now, the identity is lifted by the identity morphism, and the lift of the composite lifts the composite of the lifts, again up to homotopy using Lemma 3.5.16. Call $PR : A\text{-}Mod \longrightarrow K^-(A\text{-}Proj)$ the resulting functor.

The homology functor $H_0 : K^-(A\text{-}Proj) \longrightarrow A\text{-}Mod$ has the property $H_0 \circ PR = id_{A\text{-}Mod}$ and so the functor PR is a fully faithful embedding. □

Proposition 3.5.18 is one of the main reasons for us to be particularly interested in $K^-(A\text{-}Proj)$. We shall study complexes which are isomorphic to 0 in $K^-(A\text{-}Proj)$ in the next Proposition. The Proposition 3.5.19 will be generalised in Proposition 3.5.23 below.

Proposition 3.5.19 • *A complex $X \in C^-(A\text{-}Mod)$ is isomorphic to 0 in $K^-(A\text{-}Mod)$ if and only if X is isomorphic to a direct sum of complexes of the form*

$$Z_{i,i+1}: \quad \ldots \longrightarrow 0 \longrightarrow 0 \longrightarrow Z \xrightarrow{id} Z \longrightarrow 0 \longrightarrow 0 \longrightarrow \ldots$$

which are concentrated in degree i and $i+1$.

- *A complex $X \in C^-(A\text{-}Proj)$ has the property $H(X) = 0$ if and only if X is isomorphic to a direct sum of complexes of the form*

$$Z_{i,i+1}: \quad \ldots \longrightarrow 0 \longrightarrow 0 \longrightarrow Z \xrightarrow{id} Z \longrightarrow 0 \longrightarrow 0 \longrightarrow \ldots$$

which are concentrated in degree i and $i+1$.

Proof It is clear that complexes of the form

$$\ldots \longrightarrow 0 \longrightarrow 0 \longrightarrow Z \xrightarrow{id} Z \longrightarrow 0 \longrightarrow 0 \longrightarrow \ldots$$

are isomorphic to 0 since every endomorphism of this complex is zero homotopic. Hence direct sums of complexes of this type are also zero homotopic.

Conversely let X

$$\ldots \longrightarrow X_{n+3} \xrightarrow{d_{n+3}} X_{n+2} \xrightarrow{d_{n+2}} X_{n+1} \xrightarrow{d_{n+1}} X_n \longrightarrow 0 \longrightarrow 0 \longrightarrow \ldots$$

be a complex of projective modules which we assume to be isomorphic to 0 in $K^-(A\text{-}Mod)$. The identity endomorphism of the complex X has to be zero homotopic. Hence there is a mapping h of degree 1 so that $id_X = d \circ h + h \circ d$. In particular on degree n we get $h_n \circ d_{n+1} = id_{X_n}$. Hence d_{n+1} is split. Therefore $X_{n+1} \simeq X'_n \oplus X'_{n+1}$ and d_{n+1} is an isomorphism between X_n and X'_n, which implies

$$X \simeq (X_n)_{n,n+1} \oplus X'$$

where X' is the complex

$$\ldots \longrightarrow X_{n+3} \xrightarrow{d_{n+3}} X_{n+2} \xrightarrow{d_{n+2}} X'_{n+1} \longrightarrow 0 \longrightarrow 0 \longrightarrow 0 \longrightarrow \ldots.$$

Hence X has to be isomorphic to a direct sum of complexes of the form $Z_{i,i+1}$ for projective Z since at each degree, by induction, one removes a direct factor of this type to get a residual complex which is concentrated at higher degrees. Recursively we construct the required isomorphism.

In order to prove the second statement we just mention that the fact that if X has homology 0 then the rightmost differential, say in degree m, is surjective. Hence the rightmost differential is split, and one may remove a direct factor of the form $(X_m)_{m+1,m}$ and obtain a complex concentrated in degrees $m+1$ at least. The statement follows by induction. □

Observe that we needed to work with right bounded complexes. The statement also holds for left bounded complexes, changing projectives into injectives in the second statement, but boundedness at one side is necessary.

Complexes with vanishing homology play an important role.

Definition 3.5.20 A complex X in an abelian category $\mathcal{A}$ is *acyclic* if its homology is 0, i.e. $H(X) = 0$.

In view of Proposition 3.5.19 the following Proposition 3.5.23 is particularly illuminating. Moreover, Proposition 3.5.23 improves Proposition 3.5.11, which was valid for bounded complexes only. Up to homotopy we may go to right bounded complexes with bounded homology.

First we need to introduce a very useful construction, the "intelligent cutting" or "intelligent truncation".

Let $X^\bullet$

$$\dots \to X_{n+3} \stackrel{d_{n+3}}{\to} X_{n+2} \stackrel{d_{n+2}}{\to} X_{n+1} \stackrel{d_{n+1}}{\to} X_n \to \dots$$

be a complex, then define $\tau_{\leq m}(X)$ to be the complex

$$\dots \to 0 \to 0 \to \mathrm{im}(d_{m+1}) \stackrel{incl}{\to} X_m \stackrel{d_m}{\to} X_{m-1} \stackrel{d_{m-1}}{\to} X_{m-2} \stackrel{d_{m-2}}{\to} X_{m-3} \to \dots$$

Moreover, if $\alpha : X^\bullet \longrightarrow Y^\bullet$ is a morphism of complexes, then define

$$(\tau_{\leq m}(\alpha))_k = \begin{cases} \alpha_k & \text{if } k \leq m \\ (\alpha_m)|_{\mathrm{im}(d_{m+1})} & \text{if } k = m+1 \\ 0 & \text{if } k \geq m+2 \end{cases}$$

and obtain in this way a functor

$$\tau_{\leq m} : C(A\text{-}Mod) \longrightarrow C^-(A\text{-}Mod).$$

Observe that

$$\begin{array}{ccccccccccccc}
\dots \to X_{m+2} \to & X_{m+1} & \stackrel{d_{m+1}}{\to} & X_m & \stackrel{d_m}{\to} & X_{m-1} & \stackrel{d_{m-1}}{\to} & X_{m-2} & \stackrel{d_{m-2}}{\to} & X_{m-3} \to \dots \\
\downarrow & \downarrow d_{m+1} & & \| & & \| & & \| & & \| \\
\dots \to \quad 0 \quad \to & \mathrm{im}(d_{m+1}) & \stackrel{incl}{\to} & X_m & \stackrel{d_m}{\to} & X_{m-1} & \stackrel{d_{m-1}}{\to} & X_{m-2} & \stackrel{d_{m-2}}{\to} & X_{m-3} \to \dots
\end{array}$$

is a morphism of complexes, and hence there is always a natural mapping $X \longrightarrow \tau_{\leq m}X$. It is clear that this morphism of complexes induces an isomorphism $H^k(X) \longrightarrow H^k(\tau_{\leq m}X)$ for all $k \leq m$.

Dually one obtains the functor

$$\tau_{\geq m} : C(A\text{-}Mod) \longrightarrow C^+(A\text{-}Mod)$$

and for every object X a morphism $\tau_{\geq m-2}X \longrightarrow X$ defined by the diagram

$$\begin{array}{ccccccccccccc}
\dots \to X_{m+2} & \to & X_{m+1} & \overset{d_{m+1}}{\to} & X_m & \overset{d_m}{\to} & X_{m-1} & \overset{d_{m-1}}{\to} & X_{m-2} & \overset{d_{m-2}}{\to} & X_{m-3} & \to \dots \\
\| & & \| & & \| & & \| & & \uparrow incl & & \uparrow 0 & \\
\dots \to X_{m+2} & \to & X_{m+1} & \overset{d_{m+1}}{\to} & X_m & \overset{d_m}{\to} & X_{m-1} & \overset{d_{m-1}}{\to} & \ker(d_{m-2}) & \to & 0 & \to \dots
\end{array}$$

We hence get the following lemma.

Lemma 3.5.21 *The intelligent truncation $\tau_{\le m} : C(A\text{-}Mod) \longrightarrow C^-(A\text{-}Mod)$ is a functor and there is a natural morphism $X \longrightarrow \tau_{\le m}X$ inducing an isomorphism on homology in all degrees up to m.*

The intelligent truncation $\tau_{\ge m} : C(A\text{-}Mod) \longrightarrow C^+(A\text{-}Mod)$ is a functor and there is a natural morphism $\tau_{\ge m}X \longrightarrow X$ inducing an isomorphism on homology in all degrees at least equal to m.

Both functors descend to functors of the corresponding homotopy categories.

Proof The proof for the complex categories was done above.

Concerning the homotopy categories, we observe that actually $\tau_{\le m} : K(A\text{-}Mod) \longrightarrow K^-(A\text{-}Mod)$. Indeed, a homotopy h from X to Y gives a homotopy h from $\tau_{\le m}(X)$ to $\tau_{\le m}(Y)$ in each degree up to degree $m-2$, then the homotopy h_{m-1} from X_{m-1} to Y_m will be replaced by the mapping $d_{m+1} \circ h_m$, which has image in $\text{im}(d_{m+1})$. □

Remark 3.5.22 Another operation is sometimes useful, the so-called *stupid truncation*. Given a complex X

$$\dots \longrightarrow X_n \xrightarrow{\partial_n} X_{n-1} \xrightarrow{\partial_{n-1}} X_{n-2} \xrightarrow{\partial_{n-2}} X_{n-3} \xrightarrow{\partial_{n-3}} X_{n-4} \longrightarrow \dots$$

we define $\sigma_{\ge n-1}X$ as the complex

$$\dots \longrightarrow 0 \longrightarrow 0 \xrightarrow{0} X_{n-1} \xrightarrow{\partial_{n-1}} X_{n-2} \xrightarrow{\partial_{n-2}} X_{n-3} \xrightarrow{\partial_{n-3}} X_{n-4} \longrightarrow \dots.$$

It is clear that this gives a functor on the category of complexes only. For example the complex

$$\dots \longrightarrow X \xrightarrow{id} X \longrightarrow 0 \longrightarrow \dots$$

is homotopy equivalent to 0, but stupidly truncated in the middle gives the complex

$$\dots \longrightarrow 0 \xrightarrow{0} X \longrightarrow 0 \longrightarrow \dots$$

which is not zero homotopic. Hence, stupid truncation can create any homology at the "cutting edge" if it is used in the homotopy category.

We now come to the promised generalisation of Proposition 3.5.19.

Proposition 3.5.23 *Let K be a field and let A be a finite dimensional K-algebra. Let X and Y be two objects in $K^{-,b}(A\text{-}proj)$. Then $X \simeq Y$ in $K^{-,b}(A\text{-}proj)$ if and only if there are complexes Z, X', Y' in $C^{-,b}(A\text{-}proj)$ such that*

$$X \simeq X' \oplus Z \text{ and } Y \simeq Y' \oplus Z \text{ in } C^{-,b}(A\text{-}proj)$$

and such that $X' \simeq Y' \simeq 0$ in $K^{-,b}(A\text{-}proj)$.

Proof Let X be a complex in $C^{-,b}(A\text{-}proj)$

$$\dots \longrightarrow X_{n+3} \stackrel{d_{n+3}}{\longrightarrow} X_{n+2} \stackrel{d_{n+2}}{\longrightarrow} X_{n+1} \stackrel{d_{n+1}}{\longrightarrow} X_n \longrightarrow 0 \longrightarrow 0 \longrightarrow \dots$$

and suppose $H_k(X) = 0$ for $k \geq m$.

We construct $\tau_{\leq m+1}(X)$ and observe that this complex has the same homology as X and also that this complex belongs to $C^b(A\text{-}mod)$. We do have the Krull-Schmidt theorem for $\tau_{\leq m+1}(X)$ and decompose this complex into indecomposable direct factors

$$\tau_{\leq m+1}(X) \simeq F_1^{\leq m} \oplus F_2^{\leq m} \oplus \dots \oplus F_\ell^{\leq m} \oplus F_{\ell+1}^{\leq m} \oplus F_{\ell+2}^{\leq m} \oplus \dots \oplus F_k^{\leq m}.$$

Suppose that $H(F_s^{\leq m}) = 0$ for all s with $\ell < s \leq k$ and $H(F_s^{\leq m}) \neq 0$ for all s with $0 < s \leq \ell$.

We know that

$$Z^{\leq m} := F_1^{\leq m} \oplus F_2^{\leq m} \oplus \dots \oplus F_\ell^{\leq m}$$

carries all the homology of X in the sense that $H(Z^{\leq m}) = H(X)$. The degree $m+1$ component $(Z^{\leq m})_{m+1}$ of $Z^{\leq m}$ is not necessarily projective, but we can construct a projective resolution

$$\dots \longrightarrow (Z^{\geq m})_{m+3} \longrightarrow (Z^{\geq m})_{m+2} \longrightarrow (Z^{\geq m})_{m+1} \longrightarrow (Z^{\leq m})_{m+1} \longrightarrow 0$$

and obtain, patching the complexes $Z^{\leq m}$ and $Z^{\geq m}$ together, a complex Z

$$\begin{array}{ccccccccc} \dots & \longrightarrow (Z^{\geq m})_{m+4} & \longrightarrow & (Z^{\geq m})_{m+3} & \longrightarrow & (Z^{\geq m})_{m+2} & \longrightarrow & (Z^{\geq m})_{m+1} \\ & & & & & & & \downarrow \\ \dots \longleftarrow 0 \longleftarrow & Z_n^{\leq m} & \longleftarrow & \dots & \longleftarrow & (Z^{\leq m})_{m+1} & \longleftarrow & (Z^{\leq m})_m \end{array}$$

We get a mapping $Z \longrightarrow X$, being the direct factor inclusion in degrees smaller than m, and then by the universal property of the projective resolution of $(Z^{\leq m})_{m+1}$ in all higher degrees. We also get a mapping in the opposite direction by the direct factor projection in degrees smaller than m and by the universal property of the projective resolution in the other degrees. The composition of both mappings has to be the identity on Z, and so Z is a direct factor of X.

Since $X \simeq Y$ in the homotopy category, $\tau_{\leq m}(X) \simeq \tau_{\leq m}(Y)$ in the homotopy category, using that $\tau_{\leq m}$ is a functor. We may also decompose the complex $\tau_{\leq m}(Y)$

into indecomposable direct factors, and obtain a complex Z' similar to Z, by the same construction, so that Z' is a direct factor of Y.

Now, $X \simeq Z \oplus X'$ and $Y \simeq Z' \oplus Y'$ where X' and Y' are complexes of projective modules with zero homology. Proposition 3.5.19 then shows that $X' \simeq 0$ in $K^-(A\text{-}proj)$ and $Y' \simeq 0$ in $K^-(A\text{-}proj)$. Therefore $Z \simeq Z'$ in $K^-(A\text{-}proj)$.

Let $\alpha : Z \longrightarrow Z'$ and $\beta : Z' \longrightarrow Z$ be homomorphisms of complexes so that $\alpha \circ \beta - id_{Z'}$ is zero homotopic, and so that $\beta \circ \alpha - id_Z$ is zero homotopic. Since no direct factor of Z, or of Z', is zero homotopic, a zero homotopic mapping $Z \longrightarrow Z$ has image in the radical of Z (regarded as an A-module). Therefore $\beta \circ \alpha$ is an automorphism of complexes, and likewise $\alpha \circ \beta$ is an automorphism of complexes. This shows that α is an isomorphism of complexes. This proves the statement. □

Definition 3.5.24 For each complex X in $C(A\text{-}Mod)$ define

$$\left(\bigoplus_{i\in\mathbb{Z}} X_i, (\partial_i)_{i\in\mathbb{Z}}\right)[1] := \left(\bigoplus_{i\in\mathbb{Z}} X_{i-1}, (-\partial_i)_{i\in\mathbb{Z}}\right)$$

where $\partial_i : X_i \longrightarrow X_{i-1}$. Moreover, define $X[n] := (X[n-1])[1]$ for all $n \in \mathbb{N}$. Observe that for each complex X there is a unique complex $X[-n]$ such that $X[-n][n] = X$.

Observe that ?[1] moves the complex by one degree to the left and the differential changes the sign. There is a reason behind this strange convention which will become clear later. We may extend $[n]$ to a functor between complex categories by putting $\alpha_i[n] = (-1)^n \alpha_{i-n}$ for the degree i homogeneous component α_i of α.

Proposition 3.5.25 *Let $\mathcal{A}$ be an additive category. Then $K^{x,y}(\mathcal{A})$ is a triangulated category for all $x, y \in \{\emptyset, +, -, b\}$.*

Proof The self-equivalence $T : K^{x,y}(\mathcal{A}) \longrightarrow K^{x,y}(\mathcal{A})$ needed for the definition of a triangulated category is the shift in degrees: $T(X) := X[1]$.

We shall need to construct triangles. Let $\alpha : X \longrightarrow Y$ be a morphism of complexes in $K^{x,y}(\mathcal{A})$. Then $C(\alpha)$ is the complex

$$C(\alpha) := \left(\bigoplus_{i\in\mathbb{Z}} (X_{i-1} \oplus Y_i), \begin{pmatrix} -d^X & 0 \\ \alpha & d^Y \end{pmatrix}\right)$$

or in less dense notation

$$X : \ldots X_{n+1} \stackrel{d^X_{n+1}}{\longrightarrow} X_n \stackrel{d^X_n}{\longrightarrow} X_{n-1} \stackrel{d^X_{n-1}}{\longrightarrow} X_{n-2} \stackrel{d^X_{n-2}}{\longrightarrow} X_{n-3} \longrightarrow \ldots$$

and

$$Y : \ldots Y_{n+1} \stackrel{d^Y_{n+1}}{\longrightarrow} Y_n \stackrel{d^Y_n}{\longrightarrow} Y_{n-1} \stackrel{d^Y_{n-1}}{\longrightarrow} Y_{n-2} \stackrel{d^Y_{n-2}}{\longrightarrow} Y_{n-3} \longrightarrow \ldots$$

then $C(\alpha)$ is the complex

$$\ldots \to X_n \oplus Y_{n+1} \overset{\begin{pmatrix} -d_n^X & 0 \\ \alpha_n & d_{n+1}^Y \end{pmatrix}}{\to} X_{n-1} \oplus Y_n \overset{\begin{pmatrix} -d_{n-1}^X & 0 \\ \alpha_{n-1} & d_n^Y \end{pmatrix}}{\to} X_{n-2} \oplus Y_{n-1} \to \ldots$$

where we display the degrees $n+1$, n and $n-1$.

We see that

$$\begin{pmatrix} -d_{n-1}^X & 0 \\ \alpha_{n-1} & d_n^Y \end{pmatrix} \circ \begin{pmatrix} -d_n^X & 0 \\ \alpha_n & d_{n+1}^Y \end{pmatrix} = \begin{pmatrix} d_{n-1}^X \circ d_n^X & 0 \\ d_n^Y \circ \alpha_n - \alpha_{n-1} \circ d_n^X & d_n^Y \circ d_{n+1}^Y \end{pmatrix} = \begin{pmatrix} 0 & 0 \\ 0 & 0 \end{pmatrix}$$

and hence $C(\alpha)$ is indeed a complex.

Definition 3.5.26 We call the complex $C(\alpha)$ the cone of the morphism of complexes $\alpha : X \longrightarrow Y$.

Moreover, the projection onto the second component $C(\alpha) \longrightarrow X[1]$ gives a homomorphism of complexes:

$$\begin{array}{ccccccc}
\ldots \to X_n \oplus Y_{n+1} & \overset{\begin{pmatrix} -d_n^X & 0 \\ \alpha_n & d_{n+1}^Y \end{pmatrix}}{\to} & X_{n-1} \oplus Y_n & \overset{\begin{pmatrix} -d_{n-1}^X & 0 \\ \alpha_{n-1} & d_n^Y \end{pmatrix}}{\to} & X_{n-2} \oplus Y_{n-1} \to \ldots \\
\downarrow & & \downarrow & & \downarrow \\
\ldots \to \quad X_n & \overset{-d_n^X}{\to} & X_{n-1} & \overset{-d_{n-1}^X}{\to} & X_{n-2} \quad \to \ldots
\end{array}$$

and the embedding into the first component $Y \longrightarrow C(\alpha)$ gives a homomorphism of complexes:

$$\begin{array}{ccccccc}
\ldots \to \quad Y_{n+1} & \overset{d_n^Y}{\to} & Y_n & \overset{d_{n-1}^Y}{\to} & Y_{n-1} \quad \to \ldots \\
\downarrow & & \downarrow & & \downarrow \\
\ldots \to X_n \oplus Y_{n+1} & \overset{\begin{pmatrix} -d_n^X & 0 \\ \alpha_n & d_{n+1}^Y \end{pmatrix}}{\to} & X_{n-1} \oplus Y_n & \overset{\begin{pmatrix} -d_{n-1}^X & 0 \\ \alpha_{n-1} & d_n^Y \end{pmatrix}}{\to} & X_{n-2} \oplus Y_{n-1} \to \ldots
\end{array}$$

In this way, we obtain a triangle

$$X \xrightarrow{\alpha} Y \longrightarrow C(\alpha) \longrightarrow X[1]$$

and we define a distinguished triangle to be any triangle isomorphic to a triangle of the form

$$X \xrightarrow{\alpha} Y \longrightarrow C(\alpha) \longrightarrow X[1].$$

By definition of the cone every morphism $\alpha : X \longrightarrow Y$ can be completed to a distinguished triangle

$$X \xrightarrow{\alpha} Y \longrightarrow C(\alpha) \longrightarrow X[1].$$

This proves the second half of the axiom **Tr1**. We shall verify the first half of **Tr1**, namely that $C(id_X) \simeq 0$. We shall show that the identity on $C(id_X)$ is zero homotopic which proves $C(id_X) \simeq 0$ as claimed. The degree n homogeneous component of $C(id_X)$ is $X_{n-1} \oplus X_n$ and the degree $n+1$ homogeneous component of $C(id_X)$ is $X_{n-1} \oplus X_{n-2}$. The differential of $C(id_X)$ is defined by

$$d_{C(id_X)n} = \begin{pmatrix} -d_{n-1}^X & 0 \\ id_{X_{n-1}} & d_n^X \end{pmatrix} : X_{n-1} \oplus X_n \longrightarrow X_{n-2} \oplus X_{n-1}.$$

Define a homotopy mapping $h_n : X_{n-1} \oplus X_n \longrightarrow X_n \oplus X_{n+1}$ by

$$h_n := \begin{pmatrix} 0 & id_{X_n} \\ 0 & 0 \end{pmatrix}$$

and compute

$$\begin{aligned} & d_{C(id)_{n+1}} \circ h_n + h_{n-1} \circ d_{C(id)_n} \\ &= \begin{pmatrix} -d_n^X & 0 \\ id_{X_n} & d_{n+1}^X \end{pmatrix} \circ \begin{pmatrix} 0 & id_{X_n} \\ 0 & 0 \end{pmatrix} + \begin{pmatrix} 0 & id_{X_{n-1}} \\ 0 & 0 \end{pmatrix} \circ \begin{pmatrix} -d_{n-1}^X & 0 \\ id_{X_{n-1}} & d_n^X \end{pmatrix} \\ &= \begin{pmatrix} 0 & -d_n^X \\ 0 & id_{X_n} \end{pmatrix} + \begin{pmatrix} id_{X_{n-1}} & d_n^X \\ 0 & 0 \end{pmatrix} = \begin{pmatrix} id_{X_{n-1}} & 0 \\ 0 & id_{X_n} \end{pmatrix} \end{aligned}$$

so that the identity on $C(id_X)$ is zero-homotopic.

We need to verify **Tr2**. Let

$$X \xrightarrow{\alpha} Y \longrightarrow C(\alpha) \longrightarrow X[1]$$

be a distinguished triangle. We need to show that

$$Y \xrightarrow{incl} C(\alpha) \longrightarrow X[1] \xrightarrow{-\alpha[1]} Y[1]$$

is a distinguished triangle, or in other words that there is an isomorphism of complexes $X[1] \longrightarrow C(incl)$ making the diagram

$$\begin{array}{ccccccc} Y & \xrightarrow{incl} & C(\alpha) & \longrightarrow & X[1] & \xrightarrow{-\alpha[1]} & Y[1] \\ \| & & \| & & \downarrow & & \| \\ Y & \xrightarrow{incl} & C(\alpha) & \longrightarrow & C(incl) & \longrightarrow & Y[1] \end{array}$$

commutative.

For the proof we shall need to make $C(incl)$ explicit. The complex $C(incl)$ is the cone of the morphism

$$\begin{array}{ccccccc}
\ldots \to & Y_{n+1} & \stackrel{d_n^Y}{\to} & Y_n & \stackrel{d_{n-1}^Y}{\to} & Y_{n-1} & \to \ldots \\
& \downarrow & & \downarrow & & \downarrow & \\
\ldots \to & X_n \oplus Y_{n+1} & \stackrel{\begin{pmatrix} -d_n^X & 0 \\ \alpha_n & d_{n+1}^Y \end{pmatrix}}{\to} & X_{n-1} \oplus Y_n & \stackrel{\begin{pmatrix} -d_{n-1}^X & 0 \\ \alpha_{n-1} & d_n^Y \end{pmatrix}}{\to} & X_{n-2} \oplus Y_{n-1} & \to \ldots
\end{array}$$

which is

$$\ldots \to X_n \oplus Y_{n+1} \oplus Y_n \stackrel{\partial_{n+1}}{\to} X_{n-1} \oplus Y_n \oplus Y_{n-1} \stackrel{\partial_n}{\to} X_{n-2} \oplus Y_{n-1} \oplus Y_{n-2} \to \ldots$$

where we display the degrees $n+1$, n and $n-1$. For the differential ∂_n we get

$$\partial_n = \begin{pmatrix} -d_n^X & 0 & 0 \\ \alpha_n & d_{n+1}^Y & id_{Y_n} \\ 0 & 0 & -d_n^Y \end{pmatrix}.$$

Recalling that the differential of $X[1]$ is -1 times the differential of X, it is now obvious that the projection onto the first component is a morphism of complexes $C(incl) \longrightarrow X[1]$. The kernel of this morphism is equal to the cone of the identity morphism (up to a sign) on Y. Hence, the kernel of this morphism, taken in the category of complexes, is isomorphic to 0 in the homotopy category. Further, $C(incl) \longrightarrow X[1]$ is obviously surjective. Moreover, it is clear that the diagram

$$\begin{array}{ccc}
C(\alpha) & \longrightarrow & X[1] \\
\| & & \uparrow \\
C(\alpha) & \longrightarrow & C(incl)
\end{array}$$

is commutative. Furthermore the diagram

$$\begin{array}{ccc}
X[1] & \stackrel{\alpha}{\longrightarrow} & Y[1] \\
\downarrow & & \| \\
C(incl) & \longrightarrow & Y[1]
\end{array}$$

is commutative as is seen by direct inspection. The shift in the other direction is similar. Alternatively one may apply the previous step five times, and then apply $[-2]$. This proves **Tr2**.

We need to show **Tr3**. Given a diagram

$$\begin{array}{ccccccc}
X & \stackrel{\alpha}{\longrightarrow} & Y & \longrightarrow & C(\alpha) & \longrightarrow & X[1] \\
\downarrow \xi & & \downarrow \eta & & & & \downarrow \xi[1] \\
X' & \stackrel{\alpha'}{\longrightarrow} & Y' & \longrightarrow & C(\alpha') & \longrightarrow & X'[1]
\end{array}$$

with commuting right square, we need to find $\zeta : C(\alpha) \longrightarrow C(\alpha')$ making the two right squares

$$\begin{array}{ccccccc} X & \stackrel{\alpha}{\longrightarrow} & Y & \longrightarrow & C(\alpha) & \longrightarrow & X[1] \\ \downarrow \xi & & \downarrow \eta & & \downarrow \zeta & & \downarrow \xi[1] \\ X' & \stackrel{\alpha'}{\longrightarrow} & Y' & \longrightarrow & C(\alpha') & \longrightarrow & X'[1] \end{array}$$

involving ζ commutative. Define

$$\zeta_n := \begin{pmatrix} \xi_{n-1} & 0 \\ 0 & \eta_n \end{pmatrix} : X_{n-1} \oplus Y_n \longrightarrow X'_{n-1} \oplus Y'_n$$

and verify first that ζ is a morphism of complexes. Indeed,

$$\begin{aligned} \begin{pmatrix} \xi_{n-2} & 0 \\ 0 & \eta_{n-1} \end{pmatrix} \circ \begin{pmatrix} -d^X_{n-1} & 0 \\ \alpha_{n-1} & d^Y_n \end{pmatrix} &= \begin{pmatrix} -\xi_{n-2} \circ d^X_{n-1} & 0 \\ \eta_{n-1} \circ \alpha_{n-1} & \eta_{n-1} \circ d^Y_n \end{pmatrix} \\ &= \begin{pmatrix} -d^{X'}_{n-1} \circ \xi_{n-1} & 0 \\ \alpha'_{n-1} \circ \xi_{n-1} & d^{Y'}_n \circ \eta_n \end{pmatrix} \\ &= \begin{pmatrix} -d^{X'}_{n-1} & 0 \\ \alpha'_{n-1} & d^{Y'}_n \end{pmatrix} \circ \begin{pmatrix} \xi_{n-1} & 0 \\ 0 & \eta_n \end{pmatrix}. \end{aligned}$$

Moreover, ζ makes the two right squares commutative since it is just a diagonal mapping and the outer squares are given by the projection mappings. This shows **Tr3**.

We need to verify the octahedral axiom **Tr4**. We start with morphisms

$$\alpha_1 : X^2 \longrightarrow X^3$$

and

$$\alpha_3 : X^1 \longrightarrow X^2.$$

We shall show that then there are morphisms of complexes

$$\delta_1 : C(\alpha_3) \longrightarrow C(\alpha_1 \circ \alpha_3)$$

and

$$\delta_3 : C(\alpha_1 \circ \alpha_3) \longrightarrow C(\alpha_1)$$

so that

$$C(\alpha_3) \longrightarrow C(\alpha_1 \circ \alpha_3) \longrightarrow C(\alpha_1) \longrightarrow C(\alpha_3)[1]$$

is a distinguished triangle and so that the relevant triangles in the axiom are commutative.

Now,

$$C(\alpha_3) = ({X^1}_{n-1} \oplus {X^2}_n, \begin{pmatrix} -d^1_{n-1} & 0 \\ {\alpha_3}_{n-1} & d^2_n \end{pmatrix})$$

and

$$C(\alpha_1 \circ \alpha_3) = ({X^1}_{n-1} \oplus {X^3}_n, \begin{pmatrix} -d^1_{n-1} & 0 \\ (\alpha_1 \circ \alpha_3)_{n-1} & d^3_n \end{pmatrix})$$

so that

$$\delta_1 := \begin{pmatrix} id_{X^1} & 0 \\ 0 & \alpha_1 \end{pmatrix}$$

is a morphism of complexes: Indeed,

$$\begin{aligned} \begin{pmatrix} id_{X^1} & 0 \\ 0 & \alpha_{1n-1} \end{pmatrix} \circ \begin{pmatrix} -d^1_{n-1} & 0 \\ \alpha_{3n-1} & d^2_n \end{pmatrix} &= \begin{pmatrix} -d^1_{n-1} & 0 \\ (\alpha_1 \circ \alpha_3)_{n-1} & \alpha_{1n-1} \circ d^2_n \end{pmatrix} \\ &= \begin{pmatrix} -d^1_{n-1} & 0 \\ (\alpha_1 \circ \alpha_3)_{n-1} & d^3_n \circ \alpha_{1n} \end{pmatrix} \\ &= \begin{pmatrix} -d^1_{n-1} & 0 \\ (\alpha_1 \circ \alpha_3)_{n-1} & d^3_n \end{pmatrix} \circ \begin{pmatrix} id_{X^1} & 0 \\ 0 & \alpha_{1n} \end{pmatrix}. \end{aligned}$$

We compute the cone of the mapping δ_1 and show that this cone is isomorphic in the homotopy category to $C(\alpha_1)$. The degree n homogeneous component of $C(\delta_1)$ is

$${X^1}_{n-2} \oplus {X^2}_{n-1} \oplus {X^1}_{n-1} \oplus {X^3}_n$$

and the differential is

$$\partial_n := \begin{pmatrix} d^1_{n-2} & 0 & 0 & 0 \\ -\alpha_{3n-2} & -d^2_{n-1} & 0 & 0 \\ id_{X^1} & 0 & -d^1_{n-1} & 0 \\ 0 & \alpha_{1n-1} & (\alpha_1 \circ \alpha_3)_{n-1} & d^3_n \end{pmatrix}.$$

We see that the complex $C(id_{X^1})[1]$ is a direct factor of $C(\delta_1)$. Indeed the projection onto the first and the third component is a split morphism of complexes.

The kernel of this morphism is the complex with degree n homogeneous component ${X^2}_{n-1} \oplus {X^3}_n$ and differential the restriction of ∂ onto ${X^2}_{n-1} \oplus {X^3}_n$. This restriction is obtained by erasing the first and the third row and column, whence it is

$$\begin{pmatrix} -d^2_{n-1} & 0 \\ \alpha_{1n-1} & d^3_n \end{pmatrix}.$$

We see that this is precisely the complex $C(\alpha_1)$. We obtained from the distinguished triangles

$$X^2 \xrightarrow{\alpha_1} X^3 \xrightarrow{\beta_1} C(\alpha_1) \xrightarrow{\gamma_1} X^2[1]$$

$$X^1 \xrightarrow{\alpha_3} X^2 \xrightarrow{\beta_3} C(\alpha_3) \xrightarrow{\gamma_3} X^1[1]$$

$$X^1 \stackrel{\alpha_1\circ\alpha_3}{\longrightarrow} X^3 \stackrel{\beta_2}{\longrightarrow} C(\alpha_1\circ\alpha_3) \stackrel{\gamma_2}{\longrightarrow} X^1[1]$$

the distinguished triangle

$$C(\alpha_3) \stackrel{\delta_1}{\longrightarrow} C(\alpha_1\circ\alpha_3) \stackrel{\delta_3}{\longrightarrow} C(\alpha_1) \stackrel{\delta_2}{\longrightarrow} C(\alpha_3)[1]$$

and we still need to show that $\gamma_2\circ\delta_1 = \gamma_3$, $\delta_3\circ\beta_2 = \beta_1$, $\delta_2 = \beta_3[1]\circ\gamma_1$ and $\gamma_1\circ\delta_3 = \alpha_3[1]\circ\gamma_2$ as well as $\beta_2\circ\alpha_1 = \delta_1\circ\beta_3$.

In order to show the first equation we observe that

$$\delta_1 = \begin{pmatrix} id_{X^1} & 0 \\ 0 & \alpha_1 \end{pmatrix},\ \delta_3 = \begin{pmatrix} \alpha_3[1] & 0 \\ 0 & id_{X^3} \end{pmatrix} \text{ and } \delta_2 = \begin{pmatrix} 0 & 0 \\ id_{X^2} & 0 \end{pmatrix}.$$

Moreover, γ_2 is the projection on the first component, just as γ_3. γ_1 is the projection on the second component. Hence

$$\gamma_2\circ\delta_1 = (id_{X^1}\ 0)\cdot\begin{pmatrix} id_{X^1} & 0 \\ 0 & \alpha_1 \end{pmatrix} = (id_{X^1}\ 0) = \gamma_3.$$

For the second equation we see that β_1 and β_2 are injections into the second component. Whence we obtain

$$\delta_3\circ\beta_2 = \begin{pmatrix} \alpha_3[1] & 0 \\ 0 & id_{X_3} \end{pmatrix}\cdot\begin{pmatrix} 0 \\ id_{X^3} \end{pmatrix} = \begin{pmatrix} 0 \\ id_{X^3} \end{pmatrix} = \beta_1.$$

We still need to show the third equation. The mapping γ_1 is the projection onto the first component $X^2[1]\oplus X^3 \longrightarrow X^2[1]$, $\beta_3 : X^2 \longrightarrow X^1[1]\oplus X^2$ is an injection into the second component and $\delta_2 : X^2[1]\oplus X^3 \longrightarrow X^1[2]\oplus X^2[1]$ is the composition of the projection onto the first followed by the injection into the last component. Hence $\delta_2 = \beta_3[1]\circ\gamma_1$ as claimed.

The fourth equation follows from

$$\delta_1\circ\beta_3 = \begin{pmatrix} id_{X^1} & 0 \\ 0 & \alpha_1 \end{pmatrix}\cdot\begin{pmatrix} 0 \\ id_{X^2} \end{pmatrix} = \begin{pmatrix} 0 \\ \alpha_1 \end{pmatrix} = \begin{pmatrix} 0 \\ id_{X^3} \end{pmatrix}\cdot\alpha_1 = \beta_2\circ\alpha_1.$$

The fifth equation comes analogously from

$$\gamma_1\circ\delta_3 = (id_{X^2}\ 0)\cdot\begin{pmatrix} \alpha_3[1] & 0 \\ 0 & id_{X_3} \end{pmatrix} = (\alpha_3[1]\ \ 0) = \alpha_3[1]\cdot(id_{X^1[1]}\ \ 0) = \alpha_3[1]\circ\gamma_2.$$

This proves the statement.

The homotopy category of right bounded complexes of projective modules is already the context that is largely sufficient for our later studies. However, in some cases it is more convenient to work in a slightly different category, the derived category of complexes. This category is the subject of the next Sect. 3.5.3.

3.5.3 The Derived Category of Complexes

Example 3.5.27 For unbounded complexes of projective modules, or for complexes of non-projective modules, complexes with vanishing homology are not necessarily homotopic to 0.

- Let $A = \mathbb{Z}/4\mathbb{Z}$. Form the complex with $\mathbb{Z}/4\mathbb{Z}$ in each degree and differential given by multiplication by 2:

$$\dots \longrightarrow \mathbb{Z}/4\mathbb{Z} \xrightarrow{\cdot 2} \mathbb{Z}/4\mathbb{Z} \xrightarrow{\cdot 2} \mathbb{Z}/4\mathbb{Z} \xrightarrow{\cdot 2} \mathbb{Z}/4\mathbb{Z} \xrightarrow{\cdot 2} \mathbb{Z}/4\mathbb{Z} \longrightarrow \dots$$

  Then this complex has vanishing homology, as is immediately seen. However the complex is not homotopy equivalent to 0. Indeed, the identity is not homotopy equivalent to 0 since

$$\mathrm{im}(\partial h + h\partial) \subseteq 2 \cdot \bigoplus_{n \in \mathbb{Z}} \mathbb{Z}/4\mathbb{Z}$$

  for all mappings h being a candidate for a homotopy equivalence.
- Let $A = \mathbb{Z}$. Then the complex

$$\dots \longrightarrow 0 \longrightarrow 2\mathbb{Z}/4\mathbb{Z} \longrightarrow \mathbb{Z}/4\mathbb{Z} \longrightarrow \mathbb{Z}/2\mathbb{Z} \longrightarrow 0 \longrightarrow \dots$$

  has vanishing homology but is not homotopy equivalent to 0 by Proposition 3.5.19.

The above Examples 3.5.27 are a particular case of the phenomenon that a mapping in the homotopy category inducing an isomorphism in homology may be non-invertible. The derived category is introduced in order to circumvent this problem. The price we have to pay for this achievement is that this is a far more complicated category. In particular dealing with morphisms explicitly in the derived category is a highly challenging enterprise.

We start with an abelian category $\mathcal{A}$. Let $x \in \{\emptyset, +, -, b\}$ as we did for the homotopy category. In order to obtain the derived category $D^x(\mathcal{A})$ we shall formally invert the morphisms in $K^x(\mathcal{A})$. This procedure is done in the following way.

- The objects in $D^{\emptyset}(\mathcal{A})$ are complexes of objects in $\mathcal{A}$. We shall write $D(\mathcal{A})$ for $D^{\emptyset}(\mathcal{A})$ to simplify the notation.
- The objects in $D^-(\mathcal{A})$ are right bounded complexes in $\mathcal{A}$

$$\ldots \longrightarrow X_{n+2} \longrightarrow X_{n+1} \longrightarrow X_n \longrightarrow 0 \longrightarrow 0 \longrightarrow \ldots.$$

- The objects in $D^+(\mathcal{A})$ are left bounded complexes in $\mathcal{A}$

$$\ldots \longrightarrow 0 \longrightarrow X_n \longrightarrow X_{n-1} \longrightarrow X_{n-2} \longrightarrow X_{n-3} \longrightarrow \ldots.$$

- The objects in $D^b(\mathcal{A})$ are left and right bounded complexes in $\mathcal{A}$, that is complexes which are both in $D^-(\mathcal{A})$ and in $D^+(\mathcal{A})$.

We shall need to describe the morphisms. We shall do this for $D(\mathcal{A})$, and convene that $D^+(\mathcal{A})$, $D^-(\mathcal{A})$ and $D^b(\mathcal{A})$ are full subcategories of $D(\mathcal{A})$.

Recall from Remark 3.5.7 first that given a complex X with differential d_X and a complex Y with differential d_Y, then define $H(X) := \ker(d_X)/\text{im}(d_X)$ and likewise $H(Y) := \ker(d_Y)/\text{im}(d_Y)$. A homomorphism of complexes $\sigma : X \longrightarrow Y$ is a graded module homomorphism of degree 0 satisfying $d_Y \circ \sigma = \sigma \circ d_X$. Hence σ induces a morphism

$$H(\sigma) : H(X) \longrightarrow H(Y).$$

Definition 3.5.28 Let X and Y be complexes in $\mathcal{A}$ and let $\sigma : X \longrightarrow Y$ be a morphism of complexes. If $H(\sigma) : H(X) \longrightarrow H(Y)$ is an isomorphism, then σ is called a *quasi-isomorphism*.

Proposition 3.5.29 *Let (A, d_A), (B, d_B) and (C, d_C) be complexes in $\mathcal{A}$ and let*

$$0 \longrightarrow A \stackrel{\alpha}{\longrightarrow} B \stackrel{\beta}{\longrightarrow} C \longrightarrow 0$$

be a short exact sequence of complexes. Then there is a long exact sequence

$$\ldots \longrightarrow H_{i+1}(C) \stackrel{\delta_{i+1}}{\longrightarrow} H_i(A) \stackrel{H_i(\alpha)}{\longrightarrow} H_i(B) \stackrel{H_i(\beta)}{\longrightarrow} H_i(C) \stackrel{\delta_i}{\longrightarrow} H_{i-1}(A) \longrightarrow \ldots$$

for all $i \in \mathbb{Z}$.

Proof We get a commutative diagram

$$\begin{array}{ccccccc} 0 \longrightarrow & A & \longrightarrow & B & \longrightarrow & C & \longrightarrow 0 \\ & \downarrow d_A & & \downarrow d_B & & \downarrow d_C & \\ 0 \longrightarrow & A & \longrightarrow & B & \longrightarrow & C & \longrightarrow 0 \end{array}$$

is a morphism of short exact sequences of complexes. The category of complexes is abelian, so that we can complete the diagram with kernels and cokernels to

$$\begin{array}{ccccccccc}
 & & 0 & & 0 & & 0 & & \\
 & & \downarrow & & \downarrow & & \downarrow & & \\
0 & \longrightarrow & \ker(d_A) & \longrightarrow & \ker(d_B) & \longrightarrow & \ker(d_C) & & \\
 & & \downarrow & & \downarrow & & \downarrow & & \\
0 & \longrightarrow & A & \longrightarrow & B & \longrightarrow & C & \longrightarrow & 0 \\
 & & \downarrow d_A & & \downarrow d_B & & \downarrow d_C & & \\
0 & \longrightarrow & A & \longrightarrow & B & \longrightarrow & C & \longrightarrow & 0 \\
 & & \downarrow & & \downarrow & & \downarrow & & \\
 & & \mathrm{coker}(d_A) & \longrightarrow & \mathrm{coker}(d_B) & \longrightarrow & \mathrm{coker}(d_C) & \longrightarrow & 0 \\
 & & \downarrow & & \downarrow & & \downarrow & & \\
 & & 0 & & 0 & & 0 & &
\end{array}$$

Since $\mathrm{coker}(d_A) = A/\mathrm{im}(d_A)$, $\mathrm{coker}(d_B) = B/\mathrm{im}(d_B)$ and $\mathrm{coker}(d_C) = C/\mathrm{im}(d_C)$ this then induces a commutative diagram in the graded objects of $\mathcal{A}$ with exact lines and columns as follows

$$\begin{array}{ccccccccc}
 & & 0 & & 0 & & 0 & & \\
 & & \downarrow & & \downarrow & & \downarrow & & \\
 & & \ker(d_A)/\mathrm{im}(d_A) & \longrightarrow & \ker(d_B)/\mathrm{im}(d_B) & \longrightarrow & \ker(d_C)/\mathrm{im}(d_C) & & \\
 & & \downarrow & & \downarrow & & \downarrow & & \\
 & & A/\mathrm{im}(d_A) & \longrightarrow & B/\mathrm{im}(d_B) & \longrightarrow & C/\mathrm{im}(d_C) & \longrightarrow & 0 \\
 & & \downarrow d_A & & \downarrow d_B & & \downarrow d_C & & \\
0 & \longrightarrow & \ker(d_A) & \longrightarrow & \ker(d_B) & \longrightarrow & \ker(d_C) & & \\
 & & \downarrow & & \downarrow & & \downarrow & & \\
 & & \ker(d_A)/\mathrm{im}(d_A) & \longrightarrow & \ker(d_B)/\mathrm{im}(d_B) & \longrightarrow & \ker(d_C)/\mathrm{im}(d_C) & & \\
 & & \downarrow & & \downarrow & & \downarrow & & \\
 & & 0 & & 0 & & 0 & &
\end{array}$$

The snake lemma, Lemma 3.3.9, gives a connecting morphism and an exact sequence

$$\ldots \longrightarrow H(A) \longrightarrow H(B) \longrightarrow H(C) \xrightarrow{\delta} H(A) \longrightarrow H(B) \longrightarrow H(C) \longrightarrow \ldots$$

Recalling the construction of the connecting homomorphism, we see that δ is constructed using at some point the image by d_B, which is of degree -1, and we obtain that δ is graded of degree -1. □

Remark 3.5.30 This result is going to be shown in a more general context by more general means as a consequence of the triangulated category axioms in Corollary 3.5.52. However, since the above elementary proof is possible, it might be of interest too.

Corollary 3.5.31 *Let $\varphi : A \longrightarrow B$ be a homomorphism of complexes with $C = C(\varphi)$. Then there is a long exact sequence*

$$\ldots \longrightarrow H_{i+1}(C) \longrightarrow H_i(A) \longrightarrow H_i(B) \longrightarrow H_i(C) \longrightarrow H_{i-1}(A) \longrightarrow \ldots$$

for all $i \in \mathbb{Z}$.

Proof The sequence of complexes

$$0 \longrightarrow B \longrightarrow C \longrightarrow A \longrightarrow 0$$

is exact. Then apply Proposition 3.5.29. □

Just as in the homotopy category we use cones of morphisms.

Lemma 3.5.32 *Let $\varphi : A \longrightarrow B$ be a morphism of complexes. Then $C(\varphi)$ is acyclic if and only if φ is a quasi-isomorphism.*

Proof This is immediate from Corollary 3.5.31. □

Lemma 3.5.33 *Given objects Z, Y, Z' in $K(\mathcal{A})$ and morphisms $Z \longrightarrow Y$ and a quasi-isomorphism $Z' \longrightarrow Y$, then there is an object Z'' in $K(\mathcal{A})$ and a quasi-isomorphism $Z'' \longrightarrow Z$ as well as a morphism $Z'' \longrightarrow Z'$ making the square*

$$\begin{array}{ccc} Z'' & \stackrel{\alpha''}{\longrightarrow} & Z' \\ \downarrow \nu'' & & \downarrow \nu' \\ Z & \stackrel{\alpha}{\longrightarrow} & Y \end{array}$$

commutative.

Proof We have the following situation:

$$\begin{array}{ccccc} Z' & \stackrel{\nu'}{\longrightarrow} & Y & \stackrel{\pi}{\longrightarrow} & C(\nu') \\ & & \uparrow \alpha & & \| \\ & & Z & \stackrel{\pi\circ\alpha}{\longrightarrow} & C(\nu') \end{array}$$

Forming cones we may complete this to a diagram

$$\begin{array}{ccccccccc} C(\nu')[-1] & \longrightarrow & Z' & \stackrel{\nu'}{\longrightarrow} & Y & \stackrel{\pi}{\longrightarrow} & C(\nu') & \longrightarrow & Z'[1] \\ & & & & \uparrow \alpha & & \| & & \\ & & C(\pi\circ\alpha)[-1] & \stackrel{\nu''}{\longrightarrow} & Z & \stackrel{\pi\circ\alpha}{\longrightarrow} & C(\nu') & \longrightarrow & C(\pi\circ\alpha) \end{array}$$

of distinguished triangles where the middle square is commutative. The verification of **Tr3** in Proposition 3.5.25 establishes the existence of a mapping α'' making the diagram

$$\begin{array}{ccccccccc} C(\nu')[-1] & \longrightarrow & Z' & \stackrel{\nu'}{\longrightarrow} & Y & \stackrel{\pi}{\longrightarrow} & C(\nu') & \longrightarrow & Z'[1] \\ \| & & \uparrow \alpha'' & & \uparrow \alpha & & \| & & \uparrow \alpha''[1] \\ C(\nu')[-1] & \longrightarrow & C(\pi\circ\alpha)[-1] & \stackrel{\nu''}{\longrightarrow} & Z & \stackrel{\pi\circ\alpha}{\longrightarrow} & C(\nu') & \longrightarrow & C(\pi\circ\alpha) \end{array}$$

commutative. Define

$$Z'' := C(\pi \circ \alpha)[-1] \text{ and } \nu'' : Z'' \longrightarrow Z \text{ and } \alpha'' : Z'' \longrightarrow Z'$$

as given in the above diagram. Now, the cone of ν'' is isomorphic to $C(\nu')$ in the homotopy category, and since ν' is a quasi-isomorphism, $C(\nu')$ is acyclic by Lemma 3.5.32. Again by Lemma 3.5.32 we get that ν'' is a quasi-isomorphism as well. □

Given complexes X and Y in $\mathcal{A}$ we first consider triples

$$X \xleftarrow{\nu} Z \xrightarrow{\alpha} Y$$

where Z is any complex in $\mathcal{A}$ and α and ν are homomorphisms of complexes. We assume furthermore that ν is a quasi-isomorphism.

Given two triples

$$X \xleftarrow{\nu} Z \xrightarrow{\alpha} Y \text{ and } X \xleftarrow{\nu'} Z' \xrightarrow{\alpha'} Y$$

we say that

$$X \xleftarrow{\nu} Z \xrightarrow{\alpha} Y \text{ is covered by } X \xleftarrow{\nu'} Z' \xrightarrow{\alpha'} Y$$

if there is a morphism of complexes $\gamma : Z' \longrightarrow Z$ so that the diagram

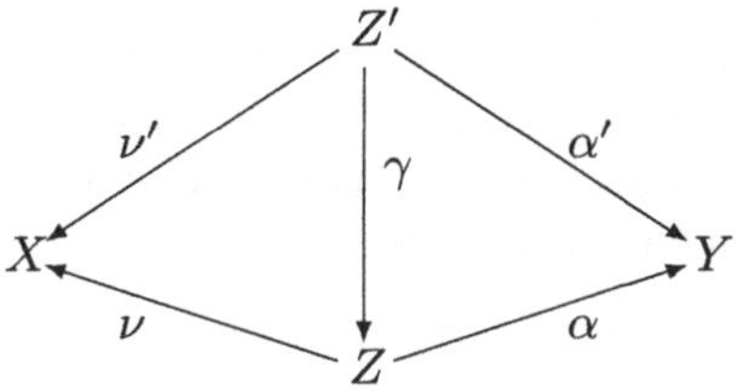

commutes, in the sense that $\nu \circ \gamma = \nu'$ and $\alpha \circ \gamma = \alpha'$.

Observe now that

$$H(\nu') = H(\nu \circ \gamma) = H(\nu) \circ H(\gamma)$$

and therefore the fact that ν and ν' are quasi-isomorphisms implies that $H(\nu)$ and $H(\nu')$ are isomorphisms, whence $H(\gamma)$ is an isomorphism. Therefore γ is necessarily a quasi-isomorphism.

Call two triples

$$X \xleftarrow{\nu} Z \xrightarrow{\alpha} Y \text{ and } X \xleftarrow{\nu'} Z' \xrightarrow{\alpha'} Y$$

equivalent if there is a triple

$$X \xleftarrow{\nu''} Z'' \xrightarrow{\alpha''} Y$$

that covers both triples

$$X \xleftarrow{\nu} Z \xrightarrow{\alpha} Y \text{ and } X \xleftarrow{\nu'} Z' \xrightarrow{\alpha'} Y.$$

This relation is transitive. Indeed, if Z_{12} covers $X \leftarrow Z_1 \rightarrow Y$ and $X \leftarrow Z_2 \rightarrow Y$, and if Z_{23} covers $X \leftarrow Z_2 \rightarrow Y$ and $X \leftarrow Z_3 \rightarrow Y$, then using Lemma 3.5.33 we obtain an object Z_{123} and quasi-isomorphisms $Z_{12} \leftarrow Z_{123} \rightarrow Z_{23}$ so that the diagram

$$\begin{array}{ccc} Z_{123} & \longrightarrow & Z_{23} \\ \downarrow & & \downarrow \\ Z_{12} & \longrightarrow & Z_2 \end{array}$$

is commutative. This gives that Z_{123} covers $X \leftarrow Z_1 \rightarrow Y$ and $X \leftarrow Z_3 \rightarrow Y$. Reflexivity and symmetricity are clear by definition.

Remark 3.5.34 We have to be careful when speaking of equivalence classes. The triples

$$X \xleftarrow{\nu} Z \xrightarrow{\alpha} Y$$

with fixed X and Y do *not* form a set since the possible Z do not form a set. We need to pass to isomorphism classes in order to be able to speak about equivalence classes. Of course any isomorphism $Z \simeq Z'$ induces a covering triple. This shows that we may pass to equivalence classes of modules. The set theoretical problems in this case are not easy to resolve, but they can be solved. The interested reader may consult Weibel [3, 10.3.3 and 10.3.6]. Having alerted the reader to these problems, we shall not elaborate on them any further.

Definition 3.5.35 Let X and Y be two objects of $D(\mathcal{A})$. Then we define $Hom_{D(\mathcal{A})}(X, Y)$ as the set of equivalence classes of triples $X \xleftarrow{\nu} Z \xrightarrow{\alpha} Y$ where Z is an object of $D(\mathcal{A})$, ν is a quasi-isomorphism and α a morphism of complexes.

Note that by Remark 3.5.34 we know that $Hom_{D(\mathcal{A})}(X, Y)$ is indeed a set, as needed for the definition of a category.

The identity morphism is represented by the triple

$$X \xleftarrow{id} X \xrightarrow{id} X.$$

We need to define composition of morphisms and verify associativity.

Recall that by Proposition 3.5.10 the category of complexes is abelian, so that pullbacks exist in the category of complexes.

Given two triples

$$X \xleftarrow{\nu} Z \xrightarrow{\alpha} Y \text{ and } Y \xleftarrow{\nu'} Z' \xrightarrow{\alpha'} W$$

we need to find an object Z'' and morphisms

$$Z \longleftarrow Z'' \longrightarrow Z'$$

giving the following diagram

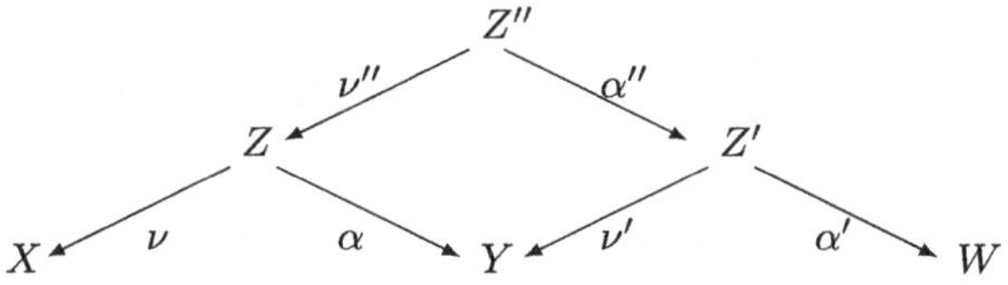

so that ν'' is a quasi-isomorphism.
Then we shall define the triple

$$X \xleftarrow{\nu \circ \nu''} Z'' \xrightarrow{\alpha' \circ \alpha''} W$$

to be the composition of the triples

$$X \xleftarrow{\nu} Z \xrightarrow{\alpha} Y \text{ and } Y \xleftarrow{\nu'} Z' \xrightarrow{\alpha'} W.$$

Lemma 3.5.33 allows us to find Z'' so that ν'' is a quasi-isomorphism. We can now define composition as we intended to do.

Moreover, we need to show that this composition is well-defined with respect to covering of triples: If

$$X \xleftarrow{\nu} Z \xrightarrow{\alpha} Y \text{ and } Y \xleftarrow{\nu'} Z' \xrightarrow{\alpha'} W$$

are triples, if

$$X \xleftarrow{\hat{\nu}} \hat{Z} \xrightarrow{\hat{\alpha}} Y \quad \text{covers} \quad X \xleftarrow{\nu} Z \xrightarrow{\alpha} Y$$

and if

$$Y \xleftarrow{\hat{\nu}'} \hat{Z}' \xrightarrow{\hat{\alpha}'} W \quad \text{covers} \quad Y \xleftarrow{\nu'} Z' \xrightarrow{\alpha'} Y,$$

then the composition of the covers defined above has a common cover with the composition of the original triples. The following scheme describes the situation:

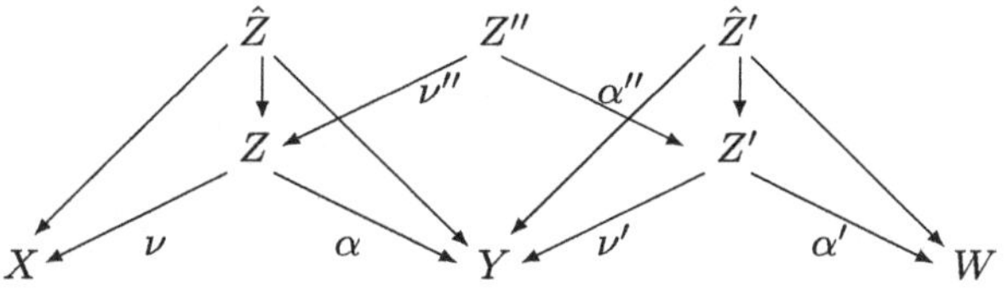

Composing the triples $(X \longleftarrow \hat{Z} \longrightarrow Y)$ and $(Y \longleftarrow \hat{Z}' \longrightarrow W)$, does this give the same result as composing $(X \longleftarrow Z \longrightarrow Y)$ and $(Y \longleftarrow Z' \longrightarrow W)$? We know by Lemma 3.5.33 that there is an object $\tilde{Z}$ and quasi-isomorphisms $\hat{Z} \longleftarrow \tilde{Z} \longrightarrow Z''$ as well as an object $\tilde{\tilde{Z}}$ with quasi-isomorphisms $Z'' \longleftarrow \tilde{\tilde{Z}} \longrightarrow \hat{Z}'$.

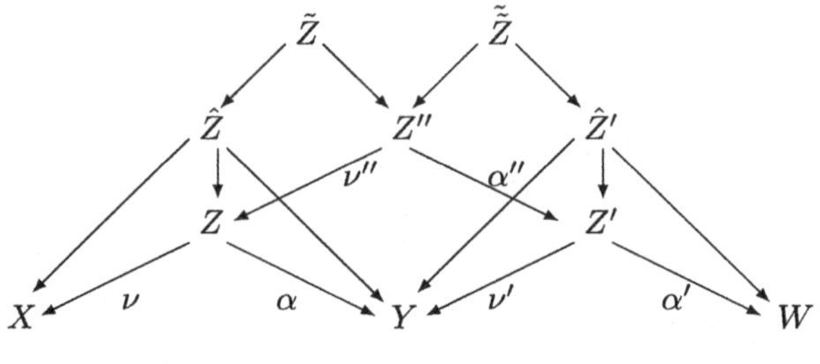

Applying Lemma 3.5.33 again shows that there is an object $\hat{\hat{Z}}$ and quasi-isomorphisms $\tilde{Z} \longrightarrow \hat{\hat{Z}} \longrightarrow \tilde{\tilde{Z}}$ so that the following diagram commutes:

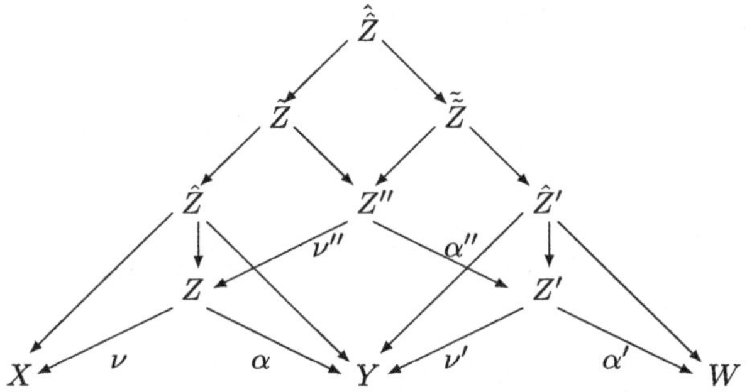

Hence $X \longleftarrow Z'' \longrightarrow W$ is covered by $X \longleftarrow \hat{\hat{Z}} \longrightarrow W$ as claimed.

Finally, associativity is clear by the construction, using Lemma 3.5.33. Indeed, express $(\nu_1, Z_1, \alpha_1) \circ (\nu_2, Z_2, \alpha_2)$ and $(\nu_2, Z_2, \alpha_2) \circ (\nu_3, Z_3, \alpha_3)$ as described in the definition of the composition, and then construct a cover (ν_4, Z_4, α_4) for the result of the compositions. Associativity can then be read off the diagram. Moreover, all quasi-isomorphisms become invertible. Indeed, let $\nu : X \longrightarrow Y$ be a quasi-isomorphism. Then $Y \overset{\nu}{\longleftarrow} X \overset{id}{\longrightarrow} X$ is a triple defining a morphism. Compose it on the left with the triple $X \overset{id}{\longleftarrow} X \overset{\nu}{\longrightarrow} Y$. Then the triple $X \overset{id}{\longleftarrow} X \overset{id}{\longrightarrow} X$ covers the composition:

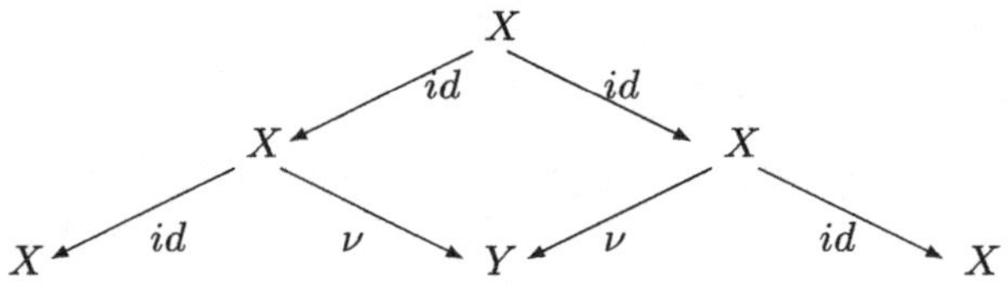

This shows that ν has a right inverse. Composing the other way round we get a scheme

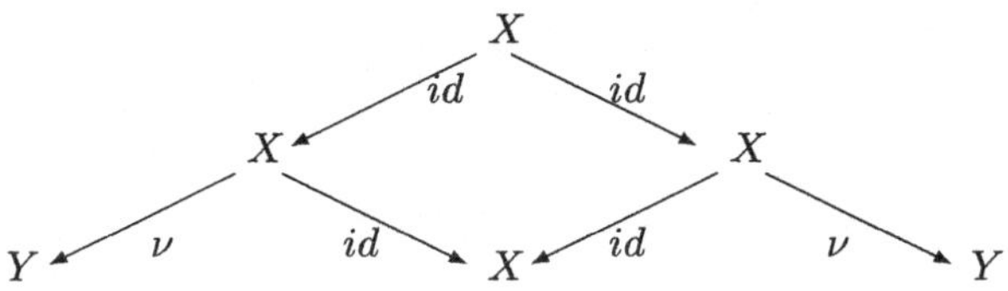

and this covers the identity on Y:

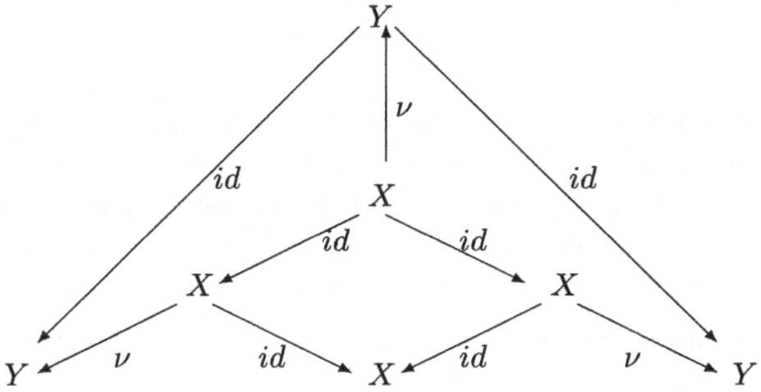

Therefore each quasi-isomorphism becomes invertible when considering triples as morphisms in a larger category.

Finally we shall need to define the abelian group structure of the morphisms. We shall again use Lemma 3.5.33. Indeed, let

$$X \stackrel{\nu}{\longleftarrow} Z \stackrel{\alpha}{\longrightarrow} Y$$

and

$$X \stackrel{\nu'}{\longleftarrow} Z' \stackrel{\alpha'}{\longrightarrow} Y$$

be two triples representing morphisms from X to Y. Then Lemma 3.5.33 shows that there is an object $\hat{Z}$ and morphisms $\mu : \hat{Z} \longrightarrow Z$ and $\mu' : \hat{Z} \longrightarrow Z'$ such that the diagram

$$\begin{array}{ccc} \hat{Z} & \stackrel{\mu}{\longrightarrow} & Z \\ \downarrow \mu' & & \downarrow \nu \\ Z' & \stackrel{\nu'}{\longrightarrow} & X \end{array}$$

is commutative and such that μ as well as μ' are quasi-isomorphisms. Hence the triple

$$X \stackrel{\nu\circ\mu}{\longleftarrow} \hat{Z} \stackrel{\alpha\circ\mu}{\longrightarrow} Y$$

covers

$$X \stackrel{\nu}{\longleftarrow} Z \stackrel{\alpha}{\longrightarrow} Y$$

by means of the morphism $\mu : \hat{Z} \longrightarrow Z$ and the triple

$$X \overset{\nu'\circ\mu'}{\longleftarrow} \hat{Z} \overset{\alpha'\circ\mu'}{\longrightarrow} Y$$

covers

$$X \overset{\nu'}{\longleftarrow} Z' \overset{\alpha'}{\longrightarrow} Y$$

by means of the morphism $\mu' : \hat{Z} \longrightarrow Z'$. Define

$$(\mu, Z, \alpha) + (\mu', Z', \alpha') := (\nu \circ \mu, \hat{Z}, \alpha \circ \mu + \alpha' \circ \mu').$$

Once the so-defined law $+$ is shown to be well-defined, it is an abelian group law by the usual arguments for fraction calculus for commutative rings. Let us show that the law $+$ is well-defined. This is again an immediate consequence of Lemma 3.5.33 since if one takes triples $(\tilde{\mu}, \tilde{Z}, \tilde{\alpha})$ and $(\tilde{\mu}', \tilde{Z}', \tilde{\alpha}')$ covering the original triples (μ, Z, α) and (μ', Z', α'), then the common cover $\hat{Z}$ is covered by the common cover $\tilde{\hat{Z}}$ of the covering triples $(\tilde{\mu}, \tilde{Z}, \tilde{\alpha})$ and $(\tilde{\mu}', \tilde{Z}', \tilde{\alpha}')$. The law then is well-defined as is immediately seen.

Definition 3.5.36 Let $\mathcal{A}$ be an abelian category. Then the *derived category* $D(\mathcal{A})$ of $\mathcal{A}$ has objects the complexes in $\mathcal{A}$ and morphisms the equivalence classes of triples (ν, Z, α) where ν is a quasi-isomorphism and where α is a morphism of complexes in $\mathcal{A}$.

For $x, y \in \{\emptyset, +, -, b\}$ let $D^{x,y}(\mathcal{A})$ be the full subcategory of $D(\mathcal{A})$ formed by objects in $K^{x,y}(\mathcal{A})$.

If A is an algebra we sometimes write $D^{x,y}(A)$ for $D^{x,y}(A\text{-}Mod)$ or for $D^{x,y}(A\text{-}mod)$ when no confusion may occur.

Remark 3.5.37 We need to have an abelian category $\mathcal{A}$ to construct the derived category rather than just an additive category since we need to be able to construct homologies of a complex. This involves kernels and cokernels, hence is a natural playground for additive categories. Nevertheless, if $\mathcal{A}$ is additive and embeds into an abelian category $\mathcal{A}'$, we may speak of the derived category with objects in $\mathcal{A}$ and homology in $\mathcal{A}'$, which is denoted by $D_{\mathcal{A}'}(\mathcal{A})$.

Remark 3.5.38 Here we must issue the following warning. Two objects X and Y in $D(\mathcal{A})$ are isomorphic if and only if there is a complex Z and there are morphisms of complexes $X \overset{\varphi}{\longleftarrow} Z \overset{\psi}{\longrightarrow} Y$ so that $H(\varphi) : H(Z) \longrightarrow H(X)$ and $H(\psi) : H(Z) \longrightarrow H(Y)$ are isomorphisms. *This is not the same* as saying that $H(X)$ and $H(Y)$ are isomorphic, in fact it is stronger. The isomorphism has to be induced by a morphism of complexes in the way described before.

Remark 3.5.39 Note that we needed to define composition in a rather complicated fashion, but now the derived category has some spectacular features.

Note further that if X and Y have left (resp. right) bounded homology, and if (μ, Z, α) is a triple representing a homomorphism in $Hom_{D(\mathcal{A})}(X, Y)$, then Z also has left (resp. right) bounded homology since μ is a quasi-isomorphism.

Proposition 3.5.40 *For any abelian category $\mathcal{A}$ and $x, y \in \{\emptyset, +, -, b\}$ the category $D^{x,y}(\mathcal{A})$ is triangulated with suspension functor inherited by the suspension functor of the homotopy category $K^{x,y}(\mathcal{A})$ and distinguished triangles being all triangles in $D^{x,y}(\mathcal{A})$ which are isomorphic to triangles of the form*

$$X \xrightarrow{\alpha} Y \longrightarrow C(\alpha) \longrightarrow X[1]$$

for $\alpha \in Hom_{K^{x,y}(\mathcal{A})}(X, Y)$.

Proof We shall first need to see what it means to be an isomorphism in the derived category. Actually $X \simeq Y$ if there is a sequence of triples $(X_i \longleftarrow Z_i \longrightarrow X_{i+1})$ for $i \in \{1, \dots, n\}$ with $X_1 = X$ and $X_{n+1} = Y$ so that each of the morphisms $Z_i \longrightarrow X_i$ and $Z_i \longrightarrow X_{i+1}$ is a quasi-isomorphism. But then, using Lemma 3.5.33 n times, one obtains that there is a triple $X \longleftarrow Z \longrightarrow Y$ where $Z \longrightarrow X$ and $Z \longrightarrow Y$ are quasi-isomorphisms.

The second half of the axiom **Tr1** is trivial since the identity is already a mapping of the homotopy category, and the image of a triangle in the homotopy category is a triangle in the derived category. For the first half of the axiom we need to define a triangle corresponding to a triple $\psi := (\nu, Z, \alpha)$ where $\nu : Z \longrightarrow X$ is a quasi-isomorphism in the homotopy category, and $\alpha : Z \longrightarrow X$ is any homomorphism in the homotopy category. But now

$$Z \xrightarrow{\alpha} Y \longrightarrow C(\alpha) \longrightarrow Z[1]$$

is a genuine triangle in the homotopy category, and hence also in the derived category. Since $X \simeq Z$ in the derived category we get an isomorphism of triangles

$$\begin{array}{ccccccc} Z & \xrightarrow{\alpha} & Y & \longrightarrow & C(\alpha) & \longrightarrow & Z[1] \\ \downarrow \nu & & \| & & \| & & \downarrow \nu[1] \\ X & \xrightarrow{\psi} & Y & \longrightarrow & C(\alpha) & \longrightarrow & Z[1] \end{array}$$

and hence ψ appears in a triangle isomorphic to the genuine triangle

$$Z \xrightarrow{\alpha} Y \longrightarrow C(\alpha) \longrightarrow Z[1].$$

The proof of **Tr2** goes the same way. Given a triangle

$$X \xrightarrow{\psi} Y \longrightarrow C(\alpha) \longrightarrow X[1]$$

then we may replace X by Z and obtain a triangle

$$Z \xrightarrow{\alpha} Y \longrightarrow C(\alpha) \longrightarrow Z[1]$$

in the homotopy category. Now, knowing that the homotopy category is indeed a triangulated category, applying **Tr2** to the homotopy triangle gives a triangle

$$Y \longrightarrow C(\alpha) \longrightarrow Z[1] \longrightarrow Y[1]$$

and using again that $Z \simeq X$ gives the required triangle. Now apply this step five times and then $[-2]$ to get that the triangle shifted in the other direction is again distinguished.

The proof of **Tr3** again uses in an essential way that the homotopy category is triangulated. Of course, this is not too surprising. A triangle in the derived category

$$X \stackrel{\psi}{\longrightarrow} Y \longrightarrow C(\psi) \longrightarrow X[1]$$

induces a commutative diagram

$$\begin{array}{ccccccc}
Z & \stackrel{\alpha}{\longrightarrow} & Y & \longrightarrow & C(\alpha) & \longrightarrow & Z[1] \\
\downarrow \nu & & \| & & \| & & \downarrow \nu[1] \\
X & \stackrel{\psi}{\longrightarrow} & Y & \longrightarrow & C(\alpha) & \longrightarrow & Z[1]
\end{array}$$

in the homotopy category. Hence the configuration scheme for the axiom **Tr3** is

$$\begin{array}{ccccccc}
X & \stackrel{\psi}{\longrightarrow} & Y & \longrightarrow & C(\psi) & \longrightarrow & X[1] \\
\downarrow \psi_X & & \downarrow \psi_Y & & & & \downarrow \psi_X[1] \\
X' & \stackrel{\psi'}{\longrightarrow} & Y' & \longrightarrow & C(\psi') & \longrightarrow & X'[1]
\end{array}$$

with $\psi = (\nu, Z, \alpha)$ and $\psi' = (\nu', Z', \alpha')$ and with $\psi_X = (\nu_X, Z_X, \alpha_X)$ and $\psi_Y = (\nu_Y, Z_Y, \alpha_Y)$. Hence this actually decomposes again as a diagram

$$\begin{array}{ccccccc}
Z & \stackrel{\alpha}{\longrightarrow} & Y & \longrightarrow & C(\alpha) & \longrightarrow & Z[1] \\
\downarrow \nu & & \| & & \| & & \downarrow \nu[1] \\
X & \stackrel{\psi}{\longrightarrow} & Y & \longrightarrow & C(\psi) & \longrightarrow & X[1] \\
\uparrow \nu_X & & \uparrow \nu_Y & & & & \uparrow \nu_X[1] \\
Z_X & & Z_Y & & & & Z_X[1] \\
\downarrow \alpha_X & & \downarrow \alpha_Y & & & & \downarrow \alpha_X[1] \\
X' & \stackrel{\psi'}{\longrightarrow} & Y' & \longrightarrow & C(\psi') & \longrightarrow & X'[1] \\
\uparrow \nu' & & \| & & \| & & \uparrow \nu'[1] \\
Z' & \stackrel{\alpha'}{\longrightarrow} & Y' & \longrightarrow & C(\alpha') & \longrightarrow & Z'[1].
\end{array}$$

Now Lemma 3.5.33 applied to the morphisms $Z_Y \longrightarrow Y \longleftarrow X$ gives a morphism $\tilde{Z} \longrightarrow Z_Y$ and Lemma 3.5.33 applied again to the morphisms $Z_Y \longrightarrow X \longleftarrow \tilde{Z}$ gives a morphism $\hat{Z}_X \stackrel{\zeta}{\longrightarrow} Z_Y$ in the homotopy category making the diagram

$$\begin{array}{ccccccc}
Z & \xrightarrow{\alpha} & Y & \longrightarrow & C(\alpha) & \longrightarrow & Z[1] \\
\downarrow \nu & & \| & & \| & & \downarrow \nu[1] \\
X & \xrightarrow{\psi} & Y & \longrightarrow & C(\psi) & \longrightarrow & X[1] \\
\uparrow \nu_X & & \uparrow \nu_Y & & & & \uparrow \nu_X[1] \\
Z_X & \xrightarrow{\zeta} & Z_Y & & & & Z_X[1] \\
\downarrow \alpha_X & & \downarrow \alpha_Y & & & & \downarrow \alpha_X[1] \\
X' & \xrightarrow{\psi'} & Y' & \longrightarrow & C(\psi') & \longrightarrow & X'[1] \\
\uparrow \nu' & & \| & & \| & & \uparrow \nu'[1] \\
Z' & \xrightarrow{\alpha'} & Y' & \longrightarrow & C(\alpha') & \longrightarrow & Z'[1]
\end{array}$$

commutative. Completing the morphism $\hat{Z}_X \xrightarrow{\zeta} Z_Y$ in the homotopy category by a cone gives a triangle

$$\hat{Z}_X \longrightarrow Z_Y \longrightarrow C(\zeta) \longrightarrow \hat{Z}_X[1].$$

Finally we need to get rid of ψ and ψ', which are not yet in the homotopy category. For this purpose we apply Lemma 3.5.33 to the morphisms $Z \longrightarrow X \longleftarrow \hat{Z}_X$ and $Z' \longrightarrow X' \longleftarrow \hat{Z}_X$ to obtain objects Z^X and $Z^{X'}$ and corresponding morphisms which replace X and X' and the corresponding triangles in the sense displayed in the diagram

$$\begin{array}{ccccccc}
Z^X & \xrightarrow{\alpha^X} & Y & \longrightarrow & C(\alpha^X) & \longrightarrow & Z^X[1] \\
\uparrow \tilde{\nu_X} & & \uparrow \nu_Y & & & & \uparrow \nu_X[1] \\
\hat{Z}_X & \longrightarrow & Z_Y & \longrightarrow & C(\zeta) & \longrightarrow & Z_X[1] \\
\downarrow \tilde{\alpha}_X & & \downarrow \alpha_Y & & & & \downarrow \alpha_X[1] \\
Z^{X'} & \xrightarrow{\alpha^{X'}} & Y' & \longrightarrow & C(\alpha^{X'}) & \longrightarrow & Z^{X'}[1].
\end{array}$$

We apply the axiom **Tr3** to the two configurations in the centre to obtain

$$\begin{array}{ccccccc}
Z^X & \xrightarrow{\alpha^X} & Y & \longrightarrow & C(\alpha^X) & \longrightarrow & Z^X[1] \\
\uparrow \tilde{\nu_X} & & \uparrow \nu_Y & & \uparrow \xi & & \uparrow \nu_X[1] \\
\hat{Z}_X & \longrightarrow & Z_Y & \longrightarrow & C(\zeta) & \longrightarrow & Z_X[1] \\
\downarrow \tilde{\alpha}_X & & \downarrow \alpha_Y & & \downarrow \eta & & \downarrow \alpha_X[1] \\
Z^{X'} & \xrightarrow{\alpha^{X'}} & Y' & \longrightarrow & C(\alpha^{X'}) & \longrightarrow & Z^{X'}[1].
\end{array}$$

Now, ξ is a quasi-isomorphism since $\tilde{\nu_X}$ and ν_Y are quasi-isomorphisms using the homology sequence Corollary 3.5.31. Hence $(\xi, C(\zeta), \eta)$ gives a morphism in the derived category inducing a morphism of triangles. The upper and lower triangles are isomorphic to the original triangles we started with and this shows **Tr3**.

The proof of **Tr4** follows the same principles and methods, repeatedly using Lemma 3.5.33 and Corollary 3.5.31. Only the context is slightly more involved and we can leave this without harm to the diligent reader, since in particular the validity of the equations required in the axiom are actually properties of the embeddings and projections into and from the cone. This indicates why they still are satisfied. □

Corollary 3.5.41 *The natural functor $K^{x,y}(\mathcal{A}) \xrightarrow{N} D^{x,y}(\mathcal{A})$ given by sending a complex X to the same complex X and a morphism $\alpha : X \longrightarrow Y$ to the triple (id_X, X, α) is a functor of triangulated categories. The subcategory of objects of $K^{x,y}(\mathcal{A})$ mapped to complexes isomorphic to 0 in $D^{x,y}(\mathcal{A})$ coincides with the full subcategory of $K^{x,y}(\mathcal{A})$ formed by acyclic complexes.*

Moreover, if $\mathcal{C}$ is a triangulated category and if $F : K^{x,y}(\mathcal{A}) \longrightarrow \mathcal{C}$ is a functor that has the following properties:

- *F is a functor of triangulated categories,*
- *F has the property that for every quasi-isomorphism ν in $K^{x,y}(\mathcal{A})$ we get $F(\nu)$ is invertible in $\mathcal{C}$,*

then there is a functor $G : D^{x,y}(\mathcal{A}) \longrightarrow \mathcal{C}$ such that $F = G \circ N$.

Proof The fact that the above defines a functor N is immediate. The statement on the subcategory sent to 0 comes from the fact that if a complex X is sent to 0, then the morphism $X \longrightarrow 0$ can be completed to a triangle which is mapped to the zero triangle $0 \longrightarrow 0 \longrightarrow 0$. Hence the cone of the morphism $X \longrightarrow 0$ is acyclic. But the cone of the morphism $X \longrightarrow 0$ is X. Hence X is acyclic. It is clear that acyclic complexes are isomorphic to 0 in the derived category since the 0 map induces an isomorphism on the homology of the complexes.

If $F : K^{x,y}(\mathcal{A}) \longrightarrow \mathcal{C}$ is a functor of triangulated categories, then set $G(X) := F(X)$, and $G((\nu, Z, \alpha)) := F(\alpha) \circ F(\nu)^{-1}$. It is easy to see that this defines a functor and by definition we get $F = G \circ N$. □

Remark 3.5.42 The concept behind the construction of a derived category out of the homotopy category is called the Verdier localisation. We refer to Gabriel and Zisman [11] for a detailed exposition of this concept. We wanted to construct a category in which the quasi-isomorphisms become invertible. Accordingly we defined the set of morphisms admissible as the first component in a triple (ν, Z, α) in which the cone of ν is acyclic. We were "localising at acyclic complexes".

In Definition 6.9.1 we shall carry out the very same procedure for morphisms with cone being in $K^b(A\text{-}proj)$ for some algebra A. The constructions go through word for word, in particular the construction leading to Lemma 3.5.33. Actually, this is the real reason why we needed Lemma 3.5.32. For details we refer to Sect. 6.9.

Proposition 3.5.43 *Let A be an algebra. Then the functor from Corollary 3.5.41 induces equivalences*

- $D^-(A\text{-}Mod) \simeq K^-(A\text{-}Proj)$
- $D^b(A\text{-}Mod) \simeq K^{-,b}(A\text{-}Proj)$

and if A is a Noetherian K-algebra over a commutative ring K we get

- $D^-(A\text{-}mod) \simeq K^-(A\text{-}proj)$
- $D^b(A\text{-}mod) \simeq K^{-,b}(A\text{-}proj)$

as triangulated categories.

Proof We shall construct for each object X in $D^-(A\text{-}Mod)$ an object P_X in $K^-(A\text{-}Proj)$ with $P_X \simeq X$ in $D^-(A\text{-}Mod)$. We call P_X a projective resolution of X.

Without loss of generality, shifting X if necessary, we may assume that X is isomorphic to a complex whose negative degree components are 0, and such that $H_0(X) \neq 0$. Indeed, if

$$X : \ldots \longrightarrow X_2 \longrightarrow X_1 \longrightarrow X_0 \longrightarrow X_{-1} \longrightarrow \ldots$$

such that $H_i(X) = 0$ for $i < 0$, then $\tau_{\geq 0} X \simeq X$ by Lemma 3.5.21, i.e.

$$\begin{array}{ccccccccc}
\ldots \longrightarrow & X_2 & \longrightarrow & X_1 & \stackrel{d_1}{\longrightarrow} & X_0 & \stackrel{d_0}{\longrightarrow} & X_{-1} & \stackrel{d_{-1}}{\longrightarrow} \ldots \\
& \| & & \| & & \uparrow & & \uparrow & \\
\ldots \longrightarrow & X_2 & \longrightarrow & X_1 & \stackrel{d_1}{\longrightarrow} & \ker(d_0) & \longrightarrow & 0 & \longrightarrow \ldots
\end{array}$$

is a quasi-isomorphism.

Hence assume that

$$X : \ldots \longrightarrow X_2 \longrightarrow X_1 \stackrel{d_1}{\longrightarrow} X_0 \longrightarrow 0 \longrightarrow \ldots$$

where d_1 is not surjective. Let $P_0 \longrightarrow X_0$ be a projective cover of X_0 and form the pullback with d_1:

$$\begin{array}{ccccccc}
\ldots \longrightarrow X_2 & \stackrel{d_2}{\longrightarrow} & X_1 & \stackrel{d_1}{\longrightarrow} & X_0 & \longrightarrow & 0 \longrightarrow \ldots \\
& & \uparrow & & \uparrow & & \uparrow \\
& & Q_1 & \stackrel{\delta_1'}{\longrightarrow} & P_0 & \longrightarrow & 0 \longrightarrow \ldots
\end{array}$$

In general Q_1 will not be projective, but take a projective cover $P_1 \stackrel{\rho_1}{\longrightarrow} Q_1$ and define $\delta_1 := \delta_1' \circ \rho_1$ to get $P_1 \stackrel{\delta_1}{\longrightarrow} P_0$. Then form the following pullback with d_2:

$$\begin{array}{ccccccccc}
\ldots \longrightarrow & X_2 & \stackrel{d_2}{\longrightarrow} & X_1 & \stackrel{d_1}{\longrightarrow} & X_0 & \longrightarrow & 0 & \longrightarrow \ldots \\
& \| & & \uparrow & & \uparrow & & \uparrow & \\
& X_2 & & P_1 & \stackrel{\delta_1}{\longrightarrow} & P_0 & \longrightarrow & 0 & \longrightarrow \ldots \\
& \uparrow & & \uparrow & & & & & \\
& Q_2 & \longrightarrow & \ker \delta_1 & & & & &
\end{array}$$

which gives a complex

$$\begin{array}{ccccccccc}
\dots \xrightarrow{d_3} & X_2 & \xrightarrow{d_2} & X_1 & \xrightarrow{d_1} & X_0 & \longrightarrow & 0 & \longrightarrow \dots \\
 & \uparrow & & \uparrow & & \uparrow & & \uparrow & \\
 & Q_2 & \xrightarrow{\delta_2'} & P_1 & \xrightarrow{\delta_1} & P_0 & \longrightarrow & 0 & \longrightarrow \dots
\end{array}$$

for a certain mapping δ_2'. We continue constructing a projective cover $P_2 \xrightarrow{\rho_2} Q_2$, define $\delta_2 := \delta_2' \circ \rho_2$ and take the pullback along d_3 to obtain

$$\begin{array}{ccccccccccc}
X_3 & \xrightarrow{d_3} & X_2 & \xrightarrow{d_2} & X_1 & \xrightarrow{d_1} & X_0 & \longrightarrow & 0 & \longrightarrow \dots \\
\| & & \uparrow & & \uparrow & & \uparrow & & \uparrow & \\
X_3 & & P_2 & \xrightarrow{\delta_2'} & P_1 & \xrightarrow{\delta_1} & P_0 & \longrightarrow & 0 & \longrightarrow \dots \\
\uparrow & & \uparrow & & & & & & & \\
Q_3 & \xrightarrow{\delta_3'} & \ker(\delta_2) & & & & & & &
\end{array}$$

which then gives a diagram

$$\begin{array}{ccccccccccc}
\dots \longrightarrow & X_3 & \xrightarrow{d_3} & X_2 & \xrightarrow{d_2} & X_1 & \xrightarrow{d_1} & X_0 & \longrightarrow & 0 & \longrightarrow \dots \\
 & \uparrow & & \uparrow & & \uparrow & & \uparrow & & & \\
\dots \longrightarrow & Q_3 & \xrightarrow{\delta_3'} & P_2 & \xrightarrow{\delta_2} & P_1 & \xrightarrow{\delta_1} & P_0 & \longrightarrow & 0 & \longrightarrow \dots
\end{array}$$

Inductively we obtain a complex (P, δ) of projective modules and a morphism of complexes $\varphi : (P, \delta) \longrightarrow (X, d)$

$$\begin{array}{ccccccccccc}
\dots \longrightarrow & X_3 & \xrightarrow{d_3} & X_2 & \xrightarrow{d_2} & X_1 & \xrightarrow{d_1} & X_0 & \longrightarrow & 0 & \longrightarrow \dots \\
 & \uparrow \varphi_3 & & \uparrow \varphi_2 & & \uparrow \varphi_1 & & \uparrow \varphi_0 & & & \\
\dots \longrightarrow & P_3 & \xrightarrow{\delta_3} & P_2 & \xrightarrow{\delta_2} & P_1 & \xrightarrow{\delta_1} & P_0 & \longrightarrow & 0 & \longrightarrow \dots
\end{array}$$

Observe that if X is a complex of finitely generated modules, then (P, δ) is a complex of finitely generated projective modules.

We shall show that the morphism $\varphi : (P, \delta) \longrightarrow (X, d)$ induces an isomorphism $H(\varphi) : H(P) \longrightarrow H(X)$.

First φ_0 induces an isomorphism

$$H_0(P) = \text{coker}(\delta_1) \longrightarrow \text{coker}(d_1) = H_0(X).$$

Indeed, since $d_1 \circ \varphi_1 = \varphi_0 \circ \delta_1$, the mapping φ_0 induces a morphism $\text{coker}(\delta_1) \longrightarrow \text{coker}(d_1)$. Since φ_0 is surjective and since $X_0 \longrightarrow H_0(X)$ is surjective $H_0(\varphi)$ is also surjective. By Lemma 1.8.27 and the snake lemma we obtain that $H_0(\varphi)$ is an isomorphism. Indeed,

$$\begin{array}{ccc} X_1 & \longrightarrow & X_0 \\ \uparrow & & \uparrow \\ Q_1 & \longrightarrow & P_0 \end{array}$$

is a pullback diagram and therefore Lemma 1.8.27 implies that the morphisms $Q_1 \longrightarrow X_1$ induce an isomorphism on the kernels K of $X_1 \longrightarrow X_0$ and $Q_1 \longrightarrow Q_0$. Hence we obtain a commutative diagram

$$\begin{array}{ccc} X_1/K & \hookrightarrow & X_0 \\ \uparrow & & \uparrow \\ Q_1/K & \hookrightarrow & P_0 \end{array}$$

Lemma 1.8.27 implies that the morphism $Q_1 \longrightarrow P_0$ induces an isomorphism on the kernels $Q_1 \longrightarrow X_1$ and $P_0 \longrightarrow X_0$ and therefore the induced morphism on the kernels of the vertical morphisms of

$$\begin{array}{ccc} \mathrm{im}(d_1) & \hookrightarrow & X_0 \\ \uparrow & & \uparrow \\ \mathrm{im}(\delta_1) & \hookrightarrow & P_0 \end{array}$$

is surjective. Since the image of δ_1 equals the image of $Q_1 \longrightarrow P_0$ we obtain a commutative diagram

$$\begin{array}{ccccccccc} & & 0 & & & & & & \\ & & \uparrow & & & & & & \\ & & C''' & \longrightarrow & 0 & \longrightarrow & 0 & & \\ & & \uparrow & & \uparrow & & \uparrow & & \\ 0 & \longrightarrow & \mathrm{im}(d_1) & \longrightarrow & X_0 & \longrightarrow & H_0(X) & \longrightarrow & 0 \\ & & \uparrow & & \uparrow & & \uparrow & & \\ 0 & \longrightarrow & \mathrm{im}(\delta_1) & \longrightarrow & P_0 & \longrightarrow & H_0(P) & \longrightarrow & 0 \\ & & \uparrow & & \uparrow & & \uparrow & & \\ 0 & \longrightarrow & C & \longrightarrow & C' & \longrightarrow & C'' & & \\ & & \uparrow & & \uparrow & & \uparrow & & \\ & & 0 & & 0 & & 0 & & \end{array}$$

with exact rows and columns. Since by construction

$$\begin{array}{ccc} X_1 & \longrightarrow & X_0 \\ \uparrow & & \uparrow \\ Q_1 & \longrightarrow & P_0 \end{array}$$

is a pullback and $P_1 \longrightarrow Q_1$ is a projective cover mapping, Lemma 1.8.27 shows that on the cokernels $H_0(\varphi) : H_0(P) \longrightarrow H_0(X)$ is an isomorphism, and hence $C'' = 0$. We have seen that $C \longrightarrow C'$ is surjective and therefore $C'' = 0$. The snake lemma implies $C''' = 0$.

Suppose by induction that $H_s(\varphi)$ is an isomorphism for all $s < n$. Then consider the complexes and morphisms of complexes

$$\begin{array}{ccccccccccc}
\dots \longrightarrow X_{n+3} & \overset{d_{n+3}}{\longrightarrow} & X_{n+2} & \overset{d_{n+2}}{\longrightarrow} & X_{n+1} & \overset{d_{n+1}}{\longrightarrow} & \ker(d_n) & \longrightarrow 0 \longrightarrow \dots \\
\uparrow \varphi_{n+3} & & \uparrow \varphi_{n+2} & & \uparrow \varphi_{n+1} & & \uparrow \varphi_n & \\
\dots \longrightarrow P_{n+3} & \overset{\delta_{n+3}}{\longrightarrow} & P_{n+2} & \overset{\delta_{n+2}}{\longrightarrow} & P_{n+1} & \overset{\delta_{n+1}}{\longrightarrow} & \ker(\delta_n) & \longrightarrow 0 \longrightarrow \dots
\end{array}$$

We may apply the discussion of the case $H_0(\varphi)$ studied above to this morphism of complexes to obtain that $H_n(\varphi)$ is an isomorphism. Hence the natural functor

$$K^-(A\text{-}Proj) \longrightarrow D^-(A\text{-}Mod)$$

is dense.

We shall need to show that the functor is fully faithful. For this purpose we use the following lemma.

Lemma 3.5.44 *Any quasi-isomorphism ν from a right bounded complex of projective A-modules to a right bounded complex of projective A-modules is invertible in the homotopy category of complexes of A-modules.*

Proof Let $\nu : (X, d^X) \longrightarrow (Y, d^Y)$ be a quasi-isomorphism in the category $K^-(A\text{-}Proj)$. By Lemma 3.5.32 the cone $C(\nu)$ is acyclic. Any right bounded acyclic complex of projective modules is zero-homotopic by Lemma 3.5.23. The structure of the triangulated category gives a distinguished triangle

$$X \longrightarrow Y \longrightarrow C(\nu) \longrightarrow X[1]$$

in the homotopy category of complexes with projective objects. Since $C(\nu)$ is an exact complex of projective modules, the map

$$\psi : C(\nu) \longrightarrow X[1]$$

is zero homotopic by Lemma 3.5.19. Hence there is a homotopy

$$h : C(\nu) \longrightarrow X[1]$$

of degree 1 so that

$$h \circ d^C + d^X \circ h = \psi.$$

Recall that $C(\nu) = X[1] \oplus Y$ as modules and

$$d^C = \begin{pmatrix} -d^X & 0 \\ \nu & d^Y \end{pmatrix}$$

such that with $h = (f\ g)$ for $f : X[1] \longrightarrow X[2]$ and $g : Y \longrightarrow X$ we obtain

$$(id_X\ 0) = \psi = (f\ g) \circ \begin{pmatrix} -d^X & 0 \\ \nu & d^Y \end{pmatrix} + (-d^X) \circ (f\ g).$$

This gives

$$(id_X\ 0) = (d^X \circ f - f \circ d^X + \nu \circ g)$$

and so g is an inverse to ν in the homotopy category. □

We continue and finish the proof of Proposition 3.5.43. Now, let X and Y be right bounded complexes of projective A-modules and let (ν, Z, α) be a triple representing a homomorphism in the derived category. Using the first part of the proof of Proposition 3.5.43 we may replace Z by its projective resolution, and so X, Y and Z are all right bounded complexes of projective A-modules. Then, by Lemma 3.5.44, ν is invertible and $\alpha \circ \nu^{-1}$ is a morphism in the homotopy category of projective A-modules. Therefore the functor $K^-(A\text{-}Proj) \longrightarrow D^-(A\text{-}Mod)$ is fully faithful and likewise for its restrictions to $K^{-,b}(A\text{-}Proj)$, $K^-(A\text{-}proj)$, $K^{-,b}(A\text{-}proj)$, and related subcategories. This proves the statement. □

Remark 3.5.45 Observe that we used in the proof of Proposition 3.5.43 only the fact that projective covers exist in the module category in the sense that every object is the cokernel of a morphism between projectives. An abelian category in which every object is the cokernel of morphisms between projective modules is said to have *sufficiently many projectives*. Analogously an abelian category in which every object is the kernel of morphisms between injective modules is said to have *sufficiently many injectives*. We use the expression "enough projectives/injectives" synonymously to "sufficiently many projectives/injective".

Formally, if an abelian category has enough injectives, one may prove that

$$D^+(\mathcal{A}) \simeq K^+(\mathcal{A} - Inj)$$

by replacing a complex by its injective resolution, defined completely dually to the projective resolution used in Proposition 3.5.43. Moreover, the dual version of Lemma 3.5.44 is true in this case.

Lemma 3.5.46 *Any quasi-isomorphism ν from a left bounded complex of A-modules to a left bounded complex of injective A-modules is invertible in the homotopy category of complexes of A-modules.*

The proof is dual to the proof of Lemma 3.5.44. □

Corollary 3.5.47 *Let $\mathcal{A}$ be an abelian category with enough projective objects. Let $pX \to X$ be a quasi-isomorphism in the homotopy category $K^-(\mathcal{A})$ so that pX is an object in the homotopy category $K^-(\mathcal{A}\text{-}Proj)$ of right bounded complexes of projective objects. Then there is a natural isomorphism $Hom_{D(\mathcal{A})}(X, Y) \simeq Hom_{K^-(\mathcal{A}\text{-}Proj)}(pX, Y)$ for every right bounded complex Y of objects in $\mathcal{A}$.*

Let $\mathcal{A}$ be an abelian category with enough injective objects. Let $Y \to \iota Y$ be a quasi-isomorphism in the homotopy category $K^+(\mathcal{A})$ so that ιY is an object in $K^+(\mathcal{A} - Inj)$ of left bounded complex of injective objects. Then there is a natural isomorphism $Hom_{D(\mathcal{A})}(X, Y) \simeq Hom_{K^+(\mathcal{A}-Inj)}(X, \iota Y)$ for every left bounded complex X of objects in $\mathcal{A}$.

Proof Let (ν, α) be a homomorphism from X to Y in $D^-(A)$, where $\nu : Z \to X$ is a quasi-isomorphism and $\alpha : Z \to Y$ is a morphism in the homotopy category, and suppose that $pX = X$. Then we obtain a quasi-isomorphism $\mu : pZ \to Z$ and get that $(\nu \circ \mu, \alpha \circ \mu)$ covers (ν, α). But by Lemma 3.5.44 we get that $\nu \circ \mu$ is actually invertible in the homotopy category. Hence (ν, α) is covered by $(id, \alpha \circ \mu \circ (\nu \circ \mu)^{-1})$. This proves the corollary for the first case. The second case is dual using Lemma 3.5.46. □

Remark 3.5.48 We may also define the intelligent truncation with the properties from Lemma 3.5.21 for derived categories, using Proposition 3.5.43.

Lemma 3.5.49 *Let A be a K-algebra for a commutative ring K and let $D^b(A\text{-}Mod)$ be the derived category of $A\text{-}Mod$. Then $A\text{-}Mod$ is a full subcategory of $D^b(A\text{-}Mod)$.*

Proof Proposition 3.5.18 shows that $A\text{-}Mod$ is a full subcategory of $K^-(A\text{-}Proj)$ by identifying a module with its projective resolution. Moreover $D^b(A\text{-}Mod) \simeq K^{-,b}(A\text{-}Proj)$ and the projective resolution of a module has bounded homology. This proves the statement. □

A nice application is the following.

Lemma 3.5.50 (Horseshoe lemma) *Let A be an algebra and let*

$$0 \longrightarrow L \longrightarrow M \longrightarrow N \longrightarrow 0$$

be a short exact sequence of A-modules. Let $P_L^\bullet$ be a projective resolution of L, and let $P_N^\bullet$ be a projective resolution of N. Then there is a projective resolution $P_M^\bullet$ of M such that $P_M^i = P_N^i \oplus P_L^i$, i.e. the homogeneous component of degree i in the projective resolution of M is the direct sum of the homogeneous components of degree i in the projective resolutions of L and of N.

Proof The short exact sequence gives rise to a distinguished triangle

$$P_N[-1] \xrightarrow{\delta} P_L \longrightarrow P_M \longrightarrow P_N$$

and so P_M is isomorphic to the cone of δ. The homogeneous components of the cone are exactly as claimed. □

Sometimes it is useful to link exact sequences in the complex category and triangles in the homotopy category.

Proposition 3.5.51 *Let*

$$0 \longrightarrow X \xrightarrow{\alpha} Y \xrightarrow{\beta} Z \longrightarrow 0$$

be an exact sequence of right bounded complexes of projective modules over an algebra A. Then

$$X \xrightarrow{\alpha} Y \xrightarrow{\beta} Z \longrightarrow X[1]$$

is a distinguished triangle in $K^-(A\text{-}Proj)$.

Proof Since β is surjective and since each homogeneous component of Y is a projective module, the sequence is 'locally' split, in the sense that it splits in each degree with a map not compatible with the differentials in general. Hence $Y \simeq Z \oplus X$ as a $\mathbb{Z}$-graded module, α is transformed into the canonical injection and β is transformed into the canonical surjection. But then, assuming $Y = Z \oplus X$ and α, resp. β are the canonical injections, resp. projections, we obtain that the differential d^Y is of the form

$$d_n^Y = \begin{pmatrix} d_n^Z & 0 \\ \rho_n & d_n^X \end{pmatrix}$$

where $\rho : Z \longrightarrow X[1]$ is a graded morphism. But since $\left(d^Y\right)^2 = 0$, we have a homomorphism of complexes. We apply TR2 to conclude the statement. □

We have seen in Proposition 3.4.11 that for every object X of the triangulated category $\mathcal{T}$ the functor

$$Hom_{\mathcal{T}}(X, -) : \mathcal{T} \longrightarrow \mathbb{Z}\text{-}Mod$$

is homological and

$$Hom_{\mathcal{T}}(-, X) : \mathcal{T} \longrightarrow \mathbb{Z}\text{-}Mod$$

is cohomological.

The reason why we call these functors homological comes from the following corollary. This corollary also gives a different approach to the statement of Proposition 3.5.29.

Corollary 3.5.52 *Let K be a commutative ring and let A be a K-algebra. Then the homology functor $H_n : D(A\text{-}Mod) \to A\text{-}Mod$ is homological.*

Proof Indeed,

$$Hom_{D(A)}(A[n], -) = H_n(-)$$

as is seen as follows: First, we may look at morphisms up to homotopy since A is projective. Then we consider the morphisms of complexes

$$\begin{array}{ccccccc} \dots \longrightarrow & X_{n+1} & \stackrel{\partial_{n+1}}{\longrightarrow} & X_n & \stackrel{\partial_n}{\longrightarrow} & X_{n-1} & \longrightarrow \dots \\ & \uparrow & & \uparrow \varphi & & \uparrow & \\ & 0 & \longrightarrow & A & \longrightarrow & 0 & \end{array}$$

and see that φ has to have image in $\ker(\partial_n)$ and the set of possible φ coincides with $\ker(\partial_n)$ since φ is A-linear and hence is fixed by the image of $1 \in A$. Moreover, we consider morphisms up to homotopy, which is the same as considering $\ker(\partial_n)/\mathrm{im}(\partial_{n+1})$. □

Remark 3.5.53 Since we did not use in the proof of Proposition 3.5.43 that the categories are triangulated, and since the triangles in the derived category are precisely the images of the triangles in the homotopy category, Proposition 3.5.43 and Proposition 3.5.25 actually show directly that $D^x(A\text{-}Mod)$ and $D^x(A\text{-}mod)$ are triangulated for each $x \in \{-, b\}$. However Proposition 3.5.40 is more general since it shows that the derived category of any abelian category is triangulated.

Remark 3.5.54 It can be shown that there is a projective resolution of unbounded complexes as well. This is a result due to Spaltenstein [12].

3.6 An Application and a Useful Tool: Hochschild Resolutions

3.6.1 The Bar Resolution

Let K be a commutative ring. We shall give an explicit free resolution over a K-algebra A which is far from minimal in general, but which is completely general, explicit and useful for many purposes. The resolution is called the bar-resolution. We shall produce a free resolution $\mathbb{B}A$ of A as an $A \otimes_K A^{op}$-module. To shorten the notation we write $\otimes$ instead of $\otimes_K$ when no confusion can occur.

The degree n component of $\mathbb{B}A$ is

$$(\mathbb{B}A)_n := \underbrace{A \otimes A \otimes \cdots \otimes A}_{n+2 \text{ factors}}$$

so that

$$\mathbb{B}A : \quad \cdots \stackrel{\partial_3}{\longrightarrow} A \otimes A \otimes A \otimes A \stackrel{\partial_2}{\longrightarrow} A \otimes A \otimes A \stackrel{\partial_1}{\longrightarrow} A \otimes A$$

with

$$\begin{aligned} \partial_n(a_0 \otimes \cdots \otimes a_{n+1}) \\ &:= a_0 a_1 \otimes a_2 \otimes \cdots \otimes a_{n+1} \\ &\quad + \sum_{i=1}^{n-1} (-1)^i a_0 \otimes \cdots \otimes a_{i-1} \otimes a_i a_{i+1} \otimes a_{i+2} \otimes \cdots \otimes a_{n+1} \\ &\quad + (-1)^n a_0 \otimes \cdots \otimes a_{n-1} \otimes a_n a_{n+1}. \end{aligned}$$

Proposition 3.6.1 *Let K be a commutative ring and let A be a K-projective K-algebra. Then $(\mathbb{B}A, \partial_\bullet)$ is a free resolution of the $A \otimes_K A^{op}$ module A.*

Proof First,

$$\text{coker}(\partial_1) = \text{coker}(a_0 \otimes a_1 \otimes a_2 \mapsto a_0a_1 \otimes a_2 - a_0 \otimes a_1a_2)$$

and the surjective map

$$\begin{array}{ccc} A \otimes A & \xrightarrow{\mu} & A \\ a_0 \otimes a_1 & \mapsto & a_0a_1 \end{array}$$

composes to 0 with ∂_1. Define the "augmented" complex $(\widehat{\mathbb{B}A}, \widehat{\partial})$ which is $\mathbb{B}A$ in non-negative degrees and which has a degree -1 component $(\widehat{\mathbb{B}A})_{-1} := A$ and which has $\partial_0 := \mu$. We shall show that $(\widehat{\mathbb{B}A}, \widehat{\partial})$ is exact. In order to do so define a K-linear map (actually it is A^{op}-linear, but for the argument the K-linearity is sufficient)

$$h_n : (\mathbb{B}A)_{n-1} \longrightarrow (\mathbb{B}A)_n$$

by putting

$$h_n(a_0 \otimes \cdots \otimes a_n) := 1 \otimes a_0 \otimes \cdots \otimes a_n.$$

Then

$$(\partial_0 \circ h_0)(a_0) = \partial_0(1 \otimes a_0) = a_0$$

and so

$$\partial_0 \circ h_0 = id_{\mathbb{B}A_{-1}}.$$

Further

$$\begin{aligned}
(h_n \circ \partial_n + &\partial_{n+1} \circ h_{n+1})(a_0 \otimes \cdots \otimes a_{n+1}) \\
&= h_n(a_0a_1 \otimes a_2 \otimes \cdots \otimes a_{n+1}) \\
&\quad + \sum_{i=1}^{n-1}(-1)^i h_n(a_0 \otimes \cdots \otimes a_{i-1} \otimes a_ia_{i+1} \otimes a_{i+1} \otimes \cdots \otimes a_{n+1}) \\
&\quad + (-1)^n h_n(a_0 \otimes \cdots \otimes a_{n-1} \otimes a_na_{n+1}) \\
&\quad + \partial_{n+1}(1 \otimes a_0 \otimes \cdots \otimes a_{n+1}) \\
&= 1 \otimes a_0a_1 \otimes a_2 \otimes \cdots \otimes a_{n+1} \\
&\quad + \sum_{i=1}^{n-1}(-1)^i(1 \otimes a_0 \otimes \cdots \otimes a_{i-1} \otimes a_ia_{i+1} \otimes a_{i+1} \otimes \cdots \otimes a_{n+1}) \\
&\quad + a_0 \otimes \cdots \otimes a_{n+1} \\
&\quad + \sum_{i=1}^{n}(-1)^{i+1}(1 \otimes a_0 \otimes \cdots \otimes a_{i-1} \otimes a_ia_{i+1} \otimes a_{i+2} \otimes \cdots \otimes a_{n+1}) \\
&= a_0 \otimes \cdots \otimes a_{n+1}
\end{aligned}$$

and therefore

$$h_n \circ \partial_n + \partial_{n+1} \circ h_{n+1} = id_{\widehat{\mathbb{B}A}}.$$

This implies that in the category of vector spaces $(\widehat{\mathbb{B}A}, \widehat{\partial})$ is homotopy equivalent to the zero complex. Hence, in particular, the complex has 0 homology as vector spaces, and therefore also as $A \otimes A^{op}$-modules. □

Corollary 3.6.2 *Let K be a commutative ring and let A be a K-projective K-algebra. Then for every A-module L the complex $(\mathbb{B}A \otimes_A L, \partial_\bullet \otimes_A id_L)$ is a projective resolution of L as an A-module.*

Proof Indeed, the completed resolution $(\widehat{\mathbb{B}A}, \widehat{\partial})$ is 0-homotopic as a complex of A-modules, as we observed in the proof of Proposition 3.6.1 above. Therefore $(\widehat{\mathbb{B}A} \otimes_A L, \widehat{\partial} \otimes id_L)$ is still exact. □

Remark 3.6.3 I thank Guodong Zhou for communicating this argument to me.

Definition 3.6.4 Let K be a commutative ring, let A be a K-projective K-algebra and let L be an A-module. Then $(\mathbb{B}A \otimes_A L, \partial_\bullet \otimes_A id_L)$ is the *bar resolution* of L. The resolution $(\mathbb{B}A, \partial_\bullet)$ is the *bar resolution* of A. Let M be an $A \otimes_K A^{op}$-module. The degree n cycles of $(Hom_{A\otimes_K A^{op}}((\mathbb{B}A, \partial_\bullet), M))$ are called *degree n Hochschild cocycles*, and the degree n boundaries of $(Hom_{A\otimes_K A^{op}}((\mathbb{B}A, \partial_\bullet), M))$ are called *degree n Hochschild coboundaries*.

Remark 3.6.5 If (Y, d_Y) is a complex, then it is possible to replace each homogeneous component by its bar resolution and lift the differential to a complex of bar resolutions. The differential of this complex of complexes is then $id_{\mathbb{B}A} \otimes d^Y$ and the total complex of this double complex is quasi-isomorphic to Y. It is an easy, though tedious, exercise to write down all the homogeneous components and the differential of the total complex explicitly.

We mention another important feature of this construction. For this we need some notation.

Definition 3.6.6 Let A be an algebra and let M be a right A-module, and let N be a left A-module. Then let $P^\bullet \to N$ be a projective resolution of N as an A-module. We form the complex $M \otimes_A P^\bullet$ and define $Tor_n^A(M, N)$ to be the degree n homology of the resulting complex.

Since two projective resolutions differ by zero homotopic direct factors, it is immediate to see that $Tor_n^A(M, N)$ does not depend on the projective resolution. A more general and less ad-hoc definition will be given in Definition 3.7.5 below.

Definition 3.6.7 Let K be a field and let A be a finite dimensional K-algebra. The degree n *Hochschild cohomology* of A is

$$HH^n(A) := Ext^n_{A\otimes_K A^{op}}(A, A).$$

The degree *n Hochschild homology* of A is

$$HH_n(A) := Tor_n^{A\otimes_K A^{op}}(A, A).$$

Since the bar resolution is a resolution of A as $A \otimes_K A^{op}$-modules we can use the bar resolution to give an explicit description of the Hochschild cohomology and the Hochschild homology, and in some rare cases compute it explicitly. Hochschild homology and Hochschild cohomology are very important invariants of an algebra. If A is not finite dimensional over the field K, then the Hochschild homology and cohomology is defined differently.

Note that by definition

$$HH_0(A) = A/[A, A] \text{ and } HH^0(A) = Z(A).$$

This follows directly from the definition, or perhaps more directly by the first terms of the bar resolution.

3.6.2 Second Hochschild Cohomology, Extensions and The Wedderburn-Malcev Theorem

We shall give an interpretation of the second Hochschild cohomology in terms of extensions and apply this result to the Wedderburn-Malcev Theorem which identifies the radical quotient of an algebra A inside A.

Proposition 3.6.8 *Let K be a field and let $\overline{A}$ be a K-algebra. Then for every $\overline{A} \otimes_K \overline{A}^{op}$-module M the group $Ext^2_{\overline{A}\otimes_K \overline{A}^{op}}(\overline{A}, M)$ parameterises equivalence classes of short exact sequences*

$$0 \longrightarrow M \longrightarrow A \longrightarrow \overline{A} \longrightarrow 0$$

where A is a K-algebra, and where M is a two-sided ideal of A with $M \cdot M = 0$. The equivalence relation is defined as usual analogously to the one used in Lemma 1.8.17.

For a Hochschild 2-cocycle g the algebra A is defined by $M \oplus \overline{A}$ as a K-module, and multiplication $(m_1, a_1) \cdot (m_2, a_2) := (m_1 + m_2 + g(a_1, a_2), a_1 a_2)$. Under this construction g is a Hochschild 2-coboundary if and only if the extension splits.

Proof Let

$$f : \overline{A} \otimes_K \overline{A} \otimes_K \overline{A} \otimes_K \overline{A} \longrightarrow M$$

be an $\overline{A} \otimes_K \overline{A}^{op}$-linear mapping so that $f \circ \partial_3 = 0$ such that f then gives an element in $Ext^2_{\overline{A}\otimes_K \overline{A}}(\overline{A}, M)$. The condition $f \circ \partial_3 = 0$ is equivalent to

$$\begin{aligned} 0 &= f(ab \otimes c \otimes d \otimes e - a \otimes bc \otimes d \otimes e + a \otimes b \otimes cd \otimes e - a \otimes b \otimes c \otimes de) \\ &= ab \cdot f(1 \otimes c \otimes d \otimes 1) \cdot e - a \cdot f(1 \otimes bc \otimes d \otimes 1) \cdot e \\ &\quad + a \cdot f(1 \otimes b \otimes cd \otimes 1) \cdot e - a \cdot f(1 \otimes b \otimes c \otimes 1) \cdot de \\ &= ab \cdot g(c \otimes d) \cdot e - a \cdot g(bc \otimes d) \cdot e + a \cdot g(b \otimes cd) \cdot e - a \cdot g(b \otimes c) \cdot de \end{aligned}$$

for $g(b \otimes c) := f(1 \otimes b \otimes c \otimes 1)$. We see that we may put $a = e = 1$ and ask the mapping g to be K-linear. We obtain that a K-linear map $g : \overline{A}^{\otimes 2} \longrightarrow M$ gives an element in $Ext^2_{\overline{A} \otimes_K \overline{A}}(\overline{A}, M)$ if and only if

$$(*) \quad b \cdot g(c \otimes d) - g(bc \otimes d) + g(b \otimes cd) - g(b \otimes c) \cdot d = 0$$

for all $b, c, d \in \overline{A}$. We say that g is a 2-cocycle of $\overline{A}$ with values in M in this case. Such a mapping g gives the 0 element in $Ext^2_{\overline{A} \otimes_K \overline{A}}(\overline{A}, M)$ if and only if there is an $\overline{A} \otimes \overline{A}^{op}$-linear mapping $h : \overline{A} \otimes_K \overline{A} \otimes_K \overline{A} \longrightarrow M$ so that

$$ag(b \otimes c)d = h \circ \partial_2(a \otimes b \otimes c \otimes d)$$

for all $a, b, c, d \in \overline{A}$. Again, we may consider the case $a = d = 1$ and K-linear maps h. Hence $g = 0$ in $Ext^2_{\overline{A} \otimes_K \overline{A}}(\overline{A}, M)$ if and only if there is an $h : A \longrightarrow M$ so that

$$g(b \otimes c) = bh(c) - h(bc) + h(b)c$$

for all $b, c \in \overline{A}$. In this case we say that g is a Hochschild 2-coboundary.

We shall see that $Ext^2_{\overline{A} \otimes_K \overline{A}}(\overline{A}, M)$ parameterises short exact sequences of $\overline{A} \otimes \overline{A}$-modules, in the same way as $H^2(G, M)$ parameterises short exact sequences of groups, as explained in Sect. 1.8.4.

On the K-vector space $M \oplus \overline{A}$ we define a multiplication by

$$(m, a) \cdot (n, b) := (mb + an + g(a \otimes b), ab)$$

for all $a, b \in \overline{A}$ and $m, n \in M$. We check if this is associative and will see that g satisfies the condition $(*)$ of being a Hochschild 2-cocycle if and only if the multiplication is associative. Indeed,

$$\begin{aligned} ((m, a) \cdot (n, b)) \cdot (\ell, c) &= (mb + an + g(a \otimes b), ab) \cdot (\ell, c) \\ &= (mbc + anc + ab\ell + g(a \otimes b)c + g(ab \otimes c), abc) \\ &= (mbc + anc + ag(b \otimes c) + g(a, bc), abc) \text{by eqtn } (*) \\ &= (m, a) \cdot (nc + b\ell + g(b \otimes c), bc) \\ &= (m, a) \cdot ((n, b) \cdot (\ell, c)) \end{aligned}$$

for all $m, n, \ell \in M$ and $a, b, c \in \overline{A}$. Denote by $M \oplus_g \overline{A}$ the algebra obtained by the 2-cocycle g.

Now, the natural projection $M \oplus_g \overline{A} \longrightarrow\!\!\!\!\rightarrow \overline{A}$ onto the second component admits a splitting $\overline{A} \longrightarrow M \oplus_g \overline{A}$ if and only if g is a 2-coboundary. Indeed, suppose g is a coboundary defined by h, then

$$\begin{array}{rcl} \overline{A} & \stackrel{\varphi}{\longrightarrow} & M \oplus_g \overline{A} \\ a & \mapsto & (-h(a), a) \end{array}$$

is a splitting. Indeed,

$$\begin{aligned} \varphi(a) \cdot \varphi(b) &= (-h(a), a) \cdot (-h(b), b) = (-h(a)b - ah(b) + g(a \otimes b), ab) \\ &= (-h(ab), ab) = \varphi(ab) \end{aligned}$$

for all $a, b \in \overline{A}$. Conversely, any splitting of the projection onto the second component is of the form φ for some h. The rest is immediate.

Observe that M is an ideal of $M \oplus_g \overline{A}$ with $M^2 = 0$. □

In the proof of Proposition 3.6.8 we used a K-linear splitting $\overline{A} \stackrel{\rho}{\to} A$ of the natural projection $A \to \overline{A}$. An easy verification shows that a Hochschild 1-cocycle is a K-linear mapping $\overline{A} \stackrel{\delta}{\to} M$ such that $\delta(a_1a_2) = a_1\delta(a_2) + \delta(a_1)a_2$. If one chooses a Hochschild 1-cocycle δ, then $\rho + \delta$ is again a splitting giving the same Hochschild 2-cycle for the extension.

Theorem 3.6.9 (Wedderburn-Malcev) *Let K be an algebraically closed field and let A be a finite dimensional K-algebra. Then there is a semisimple subalgebra $S \simeq A/\mathrm{rad}(A)$ of A such that $A \simeq S \oplus \mathrm{rad}(A)$, where the direct sum is a direct sum of vector spaces, and where S is a subalgebra of A.*

Proof We first claim that $A/\mathrm{rad}(A) \otimes_K A/\mathrm{rad}(A)$ is semisimple. Indeed, $A/\mathrm{rad}(A) \simeq \prod_{i=1}^{s} Mat_{n_i}(K)$ by Wedderburn's theorem 1.4.16. But then we see that

$$Mat_{n_i}(K) \otimes_K Mat_{n_j}(K) \simeq Mat_{n_j}(Mat_{n_i}(K) \otimes_K K) \simeq Mat_{n_j \cdot n_i}(K).$$

Now define $\overline{A} := A/\mathrm{rad}(A)$ and observe that $Ext^2_{\overline{A} \otimes_K \overline{A}^{op}}(\overline{A}, M) = 0$ for all $\overline{A} \otimes_K \overline{A}^{op}$-modules M.

Now, $\mathrm{rad}(A)$ is a nilpotent ideal of A. Suppose $\mathrm{rad}(A)^n = 0$ and suppose n is minimal with respect to this property. We proceed by induction on n. If $n = 1$, then A is semisimple, and there is nothing to prove.

Suppose the result proven for $n - 1$. Then consider $\check{A} := A/\mathrm{rad}^{n-1}(A)$ and observe that $\mathrm{rad}(A)/\mathrm{rad}^{n-1}(A) = \mathrm{rad}(\check{A})$. Hence

$$\overline{A} = A/\mathrm{rad}(A) = \check{A}/\mathrm{rad}(\check{A}).$$

By the induction hypothesis we have that

$$\check{A} = \check{S} \oplus \mathrm{rad}(\check{A})$$

for some algebra $\check{S} \simeq \overline{A}$.

Consider the projection

$$A \xrightarrow{\pi} \check{A}.$$

Then

$$0 \longrightarrow \mathrm{rad}^{n-1}(A) \longrightarrow \pi^{-1}(\check{S}) \longrightarrow \check{S} \longrightarrow 0$$

is an exact sequence, where the right-hand mapping is a morphism of algebras and where the kernel $\mathrm{rad}^{n-1}(A)$ is nilpotent of degree 2. Since $\check{S} \simeq \overline{A}$ we may again use the fact that $Ext^2_{\overline{A}\otimes\overline{A}^{op}}(\overline{A}, \mathrm{rad}^{n-1}(A)) = 0$ to conclude by Proposition 3.6.8 that $\pi^{-1}(\check{S})$ contains a subalgebra S isomorphic to $\overline{A}$ complementing $\mathrm{rad}^{n-1}(A)$. Hence S complements $\mathrm{rad}(A)$ in A. This proves the result. □

Remark 3.6.10 The hypothesis that K is algebraically closed is too strong. It is sufficient to assume that $A/\mathrm{rad}(A)$ is a separable algebra, i.e. $L \otimes_K A/\mathrm{rad}(A)$ is semisimple for all extension fields L of K.

Similar arguments and a similar statement can be used for the following result. The proof closely follows Külshammer [13].

Recall from Definition 2.1.9 that an algebra A contained as subalgebra in the algebra B is said to be a separable extension if the multiplication mapping $B \otimes_A B \longrightarrow B$ is split as a $B \otimes_A B^{op}$-module mapping.

We extend the definition of a separable extension in the following way.

Definition 3.6.11 Let $\beta : A \longrightarrow B$ be a homomorphism of K-algebras. Then we say that β is a *separable extension* of K-algebras if the subalgebra $\beta(A)$ of B is separable in the sense of Definition 2.1.9.

Note that if $\beta : A \longrightarrow B$ is an algebra homomorphism, then B is an A-A-bimodule via the action of $a \in A$ on $b \in B$ by multiplication with $\beta(a)$.

Proposition 3.6.12 *Let $\beta : A \longrightarrow B$ be a separable extension of K-algebras and let $\gamma : A \longrightarrow C$ be a homomorphism of K-algebras. Let I be a nilpotent two-sided ideal of C, denote by $\pi : C \longrightarrow C/I$ the natural homomorphism and let $\rho : B \longrightarrow C/I$ be a homomorphism of K-algebras so that $\rho \circ \beta = \pi \circ \gamma$. Suppose that there is a homomorphism $\tau_0 : B \longrightarrow C$ of A-A-bimodules, then there is a homomorphism of K-algebras $\tau : B \longrightarrow C$ satisfying $\tau \circ \beta = \gamma$.*

Proof We first suppose that $I^2 = 0$.

Since $A \longrightarrow B$ is separable, by Proposition 2.1.8 there is an $\omega \in B \otimes_A B$ such that $b\omega = \omega b$ for all $b \in B$ and such that for the multiplication map $\mu_B : B \otimes_A B \longrightarrow B$ we get $\mu_B(\omega) = 1_B$.

Put

$$\begin{array}{rcl} B\otimes_A B & \xrightarrow{\theta} & I \\ x\otimes y & \mapsto & \tau_0(xy)-\tau_0(x)\tau_0(y) \end{array} \qquad \begin{array}{rcl} B\otimes_A B\otimes_A B & \xrightarrow{\lambda} & I \\ x\otimes y\otimes z & \mapsto & \theta(x\otimes y)\tau_0(z) \end{array}$$

and

$$\begin{array}{rcl} B & \xrightarrow{\eta} & I \\ x & \mapsto & \lambda(x\otimes\omega). \end{array}$$

Then

$$\begin{aligned}
\tau_0(x)\theta(y\otimes z) &+ \theta(x\otimes yz) - \theta(xy\otimes z) - \theta(x\otimes y)\tau_0(z) \\
&= \tau_0(x)(\tau_0(yz)-\tau_0(y)\tau_0(z)) + \tau_0(xyz) - \tau_0(x)\tau_0(yz) \\
&\quad - (\tau_0(xyz)-\tau_0(xy)\tau_0(z)) - (\tau_0(xy)-\tau_0(x)\tau_0(y))\tau_0(z) \\
&= \tau_0(x)\tau_0(yz) - \tau_0(x)\tau_0(y)\tau_0(z) + \tau_0(xyz) - \tau_0(x)\tau_0(yz) \\
&\quad - \tau_0(xyz) + \tau_0(xy)\tau_0(z) - \tau_0(xy)\tau_0(z) + \tau_0(x)\tau_0(y)\tau_0(z) \\
&= 0
\end{aligned}$$

which can be written as

$$(\dagger)\quad \tau_0(x)\theta(y\otimes z) + \theta(x\otimes yz) = \theta(xy\otimes z) + \theta(x\otimes y)\tau_0(z).$$

Now, write $\omega = \sum_{j=1}^{\ell} x_j\otimes y_j$ for $x_j, y_j\in B$ for all $j\in\{1,\ldots,\ell\}$. Then

$$\eta(x) = \lambda(x\otimes\omega) = \sum_{j=1}^{\ell}\theta(x\otimes x_j)\tau_0(y_j)$$

and hence,

$$\begin{aligned}
\tau_0(x)\eta(y) &- \eta(xy) + \eta(x)\tau_0(y) \\
&= \tau_0(x)\lambda(y\otimes\omega) - \lambda(xy\otimes\omega) + \lambda(x\otimes\omega)\tau_0(y) \\
&= \sum_{j=1}^{\ell}\big(\tau_0(x)\theta(y\otimes x_j)\tau_0(y_j) - \theta(xy\otimes x_j)\tau_0(y_j) + \theta(x\otimes x_j)\tau_0(y_j)\tau_0(y)\big) \\
&= \sum_{j=1}^{\ell}\big(\theta(x\otimes y)\tau_0(x_j)\tau_0(y_j) - \theta(x\otimes yx_j)\tau_0(y_j) + \theta(x\otimes x_j)\tau_0(y_jy)\big) \\
&= \theta(x\otimes y)\sum_{j=1}^{\ell}\tau_0(x_jy_j) - \lambda(x\otimes y\omega) + \lambda(x\otimes\omega y) \\
&= \theta(x\otimes y)\tau_0(1) \\
&= \theta(x\otimes y) \\
&= \tau_0(xy) - \tau_0(x)\tau_0(y)
\end{aligned}$$

where passing from the second to the third equation one uses the equality (†). Therefore

$$\begin{aligned}(\tau_0(x) + \eta(x)) \cdot (\tau_0(y) + \eta(y)) \\ &= \tau_0(x)\tau_0(y) + \tau_0(x)\eta(y) + \tau_0(y)\eta(x) + \eta(x)\eta(y) \\ &= \tau_0(xy) + \eta(xy).\end{aligned}$$

This gives that $\tau := \tau_0 + \eta$ is a morphism of A-A-bimodules satisfying $\tau(x) - \tau_0(x) \in I$ and $\tau(xy) = \tau(x)\tau(y)$.

Hence we may lift $\rho : B \longrightarrow C/I$ to $\tau =: \rho_1 : B \longrightarrow C/I^2$.

By induction, suppose $\rho : B \longrightarrow C/I$ can be lifted to $\rho_{n-1} : B \longrightarrow C/I^{2^n}$, then the above proof shows that we may lift $\rho_{n-1} : B \longrightarrow C/I^{2^n}$ to $\rho_{n-1} : B \longrightarrow C/I^{2^{n+1}}$. Since I is nilpotent, there is an N such that $I^{2^N} = 0$, and the proposition is proved. □

Remark 3.6.13 Here we again encounter an example where lifting objects from a quotient of an object to the object itself involves cohomology in some sense. We have observed this phenomenon before, in the context of group cohomology in Proposition 1.8.39.

3.6.3 Unicity of the Form on Frobenius Algebras

We continue with a promised application. Recall that Proposition 1.10.18 gives a general method to find a bilinear form giving the structure of a self-injective algebra to a finite dimensional K-algebra.

We shall use the Wedderburn-Malcev theorem to show that this is the only possibility, at least for algebraically closed fields.

Proposition 3.6.14 [14] *Let A be a finite dimensional Frobenius K-algebra and suppose $A = KQ/I$ for a quiver Q and an admissible ideal I and an algebraically closed field K. Then for every non-degenerate associative bilinear form $\langle \cdot, \cdot \rangle$: $A \times A \longrightarrow K$ there is a K-basis $\mathcal{B}$ containing a K-basis of the socle such that $\langle x, y\rangle = \psi(xy)$, where ψ is defined by*

$$\psi(b) = \begin{cases} 1 & \text{if } b \in \mathrm{soc}(A) \cap \mathcal{B} \\ 0 & \text{if } b \in \mathcal{B} \setminus \mathrm{soc}(A). \end{cases}$$

Proof Denote by ν the Nakayama automorphism of A. Given an associative bilinear form $\langle \cdot, \cdot \rangle : A \times A \longrightarrow K$ there is a linear map $\psi : A \longrightarrow K$ defined by $\psi(x) := \langle 1, x\rangle$ and for any $x, y \in A$ one gets $\langle x, y\rangle = \langle 1, xy\rangle = \psi(xy)$. Hence ψ determines the associative bilinear map and the associative bilinear map determines ψ.

Since $A = KQ/I$ the socle of A is a direct sum of pairwise non-isomorphic one-dimensional simple A-modules. Let $\{s_1, \ldots, s_n\} \in A$ so that $s_i A$ is simple for every $i \in \{1, 2, \ldots, n\}$ and so that $\mathrm{soc}(A) =< s_1, \ldots, s_n >_K$.

Given an associative non-degenerate bilinear form $\langle \cdot, \cdot \rangle$ then $\langle \cdot, s_i \rangle : A \longrightarrow K$ is a non-zero linear form on A, since the bilinear form is non-degenerate. Hence there is an element $a \in A$ such that $\langle a, s_i \rangle \neq 0$. Now, by the Wedderburn-Malcev theorem 3.6.9 there is an element $\rho \in \mathrm{rad}(A)$ so that $a = \sum_{i=1}^n \lambda_i e_i + \rho$ for scalars $\lambda_i \in K$, and where $e_i^2 = e_i$ is an indecomposable idempotent of A, where $e_{\nu^{-1}(i)} s_i = s_i$, and where $e_{\nu^{-1}(j)} s_i = 0$ for $j \neq i$. Hence,

$$\langle a, s_i \rangle = \langle 1, a s_i \rangle = \langle 1, \lambda_{\nu^{-1}(i)} s_i \rangle = \lambda_{\nu^{-1}}(i).$$

We replace s_i by $\lambda_{\nu^{-1}(i)}^{-1} s_i$ and get $\langle 1, s_i \rangle = 1$. Take a K-basis $\mathcal{B}_i$ of $\ker(\langle \cdot, s_i \rangle)$ in $Ae_{\nu^{-1}(i)}$. Then, since $A = \bigoplus_{j=1}^n Ae_j$,

$$\mathcal{B} := \bigcup_{i=1}^n \mathcal{B}_i \cup \{s_1, s_2, \ldots, s_n\}$$

is a K-basis of A satisfying the hypotheses of Proposition 1.10.18. Moreover, if $xy \in \mathcal{B}$, then there is a unique $e_i^2 = e_i$ so that $xye_i = xy$, and so

$$\langle x, y \rangle = \langle 1, xy \rangle = \sum_{i=1}^n \langle 1, xye_i \rangle = \begin{cases} 1 & \text{if } xy \in \mathrm{soc}(A) \\ 0 & \text{otherwise} \end{cases}$$

This proves the statement. □

3.7 Hypercohomology and Derived Functors

Let $\mathcal{A}$ and $\mathcal{B}$ be additive categories and let

$$F : \mathcal{A} \longrightarrow \mathcal{B}$$

be an additive functor between these categories, i.e. F preserves finite direct sums and induces a homomorphism of abelian groups between the morphism sets.

Then, since $F(0) = 0$, it is clear that F induces a functor

$$F : C(\mathcal{A}) \longrightarrow C(\mathcal{B})$$

between the category of complexes over $\mathcal{A}$ and $\mathcal{B}$. Moreover, homotopy maps $hd^X + d^Y h$ between complexes (X, d^X) and (Y, d^Y) are mapped to homotopy maps between the image complexes since

$$F(hd^X + d^Y h) = F(h)F(d^X) + F(d^Y)F(h)$$

where we use that F is additive. Hence F induces a functor between the homotopy categories

$$F : K(\mathcal{A}) \longrightarrow K(\mathcal{B}).$$

Suppose $\mathcal{A}$ and $\mathcal{B}$ are abelian categories, then we may compute homologies and construct the derived category. However, going further to obtain a functor between the derived categories is more difficult. If we want to extend F to a functor between the derived categories we need to define functorially a quasi-isomorphism as the image of a quasi-isomorphism. If $\mathcal{A}$ and $\mathcal{B}$ are abelian and F is exact, i.e. sends exact sequences to exact sequences, then of course a quasi-isomorphism is sent to a quasi-isomorphism. However, the most natural functors we would like to apply, such as representable functors $Hom_{C(\mathcal{A})}(X, -)$ for a complex X, are not exact in general.

3.7.1 Left and Right Derived Functors

We shall need to develop the concept of left and right derived functors which are used for left and right exact functors, where a construction is possible. However, there are drawbacks. For example, we will not be able to prove associativity of composition of derived functors.

The treatment which we give here follows Verdier [15, §2 Définition 1.6, Proposition 1.6].

Definition 3.7.1 Let $\mathcal{A}$ and $\mathcal{B}$ be a abelian categories. Denote by $N_{\mathcal{A}} : K(\mathcal{A}) \longrightarrow D(\mathcal{A})$ and $N_{\mathcal{B}} : K(\mathcal{B}) \longrightarrow D(\mathcal{B})$ the natural functors from the homotopy category to the derived category.

- Then for a functor

  $$F : K(\mathcal{A}) \longrightarrow K(\mathcal{B})$$

  the *(total) right derived functor* $\mathbb{R}F$ is a functor

  $$\mathbb{R}F : D(\mathcal{A}) \longrightarrow D(\mathcal{B})$$

  together with a natural transformation

  $$\xi_F : N_{\mathcal{B}} \circ F \longrightarrow \mathbb{R}F \circ N_{\mathcal{A}}$$

  so that for all functors $G : D(\mathcal{A}) \longrightarrow D(\mathcal{B})$ with a natural transformation $\zeta : N_{\mathcal{B}} \circ F \longrightarrow G \circ N_{\mathcal{A}}$ there is a unique natural transformation

  $$\eta : \mathbb{R}F \longrightarrow G \text{ with } \zeta = (\eta_{N_{\mathcal{A}}(-)}) \circ \xi$$

in the sense that $\zeta_X = \eta_{N_{\mathcal{A}}(X)} \circ \xi_X$ for all objects X of $K(\mathcal{A})$.

- For a functor
$$F : K(\mathcal{A}) \longrightarrow K(\mathcal{B})$$
the *(total) left derived functor* $\mathbb{L}F$ is a functor
$$\mathbb{L}F : D(\mathcal{A}) \longrightarrow D(\mathcal{B})$$
together with a natural transformation
$$\xi_F : \mathbb{L}F \circ N_{\mathcal{A}} \longrightarrow N_{\mathcal{B}} \circ F$$
so that for all functors $G : D(\mathcal{A}) \longrightarrow D(\mathcal{B})$ with a natural transformation $\zeta : G \circ N_{\mathcal{A}} \longrightarrow N_{\mathcal{B}} \circ F$ there is a unique natural transformation
$$\eta : \mathbb{L}F \longrightarrow G \text{ with } \zeta = \xi \circ \eta_{N_{\mathcal{A}}}.$$

Remark 3.7.2 The universal property ensures, as usual, that whenever a right (resp. left) derived functor exists, then it is unique up to isomorphism.

As we have seen, exact functors $F : \mathcal{A} \longrightarrow \mathcal{B}$ admit left and right derived functors, which both coincide with F.

Remark 3.7.3 Recall Corollary 3.5.47. If X is a right bounded complex of projective modules and Y is a right bounded complex of (not necessarily projective) modules, then
$$Hom_{D^-(A\text{-}Mod)}(X, Y) = Hom_{K^-(A\text{-}Mod)}(X, Y).$$

Let $X \in K^-(A\text{-}Proj)$. If $Z \longrightarrow X$ is a quasi-isomorphism, replacing Z by its projective resolution, then using the equivalence of $D^-(A\text{-}Mod) \simeq K^-(A\text{-}Proj)$ and the fact that Z is a complex of projective modules, we see by Lemma 3.5.44 that $Z \longrightarrow X$ is actually invertible in $K^-(A\text{-}Proj)$. This shows that
$$Hom_{K^-(A\text{-}mod)}(X, Y[k]) = Hom_{D^-(A\text{-}mod)}(X, Y[k]).$$

Similarly, if Y is a left bounded complex of injective modules, then
$$Hom_{K^+(A\text{-}mod)}(X, Y[k]) = Hom_{D^+(A\text{-}mod)}(X, Y[k]).$$

This fact can be interpreted as saying that taking projective resolutions is left adjoint to the natural functor $K^-(A\text{-}Mod) \to D^-(A\text{-}Mod)$, and taking injective coresolutions is right adjoint to the natural functor $K^+(A\text{-}Mod) \to D^+(A\text{-}Mod)$.

Proposition 3.7.4 *Let $\mathcal{A}$ be an abelian category which admits sufficiently many projective objects. Let $\mathcal{B}$ be any abelian category.*

- *Then every additive covariant functor $F : \mathcal{A} \longrightarrow \mathcal{B}$ which induces $F : K^-(\mathcal{A}) \longrightarrow K(\mathcal{B})$ admits a left derived functor $\mathbb{L}F : D^-(\mathcal{A}) \longrightarrow D(\mathcal{B})$ given by* *the composition with the equivalence* $K^-(\mathcal{A}\text{-}Proj) \longrightarrow D^-(\mathcal{A})$ *of Proposition 3.5.43.*

- *Every additive contravariant functor* $F : \mathcal{A} \longrightarrow \mathcal{B}$ *which induces* $F : K^-(\mathcal{A}) \longrightarrow K(\mathcal{B})$ *admits a right derived functor* $\mathbb{R}F : D^-(\mathcal{A}) \longrightarrow D(\mathcal{B})$.

Proof Since $\mathcal{A}$ admits sufficiently many projectives, the restriction of the functor $N_{\mathcal{A}}$ to $K^-(\mathcal{A}\text{-}Proj)$ is an equivalence

$$N_{\mathcal{A}} : K^-(\mathcal{A}\text{-}Proj) \longrightarrow D^-(\mathcal{A}).$$

Hence

$$\mathbb{L}F := N_{\mathcal{B}} \circ F \circ N_{\mathcal{A}}^{-1}$$

is a left derived functor. The universal property comes from the adjointness formula of Remark 3.7.3.

Using Remark 3.5.45 one shows that a functor $F : K^+(\mathcal{A}) \longrightarrow K(\mathcal{B})$ from the homotopy category of left bounded complexes over an abelian category admitting enough injective objects to any homotopy category of complexes over an abelian category $\mathcal{B}$ admits a right derived functor.

For contravariant functors we may replace $\mathcal{A}$ by its opposite category $\mathcal{A}$ which consists of the same objects as $\mathcal{A}$ and morphisms are inverted: $Hom_{\mathcal{C}^{op}}(X, Y) := Hom_{\mathcal{C}}(Y, X)$ and composition is inverted. Of course, injective objects of $\mathcal{A}^{op}$ are precisely the projective objects of $\mathcal{A}$ and vice versa. A contravariant functor is hence a covariant functor from the opposite category. Therefore the existence of the right derived functor is implied by the existence of the left derived functors. □

3.7.2 Derived Tensor Product, Derived Hom Space, Hyper Tor and Hyper Ext

We shall mainly be interested in derived functors of Hom-spaces and of tensor products.

We shall start with the tensor product. Let A be an algebra and let (M, d^M) be a complex of left A-modules and let (N, d^N) be a complex of right A-modules. Then we may form $N \otimes_A M$, which is a bicomplex. A bicomplex is a $\mathbb{Z} \times \mathbb{Z}$-graded module together with two differentials d^N of degree $(-1, 0)$ and d^M of degree $(0, -1)$ so that

$$d^M \circ d^N = d^N \circ d^M.$$

Of course, this is the case for

$$N \otimes_A M = \bigoplus_{i \in \mathbb{Z}; j \in \mathbb{Z}} N_i \otimes_A M_j$$

and differentials $d^N \otimes_A id_M$ and $id_M \otimes_A d^M$.

Define the total complex $Tot(N \otimes_A M)$ of $N \otimes_A M$ by the complex formed by the codiagonals; i.e. the degree k homogeneous component is

$$Tot(N \otimes_A M)_k := \bigoplus_{i=-\infty}^{\infty} N_i \otimes_A M_{k-i}.$$

The differential of the total complex is

$$(d_{Tot(N \otimes_A M)})_{i,j} = (-1)^j d_i^N \otimes id_{M_j} + id_{N_i} \otimes d_j^M$$

where $d_i^N : N_i \longrightarrow N_{i-1}$ and $d_j^M : M_j \longrightarrow M_{j-1}$ are the usual differentials using our convention on the degrees. Hence

$$(d_{Tot(N \otimes_A M)})_{i,j} : N_i \otimes_A M_j \longrightarrow (N_{i-1} \otimes M_j) \oplus (N_i \otimes M_{j-1})$$

which is displayed in the following diagram (using d^N for $d^N \otimes id_M$ and d^M for $id_N \otimes d^M$ in order to slightly shorten the already quite cumbersome notation):

$$\begin{array}{ccccccc}
N_{i-2} \otimes M_{j+1} & \overset{(-1)^{j+1} d_{i-1}^N}{\longleftarrow} & N_{i-1} \otimes M_{j+1} & \overset{(-1)^{j+1} d_i^N}{\longleftarrow} & N_i \otimes M_{j+1} & \overset{(-1)^{j+1} d_{i+1}^N}{\longleftarrow} & N_{i+1} \otimes M_{j+1} \\
\downarrow d_{j+1}^M & & \downarrow d_{j+1}^M & & \downarrow d_{j+1}^M & & \downarrow d_{j+1}^M \\
N_{i-2} \otimes M_j & \overset{(-1)^j d_{i-1}^N}{\longleftarrow} & N_{i-1} \otimes M_j & \overset{(-1)^j d_i^N}{\longleftarrow} & N_i \otimes M_j & \overset{(-1)^j d_{i+1}^N}{\longleftarrow} & N_{i+1} \otimes M_j \\
\downarrow d_j^M & & \downarrow d_j^M & & \downarrow d_j^M & & \downarrow d_j^M \\
N_{i-2} \otimes M_{j-1} & \overset{(-1)^{j-1} d_{i-1}^N}{\longleftarrow} & N_{i-1} \otimes_A M_{j-1} & \overset{(-1)^{j-1} d_i^N}{\longleftarrow} & N_i \otimes M_{j-1} & \overset{(-1)^{j-1} d_{i+1}^N}{\longleftarrow} & N_{i+1} \otimes M_{j-1} \\
\downarrow d_{j-1}^M & & \downarrow d_{j-1}^M & & \downarrow d_{j-1}^M & & \downarrow d_{j-1}^M \\
N_{i-2} \otimes M_{j-2} & \overset{(-1)^{j-2} d_{i-1}^N}{\longleftarrow} & N_{i-1} \otimes M_{j-2} & \overset{(-1)^{j-2} d_i^N}{\longleftarrow} & N_i \otimes M_{j-2} & \overset{(-1)^{j-2} d_{i+1}^N}{\longleftarrow} & N_{i+1} \otimes M_{j-2}
\end{array}$$

Therefore the horizontal arrows have constant sign in each row, but alternating from one line to the next. Hence the restriction to $N_i \otimes_A M_j$ of the square of this differential is a mapping

$$N_i \otimes_A M_j \longrightarrow \left(N_{i-2} \otimes_A M_j\right) \oplus \left(N_{i-1} \otimes M_{j-1}\right) \oplus \left(N_i \otimes M_{j-2}\right).$$

The image in $N_{i-2} \otimes_A M_j$ is given by the mapping $(-1)^{2j} d_{i-1}^N \circ d_i^N = 0$ since d^N is a differential. Likewise the image in $N_i \otimes M_{j-2}$ is given by $d_{j-1}^M \circ d_j^M = 0$ since d^M is a differential. Finally, the image in $N_{i-1} \otimes M_{j-1}$ is given by

$$\left((-1)^{j-1} d_i^N\right) \circ d_j^M + d_j^M \circ \left((-1)^j d_i^N\right) = (-1)^{j-1} \left(d_i^N \circ d_j^M - d_j^M \circ d_i^N\right) = 0$$

since $d^N \otimes id_M$ commutes with $id_N \otimes d^M$.

This shows that $Tot(N \otimes_A M)$ is a complex with the above differential.

Definition 3.7.5 Let A be a K-algebra and let N be a complex of right A-modules. Then the left derived functor of

$$Tot(N \otimes_A -) : K^-(A\text{-}Mod) \longrightarrow K(K\text{-}Mod)$$

is the *left derived tensor product*, denoted

$$N \otimes_A^{\mathbb{L}} - : D^-(A\text{-}Mod) \longrightarrow D(K\text{-}Mod).$$

Its image on a complex M is given by taking a projective resolution of M as described in Proposition 3.5.43, forming the tensor product with N over A, and then taking the total complex.

Further the homology of the evaluation on M of the left derived tensor product gives the *higher torsion groups*

$$H_k(N \otimes_A^{\mathbb{L}} M) =: Tor_k^A(N, M).$$

Remark 3.7.6 Note that the objects M and N may well be modules. Then we obtain the groups $Tor_k^A(N, M)$, which are classical objects to study and were introduced in Definition 3.6.6. It is obvious that the definition given there coincides with the above more general concept in this special case.

Remark 3.7.7 Let A be a K-algebra, let M be an object in $K^-(A^{op}\text{-}Mod)$ and let N be an object in $K^-(A^{op}\text{-}Mod)$. Then $Tor_k^A(M, N)$ can be computed by resolving M or by resolving N. Indeed, $Tor(M \otimes_A pN) \simeq Tot(pM \otimes_A pN)) \simeq Tot(pM \otimes_A N)$ where we denote by pM a resolution of M by a complex of projective modules, and likewise for pN.

The analogous definition for the right derived functor of the Hom functor $Hom_{A\text{-}Mod}(-, X)$ is slightly more complicated and goes along the following lines. The main difficulty comes from the fact that even if X and Y are bounded complexes, the total complex of $Hom_A(X, Y)$ may be unbounded.

Take a complex (X, d^X) in the homotopy category $K(A\text{-}Mod)$ and a complex (Y, d^Y) in $K(A\text{-}Mod)$. Then $Hom_A(X, Y)$ is a bi-complex with differential $Hom_A(d^X, Y) =: (d^X)^*$ of degree $(1, 0)$ and differential $Hom_A(X, d^Y) =: (d^Y)_*$ of degree $(0, -1)$. More explicitly the differentials are given by $(d^Y)_*(\alpha) = d^Y \circ \alpha$ for $\alpha : X \longrightarrow Y$ and by $(d^X)^*(\beta) = \beta \circ d^X$ for $\beta : X \longrightarrow Y$.

Consider the complex $Hom_A^\bullet(X, Y)$ given by the degree k homogeneous component

$$\left(Hom_A^\bullet(X, Y)\right)_k := \prod_{i-j=k} Hom_A(X_i, Y_j)$$

and by the differential given by

$$\left(d_{Hom^\bullet_A(X,Y)}\right)_k := \prod_{i-j=k} (d_j^Y)_* + (-1)^j (d_{i+1}^X)^*$$
$$= \prod_{j=-\infty}^{\infty} (d_j^Y)_* + (-1)^j (d_{j+k+1}^X)^* .$$

Then again by the same argument as in the case of the left derived tensor product the so-defined $d_{Hom^\bullet_A(X,Y)}$ is actually a differential.

Suppose that X and Y are just A-modules concentrated in degree 0. Then the bi-complex $Hom^\bullet_A(X, Y)$ in this case is just $Hom_A(X, Y)$. Still assuming Y to be an A-module, then the extension of the functor $Hom_A(-, Y) : A\text{-}Mod \longrightarrow K\text{-}Mod$ to a functor $K^-(A\text{-}Mod) \longrightarrow K(K\text{-}Mod)$ is given by the functor $Hom^\bullet_A(-, Y)$. The extension of this functor $Hom^\bullet_A(-, Y)$ for a module Y to the functor $Hom^\bullet_A(-, Y)$ for a complex Y is then given by the above formula.

Since $Hom^\bullet_A(-, Y)$ is contravariant, we can form the right derived functor on right bounded complexes.

Definition 3.7.8 For any complex Y we define the *right derived functor of* $Hom^\bullet_A(-, Y)$ as

$$\mathbb{R}Hom_A(-, Y) : D^-(A\text{-}Mod) \longrightarrow D(K\text{-}Mod).$$

Further we define its higher extension groups

$$H_k(\mathbb{R}Hom_A(X, Y)) =: Ext^k_A(X, Y).$$

Remark 3.7.9 Recall how the right derived functor is defined. Replace X by a projective resolution P_X, and then apply $Hom_A(-, Y)$. If Y is a module, then this is precisely the way to obtain $Ext^k_A(X, Y)$. Hence the groups $Ext^k_A(X, Y)$ from Definition 3.7.8 and from Definition 1.8.29 coincide if X and Y are A-modules.

Lemma 3.7.10 *Let A be a K-algebra and let X and Y be complexes in $D^-(A)$. Then*

$$H_n(\mathbb{R}Hom_A(X, Y)) \simeq Hom_{D^-(A)}(X, Y[n])$$

where $[n]$ denotes the n-th iterate of the shift functor $[1]$.

Proof In order to evaluate $\mathbb{R}Hom_A(-, Y)$ on X one needs to replace X by its projective resolution, that is a right bounded complex of projective modules quasi-isomorphic to X. Hence we may consider X to be a right bounded complex of projective A-modules. Then

$$(\mathbb{R}Hom_A(X, Y))_k = \prod_{i-j=k} Hom_A(X_i, Y_j) = \prod_{j=-\infty}^{\infty} Hom_A(X_{j+k}, Y_j)$$

and

$$\left(d_{Hom^\bullet_A(X,Y)}\right)_k = \prod_{j=-\infty}^{\infty} \left(d_j^Y\right)_* + (-1)^j \left(d_{j+k+1}^X\right)^*.$$

Therefore an element in the kernel of $\left(d_{Hom^\bullet_A(X,Y)}\right)$ is a sequence of mappings $\alpha_j : X_{j+k} \longrightarrow Y_j$ such that

$$\left(d_{Hom^\bullet_A(X,Y)}\right)_k \left((\alpha_j)_{j\in\mathbb{Z}}\right) = 0.$$

But

$$\begin{aligned}\left(d_{Hom^\bullet_A(X,Y)}\right)_k &\left((\alpha_j)_{j\in\mathbb{Z}}\right)\\ &= \left(\prod_{j=-\infty}^{\infty} \left(d_j^Y\right)_* + (-1)^j \left(d_{j+k+1}^X\right)^*\right)\left((\alpha_j)_{j\in\mathbb{Z}}\right)\\ &= \prod_{j=-\infty}^{\infty} d_j^Y \circ \alpha_j + (-1)^j \alpha_{j+k} \circ d_{j+k+1}^X.\end{aligned}$$

Hence $\left(\alpha_j\right)_{j\in\mathbb{Z}}$ is in the kernel of the differential $\left(d_{Hom^\bullet_A(X,Y)}\right)_k$ if and only if $\left(\alpha_j\right)_{j\in\mathbb{Z}}$ is a homomorphism of complexes $X \longrightarrow Y[k]$.

We need to consider im $\left(\left(d_{Hom^\bullet_A(X,Y)}\right)_{k+1}\right)$.

$$\begin{aligned}\left(\left(d_{Hom^\bullet_A(X,Y)}\right)_{k+1}\right)\left((\alpha_j)_{j\in\mathbb{Z}}\right) &= \prod_{j\in\mathbb{Z}} \left(\left(d_j^Y\right)_* + (-1)^j \left(d_{j+k+2}^X\right)^*\right)\left((\alpha_j)_{j\in\mathbb{Z}}\right)\\ &= \prod_{j\in\mathbb{Z}} d_j^Y \circ \alpha_j + (-1)^j \alpha_{j+k+1} \circ d_{j+k+2}^X.\end{aligned}$$

Hence the image of $\left((\alpha_j)_{j\in\mathbb{Z}}\right)$ under $\left(d_{Hom^\bullet_A(X,Y)}\right)_{k+1}$ is a homotopy. Therefore

$$H_k(\mathbb{R}Hom_A(X,Y)) = Hom_{K^-(A\text{-}mod)}(X, Y[k]).$$

Now, using Remark 3.7.3, we have finished the proof of the lemma. □

Remark 3.7.11 We can construct the right derived functor of $Hom_A(X,-)$ by taking its evaluation on injective coresolutions of the second argument: Let X be a left bounded complex of A-modules. Then $\mathbb{R}Hom_A(X,Y) = Hom^\bullet_A(X,\iota Y)$ where $Y \to \iota Y$ is an injective coresolution of the left bounded complex Y.

The result is the same for bounded complexes X and Y. Indeed, by Remark 3.7.3

$$Hom_{K^+(A-Inj)}(X, \iota Y) \simeq Hom_{D^b(A)}(X, Y) \simeq Hom_{K^-(A\text{-}Proj)}(pX, Y),$$

where $pX \to X$ is a projective resolution of X, and so the resulting complexes are quasi-isomorphic by Lemma 3.7.10 and its dual construction for the construction via injective coresolutions.

Proposition 3.7.12 *Let A be a K-algebra for a commutative ring K and let M be an A-module. Then*

$$Ext_A^*(M, M) := \bigoplus_{n=0}^{\infty} Ext_A^n(M, M)$$

is a $\mathbb{Z}$-graded K-algebra.

Proof $Ext_A^n(M, M) = Hom_{D^-(A)}(M, M[n])$ and the K-module structure is clear from there. Hence for $\alpha \in Hom_{D^-(A)}(M, M[n])$ and $\beta \in Hom_{D^-(A)}(M, M[m])$ we may define

$$\alpha \cup \beta := (\beta)[n] \circ \alpha \in Hom_{D^-(A)}(M, M[n+m]) = Ext_A^{n+m}(M, M)$$

and obtain an associative algebra structure $\cup$. Actually associativity comes from the associativity of the composition of mappings in the derived category. Let $\alpha \in Ext_A^n(M, M)$, $\beta \in Ext_A^m(M, M)$ and $\gamma \in Ext_A^k(M, M)$. Then

$$\begin{aligned}\alpha \cup (\beta \cup \gamma) &= \alpha \cup (\gamma[m] \circ \beta) = (\gamma[m] \circ \beta)[n] \circ \alpha = \gamma[m+n] \circ \beta[n] \circ \alpha \\ &= (\beta[n] \circ \alpha) \cup \gamma = (\alpha \cup \beta) \cup \gamma\end{aligned}$$

and so $\cup$ is associative. □

Definition 3.7.13 The multiplication in $Ext_A^*(M, M)$ is called the *cup product*.

We defined the functor $\mathbb{R}Hom_A(-, Y)$ for any complex Y. We shall show that this functor is actually natural with respect to Y, in the following sense.

Lemma 3.7.14 *Let A be an algebra and let Y, Y' be right bounded complexes of A-modules. Then any morphism $(\nu, \alpha) : Y \longrightarrow Y'$ in the derived category induces a natural transformation*

$$\mathbb{R}Hom_A(-, Y) \longrightarrow \mathbb{R}Hom_A(-, Y')$$

and if (ν, α) is an isomorphism in the derived category, then the natural transformation is an isomorphism of functors.

Proof This is clear for an actual morphism α, since the morphism α induces morphisms on the homogeneous components of a complex. It is less clear for a quasi-isomorphism ν. So, we may suppose that $\nu : Y' \longrightarrow Y$ is a quasi-isomorphism, and we need to show that this induces an isomorphism

$$\mathbb{R}Hom_A(-, Y') \longrightarrow \mathbb{R}Hom_A(-, Y)$$

of functors. For this we shall use Remark 3.7.3 and Lemma 3.7.10.

Indeed, for a complex X denote by pX its projective resolution. Then

$$\begin{aligned}
H_k(\mathbb{R}Hom_A(-, Y)) &= H_k(Hom_A^\bullet(p-, Y)) \\
&= Hom_{K^-(A\text{-}Mod)}(p-, Y[k]) \text{ by Lemma 3.7.10} \\
&= Hom_{D^-(A\text{-}Mod)}(p-, Y[k]) \text{ by Remark 3.7.3} \\
&= Hom_{D^-(A\text{-}Mod)}(p-, Y'[k]) \text{ since Y} \simeq Y' \text{ in D}^-(A) \\
&= Hom_{K^-(A\text{-}Mod)}(p-, Y'[k]) \text{ by Remark 3.7.3} \\
&= H_k(\mathbb{R}Hom_A^\bullet(-, Y')) \\
&= H_k(Hom_A(p-, Y'))\text{by Lemma 3.7.10,}
\end{aligned}$$

and since ν induces a natural transformation

$$\mathbb{R}Hom_A(-, Y) \longrightarrow \mathbb{R}Hom_A(-, Y')$$

which gives an isomorphism on homology, the functors are actually isomorphic as functors in the derived category. □

Remark 3.7.15 Since for an isomorphism $\nu : Y \simeq Y'$ in the derived category we obtained that $\mathbb{R}Hom_A(-, Y) \simeq \mathbb{R}Hom_A(-, Y')$ and since for two morphisms $\alpha : Y \longrightarrow Y'$ and $\beta : Y' \longrightarrow Y''$ in the derived category we obtain that the composition of the induced natural transformations is clearly the natural transformation induced by the composition, we actually obtain a bi-functor

$$\mathbb{R}Hom_A(-, ?) : D^-(A\text{-}Mod) \times D^-(A\text{-}Mod) \longrightarrow D(\mathbb{Z}\text{-}Mod).$$

We know that $Hom_B(X, -)$ is right adjoint to $X \otimes_A -$. We shall show that under some additional hypotheses $\mathbb{R}Hom_B(X, -)$ is right adjoint to $X \otimes_A^{\mathbb{L}} -$. Observe first that if X is a complex of B-A-bimodules, then both functors $X \otimes_A^{\mathbb{L}} -$ take values in the derived category of B-modules, and $\mathbb{R}Hom_B(X, -)$ takes values in the derived category of A-modules. For this we need in general to use the definition of the right derived Hom by injective coresolutions, since for a $B \otimes_K A^{op}$-module M its projective cover M' as B-module will in general not be an A-module anymore. Take for example $B = \mathbb{Z}$ and $A = \mathbb{Z}/3\mathbb{Z}$. Then no projective B-module is an A-module. However, if A is projective as a K-module, then a projective $A \otimes_K B^{op}$-module is projective as right a B-module.

Proposition 3.7.16 *Let K be a commutative ring, let A and B be K-algebras and let X be a bounded complex of B-A-bimodules. Suppose A is projective as a K-module. Then*

$$X \otimes_A^{\mathbb{L}} - : D^-(A\text{-}Mod) \longrightarrow D^-(B\text{-}mod)$$

has a right adjoint

$$\mathbb{R}Hom_B(X,-) : D^-(B\text{-}Mod) \longrightarrow D^-(A\text{-}mod).$$

Proof By Proposition 3.5.43 we may replace X by a resolution so that all homogeneous components are projective. Now, since A is projective as a K-module, $A \otimes_K B$ is projective as a right B-module and $B \otimes_K A$ is projective as a left B-module. Therefore, $\mathbb{R}Hom(X,-)$ is just $Hom_B^\bullet(X,-)$. Since X is bounded, $Hom_B^\bullet(X,-)$ maps right bounded complexes to right bounded complexes. In order to evaluate the left derived tensor product one needs to replace the second variable Y by its projective resolution pY:

$$\begin{aligned} Hom_{D(A)}(-,\mathbb{R}Hom_B(X,?)) &= Hom_{K(A)}(p-, Hom_B^\bullet(X,?)) \\ &= Hom_{K(B)}(X \otimes_A p-, ?) \\ &= Hom_{K(B)}(X \otimes_A^{\mathbb{L}} p-, ?) \\ &= Hom_{D(B)}(X \otimes_A^{\mathbb{L}} -, ?). \end{aligned}$$

The second adjointness comes from the question of whether

$$Hom_B^\bullet(X,-) : K^-(B\text{-}Proj) \longrightarrow K^-(A\text{-}Proj)$$

is right adjoint to

$$Tot(X \otimes_A -) : K^-(A\text{-}Proj) \longrightarrow K^-(B\text{-}Proj).$$

But on the homotopy category this is clear since the adjointness holds on the module level. The third equation holds since A is K-projective, hence $X \otimes_A p-$ is a complex of projective B-modules. □

Remark 3.7.17 It is interesting to observe that Proposition 3.7.16 implies Lemma 3.7.10. Indeed

$$\begin{aligned} H_n(\mathbb{R}Hom_B(X,-)) &= Hom_{D^-(A)}(A[n], \mathbb{R}Hom_B(X,-)) \\ &= Hom_{D^-(B)}(X \otimes_A^{\mathbb{L}} A[n], -) \\ &= Hom_{D^-(B)}(X[n], -) \end{aligned}$$

where the first isomorphisms follows by Corollary 3.5.52. We observe that Lemma 3.7.10 was the main tool in the proof of Proposition 3.7.16.

We shall prove a useful statement which follows from the fact that

$$(X \otimes_B^{\mathbb{L}} -, \mathbb{R}Hom_A(X,-))$$

is an adjoint pair for every complex $X \in D^-(A \otimes_K B^{op})$.

Corollary 3.7.18 *Let K be a commutative ring and let A, B and C be K-projective K-algebras. Then for all complexes $X \in D^b(C \otimes_K A^{op})$, $Y \in D^b(A \otimes_K B^{op})$ and $Z \in D^b(C \otimes_K B^{op})$ we get isomorphisms*

$$\begin{aligned} Hom_{D^-(C\otimes_K B^{op})}(X \otimes_A^{\mathbb{L}} Y, Z) &\simeq Hom_{D^-(C\otimes_K A^{op})}(X, \mathbb{R}Hom_{B^{op}}(Y, Z)) \\ &\simeq Hom_{D^-(A\otimes_K B^{op})}(Y, \mathbb{R}Hom_C(X, Z)). \end{aligned}$$

In particular

$$Ext^n_{C\otimes_K A^{op}}(X, \mathbb{R}Hom_{B^{op}}(Y, Z)) \simeq Ext^n_{A\otimes_K B^{op}}(Y, \mathbb{R}Hom_C(X, Z))$$

for all $n \in \mathbb{N}$.

Proof As we have seen

$$(X \otimes_B^{\mathbb{L}} -, \mathbb{R}Hom_A(X, -))$$

is an adjoint pair of functors $D^-(B \otimes_K C^{op}) \longrightarrow D^-(A \otimes_K B^{op})$, but also, with the same arguments and changing left and right modules

$$(- \otimes_B^{\mathbb{L}} Y, \mathbb{R}Hom_C(Y, -))$$

is an adjoint pair of functors $D^-(A \otimes_K B^{op}) \longrightarrow D^-(B \otimes_K C^{op})$. This establishes the first isomorphism. The second isomorphism is an immediate consequence of the definition of the hyper-Ext, applying shift functors at strategic places.

This proves the statement. □

3.8 Applications: Flat and Related Objects

We obtained for any commutative ring K and any K-algebra A the bi-functors $Ext^1_A(-, ?)$ and $Tor^A_1(-, ?)$. We shall determine when $Ext^1_A(X, -) = 0$, when $Ext^1_A(-, X) = 0$ and when $Tor^A_1(X, -) = 0$ as functors $A\text{-}mod \longrightarrow \mathbb{Z}\text{-}Mod$.

Proposition 3.8.1 *Let K be a commutative ring and let A be a K-algebra. Let X be an A-module. Then*

- $Ext^1_A(X, -) = 0$ *if and only if X is a projective A-module.*
- $Ext^1_A(-, X) = 0$ *if and only if X is an injective A-module.*

Proof For any short exact sequence

$$0 \longrightarrow L \longrightarrow M \longrightarrow N \longrightarrow O$$

of A-modules we get an exact sequence

$$0 \to Hom_A(X, L) \to Hom_A(X, M) \to Hom_A(X, N) \to Ext^1_A(X, L) \to \dots$$

by Proposition 3.4.11 and so, $Ext^1_A(X, -) = 0$ if and only if

$$Hom_A(X, M) \longrightarrow Hom_A(X, N)$$

is surjective for every epimorphism $M \longrightarrow N$. But this means that if $\psi : M \longrightarrow\!\!\!\!\!\rightarrow N$ is an epimorphism, then every morphism $\alpha : X \longrightarrow N$ can be lifted to a morphism $\beta : X \longrightarrow M$ so that we get the equality $\psi \circ \beta = \alpha$. This is precisely the universal property for X to be a projective module. This proves the first statement.

We apply $Hom_A(-, X)$ to the exact sequence and get again by Proposition 3.4.11 an exact sequence

$$0 \to Hom_A(N, X) \to Hom_A(M, X) \to Hom_A(L, X) \to Ext^1_A(N, X) \to \dots$$

and so $Ext^1_A(-, X) = 0$ if and only if for every monomorphism $\varphi : L \longrightarrow M$ and every $\alpha : L \longrightarrow X$ there is a morphism $\beta : M \longrightarrow X$ with $\beta \circ \varphi = \alpha$. This is precisely the universal property of X being injective. □

What happens with Tor_1^A ? There the situation is slightly different. First the functor $Tor_1^A(-, ?)$ is covariant in both variables. Moreover, the two variables play the same role, so that it is not useful to distinguish between them.

First recall from Definition 2.5.22 the notion of flatness. Let K be a commutative ring and let A be a K-algebra. A right A-module S is said to be flat if for every monomorphism $L \hookrightarrow M$ of left A-modules the induced morphism of abelian groups $S \otimes_A L \longrightarrow S \otimes_A M$ is a monomorphism as well. The right A-module S is faithfully flat if S is flat and if moreover whenever a sequence

$$0 \longrightarrow S \otimes_A L \overset{id_S \otimes \alpha}{\longrightarrow} S \otimes_A M \overset{id_S \otimes \beta}{\longrightarrow} S \otimes_A N \longrightarrow 0$$

is exact, then the sequence

$$0 \longrightarrow L \overset{\alpha}{\longrightarrow} M \overset{\beta}{\longrightarrow} N \longrightarrow 0$$

is also exact.

We shall give a characterisation of flat modules via $Tor_n^A(-, ?)$. For this we shall need the following lemma.

Lemma 3.8.2 *Let K be a commutative ring and let A be a K-algebra. Let*

$$0 \longrightarrow L_1 \longrightarrow L_2 \longrightarrow L_3 \longrightarrow 0$$

be an exact sequence of A-modules. Then there is a long exact sequence

$$
\begin{array}{ccccccc}
 & & \dots & & \dots & & \\
 & & \downarrow & & & & \\
 & & Tor_3^A(M, L_1) & \longrightarrow & Tor_3^A(M, L_2) & \longrightarrow & Tor_3^A(M, L_3) \\
 & & & & & & \downarrow \\
 & & Tor_2^A(M, L_3) & \longleftarrow & Tor_2^A(M, L_2) & \longleftarrow & Tor_2^A(M, L_3) \\
 & & \downarrow & & & & \\
 & & Tor_1^A(M, L_1) & \longrightarrow & Tor_1^A(M, L_2) & \longrightarrow & Tor_1^A(M, L_3) \\
 & & & & & & \downarrow \\
0 & \longleftarrow & M \otimes_A L_3 & \longleftarrow & M \otimes_A L_2 & \longleftarrow & M \otimes_A L_1
\end{array}
$$

Proof We shall apply the functor $M \otimes_A^{\mathbb{L}} -$ to the short exact sequence

$$0 \longrightarrow L_1 \longrightarrow L_2 \longrightarrow L_3 \longrightarrow 0,$$

which induces a short exact sequence of the projective resolutions

$$0 \longrightarrow P_{L_1} \longrightarrow P_{L_2} \longrightarrow P_{L_3} \longrightarrow 0$$

of the modules L_1, L_2 and L_3 using the Horseshoe Lemma 3.5.50. The short exact sequence of complexes of projective modules gives an exact triangle in the homotopy category by Proposition 3.5.51. Now, apply $M \otimes_A -$ to the triangle, which gives a triangle in the category of K-modules again and now apply the homological functor "homology" (cf Corollary 3.5.52) to obtain the statement. □

Lemma 3.8.3 *Let K be a commutative ring and let A be a K-algebra. Then a right A-module M is flat if and only if $Tor_1^A(M, -) = 0$.*

Proof Recall

$$Tor_n^A(M, N) = H_n(M \otimes_A^{\mathbb{L}} N) = H_n(M \otimes_A P_N)$$

where

$$P_N : \ \dots \longrightarrow P_2 \longrightarrow P_1 \longrightarrow P_0 \longrightarrow N \longrightarrow 0$$

is a projective resolution of N. Let

$$0 \longrightarrow \Omega^2(N) \longrightarrow P_1 \longrightarrow P_0 \longrightarrow N \longrightarrow 0$$

be an exact sequence where P_0 and P_1 are projective and define

$$\Omega(N) := \mathrm{im}(P_1 \longrightarrow P_0).$$

Then

$$M \otimes_A \Omega(N) \longrightarrow M \otimes_A P_0 \longrightarrow M \otimes_A N \longrightarrow 0$$

is exact. If M is flat, then

$$0 \longrightarrow M \otimes_A \Omega(N) \longrightarrow M \otimes_A P_0 \longrightarrow M \otimes_A N \longrightarrow 0$$

is exact and therefore $H_1(M \otimes_A^{\mathbb{L}} N) = 0$.

Conversely if

$$Tor_1^A(M, -) = H_1(M \otimes_A^{\mathbb{L}} -) = 0,$$

and if

$$0 \longrightarrow L \longrightarrow N \longrightarrow N/L \longrightarrow 0$$

is an exact sequence of A-modules, then Lemma 3.8.2 shows that

$$0 \longrightarrow M \otimes_A L \longrightarrow M \otimes_A N \longrightarrow M \otimes_A N/L \longrightarrow 0$$

is exact and we obtain the statement. □

Remark 3.8.4 Flatness is related to projectivity.

1. Projective modules are flat. Indeed, free modules are flat since $A^n \otimes_A M \simeq M^n$ via $(a_1, \ldots, a_n) \otimes m \mapsto (a_1 m, \ldots, a_n m)$. If P is a projective A-module, then there is an A-module Q, such that $P \oplus Q$ is free. We get that

$$Tor_1^A(P \oplus Q, -) = Tor_1^A(P, -) \oplus Tor_1^A(Q, -)$$

since

$$M \otimes_A (P \oplus Q) = (M \otimes_A P) \oplus (M \otimes_A Q).$$

2. Not all flat modules are projective. However it is not easy to find counterexamples. Lam [16, Theorem 4.30] gives an explicit example of a flat but not projective module. Indeed the ring $\mathbb{Z} \oplus \bigoplus_{i \in \mathbb{N}} \mathbb{Z}/2\mathbb{Z}$, with appropriately defined ring structure, contains $(2, 0, 0, 0, \ldots) =: a$ and the principal ideal generated by a is flat but not projective.

We shall give a proof that flat modules are projective in the case of finite-dimensional algebras. The proof of this fact is relatively easy. I thank Zhengfang Wang for showing me this argument.

Lemma 3.8.5 *Let K be a field and let A be a finite dimensional K-algebra. Then a finitely generated A-module M is flat if and only if M is projective.*

Proof If P is projective, then P is a direct factor of a free module. Free modules are clearly flat and hence so are their direct summands.

Conversely suppose P is a finitely generated flat A-module. We need to show that P is projective, whence we have to show that

$$Hom_A(P, B) \xrightarrow{f^*} Hom_A(P, C)$$

is surjective for every A-linear epimorphism $f : B \twoheadrightarrow C$ of finitely generated A-modules. Define

$$\begin{aligned} Hom_K(M,K)\otimes_A P &\xrightarrow{\sigma_P^M} Hom_K(Hom_A(P,M),K) \\ \varphi\otimes p &\mapsto (\psi\mapsto(\varphi\circ\psi)(p)). \end{aligned}$$

We shall show that σ_P^M is an isomorphism. Clearly σ_A^M is an isomorphism. Since $\sigma_{P_1\oplus P_2}^M = \sigma_{P_1}^M \oplus \sigma_{P_2}^M$, we get that σ_F^M is an isomorphism for every finitely generated free A-module, and hence also for every projective A-module. Let

$$P_2 \longrightarrow P_1 \longrightarrow P \longrightarrow 0$$

be the first terms of a projective resolution of P. Then we get the following commutative diagram.

$$\begin{array}{ccccc} Hom_K(M,K)\otimes_A P_2 & \rightarrow & Hom_K(M,K)\otimes_A P_1 & \twoheadrightarrow & Hom_K(M,K)\otimes_A P \\ \downarrow \sigma_{P_2}^M & & \downarrow \sigma_{P_1}^M & & \downarrow \sigma_P^M \\ Hom_A(P_2,M)^* & \rightarrow & Hom_A(P_1,M)^* & \twoheadrightarrow & Hom_A(P,M)^* \end{array}$$

where we define $N^* := Hom_K(N,K)$ for all modules N. The rows of the diagram are exact since the tensor product $Hom_K(M,K)\otimes_A -$ is a right exact functor, since $Hom_A(-,M)$ is a left exact functor and since $Hom_K(-,K)$ is exact. Now, since $\sigma_{P_1}^M$ and $\sigma_{P_2}^M$ are isomorphisms, σ_P^M is also an isomorphism. If $f : B \twoheadrightarrow C$ is an epimorphism, then the K-linear dual $Hom_K(C,K) \longrightarrow Hom_K(B,K)$ is a monomorphism, and since P is flat,

$$Hom_K(C,K)\otimes_A P \longrightarrow Hom_K(B,K)\otimes_A P$$

is a monomorphism as well. Apply the isomorphism σ_P^B and σ_P^C to show that

$$Hom_K(Hom_A(P,C),K) \longrightarrow Hom_K(Hom_A(P,B),K)$$

is a monomorphism. Since all modules are finite dimensional, we may dualise again to see that

$$Hom_A(P,B) \xrightarrow{f^*} Hom_A(P,C)$$

is surjective. □

Lemma 3.8.6 (Change of rings) *Let R be a commutative ring and let S be a commutative R-algebra. If S is a faithfully flat R-module and Λ is a Noetherian R-algebra, then for all objects X and Y of $D^b(\Lambda\text{-}mod)$ we get*

$$Hom_{D^b(S\otimes_R\Lambda\text{-}mod)}(S\otimes_R X, S\otimes_R Y) \simeq S\otimes_R Hom_{D^b(\Lambda\text{-}mod)}(X,Y).$$

Proof Since S is flat over R, the functor $S\otimes_R -$ preserves quasi-isomorphisms and therefore we get a morphism

$$S\otimes_R Hom_{D^b(\Lambda\text{-}mod)}(U,V)\longrightarrow Hom_{D^b(S\otimes_R\Lambda\text{-}mod)}(S\otimes_R U, S\otimes_R V)$$

in the following way. Given a morphism ρ in $Hom_{D^b(\Lambda\text{-}mod)}(U,V)$ represented by the triple $\left(U\stackrel{\alpha}{\longleftarrow}W\stackrel{\beta}{\longrightarrow}V\right)$, for a quasi-isomorphism α and a morphism of complexes β, and $s\in S$, we map $s\otimes\rho$ to

$$\left(S\otimes_R U\stackrel{id_S\otimes\alpha}{\longleftarrow}S\otimes_R W\stackrel{s\otimes\beta}{\longrightarrow}S\otimes_R V\right).$$

This is natural in U and V.

We use the equivalence of categories $K^{-,b}(\Lambda\text{-}proj)\simeq D^b(\Lambda\text{-}mod)$ and suppose therefore that X and Y are right bounded complexes of finitely generated projective Λ-modules. But

$$S\otimes_R Hom_\Lambda(\Lambda^n,U)=S\otimes_R U^n=(S\otimes_R U)^n=Hom_{S\otimes_R\Lambda}((S\otimes_R\Lambda)^n,S\otimes_R U)$$

which proves the statement if X or Y is in $K^b(A\text{-}proj)$ since then a homomorphism is given by a direct sum of finitely many homogeneous mappings in those degrees where the complexes both have non-zero components. Now, taking the tensor product commutes with direct sums.

We come to the general case. Recall from Remark 3.5.22 the so-called stupid truncation σ_N of a complex. Let Z be a complex in $K^{-,b}(\Lambda\text{-}proj)$, denote by ∂ its differential and let $N\in\mathbb{N}$ so that $H_n(Z)=0$ for all $n\geq N$. We denote the homogeneous components of ∂ in such a way that $\partial_n:Z_n\longrightarrow Z_{n-1}$ for all n. Let $\sigma_N Z$ be the complex given by $(\sigma_N Z)_n=Z_n$ if $n\leq N$ and $(\sigma_N Z)_n=0$ otherwise. The differential δ on $\sigma_N Z$ is defined to be $\delta_n=\partial_n$ if $n\leq N$ and $\delta_n=0$ otherwise. Now, $\ker(\partial_N)=:C_N(Z)$ is a finitely generated Λ-module. Therefore we get an exact triangle, called in the sequel the truncation triangle for Z,

$$\sigma Z\longrightarrow Z\longrightarrow C_N(Z)[N+1]\longrightarrow(\sigma Z)[1]$$

for all objects Z in $K^{-,b}(A\text{-}proj)$. Obviously $\sigma(S\otimes_R Z)=S\otimes_R\sigma Z$ and since S is flat over R also $C_N(S\otimes_R Z)=S\otimes_R C_N(Z)$.

We choose N so that $H_n(X)=H_n(Y)=0$ for all $n\geq N$. To simplify the notation denote for the moment the bifunctor $Hom_{K^{-,b}(\Lambda\text{-}proj)}(-,-)$ by $(-,-)$, the bifunctor $Hom_{K^{-,b}(S\otimes_R\Lambda\text{-}proj)}(-,-)$ by $(-,-)_S$ and the bifunctor $S\otimes_R Hom_{K^{-,b}(\Lambda\text{-}proj)}(-,-)$ by $S(-,-)$. Further, put $S\otimes_R X=:X_S$ and $S\otimes_R Y=:Y_S$. From the long exact sequence obtained by applying $(X_S,-)_S$ to the truncation triangle of Y_S, using Proposition 3.4.11 we get a commutative diagram with exact lines denoted in the sequel by (†)

$$\begin{array}{ccccccccc}
(X_S, C_N(Y_S)[N])_S & \to & (X_S, \sigma_N Y_S)_S & \to & (X_S, Y_S)_S & \to & (X_S, C_N(Y_S)[M])_S & \to & (X_S, \sigma_N Y_S[1])_S \\
\uparrow & & \uparrow & & \uparrow & & \uparrow & & \uparrow \\
S(X, C_N(Y)[N]) & \to & S(X, \sigma_N Y) & \to & S(X, Y) & \to & S(X, C_N(Y)[M]) & \to & S(X, \sigma_N Y[1])
\end{array}$$

where we put $M := N + 1$. Since $\sigma_N(Y_S)$ is a bounded complex of projectives,

$$(X_S, \sigma_N Y_S)_S = S \otimes_R (X, \sigma_N Y) \text{ and } (X_S, \sigma_N Y_S[1])_S = S \otimes_R (X, \sigma_N Y[1]).$$

We apply $(-, C_N(Y_S)[k])_S$, for a fixed integer k, to the truncation triangle for X_S and obtain an exact sequence

$$\begin{array}{ccc}
(\sigma_N X_S[1], C_N(Y_S)[k])_S & \to & (C_N(X_S)[N+1], C_N(Y_S)[k])_S \\
 & & \downarrow \\
 & & (X_S, C_N(Y_S)[k])_S \\
 & & \downarrow \\
(C_N(X_S)[N], C_N(Y_S)[k])_S & \leftarrow & (\sigma_N X_S, C_N Y_S[k])_S
\end{array}$$

and a commutative diagram analogous to the diagram (†).

Now, for morphisms between finitely presented Λ-modules M and N we have that the natural map

$$S \otimes_R Hom_\Lambda(M, N) \longrightarrow Hom_{S \otimes_R \Lambda}(S \otimes_R M, S \otimes_R N)$$

is an isomorphism. Indeed, let

$$P_1 \longrightarrow P_0 \longrightarrow M \longrightarrow 0$$

be the first terms of a projective resolution of M as a Λ-module. Then

$$S \otimes_R P_1 \longrightarrow S \otimes_R P_0 \longrightarrow S \otimes_R M \longrightarrow 0$$

are the first terms of a projective resolution of $S \otimes_R M$ as an $S \otimes_R \Lambda$-module, and, defining $M_S := S \otimes_R M$, $N_S := S \otimes_R N$, $(P_i)_S := S \otimes_R P_i$ for $i \in \{0, 1\}$, and $\Lambda_S := S \otimes_R \Lambda$, we see that

$$\begin{array}{ccccc}
Hom_{\Lambda_S}(M_S, N_S) & \hookrightarrow & Hom_{\Lambda_S}((P_0)_S, N_S) & \to & Hom_{\Lambda_S}((P_1)_S, N_S) \\
\uparrow & & \uparrow & & \uparrow \\
S \otimes_R Hom_\Lambda(M, N) & \hookrightarrow & S \otimes_R Hom_\Lambda(P_0, N) & \to & S \otimes_R Hom_\Lambda(P_1, N)
\end{array}$$

is a commutative diagram with exact lines. The second and the third vertical morphisms are isomorphisms. Indeed,

$$\begin{aligned}
S \otimes_R Hom_\Lambda(\Lambda^n, N) &= S \otimes_R N^n \\
&= (S \otimes_R N)^n \\
&= Hom_{S \otimes_R \Lambda}(S \otimes_R \Lambda^n, S \otimes_R N)
\end{aligned}$$

and since P_0 and P_1 are direct factors of Λ^n, for certain n, we have the result. Therefore the leftmost vertical homomorphism is an isomorphism as well.

Given a projective resolution $P_\bullet \longrightarrow M$ of M, denote by $\partial_n : \Omega^n M \hookrightarrow P_{n-1}$ the embedding of the n-th syzygy of M into the degree $n-1$ homogeneous component of the projective resolution. Then

$$S \otimes_R \mathbb{R}Hom_\Lambda(X, Y) \simeq \mathbb{R}Hom_{S\otimes_R\Lambda}(S \otimes_R X, S \otimes_R Y)$$

by the definition of $\mathbb{R}Hom$ as a double complex of Hom-spaces, and the fact that S is flat over R. Taking homology and using that S is flat over R gives

$$\begin{aligned} S \otimes_R Ext^n_\Lambda(M, N) &= S \otimes_R H_n(\mathbb{R}Hom_\Lambda(M, N)) \\ &= H_n(S \otimes_R \mathbb{R}Hom_\Lambda(M, N)) \\ &= H_n(\mathbb{R}Hom_{S\otimes_R\Lambda}(S \otimes_R M, S \otimes_R N)) \\ &= Ext^n_{S\otimes_R\Lambda}(S \otimes_R M, S \otimes_R N) \end{aligned}$$

for all $n \in \mathbb{N}$, natural in M and N.

The case $k = N + 1$ then shows that

$$(C_N(X_S)[N+1], C_N(Y_S)[N+1])_S = S \otimes_R (C_N(X)[N+1], C_N(Y)[N+1])$$

and

$$(C_N(X_S)[N], C_N(Y_S)[N+1])_S = S \otimes_R (C_N(X)[N], C_N(Y)[N+1]).$$

By the case for bounded complex of projectives we get that the natural morphism is an isomorphism for

$$(\sigma_N X_S[1], C_N(Y_S)[N+1])_S \simeq S \otimes_R (\sigma_N X[1], C_N(Y)[N+1])$$

and

$$(\sigma_N X_S, C_N(Y_S)[N+1])_S \simeq S \otimes_R (\sigma_N X, C_N(Y)[N+1]).$$

Therefore also

$$(X_S, C_N(Y_S)[N+1])_S \simeq S \otimes_R (X, C_N(Y)[N+1])$$

and by the very same arguments, using $k = N$, we get

$$(X_S, C_N(Y_S)[N])_S \simeq S \otimes_R (X, C_N(Y)[N]).$$

This shows that we get isomorphisms in the two left and the two right vertical morphisms of (†) and hence the central vertical morphism is also an isomorphism.

Hence

$$(X_S, Y_S)_S \simeq S \otimes_R (X, Y)$$

and the lemma is proved. □

Remark 3.8.7 We should remark that the lemma is not true for $D^-(\Lambda)$. Indeed, let $M_\bullet$ be the complex with differential 0 and all homogeneous components of positive non-zero degree. Then

$$End_{D^-(\Lambda)}(M_\bullet) = \prod_{n\in\mathbb{N}} End_\Lambda(M_n)$$

is a direct product. Now, $S \otimes_R -$ commutes with direct sums, but does not commute with infinite direct products. An element of $S \otimes_R \left(\prod_{n\in\mathbb{N}} End_\Lambda(M_n)\right)$ is a sum of *finitely many* elements $s_i \otimes \varphi_i$ for $\varphi_i \in End_\Lambda(M_{n_i})$ and $s_i \in S$. However, an element of $\prod_{n\in\mathbb{N}} End_\Lambda(S \otimes_R M_n)$ is an *infinite* sequence $\left(\sum_{i=1}^{n_i(j)} s_i^{(j)} \otimes \varphi_i^{(j)}\right)_{j\in\mathbb{N}}$.

We could work with the so-called completed tensor product though. We will not go into these technical details.

We obtain the following generalisation of Theorem 2.5.28.

Theorem 3.8.8 [17] *Let R be a commutative Noetherian ring, let S be a commutative Noetherian R-algebra and suppose that S is a faithfully flat R-module. Suppose $R/\mathrm{rad}(R) = S/\mathrm{rad}(S)$. Let Λ be a Noetherian R-algebra, let X and Y be two objects of $D^b(\Lambda\text{-}mod)$ and suppose that $End_{D^b(\Lambda\text{-}mod)}(X)$ is a finitely generated R-module. Then*

$$S \otimes_R^{\mathbb{L}} X \simeq S \otimes_R^{\mathbb{L}} Y \Leftrightarrow X \simeq Y.$$

Remark 3.8.9 We observe that if R is local and $S = \hat{R}$ is the $\mathrm{rad}(R)$-adic completion, then S is faithfully flat as an R-module and $R/\mathrm{rad}(R) = S/\mathrm{rad}(S)$.

Proof of Theorem 3.8.8 According to the hypotheses we now suppose that $End_{D^b(\Lambda\text{-}mod)}(X)$ and $End_{D^-(\Lambda\text{-}mod)}(Y)$ are finitely generated R-modules and that $R/\mathrm{rad}(R) = S/\mathrm{rad}(S)$. Since S is flat over R, the tensor product of S over R is exact and we may replace the left derived tensor product by the ordinary tensor product. We only need to show "$\Rightarrow$" and assume therefore that X and Y are in $K^{-,b}(\Lambda\text{-}proj)$, and that $S \otimes_R X$ and $S \otimes_R Y$ are isomorphic.

Let $X_S := S \otimes_R X$ and $S \otimes_R Y =: Y_S$ in $D^b(S \otimes_R \Lambda\text{-}mod)$ to shorten the notation and denote by φ_S the isomorphism $X_S \longrightarrow Y_S$. Since then X_S is a direct factor of Y_S by means of φ_S, the mapping

$$\varphi_S = \sum_{i=1}^{n} s_i \otimes \varphi_i : X_S \longrightarrow Y_S$$

for $s_i \in S$ and $\varphi_i \in Hom_{D^b(\Lambda\text{-}mod)}(X, Y)$ has a left inverse $\psi : Y_S \longrightarrow X_S$ so that

$$\psi \circ \varphi_S = id_{X_S}.$$

Since $R/\text{rad}(R) = S/\text{rad}(S)$ there are $r_i \in R$ so that $1_S \otimes r_i - s_i \in \text{rad}(S)$ for all $i \in \{1, \ldots, n\}$.

Put

$$\varphi := \sum_{i=1}^{n} r_i \varphi_i \in Hom_{D^b(\Lambda\text{-}mod)}(X, Y).$$

Then

$$\begin{aligned}\sum_{i=1}^{n} \psi \circ (1_S \otimes (r_i\varphi_i)) - 1_S \otimes id_X &= \sum_{i=1}^{n} (\psi \circ (1_S \otimes r_i\varphi_i) - \psi \circ (s_i \otimes \varphi_i)) \\ &= \sum_{i=1}^{n} (1_S \otimes r_i - s_i) \cdot (\psi \circ (id_S \otimes \varphi_i)) \\ &\in \big(\text{rad}(S) \otimes_R End_{D^b(\Lambda\text{-}mod)}(X)\big)\end{aligned}$$

and since $End_{D^b(\Lambda\text{-}mod)}(X)$ is a Noetherian R-module, using Nakayama's lemma we obtain that the endomorphism $\psi \circ (\sum_{i=1}^{n} 1_S \otimes r_i\varphi_i)$ is invertible in $S \otimes_R End_{D^b(\Lambda\text{-}mod)}(X)$. Hence $id_S \otimes_R \varphi$ is left split and therefore by Lemma 3.4.9

$$X_S \stackrel{id_S \otimes_R \varphi}{\longrightarrow} Y_S \longrightarrow C(id_S \otimes_R \varphi) \stackrel{0}{\longrightarrow} X_S[1]$$

is a distinguished triangle, where $C(id_S \otimes_R \varphi)$ is the cone of $id_S \otimes_R \varphi$. However,

$$C(id_S \otimes_R \varphi) = S \otimes_R C(\varphi)$$

and hence

$$X_S \stackrel{id_S \otimes_R \varphi}{\longrightarrow} Y_S \longrightarrow S \otimes_R C(\varphi) \stackrel{0}{\longrightarrow} X_S[1]$$

is a distinguished triangle.

Since φ_S is an isomorphism, φ_S has a right inverse $\chi : Y_S \longrightarrow X_S$ as well. Now, since $X_S \simeq Y_S$, since S is faithfully flat over R, and since $End_{D^b(\Lambda\text{-}mod)}(X)$ is finitely generated as an R-module, using Lemma 3.8.6 we obtain that $End_{D^b(\Lambda\text{-}mod)}(Y)$ is finitely generated as an R-module as well. The same argument as for the left inverse ψ shows that $(id_S \otimes \varphi) \circ \chi$ is invertible in $S \otimes_R End_{D^b(\Lambda\text{-}mod)}(Y)$. Hence

$$X_S \stackrel{id_S \otimes_R \varphi}{\longrightarrow} Y_S \stackrel{0}{\longrightarrow} S \otimes_R C(\varphi) \stackrel{0}{\longrightarrow} X_S[1]$$

is a distinguished triangle, again using Lemma 3.4.9. This shows that $S \otimes_R C(\varphi)$ is acyclic, and hence

$$0 = H(S \otimes_R C(\varphi)) = S \otimes_R H(C(\varphi)).$$

Since S is faithfully flat over R, we have $H(C(\varphi)) = 0$, which implies that $C(\varphi)$ is acyclic and therefore φ is an isomorphism. This proves the theorem. □

Let A be an algebra over a complete discrete valuation ring R which is finitely generated as a module over R. The importance of a Krull-Schmidt theorem for the derived category of bounded complexes over A is evident from its importance in module categories. Different proofs exist in the literature. Another slightly more abstract approach than that of the proof presented below is proposed by Xiao-Wu Chen, Yu Ye and Pu Zhang in [18, Appendix A]. The authors show that an additive category is Krull-Schmidt if and only if all idempotents split (cf Definition 6.9.19 below) and the endomorphism ring of each object is semi-perfect, i.e. for every object X with endomorphism ring R, $R/\mathrm{rad}(R)$ is semisimple and every non-zero R-module contains a maximal submodule.

Proposition 3.8.10 *Let R be a complete discrete valuation ring and let A be an R-algebra, finitely generated as an R-module. Then the Krull-Schmidt theorem holds for $K^{-,b}(A\text{-}proj)$.*

Proof We first prove a Fitting lemma for $K^{-,b}(A\text{-}proj)$.

Let X be a complex in $K^{-,b}(A\text{-}proj)$ and let u be an endomorphism of the complex X. Then $X = X' \oplus X''$ as graded modules, by Fitting's lemma in the version for algebras over complete discrete valuation rings as in Lemma 2.5.23 so that the restriction of u on X' is an automorphism in each degree and the restriction of u on X'' is nilpotent modulo $\mathrm{rad}(R)^m$ for each m. Therefore u is a diagonal matrix $\begin{pmatrix} \iota & 0 \\ 0 & \nu \end{pmatrix}$ in each degree where $\iota : X' \longrightarrow X'$ is invertible, and $\nu : X'' \longrightarrow X''$ is nilpotent modulo $\mathrm{rad}(R)^m$ for each m in each degree. The differential ∂ on X is given by $\begin{pmatrix} \partial_1 & \partial_2 \\ \partial_3 & \partial_4 \end{pmatrix}$ and the fact that u commutes with ∂ shows that $\partial_3 \iota = \nu \partial_3$ and $\partial_2 \nu = \iota \partial_2$. Therefore, $\partial_3 \iota^s = \nu^s \partial_3$ and $\partial_2 \nu^s = \iota^s \partial_2$ for all s. Since ν is nilpotent modulo $\mathrm{rad}(R)^m$ for each m in each degree, and ι is invertible, $\partial_2 = \partial_3 = 0$. Hence the differential of X restricts to a differential on X' and a differential on X''. Moreover, X' and X'' are both projective modules, since X is projective.

Now, X, and therefore also X'', is exact in degrees higher than N, say. We fix $m \in \mathbb{N}$ and obtain therefore that u is nilpotent modulo $\mathrm{rad}(R)^m$ in each degree lower than N. Let M_m be the nilpotency degree. Then, since X'' is exact in degrees higher than N, modulo $\mathrm{rad}(R)^m$ the restriction of the endomorphism u^{M_m} to X'' is homotopy equivalent to 0 in degrees higher than N. We get therefore that the restriction of u to X'' is actually nilpotent modulo $\mathrm{rad}(R)^m$ for each m.

Hence, the endomorphism ring of an indecomposable object is local and the Krull-Schmidt theorem is an easy consequence by the classical proof, as in Corollary 2.5.24 or Theorem 1.4.3.

This proves the proposition. □

Remark 3.8.11 If R is a field and A is a finite dimensional R-algebra, then we can to argue more directly. Indeed, $X' = \mathrm{im}(u^N)$ and $X'' = \ker(u^N)$ for large enough N. Then it is obvious that X' and X'' are both subcomplexes of X. Observe that R may be a field in Proposition 3.8.10.

For Corollary 3.8.12 we closely follow Roggenkamp's paper [19]. Moreover, Corollary 3.8.12 generalises Theorem 2.5.28. Recall that a commutative ring R is *semilocal* if R has only a finite number of maximal ideals.

Corollary 3.8.12 *Let R be a commutative semilocal Noetherian ring, let $\hat{R}$ be its* $\mathrm{rad}(R)$*-adic completion and let S be a commutative R-algebra so that $\hat{S} := \hat{R}\otimes_R S$ is a faithful projective $\hat{R}$-module of finite type. Let Λ be a Noetherian R-algebra, finitely generated as an R-module, and let X and Y be two objects of $D^b(\Lambda)$ and suppose that $End_{D^b(\Lambda)}(X)$ and $End_{D^b(\Lambda)}(Y)$ are finitely generated R-modules. Then*

$$S \otimes_R^{\mathbb{L}} X \simeq S \otimes_R^{\mathbb{L}} Y \Leftrightarrow X \simeq Y.$$

Proof If $S\otimes_R^{\mathbb{L}} X \simeq S\otimes_R^{\mathbb{L}} Y$ in $D^b(S\otimes_R \Lambda)$, we get $\hat{S}\otimes_R^{\mathbb{L}} X \simeq \hat{S}\otimes_R^{\mathbb{L}} Y$ in $D^b(\hat{S}\otimes_R \Lambda)$. Since R is semilocal with maximal ideals $m_1, \dots, m_s$ we get $\hat{R} = \prod_{i=1}^s \hat{R}_{m_i}$ for the completion $\hat{R}_{m_i}$ of R at m_i. Now, $\hat{S}$ is projective faithful of finite type, and so there are $n_1, \dots, n_s$ with

$$\hat{S} \simeq \prod_{i=1}^{s} (\hat{R}_{m_i})^{n_i}$$

and therefore $\hat{S} \otimes_R^{\mathbb{L}} X \simeq \hat{S} \otimes_R^{\mathbb{L}} Y$ implies

$$\prod_{i=1}^{s} (\hat{R}_{m_i})^{n_i} \otimes_R^{\mathbb{L}} X \simeq \prod_{i=1}^{s} (\hat{R}_{m_i})^{n_i} \otimes_R^{\mathbb{L}} Y.$$

Hence

$$(\hat{R}_{m_i} \otimes_R^{\mathbb{L}} X)^{n_i} \simeq (\hat{R}_{m_i} \otimes_R^{\mathbb{L}} Y)^{n_i}$$

for each i, and therefore by Proposition 3.8.10

$$\hat{R}_{m_i} \otimes_R^{\mathbb{L}} X \simeq \hat{R}_{m_i} \otimes_R^{\mathbb{L}} Y$$

for each i. By Theorem 3.8.8 we obtain $X \simeq Y$. □

3.9 Spectral Sequences

We shall need at two points a rather sophisticated tool from homological algebra, which sometimes provides deep insight and surprising statements, but is somewhat technical to introduce. We shall use the most elegant approach of Godement [20, Section 4] and we shall closely follow the presentation given there.

3.9.1 Spectral Sequence of a Filtered Differential Module

Definition 3.9.1 Let A be an algebra.

- A *filtration* on an A-module M is a sequence of submodules

$$\cdots \subseteq F^p M \subseteq F^{p-1}M \subseteq \cdots \subseteq F^q M \subseteq \cdots \subseteq M$$

  such that

$$\bigcup_{p\in\mathbb{Z}} F^p M = M.$$

  The associated graded module is

$$Gr_F(M) := \bigoplus_{p=-\infty}^{\infty} F^p M / F^{p+1} M.$$

- If M is equipped with a differential d and if for the restriction of the differential we get $d|_{F^p M} \in End_A(F^p M)$, then (M, F, d) is a *filtered differential module.*
- If (M, d) is a complex, with a differential d of degree -1, then the *complex is filtered* by a filtration F of the module M if $d|_{F^p M} \in End_A(F^p M)$. In other words, a complex is filtered if there is a filtration of the module such that d preserves the filtration.

The associated graded complex of a filtered complex is a bigraded module by the double grading

$$Gr_F(M)^{p,q} := F^p M_{p+q} / F^{p+1} M_{p+q}$$

where $F^p M_n := (F^p M) \cap M_{-n}$. The change of sign is necessary since in most cases we have used differentials of degree -1 for a complex. Reversing the index of the homogeneous components will give a differential of degree 1 in this opposite notation.

Remark 3.9.2 Suppose $M = \bigoplus_{i\in\mathbb{Z}} M_i$ is graded and define $F^p M := \bigoplus_{i\geq p} M_i$. Then

$$M \supseteq \cdots \supseteq F^{-1}M \supseteq F^0 M \supseteq F^1 M \supseteq F^2 M \supseteq \ldots$$

is a filtration with

$$M = \bigcup_{p\in\mathbb{Z}} F^p M.$$

Moreover $Gr_F M = M$ in this case.

Let

$$M \supseteq \cdots \supseteq F^p M \supseteq F^{p+1} M \supseteq \ldots$$

be a filtered module and suppose

$$M = \bigoplus_{i \in \mathbb{Z}} M_i$$

is in addition graded.

Definition 3.9.3 The *grading and the filtration are compatible* if each $F^p M$ is homogeneous, or in other words

$$F^p M = \sum_{i \in \mathbb{Z}} \left(F^p M \cap M_{-i} \right).$$

If a grading and a filtration of a module are compatible we say for short that we have a filtration of a graded module, or a grading of a filtered module. A graded module with compatible filtration is *regular* if for all $i \in \mathbb{Z}$ there is an $n_i \in \mathbb{Z}$ so that $F^p M \cap M_{-i} = 0$ for all $p > n_i$.

Example 3.9.4 Let $M = \bigoplus_{(i,j) \in \mathbb{Z}^2} M^{i,j}$ for submodules $M^{i,j}$ of M. Then we obtain two different filtrations

$$'F^p M := \sum_{i \geq p} M^{i,j} \text{and } ''F^p M := \sum_{j \geq p} M^{i,j}.$$

The filtration $'F$ is compatible with respect to the grading $M^j := \bigoplus_{i \in \mathbb{Z}} M^{i,j}$, i.e. with respect to the second variable, and the second filtration $''F$ is compatible with the grading with respect to the first variable.

Lemma 3.9.5 *Let $M = \bigoplus_{i \in \mathbb{Z}} M_i$ and $N = \bigoplus_{i \in \mathbb{Z}} N_i$ be two graded filtered modules, and let $f : M \longrightarrow N$ be a homomorphism so that $f(F^p M) \subseteq F^p N$ for all $p \in \mathbb{Z}$, and so that $f(M_i) \subseteq N_i$ for all $i \in \mathbb{Z}$, then f defines a homogeneous homomorphism $Gr(f) : Gr(M) \longrightarrow Gr(N)$ of bidegree* (0, 0).

- *If the filtration on M is regular, then $Gr(f)$ is injective implies that f is injective.*
- *If the filtration on N is regular, then $Gr(f)$ is surjective implies that f is surjective.*

Proof Let $x \in \ker(f)$. Since f preserves the grading, $\ker(f)$ is graded and we may suppose that $x \in M_{-i}$ for some i. Moreover, there is a $p \in \mathbb{Z}$ such that $x \in F^p M$. Since $f(x) = 0$, $Gr(f)(x) = 0$, and since $Gr(f)$ is injective, we get that $x \in F^{p+1} M$. Hence $x \in F^n M$ for all n. But if n is sufficiently big, $F^n M \cap M_{-i} = 0$ since the filtration on M is regular. Hence if $Gr(f)$ is injective then f injective.

Suppose now $Gr(f)$ is surjective and the filtration on N is regular. Let $y \in N$. We may assume that $y \in N_{-i}$ is homogeneous. Then there is a $p \in \mathbb{Z}$ such that $y \in F^p N$. Since $Gr(f)$ is surjective, there is an $x_p \in F^p M_i$ such that $f(x_p) \in y + F^{p+1} N$, or what is equivalent, $f(x_p) - y \in F^{p+1} N$. Recursively we obtain a sequence $x_q \in F^q M_i$ for $q \geq p$, so that $f(x_p + x_{p+1} + \cdots + x_q) - y \in F^{q+1} N$. But since the filtration on N is regular, there is a q_0 such that $F^{q_0}N \cap N_{-i} = 0$. Hence $y = f(x_p + x_{p+1} + \cdots + x_{q_0})$. This proves the statement. □

The following proposition is crucial for the construction of spectral sequences.

Proposition 3.9.6 *Let (M, d) be a filtered differential module, i.e. the module is filtered, and has a differential d, so that the differential respects the filtration.*
Putting for all $r \in \mathbb{Z}$

$$Z_r^p := \{x \in F^pM \mid d(x) \in F^{p+r}M\}$$

we get that $d(Z_{r-1}^{p-r+1}) + Z_{r-1}^{p+1} \subseteq Z_r^p$ and putting

$$E_r^p := Z_r^p / \left(d(Z_{r-1}^{p-r+1}) + Z_{r-1}^{p+1}\right)$$

then $E_r := \sum_{p\in\mathbb{Z}} E_r^p$ admits a differential d_r of degree r so that

$$E_{r+1} \simeq H(E_r, d_r).$$

Proof By definition of Z_r^p we see that $d(Z_{r-1}^{p-r+1}) \subseteq F^pM$ and that

$$\begin{aligned} Z_{r-1}^{p+1} &= \{x \in F^{p+1}M \mid d(x) \in F^{p+r}M\} \\ &= \{x \in F^pM \mid d(x) \in F^{p+r}M\} \cap F^{p+1}M \\ &= Z_r^p \cap F^{p+1}M \subseteq \{x \in F^pM \mid d(x) \in F^{p+r}M\} = Z_r^p \end{aligned}$$

since $F^pM \supseteq F^{p+1}M$. Now,

$$d(d(Z_{r-1}^{p-r+1})) = 0 \subseteq F^{p+r}M$$

and hence $d(Z_{r-1}^{p-r+1}) \subseteq Z_r^p$ by definition of Z_r^p. Moreover,

$$d|_{Z_r^p} : Z_r^p \longrightarrow Z_r^{p+r}$$

again by definition and

$$d|_{Z_r^p}\left(d(Z_{r-1}^{p-r+1}) + Z_{r-1}^{p+1}\right) = d(Z_{r-1}^{p+1}) \subseteq d(Z_{r+1}^{p+1}) + Z_{r-1}^{p+r+1}.$$

Hence, since

$$E_r^p := Z_r^p / \left(d(Z_{r-1}^{p-r+1}) + Z_{r-1}^{p+1}\right) \text{ and } E_r^{p+r} = Z_r^{p+r} / (d(Z_{r-1}^{p+1}) + Z_{r-1}^{p+r+1}),$$

the mapping d induces a morphism

$$d_r : E_r^p \longrightarrow E_r^{p+r}$$

on the quotients.

Let $x + \left(d(Z_{r-1}^{p-r+1}) + Z_{r-1}^{p+1}\right) \in \ker(d_r)$. Then $d(x) \in d(Z_{r-1}^{p-r+1}) + Z_{r-1}^{p+r+1}$ and hence there are $y \in Z_{r-1}^{p-r+1}$ and $z \in Z_{r-1}^{p+r+1}$ such that

$$d(x) = d(y) + z.$$

This shows $d(x - y) = z \in Z_{r-1}^{p+r+1}$. Hence

$$x - y \in F^p M \text{ and } d(x - y) \in Z_{r-1}^{p+r+1} \subseteq F^{p+r+1}M.$$

By definition this implies $x - y \in Z_{r+1}^p$. Therefore

$$\ker(d_r) \cap E_r^p = \left(Z_{r+1}^p + Z_{r-1}^{p+1}\right) / \left(d(Z_{r-1}^{p-r+1}) + Z_{r-1}^{p+1}\right).$$

Moreover, the classes of $d(Z_r^{p-r})$ in E_r^p generate $\mathrm{im}(d_r) \cap E_r^p$ and therefore

$$d_r(E_{r-1}^{p-r+1}) = \left(d(Z_r^{p-r}) + Z_{r-1}^{p+1}\right) / \left(d(Z_{r-1}^{p-r+1}) + Z_{r-1}^{p+1}\right).$$

This shows

$$H^p(E_r, d_r) = \left(Z_{r+1}^p + Z_{r-1}^{p+1}\right) / \left(d(Z_r^{p-r}) + Z_{r-1}^{p+1}\right).$$

Now $d_r(Z_r^{p-r}) \subseteq Z_{r+1}^p$ and $Z_{r+1}^p \cap Z_{r-1}^{p+1} = Z_{r-1}^p$ by definition. This shows

$$\begin{aligned} H^p(E_r, d_r) &= Z_{r+1}^p / \left(Z_{r+1}^p \cap \left(d(Z_r^{p-r}) + Z_{r-1}^{p+1}\right)\right) \\ &= Z_{r+1}^p / \left(d(Z_r^{p-r}) + Z_r^{p+1}\right) \\ &= E_{r+1}^p. \end{aligned}$$

This proves the proposition. □

Definition 3.9.7 The *spectral sequence* associated to the differential filtered module (M, d) is the sequence E_r constructed in Proposition 3.9.6. The differential module E_r is called the r-th *page of the spectral sequence.*

Now, put

$$Z_\infty^p := \{x \in F^p M \mid dx = 0\}$$

and

$$B_\infty^p := (F^p M) \cap \mathrm{im}(d)$$

as well as

$$E_\infty^p := Z_\infty^p/(B_\infty^p + Z_\infty^{p+1}).$$

For $r \in \mathbb{Z}$ we may put

$$B_r^p := F^pM \cap d(F^{p-r}M)$$

and obtain in this way

$$E_r^p = Z_r^p/(B_{r-1}^p + Z_{r-1}^{p+1}).$$

Definition 3.9.8 We call $E_\infty = \sum_{p\in\mathbb{Z}} E_\infty^p$ the *infinite page* of the spectral sequence.

Since the differential d preserves the filtration, F^pM is again a differential module. Let $H(F^pM)$ be its homology. Then the inclusion of the differential module F^pM into M induces a morphism $H(F^pM) \longrightarrow H(M)$. Let $F^pH(M)$ be the image of this morphism. It is clear that $H(M)$ is filtered by the submodules $F^pH(M)$.

Theorem 3.9.9 *Let M be a differential filtered module. Then*

$$E_\infty = Gr(H(M)).$$

Proof Since $H(F^pM)$ is a quotient of Z_∞^p, and since $F^pH(M)$ is the image of $H(F^pM) \longrightarrow H(M)$, there is a surjective morphism

$$\alpha : Z_\infty^p \longrightarrow F^pH(M).$$

Now $z \in Z_\infty^p$ so that $\alpha(z) \in F^{p+1}H(M)$ is equivalent to $z \in B_\infty^p + Z_\infty^{p+1}$. This proves the statement. □

3.9.2 Spectral Sequences of Filtered Complexes

Suppose that M is differential filtered and has a grading compatible with the filtration and such that the differential is homogeneous of degree -1. In other words, suppose that (M, d) is a complex with a filtration in the category of complexes. The pages E_r of the spectral sequence are then bigraded if one puts

$$\begin{aligned} Z_r^{p,q} &:= Z_r^p \cap M_{-p-q} \\ B_r^{p,q} &:= B_r^p \cap M_{-p-q} \\ E_r^{p,q} &:= Z_r^{p,q}/(B_{r-1}^{p,q} + Z_{r-1}^{p+1,q-1}). \end{aligned}$$

Then

$$d_r : E_r^{p,q} \longrightarrow E_r^{p+r,q-r+1}$$

and

$$E_\infty^{p,q} = F^p H^{p+q}(M)/F^{p+1}H^{p+q}(M).$$

Suppose moreover that the filtration F of (M,d) is regular with respect to the grading given by the structure of the complex. This means that in each degree q of (M,d) there is an $n(q)$ such that $F^p(M_q) = 0$ whenever $p > n(q)$. Then for all $z \in Z_r^{p,q} \subseteq M_{-p-q}$ we have

$$d(z) \in M_{-p-q-1} \cap F^{p+r}M = F^{p+r}M_{(p+q+1)}.$$

But now $p+r > n(p+q+1)$ is equivalent to $r > n(p+q+1) - p$. This implies that

$$Z_r^{p,q} = Z_\infty^{p,q} \;\; \forall\, r > n(p+q+1) - p.$$

If $r > n(p+q+1) - p$, then $Z_r^{p+r,q-r+1} = 0$. Now

$$d_r(E_r^{p,q}) \subseteq Z_r^{p+r,q-r+1} = 0$$

and therefore $d_r|_{E_r^{p,q}} = 0$ if r is sufficiently big. By the definition of a spectral sequence we may identify $H(E_r) = E_{r+1}$ for all r and therefore obtain a morphism

$$E_r^{p,q} \longrightarrow E_{r+1}^{p,q}.$$

Composing these morphisms we hence obtain a morphism

$$\theta_s^r : E_r^{p,q} \longrightarrow E_s^{p,q} \;\; \forall\, s > r > n(p+q+1) - p$$

so that $\theta_t^s \circ \theta_s^r = \theta_t^r$ for all $t > s > r > n(p+q+1) - p$.

Lemma 3.9.10 *If (M,d) is a filtered complex (so that the filtration is compatible with the structure of a complex), and so that the filtration is regular, then the inductive limit of the system given by the $(\theta_r)_{r > n(p+q+1)-p}$ is the infinite page $E_\infty^{p,q}$:*

$$\mathrm{colim}_r E_r^{p,q} = E_\infty^{p,q}$$

Proof Since we have to consider the inductive limit, we may take r to be very large. If r is sufficiently big,

$$Z_r^{p,q} = Z_\infty^{p,q}, \;\; Z_{r-1}^{p+1,q-1} = Z_\infty^{p+1,q-1} \text{and } B_r^{p,q} \subseteq B_\infty^{p,q}.$$

Therefore we get a natural morphism

$$E_r^{p,q} = Z_r^{p,q}/(B_{r-1}^{p,q} + Z_{r-1}^{p+1,q-1}) \xrightarrow{\theta_\infty^r} Z_\infty^{p,q}/(B_\infty^{p,q} + Z_\infty^{p+1,q-1}) = E_\infty^{p,q}$$

By the above relations between Z_r and Z_∞, and between B_r and B_∞, we obtain that θ^r_∞ is surjective. Moreover

$$\theta^r_\infty = \theta^s_\infty \circ \theta^r_s$$

as soon as $s > r > n(p+q+1) - p$. Since $M = \bigcup_p F^pM$, we get that $B^{p,q}_\infty = \bigcup B^{p,q}_r$ and therefore we get that $E^{p,q}_\infty$ is the inductive limit of the pages $E^{p,q}_r$. □

Proposition 3.9.11 *Let (M, d^M) and (N, d^N) be two filtered complexes and suppose the filtration for each complex is regular. Let $f : (M, d^M) \longrightarrow (N, d^N)$ be a morphism of complexes and suppose that f is compatible with the filtrations. Then f induces a morphism*

$$f^{(r)} : E_r(M) \longrightarrow E_r(N).$$

If $f^{(r)}$ is an isomorphism then

$$H(f) : H(M) \longrightarrow H(N)$$

is an isomorphism as well.

Proof The facts that f induces $f^{(r)}$ and that $d^N_r \circ f^{(r)} = f^{(r)} \circ d^M_r$ are clear by an easy induction. Moreover f induces $f^{(\infty)} : E^M_\infty \longrightarrow E^N_\infty$ since f commutes with the differentials, and hence induces a morphism $H(f) : H(M) \longrightarrow H(N)$, and therefore induces $f^{(\infty)} : Gr(H(M)) \longrightarrow Gr(H(N))$. Theorem 3.9.9 then establishes the existence of $f^{(\infty)}$ on the infinite pages.

Since the filtrations on M and N are regular, the filtrations are also regular, regarded as filtrations of $H(M)$ and of $H(N)$. Now, if $f^{(r)}$ is an isomorphism in a specific degree (p,q), then $f^{(r+1)} = H(f^{(r)})$ is an isomorphism in the degree (p,q) as well, and therefore $f^{(s)}$ is an isomorphism for all $s \geq r$ in each degree. Since in each fixed degree the infinite page coincides with the degree r page for large enough r (depending on the degree), we get that $H(f)$ is an isomorphism. □

3.9.3 Spectral Sequence of a Double Complex

Very often we deal with a double complex, such as $X \otimes_A Y$ or $Hom_A(X, Y)$ where X and Y are two complexes of A-modules. More generally, a double complex (sometimes called bicomplex) is a $\mathbb{Z}^2$-graded A-module together with two differentials d' and d'' where d' is of degree $(-1, 0)$ and d'' is of degree $(0, -1)$. Further we suppose that $d' \circ d'' + d'' \circ d' = 0$. In the above example $X \otimes_A Y$ we obtain this by introducing an additional sign, as made explicit in the definition of the total complex in Definition 3.7.5, and in the case of $Hom_A(X, Y)$ one needs to define the degree of $Hom_A(X_n, X_m)$ as $(-n, m)$ and then an additional sign for the differential as made explicit in Definition 3.7.8.

Let $(M^{p,q}, d', d'')$ be a double complex. We can produce a simple complex from this by putting $Tot(M)^n := \sum_{k\in\mathbb{Z}} M^{n-k,k}$ and the differential $d = d' + d''$ induced by the differentials on the double complex. We obtain two different filtrations on $Tot(M)$

$$'F^pTot(M) := \sum_{i=p}^{\infty}\sum_{j\in\mathbb{Z}} M^{-i,-j}$$

and

$$''F^qTot(M) := \sum_{i\in\mathbb{Z}}\sum_{j=q}^{\infty} M^{-i,-j}.$$

Thus $Tot(M)$ is a filtered complex in two ways:

- First with respect to the filtration $'F^pTot(M)$ and differential d''. Call this the first filtration.
- Second with respect to the filtration $''F^qTot(M)$ and differential d'. Call this the second filtration.

The pages of the spectral sequence obtained by the first filtration are denoted by $'E_r^p$ and the pages of the spectral sequence obtained by the second filtration are denoted by $''E_r^q$.

Remark 3.9.12 We shall need to find the first pages of these spectral sequences. This is a general procedure. Let (V, d, F) be a differential filtered module, with differential d and filtration F. Then

$$E_0^p = F^pV/F^{p+1}V$$

and the differential d_0 on E_0 is still the differential on V. Hence

$$E_1^p = H(F^pV/F^{p+1}V, d).$$

What is the differential d_1 on E_1^p? We have the short exact sequence

$$0 \longrightarrow F^{p+1}V/F^{p+2}V \longrightarrow F^pV/F^{p+2}V \longrightarrow F^pV/F^{p+1}V \longrightarrow 0$$

and apply the homology functor. This yields a long exact sequence in homology (cf Proposition 3.5.29), obtained essentially by the snake lemma, part of which is

$$H(F^pV/F^{p+2}V) \to H(F^pV/F^{p+1}V) \xrightarrow{\delta} H(F^{p+1}V/F^{p+2}V) \to H(F^pV/F^{p+2}V)$$

where δ decreases the degree by 1. The construction of the snake lemma is precisely the way the differential d_1 is constructed, and therefore $\delta = d_1$.

We first obtain by Remark 3.9.12 that

$$'E_1^p = H('F^pTot(M)/'F^{p+1}Tot(M)) \simeq H(\bigoplus_{j\in\mathbb{Z}} M^{-p,-j}).$$

Here we observe the following convention. Since $d'(M^{-p,-j}) \subseteq M^{-p-1,-j}$, the differential $d' + d''$ of the total complex becomes d'', using the fact that, by the above, d' maps any term into the quotient, and hence is 0. Therefore

$$'E_1^p = H(\bigoplus_{j\in\mathbb{Z}} M^{-p,-j}, d'').$$

Similarly, by symmetry,

$$''E_1^q = H(\bigoplus_{i\in\mathbb{Z}} M^{-i,-q}, d').$$

We shall now compute the second page, again using Remark 3.9.12. We need to start with the short exact sequence

$$0 \to \frac{'F^{p+1}Tot(M)}{'F^{p+2}Tot(M)} \to \frac{'F^pTot(M)}{'F^{p+2}Tot(M)} \to \frac{'F^pTot(M)}{'F^{p+1}Tot(M)} \to 0,$$

and obtain from the connecting morphism δ of the induced homology sequence the differential $\delta = d_1$ on $'E_1^p$. This exact sequence actually comes in our case from the short exact sequence

$$0 \to \bigoplus_{j\in\mathbb{Z}} M^{(p+1),j} \to \left(\bigoplus_{j\in\mathbb{Z}} M^{(p+1),j} \oplus \bigoplus_{j\in\mathbb{Z}} M^{p,j}\right) \xrightarrow{\delta'} \bigoplus_{j\in\mathbb{Z}} M^{p,j} \to 0.$$

The leftmost and the rightmost terms carry the differential d'' and the term in the middle still carries the differential $d = d' + d''$. Taking homology of this short exact sequence we will obtain the connecting homomorphism

$$H(\bigoplus_{j\in\mathbb{Z}} M^{p,j}) \xrightarrow{\hat{\delta}} \bigoplus_{j\in\mathbb{Z}} M^{(p+1),j}$$

obtained by the snake lemma, which will give the differential d_1 on $'E_1^p$. Since $d = d' + d''$ on the middle term, and since $d'' = 0$ on the cycles (i.e. the elements of the kernel of the differential d'') of the right-hand term, we get that $\hat{\delta}$ is induced by d'. This shows that

$$'E_2^p = H^p(H(M, d''), d').$$

Since M is bigraded, $'E_2$ is bigraded as well. The first grading has already been obtained. The second grading comes from the the fact that the elements in $'E_2^{p,q}$ are presented by those elements in $'E_2^p$ coming from elements of degree $p+q$ in $Tot(M)$. The elements of $\bigoplus_{j\in\mathbb{Z}} M^{p,j}$ giving total degree $p+q$ are precisely the elements in $M^{p,q}$. This gives

$$'E_2^{p,q} = H^p(H^q(M,d''),d').$$

Likewise, by symmetry, we get

$$''E_2^{p,q} = H^q(H^p(M,d'),d'').$$

Definition 3.9.13 A spectral sequence $E_2^{p,q}$ of a double complex *converges* to some double graded module $E_\infty^{p,q}$ if for all p and q there is an $r(p,q)$ such that for all $r \geq r(p,q)$ we get $d_r^{p,q} = 0$ and $d_r^{p-r,q+r-1} = 0$. In particular, if the spectral sequence converges, then for all p and q we get $E_r^{p,q} = E_\infty^{p,q}$ whenever $r > r(p,q)$. If $E_r^{p,q}$ converges, we write

$$E_r^{p,q} \Rightarrow E_\infty^{p,q}.$$

Remark 3.9.14 We observe that this may happen frequently in the case of a double complex. In particular this happens if the complex obtained from the grading and differential in the second variable is bounded when one fixes the first index p. The same holds if one fixes the second index q and assumes that for each index q the complex in the first index and the first differential is bounded.

3.9.4 Short Exact Sequences Associated to Spectral Sequences

When one gets a spectral sequence $E_r^{p,q}$ associated to a filtered complex, then various short exact sequences relate the homology of the complex and the pages of the spectral sequences.

Proposition 3.9.15 *Let (M, d, F) be a filtered complex and suppose that there are integers $n \geq r \geq 1$ so that $E_r^{p,q} = 0$ for all $p \notin \{0, n\}$. Suppose that the filtration F is regular. Then there is an exact sequence*

$$\dots \to E_r^{n,i-n} \to H^i(M) \to E_r^{0,i} \to E_r^{n,i+1-n} \to H^{i+1}(M) \to \dots.$$

Proof As $E_\infty^{p,q} = \operatorname{colim}_r E_r^{p,q}$ by Lemma 3.9.10, we have $E_\infty^{p,q} = 0$ if $p \notin \{0, n\}$. By Theorem 3.9.9 we get that $E_\infty = Gr(H(M))$. This shows that there is an exact sequence

$$0 \longrightarrow E_\infty^{n,i-n} \longrightarrow H^i(M) \longrightarrow E_\infty^{0,i} \longrightarrow 0$$

for each degree i. Moreover, the only possibly non-zero differential is $d_n : E_n^{0,i} \longrightarrow E_n^{n,i-n+1}$ for each i. Hence no element in $E_n^{0,i}$ is in the image of a differential, whereas all elements of $E_n^{n,i-n+1}$ are in the kernel of the differential. Hence, using that $E_{n+1} = H(E_n)$, we obtain

$$E_\infty^{0,i} = E_{n+1}^{0,i} = \ker(E_n^{0,i} \xrightarrow{d_n} E_n^{n,i-n+1})$$

and

$$E_\infty^{n,i-n+1} = E_{n+1}^{n,n-i+1} = E_n^{n,i-n+1}/d_n(E_n^{0,i-1}).$$

This shows that we can compose

$$E_n^{n,i-n} \longrightarrow\!\!\!\!\to E_\infty^{n,i-n} \hookrightarrow H^i(M)$$

and similarly

$$H^i(M) \longrightarrow\!\!\!\!\to E_\infty^{0,i} \hookrightarrow E_n^{0,i}.$$

Composing these maps gives the exact sequence

$$E_n^{n,i-n} \to H^i(M) \to E_n^{0,i} \to E_n^{n,i+1-n}.$$

Since $E_r^{p,q} = E_n^{p,q}$ for $r \le n$ we obtain the sequence as requested. □

Remark 3.9.16 By the dual argument we get the dual statement. Let (M, d, F) be a filtered complex and suppose that there are integers $n \ge r \ge 1$ such that $E_r^{p,q} = 0$ for all $p \notin \{0, n\}$. Suppose that the filtration F is regular. Then there is an exact sequence

$$\ldots \to E_r^{i,0} \to H^i(M) \to E_r^{i-n,n} \to E_r^{i+1,0} \to H^{i+1}(M) \to \ldots.$$

3.9.5 Application: Homomorphisms in the Derived Category

We shall use these considerations to compute the homomorphisms in the homotopy category between two bounded complexes of projective modules as a spectral sequence. The arguments are due to Keller and are published in an appendix of [21].

Recall the intelligently truncated complex $\tau_{\ge m}X$ which is defined by the following construction.

$$\begin{array}{rcccccccc}
X: & \ldots \longrightarrow & X_{p+1} & \xrightarrow{d_{p+1}} & X_p & \xrightarrow{d_p} & X_{p-1} & \xrightarrow{d_{p-1}} & X_{p-2} \longrightarrow \ldots \\
 & & \| & & \| & & \uparrow \text{incl} & & \uparrow \\
\tau_{\ge p-1}X: & \ldots \longrightarrow & X_{p+1} & \xrightarrow{d_{p+1}} & X_p & \xrightarrow{d_p} & \ker(d_{p-1}) & \xrightarrow{0} & 0 \longrightarrow \ldots
\end{array}$$

We see that there is a sequence of natural morphisms

$$\tau_{\geq p}X \xrightarrow{\lambda_p^X} \tau_{\geq p-1}X \xrightarrow{\lambda_{p-1}^X} \tau_{\geq p-2}X \longrightarrow \ldots \longrightarrow X$$

and by composition

$$\lambda_p^\infty : \tau_{\geq p}X \longrightarrow X.$$

Moreover, we get an exact sequence of complexes

$$0 \longrightarrow \tau_{\geq p}X \xrightarrow{\lambda_p^X} \tau_{\geq p-1}X \longrightarrow (H_{p-1}(X))[p-1] \longrightarrow 0.$$

Now, $\mathbb{R}Hom_{D^b(A)}(X, Y)$ is a filtered complex, where the filtration is induced by the maps

$$\mathbb{R}Hom_{D^b(A)}(X, \lambda_p^Y) : \mathbb{R}Hom_{D^b(A)}(X, \tau_{\geq p}Y) \longrightarrow \mathbb{R}Hom_{D^b(A)}(X, Y).$$

We know how to compute the pages $E_1^{p,q}$ and $E_2^{p,q}$ of the spectral sequence. We get

$$E_1^{p,q} = \prod_{\ell\in\mathbb{Z}} Ext_A^{2p+q}(H_\ell(X), H_{\ell+p}(Y)).$$

It is clear that if X and Y have bounded homology, then $E_1^{p,q}$ will be non-zero in only finitely many positions. Therefore the spectral sequence converges. We have proved the following statement.

Proposition 3.9.17 *Let K be a commutative ring and let A be a K-algebra. Then for every two objects X and Y in $D^b(A\text{-}Mod)$ there is a converging spectral sequence*

$$E_1^{p,q} = \prod_{\ell\in\mathbb{Z}} Ext_A^{2p+q}(H_\ell(X), H_{\ell+p}(Y)) \Rightarrow Hom_{D^b(A)}(X, Y[p+q]).$$

Proof The result follows directly from the above discussion and Definition 3.9.13. □

Another consideration comes to a different spectral sequence for homomorphisms. This is a result due from Verdier [2, Section 4.6].

Proposition 3.9.18 *Let K be a commutative ring, let A be a K-algebra, and let X and Y be two complexes in $K^b(A\text{-}proj)$. Then there is a spectral sequence $E_r^{p,q}$ with $E_1^{p,q} = Hom_A(X_p, H^q(Y))$ and*

$$E_2^{p,q} = \prod_{\ell\in\mathbb{Z}} Ext_A^p(H^{\ell-q}(X), H^\ell(Y)) \Rightarrow E_\infty^{p,q} = Hom_{K^b(A\text{-}proj)}(X, Y[p+q]).$$

Proof We consider the double complex $Hom_A(X, Y)$ with

$$deg(Hom_A(X_n, X_m)) = (-n, m)$$

and the two differentials induced by the differential on X and the differential on Y, equipped with appropriate alternating signs on d^X. Then the above discussion on double complexes shows that the spectral sequence of the filtration of $Tot(Hom_A(X, Y))$ by the complex X has term

$$\begin{aligned} E_2^{p,q} &= H^p(H^q(Hom_A(X, Y), d^Y), d^X) \\ &= H^p(Hom_A(X, H^q(Y)), d^X) \\ &= \prod_{\ell \in \mathbb{Z}} Ext_A^p(H^{\ell-q}(X), H^{\ell}(Y)) \end{aligned}$$

and term $E_1^{p,q} = Hom_A(X_p, H^q(Y))$.

Since X and Y are both bounded complexes of projective modules, the bicomplex $Hom_A(X, Y)$ is actually bounded in finitely many degrees. Therefore by Remark 3.9.14 the spectral sequence converges in the sense of Definition 3.9.13.

The degree $p + q$ homology of the total complex is

$$Hom_{K^b(A\text{-}proj)}(X, Y[p+q])$$

by Lemma 3.7.10. By Theorem 3.9.9 we know that E_∞ can be identified with the homology of the total complex and

$$E_\infty^{p,q} = H^{p+q}(Hom_{K^b(A\text{-}proj)}(X, Y)) = Hom_{K^b}(A\text{-}proj)(X, Y[p+q]).$$

This proves the result. □

References

1. Mac Lane, S.: Categories for the Working Mathematician, 2nd edn. Springer, New York (1996)
2. Verdier, J.-L.: Des catégories dérivées des catégories abéliennes. Astérisque 239. Societe Mathematique de France, Paris (1996)
3. Weibel, C.: An Introduction to Homological Algebra. Cambridge University Press, Cambridge (1995)
4. Bourbaki, N.: Univers, appendix to Exposé I. In: Artin, M., Grothendieck, A., Verdier, J.L. (eds.) Théorie des Topos et Cohomologie Étale des Schémas (SGA 4), 2nd edn. Springer, Heidelberg (1972)
5. Keller, B.: Chain complexes and stable categories. Manuscripta Math. **67**, 379–417 (1990)
6. Watts, C.E.: Intrinsic characterizations of some additive functors. Proc. Am. Math. Soc. **11**, 5–8 (1960)
7. Eilenberg, S.: Abstract description of some basic functors. J. Indian Math. Soc. (New Series) **24**, 231–234 (1960)

8. Bourbaki, N.: Éléments de mathématique. Algèbre Commutative. Springer, Berlin (2006). Chapitres 1–4
9. May, J.P.: The additivity of traces in triangulated categories. Adv. Math. **163**, 34–73 (2001)
10. Spanier, E.H.: Algebraic Topology. Springer, New York (1981)
11. Gabriel, P., Zisman, M.: Calculus of Fractions and Homotopy Theory. Springer, Heidelberg (1967)
12. Spaltenstein, N.: Resolutions of unbounded complexes. Compositio Math. **65**, 121–154 (1988)
13. Külshammer, B.: Nilpotent Blocks Revisited in "Groups, Rings and Group Rings", pp. 263–274. Chapman and Hall, Boca Raton (2006)
14. Holm, T., Zimmermann, A.: Deformed preprojective algebras of type L: Külshammer spaces and derived equivalences. J. Algebra **346**, 116–146 (2011)
15. Verdier, J.-L.: Catégories dérivées; Quelques résultats (Etat 0), SGA 4 1/2. Lect. Notes Math. **569**, 262–311 (1977)
16. Lam, T.-Y.: Lectures on Modules and Rings. Springer, New York (1998)
17. Zimmermann, A.: A Noether-Deuring theorem for derived categories. Glasgow Math. J. **54**(3), 647–654 (2012)
18. Chen, X.-W., Ye, Y., Zhang, P.: Algebras of derived dimension zero. Commun. Algebra **36**, 1–10 (2008)
19. Roggenkamp, K.W.: An extension of the Noether-Deuring theorem. Proc. Am. Math. Soc. **31**, 423–426 (1972)
20. Godement, R.: Théorie des faisceaux. Hermann, Paris (1958)
21. Muchtadi-Alamsyah, I.: Homomorphisms of complexes via homologies. J. Algebra **294**, 321–345 (2005)

Chapter 4
Morita Theory

We ask for practical conditions for two rings A and B to have equivalent module categories A-*Mod* $\simeq$ B-*Mod*. This question was completely solved in the 1950s by K. Morita. The treatment which was given at that time does not clearly correspond to modern considerations. However, as is now clear, that theory is the archetype for weaker and more general classes of equivalences between representation theoretic categories. Most surprisingly, with respect to certain features, Example 3.1.25 (2) already describes the most general setting.

4.1 Progenerators

The crucial objects in Morita theory are progenerators.

Definition 4.1.1 Let $\mathcal{C}$ be a category. An object X in $\mathcal{C}$ is called a *generator* if the functor $Mor_{\mathcal{C}}(X, -) : \mathcal{C} \longrightarrow \mathcal{E}ns$ is faithful.

An object Y in $\mathcal{C}$ is called a *cogenerator* if the functor $Mor_{\mathcal{C}}(-, X) : \mathcal{C}^{op} \longrightarrow \mathcal{E}ns$ is faithful.

Example 4.1.2 If R is a ring, then the regular left module is a generator for the category R-*Mod* of R-modules. Indeed, $Hom_R(R, -) = id_{R\text{-}Mod}$, which is clearly faithful. However, in general R is not a cogenerator for R-*Mod*. For example, for $R = \mathbb{Z}$, the functor $Hom_{\mathbb{Z}}(-, \mathbb{Z})$ sends each of the $\mathbb{Z}$-modules $\mathbb{Z}/n\mathbb{Z}$ to 0.

Proposition 4.1.3 *Let $\mathcal{A} = R$-Mod for a ring R. An R-module X is a generator if and only if each R-module A is a quotient of a direct sum of copies of X.*

Proof Let X be a generator. We shall prove that each R-module is a quotient of a direct sum of copies of X.

A. Zimmermann, *Representation Theory*,
Algebra and Applications 19, DOI: 10.1007/978-3-319-07968-4_4,

Put

$$Y := \bigoplus_{f \in Hom_R(X,A)} X_f$$

where $X_f := X$. This is a direct sum of copies of X, one for each morphism. Hence, each component of the sum comes with a morphism $X \longrightarrow A$. Define $\eta \in Hom_R(X, A)$ by

$$\begin{aligned} \eta : Y &\longrightarrow A \\ \sum_{f \in Hom_R(X,A)} x_f &\mapsto \sum_{f \in Hom_R(X,A)} f(x_f) \end{aligned}$$

Observe that as Y is defined as a direct sum, all but a finite number of x_f are 0. Hence the sum on the right is well-defined. We claim that η is surjective. If this were not be the case then there would be a non-zero morphism $\alpha : A \longrightarrow A/\mathrm{im}(\eta)$. Since X is a generator there is then an $f : X \longrightarrow A$ such that $\alpha \circ f \neq 0$. But $f(X) \subseteq \eta(Y)$ and so $(\alpha \circ f)(X) = \alpha(f(X)) \subseteq \alpha(\eta(Y)) = 0$. This is a contradiction, proving that A is a quotient of a direct sum of copies of X.

Suppose conversely that for each R-module A there is an epimorphism

$$\bigoplus_{i \in I_A} X_i \xrightarrow{\xi_A} A$$

where $X_i = X$ for all i. Let $\alpha \in Hom_R(A, A') \setminus \{0\}$. Since ξ_A is surjective we get $\alpha \circ \xi_A \neq 0$. Since ξ_A has as source a direct sum we may write

$$\xi_A = \sum_{i \in I_A} \xi_A^{(i)}$$

where $\xi_A^{(i)} \in Hom_R(X, A)$. Therefore

$$\alpha \circ \xi_A = \sum_{i \in I_A} \alpha \circ \xi_A^{(i)}.$$

Hence there is an $i_0 \in I_A$ such that $\alpha \circ \xi_A^{(i_0)} \neq 0$. This proves the statement. □

Being a generator is a functorial property, as is seen by the definition.

Lemma 4.1.4 *Let $\mathcal{C}$ and $\mathcal{D}$ be categories and let X be a generator (resp. cogenerator) of $\mathcal{C}$. If $F : \mathcal{C} \longrightarrow \mathcal{D}$ is an equivalence, then $F(X)$ is a generator (resp. cogenerator) as well.*

Proof X is a generator if and only if $Mor_{\mathcal{C}}(X, -)$ is faithful. Applying F gives a bijection $Mor_{\mathcal{C}}(-, ?) \longrightarrow Mor_{\mathcal{D}}(F-, F?)$. Since F is essentially surjective, every object in $\mathcal{D}$ is isomorphic to an object of the form $F?$. Therefore $Mor_{\mathcal{D}}(FX, -)$ is faithful, and therefore X is a generator. The case for cogenerators is proved in a completely analogous fashion and is left as an exercise to the reader. □

Given a commutative ring K and K-algebras A and B we suppose there is an equivalence $F : B\text{-}Mod \longrightarrow A\text{-}Mod$. Clearly B is a generator of B-Mod. By Lemma 4.1.4 $M := F(B)$ is also a generator of A-Mod. Since F is an equivalence we get (cf Lemma 1.4.14)

$$End_A(M) \simeq End_B(B) \simeq B^{op}.$$

Hence M is an $A - B^{op}$-bimodule by putting

$$a \cdot m \cdot f = f(am)$$

for all $m \in M$, $a \in A$, $f \in End_A(M)$ and $Hom_A(M, A)$ is a B^{op}-A-bimodule by putting

$$(f \cdot g \cdot a)(m) = g(f(am))$$

for all $m \in M$, $g \in Hom_A(M, A)$, $a \in A$, $f \in End_A(M) = B^{op}$.

Proposition 4.1.5 *Let K be a commutative ring and let A be a K-algebra. The A-module M is a generator of A-mod if and only if*

$$\begin{array}{rcl} M \otimes_{B^{op}} Hom_A(M, A) & \stackrel{\epsilon}{\longrightarrow} & A \\ m \otimes f & \mapsto & f(m) \end{array}$$

is an epimorphism of A-A-bimodules for $B = End_A(M)^{op}$.

Proof We shall first show that ϵ is well-defined. Indeed, for any $b \in B^{op}$ we get

$$\epsilon(mb \otimes f) = f(mb) = (bf)(m) = \epsilon(m \otimes bf)$$

and ϵ is A-A-bilinear by

$$a_1\epsilon(m \otimes f)a_2 = a_1(f(m))a_2 = f(a_1m)a_2 = \epsilon(a_1m \otimes fa_2)$$

for all $a_1, a_2 \in A$.

Suppose that M is a generator. Then there is an $r \in \mathbb{N}$ such that the regular module A is a quotient of M^r. Let

$$M^r \stackrel{(f_1,\dots,f_r)}{\longrightarrow} A$$

be the epimorphism. Now, there are elements $m_1, \dots, m_r$ with $\sum_{i=1}^r f_i(m_i) = 1_A$. But then $\epsilon(\sum_{i=1}^r m_i \otimes f_i) = 1_A$. This proves that ϵ is surjective.

Suppose ϵ is surjective. Then there are elements $f_i \in Hom_A(M, A)$ and $m_i \in M$ such that

$$1_B = \epsilon\left(\sum_{i=1}^{s} m_i \otimes f_i\right)$$

which implies that the mapping

$$\begin{array}{rcl} M^s & \longrightarrow & A \\ (n_1, n_2, \dots, n_s) & \mapsto & \sum_{i=1}^{s} f_i(n_i) \end{array}$$

is surjective, since it is A-linear and maps $(m_1, \dots, m_s)$ to 1_A. Since for every A-module X there is an $r(X) \in \mathbb{N}$ and an epimorphism $A^{r(X)} \longrightarrow X$ there is also an epimorphism

$$M^{r(X) \cdot s} \longrightarrow A^{r(X)} \longrightarrow X.$$

Lemma 4.1.3 implies that M is a generator. □

Proposition 4.1.6 *Let K be a commutative ring and let A be a K-algebra. Suppose that M is a generator of A-mod with endomorphism ring B^{op}. Put $N := Hom_A(M, A)$. Then*

$$\epsilon : M \otimes_{B^{op}} N \longrightarrow A$$

given by $\epsilon(m \otimes f) = f(m)$ is bijective.

Proof By Proposition 4.1.5 the mapping ϵ is surjective. We need to show that ϵ is injective. Since M is a generator of A-*mod*, there are $f_i \in Hom_A(M, A)$ and $m_i \in M$ for $i \in \{1, \dots, r\}$ such that

$$1_A = \sum_{i=1}^{s} f_i(m_i) = \epsilon\left(\sum_{i=1}^{r} m_i \otimes f_i\right).$$

Let $\sum_{j=1}^{t} n_j \otimes g_j \in \ker(\epsilon)$. Then any element $m \otimes f \in M \otimes_{B^{op}} N$ induces an A-module endomorphism $f^{(m)}$ of M by putting $n \mapsto f(n) \cdot m = f^{(m)}(n)$. Observe that this shows that

$$f(m) \cdot g(n) = g(f(m)(n)) = (g \circ f^{(n)})(m) = \left(f^{(n)} \cdot g\right)(m)$$

for all $n, m \in M$, all $g \in Hom_A(M, A)$ and $f \in End_A(M)$, where we have in mind that $B^{op} = End_A(M)$. Hence

$$f^{(n)} \cdot g = f \cdot g(n).$$

We obtain for all $f \in End_A(M)$, all $m, n \in M$ and all $g \in Hom_A(M, A)$ that, using the action of the different objects on each other, and denoting this action by "$\cdot$",

$$\begin{aligned} f(m) \cdot (n \otimes g) &= (f(m) \cdot n) \otimes g = f^{(n)}(m) \otimes g = (m \cdot f^{(n)}) \otimes g \\ &= m \otimes (f^{(n)} \cdot g) = m \otimes f \cdot g(n). \end{aligned}$$

With this we get

$$\begin{aligned} \sum_{j=1}^{t} n_j \otimes g_j &= \left(\sum_{i=1}^{r} f_i(m_i)\right) \cdot \left(\sum_{j=1}^{t} (n_j \otimes g_j)\right) = \sum_{i=1}^{r} \sum_{j=1}^{t} f_i(m_i) \cdot (n_j \otimes g_j) \\ &= \sum_{i=1}^{r} \sum_{j=1}^{t} m_i \otimes f_i \cdot g_j(n_j) = \sum_{i=1}^{r} \left(m_i \otimes f_i \cdot \left(\sum_{j=1}^{t} g_j(n_j) \right) \right) \\ &= \sum_{i=1}^{r} \left(m_i \otimes f_i \cdot \epsilon \left(\sum_{j=1}^{t} n_j \otimes g_j \right) \right) = 0 \end{aligned}$$

since $\epsilon \left(\sum_{j=1}^{t} n_j \otimes g_j \right) = 0$, and where the third equation comes from the equation just above this set of equalities. Hence ϵ is injective as well. □

The following concept is of importance beyond the discussion here.

Definition 4.1.7 Let R and S be two rings. A functor $F : R\text{-}Mod \longrightarrow S\text{-}Mod$ *preserves direct sums* if $F(\oplus_{i \in I} X_i) \simeq \oplus_{i \in I} F(X_i)$ for all modules X_i and $i \in I$.

Remark 4.1.8 In general functors $F : R\text{-}mod \longrightarrow S\text{-}mod$ do not preserve direct sums. When they do, we know quite a number of things concerning their behaviour. An example of a functor which does not preserve direct sums is the functor $F : K\text{-}mod \longrightarrow K\text{-}mod$, for a fixed field K, defined by $F : X \mapsto X \otimes_K X$. Then

$$F(X \oplus Y) \simeq F(X) \oplus F(Y) \oplus (X \otimes_K Y) \oplus (Y \otimes_K X).$$

This fact can be studied in a more systematic way. Indeed, let $\pi_X : X \oplus Y \longrightarrow X$ be the projection onto the first component, and π_Y the projection onto the second component. Then $F(\pi_X) \oplus F(\pi_Y)$ induces a mapping $F(X \oplus Y) \longrightarrow F(X) \oplus F(Y)$, which is moreover functorial in X and Y. Let

$$F^{(2)}(X|Y) := \ker(F(\pi_X) \oplus F(\pi_Y))$$

which defines a functor

$$F^{(2)}(-_1, -_2) : K\text{-}mod \times K\text{-}mod \longrightarrow K\text{-}mod.$$

This functor is the *cross effect* of F and it is obvious that F preserves direct sums if and only if its cross effect is 0. It is possible to iterate this, with respect to the first variable, to obtain a functor $F^{(3)}(-_1, -_2, -_3) : K\text{-}mod \times K\text{-}mod \times K\text{-}mod \longrightarrow K\text{-}mod$, the third cross effect. Iterating, one defines a functor $F^{(n)}$ in n variables and we say that F is *polynomial of degree at most n* if $F^{(n+1)} = 0$. The theory of polynomial functors is closely related to the representation theory of the symmetric group and has been intensively studied by various authors, such as Kuhn, Franjou, Friedlander, Pirashvili, Schwartz, Suslin and others.

Recall from Definition 3.3.15 the concept of an exact functor, a right exact functor and of a left exact functor. Exact functors play an important role and in particular for module categories this is a very desirable property. For a ring A and an A-module M the functor $Hom_A(M, -) : A\text{-}mod \longrightarrow \mathcal{A}b$ is left exact. The functor $- \otimes_A M : mod\text{-}A \longrightarrow \mathcal{A}b$ is right exact.

Indeed, if $\alpha : N \longrightarrow L$ is injective, then clearly the mapping $\beta \mapsto \beta \circ \alpha$ is injective as well for all $\beta : M \longrightarrow N$. If $\gamma : L \longrightarrow K$ is surjective, then clearly $id \otimes_A \gamma$ is also surjective.

Both functors are usually not exact. Proposition 1.8.5 shows that for the *Hom*-functor surjections are mapped to surjections only when we restrict the functor to the full subcategory of injective modules.

Lemma 4.1.9 *If A is an R-algebra for a commutative ring R and if P is a projective A-module, then $Hom_A(P, -)$ preserves arbitrary coproducts if and only if P is finitely generated.*

Proof This is a special case of Lemma 3.3.13. □

We shall use exactness in the following statement.

Proposition 4.1.10 *Let $\mathcal{A} = \Lambda\text{-}Mod$ and $\mathcal{B} = \Gamma\text{-}Mod$ (or more generally two abelian categories) for two algebras Λ and Γ. Let $F : \mathcal{A} \longrightarrow \mathcal{B}$ and $G : \mathcal{A} \longrightarrow \mathcal{B}$ be two right exact functors. Suppose F and G preserve direct sums and suppose $t : F \longrightarrow G$ is a natural transformation. If X is a generator of $\mathcal{A}$ and if $t_X : F(X) \longrightarrow G(X)$ is an isomorphism, then t is an isomorphism of functors.*

Proof For every $\alpha : A \longrightarrow A'$ the diagram

$$\begin{array}{ccc} FA & \xrightarrow{F\alpha} & FA' \\ t_A \downarrow & & \downarrow t_{A'} \\ GA & \xrightarrow{G\alpha} & GA' \end{array}$$

is commutative. Since X is a generator we get exact sequences

$$\bigoplus_{J_A} X \longrightarrow \bigoplus_{I_A} X \longrightarrow A \longrightarrow 0$$

and

$$\bigoplus_{J'_A} X \longrightarrow \bigoplus_{I'_A} X \longrightarrow A' \longrightarrow 0.$$

Since F and G are right exact we get a commutative diagram with exact lines

$$\begin{array}{ccccc} F\left(\bigoplus_{J_A} X\right) & \longrightarrow & F\left(\bigoplus_{I_A} X\right) & \longrightarrow F(A) \longrightarrow 0 \\ \downarrow t_{\oplus_{J_A} X} & & \downarrow t_{\oplus_{I_A} X} & \downarrow t_A \\ G\left(\bigoplus_{J_A} X\right) & \longrightarrow & G\left(\bigoplus_{I_A} X\right) & \longrightarrow G(A) \longrightarrow 0 \end{array}$$

Since F and G preserve direct sums and since t is natural, and hence t of a direct sum is the direct sum of the t's,

$$\begin{array}{ccccc} \bigoplus_{J_A} F(X) & \longrightarrow & \bigoplus_{I_A} F(X) & \longrightarrow F(A) \longrightarrow 0 \\ \downarrow \oplus_{J_A} t_X & & \downarrow \oplus_{I_A} t_X & \downarrow t_A \\ \bigoplus_{J_A} G(X) & \longrightarrow & \bigoplus_{I_A} G(X) & \longrightarrow G(A) \longrightarrow 0 \end{array}$$

is a commutative diagram with exact lines. Each t_X is an isomorphism, and therefore t_A is an isomorphism as well. □

4.2 The Morita Theorem

Lemma 4.2.1 (Gabriel [1, Proposition 13]) *Let $\mathcal{A}$ and $\mathcal{B}$ be two abelian categories. Let $F : \mathcal{A} \longrightarrow \mathcal{B}$ be an equivalence. Then F is exact. Moreover $F(P)$ is projective whenever P is projective.*

Proof We first show that F is right exact. Since F is an equivalence, F is essentially surjective and any object of $\mathcal{B}$ is isomorphic to an object of the form FX for an object X. Let $\alpha \in Hom_{\mathcal{A}}(X, Y)$ be an epimorphism and put $F\alpha = \overline{\alpha}$. We need to show that $\overline{\alpha}$ is an epimorphism as well. Let $\overline{\beta} \in Hom_{\mathcal{B}}(FY, FZ)$ and $\overline{\gamma} \in Hom_{\mathcal{B}}(FY, FZ)$ and let $\beta \in Hom_{\mathcal{A}}(Y, Z)$ and $\gamma \in Hom_{\mathcal{A}}(Y, Z)$ such that $F\beta = \overline{\beta}$ and $F\gamma = \overline{\gamma}$. Then

$$\begin{aligned} \overline{\beta} \circ \overline{\alpha} = \overline{\gamma} \circ \overline{\alpha} &\Rightarrow F\beta \circ F\alpha = F\gamma \circ F\alpha \Rightarrow F(\beta \circ \alpha) = F(\gamma \circ \alpha) \\ &\Rightarrow \beta \circ \alpha = \gamma \circ \alpha \Rightarrow \beta = \gamma \Rightarrow F\beta = \overline{\beta} = \overline{\gamma} = F\gamma \end{aligned}$$

and so $F\alpha$ is an epimorphism.

By a similar argument one gets that if α is a monomorphism then $F\alpha$ is a monomorphism. Let $0 \to X \stackrel{\alpha}{\to} Y \stackrel{\beta}{\to} Z \to 0$ be an exact sequence. Then $F\alpha$ is a monomorphism and let $FY \stackrel{\gamma_{F\alpha}}{\longrightarrow}$ coker$(F\alpha)$ be the natural mapping to the cokernel of $F\alpha$. By the universal property of the cokernel, there is a unique morphism coker$(F\alpha) \stackrel{\delta}{\to} FZ$ such that $\delta \circ \gamma_{F\alpha} = F\beta$. Apply a (quasi-)inverse G of F to the exact sequence $0 \to FX \stackrel{F\alpha}{\to} FY \stackrel{\gamma_{F\alpha}}{\to}$ coker$(F\alpha) \to 0$. By the same argument as above we obtain a unique morphism $Z \stackrel{\varepsilon}{\to} G(\text{coker}(F\alpha))$ such that $G\gamma_{F\alpha} = \varepsilon \circ \beta$. Since $F\beta$ is an epimorphism, $\delta \circ F\varepsilon = id_{FZ}$ and ε as well as δ are isomorphisms. Therefore F is exact.

Let $\overline{\alpha} \in Hom_{\mathcal{B}}(FX, FY)$ be an epimorphism. Since F is an equivalence, $\overline{\alpha} = F(\alpha)$ for a unique epimorphism $\alpha \in Hom_{\mathcal{A}}(X, Y)$. Given a projective object P in $\mathcal{A}$ and a $\overline{\delta} \in Hom_{\mathcal{B}}(FP, FY)$, there is a unique $\delta \in Hom_{\mathcal{A}}(P, Y)$ with $F\delta = \overline{\delta}$. Since P is projective, there is an $\epsilon \in Hom_{\mathcal{A}}(P, X)$ such that $\alpha \circ \epsilon = \delta$. Applying F gives $\overline{\alpha} \circ \overline{\epsilon} = \overline{\delta}$ and hence $F(P)$ is projective. □

Corollary 4.2.2 *Suppose ${}_AM_B$ is an A-B bimodule and suppose that $M \otimes_B -$: B-Mod $\longrightarrow$ A-Mod is an equivalence. Then M is flat as a B-module and projective as an A-module.*

Proof Indeed, by Lemma 4.2.1 $M \otimes_B -$: *B-Mod* $\longrightarrow$ *A-Mod* is exact. Therefore M is flat as a B-module. Since the regular B-module ${}_BB$ is projective, again by Lemma 4.2.1 $M \otimes_B B \simeq M$ is also a projective A-module. □

Definition 4.2.3 A *progenerator* in a category $\mathcal{C}$ which admits cokernels is a finitely generated projective object which is also a generator.

Observe that whenever M is an A-module, then M is actually an $A - B$-bimodule, where $B = End_A(M)^{op}$. Hence $Hom_A(M, V)$ is a B-module for each A-module V, and actually $Hom_A(M, -)$ is a functor *A-Mod* $\longrightarrow$ *B-Mod*. We can now prove the first main result.

Proposition 4.2.4 *Let A be a K-algebra and let K be a commutative ring. Let P be a progenerator. Put $B := End_A(P)^{op}$. Then*

$$Hom_A(P, -) : A\text{-}Mod \longrightarrow B\text{-}Mod$$

is an equivalence which restricts to an equivalence A-mod $\longrightarrow$ B-mod if A is Noetherian.

Proof By Lemma 4.1.9 we get that $Hom_A(P, -)$ commutes with direct sums. Since P is projective, $Hom_A(P, -)$ is exact. Indeed, the functor $Hom_A(M, -)$ is always left exact. For a projective module P the functor $Hom_A(P, -)$ is right exact as well. Indeed, for a projective module P and any surjection $\alpha : X \longrightarrow Y$ any homomorphism $\beta : P \longrightarrow Y$ "lifts" to a homomorphism $\gamma : P \longrightarrow X$ in such a way that $\beta = \alpha \circ \gamma$. This is precisely the right exactness of $Hom_A(P, -)$. Let $F := Hom_A(P, -)$.

We shall show first that F is fully faithful. If $X = \coprod_{\lambda\in I_A} P_\lambda$ for $P_\lambda = P$ for all λ, then

$$Hom_A(X,Y) = Hom_A\left(\coprod_{\lambda\in I_A} P_\lambda, Y\right) = \prod_{\lambda\in I_A} Hom_A(P_\lambda, Y) = \prod_{\lambda\in I_A} F(Y).$$

By Lemma 4.1.9 we know that F preserves direct sums and we get

$$\begin{aligned} Hom_B(FX,FY) = Hom_B\left(\coprod_{\lambda\in I_A} F(P), F(Y)\right) &= \prod_{\lambda\in I_A} Hom_B(F(P), F(Y)) \\ &= \prod_{\lambda\in I_A} Hom_B(B, F(Y)) = \prod_{\lambda\in I_A} F(Y). \end{aligned}$$

If X is any A-module. Then, using that P is a generator, there are modules U and V, each a direct sum of copies of P, and an exact sequence

$$U \longrightarrow V \longrightarrow X \longrightarrow 0.$$

Since F is exact,

$$F(U) \longrightarrow F(V) \longrightarrow F(X) \longrightarrow 0$$

is also exact. We get a commutative diagram with exact lines

$$\begin{array}{ccccccc} 0 \longrightarrow & Hom_A(X,Y) & \longrightarrow & Hom_A(V,Y) & \longrightarrow & Hom_A(U,Y) \\ & \downarrow F_1 & & \downarrow F_2 & & \downarrow F_3 \\ 0 \longrightarrow & Hom_B(FX,FY) & \longrightarrow & Hom_B(FV,FY) & \longrightarrow & Hom_B(FU,FY) \end{array}$$

where F_2 and F_3 are isomorphisms by the first case, discussed above, where we assumed X is a direct sum of copies of P. Hence, F_1 is an isomorphism as well. This shows that $Hom_A(P,-)$ is fully faithful.

We need to show that $Hom_A(P,-)$ is essentially surjective. Let M be a B-module. Then we may choose two free B-modules $L_0 = \bigoplus_{I_0} B$ and $L_1 = \bigoplus_{I_1} B$ and an exact sequence

$$L_1 \stackrel{\zeta}{\longrightarrow} L_0 \longrightarrow M \longrightarrow 0.$$

Since $F(P) = B$ we obtain A-modules U_0 and U_1, each of which is a direct sum of copies of P, so that $F(U_0) = L_0$ and $F(U_1) = L_1$. Moreover, since F is fully faithful

$$Hom_B(L_1, L_0) = Hom_B(FU_1, FU_0) \simeq Hom_A(U_1, U_0)$$

and therefore $F(\xi) = \zeta$ for some $\xi \in Hom_A(U_1, U_0)$. Putting $N := \text{coker}(\alpha)$, using the fact that F is exact, we obtain a commutative diagram with exact rows

$$\begin{array}{ccccccc} F(U_1) & \xrightarrow{F(\xi)} & F(U_0) & \longrightarrow & F(N) & \longrightarrow & 0 \\ \| & & \| & & \downarrow & & \\ L_1 & \xrightarrow{\zeta} & L_1 & \longrightarrow & M & \longrightarrow & 0 \end{array}$$

which implies that $F(N) \simeq M$.

If A is Noetherian, then the endomorphism ring B^{op} of P is Noetherian as well and hence finitely generated modules are just the same as finitely presented modules for A as well as for B. But then, by Lemma 3.3.13, the compact objects in A-*Mod* are the objects in A-*mod* and likewise the compact objects in B-*Mod* are the compact objects in B-*mod*. Being compact is preserved under F and hence we have shown that $Hom_A(P, -)$ restricts to an equivalence A-*mod* $\longrightarrow$ B-*mod*. □

We shall frequently use the following technical lemma, and due to its importance for the following arguments we will give its proof in full detail.

Lemma 4.2.5 *Let K be a commutative ring and let A and B be K-algebras. Let M be an A-B-bimodule. Then there is a natural transformation*

$$\psi : (Hom_A(M, A) \otimes_A -) \longrightarrow Hom_A(M, -)$$

of functors A-mod $\longrightarrow$ B-Mod.

If M is finitely generated projective as an A-module, then $Hom_A(M, A)$ is a finitely generated projective right A-module and ψ is an equivalence of functors A-mod $\longrightarrow$ B-Mod.

Proof We have a morphism

$$\begin{aligned} [Hom_A(M, A) \otimes_A -] &\xrightarrow{\psi} [Hom_A(M, -)] \\ [f \otimes -] &\mapsto [m \mapsto (f(m) \cdot -)] \end{aligned}$$

We get that

$$\psi_V : Hom_A(M, A) \otimes_A V \longrightarrow Hom_A(M, V)$$

is B-linear for every A-module V. Indeed, for $f \in Hom_A(M, A)$, $m \in M$ and $v \in V$, and for $b \in B$ we get

$$\begin{aligned} \psi_V(b \cdot (f \otimes v))(m) &= \psi_V((bf) \otimes v)(m) = (bf)(m) \cdot v \\ &= f(mb) \cdot v = b \cdot [m \mapsto f(m)v] = (b \cdot \psi_V(f \otimes v))(m). \end{aligned}$$

Hence ψ_V is B-linear for each A-module V. Moreover, ψ is a natural transformation of functors A-*mod* $\longrightarrow$ B-*Mod*. Indeed, for all A-modules V and W and each A-module homomorphism $\varphi : V \longrightarrow W$ we need to show that

$$\begin{array}{ccc} [Hom_A(M,A)\otimes_A V] & \xrightarrow{\psi_V} & Hom_A(M,V) \\ {\scriptstyle Hom_A(M,A)\otimes_A \varphi}\downarrow & & \downarrow {\scriptstyle Hom_A(M,\varphi)} \\ [Hom_A(M,A)\otimes_A W] & \xrightarrow{\psi_W} & Hom_A(M,W) \end{array}$$

is commutative. But

$$\begin{aligned}(Hom_A(M,\varphi)\circ\psi_V)(f\otimes v) &= Hom_A(M,\varphi)\,([m\mapsto f(m)\cdot v]) \\ &= [m\mapsto \varphi(f(m)\cdot v)] \\ &= [m\mapsto f(m)\cdot\varphi(v)] \text{ since } \varphi \text{ is } A\text{-linear} \\ &= (\psi_W\circ(Hom_A(M,A)\otimes\varphi))\,(f\otimes v)\end{aligned}$$

for all $f\in Hom_A(M,A)$, all $v\in V$ and all $m\in M$.

Hence

$$\psi : (Hom_A(M,A)\otimes_A -)\longrightarrow Hom_A(M,-)$$

is a natural transformation.

Suppose now that M is finitely generated projective. Then the A-module $Hom_A(M,A)$ is projective. Indeed, if M is finitely generated free, then

$$Hom_A(A^s,A)=Hom_A(A,A)^s$$

is a free A-module and

$$Hom_A(A^s,A)\otimes_A V=V^s=Hom_A(A^s,V).$$

Since M is a direct summand of finitely generated free A-module, and since

$$Hom_A(M_1\oplus M_2,-)\simeq Hom_A(M_1,-)\oplus Hom_A(M_2,-)$$

we get that $Hom_A(M,A)$ is projective. Since ψ depends on M, for the moment denote ψ by ψ^M. Clearly $\psi_V^{A^s}$ is an isomorphism for all V. Since M is a direct summand of a finitely generated free A-module, and since $\psi_V^{M_1}\oplus\psi_V^{M_2}=\psi_V^{M_1\oplus M_2}$, we get that ψ_V^M is an isomorphism for all A-modules V and all finitely generated projective A-modules M. This finishes the proof. □

We are part of the way already. If P is a progenerator of A-*mod* and if $B^{op}:=End_A(P)$, then P is an A-B-bimodule. Moreover,

$$Hom_A(P,-):A\text{-}Mod\longrightarrow B\text{-}Mod$$

is an equivalence. But $P\otimes_B-$ is left adjoint to $Hom_A(P,-)$. Proposition 3.2.8 shows that

$$P \otimes_B - : B\text{-}Mod \longrightarrow A\text{-}Mod$$

is an equivalence as well, a quasi-inverse to $Hom_A(P, -)$.

Moreover, since P is a finitely generated projective A-module, Lemma 4.2.5 shows

$$Hom_A(P, -) \simeq Hom_A(P, A) \otimes_A -.$$

As a whole we have the following

Proposition 4.2.6 *Let K be a commutative ring and let A be a K-algebra. Suppose P is a progenerator in A-mod, put $\check{P} := Hom_A(P, A)$ and $B^{op} := End_A(P)$. Then P is an A-B-bimodule and $\check{P}$ is a B-A-bimodule. Moreover*

$$P \otimes_B - : B\text{-}Mod \longrightarrow A\text{-}Mod$$

and

$$\check{P} \otimes_A - : A\text{-}Mod \longrightarrow B\text{-}Mod$$

are mutually quasi-inverse equivalences. Furthermore $Hom_A(P, -) \simeq \check{P} \otimes_A -$ is exact. □

Proposition 4.2.7 *Let $\check{P}$ be a B-A-module such that $\check{P} \otimes_A - : A\text{-}Mod \longrightarrow B\text{-}Mod$ is an equivalence. Then $\check{P}$ is a progenerator in B-mod.*

Proof Corollary 4.2.2 shows that $\check{P}$ is flat as an A-module and projective as a B-module. Since A is a compact A-module and since $\check{P} \otimes_A -$ preserves direct sums, $\check{P}$ is also a compact B-module. Lemma 4.1.9 shows that $\check{P}$ is a finitely generated projective B-module and Lemma 4.1.4 shows that $\check{P}$ is a generator. □

Theorem 4.2.8 (Morita theorem) *Let K be a commutative ring and let A and B be K-algebras. If $F : B\text{-}Mod \longrightarrow A\text{-}Mod$ is an equivalence, then there is a progenerator P of A-Mod with endomorphism ring B^{op} and*

$$P \otimes_B - : B\text{-}Mod \longrightarrow A\text{-}Mod$$

is an equivalence with quasi-inverse

$$Hom_A(P, A) \otimes_A - : A\text{-}Mod \longrightarrow B\text{-}Mod.$$

Moreover, if A and B are Noetherian, then $A\text{-}mod \simeq B\text{-}mod \Leftrightarrow A\text{-}Mod \simeq B\text{-}Mod$.

Proof Since F is an equivalence, Lemma 4.1.4 shows that $M := F(B)$ is a generator in A-*Mod* with endomorphism ring B^{op}. Lemma 3.2.10 shows that F preserves arbitrary coproducts. Moreover, F is exact by Lemma 4.2.1 and $F(B) = M$ is a projective A-module. Since F preserves coproducts, F sends compact objects to compact objects, and hence M is a progenerator in A-*mod*. Proposition 4.2.4 shows that

$Hom_A(M,-) \simeq Hom_A(M,A) \otimes_A -$ is an equivalence which maps A-*mod* to B-*mod* and Proposition 4.2.6 concludes the first part of the proof. If A and B are Noetherian, then A-*mod* and B-*mod* are abelian categories. If $F : B\text{-}mod \to A\text{-}mod$ is an equivalence, then F is exact and commutes with coproducts. Moreover, $F(B) =: M$ is a progenerator. This proves the statement. □

Definition 4.2.9 Let K be a commutative ring and let A and B be K-algebras. Then an A-B-bimodule M which is a progenerator as an A-module and with $End_A(M) = B^{op}$ is a *Morita bimodule*. The equivalence $M \otimes_B - : B\text{-}Mod \longrightarrow A\text{-}Mod$ is called a *Morita equivalence* and two algebras A and B are *Morita equivalent* if there is Morita equivalence between them.

Recall that progenerators are finitely generated. We get a slight improvement which is useful in certain circumstances.

Proposition 4.2.10 *Suppose K is a commutative ring and that A and B are Noetherian K-algebras. Suppose there is an A-B-bimodule M and a $B-A$-bimodule N such that there are isomorphisms*

$$M \otimes_B N \xrightarrow{\alpha} A$$

as A-A-bimodules and

$$N \otimes_A M \xrightarrow{\beta} B$$

as B-B-bimodules. Then M is a progenerator as an A-module, $B^{op} \simeq End_A(M)$, N is a progenerator as a B-module and $A^{op} \simeq End_B(N)$. Moreover,

$$Hom_A(M,A) \simeq N \simeq Hom_B(M,B) \text{ and } Hom_A(N,A) \simeq M \simeq Hom_B(N,B).$$

Proof We see that $M \otimes_B - : B\text{-}Mod \longrightarrow A\text{-}Mod$ is an equivalence with quasi-inverse $N \otimes_A - : A\text{-}Mod \longrightarrow B\text{-}Mod$. Hence both functors $M \otimes_B - : B\text{-}Mod \longrightarrow A\text{-}Mod$ and $N \otimes_A - : A\text{-}Mod \longrightarrow B\text{-}Mod$ are exact by Corollary 4.2.2.

Now $M \otimes_B -$ is left and right adjoint to $N \otimes_A -$. Since $M \otimes_B -$ is left adjoint to $Hom_A(M,-)$, by Proposition 3.2.8 $Hom_A(M,-)$ is also an equivalence, a quasi-inverse to $M \otimes_B -$.

Since the A-module M is the image of the (projective) regular module B, M is also projective as an A-module by Lemma 4.2.1. By Lemma 4.2.5

$$Hom_A(M,-) \simeq Hom_A(M,A) \otimes_A -$$

is an exact equivalence quasi-inverse to $M \otimes_B - : B\text{-}Mod \longrightarrow A\text{-}Mod$ and we get $N \simeq Hom_A(M,A)$ by Proposition 3.2.7. Moreover,

$$End_B(B) \simeq End_A(M \otimes_B B) \simeq End_A(M)$$

since $M \otimes_B - : B\text{-}Mod \longrightarrow A\text{-}Mod$ is an equivalence of categories. Lemma 4.1.4 shows that being a generator is a functorial property. Now, trivially B is a generator of B-*mod*. Hence ${}_AM \simeq {}_AM_B \otimes_B B$ is a generator of A-*mod*.

By symmetry the analogous statements for N follow.

We still need to show that $N \simeq Hom_B(M, B)$, and likewise for M. But this follows by considering right modules instead of left modules and the corresponding equivalences. □.

Definition 4.2.11 A bimodule M as in Proposition 4.2.10 is said to be *invertible*.

Finally, we obtain

Corollary 4.2.12 *Let K be a commutative ring and let A and B be two Noetherian K-algebras. Then every additive K-linear equivalence $F : A\text{-}mod \longrightarrow B\text{-}mod$ is isomorphic to $M \otimes_A -$ for a Morita bimodule M.*

Proof Indeed, Lemma 4.2.1 shows that F is exact and that the B-A-bimodule $F(A) =: M$ is a progenerator in the category of B-modules. By Lemma 3.2.10, F commutes with direct sums and Watt's Theorem 3.3.16 shows that $F \simeq M \otimes_A -$. We conclude from Theorem 4.2.8 that M is a Morita bimodule. □

4.3 Properties Invariant Under Morita Equivalences

Let K be a commutative ring and let A and B be two Morita equivalent K-algebras. Every property that can be expressed in terms of the module category is an invariant under Morita equivalence. Denote by ${}_AM_B$ a Morita bimodule and put $\check{M} := Hom_A(M, A)$.

Being a simple object in a category $\mathcal{C}$ can be expressed in purely category theoretic terms. Indeed, S is simple if any monomorphism $T \to S$ factors either through the zero object or is an isomorphism. Since by a similar argument being a direct sum of objects of a special category theoretic kind is also a property which can be expressed in category theoretic terms, semisimplicity is a category theoretic term. If K is a field and if A is finite dimensional, then there are only a finite number n of isomorphism classes of simple A-modules. Since B is Morita equivalent to A, the module categories are equivalent, and hence B also has n isomorphism classes of simple modules.

The nilpotency degree of the radical is an invariant. Indeed, the radical of A is the smallest two-sided ideal admitting a semisimple quotient. The functor $M \otimes_B - \otimes_B \check{M} : B \otimes_K B^{op}\text{-}mod \to A \otimes_K A^{op}\text{-}mod$ is an equivalence of categories with quasi-inverse $\check{M} \otimes_A - \otimes_A M$, and the bimodule B is sent to the bimodule A. Moreover, the equivalence induces a bijection between the two-sided submodules, i.e. two-sided ideals of B and of A. The property of having a semisimple quotient is invariant under Morita equivalence and therefore

$$M \otimes_B \text{rad}(B) \otimes_B \check{M} = \text{rad}(A).$$

The composition series of projective modules is an invariant under Morita equivalence.

Proposition 4.3.1 *Let K be a commutative ring and let A and B be two Morita equivalent K-algebras. Then the centres of A and of B are isomorphic as K-algebras:*

$$Z(A) \simeq Z(B).$$

More precisely, if ${}_AM_B$ is a Morita bimodule, then an isomorphism $\gamma : Z(A) \longrightarrow Z(B)$ is characterised by the fact that for all $z \in Z(A)$ and all $m \in M$ we have $zm = m\gamma(z)$.

Proof We know that $B^{op} = End_A(M)$. Let $z \in Z(A)$. Then

$$M \xrightarrow{\mu_z} M$$
$$m \mapsto zm$$

is indeed an element in $End_A(M)$, using that z is central. Hence, there is an element $\gamma(z) \in B$ such that $zm = m\gamma(z)$ for all $m \in M$. Now, γ is obviously additive. γ is a ring homomorphism. Indeed, $\gamma(1) = 1$ is clear and for $z_1, z_2 \in Z(A)$ we have

$$m\gamma(z_1z_2) = (z_1z_2)m = z_1(z_2m) = z_1(m\gamma(z_2)) = m\gamma(z_2)\gamma(z_1)$$

for all $m \in M$. Therefore, since the elements in B are determined by their action on M, we have $\gamma(z_1z_2) = \gamma(z_1)\gamma(z_2)$ for all $z_1, z_2 \in Z(A)$. We see that $\gamma : Z(A) \longrightarrow B$ is a ring homomorphism. Since M is a generator in A-*mod*, the A-module M is faithful, i.e. $am = 0$ for all $m \in M$ implies $a = 0$. Hence γ is injective. Now, $\text{im}(\gamma) \subseteq Z(B)$. Indeed, for all $b \in B$, $m \in M$ and $z \in Z(A)$ we have

$$m(b\gamma(z)) = (mb)\gamma(z) = z(mb) = (zm)b = (m\gamma(z))b = m(\gamma(z)b)$$

and so

$$b\gamma(z) = \gamma(z)b$$

for all $z \in Z(A)$ and $b \in B$. γ is surjective since M is a progenerator in $mod - B$ as well and therefore the same argument defines a mapping $\lambda : Z(B) \longrightarrow Z(A)$ which is clearly inverse to γ. □

Remark 4.3.2 Let K be a commutative ring and let A be a K-algebra. Suppose that A satisfies the Krull-Schmidt theorem for projective modules. Then the regular module A is projective and

$$A \simeq P_1^{n_1} \oplus P_2^{n_2} \cdots \oplus P_m^{n_m}$$

for indecomposable projective modules $P_1, P_2, \ldots, P_m$ so that $P_i \simeq P_j \Rightarrow i = j$.

We claim that $M := P_1 \oplus P_2 \oplus \cdots \oplus P_m$ is a progenerator for A-*mod*. Indeed, for $k := \max(n_i |\ i \in \{1, 2, \ldots, m\})$ one has $A|M^k$. Since A is a generator, M is a generator as well. Hence $B := End_A(M)^{op}$ is Morita equivalent to A. In particular, if $k \geq 2$, the algebra A is always non-commutative, whereas $End_A(M)$ can be commutative. But, with the construction above $B/\mathrm{rad}(B)$ is a direct sum of simple modules which are pairwise non-isomorphic. Algebras of this type are called basic.

Definition 4.3.3 Let A be an artinian algebra. If for every two simple submodules S_1 and S_2 of $A/\mathrm{rad}(A)$ one has $S_1 \simeq S_2 \Rightarrow S_1 = S_2$, then A is called a *basic algebra.*

Lemma 4.3.4 *Let K be a field and let A be a finite dimensional K-algebra. Then A is Morita equivalent to a basic algebra.*

Proof The basic algebra for A is constructed above in Remark 4.3.2. □

Proposition 4.3.5 *Let K be a commutative ring and let A and B be artinian K-algebras. Then A and B are Morita equivalent if and only if their basic algebras constructed in Remark 4.3.2 are isomorphic as K-algebras.*

Proof We know that

$$_AA = P_1 \oplus \cdots \oplus P_n$$

where $P_i \not\simeq P_j$ whenever $i \neq j$. Let M be a progenerator with $End_A(M) = B^{op}$. Since B is basic, each indecomposable direct factor of M has multiplicity 1, and since M is a generator, $M \simeq P_1 \oplus \cdots \oplus P_n$. Therefore

$$B \simeq End_A(M)^{op} \simeq End_A(P_1 \oplus \cdots \oplus P_n)^{op} \simeq End_A(A)^{op} = A.$$

This proves the proposition. □

Corollary 4.3.6 *Let A be a local K-algebra. Then A is basic.*

Proof Indeed, the regular module is indecomposable, hence A is basic. □

4.4 Group Theoretical Examples

We will now illustrate Morita theory on examples from the representation theory of groups.

4.4.1 Elementary Examples

Example 4.4.1 We first analyse two trivial examples.

- Let R be any ring and let $n \in \mathbb{N}$. Denote by $Mat_n(R)$ the n by n matrix ring with coefficients in R. Then $Mat_n(R)$ is Morita equivalent to R. Indeed, R^n is a progenerator in R-*mod* and $End_R(R^n) = Mat_n(R)$ by definition. This naturally generalises Example 3.1.25.2. It may be instructive to note that $Mat_2(R)$ is Morita equivalent to $Mat_3(R)$ since both are Morita equivalent to R.
- Isomorphic rings are Morita equivalent. Let $\alpha : A \longrightarrow B$ be an isomorphism of K-algebras. Then ${}_1B_\alpha$ is a Morita B-A-bimodule with inverse ${}_\alpha B_1 \simeq {}_1A_{\alpha^{-1}}$, where the isomorphism is given by α.

Example 4.4.2 We may apply Example 4.4.1 to group rings. Indeed, if G and H are finite groups and if K is a splitting field for G and H, and if the order of G and the order of H are invertible in K, then KG is Morita equivalent to KH if and only if the number of conjugacy classes of G equals the number of conjugacy classes of H. Indeed, the number of conjugacy classes γ_G of G equals the number of isomorphism classes of simple KG-modules, and likewise for H. Hence KG is Morita equivalent to $\prod_{i=1}^{\gamma_G} K$ by Wedderburn's theorem, and KH is Morita equivalent to $\prod_{i=1}^{\gamma_H} K$ which proves the statement.

Example 4.4.3 Proposition 2.12.4 shows that the basic algebra of a block B of a group algebra with cyclic normal defect group is actually a Nakayama algebra $N_s^{|D|+1}$. Indeed, the composition series are the same, and if $P_1, P_2, \ldots, P_s$ are representatives of the isomorphism classes of the projective indecomposable B-modules, then $M := \bigoplus_{i=1}^s P_i$ is a progenerator of B with endomorphism ring $N_s^{|D|+1}$.

Suppose K is a field of characteristic $p > 0$ and G is a finite group with $N \trianglelefteq G$ and $H \leq G$. Then H acts on KN by conjugation. The principal block of KN is fixed by conjugation, and so we can speak of the principal block of $(KN)^H$ being the principal block of KN intersected with $(KN)^H$.

Proposition 4.4.4 *Let K be a field of characteristic $p > 0$, let G be a finite group and let N be a normal subgroup of G containing a Sylow p-subgroup of G, and let $H := G/N$. Then the principal block of KG is Morita equivalent to the principal block of $(KN)^H$.*

Remark 4.4.5 We shall apply a method which comes from the theory of Hopf algebras. For the general theory of Hopf algebras the reader may consult Montgomery [2]. A very brief introduction is given in Sect. 6.2.1.

Proof of Proposition 4.4.4 Since N contains a Sylow p-subgroup of G we get that p does not divide $H := G/N$. The Schur Zassenhaus Theorem 1.8.47 then implies that $G \simeq N \rtimes H$. Now KH is semisimple. Let $e_1, \ldots, e_k$ be the block idempotents of KH. We may assume that $KHe_1 \simeq K$ is the trivial module. Hence

$$KGe_1 = KG \otimes_{KH} KHe_1 \simeq KG \otimes_{KH} K.$$

Put $t := \sum_{h\in H} h$. Then

$$\begin{aligned} KN &\longrightarrow (KN)^H \\ x &\mapsto t \cdot x \end{aligned}$$

is a $(KN)^H$-$(KN)^H$-bilinear mapping.

We denote by $B_0(KN)$ the principal block of KN, by $B_0(KG)$ the principal block of KG and by $B_0((KN)^H)$ the principal block of $(KN)^H$.

Then put $M := B_0(KN)$, which is a $B_0(KG)$-$B_0((KN)^H)$-bimodule, where $N \rtimes H$ acts on KN by $(n, h) \cdot m := n(m^h)$ for $n, m \in N, h \in H$, and as usual m^h is the conjugate of m by h.

The inverse bimodule of the bimodule M will be $Hom_{KG}(M, KG)$ which is again $B_0(KN)$, but the module structure is actually a $(KN)^H$-KG bimodule structure given by the group element-wise inverse.

Then

$$\begin{aligned} B_0(KN) \otimes_{B_0((KN)^H)} B_0(KN) &\longrightarrow B_0(KG) \\ x \otimes y &\mapsto x \cdot t \cdot y \end{aligned}$$

and

$$\begin{aligned} B_0(KN) \otimes_{KG} B_0(KN) &\longrightarrow B_0((KN)^H) \\ x \otimes y &\mapsto t \cdot (xy) \end{aligned}$$

are bimodule maps. We need to show that they are bijective. Actually it is sufficient to show they are surjective. Injectivity follows from Proposition 4.1.6.

What is the principal block of KG? First, since KH is semisimple the block idempotent of the principal block of KH is a scalar multiple of t. Therefore, the principal block idempotent of KG is the product of the principal block idempotent of KN multiplied by a scalar multiple of t. Hence the mapping

$$B_0(KN) \otimes_{B_0((KN)^H)} B_0(KN) \longrightarrow B_0(KG)$$

is surjective. The H-trivial elements of KN are given by $t \cdot KN$ since the order of H is invertible in the field. Therefore,

$$B_0(KN) \otimes_{KG} B_0(KN) \longrightarrow B_0((KN)^H)$$

is also surjective. Proposition 4.2.10 proves the statement. □

We can give a related result for not necessarily principal blocks.

Proposition 4.4.6 *Let G be a finite group, let K be an algebraically closed field and let B be a block of KG with defect group D. If $G = D \cdot C_G(D)$, then there is an integer t such that $B \simeq Mat_{t\times t}(KD)$, and hence B is Morita equivalent to KD.*

Proof Let S be a simple B-module. Then S is D-projective and therefore S is a direct summand of $B \otimes_{KD} U$ for some KD-module U, and choose U as small as possible. Since S is simple, U is simple as well, for if not then let V be a submodule, then S is either a direct factor of $B \otimes_{KD} U/V$ or of $B \otimes_{KD} V$. By the minimality of U we get that U has to be simple. But then, since D is a p-group, and since KD is therefore local, D acts trivially on the simple one-dimensional module U.

Recall that Proposition 2.11.14 shows that the natural map $G \stackrel{\pi}{\longrightarrow} G/D$ induces a bijection between blocks of KG of defect D and blocks of KG/D of defect 0.

Let $b = \pi(B)$ be the block of KG/D that corresponds to B. Then b admits a unique simple b-module T, and by the above observation, B also admits a unique simple B-module, which we called S.

Hence, since T belongs to a block b of defect 0 of KG/D, we get that b is semisimple by Proposition 2.3.9, which implies that T is actually projective, and therefore by Proposition 1.9.17 there is a primitive idempotent $\overline{e}$ of KG/D with $KG/D \cdot \overline{e} = T$. Now, D is a p-group and therefore $\ker(\pi)$ is nilpotent (cf Lemma 2.11.7). This shows that there is an idempotent e of KG such that $\pi(e) = \overline{e}$. Moreover, e is primitive since $\overline{e}$ is primitive and $\ker(\pi)$ is nilpotent. $P := KG \cdot e = B \cdot e$ is an indecomposable projective B-module. Since B admits only one isomorphism class of simple modules, P is an indecomposable progenerator of B. We need to compute $End_B(P)$. Recall that $G = C_G(D) \cdot D$. Hence P is actually a $KG-KD$-bimodule for the action of D on the right of P by $x \bullet d := d^{-1}x$ for $x \in P$ and $d \in D$. Since $G = C_G(D) \cdot D$ this action commutes with the action of G on the left. Hence there is a homomorphism $KD \longrightarrow End_B(P)$. We claim that this is an isomorphism. First

$$End_B(T) = End_{KG}(T) = End_{KG/D}(T) = K$$

since K is algebraically closed and since T is a simple KG/D-module. Then P is a projective cover of T as a KG-module, and so every endomorphism of T lifts to an endomorphism of P. Moreover, each endomorphism of P induces an endomorphism of $P/\text{rad}(P) = T$. Now, $\dim_K(\ker(\pi) \cdot e) = |D|$. Since the only endomorphisms of T are scalar multiplications, each of them lifts in $|D|$ different linearly independent ways, and so

$$End_B(P) \simeq KD.$$

Hence B is Morita equivalent to KD. Since the unique simple KD-module is one-dimensional, KD is a basic indecomposable algebra, and hence

$$B \simeq Mat_{t\times t}(KD)$$

as algebras, for some integer t. □

Example 4.4.7 Another incidence of a Morita equivalence between blocks of group rings and blocks of subgroups is given by Lemma 2.11.11.

4.4.2 Nilpotent Blocks

We come to a quite sophisticated notion in group representation, namely the concept of nilpotent blocks.

Recall from Definition 2.10.12 the concept of a Brauer pair. Let G be a finite group, and let k be an algebraically closed field of characteristic $p > 0$. Let $B = kG \cdot b$ be a block of kG with $b^2 = b \in Z(kG)$. Then a Brauer pair is a pair (P, e) where P is a p-subgroup of G and e is a primitive central idempotent of $kC_G(P)$. Recall from Theorem 2.10.16 that the set of Brauer pairs is partially ordered using the Brauer map β_P which is defined for each p-subgroup P of G. The Brauer pair (P, e) *belongs to* B if $(1, b) \leq (P, e)$. Let D be the defect group of B, then by Corollary 2.10.20 there is a primitive idempotent $f \in A := (kG)^D$ such that $b \in Tr_D^G(AfA) \subseteq AfA$ and $\beta_D(f)e \neq 0$. We call f a *source idempotent of* B and fBf the *source algebra* of B. It is possible to prove uniqueness of the source idempotent, but this property is not needed here. Since $b \in AfA$, we have

$$kG \cdot b = B = kG \cdot b \cdot kG = kG \cdot f \cdot kG = B \cdot f \cdot B$$

and we obtain that B is Morita equivalent to $f \cdot B \cdot f$ with Morita bimodule Bf. More precisely we shall need the following three properties:

- Let (P, e_P) be a Brauer pair which belongs to B. If $Q \leq P$, then there is a unique e_Q such that $(Q, e_Q) \leq (P, e_P)$ (cf Theorem 2.10.16).
- Let (P, e_P) and (Q, e_Q) be two Brauer pairs which belong to B, then $(Q, e_Q) \leq (P, e_P)$ if and only if $Q \leq P$ and there is a primitive idempotent e' of $(kG)^P$ such that $\beta_P(e')e_P \neq 0 \neq \beta_Q(e')e_Q$ (cf Corollary 2.10.17).
- Let f be a source idempotent of kG, then B is Morita equivalent to fBf.

Further, recall from Definition 3.6.11 the concept of a separable extension. An algebra homomorphism $\beta : A \longrightarrow B$ is a separable extension if the multiplication map $\mu_B : B \otimes_A B \longrightarrow B$ is split as a morphism of $B \otimes_K B^{op}$-modules. By Proposition 2.1.8 the extension β is separable if and only if there is an $\omega \in B \otimes_A B$ such that $b\omega = \omega b$ for all $b \in B$ and such that $\mu_B(\omega) = 1_B$.

Definition 4.4.8 Let k be an algebraically closed field of characteristic $p > 0$ and let G be a finite group. A block B of kG with defect group D is *nilpotent* if whenever there is a Brauer pair (Q, e_Q) that belongs to B and satisfies $(Q, e_Q) \leq (D, e)$, then $(Q, e_Q)^g \leq (D, e)$ implies that there is a $c \in C_G(Q)$ and $u \in D$ such that $g = cu$.

Lemma 4.4.9 [3, Lemma 3.1] *Let k be an algebraically closed field of characteristic $p > 0$, let G be a finite group and let B be a block of kG with defect group D. Then B is nilpotent if $N_G(Q, e_Q)/C_G(Q)$ is a p-group for all Brauer pairs (Q, e_Q) that belong to B.*

Proof Let (Q, e_Q) be a Brauer pair in B and let $P := D \cap N_G(Q, e_Q)$. By Lemma 2.10.10

$$e_Q \in Tr_P^{N_G(Q,e_Q)}(kC_G(Q)) \subseteq Tr_{PC_G(Q)}^{N_G(Q,e_Q)}(Z(kC_G(Q))).$$

Let $Z(kC_G(Q))e_Q \xrightarrow{\pi} Z(kC_G(Q))e_Q/\text{rad}(Z(kC_G(Q))e_Q)$ be the canonical morphism. The radical quotient is k, since k is algebraically closed. Moreover, $N_G(Q, e_Q)$ acts trivially and k-linearly on it. Hence $\pi(e_Q) \in |N_G(Q, e_Q) : P \cdot C_G(Q)| \cdot k$. This shows that p does not divide $|N_G(Q, e_Q) : PC_G(Q)|$. The hypothesis that $N_G(Q, e_Q)/C_G(Q)$ is a p-group now implies that $N_G(Q, e_Q) = P \cdot C_G(Q)$. Therefore B is nilpotent. □

The purpose of this section is to prove the following result of Puig.

Theorem 4.4.10 (Puig) *Let k be an algebraically closed field of characteristic $p > 0$ and let G be a finite group. Let B be a nilpotent block of kG with defect group D and source idempotent f.*

Then there is an extension of algebras $kD \longrightarrow Mat_n(k)$ and an isomorphism

$$f \cdot kG \cdot f \longrightarrow kD \otimes_k Mat_n(k)$$

of extensions of kD. Furthermore, B is Morita equivalent to $Mat_n(kD)$.

Remark 4.4.11 The proof we present here is due to Külshammer [4]. The original proof of Puig, and of Broué, is much more involved, gives slightly more information on the size n of the matrix ring in particular, and uses most heavily deep Brauer character theory.

The proof which we are going to present here proceeds in several steps, and we shall first need a few preliminary results.

Lemma 4.4.12 *Given k, G, B, D, $A = (kG)^D$, b, e and f as above. Then the map*

$$\begin{aligned} kD &\longrightarrow fBf \\ x &\mapsto fx = xf \end{aligned}$$

is a separable extension of k-algebras.

Proof Since $b \in Tr(AfA)$, we find $x_j, y_j \in A = (kG)^D$ for $j \in \{1, \ldots, \ell\}$ so that

$$b = \sum_{j=1}^{\ell} Tr_D^G(x_j f y_j).$$

But $Z(B)$ is local since B is indecomposable, and $Tr_D^G(x_j f y_j) \in Z(B)$ for each j. Since b is an idempotent, $b \notin \text{rad}(Z(B))$ and therefore there is a $j_0 \in \{1, \ldots, \ell\}$ with

$$c := Tr_D^G(x_{j_0} f y_{j_0}) \notin \text{rad}(Z(B)).$$

Since $Z(B)$ is local, c is invertible and

$$b = cc^{-1} = Tr_D^G(x_{j_0}fy_{j_0})c^{-1} = Tr_D^G(x_{j_0}fy_{j_0}c^{-1}).$$

Therefore there are $x, y \in A$ such that $b = Tr_D^G(xfy)$. Let

$$G = D \cup g_1 D \cup \cdots \cup g_s D$$

be a disjoint union and define

$$\omega := \sum_{i=1}^{s} g_i xf \otimes fyg_i^{-1} \in kG \otimes_{kD} kG.$$

Since the tensor product is taken over kD and since $x, y \in A$ commute with each element in D, we get that ω does not depend on the choice of the representatives g_i of the cosets G/D. Hence,

$$G = gG = gD \cup gg_1 D \cup \cdots \cup gg_s D$$

and as a consequence $g\omega = \omega g$ for each $g \in G$. Moreover, denoting by $\mu : kG \otimes_{kD} kG \longrightarrow kG$ the multiplication mapping, we clearly get $\mu(\omega) = Tr_D^G(xfy) = b$, by construction. Then $f\omega f \in (fkGf \otimes_{kD} fkGf)$ and $\mu(f\omega f) = f\mu(\omega)f = f$ is the unit element of fBf. Using Proposition 2.1.8 the algebra homomorphism $kD \longrightarrow fBf$ is a separable extension. □

Consider kG as a kD-kD-bimodule. Then, by Mackey's argument, we get

$$kG = \bigoplus_{DgD \in D\backslash G/D} kDgD$$

and each direct factor $kDgD$ is indecomposable. Indeed, if P is a p-group and $Q \leq P$, then the kP-module kP/Q has socle k by Frobenius reciprocity. Now $fkGf$ is a kD-kD-bimodule, and hence a direct factor of the kD-kD-bimodule kG. Let

$$fkGf = \bigoplus_{i=1}^{r} M_i$$

be a decomposition of the kD-kD-bimodule $fkGf$ into indecomposable direct factors. By the Krull-Schmidt theorem for each $i \in \{1, \ldots, r\}$ there is a $g_i \in G$ and an isomorphism $\phi_i : k(Dg_iD) \longrightarrow M_i$ of $kD - kD$-bimodules. For each $i \in \{1, \ldots, r\}$ define

$$x_i := \phi_i(g_i) \in \phi_i(Dg_iD) =: X_i$$

and

$$D_i := D \cap g_i D g_i^{-1}.$$

By Theorem 2.10.16 there is a unique primitive idempotent e_i in $Z(kC_G(D_i))$ such that we have an inclusion of Brauer pairs $(D_i, e_i) \leq (D, e)$.

Proposition 4.4.13 *$(D_i, e_i)^{g_i} \leq (D, e)$ for all $i \in \{1, \ldots, r\}$.*

Proof Let $u \in D_i$. Then

$$\begin{aligned} ux_ig_i^{-1}u^{-1} &= \underbrace{u}_{\in D} \phi_i(g_i) \underbrace{g_i^{-1}u^{-1}g_i}_{\in D} g_i^{-1} = \phi_i(ug_ig_i^{-1}u^{-1}g_i)g_i^{-1} \\ &= \phi_i(g_i)g_i^{-1} = x_ig_i^{-1}. \end{aligned}$$

Therefore $x_ig_i^{-1}$ is invariant under conjugation with elements of D_i. Hence

$$X = X_1 \dot\cup X_2 \dot\cup \cdots \dot\cup X_r$$

is a k-basis of $fkGf$, which is invariant under left and right multiplication by D. We may complete this basis into a k-basis Y of kG which is invariant under left and right multiplication by D. Hence Yg_i^{-1} is a k-basis of kG which is invariant under D_i-left or right multiplication. Since $x_ig_i^{-1} \in Yg_i^{-1} \cap (kG)^{D_i}$ we get $\beta_{D_i}(x_ig_i^{-1}) \neq 0$. Since

$$x_ig_i^{-1} = fx_ig_i^{-1}g_ifg_i^{-1}$$

there are primitive idempotents $f_i \in f\left((kG)^{D_i}\right)f$ and $f_i' \in f\left((kG)^{g_i^{-1}D_ig_i}\right)f$ such that

$$0 \neq \beta_{D_i}(f_ix_ig_i^{-1}g_if_i'g_i^{-1}) = \beta_{D_i}(f_i)\beta_{D_i}(x_ig_i^{-1})\beta_{D_i}(g_if_i'g_i^{-1}).$$

Therefore $\beta_{D_i}(f_i)$ and $\beta_{D_i}(g_if_i'g_i^{-1})$ belong to the same block of $kC_G(D_i)$.

Now, $(D_i, e_i) \leq (D, e)$ implies

$$\begin{aligned} \beta_{D_i}(f_i)e_i &= \beta_{D_i}(f_if)e_i = \beta_{D_i}(f_i)\beta_{D_i}(f)e_i \\ &= \beta_{D_i}(f_i)\beta_{D_i}(f) = \beta_{D_i}(f_if) = \beta_{D_i}(f_i). \end{aligned}$$

This shows that both $\beta_{D_i}(f_i)$ and $\beta_{D_i}(g_if_i'g_i^{-1})$ belong to the block $kC_G(D_i)e_i$. Since

$$0 \neq \beta_{D_i}(g_if_i'g_i^{-1})e_i = \beta_{D_i}(g_if_i'g_i^{-1})\beta_{D_i}(g_ifg_i^{-1})e_i$$

we get that

$$\beta_{D_i}(g_ifg_i^{-1})e_i \neq 0.$$

Now

$$\beta_{g_iDg_i^{-1}}(g_ifg_i^{-1})g_ie_ig_i^{-1} \neq 0$$

which implies $(D_i, e_i) \leq g_i(D, e)g_i^{-1}$ as claimed. □

Proof of Theorem 4.4.10 Let now $B = kG \cdot b$ be a nilpotent block with defect group D, and hence $(Q,f) \leq (D, e)$ and $(Q,f)^g \leq (P, e)$ implies there is a $c \in C_G(Q)$ and $u \in D$ with $g = c \cdot u$.

We use the notation introduced in Proposition 4.4.13 and Lemma 4.4.12. We first get that Proposition 4.4.13 implies $g_i \in C_G(D_i)$ for all $i \in \{1, \ldots, r\}$. Moreover,

$$\begin{aligned} kD &\longrightarrow kD \otimes_k f \cdot kG \cdot f \\ D \ni u &\mapsto u \otimes uf \end{aligned}$$

is an extension of k-algebras. Further,

$$\begin{aligned} kD \otimes_k f \cdot kG \cdot f &\xrightarrow{\sigma} f \cdot kG \cdot f \\ u \otimes x &\mapsto x \end{aligned}$$

for $u \in D$ and $x \in f \cdot kG \cdot f$, is a morphism of extensions of the algebra kD. We get that

$$I := \ker(\sigma) = \mathrm{rad}(kD) \otimes_k f \cdot kG \cdot f$$

is a nilpotent ideal of $kD \otimes_k fkGf$. Let

$$(kD \otimes_k f \cdot kG \cdot f)\,/I \xrightarrow{\overline{\sigma}} f \cdot kG \cdot f$$

be the isomorphism induced by σ. Put $\rho := \overline{\sigma}^{-1}$ and let

$$\pi : (kD \otimes_k f \cdot kG \cdot f) \longrightarrow (kD \otimes_k f \cdot kG \cdot f)\,/I$$

be the canonical morphism.

We have seen that $f \cdot kG \cdot f$ and $kD \otimes_k f \cdot kG \cdot f$ are both $k(D \times D)$-permutation modules. Let $u, v \in D$ with $ux_{jv} = x_j$. Then $ug_{jv} = g_j$ and hence

$$u = g_{jv}^{-1}g_j^{-1} \in D \cap g_jDg_j^{-1}.$$

We can assume that $g_j \in C_G(D_j)$, and so

$$u = g_j^{-1}ug_j = v^{-1}.$$

Hence

$$u(1 \otimes x_j)v = uv \otimes ux_{jv} = 1 \otimes x_j.$$

This shows that for each $j \in \{1, \ldots, r\}$ the stabilisers of the action of $D \times D$ on $x_j \in f \cdot kG \cdot f$ coincides with the stabilisers of the of the action of $D \times D$ on $1 \otimes x_j \in kD \otimes_k f \cdot kG \cdot f$.

Therefore

$$\begin{aligned} f \cdot kG \cdot f &\xrightarrow{\tau_0} kD \otimes_k f \cdot kG \cdot f \\ x_j &\mapsto 1 \otimes x_j \end{aligned}$$

is a homomorphism of kD-kD-bimodules. Since

$$\overline{\sigma}(\tau_0(x_j) + I) = \sigma(\tau_0(x_j)) = x_j$$

we get $\tau_0(x_j) + I = \rho(x_j)$ and hence τ_0 lifts ρ to a homomorphism of bimodules. Since by Lemma 4.4.12

$$kD \longrightarrow f \cdot kG \cdot f$$

is a separable extension, Proposition 3.6.12 shows that there is a homomorphism of extensions

$$\tau : f \cdot kG \cdot f \longrightarrow kD \otimes_k f \cdot kG \cdot f$$

lifting ρ. This implies that

$$\sigma \circ \tau = \overline{\sigma} \circ \pi \circ \tau = \overline{\sigma} \circ \rho = id_{fkGf}$$

and hence

$$kD \otimes_k f \cdot kG \cdot f = \tau(f \cdot kG \cdot f) \oplus I.$$

For every simple $f \cdot kG \cdot f$-module M we form the $kD \otimes_k f \cdot kG \cdot f$-module $kD \otimes_k M$. Via the morphism τ the module $kD \otimes_k M$ is a $f \cdot kG \cdot f$-module and via

$$\begin{aligned} kD &\longrightarrow f \cdot kG \cdot f \\ u &\mapsto u \cdot f \end{aligned}$$

the module $kD \otimes_k M$ becomes a kD-module. Now, for every $u, v \in D$ and $m \in M$ we get that the action of kD on $kD \otimes_k M$ is given by $u \cdot (v \otimes m) = uv \otimes fum$. Hence $kD \otimes_k M$ is a projective kD-module. Since by Lemma 4.4.12

$$kD \longrightarrow f \cdot kG \cdot f$$

is a separable extension, Proposition 2.1.8 shows that $kD \otimes_k M$ is a projective $f \cdot kG \cdot f$-module as well.

Since M is simple, and since the only simple kD-module is the trivial module k, there is a composition series of kD of length $|D|$, and hence a composition series of $kD \otimes_k M$ as a $kD \otimes_k f \cdot kG \cdot f$-module, but also as an $f \cdot kG \cdot f$-module, of length $|D|$ as well. Moreover, any two composition factors of $kD \otimes_k M$ are isomorphic, since this holds true for kD as a kD-module.

Therefore any simple $f \cdot kG \cdot f$-module is isomorphic to M and $kD \otimes_k M$ is its projective cover. Hence the $k(f \cdot kG \cdot f)$-module $kD \otimes_k M$ is faithful. Moreover, since M is simple, $kD \otimes_k \mathrm{rad}(f \cdot kG \cdot f) =: J$ annihilates $kD \otimes_k M$. Faithfulness shows that the composition

$$\varphi : f \cdot kG \cdot f \xrightarrow{\tau} kD \otimes f \cdot kG \cdot f \longrightarrow (kD \otimes f \cdot kG \cdot f)\,/J$$

is injective. Since the dimensions are equal, we get that φ is an isomorphism. Hence $f \cdot kG \cdot f$ is isomorphic to a direct sum of $\dim_k(M)$ copies of $kD \otimes_k M$. This proves the statement. □

Corollary 4.4.14 *A nilpotent block B is Morita equivalent to the group ring over its defect group.*

4.5 Some Basic Algebras

4.5.1 Basic Hereditary Algebras and Gabriel's Theorem

We shall prove that every hereditary algebra over an algebraically closed field is Morita equivalent to a quiver algebra. Recall from Proposition 1.11.11 that every quiver algebra is hereditary. We shall also show that every finite dimensional algebra is Morita equivalent to a quotient of a quiver algebra. This partially explains the importance of hereditary algebras. Detailed presentations of this subject can be found in various places. For further reading we recommend Assem-Simson-Skowroński [5].

Proposition 4.5.1 *Let K be an algebraically closed field and let A be an indecomposable, basic, finite dimensional K-algebra. Then there is a quiver Q and an epimorphism*

$$KQ \longrightarrow A$$

of K-algebras. The set of arrows of Q form a K-basis of $\mathrm{rad}(A)/\mathrm{rad}^2(A)$.

Proof The statement is trivially true for $A = K$, and so we exclude this case in the sequel.

We know that A is artinian and therefore the Jacobson radical $\mathrm{rad}(A)$ is nilpotent by Lemma 1.6.6. Since A is basic and K is algebraically closed

$$A/\mathrm{rad}(A) \simeq K \times \cdots \times K$$

by Wedderburn's Theorem 1.4.16. Hence in particular, $\mathrm{rad}(A) \neq 0$ and $A/\mathrm{rad}(A)$ has a K-basis $V := \{e_1, \ldots, e_n\}$ of primitive idempotents of A. Recall that by Proposition 2.9.6 we get that n is the number of isomorphism classes of simple A-modules and that e_i are actually primitive idempotents. Let $S_1, \ldots, S_n$ be simple A-modules such that $e_iS_i \neq 0$ and $e_iS_j = 0$ for all $i, j \in \{1, \ldots, n\}$ and all $j \neq i$.

Define the vertices of Q to be V, and denote an idempotent e_i by v_i when we consider it as a vertex of Q.

Since A is finite dimensional, $\mathrm{rad}^2(A) \neq \mathrm{rad}(A)$ by Nakayama's lemma (Lemma 1.6.5). Let

$$\{\alpha_\ell^{i,j} \mid 1 \leq \ell \leq d_{i,j}\}$$

be a K-basis of $e_i \cdot \big(\mathrm{rad}(A)/\mathrm{rad}^2(A)\big) \cdot e_j$. The arrows in the quiver Q from the vertex i to the vertex j are the elements $\alpha_\ell^{i,j}$ for $\ell \in \{1, \ldots, d_{i,j}\}$. We denote the arrow $\alpha_\ell^{i,j}$ by $\beta_\ell^{i,j}$ when we consider it as being part of Q.

Consider the quiver algebra KQ. Observe that we allowed loops or oriented cycles in Q, so that KQ is a priori an infinite dimensional K-algebra. Define a mapping

$$\Psi : KQ \longrightarrow A$$

by $\Psi(v_i) := e_i$ and define $\Psi(\beta_\ell^{i,j}) := \alpha_\ell^{i,j}$ for each arrow $\alpha = \beta_\ell^{i,j}$ of Q. Now extend Ψ K-linearly, and define by induction on the length of a path $c = c_1 \cdot c_2$ for a path c_2 of KQ and an arrow c_1 of Q the image $\Psi(c_1 \cdot c_2) := \Psi(c_1) \cdot \Psi(c_2)$. We have to show that this is actually a ring homomorphism. Indeed K-linearity is clear by definition. Also multiplicativity is clear by the inductive definition on the length of paths and the fact that the paths in Q form a basis of KQ. Hence Ψ is a homomorphism of algebras.

We need to show that Ψ is surjective. We get as a K-vector space

$$A = \bigoplus_{i=1}^{n} \bigoplus_{j=1}^{n} e_iAe_j.$$

Hence we need to show that $e_iAe_j \subseteq \mathrm{im}(\Psi)$ for all $i, j \in \{1, \ldots, n\}$.

Let $x \in e_iAe_j$ and suppose $x \in \mathrm{rad}^m A \setminus \mathrm{rad}^{m+1} A$ for some m. We shall show that x is the image under Ψ of a linear combination of paths of length at most m. We shall proceed by downward induction on m.

Since there is an ℓ with $\text{rad}^\ell(A) = 0$, if $m \geq \ell$, we are done since 0 is in the image of Ψ. Otherwise, suppose that all elements in $\text{rad}^{m+1}(A)$ are in the image of Ψ. Since $x \in \text{rad}^m(A)$, there are elements $y_1^s, y_2^s, \ldots, y_m^s \in \text{rad}(A)$ such that $x = \sum_s y_1^s y_2^s \ldots y_m^s$. Recall that the arrows in Q correspond under Ψ to a basis of $\text{rad}(A)/\text{rad}^2(A)$ and since the vertices in Q correspond under Ψ to a basis of $A/\text{rad}(A)$, we have a short exact sequence

$$0 \longrightarrow \text{rad}(A)/\text{rad}^2(A) \longrightarrow A/\text{rad}^2(A) \longrightarrow A/\text{rad}(A) \longrightarrow 0$$

which shows by the Wedderburn-Malcev theorem that there is a decomposition as vector spaces

$$A/\text{rad}^2(A) = \text{rad}(A)/\text{rad}^2(A) \oplus A/\text{rad}(A)$$

and hence the composition

$$KQ \xrightarrow{\Psi} A \longrightarrow A/\text{rad}^2(A)$$

is a surjective ring homomorphism. Choose linear combinations of idempotents and arrows γ_i^s so that

$$\Psi(\gamma_i^s) - y_i^s \in \text{rad}^2(A)$$

for all i, s. Then

$$\begin{aligned}\sum_s \Psi(\gamma_1^s\gamma_2^s \ldots \gamma_m^s) - y_1^s y_2^s \ldots y_m^s &= \sum_s \Psi(\gamma_1^s)\Psi(\gamma_2^s) \ldots \Psi(\gamma_m^s) - y_1^s y_2^s \ldots y_m^s \\ &= \sum_s y_1^s y_2^s \ldots y_m^s + \rho^s - y_1^s y_2^s \ldots y_m^s\end{aligned}$$

for some $\rho^s \in \text{rad}^{m+1}(A)$. Since by the induction hypothesis $\text{rad}^{m+1}(A) \subseteq \Psi(KQ)$ we see that $x = \sum_s y_1^s y_2^s \ldots y_m^s \in \text{im}(\Psi)$.

Induction finishes the proof. □

Recall from Definition 1.11.13 the definition of an admissible ideal in a quiver and from Definition 1.11.5 the definitions and notations used in a quiver.

Theorem 4.5.2 (Gabriel) *Let K be an algebraically closed field and let A be a finite dimensional K-algebra. Then there is a finite quiver Q and a two-sided admissible ideal I of KQ such that A is Morita equivalent to KQ/I.*

Proof Since the statement is up to Morita equivalence only, we may use Lemma 4.3.4 to replace A by its basic algebra. Proposition 4.5.1 then shows that A is a quotient of KQ by some ideal I and the arrows of Q are a K-basis of $\text{rad}(A)/\text{rad}^2(A)$. It remains to show that I is admissible. Let R be the ideal of Q generated by the arrows. Since the arrows of Q form a K-basis of $\mathrm{rad}(A)/\mathrm{rad}^2(A)$, the ideal R maps to $\mathrm{rad}(A)$. Therefore the image $\overline{R}$ of R in $A = KQ/I$ is nilpotent. Hence there is an $\ell \in \mathbb{N}$ such that $I \supseteq R^\ell$. Let $x \in I$ and write

$$x = \sum_{v \in Q_0} \lambda_v e_v + \sum_{a \in Q_1} \mu_a a + y$$

for some scalars $\lambda_v, \mu_a \in K$, almost all 0, where e_v is the lazy path corresponding to v, and where $y \in R^2$. Then, modulo I, we obtain that $x = 0$ and hence, considering the image of the equation in A,

$$\sum_{v \in Q_0} \lambda_v e_v = -\sum_{a \in Q_1} \mu_a a - y \in \mathrm{rad}(A).$$

Since $e_v^2 = e_v$ and $e_v e_w = 0$ for $v \neq w$, we get that the left-hand side is nilpotent only if $\lambda_v = 0$ for all $v \in Q_0$. But this shows that

$$\sum_{a \in Q_1} \mu_a a + I = -y + I \in \overline{R}^2 \subseteq \mathrm{rad}^2(A).$$

Since the arrows of Q form a K-basis of $\mathrm{rad}(A)/\mathrm{rad}^2(A)$ we get $\mu_a = 0$ for all $a \in Q_1$ and therefore $x = y \in R^2$ and $R^2 \supseteq I \supseteq R^\ell$. This proves the statement. □

Proposition 4.5.3 *Let K be an algebraically closed field and let A be a finite dimensional hereditary K-algebra. Then A is Morita equivalent to a quiver algebra KQ.*

Proof Gabriel's Theorem 4.5.2 shows that A is Morita equivalent to KQ/I for some admissible ideal I. Since A is finite dimensional, $\mathrm{rad}(A)$ is nilpotent, hence there is an m with $\mathrm{rad}^m(A) = 0$. Let J be the ideal of KQ generated by the m-th power of all non-trivial closed paths. Then $J \subseteq I$ and hence KQ/J still maps onto A. But KQ/J is finite dimensional and therefore artinian. Since there are only a finite number of closed paths, we get that I is finitely generated. Since I is admissible, I is generated by linear combinations of paths of length strictly bigger than 1.

We shall need to show that KQ/I is not hereditary if I is an ideal generated by a finite number of linear combinations of paths of length at least 2, and so that the paths contributing in a single linear combination of paths all start at the same vertex and end at the same vertex.

Suppose KQ/I is hereditary and let e be a primitive idempotent. Let $\alpha_1, \alpha_2, \ldots, \alpha_n$ be all the arrows in Q with $\alpha_i e \neq 0$. Then necessarily $\alpha_i e = \alpha_i$ for all i. Let e_i be the primitive idempotent such that $e_i \alpha_i = \alpha_i$. Then, since the algebra is hereditary, $P_e := (KQ/I)e$ is a projective module with

$$\mathrm{rad}(P_e) = \oplus_{i=1}^n P_{e_i}$$

where $P_{e_i} := (KQ/I)e_i$. The mapping $P_{e_i} \longrightarrow P$ is given by multiplication by α_i and this has to be an embedding. Indeed, since submodules of projectives are projective, the image of this mapping is projective, and if multiplication by α_i is not injective then we get a contradiction to P_{e_i} being indecomposable. Let μ_i be right multiplication by α_i. We need to consider $\ker(\mu_i)$.

$$\ker(\mu_i) = \{u \in KQ/I \mid u \cdot \alpha_i = 0\} = \{u \in KQ \mid u \cdot \alpha_i \in I\}/I$$

and hence $\ker(\mu_i) = 0$ if and only if

$$u \cdot \alpha_i \in I \Rightarrow u \in I.$$

Since this holds for all vertices e and all arrows, this can happen only if $I = 0$. This proves the statement. □

Remark 4.5.4 Observe that we needed to have an algebraically closed base field at the very beginning of the proof of Proposition 4.5.1 when we were using Wedderburn's theorem. Theorem 4.5.2 is actually false if the base field is not algebraically closed. Indeed, if I is admissible, then

$$(KQ/I)/\mathrm{rad}(KQ/I) \simeq \prod_{i=1}^{t} K$$

for some integer t. Hence K is a splitting field for A. There are finite dimensional K-algebras for which K is not a splitting field if K is not algebraically closed. Indeed, any finite field extension D will do. In this case $D \simeq K[X]/f(X)$ and $K[X]$ is KQ where Q is the quiver with one vertex and one arrow, a single loop. However, $f(X)$ is an irreducible polynomial in $K[X]$ and the ideal generated by $f(X)$ cannot be admissible.

Remark 4.5.5 Observe that Proposition 4.5.1 heavily uses the fact that the algebras are finite dimensional. The quiver Q one gets for $K[X]$ is a single loop, and indeed, $K[X]$ is isomorphic to the algebra KQ when Q is a single loop. The construction in the proof of Proposition 4.5.1 will also give a single loop for Q in the case of $K[[X]]$, the power series ring. However, $K[X]$ is not Morita equivalent to $K[[X]]$.

4.5.2 Frobenius Algebras and Self-injective Algebras II

We can prove a partial converse to Lemma 1.10.3. For further reading on criteria for Frobenius algebras we refer to [6, § 16C].

Let K be a field and let A be a finite dimensional K-algebra. Denote by $\mathcal{S}$ the set of isomorphism classes of simple A-modules. We have a functor $Hom_K(Hom_A(-, A), K) : A\text{-}mod \longrightarrow A\text{-}mod$ which we call $\mathcal{N}$, the Nakayama functor.

Lemma 4.5.6 *Let K be a field and let A be a finite dimensional K-algebra. Then*

- *$\mathcal{N}$ is an equivalence, with quasi-inverse $\mathcal{N}^{-1} := Hom_A(Hom_K(-, K), A)$.*
- *If A is a Frobenius algebra, then $\mathcal{N} \simeq {}_1A_\nu \otimes_A -$.*
- *For every projective indecomposable A-module P we get that*

$$\mathcal{N}(P/\mathrm{rad}(P)) \simeq \mathrm{soc}(P) \text{ and } \mathcal{N}(P_S) = I_S$$

if P_S is the projective cover of the simple module S and I_S the injective hull of the simple module S.

Proof Concerning the first statement, since A is finite dimensional, applying twice the evaluation mapping which identifies a module with its double dual, $\mathcal{N}$ has an inverse

$$\mathcal{N}^{-1} := Hom_A(Hom_K(-, K), A).$$

For the second statement let ν be the Nakayama automorphism, then by Definition 1.10.13

$$Hom_K(A, K) \simeq {}_1A_\nu \simeq {}_{\nu^{-1}}A_1$$

as A-A-bimodules, and we get isomorphisms of functors

$$\begin{aligned} Hom_K(Hom_A(-, A), K) &\simeq Hom_K(Hom_A(-, Hom_K(Hom_K(A, K), K)), K) \\ &\simeq Hom_K(Hom_K(Hom_K(A, K) \otimes_A -, K), K) \\ &\simeq Hom_K(A, K) \otimes_A - \\ &\simeq {}_{\nu^{-1}}A_1 \otimes_A -. \end{aligned}$$

For the third statement, assume $P/\mathrm{rad}(P)$ is simple. Then

$$A \simeq \bigoplus_{S \in \mathcal{S}} P_S^{n_S}$$

for certain integers n_S and we get that

$$Hom_A(P/\mathrm{rad}(P), A) = Hom_A\left(P/\mathrm{rad}(P), \bigoplus_{S \in \mathcal{S}} P_S^{n_S}\right) = Hom_A\left(S, P_{\nu^{-1}(S)}^{n_{\nu^{-1}(S)}}\right).$$

Since

$$A^{op} \simeq End_A(A) \simeq End_A\left(\bigoplus_{S \in \mathcal{S}} P_S^{n_S}\right),$$

the natural $End_A\left(P_{\nu^{-1}(S)}^{n_{\nu^{-1}(S)}}\right)$-right module structure of $Hom_A\left(S, P_{\nu^{-1}(S)}^{n_{\nu^{-1}(S)}}\right)$ gives the A-right module structure of A. It is easy to see that the corresponding module is

$Hom_K(\mathrm{soc}(P_{n_{\nu^{-1}(S)}}), K)$. The K-linear dual transports the right module structure into a left module structure, and this proves the statement. Since the K-linear dual transforms projective right modules into injective left modules, we see that $\mathcal{N}(P_S)$ is an injective module with socle S, whence isomorphic to I_S. □

Proposition 4.5.7 (Nakayama [7, 8]) *Let K be a field and let A be a finite dimensional self-injective K-algebra. Let $\mathcal{S}$ be a set of representatives of the isomorphism classes of simple A-modules. Then A is a Frobenius algebra if and only* $\dim_K(\mathrm{soc}(P)) = \dim_K(P/\mathrm{rad}(P))$ *for each projective indecomposable A-module P. If A is basic, then A is a Frobenius K-algebra.*

Proof Let A be a Frobenius algebra. Denote by P_T the projective cover of the module T. Put $S := P/\mathrm{rad}(P)$. We get by Lemma 4.5.6 that $\mathrm{soc}(P) = {}^{\nu}S$ for the Nakayama automorphism ν of A. Since the underlying vector space ${}^{\nu}S$ is just the same as S, we get that $\dim_K(\mathrm{soc}(P)) = \dim_K(P/\mathrm{rad}(P))$ for each projective indecomposable A-module P.

For the inverse direction suppose that A is a self-injective K-algebra such that $\dim_K(\mathrm{soc}(P)) = \dim_K(P/\mathrm{rad}(P))$ for each projective A-module P. Using the functors $\mathcal{N}$ and its inverse we obtain a ring homomorphism

$$End_A(S) \longrightarrow End_A(\nu(S))$$

induced by $\mathcal{N}$, and a ring homomorphism

$$End_A(\nu(S)) \longrightarrow End_A(S)$$

induced by $\mathcal{N}^{-1}$. Since both endomorphism algebras are skew-fields by Schur's lemma, both morphisms are injective. Counting the dimensions give that they are actually isomorphic.

Now, denote by P_S the projective cover of the simple module S, and by I_S the injective hull of S. For

$$A \simeq \bigoplus_{S\in\mathcal{S}} P_S^{n_S}$$

we obtain a method to compare n_S and $n_{\nu(S)}$ by

$$n_S = \frac{\dim_K(S)}{\dim_K(End_A(S))} = \frac{\dim_K(\nu(S))}{\dim_K(End_A(\nu(S)))} = n_{\nu(S)}$$

and since the K-linear dual transforms projective right modules into injective left modules, and since by Lemma 4.5.6 $\mathcal{N}(P_S) = I_S$, we get A-module isomorphisms

$$\begin{aligned} Hom_K(A_A, K) &= Hom_K(End_A(A)_A, K) \\ &= \mathcal{N}\left(\bigoplus_{S\in\mathcal{S}} P_S^{n_S}\right) = \bigoplus_{S\in\mathcal{S}} I_S^{n_S} = \bigoplus_{S\in\mathcal{S}} P_{\nu(S)}^{n_S} = {}_AA. \end{aligned}$$

Therefore $A \simeq Hom_K(A, K)$ as left modules, and hence A is a Frobenius algebra.

A special case occurs if A is basic, then $n_S = 1$ for all S, and the condition on the equality of the dimensions of the simple modules is superfluous. □

4.6 Picard Groups

As we have seen the group of automorphisms of a ring is not a Morita invariant. For example $Aut(\mathbb{Z}) = 1$ whereas $Aut(Mat_2(\mathbb{Z}))$ contains $PGL_2(\mathbb{Z})$, which itself contains the modular group $PSL_2(\mathbb{Z}) \simeq C_2 * C_3$ since conjugation by an invertible matrix is an automorphism. Moreover, the group of units of a ring is not a Morita invariant either. Again, the group of units of $\mathbb{Z}$ is $\{\pm 1\}$, whereas the group of units of $Mat_2(\mathbb{Z})$ is $GL_2(\mathbb{Z})$.

The concept in this section gives Morita invariant replacements of the group of automorphisms and the group of units. Moreover, the concepts developed here find their counterparts in more subtle equivalences studied later in Sect. 6.12. Large parts of the theory were developed by A. Fröhlich and his collaborators (cf [9]).

Denote by $[X]$ the isomorphism class of a module X.

Definition 4.6.1 Let K be a commutative ring and let A be a K-algebra. Then

$$\begin{aligned} Pic_K(A) := \{[X] \in \ & A \otimes_K A^{op}\text{-}mod \mid \exists Y \in A \otimes_K A^{op}\text{-}mod : \\ & X \otimes_A Y \simeq A \simeq Y \otimes_A X \text{ as } A\text{-}A\text{-bimodules}\} \end{aligned}$$

the *Picard group* of the K-algebra A.

The Picard group is actually a group where multiplication is given by the tensor product over A. Of course, in view of Corollary 4.2.12 the group $Pic_K(A)$ is the group of additive K-linear self-equivalences of A-*mod* inducing self-equivalences of A-*Mod*.

In particular, for every $[M]$ in $Pic_K(A)$ we get that

$$End_A(M) \simeq A^{op}.$$

The link to the automorphism group is easy.

Lemma 4.6.2 *Let K be a commutative ring and let A be a K-algebra. Then there is a group homomorphism*

$$\begin{aligned} Aut_K(A) &\xrightarrow{\Phi} Pic_K(A) \\ \alpha &\mapsto {}_1A_\alpha \end{aligned}$$

with kernel $Inn(A)$ the group of inner automorphisms.

Proof The fact that the above is a group homomorphism follows directly from Lemma 1.10.10. Lemma 1.10.9 shows that the kernel of this mapping is the group of inner automorphisms, as claimed. □

Recall that we define $Out_K(A) = Aut_K(A)/Inn(A)$.

Proposition 4.6.3 *Let K be a commutative ring and let A be a K-algebra. Two elements $[M]$ and $[N]$ of $Pic_K(A)$ are isomorphic as left A-modules if and only if there is an automorphism α of A such that*

$$M \simeq N \otimes_A {}_1A_\alpha.$$

In particular, an element $[M]$ in $Pic_K(A)$ is in the image of Φ if and only if M is free of rank 1 as an A-module.

Proof It is clear that for all automorphisms α we get that ${}_1A_\alpha$ is free as a left module, since ${}_1A_\alpha \simeq A$ as an A-module. Therefore, as a left A-module $M \otimes_A {}_1A_\alpha \simeq M$ and if $N \simeq M \otimes_A {}_1A_\alpha$, then ${}_AN \simeq {}_AM$.

Conversely, suppose $M \simeq N$ as A-modules. Then there is an isomorphism

$$N \xrightarrow{\rho} M$$

as A-left modules. Let $\check{N}$ be the inverse of N. Then ρ induces an isomorphism

$$A = \check{N} \otimes_A N \xrightarrow{id \otimes \rho} \check{N} \otimes_A M$$

and so we may assume that $N \simeq A$ as a left A-module. Suppose therefore that

$$\rho : A \longrightarrow M$$

is an isomorphism of left A-modules. Since M is invertible we get $End_A(M) \simeq A^{op}$ so that each endomorphism of M is given by right multiplication by an element in A.

$$A^{op} \simeq End_A(A) \simeq End_A(M)$$

where the left isomorphism is given by right multiplication. Hence for each $a \in A$ the A-linear endomorphism μ_a^M given by right multiplication by a on M induces an A-linear endomorphism

$$\rho^{-1} \circ \mu_a^M \circ \rho \in End_A(A).$$

But also $End_A(A) \simeq A^{op}$ by right multiplication, and so for each $a \in A$ there is an $\alpha(a) \in A$ such that

$$(\rho^{-1} \circ \mu_a^M \circ \rho)(x) = x \cdot \alpha(a)$$

for all $x \in A$. Hence

$$\rho(x) \cdot a = \rho(x \cdot \alpha(a))$$

for all $x \in A$. The mapping

$$_1A_\alpha \xrightarrow{\rho} M$$

is now an isomorphism of A-A-bimodules. Indeed, ρ is an isomorphism of A-left modules by construction, and the equation

$$\rho(x) \cdot a = \rho(x \cdot \alpha(a))$$

shows that ρ is even A-A-linear. □

Corollary 4.6.4 *Let K be a commutative ring and let A be a K-algebra. Suppose that $A - proj$, the category of finitely generated A-modules, has the Krull-Schmidt property, and suppose that the left regular A-module decomposes into a finite number of indecomposable projective A-modules. Suppose moreover that each projective indecomposable A-module P occurs exactly once as a direct factor of A. Then $Pic_K(A) \simeq Out_K(A)$. In particular, if K is a field and if A is a finite dimensional basic K-algebra, then $Pic_K(A) \simeq Out_K(A)$.*

Proof Given an element $[M]$ in $Pic_K(A)$ we see that M is a progenerator of A-*mod* with endomorphism ring isomorphic to A^{op}. Since $A - proj$ has the Krull-Schmidt property, M has to be free as an A-left module. Indeed, each projective indecomposable A-module has to be a direct factor of M, and since $End_A(M) \simeq A^{op}$, each projective indecomposable direct factor has to occur in the direct sum decomposition of M exactly once. Then Proposition 4.6.3 proves the statement. □

For the next example we shall need the following classical result.

Theorem 4.6.5 (Skolem-Noether Theorem) (cf e.g. [10, Theorem 7.21]) *Let K be a field, and let A be a finite dimensional simple K-algebra. Suppose $Z(A) = K$. Then every K-linear automorphism α of A is inner in the sense that there is an invertible element $a \in A$ such that $\alpha(b) = a \cdot b \cdot a^{-1}$.*

Proof Since $\alpha|_{Z(A)} = id_{Z(A)}$, we obtain that α fixes each primitive central idempotent of A and hence induces automorphisms of each simple direct factor of A. We may hence assume that A is simple and therefore $A \simeq Mat_{n\times n}(D)$ for an integer n and a finite dimensional skew-field D over K.

We shall prove in a first step that $A \otimes_K D$ is a simple algebra. Since $A \simeq Mat_{n\times n}(D)$ it is sufficient to show that $D \otimes_K D$ is simple. Since $Z(A) = K$, we also have $Z(D) = K$. Hence, $Z(D \otimes_K D) = K$. Fix a K-basis

$$\{d_i \in D \mid i \in \{1, \ldots, n\}\}$$

of D. Let X be a two-sided ideal of $D \otimes_K D$ and let

$$x = \sum_{j=1}^{m} d_{i_j} \otimes \delta_j \in X$$

for $\delta_j \in D \setminus \{0\}$ be chosen so that m is minimal. We may consider $x \cdot \left(1 \otimes \delta_1^{-1}\right)$ instead of x, and may therefore assume that

$$x = d_{i_1} \otimes 1 + \sum_{j=2}^{m} d_{i_j} \otimes \delta_j.$$

Let $\delta \in D \setminus K$. Then

$$y := (1 \otimes \delta) \cdot x \cdot (1 \otimes \delta)^{-1} = d_{i_1} \otimes 1 + \sum_{j=2}^{m} d_{i_j} \otimes \delta\delta_j\delta^{-1} \in X$$

and $x - y \in X$. Since m is minimal, and since the first term in x and y coincide, we get that $x - y = 0$. This shows that δ commutes with each δ_j for $j > 1$ and hence $\delta_j \in Z(D) = K$ for each j. In fact, this shows that $m = 1$, and therefore $x \in (D \otimes_K D)^\times$, the unit group of the algebra. Therefore $X = D \otimes_K D$, and hence $D \otimes_K D$ is simple.

Let V be a simple A-module. Then V is a simple $A \otimes_K D$-module by putting $(a \otimes d) \cdot v := a \cdot d \cdot v$. Hence twisting the action of A by α gives ${}^{\alpha}V$, which is a simple $A \otimes_K D$-module by the action

$$(a \otimes d) \bullet_\alpha v := \alpha(a) \cdot d \cdot v.$$

Since $\dim_K(V) = \dim_K({}^{\alpha}V)$, and since $A \otimes_K D$ is simple by the first step, we get that ${}^{\alpha}V \simeq V$ as $A \otimes_K D$-modules. Let

$$\varphi : V \longrightarrow {}^{\alpha}V$$

be an isomorphism. Then

$$\varphi(a \cdot d \cdot v) = \alpha(a) \cdot d \cdot \varphi(v)$$

for all $a \in A, d \in D, v \in V$. Putting $a = 1$ we get that $\varphi \in End_D(V) = A$ and so there is a $u \in A$ with $\varphi(v) = u \cdot v$. Since φ is an isomorphism and since V is a faithful A-module, we get that u is invertible. Hence

$$u \cdot a \cdot d \cdot v = \alpha(a) \cdot d \cdot u \cdot v$$

for all $a \in A, d \in D, v \in V$. With $d = 1$ we get that $u \cdot a - \alpha(a) \cdot u$ acts as 0 on V. However, V is a faithful A-module, and therefore

$$u \cdot a - \alpha(a) \cdot u = 0$$

which implies the statement. □

Example 4.6.6 Let K be a field and let $A = K \times Mat_{2\times 2}(K)$. Then

$$Aut_K(A) = Aut_K(K) \times Aut_K(Mat_{2\times 2}(K))$$

and all automorphisms of A as K-algebra are inner, using that

$$Aut_K(Mat_{2\times 2}(K)) = Inn(Mat_{2\times 2}(K))$$

by the Skolem Noether Theorem 4.6.5. However, A is Morita equivalent to $K \times K$ and there is a non-inner automorphism α interchanging the two components. Hence $Pic_K(A) = C_2$.

This example shows that the assumption in Corollary 4.6.4 that A is basic is necessary, but it shows as well that the outer automorphism group $Out_K(A)$ is not invariant under Morita equivalences.

Of course, $Pic_K(A)$ is invariant under Morita equivalences and therefore serves as the most suitable replacement for $Out_K(A)$ when one needs a structure close to $Out_K(A)$ but invariant under Morita equivalences.

If K is an algebraically closed field and A is a finite dimensional K-algebra then it can be shown that $Out_K(A)$ is an algebraic group. The connected component of the identity $Out_K(A)^\circ$ is then a normal subgroup which is invariant under Morita equivalences, and moreover which is invariant under the much more general concept of derived equivalences which will be introduced in Chap. 6. This fact was discovered by Huisgen-Zimmermann and Saorin [11] as well as independently by Rouquier [12]. Rouquier attributes the case of Morita equivalences to Brauer.

The idea of Proposition 4.6.3 can be used to obtain a group homomorphism of the Picard group to the automorphism group of the centre of an algebra.

Proposition 4.6.7 *Let K be a commutative ring and let A be a K-algebra. Denote by $Z(A)$ its centre. Then there is a group homomorphism*

$$\Gamma : Pic_K(A) \longrightarrow Aut_K(Z(A)).$$

Proof Let $c \in Z(A)$ and $[M] \in Pic_K(A)$. Then left multiplication by c on M is an A-linear automorphism of M, since $c \in Z(A)$. Hence there is a $\gamma_M(c) \in A$ such that

$$c \cdot m = m \cdot \gamma_M(c)$$

for all $m \in M$. Now, by Proposition 4.3.1 we get that γ_M is a ring automorphism.

We shall see that

$$\gamma_{M_1 \otimes_A M_2} = \gamma_{M_2} \circ \gamma_{M_1}.$$

Let $m_1 \in M_1$ and $m_2 \in M_2$. Then for all $z \in Z(A)$ we have

$$\begin{aligned} z \cdot (m_1 \otimes m_2) &= (m_1 \gamma_{M_1}(z)) \otimes m_2 = m_1 \otimes (\gamma_{M_1}(z) m_2) \\ &= m_1 \otimes (m_2 \gamma_{M_2}(\gamma_{M_1}(z))) = (m_1 \otimes m_2) \cdot (\gamma_{M_2} \circ \gamma_{M_1})(z) \\ &= (m_1 \otimes m_2) \cdot \gamma_{M_1 \otimes_A M_2}(z). \end{aligned}$$

Hence

$$\begin{aligned} Pic_K(A) &\xrightarrow{\Gamma} Aut_K(Z(A)) \\ [M] &\mapsto \gamma_M^{-1} \end{aligned}$$

is the desired group homomorphism. □

Definition 4.6.8 Let K be a commutative ring and let A be a K-algebra. Then define $Picent(A) := \ker(\Gamma)$.

Recall the definition of Φ from Lemma 4.6.2.

Corollary 4.6.9 *Let K be a commutative ring and let A be a commutative K-algebra. Then $\Gamma \circ \Phi = id_{Aut_K(A)}$. In particular*

$$Pic_K(A) = Picent(A) \rtimes Aut_K(A)$$

is a semidirect product.

Proof Indeed, if A is commutative, $Z(A) = A$ and Γ is an automorphism of A. The fact that $\Gamma \circ \Phi = id_{Aut_K(A)}$ is immediate by the definition. □

Remark 4.6.10 Observe that the definition of $Picent(A)$ implies that the group $Picent(A)$ is formed by the invertible bimodules over A on which the centre of A acts the same way on the left and on the right. In particular this fact does not involve K.

References

1. Gabriel, P.: Des catégories abéliennes. Bull. S.M.F., Tome **90**, 323–448 (1962)
2. Montgomery, S.: Hopf algebras and their actions on rings. CBMS Regional Conference Series in Mathematics, vol. 82. American Mathematical Society, Providence, R.I. (1993)

3. Broué, M., Puig, L.: A Frobenius theorem for blocks. Inventiones Math. **56**, 117–128 (1980)
4. Külshammer, B.: Nilpotent blocks revisited. In: Groups, Rings and Group Rings, pp. 263–274. Chapman and Hall, Boca Raton (2006)
5. Assem, I., Simson, D., Skowroński, A.: Elements of the Representation Theory of Associative Algebras: Techniques of Representation Theory, vol. 1. Cambridge University Press, New York (2006)
6. Lam, T.-Y.: Lectures on Modules and Rings. Springer, New-York (1998)
7. Nakayama, T.: On Frobeniusean algebras I. Ann. Math. **40**(3), 611–633 (1939)
8. Nakayama, T.: On Frobeniusean algebras II. Ann. Math. **42**(1), 1–21 (1941)
9. Fröhlich, A.: The Picard group of noncommutative rings, in particular of orders. Trans. Am. Math. Soc. **180**, 1–45 (1973)
10. Reiner, I.: Maximal Orders. Academic Press, London (1975)
11. Huisgen-Zimmermann, B., Saorin, M.: Geometry of chain complexes and outer automorphisms under derived equivalences. Trans. Am. Math. Soc. **353**, 4757–4777 (2001)
12. Rouquier, R.: Automorphismes, graduations et catégories triangulées. J. Inst. Math. Jussieu **10**, 713–751 (2011)

Chapter 5
Stable Module Categories

Morita equivalences provide a very strong relationship between two rings, and in particular their representation theory. However, one observes in examples similarities between module categories which are not given by a Morita equivalence. Nevertheless, a structural connection is reasonable. One of the possible connections is a stable equivalence. A stable equivalence is the weakest possible equivalence within a hierarchy. The strongest is of course isomorphism, then Morita equivalence, and the weakest will be stable equivalence. In between these two equivalences we shall put two classes, called derived equivalence, to be introduced in Chap. 6, and weaker than that, stable equivalences of Morita type, introduced in this chapter.

5.1 Definitions, Elementary Properties, Examples

5.1.1 Definition of Stable Categories

Let G be a finite group, let p be a prime and let P be a Sylow p-subgroup of G. Then Green correspondence actually gives an equivalence of certain categories. Auslander and Kleiner [1] gives a very detailed and subtle discussion of an abstract version of Green correspondence.

Recall that Green correspondence shows that given an indecomposable KG-module M with vertex D then $M\downarrow^G_{N_G(D)}$ is a direct sum of indecomposable $N_G(D)$-modules, one has vertex D and the other indecomposable factors have vertex in $N_G(D)\cap D^g$ for $g\in G\setminus N_G(D)$. Moreover, given an indecomposable $KN_G(D)$-module N with vertex D, then $N\uparrow^G_{N_G(D)}$ is a direct sum of an indecomposable module with vertex D and other indecomposable modules with vertex in $D\cap D^g$ for $g\notin N_G(D)$.

Suppose that D is a trivial intersection p-subgroup. This means that $D\cap D^g=\{1\}$ whenever $g\notin N_G(D)$. Then a module with vertex of the form $D\cap D^g$ for $g\notin N_G(D)$ has vertex 1, and is hence projective. Therefore the functor

A. Zimmermann, *Representation Theory*, 427
Algebra and Applications 19, DOI: 10.1007/978-3-319-07968-4_5,

$$\uparrow_{N_G(D)}^G : KN_G(D)\text{-}mod \longrightarrow KG\text{-}mod$$

does not really send indecomposable objects to indecomposable objects, but sends indecomposable objects to indecomposable objects "after cancelling out projective objects". This behaviour would be perfect if we could "eliminate" the projective objects. We need to construct a category whose objects are modules, and morphisms are defined in such a way that any projective module is isomorphic to the 0 module.

Definition 5.1.1 Let K be a commutative ring and let A be a K-algebra. Then the *stable category* A-$\underline{Mod}$ is the category with objects the A-modules and with morphisms, for all pairs M, N of A-modules

$$\underline{Hom}_A(M, N) := Hom_A(M, N)/PHom_A(M, N)$$

where

$$PHom_A(M, N) := \{f \in Hom_A(M, N) \mid \exists \text{ a projective } A\text{-module } P : \\ f \in Hom_A(P, N) \circ Hom_A(M, P)\}.$$

The category A-$\underline{mod}$ is the full sub-category of A-$\underline{Mod}$ formed by finitely generated A-modules.

Remark 5.1.2 The following observations should be made.

- In order to have a proper category we need to define composition of morphisms in A-$\underline{Mod}$. This is, of course, induced by the usual composition of mappings. We shall verify that this is indeed well-defined. Given four A-modules L, M, N, R, by definition

$$Hom_A(N, L) \circ PHom_A(M, N) \circ Hom_A(R, M) \subseteq PHom_A(R, L)$$

  and so the composition is well-defined.
- Projective modules are isomorphic to 0 in A-$\underline{Mod}$. Indeed, for P a projective module there is a unique A-module homomorphism $P \longrightarrow 0$ and a unique A-module homomorphism $0 \longrightarrow P$. Both are in fact isomorphisms, inverse to one another in A-$\underline{Mod}$. Indeed, the composition $0 \longrightarrow P \longrightarrow 0$ is the identity of course. Denote by α the composition $P \longrightarrow 0 \longrightarrow P$. Then $\alpha - id_P \in PHom_A(P, P)$ by definition. Hence $\alpha = id_P \in \underline{Hom}_A(P, P)$. This shows that $P \simeq 0$.

5.1.2 Syzygies

We recall from Definition 1.8.10 the very important construction of a syzygy.

Let K be a commutative ring and let A be a K-algebra. Then for every A-module M there is a projective A-module P and an epimorphism $P \stackrel{\alpha}{\longrightarrow} M$. Let $\Omega(M) := \ker(\alpha)$.

Remark 5.1.3 In Definition 1.8.10 we denoted the syzygy of M by Ω_M. In order to emphasise the functorial character we shall write $\Omega(M)$.

We need to show that $\Omega(M)$ does not depend on the choice of P and the epimorphism α. Actually, it is clear that we may modify $\Omega(M)$ by an isomorphism since kernels are only defined up to an isomorphism. The reason for this is Schanuel's Lemma 1.8.12.

Hence $\Omega(M)$ is well-defined up to projective direct factors and isomorphism. Therefore the isomorphism class of $\Omega(M)$ is well-defined in the stable category A-$\underline{Mod}$. Now, define for all $n \in \mathbb{N}$ the module $\Omega^n(M) := \Omega(\Omega^{n-1}(M))$ where $\Omega^1(M) := \Omega(M)$.

We shall show that Ω is a functor A-$\underline{Mod} \longrightarrow A$-$\underline{Mod}$. Let M and N be two A-modules and let $\alpha : P \longrightarrow M$ and $\beta : Q \longrightarrow N$ be epimorphisms. Let $\gamma : M \longrightarrow N$ be a homomorphism. Then, since P is projective there is a morphism $\delta : P \longrightarrow Q$ with $\beta \circ \delta = \gamma \circ \alpha$. Since Q is projective, there is a morphism $\delta : Q \longrightarrow P$ such that $\alpha \circ \delta = \beta$. The mapping $\epsilon := \delta|_{\Omega(M)}$ induces a morphism $\Omega(M) \longrightarrow \Omega(N)$

$$\begin{array}{ccccccc}
0 \longrightarrow & \Omega(M) & \longrightarrow & P & \stackrel{\alpha}{\longrightarrow} & M & \longrightarrow 0 \\
 & \downarrow \epsilon & & \downarrow \delta & & \downarrow \gamma & \\
0 \longrightarrow & \Omega(N) & \longrightarrow & Q & \stackrel{\beta}{\longrightarrow} & N & \longrightarrow 0
\end{array}$$

Define $\Omega(\gamma) := \epsilon$.

We need to show that $\Omega(\gamma)$ is well-defined. Suppose δ_1 and δ_2 both lift the same morphism γ. Then the difference lifts the map $0 \in Hom_A(M, N)$. But then $\delta_1 - \delta_2$ factors through $\Omega(N)$ via a map $P \longrightarrow \Omega(N)$ and so $\epsilon_1 - \epsilon_2$ is the composition $\Omega(M) \longrightarrow P \longrightarrow \Omega(N)$. This map factors through the projective module P and is hence 0 in the stable category.

The fact that $\Omega(id_M) = id_{\Omega(M)}$ is clear since the identity on P lifts the identity on M. The fact that Ω composes well with composite mappings comes from the commutativity of the diagram

$$\begin{array}{ccccccc}
0 \longrightarrow & \Omega(M) & \longrightarrow & P & \longrightarrow & M & \longrightarrow 0 \\
 & \downarrow \Omega(\gamma) & & \downarrow \delta & & \downarrow \gamma & \\
0 \longrightarrow & \Omega(N) & \longrightarrow & Q & \longrightarrow & N & \longrightarrow 0 \\
 & \downarrow \Omega(\epsilon) & & \downarrow \delta & & \downarrow \epsilon & \\
0 \longrightarrow & \Omega(L) & \longrightarrow & R & \longrightarrow & L & \longrightarrow 0
\end{array}$$

and so $\Omega(\epsilon \circ \gamma) = \Omega(\epsilon) \circ \Omega(\gamma)$. Finally, we observe that $\Omega(\gamma_1 + \gamma_2) = \Omega(\gamma_1) + \Omega(\gamma_2)$ since $\epsilon_1 + \epsilon_2$ lifts $\gamma_1 + \gamma_2$.

Proposition 5.1.4 *Let K be a commutative ring and let A be a K-algebra. Then the syzygy Ω is a functor $\Omega : A\text{-}\underline{Mod} \longrightarrow A\text{-}\underline{Mod}$. If A is Noetherian, then Ω restricts to a functor $\Omega : A\text{-}\underline{mod} \longrightarrow A\text{-}\underline{mod}$.*

Proof The first part is actually proved in the discussion above. If A is Noetherian, and if M is finitely generated, then there is a finitely generated projective A-module P and an epimorphism $\pi : P \longrightarrow M$. Since A is Noetherian, $\ker(\pi) = \Omega(M)$ is again finitely generated. □

Example 5.1.5 Usually not all modules are syzygies of some module. As a consequence, Ω will not be dense in general, even up to projective direct factors, and hence Ω is not an equivalence in general. We give the following example of this phenomenon.

Let A be the algebra given by the quiver

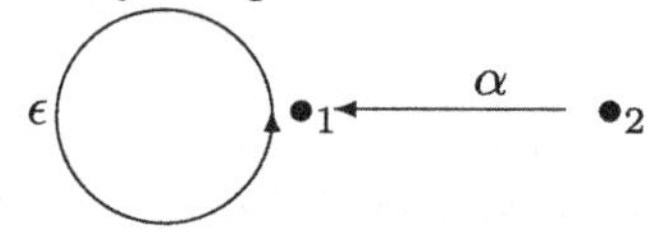

subject to the relations $\epsilon^2 = 0 = \alpha\epsilon$. Recall from Definition 1.11.6 that the algebra has dimension 4, a basis $\{e_1, e_2, \alpha, \epsilon\}$ over any fixed field K and multiplication given by the fact that $e_i^2 = e_i$, $e_i e_j = 0$ for $i \neq j$, $e_2\alpha = \alpha = \alpha e_1$, $e_1\epsilon = \epsilon = \epsilon e_1$, and all other products of basis elements are 0. Then e_1A is a projective indecomposable module of dimension 2, and e_1A is a projective indecomposable module of dimension 2 as well. Let $e_1A/\text{rad}(e_1A) = S_1$ and $e_2A/\text{rad}(e_2A) = S_2$. Then $\text{soc}(e_2A) = S_1 = \text{soc}(e_1A)$. The only non-projective indecomposable modules are S_1 and S_2 and we get

$$\Omega(e_1A) = 0 = \Omega(e_2A) \text{ and } \Omega(S_1) = S_1 = \Omega(S_2).$$

Hence S_2 does not occur as a syzygy.

5.1.3 Representation Type and Equivalences of Stable Categories

A first invariant of general equivalences between stable categories of algebras is the property of being of finite representation type. Recall that an algebra A is of finite representation type if and only if A-*mod* admits only finitely many isomorphism classes of indecomposable A-modules.

Proposition 5.1.6 *Let K be a field and let A and B be finite dimensional K-algebras. Let $F : A\text{-}\underline{mod} \longrightarrow B\text{-}\underline{mod}$ be an equivalence and let M be an indecomposable A-module.*

Then $F(M)$ is indecomposable in $B\text{-}\underline{mod}$, i.e. $F(M)$ is a B-module having precisely one non-projective indecomposable direct summand. Moreover, A is of finite representation type if and only if B is of finite representation type. Further, if M is not projective, then every endomorphism of M that factors through a projective module is nilpotent.

Proof Since A is finite dimensional, and since M is non-projective indecomposable, any endomorphism α that factors through a projective module is nilpotent. Indeed, by Fitting's Lemma 1.4.4, there are A-submodules M_0 and M_1 such that $M \simeq M_0 \oplus M_1$ and α restricts to a nilpotent endomorphism of M_0 and to an automorphism of M_1. Since M is indecomposable, either $M = M_0$ or $M = M_1$. Suppose $M = M_1$, and $\alpha = \beta \circ \gamma$ for $\gamma \in Hom_A(M, P)$ and $\beta \in Hom_A(P, M)$, for some projective A-module P. Then $id_M = (\alpha^{-1} \circ \beta) \circ \gamma$, and therefore M is a direct factor of P. Since P was assumed to be projective, M is projective as well. This proves the first part of the statement.

We shall need to show that if $M \simeq N$ in A-$\underline{mod}$ and if M and N are both indecomposable A-modules, then $M \simeq N$ as A-modules. Let $\alpha : M \longrightarrow N$ be a stable isomorphism with stable inverse α'. The same proof as above for endomorphisms, applied to α, shows that M (or N) is projective.

Since F is an equivalence,

$$\underline{End}_A(M) \simeq \underline{End}_B(FM).$$

Therefore, there is a non-trivial idempotent $e \in \underline{End}_A(M)$ if and only if there is a non-trivial idempotent $Fe \in \underline{End}_B(FM)$. If FM is decomposable, there is such an idempotent $f = Fe$. We identify the idempotent endomorphism e of M in the stable category of A-modules. Moreover, the natural projection

$$End_A(M) \longrightarrow\!\!\!\!\rightarrow \underline{End}_A(M)$$

is a surjective homomorphism of rings. The kernel of this epimorphism contains only nilpotent elements by the first step, and hence is a nilpotent ideal by Proposition 1.6.18. Now Lemma 1.9.17 shows that there is a non-trivial idempotent endomorphism $\hat{e}$ of M as an A-module. Therefore M is decomposable. Likewise, if M is decomposable, FM is also decomposable. This shows that F induces a bijection between the stable isomorphism classes of indecomposable A-modules and stable isomorphism classes of indecomposable B-modules. Since we have seen that for indecomposable non-projective modules stable isomorphism classes coincide with isomorphism classes as modules, we are done. □

Remark 5.1.7 The notion of tame and wild algebra is invariant under stable equivalences, as follows from a result of Krause [2] and of Krause and Zwara [3]. We shall not elaborate on this here.

5.1.4 The Case of Self-Injective Algebras; Triangulated Structure

The structure of a stable category is richer for self-injective algebras. Let K be a field. Recall from Sect. 1.10.1 that a finite dimensional K-algebra A is Frobenius if $Hom_K(A, K) \simeq A$ as A-modules and that A is self-injective if each projective module is injective and each injective module is projective. Hence, for a finite-dimensional A-module M let $Hom_K(M, K)$ be the K-linear dual, which is an A-right module. Then take a projective right A-module P and an epimorphism $\alpha : P \longrightarrow Hom_K(M, K)$ with kernel L.

$$0 \longrightarrow L \longrightarrow P \longrightarrow Hom_K(M, K) \longrightarrow 0.$$

Applying $Hom_K(-, K)$ gives an exact sequence

$$0 \longrightarrow Hom_K(Hom_K(M, K)) \longrightarrow Hom_K(P, K) \longrightarrow Hom_K(L, K) \longrightarrow 0.$$

Since K is a field and since M is finite dimensional over K, the evaluation mapping gives an isomorphism

$$Hom_K(Hom_K(M, K), K) \simeq M.$$

Moreover, the K-linear dual of P is an injective left A-module. Since A is self-injective, injective modules are projective. Therefore

$$0 \longrightarrow M \longrightarrow Hom_K(P, K) \longrightarrow Hom_K(L, K) \longrightarrow 0$$

is exact where $Hom_K(P, K)$ is projective. Define

$$\Omega^{-1}(M) := Hom_K(L, K),$$

$\Omega^0 := id_{A\text{-}\underline{mod}}$ and

$$\Omega^{-n}(M) := \Omega^{-1}(\Omega^{-n+1}(M))$$

for all $n \in \mathbb{N}$. Again, as before in Proposition 5.1.4 for Ω^n, for $n \in \mathbb{N}$, we get that $\Omega^{-1} : A\text{-}\underline{mod} \longrightarrow A\text{-}\underline{mod}$ is a functor. Moreover,

$$\Omega \circ \Omega^{-1} = id_{A\text{-}\underline{mod}} = \Omega^{-1} \circ \Omega$$

by definition and so Ω is a self-equivalence of A-$\underline{mod}$.

Proposition 5.1.8 *Let K be a field and let A be a finite dimensional self-injective K-algebra. Then*

$$\Omega : A\text{-}\underline{mod} \longrightarrow A\text{-}\underline{mod}$$

is a self-equivalence of categories.

Proof The proposition has already been proved above. □

Remark 5.1.9 If A is not finite dimensional or if K is not a field, then it is not possible to work over the category of finitely generated modules. Also the above double dual argument is not valid. One needs to pass to the general module category, and then again one gets invertibility of the syzygy functor using the existence of injective coresolutions.

We shall show that the stable category of self-injective algebras is triangulated. For this purpose we shall need to define an invertible endo-functor, which will be $T := \Omega^{-1}$, and a class of distinguished triangles.

Let

$$0 \longrightarrow M \longrightarrow N \longrightarrow L \longrightarrow 0$$

be an exact sequence of A-modules. Then we shall define a module homomorphism $\Omega(L) \longrightarrow M$ which is going to define a triangle

$$M \longrightarrow N \longrightarrow L \longrightarrow \Omega^{-1}(M)$$

and all sequences of this form will then be the distinguished triangles. We start with a projective cover P_L, obtain the diagram

$$\begin{array}{ccccccccc} 0 & \longrightarrow & M & \longrightarrow & N & \longrightarrow & L & \longrightarrow & 0 \\ & & & & & & \| & & \\ 0 & \longrightarrow & \Omega(L) & \longrightarrow & P_L & \longrightarrow & L & \longrightarrow & 0 \end{array}$$

and observe that P_L is projective, hence there is a morphism $P_L \longrightarrow N$ making the right-hand square commutative, thus inducing a morphism $\Omega(L) \longrightarrow M$:

$$\begin{array}{ccccccccc} 0 & \longrightarrow & M & \longrightarrow & N & \longrightarrow & L & \longrightarrow & 0 \\ & & \uparrow & & \uparrow & & \| & & \\ 0 & \longrightarrow & \Omega(L) & \longrightarrow & P_L & \longrightarrow & L & \longrightarrow & 0 \end{array}$$

We need to show that this mapping is unique in the stable module category. Suppose we have two different lifts of the identity on L to $P_L \longrightarrow N$. Then the difference δ of the two lifts lifts the 0 morphism. We get the diagram

$$\begin{array}{ccccccccc} 0 & \longrightarrow & M & \stackrel{\beta}{\longrightarrow} & N & \stackrel{\gamma}{\longrightarrow} & L & \longrightarrow & 0 \\ & & \delta' \uparrow & & \delta \uparrow & & 0 \uparrow & & \\ 0 & \longrightarrow & \Omega(L) & \longrightarrow & P_L & \longrightarrow & L & \longrightarrow & 0 \end{array}$$

But now, $\gamma \circ \delta = 0$ and so δ actually factorises through M: there is an $\alpha : P_L \longrightarrow M$ such that $\delta = \beta \circ \alpha$. Hence δ' factorises through P_L, which implies that $\delta' = 0$ in A-$\underline{Mod}$.

Proposition 5.1.10 *Let K be a field and let A be a finite dimensional self-injective K-algebra. Then A-$\underline{mod}$ is a triangulated category.*

Proof We need to verify the axioms from Definition 3.4.1.

Let us verify TR1. First

$$0 \longrightarrow X \stackrel{id}{\longrightarrow} X \longrightarrow 0 \longrightarrow 0$$

is exact. Now, given a module homomorphism $\alpha : X \longrightarrow Y$, let $X \stackrel{\iota_X}{\longrightarrow} I_X$ be the injective hull of X, then $Y \simeq Y \oplus I_X$ in $A\text{-}\underline{mod}$ since A is selfinjective, and so I_X is projective. Hence

$$0 \longrightarrow X \stackrel{(\alpha,\iota_X)}{\longrightarrow} Y \oplus I_X \longrightarrow C_X(\alpha) \longrightarrow 0$$

is exact for some A-module $C_X(\alpha)$. We obtain a distinguished triangle

$$X \stackrel{(\alpha,\iota_X)}{\longrightarrow} Y \oplus I_X \longrightarrow C_X(\alpha) \longrightarrow \Omega^{-1}(X).$$

Consider TR2. A short exact sequence

$$0 \longrightarrow X \stackrel{\alpha}{\longrightarrow} Y \stackrel{\beta}{\longrightarrow} Z \longrightarrow 0$$

of A-modules yields a distinguished triangle

$$X \stackrel{\alpha}{\longrightarrow} Y \stackrel{\beta}{\longrightarrow} Z \stackrel{\gamma[1]}{\longrightarrow} \Omega^{-1}X$$

in $A\text{-}\underline{mod}$.

We will construct a short exact sequence of A-modules

$$0 \longrightarrow \Omega Z \stackrel{\gamma'}{\longrightarrow} X' \stackrel{-\alpha'}{\longrightarrow} Y \longrightarrow 0$$

where $X' \simeq X$ in $A\text{-}\underline{mod}$, where α' becomes α and γ' becomes γ after composing with the isomorphism $X' \simeq X$. For this purpose consider the commutative diagram

$$\begin{array}{ccccccccc} 0 & \longrightarrow & X & \stackrel{\alpha}{\longrightarrow} & Y & \stackrel{\beta}{\longrightarrow} & Z & \longrightarrow & 0 \\ & & \uparrow \gamma & & \uparrow \lambda & & \| & & \\ 0 & \longrightarrow & \Omega(Z) & \stackrel{\iota_Z}{\longrightarrow} & P_Z & \stackrel{\beta}{\longrightarrow} & Z & \longrightarrow & 0 \end{array}$$

where λ exists since P_Z is projective, and where γ is induced by the universal property of the kernel. By Lemma 1.8.27 the left-hand square is a pushout diagram. The sequence

$$0 \longrightarrow \Omega(Z) \stackrel{\binom{\iota_Z}{\gamma}}{\longrightarrow} P_Z \oplus X \stackrel{(\lambda,\ -\alpha)}{\longrightarrow} Y \longrightarrow 0$$

is exact. Indeed, since ι_Z is injective, $\binom{\iota_Z}{\gamma}$ is also injective. The fact that the above left-hand square is a pushout diagram is equivalent to saying that Y is the cokernel of $\binom{\iota_Z}{\gamma}$ via $(\lambda,\ -\alpha)$, since the universal property of the pushout is precisely the same as the universal property of the cokernel in this case. Since P_Z is projective, whence 0 in the stable category, we get a short exact sequence as required. Then this induces a distinguished triangle in the stable category

$$\Omega(Z) \xrightarrow{\gamma} X \xrightarrow{-\alpha} Y \longrightarrow \xrightarrow{\gamma[1]} Z$$

as required.

We shall prove TR3. Suppose

$$0 \longrightarrow X \longrightarrow Y \longrightarrow Z \longrightarrow 0$$

and

$$0 \longrightarrow X' \longrightarrow Y' \longrightarrow Z' \longrightarrow 0$$

are short exact sequences of A-modules. Then if we have mappings

$$\begin{array}{ccccccc} X & \xrightarrow{\alpha} & Y & \xrightarrow{\beta} Z \xrightarrow{\gamma} & \Omega^{-1}X \\ \downarrow\rho & & \downarrow\sigma & & \downarrow\Omega^{-1}\rho \\ X' & \xrightarrow{\alpha'} & Y' & \xrightarrow{\beta'} Z' \xrightarrow{\gamma'} & \Omega^{-1}X' \end{array}$$

so that the left square commutes in the stable category, then we need to define a mapping $Z \longrightarrow Z'$ so that the square on the right and in the middle commute in the stable category. Suppose $\sigma\circ\alpha-\alpha'\circ\rho : X \longrightarrow Y'$ factors through a projective module P, i.e.

$$\sigma\circ\alpha-\alpha'\circ\rho=\lambda\circ\mu : X \longrightarrow Y'$$

for $\lambda : P \longrightarrow Y'$ and $\mu : X \longrightarrow P$. We shall change the short exact sequence

$$0 \longrightarrow X \xrightarrow{\alpha} Y \xrightarrow{\beta} Z \longrightarrow 0$$

into a sequence

$$0 \longrightarrow X \xrightarrow{\binom{\alpha}{-\mu}} Y\oplus P \xrightarrow{(\beta,\tilde\beta)} \tilde Z \longrightarrow 0$$

and a commutative diagram

$$\begin{array}{ccccccccc} 0 \longrightarrow & X & \xrightarrow{\binom{\alpha}{-\mu}} & Y\oplus P & \xrightarrow{(\beta,\tilde\beta)} & \tilde Z & \longrightarrow 0 \\ & \| & & \downarrow\pi_Y & & \downarrow\varphi & \\ 0 \longrightarrow & X & \xrightarrow{\alpha} & Y & \xrightarrow{\beta} & Z & \longrightarrow 0 \end{array}$$

which yields an isomorphic triangle

$$\begin{array}{ccccccc} X & \xrightarrow{\binom{\alpha}{-\mu}} & Y \oplus P & \xrightarrow{(\beta,\tilde{\beta})} & \tilde{Z} & \xrightarrow{\tilde{\gamma}} & \Omega^{-1}X \\ \| & & \downarrow \pi_Y & & \downarrow \varphi & & \| \\ X & \xrightarrow{\alpha} & Y & \xrightarrow{\beta} & Z & \xrightarrow{\gamma} & \Omega^{-1}X \end{array}$$

but in which the left-hand square really commutes, not only in the stable category. Hence the mapping φ makes the squares in the diagram

$$\begin{array}{ccccccc} X & \xrightarrow{\binom{\alpha}{-\mu}} & Y \oplus P & \xrightarrow{(\beta,\tilde{\beta})} & \tilde{Z} & \xrightarrow{\tilde{\gamma}} & \Omega^{-1}X \\ \downarrow \rho & & \downarrow (\sigma,\lambda) & & \downarrow \varphi & & \downarrow \Omega^{-1}\rho \\ X' & \xrightarrow{\alpha'} & Y' & \xrightarrow{\beta'} & Z' & \xrightarrow{\gamma'} & \Omega^{-1}X' \end{array}$$

commute since they really come from mappings of short exact sequences.

TR4 is shown as follows. Recall the diagram illustrating the octahedral axiom from Definition 3.4.1.

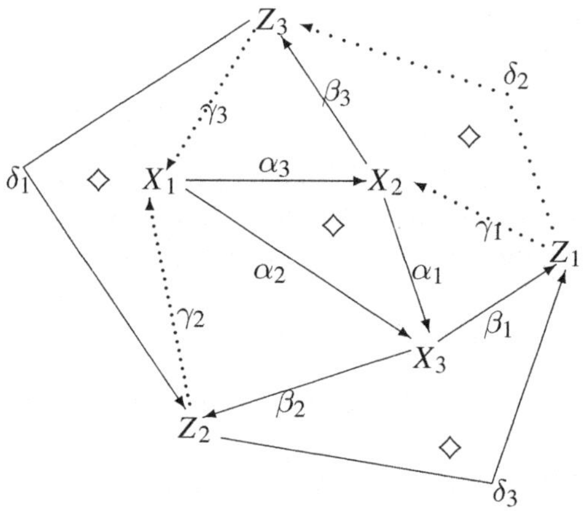

where in the present setting we may assume that the mappings α_i are kernels of β_i and the mappings β_i are cokernels of α_i.

Further, the equality $\alpha_1 \circ \alpha_3 = \alpha_2$ only holds in the stable category. Let P be a projective module such that $\alpha_1 \circ \alpha_3 - \alpha_2 = \lambda \circ \mu$ for $\lambda : P \longrightarrow X_3$ and $\mu : X_1 \longrightarrow P$. Then we may replace X_2 by $X_2 \oplus P$, α_3 by $\binom{\alpha_3}{-\mu}$ and α_1 by (α_1, λ), again changing Z_1 and Z_3 so that the kernel-cokernel property still holds, then the triangles formed by X_i, X_j and Z_k still come from short exact sequences, and the relation $\alpha_1 \circ \alpha_3 = \alpha_2$ holds in the module category.

The mapping δ_3 exists by the universal property of the cokernel. In fact

$$\begin{array}{ccccccccc} 0 \longrightarrow & X_1 & \stackrel{\alpha_2}{\longrightarrow} & X_3 & \stackrel{\beta_2}{\longrightarrow} & Z_2 & \longrightarrow 0 \\ & \downarrow \alpha_3 & & \| & & \downarrow \delta_3 & \\ 0 \longrightarrow & X_2 & \stackrel{\alpha_1}{\longrightarrow} & X_3 & \stackrel{\beta_1}{\longrightarrow} & Z_1 & \longrightarrow 0 \end{array}$$

is a diagram with exact lines and commutative squares. The right-hand mapping exists by the fact that the left-hand square is commutative.

We have short exact sequences and a morphism of short exact sequences as follows.

$$\begin{array}{ccccccc} 0 \longrightarrow X_1 & \stackrel{\alpha_3}{\longrightarrow} & X_2 & \stackrel{\beta_3}{\longrightarrow} & Z_3 & \longrightarrow 0 \\ \| & & \downarrow \alpha_1 & & \downarrow \delta_1 & \\ 0 \longrightarrow X_1 & \stackrel{\alpha_2}{\longrightarrow} & X_3 & \stackrel{\beta_2}{\longrightarrow} & Z_2 & \longrightarrow 0 \end{array}$$

and δ_1 exists, again since the left square commutes.

Further

$$Z_3 \longrightarrow Z_2 \longrightarrow Z_1 \longrightarrow \Omega^{-1} Z_3$$

is a distinguished triangle since

$$(X_3/X_1)/(X_2/X_1) \simeq X_3/X_2$$

canonically.

We need to show

$$\gamma_2 \circ \delta_1 = \gamma_3, \ \delta_3 \circ \beta_2 = \beta_1, \ \delta_2 = \Omega^{-1}(\beta_3) \circ \gamma_1$$

as well as

$$\gamma_1 \circ \delta_3 = \Omega^{-1}(\alpha_3) \circ \gamma_2 \ \text{ and } \ \beta_2 \circ \alpha_1 = \delta_1 \circ \beta_3.$$

For the equation $\delta_3 \circ \beta_2 = \beta_1$ we observe that the diagram

$$\begin{array}{ccc} X_3/X_1 & \stackrel{\delta_3}{\longrightarrow} & X_3/X_2 \\ \uparrow \beta_2 & & \uparrow \beta_1 \\ X_3 & = & X_3 \end{array}$$

with the natural maps is commutative. The equation $\beta_2 \circ \alpha_1 = \delta_1 \circ \beta_3$ is a consequence of the fact that the diagram

$$\begin{array}{ccc} X_2 & \stackrel{\alpha_1}{\longrightarrow} & X_3 \\ \downarrow \beta_3 & & \downarrow \beta_2 \\ X_2/X_1 & \stackrel{\delta_1}{\longrightarrow} & X_3/X_1 \end{array}$$

with the natural maps is clearly commutative.

The equation $\delta_2 = \Omega^{-1}(\beta_3)\circ\gamma_1$ is equivalent to the equation $\Omega(\delta_2) = \beta_3\circ\Omega(\gamma_1)$. In order to prove this we let $P \longrightarrow X_3$ be a projective cover. Then this induces an epimorphism $P \longrightarrow X_3/X_1$, which may serve to compute the syzygy mapping. Using the commutativity of the first two diagrams just proved above, the diagram

$$\begin{array}{ccccc} \Omega(X_3/X_2) & \longrightarrow & P & \longrightarrow & X_3/X_2 \\ \downarrow \Omega(\gamma_1) & & \downarrow & & \| \\ X_2 & \stackrel{\alpha_1}{\longrightarrow} & X_3 & \stackrel{\beta_1}{\longrightarrow} & X_3/X_2 \\ \downarrow \beta_3 & & \downarrow \beta_2 & & \| \\ X_2/X_1 & \stackrel{\delta_1}{\longrightarrow} & X_3/X_1 & \stackrel{\delta_3}{\longrightarrow} & X_3/X_2 \end{array}$$

is commutative, and since the projective cover $P \longrightarrow X_3$ induces an epimorphism $P \longrightarrow X_3/X_1$, the definition of $\Omega(\delta_3)$ implies that the map

$$\Omega(X_3/X_2) \stackrel{\beta_3\circ\Omega(\gamma_1)}{\longrightarrow} X_2/X_1$$

is equal to $\Omega(\delta_3)$.

The equation $\gamma_2 \circ \delta_1 = \gamma_3$ is equivalent to the equation

$$\Omega(\gamma_2) \circ \Omega(\delta_1) = \Omega(\gamma_3).$$

In order to prove this, let $P \longrightarrow\!\!\!\!\rightarrow X_2/X_1$ and $Q \longrightarrow\!\!\!\!\rightarrow X_3/X_1$ be projective cover maps. Since $X_2/X_1 \stackrel{\delta_1}{\hookrightarrow} X_3/X_1$, the universal property of the projective module P implies the existence of a map $P \longrightarrow Q$ such that the diagram

$$\begin{array}{ccc} P & \longrightarrow & X_2/X_1 \\ \downarrow & & \downarrow \delta_1 \\ Q & \longrightarrow & X_3/X_1 \end{array}$$

is commutative. This induces

$$\Omega(X_2/X_1) \stackrel{\Omega(\delta_1)}{\longrightarrow} \Omega(X_3/X_1).$$

Since the map $P \longrightarrow Q$ makes the diagram

$$\begin{array}{ccc} P & \longrightarrow & X_2 \\ \downarrow & & \downarrow \alpha_1 \\ Q & \longrightarrow & X_3 \end{array}$$

commutative, and since the diagram

$$\begin{array}{ccccc}
\Omega(X_2/X_1) & \longrightarrow & P & \longrightarrow & X_2/X_1 \\
\downarrow \Omega(\gamma_3) & & \downarrow & & \| \\
X_1 & \longrightarrow & X_2 & \longrightarrow & X_2/X_1 \\
\| & & \downarrow \alpha_1 & & \downarrow \\
X_1 & \longrightarrow & X_3 & \longrightarrow & X_3/X_1 \\
\uparrow \Omega(\gamma_2) & & \uparrow & & \| \\
\Omega(X_3/X_1) & \longrightarrow & Q & \longrightarrow & X_3/X_1
\end{array}$$

is commutative, we get

$$\Omega(\gamma_2) \circ \Omega(\delta_1) = \Omega(\gamma_3).$$

Finally, the equation $\gamma_1 \circ \delta_3 = \Omega^{-1}(\alpha_3) \circ \gamma_2$ is equivalent to

$$\Omega(\gamma_1) \circ \Omega(\delta_3) = \alpha_3 \circ \Omega(\gamma_2).$$

In order to prove this we first observe that if $Q \longrightarrow X_3/X_1$ is a projective cover, then $Q \longrightarrow X_3/X_1 \longrightarrow X_3/X_2$ is an epimorphism. By the universal property of projective modules the morphism $Q \longrightarrow X_3/X_1$ lifts to a morphism $Q \longrightarrow X_3$. We obtain a commutative diagram

$$\begin{array}{ccccc}
\Omega(X_3/X_2) & \longrightarrow & Q & \longrightarrow & X_3/X_2 \\
\downarrow \Omega(\gamma_1) & & \downarrow & & \| \\
X_2 & \stackrel{\alpha_1}{\longrightarrow} & X_3 & \stackrel{\beta_1}{\longrightarrow} & X_3/X_2 \\
\uparrow \alpha_3 & & \| & & \uparrow \delta_3 \\
X_1 & \stackrel{\alpha_2}{\longrightarrow} & X_3 & \stackrel{\beta_2}{\longrightarrow} & X_3/X_1 \\
\uparrow \Omega(\gamma_1) & & \uparrow & & \| \\
\Omega(X_3/X_1) & \longrightarrow & Q & \longrightarrow & X_3/X_1
\end{array}$$

and the map δ_3 induces the map

$$\Omega(\delta_3) : \Omega(X_3/X_1) \longrightarrow \Omega(X_3/X_2)$$

on the far left, proving the equation

$$\Omega(\gamma_1) \circ \Omega(\delta_3) = \alpha_3 \circ \Omega(\gamma_2).$$

This proves the proposition. □

We have seen in the proof that monomorphism and epimorphism do not make sense in the stable category. We even have the following strange phenomenon.

Lemma 5.1.11 *Let K be a field and let A be a self-injective K-algebra. Then for any two A-modules M and N and any A-module homomorphism $\alpha : M \longrightarrow N$ we may replace M by an isomorphic copy M' so that α is represented by a surjective* *mapping, and we may replace N by an isomorphic copy N' so that α is represented by an injective mapping.*

Proof Let $\pi_N : P_N \longrightarrow N$ be a projective cover of N and $\iota_M : M \longrightarrow I_M$ be an injective hull of M. Since A is self-injective, I_M is projective. Now $M \stackrel{\alpha}{\longrightarrow} N$ is in the same class as $M \oplus P_N \stackrel{\binom{\alpha}{0}}{\longrightarrow} N$ which is equal, in the stable category, to $M \oplus P_N \stackrel{\binom{\alpha}{\pi_N}}{\longrightarrow} N$. Now, $M' := M \oplus P_N$ is as claimed. Likewise $M \stackrel{\alpha}{\longrightarrow} N$ can be replaced by $M \stackrel{(\alpha,0)}{\longrightarrow} N \oplus I_M$ which is equal to $M \stackrel{(\alpha,\iota_M)}{\longrightarrow} N \oplus I_M$.

This proves the statement. □

5.2 Some Examples of Stable Categories

Just as Morita theory is the right answer to the ultimate homological properties of module categories, we want to get an answer to the question of when when two stable categories of two algebras are equivalent. In particular we want to determine how similar algebras really are when their stable categories are equivalent. We shall see that our ambitions have to be much more modest than for equivalences of module categories.

Definition 5.2.1 Let K be a commutative ring andlet A and B be Noetherian K-algebras. Then we say that A and B are *stably equivalent* if

$$A\text{-}\underline{mod} \simeq B\text{-}\underline{mod}$$

as K-linear categories.

In the stable category A-$\underline{mod}$ for a K-algebra A all projective A-modules P have endomorphism ring

$$End_{A\text{-}\underline{mod}}(P) = 0$$

by definition of *PHom* and so if $A = B \times C$ for a semisimple K-algebra C, then A-$\underline{mod} \simeq B$-$\underline{mod}$. Therefore B and A are stably equivalent. Semisimple direct factors cannot be discovered from the stable category.

But the situation is even worse. Abstract stable categories are rather weak invariants as is shown by the following example.

Example 5.2.2 We shall start with the following examples due to Auslander and Reiten [4]. Let K be a field.

1. Then define

$$A_1 := \begin{pmatrix} K & K \\ 0 & K \end{pmatrix}$$

and observe that A_1 is a serial ring, i.e. every projective indecomposable A_1-module is uniserial. There are only three isomorphism classes of (non-zero) indecomposable A-modules:

$$P_1^1 := \begin{pmatrix} K \\ 0 \end{pmatrix} \; ; \; P_2^1 := \begin{pmatrix} K \\ K \end{pmatrix} \; ; \; S_2^1 := P_2^1/P_1^1.$$

The modules P_1^1 and P_2^1 are projective, so that A_1-$\underline{mod}$ admits exactly one indecomposable object, namely P_2^1/P_1^1. We see that

$$End_{A_1\text{-}\underline{mod}}(P_2^1/P_1^1) = K.$$

Hence

$$A_1\text{-}\underline{mod} \simeq K\text{-}mod$$

as K-linear categories.

2. Let $A_2 := K[X]/X^2$ be the algebra of dual numbers. Then A_2 is uniserial. The indecomposable A_2-modules are given by the ideals of A_2 and again we have exactly two isomorphism classes of (non-zero) indecomposable A_2-modules, namely $XK[X]/X^2$ and the regular A_2-module A_2. Of course, the regular module is projective, whereas the module $XK[X]/X^2$ is not projective since A_2 is local. Again we get

$$End_{A_2\text{-}\underline{mod}}(XK[X]/X^2) = K$$

and we obtain

$$A_2\text{-}\underline{mod} \simeq K\text{-}mod$$

as K-linear categories.

We observe that A_1 is hereditary, whereas A_2 is symmetric.

3. Let

$$A_3 := \begin{pmatrix} K & K & K \\ 0 & K & K \\ 0 & 0 & K \end{pmatrix} \Big/ \begin{pmatrix} 0 & 0 & K \\ 0 & 0 & 0 \\ 0 & 0 & 0 \end{pmatrix}$$

be the quotient of the hereditary algebra of upper triangular 3×3-matrices modulo the matrices with coefficient in the upper right corner. Again, since the upper triangular matrix ring is isomorphic to the hereditary algebra given as quiver algebra of the quiver

$$Q_3 : \; \bullet \longrightarrow \bullet \longrightarrow \bullet$$

we can give the indecomposable A_3-modules as certain indecomposable KQ_3-modules. The indecomposable KQ_3-modules are

$$P_1^3 := \begin{pmatrix} K \\ 0 \\ 0 \end{pmatrix} \; ; \; P_2^3 := \begin{pmatrix} K \\ K \\ 0 \end{pmatrix} \; ; \; P_3^3 := \begin{pmatrix} K \\ K \\ K \end{pmatrix}$$

$$S_2^3 := P_2^3 / P_1^3 \; ; \; S_3^3 := P_3^3 / P_2^3 \; ; \; M_3^3 := P_3^3 / P_1^3 .$$

The algebra A_3 is the quotient of $K Q_3$ by the subspace of morphisms from P_1 to P_3. Hence the indecomposable non-projective A_2-modules are

$$S_2^3 \text{ and } S_3^3$$

each with stable endomorphism ring K, and there is no non-zero homomorphism from one to the other, since in one direction there is no morphism in $K Q_3$, and in the other direction the morphisms are in the quotient. Hence

$$A_3\text{-}\underline{mod} \simeq (K \times K)\text{-}mod$$

as a K-linear category. Therefore

$$A_3\text{-}\underline{mod} \simeq (A_2 \times A_2)\text{-}\underline{mod}$$

as triangulated categories. Observe that A_3 is indecomposable, whereas $A_2 \times A_2$ is not indecomposable, and each of the factors is not semisimple.

4. Let K be a field and let A be a finite dimensional K-algebra with $\text{rad}^2(A) = 0$. Then Auslander–Reiten show in [4, Theorem 2.1] that

$$A\text{-}\underline{mod} \simeq \begin{pmatrix} A/\text{rad}(A) & 0 \\ \text{rad}(A) & A/\text{rad}(A) \end{pmatrix}\text{-}\underline{mod}$$

as K-linear categories, and that $\begin{pmatrix} A/\text{rad}(A) & 0 \\ \text{rad}(A) & A/\text{rad}(A) \end{pmatrix}$ is hereditary. Moreover, they establish a necessary and sufficient criterion for when two finite dimensional K-algebras A and B with $\text{rad}^2(A) = 0 = \text{rad}^2(B)$ are stably equivalent. This is a generalisation of the statement that

$$K[X]/X^2\text{-}\underline{mod} \simeq \begin{pmatrix} K & 0 \\ K & K \end{pmatrix}\text{-}\underline{mod},$$

which was deduced above.

We observe that the equivalences of stable categories obtained in Example 5.2.2 are rather abstract, and in particular are not necessarily induced by a functor on the level of the module categories.

5.3 Stable Equivalences of Morita Type: The Definition

A much more natural class of equivalences between stable categories comes from a definition given by Broué.

5.3.1 Some Auxiliary Results on Functors Between Stable Categories

By definition we get for every commutative ring K and every K-algebra A a natural functor

$$A\text{-}mod \longrightarrow A\text{-}\underline{mod}.$$

We say that for every two K-algebras A and B a *functor*

$$\underline{F} : A\text{-}\underline{mod} \longrightarrow B\text{-}\underline{mod}$$

is induced by a functor

$$F : A\text{-}mod \longrightarrow B\text{-}mod$$

if the diagram

$$\begin{array}{ccc} A\text{-}mod & \xrightarrow{F} & B\text{-}mod \\ \downarrow & & \downarrow \\ A\text{-}\underline{mod} & \xrightarrow{\underline{F}} & B\text{-}\underline{mod} \end{array}$$

is commutative, where of course the vertical functors are the natural ones.

Lemma 5.3.1 *Let K be a Noetherian commutative ring, let A and B be K-algebras, finitely generated as K-modules, and let M be a finitely generated B–A-bimodule. Then $M \otimes_A - : A\text{-}mod \longrightarrow B\text{-}mod$ induces a functor $A\text{-}\underline{mod} \longrightarrow B\text{-}\underline{mod}$ if and only if M is a projective B-module.*

Proof Since A and B are finitely generated as K-modules, a finitely generated bimodule M is also finitely generated as an A-module, and as a B-module. In order to be well-defined a functor F induces a functor $\underline{F}$ if and only if $F(P)$ is a projective B-module for every projective A-module P. If $F = M \otimes_A -$ for a B–A-bimodule M, then we just need that $F(A)$ is a projective B-module since F commutes with finite direct sums and hence $F(P)$ is a projective B-module for every projective A-module P. Indeed, if P is projective, then there is a projective A-module Q such that $P \oplus Q \simeq A^n$. Now

$$F(A)^n \simeq F(A^n) \simeq F(P \oplus Q) \simeq F(P) \oplus F(Q)$$

and if ${}_BM = F(A)$ is projective then $F(A^n)$ is projective, and therefore $F(P)$ is a direct summand of the projective B-module $F(A)^n$. Hence,

$$M \otimes_A - : A\text{-}mod \longrightarrow B\text{-}mod$$

induces a functor

$$M \otimes_A - : A\text{-}\underline{mod} \longrightarrow B\text{-}\underline{mod}$$

if and only if M is projective as a B-module. □

Lemma 5.3.2 *Let K be a commutative Noetherian ring and let A and B be finitely generated K-algebras. Suppose that the functor*

$$\underline{F} : A\text{-}\underline{mod} \longrightarrow B\text{-}\underline{mod}$$

is induced by an additive exact functor

$$F : A\text{-}mod \longrightarrow B\text{-}mod.$$

Then there is a B–A-bimodule M with the property that M is projective as a B-module and M is flat as an A-module, so that $\underline{F}$ is isomorphic to

$$\underline{F} \simeq M \otimes_A - : A\text{-}\underline{mod} \longrightarrow B\text{-}\underline{mod}.$$

In particular, if K is a field and A and B are finite dimensional K-algebras, then M is projective as an A-module.

Proof Since F is additive F commutes with finite direct sums. By Watts' Theorem 3.3.16 we get that $F \simeq M \otimes_A -$ for a B–A-bimodule M. Now, by Lemma 5.3.1 the functor $M \otimes_A -$ induces a functor on the level of stable categories if and only if M is projective as a B-module. Now, $M \otimes_A - : A\text{-}mod \longrightarrow \mathbb{Z}\text{-}mod$ is exact if and only if M is flat as an A-module. By Remark 3.8.4 we know that for finitely presented modules M this happens if and only if M is projective. □

We shall need the following technical lemma in the sequel.

Lemma 5.3.3 *Let K be a commutative ring and let A, B and C be K-algebras. Suppose that B is finitely generated projective as a K-module.*

Suppose U is an $A \otimes_K B^{op}$-module and V is a $B \otimes_K C^{op}$-module.

1. *If U is projective of finite rank as an $A \otimes_K B^{op}$-module and V is projective of finite rank as a C^{op}-module, then $U \otimes_B V$ is projective of finite rank as $A \otimes_K C^{op}$-module.*
2. *If U is projective of finite rank as A-module and V is projective of finite rank as a $B \otimes_K C^{op}$-module, then $U \otimes_B V$ is projective of finite rank as a $A \otimes_K C^{op}$-module.*

Proof By symmetry the statements (1) and (2) are equivalent. Hence we shall prove statement (1). We suppose first that B is free as a K-module

$$\begin{aligned}(A \otimes_K B) \otimes_B (B \otimes_K C) &\simeq A \otimes_K ((B \otimes_B B) \otimes_K C)\\ &\simeq A \otimes_K (B \otimes_K C)\\ &\simeq (A \otimes_K C)^{\mathrm{rank}_K(B)}.\end{aligned}$$

If B is projective of finite rank as a K-module, then B is a direct factor of K^m for some m and

$$(A \otimes_K B) \otimes_B (B \otimes_K C) \simeq A \otimes_K (B \otimes_K C))$$

is a direct factor of

$$A \otimes_K (K^m \otimes_K C)) \simeq (A \otimes_K C)^m$$

as an $A \otimes_K C^{op}$-module.

Since U is projective as an $A \otimes_K B^{op}$-module, U is a direct factor of some $(A \otimes_K B)^{n_U}$ and since V is projective as a C^{op}-module, V is a direct factor of some C^{n_V}. Hence $U \otimes_B V$ is a direct factor of $(A \otimes_K C)^{n_U \cdot n_V \cdot m}$ and this proves the statement. □

We shall use pairs of adjoint functors in the sequel. The following statement is most useful in this context.

Lemma 5.3.4 (Auslander–Kleiner) *Let A and B be rings. If (G, F) is a pair of adjoint functors A-mod $\longrightarrow$ B-mod so that F and G map projective modules to projective modules. Then F and G induce functors A-$\underline{mod}$ $\longrightarrow$ B-$\underline{mod}$ and (G, F) is a pair of adjoint functors A-$\underline{mod}$ $\longrightarrow$ B-$\underline{mod}$.*

Proof If P is a projective B module and $f = h \circ g : U \longrightarrow FV$ is a B-linear homomorphism factoring through P

$$U \xrightarrow{g} P \xrightarrow{h} FV$$

then using the isomorphism

$$\psi : Hom_B(-, F?) \xrightarrow{\simeq} Hom_A(G-, ?)$$

we obtain that $\psi(f)$ factors into

$$GU \xrightarrow{\psi(g)} GP \xrightarrow{\psi(h)} V$$

and since G maps projectives to projectives, $\psi(f)$ factors through a projective A-module. Hence ψ induces a homomorphism

$$\underline{\psi} : \underline{Hom}_B(-, F?) \longrightarrow \underline{Hom}_A(G-, ?)$$

which is an isomorphism since the same argument also holds for ψ^{-1}. Finally

$$\begin{array}{ccc} Hom_B(U, FV) & \xrightarrow{\psi} & Hom_A(GU, V) \\ \downarrow & & \downarrow \\ \underline{Hom}_B(U, FV) & \xrightarrow{\underline{\psi}} & \underline{Hom}_A(GU, V) \end{array}$$

is commutative. This proves the statement. □

5.3.2 Broué's Original Definition

Definition 5.3.5 (Broué [5]) Let K be a commutative ring and let A and B be K-algebras. Let M be a B–A-bimodule and let N be an A–B-bimodule such that

1. (a) M is projective as a B-module and as an A^{op}-module.
   (b) N is projective as a B^{op}-module and as an A-module.
2. (a) $M \otimes_A N \simeq B \oplus Q$ as B–B-bimodules, where Q is a finitely generated projective B–B-bimodule.
   (b) $N \otimes_B M \simeq A \oplus P$ as A–A-bimodules, where P is a finitely generated projective A–A-bimodule.

Then we say that *(M, N) induces a stable equivalence of Morita type* between A and B.

Remark 5.3.6 We should emphasise that, for a commutative ring K and K-algebras A and B and an A–B-bimodule M, being projective as an A-module and as a B-module is not the same as being projective as an A–B-bimodule. If B is projective as a K-module then a projective A–B-bimodule is projective as an A-module. Indeed, M is a direct factor of $(A \otimes_K B^{op})^n$ for some n, and if B is projective as a K-module, it is a direct factor of K^m for some m, and $A \otimes_K B^{op}$ is a direct factor of $A \otimes_K K^m \simeq A^m$ for some m, as A-modules. Let us examine the case $A = B = KP$ for some p-group P and K a field of characteristic p. Then KP is local by Proposition 1.6.22 and $KP \otimes_K (KP)^{op} \simeq K(P \times P)$ is local as well. So, a projective $KP \otimes_K (KP)^{op}$-module is free, and therefore of dimension a multiple of $|P|^2$ over K. However the $KP \otimes_K (KP)^{op}$-module KP is projective as a KP-module on the left and on the right. Since the dimension of KP is $|P|$, we see that KP is not projective as $KP \otimes_K (KP)^{op}$-module, unless $P = 1$.

Remark 5.3.7 We give some important remarks on the definition of a stable equivalence of Morita type.

1. Let K be a field, let A and B be K-algebras and let (M, N) be a pair of A–B-bimodules inducing a stable equivalence of Morita type. Then

$$M \otimes_A - : A\text{-}\underline{mod} \longrightarrow B\text{-}\underline{mod}$$

is a well-defined functor by Lemma 5.3.1 and

$$N \otimes_B - : B\text{-}\underline{mod} \longrightarrow A\text{-}\underline{mod}$$

as well. Moreover, for the composition we get

$$\begin{aligned}(M \otimes_A -) \circ (N \otimes_B -) &= M \otimes_A (N \otimes_B -) = (M \otimes_A N) \otimes_B - \\ &\simeq (B \oplus Q) \otimes_B - \simeq (B \otimes_B -) \oplus (Q \otimes_B -) \\ &\simeq id \oplus (Q \otimes_B -).\end{aligned}$$

But if Q is projective as a B–B-bimodule, then Q is a direct factor of $(B \otimes_K B)^n$ for some n and therefore $Q \otimes_B -$ is a direct factor of

$$(B \otimes_K B)^n \otimes_B - \simeq ((B \otimes_K B) \otimes_B -)^n \simeq (B \otimes_K (B \otimes_B -))^n \simeq (B \otimes_K -)^n$$

and $B \otimes_K S \simeq B^{\dim_K(S)}$ is a projective B-module. Hence

$$(M \otimes_A -) \circ (N \otimes_B -) \simeq id_{B\text{-}\underline{mod}}.$$

Similarly we get

$$(N \otimes_B -) \circ (M \otimes_A -) \simeq id_{A\text{-}\underline{mod}}$$

and therefore $M \otimes_A - : A\text{-}\underline{mod} \longrightarrow B\text{-}\underline{mod}$ is an equivalence with quasi-inverse $N \otimes_B - : B\text{-}\underline{mod} \longrightarrow A\text{-}\underline{mod}$.

2. If one wants to have a pair of bimodules (M, N) such that $M \otimes_B - : B\text{-}\underline{mod} \longrightarrow A\text{-}\underline{mod}$ and $N \otimes_A - : A\text{-}\underline{mod} \longrightarrow B\text{-}\underline{mod}$ are mutually inverse equivalences, we need to have that $M \otimes_B N \simeq A \oplus P$ as A–A-bimodules so that $P \otimes_A U$ is projective for every A-module U. As we have seen, this holds if P is projective as a bimodule. But in general the converse does not hold, as was observed by Auslander–Reiten in 1990 [6]. The example provided there is the following. Let k be a non-perfect field and let $K = k(x)$ be a purely inseparable field extension of k. Then $K \otimes_k K$ is a local algebra, but not a field (cf Example 1.7.17). Hence here is a non-free K–K-bimodule M and therefore M is not projective. However, M is free as a left (and as a right) K-module since K is a field, and hence $M \otimes_K X$ is a free K-module for all K-modules X.

5.3.3 Stable Equivalences Induced by Exact Functors

We shall use an auxiliary lemma which is also useful in other situations concerning bimodules and tensor products of algebras. Recall Examples 1.7.15, 1.7.16 and 1.7.17 for possible phenomena.

Lemma 5.3.8 *If K is a field and if C and D are finite dimensional K-algebras such that $Z(C)$ and $Z(D)$ are separable field extensions of K, then for every simple C-module S and every simple D-module T, the $C \otimes_K D$-module $S \otimes_K T$ is semisimple.*

Proof As a first observation this statement is invariant under a Morita equivalence applied to C and to D. Hence we assume C and D are both basic. Then we get that S is a skew-field D_S and T is a skew-field D_T, both finite dimensional over K.

Let I be a two-sided ideal of $D_S \otimes_K D_T$ and let

$$u = \sum_{i=1}^{n} a_i \otimes b_i \in I \setminus \{0\}$$

such that n is minimal, and such that $\{b_1, \ldots, b_n\}$ is a K-linearly independent family. Then $a_1 \neq 0$ by minimality of n, and hence $D_S \cdot a_1 = D_S$. Therefore replacing u by $a_1^{-1} \cdot u$ we may suppose that $a_1 = 1$ and n is still minimal. For any $a \in D_S$ we get

$$au - ua = \sum_{i=2}^{n} (a \cdot a_i - a_i \cdot a) \otimes b_i \in I$$

and therefore $(a \cdot a_i - a_i \cdot a) = 0$ for all i. This shows that $a_i \in Z(D_T)$ for all $i \in \{1, \ldots, n\}$.

Sublemma: If $K = Z(D_S)$, then $D_S \otimes_K D_T$ is simple.

Indeed, in this case we obtain that $a_i \in K$ for all i and therefore $u = 1 \otimes b$ for some $b \in D_T \setminus \{0\}$. But therefore

$$I \supseteq (1 \otimes b) \cdot (D_S \otimes D_T) \supseteq (1 \otimes b) \cdot (1 \otimes D_T) = 1 \otimes D_T.$$

But then $D_S \otimes 1 \subseteq I$ and hence $I = D_S \otimes D_T$.

If now $K \leq L := Z(D_S)$, then

$$D_S \otimes_K D_T = (D_S \otimes_L L) \otimes_K D_T = D_S \otimes_L (L \otimes_K D_T)$$

so that, using the sublemma, we only need to show that $L \otimes_K D_T$ is semisimple. Therefore we may assume that D_S is a field.

The same arguments as for the sublemma, applied to D_T, shows that we may also assume that D_T is a field.

We therefore assume that D_T and D_S are separable field extensions of finite degree of K. Then (cf e.g. Fröhlich and Taylor [7, 1.50]) let $D_S = K[X]/f(X)$ for some irreducible polynomial $f(X)$ which has only simple roots in an algebraic closure $\overline{K}$ of K. We obtain $f(X) = g_1(X) \cdot \dots \cdot g_s(X)$ in $D_T[X]$ for irreducible polynomials $g_i(X)$, $i \in \{1, \dots, s\}$, where g_i and g_j do not have any common roots in $\overline{K}$. Again, the polynomials g_i have only simple roots in $\overline{K}$, and $D_S \otimes_K D_T = D_T[X]/f(X) = \prod_{i=1}^{s} D_T[X]/g_i(X)$. We get that $D_S \otimes_K D_T$ is a product of separable field extensions of D_T, and in particular $D_S \otimes_K D_T$ is semisimple. We have proved the lemma. □

Remark 5.3.9 As is well-known, if K is a perfect field, then every finite dimensional extension field gives a separable field extension.

Corollary 5.3.10 *If K is a perfect field and if C and D are finite dimensional K-algebras, then* $\mathrm{rad}(C \otimes_K D) = \mathrm{rad}(C) \otimes_K D + C \otimes_K \mathrm{rad}(D)$ *and every simple $C \otimes_K D$-module is a direct factor of a module $S \otimes_K T$ for some simple C-module S and a simple D-module T.*

Proof Indeed, $C/\mathrm{rad}(C) \otimes_K D/\mathrm{rad}(D)$ is semisimple by Lemma 5.3.8 and hence

$$\mathrm{rad}(C \otimes_K D) \subseteq \mathrm{rad}(C) \otimes_K D + C \otimes_K \mathrm{rad}(D).$$

However, since $\mathrm{rad}(C)$ and $\mathrm{rad}(D)$ are both nilpotent ideals of C and D respectively, $\mathrm{rad}(C) \otimes_K D + C \otimes_K \mathrm{rad}(D)$ is a nilpotent ideal of $C \otimes_K D$, and therefore

$$\mathrm{rad}(C \otimes_K D) \supseteq \mathrm{rad}(C) \otimes_K D + C \otimes_K \mathrm{rad}(D).$$

Therefore

$$(C \otimes_K D)/\mathrm{rad}(C \otimes_K D) = C/\mathrm{rad}(C) \otimes D/\mathrm{rad}(D)$$

and since every simple module is a direct factor of the radical quotient, this proves the statement. □

Proposition 5.3.11 (Auslander–Reiten [6]) *Let K be a perfect field and let A be a finite dimensional K-algebra. Let P be an $A \otimes_K A^{op}$-module.*

- *If $X \otimes_A P$ is a projective right A-module for all right A-modules X, then P is a projective $A \otimes_K A^{op}$-module.*
- *If $P \otimes_A X$ is a projective left A-module for all left A-modules X, then P is a projective $A \otimes_K A^{op}$-module.*

Proof The two statements are completely symmetric. We only need to prove one of them and suppose therefore that $X \otimes_A P$ is projective for all right A-modules X.

First step: We first note that for two algebras C and D over a commutative ring K we get that

$$\begin{array}{rcl} (U \otimes_K V) \otimes_{C \otimes_K D^{op}} W & \longrightarrow & U \otimes_C W \otimes_D V \\ (u \otimes v) \otimes w & \mapsto & u \otimes w \otimes v \end{array}$$

is an isomorphism for every right C-module U and every D-module V and every C–D^{op}-module W, and that this isomorphism extends to a functor

$$(C^{op}\text{-}mod) \times (D\text{-}mod) \times ((C \otimes_K D^{op})\text{-}mod) \longrightarrow K\text{-}mod.$$

Second step: For the given $A \otimes_K A^{op}$-module P let

$$\dots \xrightarrow{\partial_2} Q_1 \xrightarrow{\partial_1} Q_0 \longrightarrow P \longrightarrow 0$$

be a projective resolution as an $A \otimes_K A^{op}$-module. Since by assumption $X \otimes_A P$ is a projective A-module, $A \otimes_A P = P$ is projective as a right A-module and the resolution $Q^\bullet$ splits as a sequence of A-modules. Since $Q^\bullet$ is split, we may apply $X \otimes_A -$ to the resolution of P to get a resolution

$$\dots \xrightarrow{\partial_2^X} X \otimes_A Q_1 \xrightarrow{\partial_1^X} X \otimes_A Q_0 \longrightarrow X \otimes_A P \longrightarrow 0$$

for $id_X \otimes_A \partial_i =: \partial_i^X$, which is then a projective resolution of $X \otimes_A P$ as A^{op}-modules, using Lemma 5.3.3. Hence, for any A^{op}-module X and any A-module Y we get

$$\begin{aligned}
&Tor_1^{A\otimes_K A^{op}}(X \otimes_K Y, P)\\
&= \ker((id_X \otimes id_Y) \otimes_{A\otimes_K A^{op}} \partial_1)/\mathrm{im}((id_X \otimes id_Y) \otimes_{A\otimes_K A^{op}} \partial_2)\\
&= \ker(\partial_1^X \otimes_A id_Y)/\mathrm{im}(\partial_2^X \otimes_A id_Y)\\
&= Tor_1^A(X \otimes_A P, Y)
\end{aligned}$$

where the second equality follows by the first step. Since $X \otimes_A P$ is projective for all A-modules X, we get that $Tor_1^A(X \otimes_A P, Y) = 0$ for all A-modules Y, and therefore

$$Tor_1^{A\otimes_K A^{op}}(X \otimes_K Y, P) = 0$$

for all A-modules Y and all A^{op}-modules X.

Third step: Observe that we haven't used yet the hypothesis that K is perfect. By Corollary 5.3.10 we see that every simple $A \otimes_K A^{op}$-module is a direct factor of a tensor product $S \otimes T$ for a simple A-module T and a simple A^{op}-module T. But then $Tor_1^{A\otimes_K A^{op}}(U, P) = 0$ for every simple $A \otimes_K A^{op}$-module U. Hence $Tor_1^{A\otimes_K A^{op}}(-, P) = 0$. Indeed we may show by induction on the composition length of an arbitrary $A \otimes_K A^{op}$-module V, using the long exact sequence Lemma 3.8.2, that $Tor_1^{A\otimes_K A^{op}}(V, P) = 0$. Hence P is flat. By Lemma 3.8.5 we obtain that in this case a flat module P is projective.

This proves the statement. □

Remark 5.3.12 Let K be a Noetherian commutative ring and let M be a finitely generated A–B-module for finitely generated Noetherian K-algebras A and B. We know that the functor

$$M \otimes_B - : B\text{-}mod \longrightarrow A\text{-}mod$$

has a right adjoint

$$Hom_A(M, -) : A\text{-}mod \longrightarrow B\text{-}mod.$$

Hence, if M is projective as an A-module, then $M \otimes_B -$ induces a functor between the stable categories by Lemma 5.3.1. If moreover

$$M \otimes_B - : B\text{-}\underline{mod} \longrightarrow A\text{-}\underline{mod}$$

is an equivalence, then by Lemma 5.3.4 the inverse equivalence is

$$Hom_A(M, -) : A\text{-}\underline{mod} \longrightarrow B\text{-}\underline{mod}.$$

Now, if M is finitely generated projective as an A-module, we get

$$Hom_A(M, -) \simeq Hom_A(M, A) \otimes_A -$$

by Lemma 4.2.5. Moreover, M is projective as an A-left module implies that $Hom_A(M, A)$ is projective as an A-right module.

Finally, defining $N := Hom_A(M, A)$, by Remark 5.3.7 item 2 we get that

$$M \otimes_B N \simeq A \oplus P$$

as A–A-bimodules and

$$N \otimes_A M \simeq B \oplus Q$$

as B–B-bimodules for a projective A–A-bimodule P and a projective B–B-bimodule Q.

However, if M is projective as a right B-module *we do not obtain automatically yet that $Hom_A(M, A)$ is projective as a left B-module.*

Proposition 5.3.13 *Let K be a perfect field and let A and B be finite dimensional K-algebras. Suppose M is an A–B-bimodule and suppose that M is finitely generated projective as an A-module and as a B-module. If*

$$M \otimes_B - : B\text{-}\underline{mod} \longrightarrow A\text{-}\underline{mod}$$

is an equivalence and if $N := Hom_A(M, A)$ is projective as a B-module, then (M, N) induces a stable equivalence of Morita type.

Proof Remark 5.3.12 shows that

$$M \otimes_B N \simeq A \oplus P$$

as A–A-bimodules and

$$N \otimes_A M \simeq B \oplus Q$$

as B–B-bimodules for an A–A-bimodule P and a B–B-bimodule Q. Proposition 5.3.11 show that P is projective as an $A \otimes_K A^{op}$-module, and Q is projective as a $B \otimes_K B^{op}$-module. The module M is projective as an A-module and as a B-module by hypothesis. Lemma 4.2.5 shows that N is projective as an A-module and by hypothesis as a B-module.

This proves the statement. □

Remark 5.3.14 We observe that Broué's definition 5.3.5 for stable equivalences of Morita type was originally designed for finite dimensional algebras, mainly over algebraically closed fields. For general algebras over general rings, in view of Proposition 5.3.11, a more natural concept is given in Definition 5.3.15 below. Indeed, in general not all finitely generated flat modules are finitely presented flat. For finitely presented modules the concept of flatness and projectivity coincides. We refer to Lemma 3.8.5 for a proof in the case of finite dimensional algebras.

Definition 5.3.15 Let K be a commutative ring and let A and B be K-algebras. Let M be a B–A-bimodule and let N be an A–B-bimodule so that

1. (a) M is projective as a B-module and as an A^{op}-module.
   (b) N is projective as a B^{op}-module and as an A-module.
2. (a) $M \otimes_A N \simeq B \oplus Q$ as B–B-bimodules, where Q is a finitely generated flat B–B-bimodule.
   (b) $N \otimes_B M \simeq A \oplus P$ as A–A-bimodules, where P is a finitely generated flat A–A-bimodule.
3. $P \otimes_A X$ is projective for each A-module X and $Q \otimes_B Y$ is projective for every B-module Y.

Then we say that (M, N) induces a *stable equivalence of Morita type in the generalised sense* between A and B.

Observe that since M and N are projective as left modules, and projective as right modules, P and Q are projective as left modules and as right modules.

Working in the "universe" of finite dimensional algebras over perfect fields and finitely generated modules we see that a stable equivalence of Morita type is just a stable equivalence induced by an exact additive functor on the level of the module category, plus a little supplement, namely that the functor inducing the quasi-inverse comes from an exact functor as well.

Proposition 5.3.16 (Linckelmann [8, Chap. 11]) *Let K be a perfect field and let A and B be finitely generated K-algebras. Suppose A and B are symmetric K-algebras and let $F : B\text{-mod} \longrightarrow A\text{-mod}$ be an exact additive functor inducing an equivalence $\underline{F} : B\text{-}\underline{mod} \longrightarrow A\text{-}\underline{mod}$. Then $F \simeq M \otimes_B -$ for some A–B-bimodule M and (M, N) induces a stable equivalence of Morita type with $N := Hom_A(M, A)$.*

Proof The existence of M is a consequence of Lemma 5.3.2. Now M is projective as an A-module since $M \otimes_B -$ passes to the stable category. M is projective as a B-module since $M \otimes_B -$ is exact. Lemma 4.2.5 shows that $Hom_A(M, A) =: N$ is projective as an A-module. But since A and B are symmetric K-algebras and hence

$$Hom_B(-, B) \simeq Hom_K(-, K) \simeq Hom_A(-, A)$$

by Proposition 1.10.28, we get that $N \simeq Hom_B(M, B)$ and again by Proposition 5.3.13 we see that $- \otimes_A M$ is an exact functor between the corresponding categories of right modules inducing a stable equivalence and that $Hom_B(M, B)$ is projective as a B-module. Again Proposition 5.3.13 shows that

$$M \otimes_B N \simeq A \oplus P$$

as A–A-bimodules and

$$N \otimes_A M \simeq B \oplus Q$$

as B–B-bimodules for a projective A–A-bimodule P and a projective B–B-bimodule Q. This finishes the proof. □

A slightly more complicated proof given by Rickard in [9, Theorem 3.2] generalises this statement to self-injective algebras.

Proposition 5.3.17 (Rickard [9, Theorem 3.2]) *Let K be a field and let A and B be two finite dimensional indecomposable self-injective K-algebras and let $F : A\text{-}mod \longrightarrow B\text{-}mod$ be an exact additive functor. Suppose that F induces an equivalence $\underline{F} : A\text{-}\underline{mod} \longrightarrow B\text{-}\underline{mod}$. Then F is isomorphic to a stable equivalence of Morita type.*

Proof Let $M := F(A)$. By Watt's theorem 3.3.16 we see that $F \simeq M \otimes_A -$. Since F is exact, M is projective as an A-right module. Since F induces a functor on the stable category, M is projective as a B-left module. The functor

$$M \otimes_A - : A\text{-}mod \longrightarrow B\text{-}mod$$

admits a right adjoint $Hom_B(M, -)$, which is exact since M is projective as a B-module. But now

$$Hom_A(-, Hom_B(M, B)) = Hom_B(M \otimes_A -, B)$$

is exact since B is self-injective, and hence the regular module is injective, and so $Hom_B(?, B)$ exact. Therefore $Hom_B(M, B)$ is injective, and therefore projective since B is self-injective. This shows that $Hom_B(M, -)$ sends projective B-modules to projective B-modules. Therefore $Hom_B(M, -)$ induces a functor

$$Hom_B(M, -) : B\text{-}\underline{mod} \longrightarrow A\text{-}\underline{mod}.$$

By Lemma 5.3.4 $Hom_B(M,-)$ is right adjoint to $M\otimes_A-$ as functors between A-$\underline{mod}$ and B-$\underline{mod}$. Put $N := Hom_B(M,B)$ and observe that $Hom_B(M,-) = N\otimes_B -$. The unit of the adjunction gives a map

$$A \longrightarrow N\otimes_B M$$

which fits into a distinguished triangle

$$A \longrightarrow N\otimes_B M \longrightarrow \Pi \longrightarrow \Omega^{-1}A.$$

Since $M\otimes_A -$ is an equivalence of stable categories, and hence so is its adjoint $N\otimes_B -$, we see that for all A-modules U we get

$$A\otimes_A U \longrightarrow N\otimes_B M\otimes_A U$$

is an isomorphism in the stable category, and therefore $\Pi\otimes_A U$ is a projective A-module for all A-modules U. Proposition 5.3.11 implies that Π is a projective $A\otimes A^{op}$-module, and therefore $A\simeq N\otimes_B M$ in $A\otimes A^{op}$-$\underline{mod}$. Since A is indecomposable we get that there is a projective $A\otimes A^{op}$-module X such that

$$N\otimes_B M\simeq A\oplus X$$

as $A\otimes A^{op}$-modules. Interchanging the roles of M and N gives the remaining assertion. □

Proposition 5.3.18 (Liu [10, Lemma 2.2] and independently Dugas and Martinez-Villa [11]) *Let K be a field, let A and B be finite dimensional K-algebras and suppose that (M,N) induces a stable equivalence of Morita type between A and B. Then ${}_AM$ is a progenerator, M_B is a progenerator, ${}_BN$ is a progenerator and N_A is a progenerator.*

Proof By symmetry we only need to show that ${}_AM$ is a progenerator.

By definition of a stable equivalence of Morita type, ${}_AM$ is projective. Let $P_1, P_2, \dots, P_n$ be a complete set of isomorphism classes of indecomposable projective B-modules. Let E be a projective A-module. We shall show that E is in $add(\bigoplus_{i=1}^n M\otimes_B P_i)$, the additive category generated by the modules $M\otimes_B P_i$. Since this category contains all direct summands of directs sums of the A-module $M\otimes_B B = M$, we have shown that ${}_AM$ is a progenerator.

But now, since ${}_BN$ is projective, $N\otimes_A E$ is a projective B-module, and therefore

$$N\otimes_A E\in add\left(\bigoplus_{i=1}^n P_i\right).$$

Let $M \otimes_B N \simeq A \oplus Q$ for some projective $A \otimes_K A^{op}$-module Q. Then

$$A \otimes_A E \oplus Q \otimes_A E \in add(\bigoplus_{i=1}^{n} M \otimes_B P_i).$$

But since $E \simeq A \otimes_A E$ is a direct summand of $E \oplus Q \otimes_A E$, we have $E \in add(\bigoplus_{i=1}^{n} M \otimes_B P_i)$. This proves the statement. □

Remark 5.3.19 We give two remarks on the above proof.

- The proof given above is the one proposed by Yuming Liu. Dugas and Martinez-Villa use a short argument from Auslander–Reiten theory for their proof.
- Note that if M is an A–B-bimodule, and ${}_AM$ is a progenerator with $B^{op} \simeq End_A(M)$, then $M \otimes_B - : B\text{-}mod \longrightarrow A\text{-mod}$ is a Morita equivalence (cf Theorem 4.2.8). Proposition 5.3.18 shows that a stable equivalence of Morita type (M, N) has the property that M is a progenerator as an A-module. However, B is not assumed to be isomorphic to $End_A(M)^{op}$. The next corollary clarifies this situation.

Corollary 5.3.20 *Let K be a field and let A and B be finite dimensional K-algebras. Suppose that (M, N) induces a stable equivalence of Morita type between A and B. Then there is an algebra A' Morita equivalent to A such that B is a subalgebra of A', and there is an algebra B' Morita equivalent to B such that A is a subalgebra of B'.*

Proof We only need to construct A' since the other statement follows by symmetry.

Let $A' := End_A(M)^{op}$. Since M is a progenerator of A-*mod*, we get that A' is Morita equivalent to A by Theorem 4.2.8. Moreover, since M is an A–B-bimodule, there is a ring homomorphism

$$B \xrightarrow{\mu_B} End_A(M) = A'.$$

We claim that this homomorphism is injective. Indeed, if $I := \ker(\mu_B)$, then M is actually an $A - (B/I)$-bimodule and on $N \otimes_A M$ we get that I acts as 0 on the right. However $N \otimes_A M \simeq B \oplus X$ as B–B-bimodules for some B–B-bimodule X. In particular, B acts on the right as a regular module on B, and the regular action does not have a kernel. This shows that $I = 0$.

Therefore, if (M, N) induces a stable equivalence of Morita type, then A can be replaced by a Morita equivalent copy A' so that B is a subalgebra of A'. □

5.4 Stable Equivalences of Morita Type Between Indecomposable Algebras

Let now (M, N) be a stable equivalence of Morita type between two indecomposable K-algebras A and B. Then under reasonable hypotheses, M and N are essentially indecomposable as well. This is the essence of the following result, which was first formulated by Rouquier for selfinjective algebras and published in a paper by Linckelmann. The present more general result is due to Dugas and Martinez-Villa.

Proposition 5.4.1 [11] *Let K be a commutative ring, let A and B be K-algebras and assume that A is indecomposable. Suppose that the Krull-Schmidt theorem is valid for finitely generated $A \otimes_K A^{op}$-modules, for finitely generated $A \otimes_K B^{op}$-modules and for finitely generated $B \otimes_K A^{op}$-modules. Let $({}_AM_B, {}_BN_A)$ be a pair of bimodules over A and B inducing a stable equivalence of Morita type between A and B.*

Then there are uniquely determined indecomposable direct factors M' of M and N' of N such that M' is not projective as an $A \otimes_K B^{op}$-module, and N' is not projective as a $B \otimes_K A^{op}$-module. Moreover, (M', N') induces a stable equivalence of Morita type between A and B.

Proof First, it is clear that if A is indecomposable as an algebra, then A is indecomposable as an A–A-bimodule as well. We know that

$$M \otimes_B N \simeq A \oplus P$$

as A–A-bimodules and

$$N \otimes_A M \simeq B \oplus Q$$

as B–B-bimodules, where P and Q are projective. Suppose now

$$M \simeq M' \oplus M''$$

as A–B-bimodules. Then

$$A \oplus P \simeq (M' \otimes_B N) \oplus (M'' \otimes_B N)$$

as A–A-bimodules. Since A is indecomposable, by the Krull-Schmidt theorem, either $M' \otimes_B N$ or $M'' \otimes_B N$ is projective as an A–A-bimodule. If $M'' \otimes_B N$ is projective as an A–A-bimodule, we get that

$$(M'' \otimes_B N) \otimes_A M \simeq M'' \otimes_B (N \otimes_A M) \simeq M'' \oplus M'' \otimes_B Q$$

is a projective A–B-bimodule. Hence M'' is a direct factor of the projective A–B-bimodule $(M'' \otimes_B N) \otimes_A M$ and is therefore projective as an A–B-bimodule. Let

$$N \simeq N' \oplus N''$$

as B–A-bimodules. Then

$$A \oplus P \simeq (M \otimes_B N') \oplus (M \otimes_B N'')$$

as A–A-bimodules. By the Krull-Schmidt theorem either $M \otimes_B N'$ or $M \otimes_B N''$ is projective as A–A-bimodules. Suppose $M \otimes_B N''$ is projective as an A–A-bimodule. Then again by an analogous arguments $N \otimes_A (M \otimes_B N'')$ is projective as a B–A-bimodule, N'' is a direct factor of that module, and is therefore projective as a B–A-bimodule.

The Krull-Schmidt theorem for $A \otimes_K B^{op}$-modules shows that there is a unique indecomposable non-projective direct factor M' of M, and the Krull-Schmidt theorem for $B \otimes_K A^{op}$-modules shows that there is a unique indecomposable non-projective direct factor N' of N as an $A \otimes_K B^{op}$-module. Hence

$$M \simeq M' \oplus M'' \text{ and } N \simeq N' \oplus N''$$

as A–B-bimodules and as B–A-bimodules, and M'' as well as N'' are projective as bimodules. We compute

$$\begin{aligned} A \oplus P &\simeq M \otimes_B N \\ &\simeq (M' \otimes_B N') \oplus (M' \otimes_B N'') \oplus (M'' \otimes_B N') \oplus (M'' \otimes_B N'') \end{aligned}$$

and all direct factors except $M' \otimes_B N'$ are projective as bimodules. Hence

$$A \oplus P' \simeq M' \otimes_B N'$$

as A–A-bimodules for a projective bimodule P'. Likewise

$$B \oplus Q' \simeq N' \otimes_B M'$$

for a projective bimodule Q'.

Since M' and N' are direct summands of M and N respectively, and since M and N are both projective as A-modules, and as B-modules, M' and N' are also projective as A-modules and as B-modules.

This proves the statement. □

Remark 5.4.2 Note that we did not need to assume that the Krull-Schmidt theorem is valid for $B \otimes_K B^{op}$-modules, nor did we need that B is indecomposable. However we shall see that the fact that B is indecomposable is a consequence of the other hypotheses.

In a certain sense, we can improve Proposition 5.4.1 so that we only need a stable equivalence given by a tensor product with a bimodule which is projective on each side. As we have seen in Remark 5.3.12 this is not quite the same as a stable equivalence of Morita type. However below we need finite dimensional algebras over fields.

Proposition 5.4.3 (Dugas and Martinez-Villa [11]) *Let K be a field and let A and B be finite dimensional K-algebras. Let ${}_BN_A$ be a bimodule which is projective as an A-module and projective as a B-module. Suppose*

$$N \otimes_A - : A\text{-}\underline{mod} \longrightarrow B\text{-}\underline{mod}$$

is an equivalence of K-linear categories. Then there is an indecomposable direct factor N' of N such that

$$(N' \otimes_A -) \simeq (N \otimes_A -)$$

as functors A-$\underline{mod} \longrightarrow B$-$\underline{mod}$.

Proof Let ${}_BN_A \simeq {}_BX_A \oplus {}_BY_A$ as B–A-bimodules. We shall show that either $X \otimes_A U$ is projective for every indecomposable A-module U or that $Y \otimes_A U$ is projective for every indecomposable A-module U. Once this is done, either $N \otimes_A - \simeq Y \otimes_A -$ in the first case or $N \otimes_A - \simeq X \otimes_A -$ in the second case. Since N is finite dimensional we get the statement by induction on the dimension of N.

We observe that for every indecomposable non-projective A-module U we get $End_A(U)$ is local, and therefore also $\underline{End}_A(U)$ is local. Since $N \otimes_A -$ is an equivalence between stable categories, we get that also

$$\underline{End}_A(U) \simeq \underline{End}_B(N \otimes_A U)$$

is local. Hence U has a unique indecomposable non-projective direct summand. Therefore either $X \otimes_A U$ is projective or $Y \otimes_A U$ is projective, but not both of them, since otherwise

$$\underline{End}_A(U) \simeq \underline{End}_B(N \otimes_A U) = 0$$

and therefore U is projective. Moreover, if $U \not\simeq V$ but U and V are both indecomposable non-projective A-modules, then $N \otimes_A U \not\simeq N \otimes_A V$ as B-modules. Indeed, $N \otimes_A U$ has a unique non-projective indecomposable direct factor F_U as B-modules, and likewise there is a unique non-projective indecomposable B-module F_V which is a direct factor of $N \otimes_A V$. In the stable category B-$\underline{mod}$ one has

$$F_U \simeq N \otimes_A U \text{ and } F_V \simeq N \otimes_A V$$

so that $F_U \simeq F_V$ contradicts the property of $N \otimes_A -$ being an equivalence.

Since N is projective as an A-module and as a B-module, Lemma 5.3.2 implies that the functor $N \otimes_A - : A\text{-}mod \longrightarrow B\text{-}mod$ is exact and induces a functor between the stable categories. Hence $X \otimes_A -$ is also exact and $Y \otimes_A -$ is exact. Suppose S and T are simple A-modules and that there is a non-split exact sequence

$$(*) \quad 0 \longrightarrow T \longrightarrow U \longrightarrow S \longrightarrow 0$$

of A-modules. In particular S is not projective. Suppose $X \otimes_A S$ is non-projective. Then $Y \otimes_A S$ has to be projective (or 0) and

$$0 \longrightarrow Y \otimes_A T \longrightarrow Y \otimes_A U \longrightarrow Y \otimes_A S \longrightarrow 0$$

is split exact since it is exact by the above, and since $Y \otimes_A S$ is projective or 0. Since the exact sequence $(*)$ above is non-split, U is indecomposable and therefore $N \otimes_A U$ has a unique non-projective indecomposable direct factor. Now, since $T \not\simeq U$, we get that $N \otimes_A T \not\simeq N \otimes_A U$. If $Y \otimes_A T$ is not projective, $Y \otimes_A U$ is not projective either, but since $Y \otimes_A S$ is projective, we would get that

$$N \otimes_A T \simeq Y \otimes_A T \simeq Y \otimes_A U \simeq N \otimes_A U$$

in the stable category $B\text{-}\underline{mod}$. This contradiction shows that if $Y \otimes_A S$ is projective then $Y \otimes_A T$ and $Y \otimes_A U$ are projective. Now, let

$$0 \longrightarrow T \longrightarrow U' \longrightarrow S' \longrightarrow 0$$

be a non-split exact sequence of A-modules, so that S' is simple. In particular S' is not projective. We maintain our hypothesis that $Y \otimes_A S$ is projective, and hence $Y \otimes_A T$ is projective. We shall show that $Y \otimes_A S'$ is projective as well. Suppose $Y \otimes_A S'$ is not projective. Then $X \otimes_A S'$ is projective, and by the first step, we get that $X \otimes_A T$ is projective. Therefore $X \otimes_A T$ is projective and $Y \otimes_A T$ is projective, which implies that T is projective, as we have seen above. We shall form the pushout along the embeddings of T into U and into U'

$$\begin{array}{ccccccccc}
 & & 0 & & 0 & & 0 & & \\
 & & \downarrow & & \downarrow & & \downarrow & & \\
0 & \longrightarrow & T & \longrightarrow & U & \longrightarrow & S & \longrightarrow & 0 \\
 & & \downarrow & & \downarrow & & \| & & \\
0 & \longrightarrow & U' & \longrightarrow & V & \longrightarrow & S & \longrightarrow & 0 \\
 & & \downarrow & & \downarrow & & \downarrow & & \\
0 & \longrightarrow & S' & = & S' & \longrightarrow & 0 & & \\
 & & \downarrow & & \downarrow & & & & \\
 & & 0 & & 0 & & & &
\end{array}$$

where we use Lemma 1.8.27 to describe the cokernels of the embeddings $U \longrightarrow V$ and $U' \longrightarrow V$.

We shall show that V is non-projective. Indeed, the construction of the pushout gives a short exact sequence

$$0 \longrightarrow T \longrightarrow U \oplus U' \longrightarrow V \longrightarrow 0.$$

If V is projective, $U \oplus U' \simeq T \oplus V$ and then $U \simeq T \oplus S$ or $U' \simeq T \oplus S'$. This contradicts the fact that the sequences

$$0 \longrightarrow T \longrightarrow U \longrightarrow S \longrightarrow 0$$

and

$$0 \longrightarrow T \longrightarrow U' \longrightarrow S' \longrightarrow 0$$

are non-split. Hence V is non projective.

Since T is projective, and since we assumed $Y \otimes_A S$ is projective,

$$N \otimes_A S \simeq X \otimes_A S \simeq X \otimes_A U \simeq N \otimes_A U$$

in B-$\underline{mod}$ and since we assumed $X \otimes_A S'$ projective,

$$N \otimes_A S' \simeq Y \otimes_A S' \simeq Y \otimes_A U' \simeq N \otimes_A U' \text{ in } B\text{-}\underline{mod}.$$

In particular, U and U' are both non-projective. Moreover,

$$N \otimes_A U' \simeq Y \otimes_A U' \simeq Y \otimes_A V \text{ in } B\text{-}\underline{mod}$$

and

$$N \otimes_A U \simeq X \otimes_A U \simeq X \otimes_A V \text{ in } B\text{-}\underline{mod}.$$

Since $N \simeq X \oplus Y$ we get

$$N \otimes_A V \simeq (X \otimes_A V) \oplus (Y \otimes_A V) \simeq (N \otimes_A U') \oplus (N \otimes_A U) \simeq N \otimes_A (U' \oplus U)$$

in B-$\underline{mod}$. Since U and U' are both non-projective indecomposable, V has to be decomposable as well, and indeed $V \simeq U \oplus U'$ in A-$\underline{mod}$. Therefore, there are projective A-modules R_1 and R_2 so that

$$V \oplus R_1 \simeq U \oplus U' \oplus R_2.$$

The modules U and U' are both indecomposable non-projective. Hence $U \oplus U'$ is a direct factor of V by the Krull-Schmidt theorem. However, the composition factors of V are T, S and S', whereas the composition factors of U are S and T, and the composition factors of U' are T and S'. This is a contradiction and therefore $Y \otimes_A S'$ is projective.

Since A is indecomposable, Proposition 2.7.4 shows that for all simple A-modules L we get $Y \otimes_A L$ is projective.

Now, let U be an indecomposable A-module of finite composition length at least 2. Then

$$0 \longrightarrow \mathrm{rad}(U) \longrightarrow U \longrightarrow U/\mathrm{rad}(U) \longrightarrow 0$$

is a non-split exact sequence. We shall show by induction on the composition length that $Y \otimes_A U$ is projective. The case of composition length at most 2 has already been done. We see that

$$0 \longrightarrow Y \otimes_A \mathrm{rad}(U) \longrightarrow Y \otimes_A U \longrightarrow Y \otimes_A (U/\mathrm{rad}(U)) \longrightarrow 0$$

is exact. By induction hypothesis $Y \otimes_A \mathrm{rad}(U)$ is projective, and since $U/\mathrm{rad}(U)$ is semisimple, whence a direct sum of simple modules, using that simple modules are of composition length 1, we obtain that $Y \otimes_A (U/\mathrm{rad}(U))$ is projective as well. Hence the sequence

$$0 \longrightarrow Y \otimes_A \mathrm{rad}(U) \longrightarrow Y \otimes_A U \longrightarrow Y \otimes_A (U/\mathrm{rad}(U)) \longrightarrow 0$$

splits and $Y \otimes_A U$ is projective. This proves the statement by induction on the number of indecomposable direct factors of N. □

Let K be a commutative ring and let A and B be K-algebras. Let $A = A_1 \times A_2$ and let (M, N) induce a stable equivalence of Morita type between A and B. Can we show that $B = B_1 \times B_2$ and that we can decompose $M = M_1 \oplus M_2$ and $N = N_1 \oplus N_2$ so that (M_1, N_1) induces a stable equivalence of Morita type between A_1 and B_1 and that (M_2, N_2) induces a stable equivalence of Morita type between A_2 and B_2? The answer is "yes". This is a theorem of Yuming Liu.

Proposition 5.4.4 (Liu [12]) *Let K be a commutative ring and let A and B be K-algebras. Suppose that the Krull-Schmidt theorem is valid for $A \otimes_K A^{op}$, for $A \otimes_K B^{op}$, for $B \otimes_K A^{op}$-modules and for $B \otimes_K B^{op}$-modules. Suppose that (M, N) induces a stable equivalence of Morita type between A and B and suppose that $A = A_1 \times A_2$.*

Then $B = B_1 \times B_2$ and there are indecomposable direct factors M_1 and M_2 of M as well as N_1 and N_2 of N such that (M_1, N_1) induces a stable equivalence of Morita type between A_1 and B_1 and such that (M_2, N_2) induces a stable equivalence of Morita type between A_2 and B_2. In particular A is indecomposable if and only if B is indecomposable.

Proof of Proposition 5.4.4 Let e_1 be a central primitive idempotent of A and let $e_2 := 1 - e_1$. Then

$$M \otimes_B N \simeq A \oplus P$$

for a projective A–A-bimodule P implies that

$$e_1 M \otimes_B N e_1 \simeq e_1 A e_1 \oplus e_1 P e_1.$$

We shall show that $e_1 P e_1$ is a projective $e_1 A \otimes_K (e_1 A)^{op}$-module. Indeed, P is a direct factor of $(A \otimes_K A^{op})^n$ and so $e_1 P e_1$ is a direct factor of

$$e_1 \left(A \otimes_K A^{op}\right)^n e_1 = \left(e_1 A \otimes_K A^{op} e_1\right)^n = \left(e_1 A \otimes_K (e_1 A^{op})\right)^n$$

and therefore $e_1 P e_1$ is projective as an $e_1 A \otimes_K (e_1 A)^{op}$-module. Let

$$1_B = f_1 + \cdots + f_m$$

for primitive central idempotents $f_1, \ldots, f_m$, then

$$\begin{aligned} e_1 M \otimes_B N e_1 &= e_1 M(f_1 + \cdots + f_m) \otimes_B (f_1 + \cdots + f_m) N e_1 \\ &= \bigoplus_{j=1}^{m} e_1 M f_j \otimes_{(f_j B)} f_j N e_1 \\ &= e_1 A e_1 \oplus e_1 P e_1, \end{aligned}$$

since $f_j \cdot f_{j'} = 0$ if $j \neq j'$. Now, e_1 is primitive, and so $e_1 A = e_1 A e_1$ is an indecomposable A–A-bimodule, and even an indecomposable $e_1 A \otimes_K (e_1 A)^{op}$-module since the other direct factors of $A \otimes_K A^{op}$ act as 0 on $e_1 A$. Hence by the Krull-Schmidt theorem there is a unique j_0 such that $e_1 A$ is a direct factor of $e_1 M f_{j_0} \otimes_{(f_{j_0} B)} f_{j_0} N e_1$, and that $e_1 M f_j \otimes_{(f_j B)} f_j N e_1$ is a direct factor of $e_1 P e_1$ for all $j \neq j_0$. Without loss of generality, renumbering the central primitive idempotents of B if necessary, we may assume $j_0 = 1$. Moreover, since

$$e_1 M f_1 \otimes_{(f_1 B)} f_1 N e_1$$

is a direct factor of

$$e_1 (A \oplus P) e_1 = e_1 A \oplus e_1 P e_1,$$

we get that one obtains an isomorphism of $e_1 A \otimes_K (e_1 A)^{op}$-modules

$$e_1 M f_1 \otimes_{(f_1 B)} f_1 N e_1 \simeq e_1 A \oplus P_1$$

for some direct factor P_1 of P. We have seen that $e_1 P e_1$ is a projective $e_1 A \otimes_K (e_1 A)^{op}$-module, and since P_1 is a direct factor of $e_1 P e_1$, P_1 is also a projective $e_1 A \otimes_K (e_1 A)^{op}$-module.

Now, suppose U is a projective $e_1 A \otimes_K f_1 B$-module and V is an $f_1 B \otimes_K e_1 A$-module, which is assumed to be projective as an $e_1 A$-module. Then $U \otimes_{f_1 B} V$ is a projective $e_1 A \otimes_K f_1 B$-module by Lemma 5.3.3.

In particular since $e_1 A$ is not a projective $e_1 A \otimes e_1 A$-module, $e_1 M f_1$ is not projective as an $e_1 A \otimes_K (f_1 B)^{op}$-module.

We need to show that

$$f_1 N e_1 \otimes_{(e_1 A)} e_1 M f_1 \simeq f_1 B \oplus Q_1$$

for some projective $f_1 B \otimes_K (f_1 B)^{op}$-module Q_1. By the first step there is a unique primitive central idempotent $\hat{e}$ of A such that

$$f_1 N \hat{e} \otimes_{(\hat{e} A)} \hat{e} M f_1 \simeq f_1 B \oplus Q_1$$

for some direct factor Q_1 of $f_1 Q f_1$, and such that $f_1 N e \otimes_{(eA)} e M f_1$ is a direct factor of $f_1 Q f_1$ for all $e \neq \hat{e}$. If $e_1 \neq \hat{e}$, we therefore obtain that $f_1 N e_1 \otimes_{(e_1 A)} e_1 M f_1$ is a direct factor of f_1Qf_1, and hence a projective $f_1B \otimes_K f_1B$-module. We see that

$$\begin{aligned} e_1Mf_1 \otimes_{f_1B}&(f_1Ne_1 \otimes_{(e_1A)} e_1Mf_1) \\ &\simeq (e_1Mf_1 \otimes_{f_1B} f_1Ne_1) \otimes_{(e_1A)} e_1Mf_1 \\ &\simeq (e_1A \oplus P_1) \otimes_{(e_1A)} e_1Mf_1 \\ &\simeq (e_1A \otimes_{(e_1A)} e_1Mf_1) \oplus (P_1 \otimes_{(e_1A)} e_1Mf_1) \\ &\simeq e_1Mf_1 \oplus (P_1 \otimes_{(e_1A)} e_1Mf_1). \end{aligned}$$

Since $f_1Ne_1 \otimes_{(e_1A)} e_1Mf_1$ is a projective $f_1B \otimes_K f_1B$-module and since e_1Mf_1 is projective as an A-module, Lemma 5.3.3 shows that

$$e_1Mf_1 \otimes_{f_1B} (f_1Ne_1 \otimes_{(e_1A)} e_1Mf_1) \simeq e_1Mf_1 \oplus (P_1 \otimes_{(e_1A)} e_1Mf_1)$$

is a projective $e_1A \otimes_K f_1B$-module. However, e_1Mf_1 is not a projective $e_1A \otimes_K f_1 B$-module. This contradiction shows that $e_1 = \hat{e}$.

Of course, e_1Mf_1 is projective as an e_1A-module and as an f_1B-module since M is projective as an A-module and as a B-module. Likewise f_1Ne_1 is projective on either side.

An induction on the number of direct factors of A and of B shows that (e_2Mf_2, f_2Ne_2) also induces a stable equivalence of Morita type between e_2A and f_2B. This proves the statement. □

Example 5.4.5 Example 5.2.2 part 3 gives an example of a stable equivalence between an indecomposable finite dimensional K-algebra and a decomposable finite dimensional K-algebra. By Proposition 5.4.4 this stable equivalence cannot be of Morita type.

5.5 Stable Equivalences of Morita Type: Symmetricity and Further Properties

We shall see that if (M, N) induces a stable equivalence of Morita type, then M determines N and N determines M.

Proposition 5.5.1 *Let K be a commutative ring and let A and B be two K-algebras without any semisimple direct factors. Suppose that $({}_AM_B, {}_BN_A)$ and $({}_AM_B, {}_BN'_A)$ induce stable equivalences of Morita type between A and B, and suppose that M, N and N' are all indecomposable. Assume that the Krull-Schmidt theorem is valid for $A \otimes_K A^{op}$, for $A \otimes_K B^{op}$, for $B \otimes_K A^{op}$-modules and for $B \otimes_K B^{op}$-modules. Then $N \simeq N'$ as $A \otimes_K B^{op}$-modules.*

Proof By Proposition 5.4.4 we may assume that A and B are indecomposable. Then

$$M \otimes_B N \simeq A \oplus P \text{ and } N \otimes_A M \simeq B \oplus Q$$

as well as

$$M \otimes_B N' \simeq A \oplus P' \text{ and } N' \otimes_A M \simeq B \oplus Q'$$

as bimodules and for projective A–A-bimodules P and P' and projective B–B-bimodules Q and Q'. Hence

$$N \oplus (Q' \otimes_B N) \simeq (N' \otimes_A M) \otimes_B N \simeq N' \otimes_A (M \otimes_B N) \simeq N' \oplus (N \otimes_A P)$$

as bimodules. Lemma 5.3.3 shows that $(Q' \otimes_B N)$ is a projective B–A-bimodule and $(N \otimes_A P)$ is a projective A–B-bimodule. Proposition 5.4.1 shows that there is a unique indecomposable non-projective direct factor in N and in N'. This proves the statement. □

Remark 5.5.2 Changing the roles of A and B, N also determines M in Proposition 5.5.1.

Proposition 5.5.3 *Let K be a perfect field and let A and B be finite dimensional K-algebras without semisimple direct factors. If (M, N) induces a stable equivalence of Morita type between A and B, and if M and N are indecomposable as bimodules, then*

$$N \simeq Hom_B(M, B) \simeq Hom_A(M, A)$$

and

$$M \simeq Hom_A(N, A) \simeq Hom_B(N, B).$$

Proof Since

$$M \otimes_B - : B\text{-}\underline{mod} \longrightarrow A\text{-}\underline{mod}$$

is an equivalence with quasi-inverse

$$N \otimes_A - : A\text{-}\underline{mod} \longrightarrow B\text{-}\underline{mod},$$

the functors

$$(M \otimes_B -, N \otimes_A -)$$

form an adjoint pair of functors between the stable module categories. However,

$$(M \otimes_B -, Hom_A(M, -))$$

is an adjoint pair of functors between the module categories.

Hence we get isomorphisms of functors $A\text{-}mod \longrightarrow B\text{-}mod$

$$N \otimes_A - \simeq Hom_A(M, -) \simeq Hom_A(M, A) \otimes_A -$$

where the first isomorphism comes from Proposition 3.2.7 and the second isomorphism comes from Lemma 4.2.5 and the fact that M is projective as an A-module.

Since (M, N) induces a stable equivalence of Morita type, $N \otimes_A -$ induces a functor $A\text{-}\underline{mod} \longrightarrow B\text{-}\underline{mod}$. Therefore $N \otimes_A -$ sends projective A-modules to projective B-modules. This shows that $Hom_A(M, A) \otimes_A -$ also sends projective A-modules to projective B-modules, and this implies that $Hom_A(M, A)$ is projective as a B-module. By Proposition 5.3.13 we get that $(M, Hom_A(M, A))$ also induce a stable equivalence of Morita type. Clearly $Hom_A(M, A)$ is indecomposable since M is indecomposable. Proposition 5.5.1 then shows that $N \simeq Hom_A(M, A)$ as B–A-bimodules. By the very same argument we get $M \simeq Hom_B(N, B)$. Now, dealing with equivalences of stable categories of right modules we obtain $M \simeq Hom_A(N, A)$ and $N \simeq Hom_B(M, B)$. □

The above can be used to show that if A and B are stably equivalent of Morita type, then A is symmetric if and only if B is symmetric in the sense of Definition 1.10.20.

Proposition 5.5.4 *Let K be a perfect field and let A and B be finite dimensional K-algebras without any semisimple direct factors. Suppose that $({}_AM_B, {}_BN_A)$ induces a stable equivalence of Morita type. Then A is a symmetric algebra if and only if B is a symmetric algebra.*

Proof We first see that

$$\begin{aligned} Hom_K(M \otimes_B Hom_K(B, K), K) &\simeq Hom_B(M, Hom_K(Hom_K(B, K), K)) \\ &\simeq Hom_B(M, B) \end{aligned}$$

by the usual Frobenius reciprocity formula, and the fact that B is finite dimensional over K. Dualising again gives

$$M \otimes_B Hom_K(B, K) \simeq Hom_K(Hom_B(M, B), K).$$

Similarly

$$\begin{aligned} &Hom_K(Hom_K(Hom_B(M, B), K) \otimes_B N, K) \\ &\qquad \simeq Hom_B(N, Hom_K(Hom_K(Hom_B(M, B), K), K)) \\ &\qquad \simeq Hom_B(N, Hom_B(M, B)). \end{aligned}$$

We then compute

$$\begin{aligned} M \otimes_B Hom_K(B, K) \otimes_B N &\simeq Hom_K(Hom_B(M, B), K) \otimes_B N \\ &\simeq Hom_K(Hom_B(N, Hom_B(M, B)), K) \\ &\simeq Hom_K(Hom_B(N, B) \otimes_B Hom_B(M, B)), K) \\ &\simeq Hom_K(M \otimes_B N, K) \end{aligned}$$

where the last isomorphism is a consequence of Proposition 5.5.3.

Now, (M, N) induces a stable equivalence of Morita type, and therefore $M \otimes_B N \simeq A \oplus P$ in $A \otimes_K A^{op}\text{-}mod$ for some projective $A \otimes_K A^{op}$-module P.

Suppose now that B is symmetric. Then $B \simeq Hom_K(B, K)$ as $B \otimes_K B^{op}$-modules. Therefore

$$A \oplus P \simeq M \otimes_B N \simeq Hom_K(M \otimes_B N, K) \simeq Hom_K(A, K) \oplus Hom_K(P, K).$$

Denote by $\sigma : A \longrightarrow Hom_K(A, K)$ the composition of the inclusion $A \longrightarrow M \otimes_B N$, the above isomorphism, and the projection $Hom_K(M \otimes_B N, K) \longrightarrow Hom_K(A, K)$. Further $Hom_K(P, K)$ is an injective $A \otimes_K A^{op}$-module, since P is a projective $A \otimes_K A^{op}$-module. Since all indecomposable direct factors of A and $Hom_K(A, K)$ are definitively neither projective nor injective, σ is an isomorphism. Indeed σ is one component of an isomorphism $A \oplus P \simeq Hom_K(A, K) \oplus Hom_K(P, K)$, and P is projective, $Hom_K(P, K)$ is injective, and none of the direct factors of A or of $Hom_K(A, K)$ are injective or projective. This shows that A is symmetric as well. □

Remark 5.5.5 Proposition 5.5.4 actually gives an explicit symmetrising form, using Propositions 1.10.23, 1.10.6 and Definition 1.10.13.

Hence, we may say that a stable equivalence of Morita type between the symmetric algebras A and B induces for any given symmetrising form $\langle\ ,\ \rangle_A$ on A a well-defined symmetrising form $\langle\ ,\ \rangle_{B,M}$.

5.6 Image of Simple Modules Under Stable Equivalences of Morita Type

We shall discuss a particular nice situation discovered by Linckelmann which relies on the behaviour of simple modules under a stable equivalence of Morita type. We shall closely follow Linckelmann [8].

We start with a technical lemma.

Lemma 5.6.1 *Let K be a field and let C be a finite dimensional self-injective K-algebra. Then* $\mathrm{soc}(C)\cdot U = 0$ *for every non-projective indecomposable C-module U, and* $\mathrm{soc}(C)\cdot P = \mathrm{soc}(P)$ *for every projective indecomposable C-module P.*

Proof We need to show that the indecomposable module U is projective if and only if $\mathrm{soc}(C)\cdot U \neq 0$. First, if U is free (not necessarily indecomposable), then

$$\mathrm{soc}(C)\cdot U \simeq \mathrm{soc}(C)\cdot C^n = \mathrm{soc}(C)^n$$

and therefore if U is projective, U is a direct factor of C^n for some n, and hence

$$\mathrm{soc}(C)\cdot U = \mathrm{soc}(U).$$

If U is indecomposable non-projective, then let

$$P \stackrel{\alpha}{\longrightarrow\!\!\!\!\rightarrow} U$$

be a projective cover. We claim that $\mathrm{soc}(P) \subseteq \ker(\alpha)$. Suppose this is not true. Then there is a simple direct factor S of $\mathrm{soc}(P)$ such that $\alpha(S) \neq 0$. Hence the restriction $\alpha|_S$ of α to S is injective. Let I_S be the injective hull of S. Then, since C is self-injective, I_S is projective and a direct factor of the injective module P. Since $\mathrm{soc}(I_S) = S$, $\alpha|_{I_S}$ is injective as well and therefore $\alpha(I_S) \simeq I_S$ is a submodule of U. Since I_S is injective, I_S is a direct factor of U. Since U is indecomposable, $I_S \simeq U$ and hence U is projective. The case of projective modules has already been shown. This proves the statement. □

Remark 5.6.2 We need the self-injectivity here. For general finite dimensional algebras the statement of Lemma 5.6.1 is false.

Consider the quiver Q

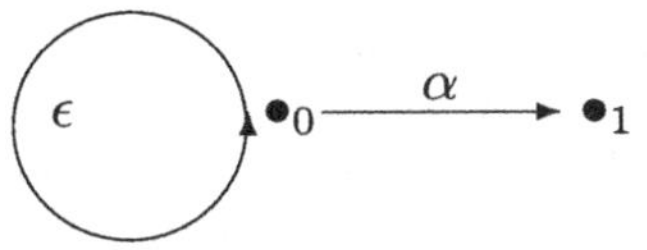

and the algebra A given by the quiver Q subject to the relations $\epsilon^2 = 0$.

Then $\mathrm{soc}(A) = K \cdot \epsilon\alpha + K \cdot \alpha$. We denote by S_0 and by S_1 the simple modules associated to the vertices 0 and 1 respectively. Consider the module $U = e_0A/\alpha A$ which is uniserial with top S_0, socle S_1 and $\mathrm{rad}(U)/\mathrm{soc}(U)$ equal to S_0. Then $\mathrm{soc}(U) = K \cdot \epsilon\alpha = \mathrm{soc}(A) \cdot U$. However, U is not projective.

We now come to a slight refinement of a result due to Linckelmann

Lemma 5.6.3 (Linckelmann [8, Theorem 11.2.3] for stable equivalences of Morita type) *Let K be a perfect field and let A and B be indecomposable self-injective finite-dimensional K-algebras. Suppose ${}_AM_B$ is an A–B-bimodule such that $M \otimes_B - : B\text{-}\underline{mod} \longrightarrow A\text{-}\underline{mod}$ is an equivalence and suppose that M is indecomposable. Then $M \otimes_B S$ is indecomposable for each simple B-module S.*

Proof Since K is perfect, Lemma 5.3.8 and Corollary 5.3.10 show that $\mathrm{soc}(A \otimes_K B^{op}) = \mathrm{soc}(A) \otimes_K \mathrm{soc}(B^{op})$. Now, if A and B are self-injective, $A \otimes_K B^{op}$ is also self-injective, as is immediate by the definition. Using Lemma 5.6.1, since M has no non-zero $A \otimes_K B^{op}$-projective direct factors, we get

$$\begin{aligned} 0 = \mathrm{soc}(A \otimes_K B^{op}) \cdot M &= \big(\mathrm{soc}(A) \otimes_K \mathrm{soc}(B^{op})\big) \cdot M \\ &= \mathrm{soc}(A) \cdot M \cdot \mathrm{soc}(B) = \mathrm{soc}(A) \cdot (M \cdot \mathrm{soc}(B)) \end{aligned}$$

and therefore $M \otimes_B \mathrm{soc}(B) = M \cdot \mathrm{soc}(B)$ is annihilated by $\mathrm{soc}(A)$. Since A is self-injective, $M \otimes_B \mathrm{soc}(B)$ does not have any A-projective direct factors using Lemma 5.6.1 again. Every simple B-module is a direct factor of $\mathrm{soc}(B)$, using again that B is self-injective. Hence for every simple B-module S the module $M \otimes_B S$ does not have any non-zero A-projective indecomposable direct factors. But now,

$$D = \underline{End}_B(S) \simeq \underline{End}_A(M \otimes_B S)$$

is a skew-field since S is simple. If $M \otimes_B S$ decomposed into two non-trivial direct factors, both of them would be non-projective, and hence the projection onto each of these direct factors would be non-trivial idempotents, which do not factor through a projective module, using again that A is self-injective. This proves the statement. □

Remark 5.6.4 Proposition 5.4.4 shows that the indecomposability condition on M in the above Lemma 5.6.3 is not essential. By Proposition 5.4.3 any M inducing a stable equivalence of Morita type has a unique indecomposable direct factor M' of M so that M' can be used instead to obtain the same equivalence of stable categories. König and Liu [13, Lemma 4.4] showed that the hypothesis on A and B to be self-injective is not necessary.

Proposition 5.6.5 (Linckelmann [8, Theorem 11.2.3]) *Let K be a perfect field and let A and B be indecomposable self-injective K-algebras. Let M be an indecomposable A–B-bimodule and let N be an indecomposable B–A-bimodule such that (M, N) induces a stable equivalence of Morita type. If $M \otimes_B S$ is a simple A-module for each simple B-module S, then M is a Morita bimodule, and hence A and B are Morita equivalent.*

Proof We get

$$N \otimes_A M \simeq B \oplus Q$$

as B–B-bimodules for a projective bimodule Q. Then $M \otimes_B S \simeq T$ is a simple A-module T by hypothesis. Lemma 5.6.3 shows that $N \otimes_A T$ is indecomposable. Hence

$$S \oplus (Q \otimes_B S) \simeq (N \otimes_A M) \otimes_B S \simeq N \otimes_A (M \otimes_B S) \simeq N \otimes_A T$$

and therefore $Q \otimes_B S = 0$ for all simple B-modules S. This shows

$$Q \otimes_B B/\mathrm{rad}(B) = Q/\mathrm{rad}(Q) = 0.$$

But this implies $Q = 0$ and hence $N \otimes_B M \simeq A$ as A–A-bimodules. By the dual argument $M \otimes_A M \simeq B$ as B–B-bimodules. Therefore (M, N) induces a Morita equivalence. This proves the statement. □

Remark 5.6.6 Proposition 5.6.5 is very important in Sect. 6.10.2 where it is used in a most ingenious way, employing a method due to Okuyama, to give explicit examples of equivalences between derived categories of blocks of group rings.

5.7 The Stable Grothendieck Group

Usually the Grothendieck group of an algebra is not invariant under equivalences of stable categories. An easy example is given in Example 5.2.2. Indeed, the algebra $\begin{pmatrix} K & K \\ 0 & K \end{pmatrix}$ has two simple modules, whereas $K[X]/X^2$ has only one simple module. Nevertheless they have equivalent stable module categories. Tachikawa-Wakamatsu [14] proposed a replacement for the Grothendieck group, called the stable Grothendieck group.

5.7.1 The Definition

Definition 5.7.1 [14] Let A be a K-algebra and let K be a field. Then let $c_A : K_0(A\text{-}proj) \longrightarrow G_0(A)$ be the Cartan mapping defined by sending the class $[P]$ of a projective A-module P to its class $[P]$ in the Grothendieck group of all finitely generated A-modules. The *stable Grothendieck group* of A is defined to be $G_0^{st}(A) := \text{coker}(c_A)$.

We may define the Grothendieck group $G_0(\mathcal{T})$ of a triangulated category $\mathcal{T}$ as the quotient of the free abelian group on isomorphism classes of objects of $\mathcal{T}$ modulo all relations given by

$$[X] - [Y] - [Z]$$

whenever $X \longrightarrow Y \longrightarrow Z \longrightarrow TX$ is a distinguished triangle. We shall study this concept in more detail later in Sect. 6.8.1.

The first thing to see is that the stable Grothendieck group is the same as the Grothendieck group of the stable category if A is self-injective. Recall from Proposition 5.1.10 that if A is a self-injective K-algebra over a field K, then $A\text{-}\underline{mod}$ is triangulated.

Lemma 5.7.2 [14] *Let K be a field and let A be a finite dimensional self-injective K-algebra. Then*

$$G_0(A\text{-}\underline{mod}) \simeq G_0^{st}(A).$$

Proof If

$$0 \longrightarrow X \longrightarrow Y \longrightarrow Z \longrightarrow 0$$

is an exact sequence of A-modules, then by definition this sequence induces a distinguished triangle

$$X \longrightarrow Y \longrightarrow Z \longrightarrow \Omega^{-1}X$$

in $A\text{-}\underline{mod}$ and vice versa. Hence the natural map

$$G_0(A\text{-}\underline{mod}) \longrightarrow G_0^{st}(A)$$

given by $[X] \mapsto [X]$ is well-defined. It is clear that the map

$$\begin{array}{ccc} G_0(A\text{-}mod) & \stackrel{\rho}{\longrightarrow} & G_0(A\text{-}\underline{mod}) \\ [M] & \mapsto & [M] \end{array}$$

is well-defined since any short exact sequence of modules gives a corresponding triangle. The image of c_A is generated by $[P]$ for two projective A-modules P. Since $P \simeq 0$ in $G_0(A\text{-}\underline{mod})$ we obtain that ρ induces a morphism

$$G_0^{st}(A) \stackrel{\overline{\rho}}{\longrightarrow} G_0(A\text{-}\underline{mod})$$

as is made clear in the commutative diagram below:

$$\begin{array}{ccccc} K_0(A-proj) & \stackrel{c_A}{\longrightarrow} & G_0(A) & \longrightarrow\!\!\!\!\!\rightarrow & G_0^{st}(A) \\ \downarrow & & \downarrow \rho & & \downarrow \overline{\rho} \\ 0 & \longrightarrow & G_0(A\text{-}\underline{mod}) & = & G_0(A\text{-}\underline{mod}) \end{array}$$

The morphisms are obviously inverse to one another. This proves the statement. □

Remark 5.7.3 Let K be a field and let A and B be finite dimensional self-injective K-algebras. Lemma 5.7.2 shows that any equivalence of triangulated categories $A\text{-}\underline{mod} \longrightarrow B\text{-}\underline{mod}$ induces an isomorphism $G_0^{st}(A) \simeq G_0^{st}(B)$. We do not need that the stable equivalence is of Morita type. For stable equivalences of Morita type we can also prove this last fact for more general algebras.

Proposition 5.7.4 (Xi [15]) *Let K be a field, let A and B be finite dimensional K-algebras, let M be an $A \otimes_K B^{op}$-module and let N be a $B \otimes_K A^{op}$-module. Suppose (M, N) induces a stable equivalence of Morita type between A and B. Then $G_0^{st}(A) \simeq G_0^{st}(B)$ as abelian groups.*

Proof Since $M \otimes_B -$ is additive, and since M is projective as an A-module, we get a homomorphism of abelian groups

$$\begin{array}{ccc} K_0(B-proj) & \stackrel{K_0(M)}{\longrightarrow} & K_0(A-proj) \\ [P] & \mapsto & [M \otimes_B P] \end{array}$$

and since M is projective as a B-module, $M \otimes_B -$ is exact and hence induces a homomorphism of abelian groups

$$\begin{array}{ccc} G_0(B) & \stackrel{G_0(M)}{\longrightarrow} & G_0(A) \\ [U] & \mapsto & [M \otimes_B U]. \end{array}$$

Moreover, it is easy to see that the diagram

$$\begin{array}{ccc} K_0(B-proj) & \stackrel{K_0(M)}{\longrightarrow} & K_0(A-proj) \\ \downarrow c_B & & \downarrow c_A \\ G_0(B) & \stackrel{G_0(M)}{\longrightarrow} & G_0(A) \end{array}$$

is commutative. Hence we get an induced map

$$G_0^{st}(B) \stackrel{G_0^{st}(M)}{\longrightarrow} G_0^{st}(A)$$

on the cokernels. Finally, for the same reason, we obtain the same map for N and get a commutative diagram

$$\begin{array}{ccccc} K_0(B-proj) & \stackrel{K_0(M)}{\longrightarrow} & K_0(A-proj) & \stackrel{K_0(N)}{\longrightarrow} & K_0(B-proj) \\ \downarrow c_B & & \downarrow c_A & & \downarrow c_B \\ G_0(B) & \stackrel{G_0(M)}{\longrightarrow} & G_0(A) & \stackrel{G_0(N)}{\longrightarrow} & G_0(B). \end{array}$$

Let $N \otimes_A M \simeq B \oplus Q$ for a projective $B \otimes_K B^{op}$-module Q. The composition gives a square

$$\begin{array}{ccc} K_0(B-proj) & \stackrel{(B\oplus Q)\otimes_B -}{\longrightarrow} & K_0(B-proj) \\ \downarrow c_B & & \downarrow c_A \\ G_0(B) & \stackrel{(B\oplus Q)\otimes_B -}{\longrightarrow} & G_0(B) \end{array}$$

Since Lemma 5.3.3 shows that $Q \otimes_B U$ is projective for every B-module U, we get that the diagram

$$\begin{array}{ccc} K_0(B-proj) & \stackrel{(B\oplus Q)\otimes_B -}{\longrightarrow} & K_0(B-proj) \\ \downarrow c_B & & \downarrow c_A \\ G_0(B) & \stackrel{(B\oplus Q)\otimes_B -}{\longrightarrow} & G_0(B) \\ \downarrow & & \downarrow \\ G_0^{st}(B) & \stackrel{B\otimes_B -}{\longrightarrow} & G_0^{st}(B) \\ \downarrow & & \downarrow \\ 0 & & 0 \end{array}$$

is commutative with exact columns. Since $B \otimes_B - \simeq id$ and since we may as well interchange the roles of M and N, we have proved the statement. □

Remark 5.7.5 Let K be a field and let A and B be two finite dimensional K-algebras. Recall that in this situation we get that $G_0(A)$ is a free abelian group of finite rank n_A. Hence $G_0^{st}(A)$ is a finitely generated abelian group and its structure is determined by the elementary divisors of c_A. Recall that the elementary divisors of c_A are positive integers $\delta_1, \delta_2, \ldots, \delta_s$ such that $s \le n_A$ and such that δ_i divides δ_{i+1} for all $i \in \{1, \ldots, s-1\}$. The elementary divisors depend only on c_A and we get

$$G_0^{st}(A) \simeq \mathbb{Z}/\delta_1\mathbb{Z} \times \cdots \times \mathbb{Z}/\delta_s\mathbb{Z} \times \mathbb{Z}^{n_A - s}.$$

In order to simplify the notation we define $\mathbb{Z}^0 := \{0\}$. Hence if the two K-algebras A and B are stably equivalent of Morita type, then the elementary divisors different from 1 of c_A and the elementary divisors different from 1 of c_B coincide, including their multiplicities. Observe, however, that since the elementary divisor 1 can occur with some multiplicity, it may happen a priori that $n_A \neq n_B$. The absolute value $|\det(c_A)|$ of the determinant of the Cartan mapping c_A of A is $\delta_1 \cdot \dots \cdot \delta_s$ if $s = n_A$, and 0 otherwise.

This is a very strong invariant. Moreover, a slight modification of Gauss elimination allows us to effectively compute the elementary divisors of square integer matrices.

5.7.2 Application to Nakayama Algebras

Recall from Example 1.10.7 that a Nakayama algebra N_n^m has n isomorphism classes of simple modules, represented by the simple modules $S_1, \dots, S_n$, and the projective cover P_i of S_i is uniserial of composition length m. The Cartan matrix $C_{n,m}$ of N_n^m is quite easy to determine. Put $m = kn + r$ for some non-negative integer $r < n$ and some non-negative integer k. Then

$$C_{n,m} = \begin{pmatrix} k+1 & k & \dots & \dots & k & k+1 & \dots & k+1 \\ \vdots & k+1 & k & \ddots & \ddots & k & k+1 & \vdots \\ \vdots & \ddots & \ddots & \ddots & \ddots & & \ddots & k+1 \\ k+1 & \ddots & \ddots & k+1 & k & \ddots & \ddots & k \\ k & k+1 & \ddots & \ddots & k+1 & k & \ddots & k \\ \vdots & \ddots & \ddots & \ddots & \ddots & \ddots & \ddots & \vdots \\ \vdots & & \ddots & \ddots & \ddots & \ddots & k+1 & k \\ k & \dots & \dots & k & k+1 & k+1 & \dots & k+1 \end{pmatrix}$$

so that in each column and in each row there are exactly $n-r$ coefficients k and r coefficients $k+1$.

This kind of matrix C is called a circulant matrix, i.e. starting from a line vector $(c_1\ c_2\ \dots\ c_n)$ which forms the first line of C, the second line is given by the vector $(c_n\ c_1\ \dots\ c_{n-1})$, the third by $(c_{n-1}\ c_n\ c_1\ \dots\ c_{n-2})$, and so on. Let ζ be a primitive n-th root of unity in $\mathbb{C}$. The eigenvectors of C are always the vectors

$$v_j := (1\ \zeta^j\ \zeta^{2j}\ \dots\ \zeta^{(n-1)j}).$$

The eigenvalue of C associated to the eigenvector v_j is

$$\lambda_j := \sum_{\ell=0}^{n-1} c_{n-\ell}\zeta^{\ell j}.$$

The determinant of C is therefore

$$\det(C) = \prod_{j=0}^{n-1} \lambda_j.$$

If $r = 0$, then $\det(C_{n,m}) = 0$ and there is only one elementary divisor of $C_{n,m}$, namely $\delta_1 = k = m/n$.

In our special case we get

$$c_1 = \cdots = c_{n-r} = k+1 \text{ and } c_{n-r+1} = \cdots = c_n = k$$

so that

$$\lambda_0 = r(k+1) + (n-r)k = nk + r = m$$

and

$$\lambda_j = \sum_{\ell=0}^{n-1} c_{n-\ell}\zeta^{\ell j} = k \cdot \sum_{\ell=0}^{n-1} \zeta^{\ell j} + \sum_{\ell=0}^{r-1} \zeta^{\ell j} = \frac{\zeta^{rj}-1}{\zeta^j - 1}.$$

Therefore

$$\det(C) = \prod_{j=0}^{n-1} \lambda_j = \lambda_0 \cdot \underbrace{\left(\prod_{j=1}^{n-1} \lambda_j\right)}_{=1} = m.$$

Lemma 5.7.6 *Suppose the algebras N_n^m and $N_{n'}^{m'}$ are stably equivalent of Morita type. Then $m = m'$ and if $n|m$, then also $n'|m'$.*

Proof If m does not divide n, the Cartan matrix of the Nakayama algebra N_n^m is non-singular of determinant m. Hence $N_{n'}^{m'}$ has non-singular Cartan matrix with determinant m' and $m = m'$ by Proposition 5.7.4 and Remark 5.7.5. If $n|m$, we get that the Cartan matrix of N_n^m is singular and the only elementary divisor is m. Hence by Proposition 5.7.4 and Remark 5.7.5 again the Cartan matrix of $N_{n'}^{m'}$ is singular and n' divides m'. Moreover, $m = m'$ in this case. This proves the statement. □

Remark 5.7.7 Proposition 5.7.4 sometimes allows us to prove that a stable equivalence of Morita type fixes the Morita type of the algebra.

- A special case is the Nakayama algebra N_p^p for some prime p. Lemma 5.7.6 shows that any Nakayama algebra stably equivalent to N_p^p is actually isomorphic to N_p^p.
- Another case is given by a symmetric Nakayama algebra N_n^{kn+1} and another symmetric Nakayama algebra $N_{n'}^{k'n'+1}$. If $k = 1$ and $n = p$ is a prime number, and

if the algebra N_p^{p+1} is stably equivalent to another symmetric non-local Nakayama algebra $N_{n'}^{k'n'+1}$, then $n' = p = n$ and $k' = k = 1$. Indeed, $p+1 = n+1 = k'n'+1$ and therefore $k'n' = p$. Since $N_{n'}^{k'n'+1}$ was assumed to be non-local, $n' > 1$ and hence $n' = p$ and $k' = 1$.

Remark 5.7.8 We shall see in Sect. 6.10.1 that there are algebras which are stably equivalent of Morita type to Nakayama algebras, but which are not Nakayama algebras themselves.

5.8 Stable Equivalences and Hochschild (Co-)Homology

In this section we shall prove that Hochschild homology and Hochschild cohomology are invariant under stable equivalences of Morita type, at least in strictly positive degrees. In degree 0 a quotient of Hochschild cohomology and a submodule of Hochschild homology is invariant under stable equivalences of Morita type.

5.8.1 Stable Equivalence of Morita Type and Hochschild Homology Using Bouc's Trace

We present here an approach due to Serge Bouc. Unfortunately Bouc's approach [16] is unpublished and I am very grateful for Bouc's permission to include his results in this section.

We start with a characterisation of projective modules.

Lemma 5.8.1 *Let K be a commutative ring and let A be a K-algebra. If P is a finitely generated projective A-module then there are elements $\{p_i \mid i \in \{1, \dots, n\}\}$ in P and elements $\{\varphi_i \mid i \in \{1, \dots, n\}\}$ in $Hom_A(P, A)$ such that $id_P = \left(p \mapsto \sum_{i=1}^n \varphi_i(p)p_i\right)$.*

If Q is a finitely generated projective A-right module, then there are elements $\{q_i \mid i \in \{1, \dots, m\}\}$ in Q and elements $\{\psi_i \mid i \in \{1, \dots, m\}\}$ in $Hom_A(Q, A)$ such that $id_Q = \left(q \mapsto \sum_{i=1}^m q_i\psi_i(q)\right)$.

Proof We know that if P is projective, then by Lemma 4.2.5 we obtain an isomorphism of functors

$$Hom_A(P, A) \otimes_A - \simeq Hom_A(P, -).$$

We may evaluate on P to get an isomorphism

$$Hom_A(P, A) \otimes_A P \simeq Hom_A(P, P)$$

which is actually linear with respect to the $End_A(P)$-module structure from the right on P, since it is obtained from the isomorphism of functors $Hom_A(P, A) \otimes_A$

$- \simeq Hom_A(P, -)$. The element $id_P \in Hom_A(P, P)$ corresponds to some element $\sum_{i=1}^{n} \varphi_i \otimes_A p_i$. Recall that the isomorphism from Lemma 4.2.5 maps $\psi \otimes q$ to $(p \mapsto \psi(p)q)$. Hence

$$id_P = \left(p \mapsto \sum_{i=1}^{n} \varphi_i(p) p_i \right).$$

The proof of the statement on right modules is analogous. This proves the lemma. □

Given a commutative ring K and two K-algebras A and B for an A–B-module ${}_AM_B$ which is finitely generated projective as a B-module, we define $\check{M} := Hom_B(M, B)$. As usual $\check{M}$ is a B–A-bimodule ${}_B\check{M}_A$. As in Lemma 5.8.1 we obtain elements $m_i \in M$ and $\varphi_i \in \check{M}$ for $i \in \{1, \ldots, n\}$ such that

$$id_M = \left(m \mapsto \sum_{i=1}^{n} m_i \varphi_i(m) \right).$$

In order to abbreviate the unavoidably cumbersome notation, for every K-algebra C we write

$$C^e := C \otimes_K C^{op}.$$

We shall define for every B^e-module U a morphism of complexes

$$tr_M : M \otimes_B U \otimes_B \check{M} \otimes_{A^e} \mathbb{B}A \longrightarrow U \otimes_{B^e} \mathbb{B}B,$$

between the Hochschild complexes of A and of B. Recall from Proposition 3.6.1 that the degree s homogeneous component of $\mathbb{B}A$ is given as $A^{\otimes(s+1)}$ and that the differential on $\mathbb{B}A$ is given by

$$\partial_s : A^{\otimes(s+2)} \longrightarrow A^{\otimes(s+1)}$$

$$\begin{aligned}\partial_s(a_0 \otimes \cdots \otimes a_{s+1}) := & \; a_0 a_1 \otimes a_2 \otimes \cdots \otimes a_{s+1} \\ & + \sum_{i=1}^{s-1} (-1)^i a_0 \otimes \cdots \otimes a_{i-1} \otimes a_i a_{i+1} \otimes a_{i+2} \otimes \cdots \otimes a_{s+1} \\ & + (-1)^s a_0 \otimes \cdots \otimes a_{s-1} \otimes a_s a_{s+1}.\end{aligned}$$

In the sequel, to save space we will write $\otimes$ for $\otimes_K$ when no confusion is likely to occur. We first observe that

$$(*) \quad V \otimes_{A^e} A^{\otimes(s+1)} \simeq V \otimes A^{\otimes(s-1)}$$

for all $s \geq 1$ by the following map:

$$(v \otimes_{A^e} (a_0 \otimes a_1 \otimes \cdots \otimes a_{s-1} \otimes a_s)) \mapsto ((a_s v a_0) \otimes a_1 \otimes \cdots \otimes a_{s-1}) .$$

This fact is immediate by the definition of the A^e-action on $A^{\otimes(s+1)}$. The formula for the complex for B is similar. Observe what happens with the differential $id_V \otimes \partial_s$ of the Hochschild complex when the Hochschild complex $V \otimes_{A^e} \mathbb{B}A$ is replaced by the corresponding complex under the above isomorphism. The resulting differential is denoted ∂_s^V.

$$\begin{array}{ccc} V \otimes_{A^e} A^{\otimes(s+2)} & \stackrel{id_V \otimes \partial_s}{\longrightarrow} & V \otimes_{A^e} A^{\otimes(s+2)} \\ \downarrow\simeq & & \downarrow\simeq \\ V \otimes_K A^{\otimes(s)} & \stackrel{\partial_s^V}{\longrightarrow} & V \otimes_K A^{\otimes(s)} \end{array}$$

is commutative if and only if

$$\begin{aligned} \partial_s^V (v \otimes a_1 \otimes \cdots \otimes a_s) &= v a_1 \otimes a_2 \otimes \cdots \otimes a_s \\ &\quad + \sum_{i=1}^{s-1} (-1)^i v \otimes \cdots \otimes a_{i-1} \otimes a_i a_{i+1} \otimes a_{i+1} \otimes \cdots \otimes a_s \\ &\quad + (-1)^s a_s v \otimes \cdots \otimes a_{s-1}. \end{aligned}$$

The next definition prepares the definition of Bouc's trace in each homogeneous component of the Hochschild complex.

Definition 5.8.2 We define mappings

$$M \otimes_B U \otimes_B \check{M} \otimes_K A^{\otimes s} \longrightarrow U \otimes_K B^{\otimes s}$$

by

$$\begin{aligned} tr_M^s(m \otimes u \otimes \varphi \otimes a_1 \otimes \cdots \otimes a_s) := \sum_{1 \leq i_0, \ldots, i_s \leq n} & (\varphi_{i_0}(m) u \varphi(m_{i_1})) \otimes \varphi_{i_1}(a_1 m_{i_2}) \otimes \ldots \\ & \otimes \varphi_{i_{s-1}}(a_{i_{s-1}} m_{i_s}) \otimes \varphi_{i_s}(a_{i_s} m_{i_0}). \end{aligned}$$

Lemma 5.8.3 *The map tr_M commutes with the differentials on the Hochschild complexes:*

$$tr_M^{s-1} \circ \partial_{s-1}^{M \otimes_B U \otimes_B \check{M}} = \partial_s^U \circ tr_M^s$$

for each $s \in \mathbb{N}$.

Remark 5.8.4 Unfortunately the proof of this lemma is somewhat technical. There is nothing difficult in the proof. We just apply the definitions of the mappings, and then apply the fact that $\sum_{i=1}^n m_i \varphi_i = id_M$. For completeness we display the cumbersome calculation.

Proof of Lemma 5.8.3 We first compute

$$
\begin{aligned}
&\partial_s^U(tr_M^s(m\otimes u\otimes\varphi\otimes a_1\otimes\cdots\otimes a_s))\\
&=\partial_s^U\Bigg(\sum_{1\le i_0,\dots,i_s\le n}(\varphi_{i_0}(m)u\varphi(m_{i_1}))\otimes\varphi_{i_1}(a_1m_{i_2})\otimes\cdots\\
&\quad\otimes\varphi_{i_{s-1}}(a_{i_{s-1}}m_{i_s})\otimes\varphi_{i_s}(a_{i_s}m_{i_0})\Bigg)\\
&=\sum_{1\le i_0,\dots,i_s\le n}(\varphi_{i_0}(m)u\varphi(m_{i_1}))\varphi_{i_1}(a_1m_{i_2})\otimes\cdots\\
&\quad\otimes\varphi_{i_{s-1}}(a_{i_{s-1}}m_{i_s})\otimes\varphi_{i_s}(a_{i_s}m_{i_0})\\
&\quad+\sum_{j=1}^{s-1}(-1)^j\Bigg(\sum_{1\le i_0,\dots,i_s\le n}(\varphi_{i_0}(m)u\varphi(m_{i_1}))\otimes\varphi_{i_1}(a_1m_{i_2})\otimes\dots\\
&\quad\otimes\varphi_{i_{j-1}}(a_{j-1}m_{i_j})\otimes\varphi_{i_j}(a_jm_{i_{j+1}})\varphi_{i_{j+1}}(a_{j+1}m_{i_{j+2}})\otimes\varphi_{i_{j+2}}(a_{j+2}m_{i_{j+3}})\otimes\cdots\\
&\quad\otimes\varphi_{i_{s-1}}(a_{i_{s-1}}m_{i_s})\otimes\varphi_{i_s}(a_{i_s}m_{i_0})\Bigg)\\
&\quad+(-1)^s\sum_{1\le i_0,\dots,i_s\le n}(\varphi_{i_0}(m)u\varphi(m_{i_1}))\otimes\varphi_{i_1}(a_1m_{i_2})\otimes\dots\\
&\quad\otimes\varphi_{i_{s-2}}(a_{i_{s-2}}m_{i_{s-1}})\otimes\varphi_{i_{s-1}}(a_{i_{s-1}}m_{i_s})\varphi_{i_s}(a_{i_s}m_{i_0})\\
&=\sum_{1\le i_0,\dots,i_s\le n}(\varphi_{i_0}(m)u\varphi(m_{i_1}\varphi_{i_1}(a_1m_{i_2})))\otimes\cdots\\
&\quad\otimes\varphi_{i_{s-1}}(a_{i_{s-1}}m_{i_s})\otimes\varphi_{i_s}(a_{i_s}m_{i_0})\\
&\quad+\sum_{j=1}^{s-1}(-1)^j\Bigg(\sum_{1\le i_0,\dots,i_s\le n}(\varphi_{i_0}(m)u\varphi(m_{i_1}))\otimes\varphi_{i_1}(a_1m_{i_2})\otimes\dots\\
&\quad\otimes\varphi_{i_{j-1}}(a_{j-1}m_{i_j})\otimes\varphi_{i_j}(a_jm_{i_{j+1}}\varphi_{i_{j+1}}(a_{j+1}m_{i_{j+2}}))\otimes\varphi_{i_{j+2}}(a_{j+2}m_{i_{j+3}})\otimes\cdots\\
&\quad\otimes\varphi_{i_{s-1}}(a_{i_{s-1}}m_{i_s})\otimes\varphi_{i_s}(a_{i_s}m_{i_0})\Bigg)\\
&\quad+(-1)^s\sum_{1\le i_0,\dots,i_s\le n}(\varphi_{i_0}(m)u\varphi(m_{i_1}))\otimes\varphi_{i_1}(a_1m_{i_2})\otimes\cdots\\
&\quad\otimes\varphi_{i_{s-2}}(a_{i_{s-2}}m_{i_{s-1}})\otimes\varphi_{i_{s-1}}(a_{i_{s-1}}m_{i_s}\varphi_{i_s}(a_{i_s}m_{i_0}))
\end{aligned}
$$

where the last equality uses the B-linearity of the mappings φ and φ_j. But recall the definition of the elements φ_i and m_i, namely $\sum_i m_i\varphi_i = id_M$. Hence

$$\sum_{1\leq i_0,\dots,i_s\leq n} \varphi_{i_0}(m)u\varphi(m_{i_1}\varphi_{i_1}(a_1m_{i_2}))\otimes\dots\otimes\varphi_{i_{s-1}}(a_{i_{s-1}}m_{i_s})\otimes\varphi_{i_s}(a_{i_s}m_{i_0})$$

$$= \sum_{1\leq i_0,i_2,\dots,i_s\leq n} (\varphi_{i_0}(m)u\varphi(\sum_{i_1=1}^{n} m_{i_1}\varphi_{i_1}(a_1m_{i_2})))\otimes\dots$$

$$\otimes\varphi_{i_{s-1}}(a_{i_{s-1}}m_{i_s}\otimes\varphi_{i_s}(a_{i_s}m_{i_0})$$

$$= \sum_{1\leq i_0,i_2,\dots,i_s\leq n} \varphi_{i_0}(m)u\varphi(a_1m_{i_2})\otimes\dots$$

$$\otimes\varphi_{i_{s-1}}(a_{i_{s-1}}m_{i_s})\otimes\varphi_{i_s}(a_{i_s}m_{i_0})$$

and likewise

$$\sum_{i_{j+1}=1}^{n} \varphi_{i_j}(a_jm_{i_{j+1}}\varphi_{i_{j+1}}(a_{j+1}m_{i_{j+2}})) = \varphi_{i_j}(a_ja_{j+1}m_{i_{j+2}})$$

and

$$\sum_{1\leq i_0,\dots,i_s\leq n} (\varphi_{i_0}(m)u\varphi(m_{i_1}))\otimes\varphi_{i_1}(a_1m_{i_2})\otimes\dots$$

$$\otimes\varphi_{i_{s-2}}(a_{i_{s-2}}m_{i_{s-1}})\otimes\varphi_{i_{s-1}}(a_{i_{s-1}}m_{i_s}\varphi_{i_s}(a_{i_s}m_{i_0}))$$

$$= \sum_{1\leq i_0,\dots,i_{s-1}\leq n} (\varphi_{i_0}(m)u\varphi(m_{i_1}))\otimes\varphi_{i_1}(a_1m_{i_2})\otimes\dots$$

$$\otimes\varphi_{i_{s-2}}(a_{i_{s-2}}m_{i_{s-1}})\otimes\varphi_{i_{s-1}}(a_{i_{s-1}}\sum_{i_s=1}^{n} m_{i_s}\varphi_{i_s}(a_{i_s}m_{i_0}))$$

$$= \sum_{1\leq i_0,\dots,i_{s-1}\leq n} (\varphi_{i_0}(m)u\varphi(m_{i_1}))\otimes\varphi_{i_1}(a_1m_{i_2})\otimes\dots$$

$$\otimes\varphi_{i_{s-2}}(a_{i_{s-2}}m_{i_{s-1}})\otimes\varphi_{i_{s-1}}(a_{i_{s-1}}a_{i_s}m_{i_0}).$$

As a whole we get

$$\partial_s^U(tr_M^s(m\otimes u\otimes\varphi\otimes a_1\otimes\dots\otimes a_s))$$

$$= \sum_{1\leq i_0,i_2,\dots,i_s\leq n} \varphi_{i_0}(m)u\varphi(a_1m_{i_2})\otimes\dots\otimes\varphi_{i_{s-1}}(a_{i_{s-1}}m_{i_s})\otimes\varphi_{i_s}(a_{i_s}m_{i_0})$$

$$+\sum_{j=1}^{s-1}(-1)^j\Bigg(\sum_{1\leq i_0,\dots,\hat{i}_{j+1},\dots,i_s\leq n} (\varphi_{i_0}(m)u\varphi(m_{i_1}))\otimes\varphi_{i_1}(a_1m_{i_2})\otimes\dots$$

$$\otimes\varphi_{i_{j-1}}(a_{j-1}m_{i_j})\otimes\varphi_{i_j}(a_jm_{i_{j+1}}m_{i_{j+2}})\otimes\varphi_{i_{j+2}}(a_{j+2}m_{i_{j+3}})\otimes\dots$$

$$\otimes\varphi_{i_{s-1}}(a_{i_{s-1}}m_{i_s})\otimes\varphi_{i_s}(a_{i_s}m_{i_0})\Bigg)$$

$$+(-1)^s \sum_{1\le i_0,\ldots,i_{s-1}\le n} (\varphi_{i_0}(m)u\varphi(m_{i_1}))\otimes\varphi_{i_1}(a_1m_{i_2})\otimes\ldots$$
$$\otimes\varphi_{i_{s-2}}(a_{i_{s-2}}m_{i_{s-1}})\otimes\varphi_{i_{s-1}}(a_{i_{s-1}}a_{i_s}m_{i_0}).$$

But this last term is precisely

$$\left(tr_M^{s-1}\circ\partial_{s-1}^{M\otimes_BU\otimes_B\check{M}}\right)(m\otimes u\otimes\varphi\otimes a_1\otimes\cdots\otimes a_s)$$

as is seen by a similar, slightly easier, computation. This proves the lemma. □

Remark 5.8.5 Observe that the definition of tr_M depends on the choice of an isomorphism $M\otimes_B Hom_B(M,B)\simeq End_B(M)$. It can be shown that a different choice of an isomorphism yields a mapping tr'_M such that $tr_M-tr'_M$ is homotopic to 0. The proof of this fact is more technical than the proof of Lemma 5.8.3 and can be found in Bouc [16].

We slightly generalise the Definition 3.6.7 to the concept of Hochschild homology $HH_s(B,U)$ with values in U. This is defined as

$$HH_s(B,U):=Tor_s^{B\otimes_K B^{op}}(B,U)$$

and we note that

$$HH_s(B)=HH_s(B,B).$$

Taking homology of the Hochschild complexes, if A and B are projective as K-module, we thus obtain a morphism

$$tr_M^s: HH_s(A,M\otimes_B U\otimes_B Hom_B(M,B))\longrightarrow HH_s(B,U)$$

for every $B\otimes_K B^{op}$-module U.

Definition 5.8.6 [16] Let K be a commutative ring and let A and B be two K-algebras. Let M be an $A\otimes_K B^{op}$-module and suppose that M is finitely generated projective as a B^{op}-module. Then for all $s\in\mathbb{N}$, if A and B are projective as K-module,

$$tr_M^s: HH_s(A,M\otimes_B U\otimes_B Hom_B(M,B))\longrightarrow HH_s(B,U)$$

is called *Bouc's trace map* on homology.

Remark 5.8.7 Let K be a commutative ring and let A be a K-projective K-algebra. Then $tr_A^s=id$. Indeed, $id_A=1\cdot id_A$ and so we may take 1 and id_A as elements to compute the trace. The formula gives immediately the statement.

Example 5.8.8 Let K be a commutative ring and let A and B be K-algebras. Let M be an $A\otimes_K B^{op}$-module, let $N=Hom_B(M,B)$ and suppose that (M,N) induces a stable equivalence of Morita type (it could even be in the generalised sense). Then, if A and B are projective as K-module and if $s \geq 1$ we obtain a map

$$tr_M^s : HH_s(A, M \otimes_B B \otimes_B N) \longrightarrow HH_s(B, B).$$

Since $M \otimes_B B \otimes_B N \simeq A \oplus P$ for a flat $A \otimes_K A^{op}$-module, we get

$$\begin{aligned} HH_s(A, M \otimes_B B \otimes_B N) &\simeq HH_s(A, A \oplus P) \simeq HH_s(A) \oplus HH_s(A, P) \\ &= HH_s(A) \end{aligned}$$

since P is flat as an $A \otimes_K A^{op}$-module, and hence $HH_s(A, P) = 0$ as soon as $s \geq 1$. This is not true for $s = 0$!

Bouc's trace map is additive.

Lemma 5.8.9 *Let K be a commutative ring, let A and B be K-algebras, projective as K-modules and let M_1 and M_2 be $A \otimes_K B^{op}$-modules. Suppose that M_1 and M_2 are finitely generated projective as B^{op}-modules. Let*

$$(M_1 \oplus M_2) \otimes_B - \otimes_B (\check{M}_1 \oplus \check{M}_2) \stackrel{\pi_{1,2}}{\longrightarrow\!\!\!\!\rightarrow} M_1 \otimes_B - \otimes_B \check{M}_1 \oplus M_2 \otimes_B - \otimes_B \check{M}_2$$

be the canonical projection onto the corresponding direct factors. Then for all $s \geq 1$ we get

$$tr^s_{M_1 \oplus M_2} = tr^s_{M_1} + tr^s_{M_2}$$

in the sense that the following diagram of functors

$$\begin{array}{ccc} HH_s(A, M \otimes_B - \otimes_B \check{M}) & \stackrel{HH_s(A,\pi_{1,2})}{\longrightarrow} & HH_s(A, M_1 \otimes_B - \otimes_B \check{M}_1) \\ & & \oplus \\ \| & & HH_s(A, M_2 \otimes_B - \otimes_B \check{M}_2) \\ & & \downarrow tr^s_{M_1} + tr^s_{M_2} \\ HH_s(A, M \otimes_B - \otimes_B \check{M}) & \stackrel{tr^s_{M_1 \oplus M_2}}{\longrightarrow} & HH_s(B, -) \end{array}$$

is commutative. Here we define $M := M_1 \oplus M_2$.

Proof Let

$$id_{M_1} = \sum_{j=1}^{n_1} m^1_j \varphi^1_j \text{ and } id_{M_2} = \sum_{j=1}^{n_2} m^2_j \varphi^2_j$$

where $m^1_i \in M_1$, $\varphi^1_i \in \check{M}_1 = Hom_B(M_1, B)$, $m^2_i \in M_2$ and $\varphi^2_i \in \check{M}_2 = Hom_B(M_2, B)$. Then

$$(\dagger)\quad id_{M_1\oplus M_2}=id_{M_1}+id_{M_2}=\left(\sum_{j=1}^{n_1}m_j^1\varphi_j^1\right)+\left(\sum_{j=1}^{n_2}m_j^2\varphi_j^2\right).$$

The isomorphism

$$Hom_B(M_1\oplus M_2,M_1\oplus M_2)\simeq(M_1\oplus M_2)\otimes_B Hom_B(M_1\oplus M_2,B)$$

can be chosen so that

$$\begin{array}{rcl}Hom_B(M_1\oplus M_2,M_1\oplus M_2)&\longrightarrow&(M_1\oplus M_2)\otimes_B Hom_B(M_1\oplus M_2,B)\\ id_{M_1\oplus M_2}&\mapsto&\left(\sum_{j=1}^{n_1}m_i^1\varphi_i^1\right)+\left(\sum_{j=1}^{n_2}m_i^2\varphi_i^2\right)\end{array}$$

using equation (†) and the fact that the maps are $End_B(M_1\oplus M_2)$-linear. We may then define elements of $M_1\oplus M_2$ by putting for $j\in\{1,\dots,n_1+n_2\}$

$$m_j^3:=\begin{cases}m_j^1 & \text{if } j\le n_1\\ m_{j-n_1}^2 & \text{if } j>n_1\end{cases}$$

and

$$\varphi_j^3:=\begin{cases}\varphi_j^1 & \text{if } j\le n_1\\ \varphi_{j-n_1}^2 & \text{if } j>n_1\end{cases}$$

to obtain with $n_3:=n_1+n_2$

$$id_{M_1\oplus M_2}=\sum_{j=1}^{n_3}m_j^3\varphi_j^3.$$

For simplicity of notation we shall replace the variable "–" of the functors by a B–B-bimodule B. Functoriality is clear.

With this set of elements we may evaluate the trace $tr_{M_1\oplus M_2}$ as in Definition 5.8.2 by

$$\begin{aligned}&tr_M^s(m\otimes u\otimes\varphi\otimes a_1\otimes\cdots\otimes a_s)\\ &:=\sum_{1\le i_0,\dots,i_s\le n_3}(\varphi_{i_0}^3(m)u\varphi(m_{i_1}^3))\otimes\varphi_{i_1}^3(a_1m_{i_2}^3)\otimes\dots\\ &\quad\otimes\varphi_{i_{s-1}}^3(a_{i_{s-1}}m_{i_s}^3)\otimes\varphi_{i_s}^3(a_{i_s}m_{i_0}^3).\end{aligned}$$

Now, observe that $\varphi_k^3(am_\ell^3)=0$ as soon as $k\le n_1<\ell$ since in this case am_ℓ^3 is in the first factor M_1 of $M_1\oplus M_2$ whereas φ_k^3 evaluates the second factor M_2 of

$M_1 \oplus M_2$. Similarly, for $\ell \le n_1 < k$ one gets $\varphi_k^3(am_\ell^3) = 0$. Hence

$$
\begin{aligned}
& tr^s_M(m \otimes u \otimes \varphi \otimes a_1 \otimes \cdots \otimes a_s) \\
:= & \sum_{1 \le i_0, \ldots, i_s \le n_3} (\varphi^3_{i_0}(m) u \varphi(m^3_{i_1})) \otimes \varphi^3_{i_1}(a_1 m^3_{i_2}) \otimes \cdots \\
& \otimes \varphi^3_{i_{s-1}}(a_{i_{s-1}} m^3_{i_s}) \otimes \varphi^3_{i_s}(a_{i_s} m^3_{i_0}) \\
= & \sum_{1 \le i_0, \ldots, i_s \le n_1} (\varphi^3_{i_0}(m) u \varphi(m^3_{i_1})) \otimes \varphi^3_{i_1}(a_1 m^3_{i_2}) \otimes \cdots \\
& \otimes \varphi^3_{i_{s-1}}(a_{i_{s-1}} m^3_{i_s}) \otimes \varphi^3_{i_s}(a_{i_s} m^3_{i_0}) \\
& + \sum_{n_1+1 \le i_0, \ldots, i_s \le n_3} (\varphi^3_{i_0}(m) u \varphi(m^3_{i_1})) \otimes \varphi^3_{i_1}(a_1 m^3_{i_2}) \otimes \cdots \\
& \otimes \varphi^3_{i_{s-1}}(a_{i_{s-1}} m^3_{i_s}) \otimes \varphi^3_{i_s}(a_{i_s} m^3_{i_0}) \\
& = tr^s_{M_1}(m \otimes u \otimes \varphi \otimes a_1 \otimes \cdots \otimes a_s) + tr^s_{M_2}(m \otimes u \otimes \varphi \otimes a_1 \otimes \cdots \otimes a_s) \\
& = (tr_{M_1} + tr_{M_2})(m \otimes u \otimes \varphi \otimes a_1 \otimes \cdots \otimes a_s).
\end{aligned}
$$

Neglecting the terms for which not all terms φ_k^3 and m_ℓ^3 belong to the same part of the index set, the indices up to n_1 or those larger than $n_1 + 1$, corresponds exactly to the projection of

$$(M_1 \oplus M_2) \otimes_B - \otimes_B (\check{M}_1 \oplus \check{M}_2)$$

to

$$(M_1 \otimes_B - \otimes_B \check{M}_1) \oplus (M_2 \otimes_B - \otimes_B \check{M}_2).$$

This proves the statement. □

What is more important for us is that Bouc's trace is multiplicative.

Proposition 5.8.10 *Let K be a commutative ring and let A, B and C be K-projective K-algebras. Let M be an $A \otimes_K B^{op}$-module and let N be a $B \otimes_K C^{op}$-module. Suppose that M is finitely generated projective as a B^{op}-module and that N is finitely generated projective as a C^{op}-module. Then $M \otimes_B N$ is finitely generated projective as a C^{op}-module and*

$$tr^s_{M \otimes_B N} = tr^s_N \circ tr^s_M$$

as functors

$$HH_s(A, M \otimes_B N \otimes_C - \otimes_C Hom_C(M \otimes_B N, C)) \longrightarrow HH_s(C, -).$$

Proof Lemma 5.3.3 shows that $M \otimes_B N$ is finitely generated projective as a C^{op}-module. Hence $tr^s_{M \otimes_B N}$ is defined.

Now,

$$\begin{aligned} Hom_C(M \otimes_B N, C) &\simeq Hom_B(M, Hom_C(N, C)) \\ &\simeq Hom_C(N, C) \otimes_B Hom_B(M, B) \end{aligned}$$

using the $Hom - \otimes$-adjointness formula and Lemma 4.2.5 which can be applied since M is projective as a B^{op}-module. Observe that the isomorphism is given by

$$\begin{aligned} Hom_C(N, C) \otimes_B Hom_B(M, B) &\longrightarrow Hom_C(M \otimes_B N, C) \\ \psi \otimes \varphi &\mapsto (m \otimes n \mapsto \psi(\varphi(m)n)) \end{aligned}$$

Put $\check{N} := Hom_C(N, C)$ and $\check{M} := Hom_B(M, B)$. Let

$$id_M = \sum_{j=1}^{n_M} m_j \varphi_j \text{ and } id_N = \sum_{j=1}^{n_N} n_j \psi_j$$

so that we get

$$id_{M \otimes_B N} = \sum_{j_1=1}^{n_M} \sum_{j_2=1}^{n_N} (m_{j_1} \otimes n_{j_2})(\psi_{j_2} \varphi_{j_1}).$$

Hence we may use these elements $m_{j_1} \otimes n_{j_2}$ and $(\psi_{j_2} \varphi_{j_1})$ to evaluate $tr_{M \otimes_B N}$. Then plugging these elements into the trace formula, using

$$(M \otimes_B N) \otimes_C - \otimes_C \left(\check{N} \otimes_B \check{M}\right) = M \otimes_B \left(N \otimes_C - \otimes_C \check{N}\right) \otimes_B \check{M}$$

we obtain the statement. □

Lemma 5.8.11 *Let K be a commutative ring and let A and B be K-projective K-algebras. Let M be an $A \otimes_K B^{op}$-module and suppose M is projective as a B^{op}-module and projective as an $A \otimes_K B^{op}$-module. Then for each $s \geq 1$ we get $tr_M^s = 0$ as a functor*

$$HH_s(A, M \otimes_B - \otimes_B \check{M}) \longrightarrow HH_s(B, -).$$

Proof $M \otimes_B - \otimes_B \check{M}$ is projective as an $A \otimes_K A^{op}$-module by Lemma 5.3.3. This proves the statement. □

Theorem 5.8.12 *Let K be a commutative ring, let A and B be K-projective K-algebras, and let (M, N) induce a stable equivalence of Morita type (of generalised type) between A and B. Then for each $s \geq 1$ we get that*

$$tr_M^s : HH_s(A) \longrightarrow HH_s(B) \text{ and } tr_N^s : HH_s(B) \longrightarrow HH_s(A)$$

are mutually inverse isomorphisms.

Proof Example 5.8.8 shows that tr^s_M and tr^s_N are well-defined morphisms on the Hochschild homology. Lemmas 5.8.7 and 5.8.11 and Proposition 5.8.10 show that

$$id_{HH_s(A)} = tr^s_A = tr^s_{M\otimes_B N} = tr^s_N \circ tr^s_M$$

as well as

$$id_{HH_s(B)} = tr^s_B = tr^s_{N\otimes_A M} = tr^s_M \circ tr^s_N.$$

This completes the proof. □

Remark 5.8.13 The case $s = 0$ is much more complicated. Whether the degree 0 Hochschild homology is preserved under stable equivalences of Morita type is a difficult and open question. We shall come back to this problem in Sect. 5.9.

We shall give another more direct map on Hochschild homology. Let K be a commutative ring, let A and B be K-algebras and let M be an $A \otimes_K B^{op}$-module. Suppose that M is finitely generated projective as a B^{op}-module. Then there is a map

$$\psi : A \longrightarrow End_B(M)$$

given by left multiplication of A on M.

Lemma 5.8.14 *The map $A \longrightarrow End_B(M)$ is a map of $A \otimes_K A^{op}$-modules.*

Proof Let $a, a_1, a_2 \in A, m \in M$ and let $\psi : A \longrightarrow End_B(M)$ be the above map. Then

$$\psi(a_1 \cdot a \cdot a_2)(m) = a_1 \cdot a \cdot a_2 \cdot m = a_1 \cdot (a \cdot (a_2 \cdot m)) = (a_1 \cdot \psi(a) \cdot a_2)(m).$$

This shows the lemma. □

Remark 5.8.15 Let M be an $A \otimes_K B^{op}$-module and suppose that M is in addition projective as a B^{op}-module. Then ψ and tr_M induce a map of complexes $\widehat{tr}_M$

$$\begin{array}{ccc} \mathbb{B}A \otimes_{A\otimes_K A^{op}} A & \stackrel{id_{\mathbb{B}A}\otimes\psi}{\longrightarrow} & \mathbb{B}A \otimes_{A\otimes_K A^{op}} M \otimes_B Hom_B(M, B) \\ \downarrow \widehat{tr}_M & & \| \\ \mathbb{B}B \otimes_{B\otimes_K B^{op}} B & \stackrel{tr_M}{\longleftarrow} & \mathbb{B}A \otimes_{A\otimes_K A^{op}} M \otimes_B B \otimes_B Hom_B(M, B). \end{array}$$

Moreover, taking homology of the complexes we get a morphism

$$\widehat{tr}^n_M : HH_n(A) \longrightarrow HH_n(B)$$

for all $n \in \mathbb{N}$, including $n = 0$.

Lemma 5.8.16 *If (M, N) induces a stable equivalence of Morita type between the K-projective K-algebras A and B, and if $N = Hom_B(M, B)$ then for all $n \geq 1$ the* *map defined on Hochschild homology $\widehat{tr}^n_M$ is the same as the map tr^n_M defined in Theorem 5.8.12.*

Proof Indeed, $M \otimes_B N \simeq A \oplus P$ as $A \otimes_K A^{op}$-modules and the map

$$A \longrightarrow M \otimes_B N$$

used in the definition of $\widehat{tr}_M$ is $A \otimes_K A^{op}$-linear. Left and right multiplication by elements of A act diagonally on $A \oplus P$ and only the direct factor A will give a contribution to Hochschild homology. Hence $HH_n(id_{\mathbb{B}A} \otimes \psi) = id$, taking homology. This proves the statement. □

5.8.2 Stable Equivalences of Morita Type and Hochschild Cohomology

The invariance of Hochschild cohomology under stable equivalences of Morita type is slightly easier. We shall give a proof due to Xi [15]. There the case of Hochschild homology is also covered, but the hypotheses there are stronger than what we have using Bouc's approach and Bouc's approach seems to be more explicit.

Theorem 5.8.17 (Xi [15]) *Let K be a perfect field, let A and B be finite dimensional K-algebras and suppose that (M, N) induces a stable equivalence of Morita type between A-$\underline{mod}$ and B-$\underline{mod}$. Then*

$$HH^n(A) \simeq HH^n(B)$$

for all $n \geq 1$, and the isomorphism preserves the cup product.

Proof Let

$$M \otimes_B N \simeq A \oplus P$$

for a projective $A \otimes_K A^{op}$-module P and

$$N \otimes_A M \simeq B \oplus Q$$

for a projective $B \otimes_K B^{op}$-module Q. Then $M \simeq Hom_A(N, A)$ and $N \simeq Hom_A(M, A)$ by Proposition 5.5.3.

We shall apply Corollary 3.7.18 several times.

$$\begin{aligned} HH^n(A) &= Ext^n_{A \otimes_K A^{op}}(A, A) = Ext^n_{A \otimes_K A^{op}}(A \oplus P, A) \\ &\simeq Ext^n_{A \otimes_K A^{op}}(M \otimes_B N, A) \simeq Ext^n_{A \otimes_K B^{op}}(M, Hom_A(N, A)) \\ &\simeq Ext^n_{A \otimes_K B^{op}}(M, M). \end{aligned}$$

Moreover, by the same argument

$$HH^n(B) \simeq Ext^n_{B \otimes_K A^{op}}(N, N).$$

Finally

$$\begin{aligned} Ext^n_{A\otimes_K B^{op}}(M,M) &\simeq Ext^n_{A\otimes_K B^{op}}(M, Hom_A(N,A)) \\ &\simeq Ext^n_{B\otimes_K A^{op}}(N, Hom_A(M,A)) \text{ by Corollary 3.7.18} \\ &\simeq Ext^n_{B\otimes_K A^{op}}(N,N). \end{aligned}$$

This proves the statement for the structure as a vector space. The multiplicative structure follows by the fact that the multiplicative structure on

$$\bigoplus_{n\geq 1} Ext^n_{A\otimes_K B^{op}}(M,M)$$

is given by composing morphisms in

$$Ext^n_{A\otimes_K B^{op}}(M,M) = Hom_{D^-(A\otimes_K B^{op})}(M, M[n])$$

in the derived category (cf Proposition 3.7.12). But the same is true for

$$Ext^n_{A\otimes_K B^{op}}(N,N) = Hom_{D^-(B\otimes_K A^{op})}(N, N[n])$$

and the identification above respects this composition. This proves the statement. □

Remark 5.8.18 There are many more correspondences and compatibilities between the Hochschild homology structure of two algebras, the Hochschild cohomology structure of two algebras and stable equivalences of Morita type.

5.9 Degree Zero Hochschild (Co-)Homology

Most surprisingly, properties of degree n Hochschild homology and cohomology with respect to stable equivalences are more easily studied for $n > 0$. In degree 0 the situation is very complicated and is a source of deep open problems and conjectures.

5.9.1 Stable Centre

We haven't yet look at the invariance of the centre, although for Morita equivalence this was our first application. There is a reason for this: the result which corresponds to the invariance of the centre under Morita equivalence gives the invariance of another ring, a certain quotient of the centre, called the stable centre.

Definition 5.9.1 (Broué [5]) Let K be a commutative ring and let A be a K-algebra. The *stable centre* of A is

$$Z^{st}(A) := \underline{End}_{A\otimes_K A^{op}}(A).$$

The *projective centre* of A is the kernel $Z^{pr}(A)$ of the natural quotient of the centre to the stable centre. More precisely the sequence

$$0 \longrightarrow Z^{pr}(A) \longrightarrow End_{A\otimes_K A^{op}}(A) \longrightarrow Z^{st}(A) \longrightarrow 0$$

is exact.

Remark 5.9.2 Note that $Z(A) = End_{A\otimes_K A^{op}}(A)$. Hence $Z^{st}(A)$ is a quotient ring of the centre of A.

We want to show that the K-algebra $Z^{st}(A)$ is an invariant under stable equivalences of Morita type.

Proposition 5.9.3 *Let K be a commutative ring and let A and B be K-algebras. Let M be an $A\otimes_K B^{op}$-module and let N be a $B\otimes_K A^{op}$-module. Suppose (M,N) induces a stable equivalence of Morita type between A and B. Let $A\otimes_K A^{op}$-$\underline{mod}^\circ$ be the full subcategory of $A\otimes_K A^{op}$-$\underline{mod}$ generated by the modules which are projective as A-modules and projective as A^{op}-modules, and let $B\otimes_K B^{op}$-$\underline{mod}^\circ$ be the full subcategory of $B\otimes_K B^{op}$-$\underline{mod}$ generated by the modules which are projective as B-modules and projective as B^{op}-modules.*

Then

$$M\otimes_B - \otimes_B N : B\otimes_K B^{op}\text{-}mod \longrightarrow A\otimes_K A^{op}\text{-}mod$$

induces an equivalence

$$M\otimes_B - \otimes_B N : B\otimes_K B^{op}\text{-}\underline{mod}^\circ \longrightarrow A\otimes_K A^{op}\text{-}\underline{mod}^\circ.$$

Proof If U is a $B\otimes_K B^{op}$-module, then $M\otimes_B U\otimes_B N$ is an $A\otimes_K A^{op}$-module. If U is a projective $B\otimes_K B^{op}$-module, then U is a direct factor of $(B\otimes_K B^{op})^n$ for some $n\in\mathbb{N}$. Hence $M\otimes_B U\otimes_B N$ is a direct factor of

$$M\otimes_B (B\otimes_K B^{op})^n\otimes_B N \simeq (M\otimes_K N)^n.$$

But M is projective as an A-module and N is projective as an A^{op}-module. Hence M is a direct factor of the A-module A^n and N is a direct factor of the right A-module A^m. Therefore $M\otimes_K N$ is a direct factor of the $A\otimes_K A^{op}$-module $(A\otimes_K A^{op})^{nm}$, whence it is projective. This shows that the functor

$$M\otimes_B - \otimes_B N : B\otimes_K B^{op}\text{-}mod \longrightarrow A\otimes_K A^{op}\text{-}mod$$

induces a functor

$$\underline{F}^e_M : M\otimes_B - \otimes_B N : B\otimes_K B^{op}\text{-}\underline{mod} \longrightarrow A\otimes_K A^{op}\text{-}\underline{mod}.$$

Likewise

$$N \otimes_A - \otimes_A M : A \otimes_K A^{op}\text{-}mod \longrightarrow B \otimes_K B^{op}\text{-}mod$$

induces a functor

$$\underline{F}_N^e : N \otimes_A - \otimes_A M : A \otimes_K A^{op}\text{-}\underline{mod} \longrightarrow B \otimes_K B^{op}\text{-}\underline{mod}.$$

We shall consider the composition of these two functors in order to determine when they produce an equivalence. The composition $\underline{F}_M^e \circ \underline{F}_N^e$ is

$$\begin{aligned}
&M \otimes_B (N \otimes_A - \otimes_A M) \otimes_B N \\
&= (M \otimes_B N) \otimes_A - \otimes_A (M \otimes_B N) \\
&= (A \oplus P) \otimes_A - \otimes_A (A \oplus P) \\
&= id \oplus (- \otimes_A P) \oplus (P \otimes_A -) \oplus (P \otimes_A - \otimes_A P).
\end{aligned}$$

Now, by Lemma 5.3.3 the $A \otimes_K A^{op}$-module $(P \otimes_A U \otimes_A P)$ is projective for every $A \otimes_K A^{op}$-module U. Given an $A \otimes_K A^{op}$-module U, the module $P \otimes_A U$ is in general not a projective $A \otimes_K A^{op}$-module. The module $P \otimes_A U$ is a projective $A \otimes_K A^{op}$-module if U is projective as an A^{op}-module, using Lemma 5.3.3 again. Likewise $U \otimes_A P$ is a projective $A \otimes_K A^{op}$-module if U is projective as an A-module. Hence we may restrict $\underline{F}_M^e$ and $\underline{F}_N^e$ to $A \otimes_K A^{op}\text{-}\underline{mod}^\circ$ and to $B \otimes_K B^{op}\text{-}\underline{mod}^\circ$ to get

$$\underline{F}_M^e \circ \underline{F}_N^e|_{A \otimes_K A^{op}\text{-}\underline{mod}^\circ} = id_{A \otimes_K A^{op}\text{-}\underline{mod}^\circ}.$$

Likewise,

$$\underline{F}_N^e \circ \underline{F}_M^e|_{B \otimes_K B^{op}\text{-}\underline{mod}^\circ} = id_{B \otimes_K B^{op}\text{-}\underline{mod}^\circ}.$$

This proves the statement. □

Remark 5.9.4 Observe that $\underline{F}_M^e$ and $\underline{F}_N^e$ are well-defined functors between $A \otimes_K A^{op}\text{-}\underline{mod}$ and $B \otimes_K B^{op}\text{-}\underline{mod}$, but in general these functors will not be equivalences between $A \otimes_K A^{op}\text{-}\underline{mod}$ and $B \otimes_K B^{op}\text{-}\underline{mod}$. However the restriction to the subcategories $A \otimes_K A^{op}\text{-}\underline{mod}^\circ$ and $B \otimes_K B^{op}\text{-}\underline{mod}^\circ$ will always be equivalences.

Observe further that we needed that (M, N) induces a stable equivalence of Morita type in the strict sense according to Broué's original definition 5.3.5. The generalised Definition 5.3.15 will not produce the equivalence claimed in Proposition 5.9.3.

Proposition 5.9.5 *Let K be a commutative ring and let A and B be K-algebras. Let $M \in A \otimes_K B^{op}\text{-}mod$ and let $N \in B \otimes_K A^{op}\text{-}mod$ such that (M, N) induces a stable equivalence of Morita type in the strict sense.*

Then the functor $M \otimes_B - \otimes_B N$ induces an isomorphism $Z^{st}(B) \longrightarrow Z^{st}(A)$ as K-algebras.

Proof We observe that the $A \otimes_K A^{op}$-module A is in $A \otimes_K A^{op}\text{-}\underline{mod}^\circ$ and that the $B \otimes_K B^{op}$-module B is in $B \otimes_K B^{op}\text{-}\underline{mod}^\circ$. Moreover,

$$M \otimes_B B \otimes_B N \simeq M \otimes_B N \simeq A \oplus P$$

in $A \otimes_K A^{op}$-*mod* for some projective $A \otimes_K A^{op}$-module P. Hence, denoting the equivalence by

$$\underline{F}_M^e := M \otimes_B - \otimes_B N : B \otimes_K B^{op}\text{-}\underline{mod}^\circ \longrightarrow A \otimes_K A^{op}\text{-}\underline{mod}^\circ$$

we get $\underline{F}_M^e(B) \simeq A$ in $A \otimes_K A^{op}$-$\underline{mod}$. Therefore

$$\begin{aligned} Z^{st}(B) &= \underline{End}_{B\otimes_K B^{op}}(B) \simeq \underline{End}_{A\otimes_K A^{op}}(\underline{F}_M^e(B)) \\ &\simeq \underline{End}_{A\otimes_K A^{op}}(A) = Z^{st}(A). \end{aligned}$$

Since this isomorphism is just the isomorphism of a K-linear category equivalence on morphism maps, the isomorphism is compatible with composition of morphisms and this means that the isomorphism is an isomorphism of K-algebras. This proves the proposition. □

How big is $Z^{st}(A)$ compared to $Z(A)$? We shall give an answer to this for nice algebras A later. However we can already give a partial answer now. Recall from Definition 3.7.8 that

$$Ext^1_{A\otimes_K A^{op}}(A, -) = Hom_{D^b(A\otimes_K A^{op})}(A, -[1]).$$

But now,

$$Z(A) = End_{A\otimes_K A^{op}}(A) = End_{D^b(A\otimes_K A^{op})}(A)$$

acts on the functor $Hom_{D^b(A\otimes_K A^{op})}(A, -[1])$ by composition of morphisms. Hence for all $A \otimes_K A^{op}$-modules V the space $Ext^1_{A\otimes_K A^{op}}(A, V)$ is a $Z(A)$-module.

Definition 5.9.6 [17] Let K be a commutative ring and let A be a K-algebra. The Higman ideal $H(A)$ is the kernel of the action of $Z(A)$ on $Ext^1_{A\otimes_K A^{op}}(A, -)$, that is

$$\begin{aligned} H(A) &:= Ann_{Z(A)}(Ext^1_{A\otimes_K A^{op}}(A, -)) \\ &:= \{a \in Z(A) \mid a \cdot \zeta = 0 \ \forall \zeta \in Ext^1_{A\otimes_K A^{op}}(A, -)\} \\ &:= \{a \in Z(A) \mid a \cdot \zeta = 0 \ \forall \zeta \in Ext^1_{A\otimes_K A^{op}}(A, V) \\ &\qquad \forall V \in A \otimes_K A^{op}\text{-}mod\}. \end{aligned}$$

It should be noted that it is possible to compute $H(A)$ with this definition even if it might appear difficult at first glance.

Lemma 5.9.7 *Let K be a commutative ring, let A be a K-algebra and let $\Omega(A)$ be the kernel of the multiplication map*

$$A \otimes_K A \longrightarrow A$$

$$a_1 \otimes a_2 \mapsto a_1 a_2$$

and denote by ζ_{univ} *the exact sequence*

$$0 \longrightarrow \Omega(A) \longrightarrow A \otimes_K A \longrightarrow A \longrightarrow 0$$

regarded as an element in $Ext^1_{A\otimes_K A^{op}}(A, \Omega(A))$. *Then*

$$H(A) = \{z \in Z(A) \mid z \cdot \zeta_{univ} = 0\}.$$

Proof Since $\zeta_{univ} \in Ext^1_{A\otimes_K A^{op}}(A, \Omega(A))$ and $\Omega(A)$ is a possible module that can be used as V in the definition of $H(A)$, we get

$$H(A) \subseteq \{z \in Z(A) \mid z \cdot \zeta_{univ} = 0\}.$$

However, the first term of the bar resolution Definition 3.6.4 is precisely the exact sequence ζ_u. Hence any element in $Ext^1_{A\otimes_K A^{op}}(A, V)$ can be regarded as a morphism $\nu : \Omega(A) \longrightarrow V$ modulo those morphisms factoring through the embedding $\Omega(A) \hookrightarrow A \otimes_K A$. Hence we may form the pushout along ν to obtain

$$\begin{array}{ccccccccc}
0 & \longrightarrow & \Omega(A) & \longrightarrow & A \otimes_K A & \longrightarrow & A & \longrightarrow & 0 \\
 & & \downarrow \nu & & \downarrow & & \| & & \\
0 & \longrightarrow & V & \longrightarrow & X_\nu & \longrightarrow & A & \longrightarrow & 0
\end{array}$$

where the lower exact sequence ζ_ν is the element of $Ext^1_{A\otimes_K A^{op}}(A, V)$ corresponding to ν. The action of $z \in Z(A)$ acts on ζ_ν by the pullback

$$\begin{array}{ccccccccc}
0 & \longrightarrow & V & \longrightarrow & X_\nu^z & \longrightarrow & A & \longrightarrow & 0 \\
 & & \|\nu & & \downarrow & & \downarrow \cdot z & & \\
0 & \longrightarrow & V & \longrightarrow & X_\nu & \longrightarrow & A & \longrightarrow & 0
\end{array}$$

and likewise on ζ_{univ}. The diagram

$$\begin{array}{ccccccccc}
0 & \longrightarrow & \Omega(A) & \longrightarrow & A \otimes_K A & \longrightarrow & A & \longrightarrow & 0 \\
 & & \downarrow \nu & & \downarrow & & \| & & \\
0 & \longrightarrow & V & \longrightarrow & X_\nu & \longrightarrow & A & \longrightarrow & 0 \\
 & & \| & & \uparrow & & \uparrow \cdot z & & \\
0 & \longrightarrow & V & \longrightarrow & X_\nu^z & \longrightarrow & A & \longrightarrow & 0
\end{array}$$

is commutative with exact rows and

$$\begin{array}{ccccccccc}
0 & \longrightarrow & \Omega(A) & \longrightarrow & A \otimes_K A & \longrightarrow & A & \longrightarrow & 0 \\
 & & \| & & \uparrow & & \uparrow \cdot z & & \\
0 & \longrightarrow & \Omega(A) & \longrightarrow & Y^z & \longrightarrow & A & \longrightarrow & 0 \\
 & & \downarrow \nu & & \downarrow & & \| & & \\
0 & \longrightarrow & V & \longrightarrow & Y^z_\nu & \longrightarrow & A & \longrightarrow & 0.
\end{array}$$

However, the universal properties of the pullback and pushout give that the two operations pushout along ν and pullback along $\cdot z$ commute, so that the sequence

$$0 \longrightarrow V \longrightarrow Y^z_\nu \longrightarrow A \longrightarrow 0$$

is isomorphic to the sequence

$$0 \longrightarrow V \longrightarrow X^z_\nu \longrightarrow A \longrightarrow 0.$$

Hence if z annihilates the sequence ζ_{univ}, then

$$0 \longrightarrow \Omega(A) \longrightarrow Y^z \longrightarrow A \longrightarrow 0$$

splits and therefore so does its pushout

$$0 \longrightarrow V \longrightarrow Y^z_\nu \longrightarrow A \longrightarrow 0$$

along ν. But this is isomorphic to

$$0 \longrightarrow V \longrightarrow X^z_\nu \longrightarrow A \longrightarrow 0$$

which splits. This is the image of ζ_ν under the action of z and hence z annihilates ζ_ν. This gives the other inclusion and proves the statement. □

Example 5.9.8 This example arose in a discussion with Thorsten Holm. Let K be a field and let $A = K[X]/X^n$ for $n \geq 1$. Then

$$A \otimes_K A \simeq K[X, Y]/(X^n, Y^n)$$

and the multiplication map $A \otimes_K A \longrightarrow A$ corresponds to the map

$$K[X, Y]/(X^n, Y^n) \longrightarrow\!\!\!\!\!\rightarrow K[X, Y]/(X^n, Y^n, X - Y) \simeq K[Z]/Z^n.$$

The algebra A is commutative. Let $z \in A = Z(A)$. The pullback of ζ_{univ} along multiplication by z gives

$$Y^z = \{(p(X, Y), q(Z)) \in K[X, Y]/(X^n, Y^n) \times K[Z]/(Z^n) \mid p(Z, Z) = zq(Z)\}.$$

The sequence

$$0 \longrightarrow \Omega A \longrightarrow Y^z \longrightarrow A \longrightarrow 0$$

splits if and only if the epimorphism

$$Y^z \longrightarrow\!\!\!\!\!\rightarrow A$$

splits. Let $\sigma : A \longrightarrow Y^z$ be such a splitting morphism. Since A is $A \otimes_K A^{op}$-linear, σ is determined by $\sigma(1) = (p_1(X, Y), q_1(Z))$. But $q_1(Z) = 1$ by the fact that σ splits the projection onto the second component $Y^z \longrightarrow\!\!\!\!\!\rightarrow A$, and hence $\sigma(1) = (p_1(X, Y), 1)$ so that $z = p_1(Z, Z)$. Let $\sigma(1) = \sum_{i,j=0}^{n-1} a_{i,j} X^i Y^j$ for some $a_{i,j} \in K$. Then

$$\sigma(Z) = X \cdot \sigma(1) = \sigma(1) \cdot Y$$

implies that $a_{i,j} = 0$ for $i + j < n$. Hence there are parameters $\lambda_i \in K$ (for $i \in \{0, 1, \ldots, n-1\}$) such that

$$\sigma(1) = \sum_{i=0}^{n-1} \lambda_i X^i Y^{n-i-1} + \rho$$

for some $\rho \in \mathrm{rad}^n(K[X, Y]/(X^n, Y^n))$ and $z = (\sum_{i=0}^{n-1} \lambda_i) Z^{n-1}$. Again since $\sigma(Z^j) = X^\ell \cdot \sigma(1) \cdot Y^{n-1-\ell}$ for each $\ell \in \{0, \ldots, n-1\}$ we need to show that $\lambda_i =: \lambda$ is independent from i and hence obtain $z = n\lambda Z^{n-1}$. Any such element z gives an element in $H(A)$ and so

$$H(A) = \begin{cases} 0 & \text{if the characteristic of } K \text{ divides } n \\ \mathrm{soc}(A) & \text{if the characteristic of } K \text{ does not divide } n. \end{cases}$$

Proposition 5.9.9 [18, Proposition 2.3] *Let K be a commutative ring and let A be a K-algebra. Then $H(A) = Z^{pr}(A)$.*

Proof Put $A^e := A \otimes_K A^{op}$. A morphism $f \in End_{A^e}(A)$ factorises through a finitely generated projective module P if and only if it factorises through a free module $(A^e)^n$ so that

$$f = (f_1 \; \ldots \; f_n) \circ \begin{pmatrix} g_1 \\ \vdots \\ g_n \end{pmatrix}$$

for $f_i \in Hom_{A^e}(A^e, A)$ and $g_i \in Hom_{A^e}(A, A^e)$. But clearly the space generated by maps of this kind is the same as the space generated by maps factorising through A^e. Hence we may take $n = 1$ and obtain that

$$Z^{pr}(A) = \mathrm{im}\left(Hom_{A^e}(A^e, A) \otimes_K Hom_{A^e}(A, A^e) \xrightarrow{\psi} End_{A^e}(A)\right)$$

where we put

$$\begin{array}{rcl} Hom_{A^e}(A^e, A) \otimes_K Hom_{A^e}(A, A^e) & \xrightarrow{\psi} & End_{A^e}(A) \\ f \otimes g & \mapsto & f \circ g \end{array}$$

This map factors through the canonical map

$$Hom_{A^e}(A^e, A) \otimes_K Hom_{A^e}(A, A^e) \longrightarrow Hom_{A^e}(A^e, A) \otimes_{A^e} Hom_{A^e}(A, A^e)$$

and

$$End_{A^e}(A) \supseteq Hom_{A^e}(A^e, A) \otimes_{A^e} Hom_{A^e}(A, A^e) \simeq A \otimes_{A^e} Hom_{A^e}(A, A^e)$$

where the right-hand side is just the image $Z^{pr}(A)$ of ψ, and where the identification of the right-hand side within the left-hand side is given by $f \otimes g \mapsto f \circ g$. Precisely those A^e-linear endomorphisms of A which factor through the multiplication $A \otimes_K A \longrightarrow A$ give elements in $H(A)$. The multiplication map $A \otimes_K A \longrightarrow A$ is surjective, and A^e is A^e-free, which implies that every A^e-linear morphism to A factors through the multiplication map. Hence, precisely those endomorphisms of A which belong to $Hom_{A^e}(A^e, A) \otimes_{A^e} Hom_{A^e}(A, A^e)$ give elements in $H(A)$ and therefore $H(A) = Z^{pr}(A)$. This proves the statement. □

Particularly easy is the case of self-injective algebras. First we shall give a definition of the Higman ideal which can be found, for example, in a paper by Héthelyi et al. [19].

Proposition 5.9.10 [18] *Let K be a field and let A be a Frobenius K-algebra with associated bilinear form $\langle\ ,\ \rangle : A \times A \longrightarrow K$. Let B be a K-basis of A and let $\hat{B} = \{\hat{a} \mid a \in B\}$ be its dual basis with respect to $\langle\ ,\ \rangle$, in the sense that*

$$\langle a', \hat{a}\rangle = \begin{cases} 1 & if a = a' \in B \\ 0 & if a \neq a' \quad but\, a, a' \in B. \end{cases}$$

Then $H(A) = \mathrm{im}(\tau)$ for $\tau : A \longrightarrow A$ given by $\tau(x) := \sum_{a\in B} \hat{a}xa$, where we identify $Z(A) = End_{A\otimes_K A^{op}}(A)$ by mapping $h \in End_{A\otimes_K A^{op}}(A)$ to $h(1)$.

Proof [18] We define $f : A \longrightarrow A \otimes_K A$ by $f(x) := \sum_{a\in B} x\hat{a} \otimes a$. We observe that f is actually $A \otimes_K A^{op}$-linear. Indeed, if

$$x\hat{a} = \sum_{b\in B} \lambda_{a,b}\hat{b}$$

for some $\lambda_{a,b} \in K$ and

$$ax = \sum_{b\in B} \mu_{a,b} b$$

then

$$\mu_{a,b} = \sum_{c \in B} \mu_{a,c} \langle c, \hat{b} \rangle = \langle \sum_{c \in B} \mu_{a,c} c, \hat{b} \rangle = \langle ax, \hat{b} \rangle$$
$$= \langle a, x\hat{b} \rangle = \langle a, \sum_{c \in B} \lambda_{b,c} \hat{c} \rangle = \sum_{c \in B} \lambda_{b,c} \langle a, \hat{c} \rangle = \lambda_{b,a}$$

and so

$$ax = \sum_{b \in B} \lambda_{b,a} b \text{ if and only if } x\hat{a} = \sum_{b \in B} \lambda_{a,b} \hat{b}.$$

Hence

$$x \cdot f(1) = x \left(\sum_{a \in B} \hat{a} \otimes a \right) = \sum_{a \in B} \sum_{b \in B} \lambda_{a,b} \hat{b} \otimes a = \sum_{a \in B} \sum_{b \in B} \lambda_{b,a} \hat{a} \otimes b$$
$$= \sum_{a \in B} \hat{a} \otimes \left(\sum_{b \in B} \lambda_{b,a} b \right) = \sum_{a \in B} \hat{a} \otimes ax = f(1) \cdot x.$$

Now, B is a K-basis of A and hence $\{a \otimes b |\ a, b \in B\}$ is a K-basis of $A \otimes_K A$. Therefore

$$0 = f(x) = \sum_{a \in B} \sum_{b \in B} \lambda_{b,a} \hat{a} \otimes b$$

implies $\lambda_{a,b} = 0$ for all $a, b \in B$, which implies $x = 0$. This shows that f is injective. Since A is self-injective, every $A \otimes_K A^{op}$-linear endomorphism of A which factors through a projective, whence injective bimodule, actually factors through the injective mapping f. Thus let $h : A \longrightarrow A$ be an element in $Z^{pr}(A)$. Then $h = g_x \circ f$ for some $A \otimes_K A^{op}$-linear mapping $g_x : A \otimes_K A \longrightarrow A$. Hence $g_x(1 \otimes 1) = x$ and $g_x(u \otimes v) = uxv$. Therefore

$$h(1) = (g_x \circ f)(1) = g_x(\sum_{a \in A} \hat{a} \otimes a) = \sum_{a \in A} \hat{a}xa = \tau(x)$$

and therefore $H(A) = \text{im}(\tau)$, where we identify $End_{A \otimes_K A^{op}}(A)$ with $Z(A)$ by means of $h \mapsto h(1)$. This proves the statement. □

Corollary 5.9.11 *The image of τ is independent of the choice of the basis B.*

Proof Indeed, the Definition 5.9.6 of $H(A)$ is independent of any choice of a basis. □

The next lemma is formulated only for symmetric algebras since then left and right orthogonality with respect to the bilinear form $\langle\ ,\ \rangle$ is the same. With a little more care it is possible to obtain the result for Frobenius algebras as well. For details see [18].

Proposition 5.9.12 (Héthelyi et al. [19]) *Let K be a field and let A be a finite dimensional symmetric K-algebra. Then $H(A) \subseteq \mathrm{soc}(A) \cap Z(A) = R(A)$ where $R(A)$ is the Reynolds ideal.*

Proof By definition $H(A)$ is an ideal of $Z(A)$. Hence we only need to show that $H(A) \subseteq \mathrm{soc}(A)$. By Proposition 5.9.10 it is sufficient to show that $\mathrm{im}(\tau) \subseteq \mathrm{soc}(A)$. Recall that τ is defined by the choice of a K-basis of A. Since

$$0 \to \mathrm{rad}^i(A)/\mathrm{rad}^{i+1}(A) \to \mathrm{rad}^{i-1}(A)/\mathrm{rad}^{i+1}(A) \to \mathrm{rad}^{i-1}(A)/\mathrm{rad}^i(A) \to 0$$

is exact for every $i \in \mathbb{N}$ and since the sequences split as K-vector spaces, we choose a basis $B = \bigcup_{i\in\mathbb{N}} B_i$ of

$$\bigoplus \mathrm{rad}^i(A)/\mathrm{rad}^{i+1}(A)$$

by taking B_i to be a basis of $\mathrm{rad}^{i-1}(A)/\mathrm{rad}^i(A)$.

Recall from Lemma 2.9.10 that if B_1 is a basis of $A/\mathrm{rad}(A)$, then $\{\hat{b} \mid b \in B_1\}$ is a basis of $B_1^\perp$, which is the socle of A. If $x \in A$ and $y \in \mathrm{rad}(A)$, we get for all $a \in B_1$

$$\hat{a}xay \in \mathrm{rad}(A)^\perp \cdot A \cdot A \cdot \mathrm{rad}(A) = \mathrm{rad}(A)^\perp \cdot \mathrm{rad}(A) = 0$$

since for all $u \in \mathrm{rad}(A)^\perp$ and $v \in \mathrm{rad}(A)$, we get $su \in \mathrm{rad}(A)^\perp = \mathrm{soc}(A)$ for all $s \in A$. Hence

$$\langle s, u \cdot v\rangle = \langle s \cdot u, v\rangle = 0$$

and therefore $\tau(x)y = 0$ for all $x \in A$ and all $y \in \mathrm{rad}(A)$. But this shows $\tau(x) \in Ann_A(\mathrm{rad}(A)) = \mathrm{soc}(A)$. We have proved the statement. □

Remark 5.9.13 We should emphasise here that $\mathrm{soc}(A) \cap Z(A) \neq \mathrm{soc}(Z(A))$ in general.

Let A be a finite dimensional K-algebra and let $e_1, \ldots, e_n$ with $1 = \sum_{i=1}^n e_i$ be a complete set of primitive idempotents of A. Fix a basis $B_{i,j}$ of e_iAe_j and suppose $e_i \in B_{i,i}$. Then

$$B := \bigcup_{i,j=1}^n B_{i,j}$$

is a K-basis of A.

Lemma 5.9.14 (Héthelyi et al. [19]) *Let K be a field and let A be a finite dimensional basic symmetric K-algebra. Let B be the basis of A defined above and let $r_1, \ldots, r_n$ be the dual basis elements to $e_1, \ldots, e_n$. If we compute τ with respect to this basis, we get*

$$\tau(e_i) = \sum_{j=1}^n \dim_K(e_iAe_j)r_j.$$

Proof Let $\{\hat{b} \mid b \in B\} =: \hat{B}$ be the dual basis of B with respect to the symmetrising form $\langle\ ,\ \rangle$. By definition

$$\hat{B}_{i,j} := \{\hat{b} \mid b \in B_{i,j}\}$$

is a K-basis of

$$Hom_K((e_iAe_j), K) = e_jHom_K(A, K)e_i = e_jAe_i.$$

Hence

$$\tau(e_i)e_j = e_j\tau(e_i)e_j = \sum_{b\in B} e_j\hat{b}e_ibe_j = \sum_{b\in B_{i,j}} e_j\hat{b}e_ibe_j = \sum_{b\in B_{i,j}} \hat{b}b$$

and

$$\begin{aligned}\langle\tau(e_i),\ e_j\rangle &= \langle\tau(e_i)e_j,\ 1\rangle \\ &= \langle \sum_{b\in B_{i,j}} \hat{b}b,\ 1\rangle = \sum_{b\in B_{i,j}} \langle\hat{b}b,\ 1\rangle = \sum_{b\in B_{i,j}} \langle\hat{b},\ b\rangle \\ &= |B_{i,j}| = \dim_K(e_iAe_j) \in K.\end{aligned}$$

Now, since for all $i \in \{1, \ldots, n\}$ we have $\tau(e_i) \in \mathrm{soc}(A) \cap Z(A)$ we get

$$\tau(e_i) = \sum_{j=1}^{n}\langle\tau(e_i),\ e_j\rangle r_j = \sum_{j=1}^{n}\dim_K(e_iAe_j)r_j \in A.$$

This proves the statement. □

If K is algebraically closed and A is symmetric then the Cartan matrix determines the dimension of $H(A) = Z^{pr}(A)$. Altogether we obtain a relatively clear picture of what $Z^{pr}(A)$ might be inside $Z(A)$.

Corollary 5.9.15 *Let K be a splitting field for the symmetric finite dimensional K-algebra A. Let C_A be the Cartan matrix of A and let $C_A \otimes_{\mathbb{Z}} K$ be the Cartan matrix of A, and where we consider the coefficients of C_A as elements of K, by means of the ring homomorphism $\mathbb{Z} \longrightarrow K$. Then*

$$\dim_K(Z^{pr}(A)) = \mathrm{rank}(C_A \otimes_{\mathbb{Z}} K).$$

Proof Indeed, $\dim_K(e_iAe_j) = \dim_K(Hom_A(Ae_i, Ae_j))$ are the coefficients of the Cartan matrix, using the isomorphism classes of the projective indecomposable A-modules as basis of $K_0(A - proj)$ and the isomorphism classes of the simple A-modules as basis of $G_0(A)$.

Proposition 5.9.10 shows that $H(A)$ equals $Z^{pr}(A)$. Proposition 5.9.12 shows that the image of τ is in the socle of A. Let $E := \{e_1, \ldots, e_n\}$ be a complete set of idempotents of A with $\sum_{i_1}^{n} e_i = 1$. Then $\mathrm{soc}(A)$ has a basis given by $\hat{E} := \{r_i \mid i \in \{1, \ldots, n\}\}$. Lemma 5.9.14 shows that the coefficients of $\tau(e_i)$ in terms of the basis R are exactly the Cartan matrix coefficients. Since E is a basis of $A/\mathrm{rad}(A)$ and since $\tau(\mathrm{rad}(A)) = 0$ by Proposition 5.9.12 and its proof we showed the statement. □

5.9.2 Stable Cocentre

The stable centre $Z^{st}(A) = (HH^0)^{st}(A)$ is a quotient of the centre $Z(A) = HH^0(A)$ of an algebra. Let A be a symmetric K-algebra. Then

$$\begin{aligned} Hom_K(HH_n(A), K) &= Hom_K(H_n(\mathbb{B}A \otimes_{A\otimes_K A^{op}} A), K) \\ &= H^n(Hom_K(\mathbb{B}A \otimes_{A\otimes_K A^{op}} A, K)) \\ &\simeq H^n(Hom_{A\otimes_K A^{op}}(\mathbb{B}A, Hom_K(A, K))) \\ &\simeq H^n(Hom_{A\otimes_K A^{op}}(\mathbb{B}A, A)) \text{ since } A \text{ is symmetric} \\ &\simeq Ext^n_{A\otimes_K A^{op}}(A, A) \\ &= HH^n(A). \end{aligned}$$

Hence the dual of the Hochschild homology of a symmetric algebra is isomorphic to the Hochschild cohomology of this algebra. Since by dualising monomorphisms become epimorphisms and epimorphisms become monomorphisms, it is therefore natural to consider the stable Hochschild homology as sub-object of the usual Hochschild homology.

We use a special case of a variant of Bouc's trace from Definition 5.8.6.

Let A and B be K-algebras and let P be an $A \otimes_K B^{op}$-module. As usual we denote by $[B, B]$ the K-submodule of B generated by all elements $b_1b_2 - b_2b_1$ where $b_1, b_2 \in B$. Suppose P is projective as a B^{op}-module. Then Bouc's trace gives a mapping

$$tr_P : End_B(P) = P \otimes_B Hom_B(P, B) \longrightarrow B \longrightarrow B/[B, B].$$

Since P is an A-module, by the very definition of a module we obtain a ring homomorphism

$$A \longrightarrow End_B(P)$$

and compose it with tr_P to give a mapping

$$A \longrightarrow End_B(P) \longrightarrow B/[B, B]$$

which we call tr_P as well, in order to simplify the notation. For two square matrices X and Y we get $trace(XY) = trace(YX)$ and therefore

$$tr_P(ab - ba) = 0 \;\; \forall a, b \in A.$$

Hence $tr_P : A \longrightarrow B/[B, B]$ factors through $A \longrightarrow A/[A, A]$ and we get a map

$$tr_P : A/[A, A] \longrightarrow B/[B, B].$$

In the sequel we will always refer to this map as the Hattori-Stallings trace map.

Definition 5.9.16 Let K be a commutative ring, let A and B be K-algebras and let P be a $B \otimes_K A^{op}$-module. Suppose P is finitely generated projective as a B-module. Then the map

$$tr_P : A/[A, A] \longrightarrow B/[B, B]$$

is the *Hattori-Stallings trace.*

Remark 5.9.17 The above consideration and Remark 5.8.15 show that Bouc's trace from Definition 5.8.6 is a generalisation of the Hattori-Stallings trace in the sense that the Hattori-Stallings trace corresponds to the mapping of $A \otimes_K A^{op}$-modules $A \oplus P \longrightarrow A$ and the map induced by this on the degree zero Hochschild homology.

Lemma 5.9.18 *Let K be a field, let A, B and C be K-algebras and let U be an $A \otimes_K B^{op}$-module and let V be a $B \otimes_K C^{op}$-module. Suppose that U is projective as an A-module and as a B^{op}-module. Moreover, suppose that V is projective as a B-module and as a C^{op}-module. Then we obtain for the Hattori-Stallings trace $tr_{U \otimes_B V} = tr_V \circ tr_U$.*

Proof This is an immediate consequence of Remark 5.9.17 and Proposition 5.8.10. □

Observe that if $B = K$ is a field, then any right A-module U is a K–A-bimodule, projective as a K-module. Hence

$$tr_U : A/[A, A] \longrightarrow K$$

is just the map sending $a \in A$ first to the endomorphism $a\cdot : U \longrightarrow U$ and then to the trace of this endomorphism.

Definition 5.9.19 [18] Let K be a field and let A be a finite dimensional K-algebra. Then

$$HH_0^{st}(A) := \bigcap_{P \text{ f.g. projective } A\text{-module}} \ker(tr_P).$$

Observe that the trace is additive in the sense that

$$tr_{P_1 \oplus P_2} = tr_{P_1} + tr_{P_2}$$

and so

$$\bigcap_{P \text{ f.g. projective } A\text{-module}} \ker(tr_P) = \bigcap_{\substack{P \text{ finitely generated} \\ \text{projective indecomposable} \\ A\text{-module}}} \ker(tr_P).$$

In other words we may take the intersection of only indecomposable finitely generated projective A-modules in the definition of $HH_0^{st}(A)$.

Theorem 5.9.20 [18] *Let K be an algebraically closed field, let A and B be finite dimensional K-algebras and suppose that K is a splitting field for A and for B. Suppose that there is a stable equivalence of Morita type induced by the bimodules (M, N), so that M is an $A \otimes_K B^{op}$-module and N is a $B \otimes_K A^{op}$-module. Then $tr_M : HH_0^{st}(A) \longrightarrow HH_0^{st}(B)$ is an isomorphism.*

Proof We use the Hattori-Stallings trace. We have already shown that we obtain a morphism $tr_M : HH_0(A) \longrightarrow HH_0(B)$. We need to show that this passes to a morphism $HH_0^{st}(A) \longrightarrow HH_0^{st}(B)$.

For this, let P be a projective left B-module. Then $M \otimes_B P$ is a projective left A-module. Hence if $x \in HH^{st}(A)$, then $tr_{M\otimes_B P}(x) = 0$. Observe that

$$tr_M : A/[A, A] \longrightarrow B/[B, B]$$

and

$$tr_P : B/[B, B] \longrightarrow K.$$

We have by definition that

$$tr_M(x) \in HH^{st}(A) \Leftrightarrow tr_P(tr_M(x)) = 0$$

for each projective B-module P. But,

$$tr_{M\otimes_B P} = tr_P \circ tr_M$$

by Proposition 5.8.10 and since $x \in HH_0^{st}(A)$, we get that $tr_Q(x) = 0$ for each projective A-module Q. A special case is $Q = M \otimes_B P$ and we obtain that tr_M restricts to a morphism

$$tr_M : HH_0^{st}(A) \longrightarrow HH_0^{st}(B).$$

Similarly we obtain a morphism

$$tr_N : HH_0^{st}(B) \longrightarrow HH_0^{st}(A).$$

Let $N \otimes_A M \simeq B \oplus X$ and $M \otimes_B N \simeq A \oplus Y$ for projective bimodules X and Y. Now,

$$tr_M \circ tr_N = tr_{N \otimes_A M} = tr_{B \oplus X} = tr_B + tr_X = id + tr_X$$

and likewise

$$tr_N \circ tr_M = tr_{M \otimes_B N} = tr_{A \oplus Y} = tr_A + tr_Y = id + tr_Y.$$

We need to show that $tr_X = 0$ and $tr_Y = 0$.

Since K is algebraically closed, the projective indecomposable $A \otimes_K A^{op}$-modules are direct factors of modules $P \otimes_K Hom_K(P', K)$ where P and P' are projective indecomposable A-modules (cf Lemma 5.3.8). Indeed, we can first assume that A is basic. Then Lemma 5.3.8 shows that $A/\text{rad}(A) \otimes_K A^{op}/\text{rad}(A^{op})$ is semisimple and therefore all projective indecomposable modules are direct summands of tensor products of projective indecomposable modules of each factor. Now, if $x \in HH_0^{st}(A)$, since $HH_0^{st}(A) \subseteq \ker(tr_P)$, we get

$$\begin{aligned} tr_{P \otimes_K Hom_K(P',K)}(x) &= (tr_{Hom_K(P',K)} \circ tr_P)(x) \\ &= tr_{Hom_K(P',K)}(tr_P(x)) \\ &= tr_{Hom_K(P',K)}(0) = 0. \end{aligned}$$

The arguments for tr_X are analogous.

This proves the proposition. □

5.9.3 Külshammer Ideals and Stable Equivalences of Morita Type

We will use our knowledge of the stable centre and the stable cocentre to get a stable version of Külshammer ideals and Külshammer spaces as introduced in Definitions 2.9.4 and 2.9.12.

The usual Külshammer spaces are defined as ideals or subspaces of the degree zero Hochschild homology and cohomology. We shall imitate the classical definition inside the degree zero stable Hochschild homology and stable Hochschild cohomology.

Proposition 5.9.21 [18, 20] *Külshammer ideals are stably invariant. Let K be an algebraically closed field of characteristic $p > 0$, let A and B be finite dimensional K-algebras and suppose that*

$$F : A\text{-}\underline{mod} \longrightarrow B\text{-}\underline{mod}$$

is a stable equivalence of Morita type. Then the isomorphism

$$HH_0^{st}(A) \longrightarrow HH_0^{st}(B)$$

induced by F maps $T_n(A)/[A, A]$ to $T_n(B)/[B, B]$ for all n.

Proof Let ${}_BM_A$ be a bimodule such that $F = M \otimes_A -$ is the stable equivalence of Morita type.

We first prove the statement on the quotients $T_n(A)/[A, A]$. Let

$$\begin{array}{rcl} A/[A,A] & \stackrel{\mu_A^p}{\longrightarrow} & A/[A,A] \\ a+[A,A] & \mapsto & a^p+[A,A] \end{array}$$

be the p-power mapping. We need to show

$$tr_M \circ \mu_A^p = \mu_B^p \circ tr_M.$$

But by definition the natural morphisms

$$A \longrightarrow End_B(M) \longrightarrow B/[B,B]$$

factor through the commutator space. If $M = \bigoplus_{i=1}^n M_i$ as a B-module for indecomposable M_i, then $End_B(M)$ is isomorphic to

$$\begin{pmatrix} End_B(M_1) & Hom_B(M_2,M_1)\dots & & \dots & Hom_B(M_n,M_1) \\ Hom_B(M_1,M_2) & \ddots & \ddots & & \vdots \\ \vdots & \ddots & \ddots & \ddots & \vdots \\ \vdots & & \ddots & \ddots & Hom_B(M_n,M_{n-1}) \\ Hom_B(M_1,M_n) & \dots & \dots Hom_B(M_{n-1},M_n) & & End_B(M_n) \end{pmatrix}.$$

Since M is projective as a B-module, there are primitive idempotents e_i of B such that $M_i \simeq Be_i$. Under this isomorphism the image of $a \in A$ is a matrix $(a_{i,j})_{1\leq i,j\leq n}$ where $a_{i,j} : M_i \longrightarrow M_j$, whence $a_{i,j} \in e_jBe_i$. The trace then takes this matrix to $\sum_{i=1}^n a_{i,i} = tr_M(a)$. But now

$$tr_M(a)^p = tr(((a_{i,j})_{1\leq i,j\leq n})^p).$$

$\left(((a_{i,j})_{1\leq i,j\leq n})^p\right)$ has entries a sum of elements

$$a_{i_1,i_2}a_{i_2,i_3}\dots a_{i_{p-1},i_p}$$

and for the trace only those elements with $i_1 = i_p$ contribute. The cyclic group of order p acts on the set of all the terms occurring by cyclic permutation of the factors, giving orbits of length p, except for the fixed points $a_{i,i}^p$, which occur only once for each i. Since we consider the result only up to commutators, and since

$$a_{i_1,i_2}a_{i_2,i_3}\dots a_{i_{p-1},i_1} - a_{i_2,i_3}\dots a_{i_{p-1},i_p}a_{i_p,i_2}$$

is in the commutator, the terms corresponding to orbits of length p occur in multiples of p identical factors in $B/[B, B]$. Hence they disappear, the field being of characteristic $p > 0$. Therefore

$$tr_M(a^p + [A, A]) = (tr_M(a + [A, A]))^p .$$

This shows $tr_M \circ \mu_A^p = \mu_B^p \circ tr_M$ as required.

Moreover, for every idempotent e of A we have

$$tr_{Ae}(x^{p^n}) = (tr_{Ae}(x))^{p^n}$$

and so $T_n(A)/[A, A] \subseteq HH_0^{st}(A)$ for all $n \in \mathbb{N}$.

Therefore tr_M induces a morphism

$$T_n(A)/[A, A] \longrightarrow T_n(B)/[B, B]$$

on the kernels $T_n(A)/[A, A]$ and $T_n(B)/[B, B]$ of μ_A^p and μ_B^p. □

If A and B are stably equivalent symmetric K-algebras, then $Z^{st}(A)$ is the dual of $HH_0^{st}(A)$.

Lemma 5.9.22 [18, Proposition 1.3] *Let A be a symmetric K-algebra and let $\langle\ ,\ \rangle : A \times A \longrightarrow K$ be a symmetrising form. Then this symmetrising form induces a non-degenerate bilinear form*

$$\langle\ ,\ \rangle : Z^{st}(A) \times HH_0^{st}(A) \longrightarrow K.$$

Proof We have already seen in Lemma 2.9.10 that the symmetrising form induces a non-degenerate bilinear form

$$Z(A) \times A/[A, A] \longrightarrow K.$$

We shall show that

$$Z^{pr}(A)^{\perp} = HH_0^{st}(A)$$

which proves the result. We first remark that Proposition 5.9.10 shows that the Higman ideal $H(A)$ satisfies $H(A) = \mathrm{im}(\tau)$ for some form τ. Moreover, $Z^{pr}(A) = H(A)$ by Proposition 5.9.9. Recall that

$$\tau(x) = \sum_{i=1}^{n} b_i x a_i$$

where $x \in A$ and $\{b_1, \ldots, b_n\}$ as well as $\{a_1, \ldots, a_n\}$ are K-bases of A satisfying $\langle a_i, b_j\rangle = \delta_{i,j}$ for all i, j. Here, $\delta_{i,j}$ denotes the Kronecker delta, being 1 if $i = j$ and 0 otherwise.

Let $e^2 = e \in A$ be an idempotent, and choose $\{a_1, \ldots, a_n\}$ so that $\{a_1, \ldots, a_m\}$ is a basis of Ae, and so that $\{a_{m+1}, \ldots, a_n\}$ is a basis of $A(1-e)$. Then for $a_i^* := \langle -, a_i \rangle \in Hom_K(A, K)$ we get

$$tr_{Ae}(x) = \sum_{i=1}^{m} a_i^*(xa_i)$$

and

$$\begin{aligned}\langle x, \tau(e)\rangle &= \langle x, \sum_{i=1}^{n} b_i e a_i\rangle = \langle \sum_{i=1}^{n} b_i e a_i, x\rangle = \langle \sum_{i=1}^{n} b_i, e a_i x\rangle \\ &= \langle \sum_{i=1}^{m} b_i, a_i x\rangle = \sum_{i=1}^{m} \langle b_i, a_i x\rangle = \sum_{i=1}^{m} a_i^*(xa_i) = tr_{Ae}(x).\end{aligned}$$

But this shows for the induced form

$$\langle\ ,\ \rangle : Z(A) \times A/[A, A] \longrightarrow K$$

that

$$x + [A, A] \in HH_0^{st}(A) \Leftrightarrow x \in \mathrm{im}(\tau)^{\perp} = H(A)^{\perp} = (Z^{pr})^{\perp}.$$

This proves the lemma. □

Remark 5.9.23 The lemma can be modified so that it is true for self-injective algebras as well. Several modifications have to be made, and the arguments are more intricate. We shall not need this generalisation in the sequel, but the interested reader might like to see the arguments in [18].

Corollary 5.9.24 *Let K be an algebraically closed field and let A and B be two symmetric K-algebras. Given a stable equivalence of Morita type (M, N) between the K-algebras A and B we obtain an isomorphism*

$$tr_M^* : Z^{st}(A) \longrightarrow Z^{st}(B)$$

as K-vector spaces given by the transpose of

$$tr_M : HH_0^{st}(A) \longrightarrow HH_0^{st}(B)$$

with respect to the symmetrising form on A and the symmetrising form induced by M on B (cf Remark 5.5.5). In other words, if

$$\langle\ ,\ \rangle_A : A \times A \longrightarrow K$$

is a non-degenerate symmetric, associative bilinear form on A, then M induces a non-degenerate symmetric, associative bilinear form

$$\langle\ ,\ \rangle_{B,M} : B \times B \longrightarrow K$$

on B and

$$\langle z, tr_M(m)\rangle_{B,M} = \langle tr_M^*(z), m\rangle_A$$

for all $z \in Z^{st}(B)$ and $m \in HH_0^{st}(A)$.

Proof Since the form

$$\langle\ ,\ \rangle_A : Z^{st}(A) \times HH_0^{st}(A) \longrightarrow K$$

is non-degenerate by Lemma 5.9.22, linear algebra shows that the K-linear map

$$tr_M : HH_0^{st}(A) \longrightarrow HH_0^{st}(B)$$

induces a transpose map

$$tr_M^* : Z^{st}(B) \longrightarrow Z^{st}(A)$$

so that

$$\langle z, tr_M(m)\rangle_B = \langle tr_M^*(z), m\rangle_A$$

for all $z \in Z^{st}(B)$ and $m \in HH_0^{st}(A)$. Moreover, tr_M is an isomorphism if and only if tr_M^* is an isomorphism. □

Remark 5.9.25 König et al. show in [20] that tr_M^* is a ring homomorphism and that tr_M^* maps the ideal $T_n^{\perp}(A)/Z^{pr}(A)$ of $Z^{st}(A)$ to the ideal $T_n^{\perp}(B)/Z^{pr}(B)$ of $Z^{st}(B)$. This fact can be used to distinguish in a rather sophisticated way similar algebras up to stable equivalence of Morita type. The paper [21] gives such an example.

They show further that the so-called Gerstenhaber structure on the Hochschild cohomology is preserved under tr_M^*. The Gerstenhaber structure is a Lie algebra structure on the Hochschild cohomology ring. We will come back to this in Remarks 6.4.5 and 6.12.33.

5.10 Brauer Tree Algebras and the Structure of Blocks with Cyclic Defect Groups

Proposition 2.12.9 shows that the blocks of finite representation type are precisely the blocks with cyclic defect group. In Proposition 2.12.4 we described the precise structure of blocks in the case where the cyclic defect group is normal. The main purpose of this section is to prove Theorem 5.10.37 at the end of this section, where we shall give the precise structure in the case where the defect group is cyclic but not necessarily normal. Stable equivalences are one of the main ingredients of the proof.

5.10.1 Brauer Tree Algebras

We first need to introduce a certain class of algebras, defined by certain combinatorial data, which will occur in the structure theorem for blocks with cyclic defect groups.

Throughout this subsection let K be a field.

Recall from Definition 1.11.5 the definition of a quiver. A quiver is formed by a set of edges Γ_a, a set of vertices Γ_v and two maps $s : \Gamma_a \longrightarrow \Gamma_v$ and $t : \Gamma_a \longrightarrow \Gamma_v$, attaching to each arrow α a tail $s(\alpha)$ and a head $t(\alpha)$.

A finite graph Λ is obtained from a quiver $\Gamma = (\Gamma_v, \Gamma_a)$ by roughly speaking "forgetting the orientation". More precisely,

Definition 5.10.1 We complete our vocabulary on graphs and quivers.

- Given a quiver Γ, the *underlying finite graph* Λ is given by a vertex set $\Lambda_v = \Gamma_v$ and a set of edges $\Lambda_e = \Gamma_a$. Further we get a mapping $\epsilon_\Gamma : \Lambda_e \longrightarrow \mathcal{P}(\Lambda_v)$ from the set of edges to one- or two-element subsets of Λ_v. This is given by $\epsilon_\Lambda(e) = \{t(e), s(e)\}$ the set formed by the head and the tail of the arrow e.
  We say that the vertex v is *adjacent* to the edge e if $v \in \epsilon_\Lambda(e)$.
- A finite quiver is a *tree* if it does not allow a closed walk.

The basic combinatorial object we are going to deal with is a finite tree Λ on which we associate to every vertex $x \in \Lambda_v$ a cyclic ordering of the edges adjacent to x. By this we mean that we impose a numbering by integers of each of the sets

$$\Lambda_v(x) := \{e \in \Lambda_e \mid x \in \epsilon_\Lambda(e)\}.$$

In particular, for each vertex v and an edge e adjacent to v, we know what is "the next edge", i.e. the edge following e in the numbering, or if e carries the highest number of all edges adjacent to v, then the "next edge" is the edge with the first number. Call $\sigma_v = \sigma$ the cyclic permutation on the set of edges adjacent to v associating the "next edge" to any edge in the above ordering.

Further, we may designate an "exceptional vertex", which is just a vertex $v_0 \in \Lambda_v$, and an integer $\mu \geq 2$, called the *exceptional multiplicity.*

Definition 5.10.2 A *Brauer tree* is a finite tree Λ with a cyclic ordering on the edges at each vertex. A Brauer tree with an *exceptional vertex* is a Brauer tree together with an *exceptional vertex* and *exceptional multiplicity* μ. A Brauer tree without an exceptional vertex can be considered as a Brauer tree with an exceptional vertex and *exceptional multiplicity* 1, where the exceptional vertex is then chosen arbitrarily.

This somewhat cumbersome definition is made for historical reasons. One could define all data by quivers and relations. We shall do this, but we shall see that the language of Brauer trees has some practical advantages, and in particular the quasi-totality of the existing literature uses the language of Brauer trees.

Since the graph underlying a Brauer tree is a tree, the cyclic ordering of the edges adjacent to a vertex allows us to physically draw the tree in $\mathbb{R}^2$ by means of segments of lines for edges and small circles $\circ$ for vertices, understanding that the cyclic ordering is presented by drawing the edges in counterclockwise order, and using a black circle $\bullet$ for the exceptional vertex.

Example 5.10.3 To illustrate the procedure, consider the following example.

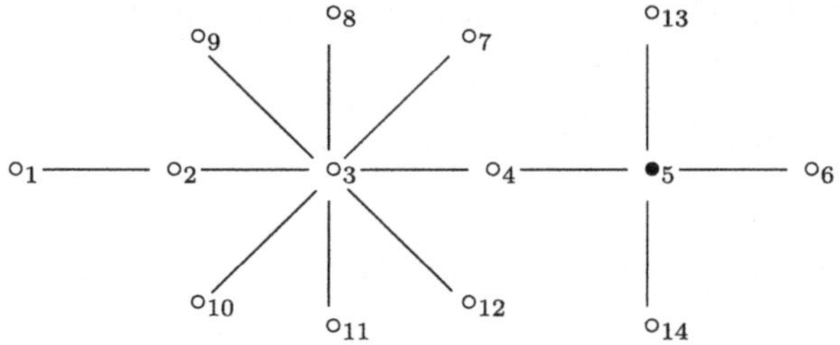

This means that the ordering at the vertices is as follows:

Vertex number	Ordering of the edges
1	$(1,2) \to (1,2)$
2	$(2,1) \to (2,3) \to (2,1)$
3	$(3,7) \to (3,8) \to (3,9) \to (3,2) \to (3,10) \to$ $\to (3,11) \to (3,12) \to (3,4) \to (3,7)$
4	$(4,5) \to (4,3) \to (4,5)$
5	$(5,4) \to (5,14) \to (5,6) \to (5,13) \to (5,4)$
6	$(6,5) \to (6,5)$
7	$(7,3) \to (7,3)$
8	$(8,3) \to (8,3)$
9	$(9,3) \to (9,3)$
10	$(10,3) \to (10,3)$
11	$(11,3) \to (11,3)$
12	$(12,3) \to (12,3)$
13	$(13,5) \to (13,5)$
14	$(14,5) \to (14,5)$

Here, we denote the edge linking the vertex i and the vertex j by the symbol $(i,\ j)$. The exceptional vertex carries the number 5.

Definition 5.10.4 Given a Brauer tree Λ, we say that a K-algebra A is a *Brauer tree algebra* if

- the projective indecomposable A-modules are indexed by the edges of Λ,

- the top $P/\mathrm{rad}(P)$ of the indecomposable A-module P is isomorphic to the socle of P,
- for each projective indecomposable A-module P_e we get $\mathrm{rad}(P_e)/\mathrm{soc}(P_e) \simeq U_{v_1}(e) \oplus U_{v_2}(e)$ for two (possibly zero) uniserial A-modules $U_{v_1}(e)$ and $U_{v_2}(e)$, where $\epsilon_\Lambda(e) = \{v_1, v_2\}$ are the vertices adjacent to e,
- if v is not the exceptional vertex and if v is adjacent to e then the composition length of $U_v(e)$ is $s(v) - 1$ where $s(v)$ is the number of edges adjacent to v, (observe that this implies $U_v(e) = 0$ if there is a unique edge adjacent to v if v is not exceptional),
- if v is the exceptional vertex with multiplicity μ and if v is adjacent to e, then the composition length of $U_v(e)$ is $\mu \cdot s(v) - 1$ where $s(v)$ is the number of edges adjacent to v,
- if v is adjacent to e then the composition series of $U_v(e)$ is given by the condition that
$$\mathrm{rad}^{i-1}(U_v(e))/\mathrm{rad}^i(U_v(e)) \simeq P_{\sigma_v^i(e)}/\mathrm{rad}(P_{\sigma_v^i(e)})$$
for all i as long as i is smaller than the composition length of $U_v(e)$.

The above combinatorics seem to be quite complicated. Actually, they are not really, and the graphical presentation of the Brauer tree gives a very convenient and practical way to get the composition series of the projective indecomposable A-modules. We shall illustrate the combinatorics with the following example.

Example 5.10.5 Recall the Brauer tree from Example 5.10.3. There are 13 projective indecomposable modules. The modules $U_1, U_6, U_7, U_8, U_9, U_{10}, U_{11}, U_{12}, U_{13}$ and U_{14} are all 0, independently from which adjacent edge one considers them.

Hence, $P_{(1,2)}, P_{(5,6)}, P_{(3,7)}, P_{(3,8)}, P_{(3,9)}, P_{(3,10)}, P_{(3,11)}, P_{(3,12)}, P_{(5,13)}$ and $P_{(5,14)}$ are uniserial. The Loewy length of $P_{(5,6)}$, of $P_{(5,13)}$ and $P_{(5,14)}$ is $4 \cdot \mu + 1$, whereas the composition length of $P_{(1,2)}$ is 3, and the composition length of $P_{(3,7)}$, $P_{(3,8)}, P_{(3,9)}, P_{(3,10)}, P_{(3,11)}$ and $P_{(3,12)}$ is 9. In the following we read a composition series following the radical quotients from top to bottom and a simple module associated to the edge $(i\ j)$ is abbreviated by the symbol $(i\ j)$. For $\mu = 3$ the composition series are as follows:

$$P_{(3,7)} = \begin{pmatrix} (3,7) \\ (3,8) \\ (3,9) \\ (3,2) \\ (3,10) \\ (3,11) \\ (3,12) \\ (3,4) \\ (3,7) \end{pmatrix};\ P_{(3,8)} = \begin{pmatrix} (3,8) \\ (3,9) \\ (3,2) \\ (3,10) \\ (3,11) \\ (3,12) \\ (3,4) \\ (3,7) \\ (3,8) \end{pmatrix};\ P_{(3,9)} = \begin{pmatrix} (3,9) \\ (3,2) \\ (3,10) \\ (3,11) \\ (3,12) \\ (3,4) \\ (3,7) \\ (3,8) \\ (3,9) \end{pmatrix}$$

$$P_{(3,10)} = \begin{pmatrix} (3,10) \\ (3,11) \\ (3,12) \\ (3,4) \\ (3,7) \\ (3,8) \\ (3,9) \\ (3,2) \\ (3,10) \end{pmatrix} ; \; P_{(3,11)} = \begin{pmatrix} (3,11) \\ (3,12) \\ (3,4) \\ (3,7) \\ (3,8) \\ (3,9) \\ (3,2) \\ (3,10) \\ (3,11) \end{pmatrix} ; \; P_{(3,12)} = \begin{pmatrix} (3,12) \\ (3,4) \\ (3,7) \\ (3,8) \\ (3,9) \\ (3,2) \\ (3,10) \\ (3,11) \\ (3,12) \end{pmatrix}$$

$$P_{(3,2)} = \begin{pmatrix} & (3,2) & \\ & & (3,10) \\ & & (3,11) \\ & & (3,12) \\ (3,1) & & (3,4) \\ & & (3,7) \\ & & (3,8) \\ & & (3,9) \\ & (3,2) & \end{pmatrix} ; \; P_{(3,4)} = \begin{pmatrix} & (3,4) & \\ & & (3,7) \\ & & (3,8) \\ & & (3,9) \\ (4,5) & & (3,2) \\ & & (3,10) \\ & & (3,11) \\ & & (3,12) \\ & (3,4) & \end{pmatrix}$$

$$P_{(1,2)} = \begin{pmatrix} (1,2) \\ (2,3) \\ (1,2) \end{pmatrix} ; \; P_{(4,5)} = \begin{pmatrix} & (4,5) & \\ & & (5,14) \\ & & (5,6) \\ & & (5,13) \\ & & (4,5) \\ & & (5,14) \\ (4,3) & & (5,6) \\ & & (5,13) \\ & & (4,5) \\ & & (5,14) \\ & & (5,6) \\ & & (5,13) \\ & (4,5) & \end{pmatrix}$$

$$P_{(5,14)} = \begin{pmatrix} (5,14) \\ (5,6) \\ (5,13) \\ (5,4) \\ (5,14) \\ (5,6) \\ (5,13) \\ (5,4) \\ (5,14) \\ (5,6) \\ (5,13) \\ (5,4) \\ (5,14) \end{pmatrix} ; \; P_{(5,6)} = \begin{pmatrix} (5,6) \\ (5,13) \\ (5,4) \\ (5,14) \\ (5,6) \\ (5,13) \\ (5,4) \\ (5,14) \\ (5,6) \\ (5,13) \\ (5,4) \\ (5,14) \\ (5,6) \end{pmatrix} ; \; P_{(5,13)} = \begin{pmatrix} (5,13) \\ (5,4) \\ (5,14) \\ (5,6) \\ (5,13) \\ (5,4) \\ (5,14) \\ (5,6) \\ (5,13) \\ (5,4) \\ (5,14) \\ (5,6) \\ (5,13) \end{pmatrix}$$

For these schemes we chose the convention that the highest symbol indicates the top, the lowest symbol indicates the socle, the left column is $U_{v_1}(e)$ and the right column is $U_{v_2}(e)$, if $U_{v_1}(e) \neq 0 \neq U_{v_2}(e)$.

Do Brauer tree algebras really exist? First of all, we observe that symmetric Nakayama algebras N_n^m are Brauer tree algebras with respect to a "star" with exceptional vertex in the centre of the star with multiplicity $\frac{m-1}{n}$ (see Remark 5.10.10 below).

We will show in Theorem 5.10.37 that blocks with cyclic defect group are Brauer tree algebras. Now we shall construct for every Brauer tree an algebra which is a Brauer tree algebra with this particular tree, this particular exceptional vertex and this particular multiplicity.

This is the purpose of the following construction.

Constructing the quiver: For a Brauer tree Λ we shall construct a quiver Γ'. The vertices of Γ' are the edges of Λ. The arrows of Γ' are obtained by the following rule. Let e be an edge of Λ, i.e. a vertex of Γ' and let $v \in \epsilon_\Lambda(e)$. Suppose that $e' \neq e$ is the arrow following e at v. Then draw an arrow from e to e', i.e. put $t(v) := e'$ and $s(v) := e$. This produces for every vertex of Γ' a cycle of length n_v where v is adjacent to n_v edges of Λ. In this way we obtain a quiver Γ'.

Example 5.10.6 Let us detail the construction for the Brauer tree from Example 5.10.3. The Brauer tree has 13 edges.

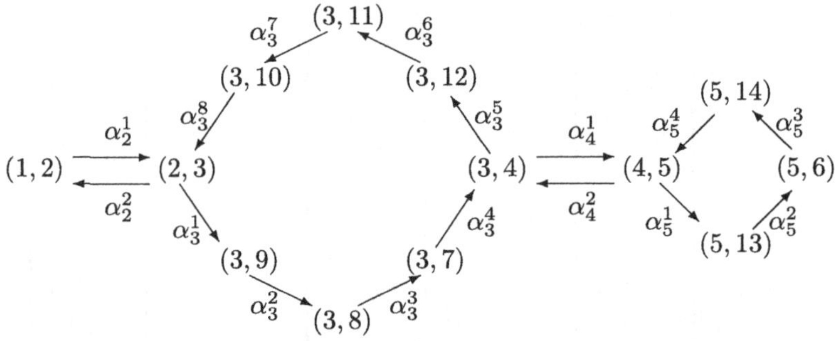

Each vertex v of Λ which has at least two edges adjacent to v therefore gives an oriented cycle. We call such a cycle irreducible if it is not part of another oriented cycle. The number of irreducible cycles in the graph is therefore equal to the number of vertices which are not "leaves" of the tree. In the above example we count four cycles. These cycles occur for the vertices 2, 3, 4 and 5 of the Brauer tree.

Constructing the relations:

Relations of the first type: Let α be an arrow belonging to the cycle of the vertex v of Λ and let β be an arrow belonging to the cycle of the vertex w of Λ. If $v \neq w$ then we impose the relation $\alpha\beta = 0$.

Relations of the second type: Let v be a vertex with exceptional multiplicity $\mu \geq 1$ (i.e. a non-exceptional vertex is included) and let w be a non-exceptional vertex so that there is an edge e with $\epsilon_\Lambda(e) = \{v, w\}$. We shall suppose that the cycle in Γ' associated to v is named $\ldots \alpha_1, \alpha_2, \ldots, \alpha_n, \alpha_1 \ldots$ and that the cycle in Γ' associated to w is named $\ldots \beta_1, \beta_2, \ldots, \beta_m, \beta_1 \ldots$ and suppose that the arrow starting at the vertex associated to e in Γ' is α_1 for the cycle associated to v, respectively β_1 for the cycle associated to w. We further suppose $n \geq 2$ and $m \geq 2$. Then we impose the relation

$$(\alpha_1\alpha_2 \ldots \alpha_n)^\mu = (\beta_1\beta_2 \ldots \beta_m).$$

Relations of the third type: Finally for vertices v of Λ such that there is only one edge e adjacent to v we impose the following relation: Let $\epsilon_\Lambda(e) = \{v, w\}$ for $w \neq v$ and let $\alpha_1, \alpha_2, \ldots, \alpha_n$ be the cycle of arrows in Γ' associated to w. Assume that α_1 is the arrow starting at the vertex of Γ' corresponding to e. Then we impose the relation

$$(\alpha_1\alpha_2 \ldots \alpha_n)^\mu\alpha_1 = 0.$$

Example 5.10.7 We continue with Example 5.10.3 and the quiver displayed in Example 5.10.6. We shall examine the relations described above for our Brauer tree of Example 5.10.3 and the associated quiver Example 5.10.6. The first set of relations is

$$\alpha_2^1\alpha_3^1 = \alpha_3^4\alpha_4^1 = \alpha_4^1\alpha_5^1 = \alpha_5^4\alpha_4^2 = \alpha_4^2\alpha_3^5 = \alpha_3^8\alpha_2^2 = 0.$$

The second set of relations is

$$\begin{aligned} \alpha_2^1\alpha_2^2 &= \alpha_3^1\alpha_3^2\alpha_3^3\alpha_3^4\alpha_3^5\alpha_3^6\alpha_3^7\alpha_3^8 \\ \alpha_4^1\alpha_4^2 &= \alpha_3^5\alpha_3^6\alpha_3^7\alpha_3^8\alpha_3^1\alpha_3^2\alpha_3^3\alpha_3^4 \\ \alpha_4^2\alpha_4^1 &= (\alpha_5^1\alpha_5^2\alpha_5^3\alpha_5^4)^\mu. \end{aligned}$$

The third set of relations is

$$\begin{aligned} \alpha_2^1\alpha_2^2 &= 0 \\ \alpha_3^2\alpha_3^3\alpha_3^4\alpha_3^5\alpha_3^5\alpha_3^6\alpha_3^7\alpha_3^8\alpha_3^1\alpha_3^2 &= 0 \\ \alpha_3^3\alpha_3^4\alpha_3^5\alpha_3^5\alpha_3^6\alpha_3^7\alpha_3^8\alpha_3^1\alpha_3^2\alpha_3^3 &= 0 \\ \alpha_3^6\alpha_3^7\alpha_3^8\alpha_3^1\alpha_3^2\alpha_3^3\alpha_3^4\alpha_3^5\alpha_3^5\alpha_3^6 &= 0 \\ \alpha_3^7\alpha_3^8\alpha_3^1\alpha_3^2\alpha_3^3\alpha_3^4\alpha_3^5\alpha_3^5\alpha_3^6\alpha_3^7 &= 0 \\ \alpha_3^8\alpha_3^1\alpha_3^2\alpha_3^3\alpha_3^4\alpha_3^5\alpha_3^5\alpha_3^6\alpha_3^7\alpha_3^8 &= 0 \\ \alpha_3^1\alpha_3^2\alpha_3^3\alpha_3^4\alpha_3^5\alpha_3^5\alpha_3^6\alpha_3^7\alpha_3^8\alpha_3^1 &= 0 \\ (\alpha_5^2\alpha_5^3\alpha_5^4\alpha_5^1)^\mu\alpha_5^2 &= 0 \\ (\alpha_5^3\alpha_5^4\alpha_5^1\alpha_5^2)^\mu\alpha_5^3 &= 0 \\ (\alpha_5^4\alpha_5^1\alpha_5^2\alpha_5^3)^\mu\alpha_5^4 &= 0. \end{aligned}$$

It is not difficult to see that the algebra with this quiver and relations has exactly the composition series of projectives as described in Example 5.10.3.

Gabriel and Riedtmann gave in [22, Sect. 1.4] an axiomatic description of the quiver obtained this way. They use the sophisticated so-called covering technique, which we do not detail here.

Structure of the projective indecomposables for this quiver and relations. We claim that the quiver algebra modulo the above-described relations is indeed a Brauer tree algebra associated to the Brauer tree (Λ, v_0, μ), with exceptional vertex v_0 with multiplicity μ. Let A be the quiver algebra modulo the ideal generated by the relations above.

Indeed, suppose first that e is an edge leading to a leaf v, i.e. $\epsilon(e) = \{v, w\}$ so that the tree one edge (namely e) is adjacent to v. Then the relation of the first type insures that the composition series of the projective indecomposable associated to e has composition factors only the simple modules in the cycle associated to the circular ordering of the vertices of the quiver (i.e. the edges of Λ adjacent to w) to which e belongs, since this type of relation ensures that a path starting at e will never leave the cycle to which e belongs. This cycle is unique since the only way that a vertex u of the quiver Γ' can belong to two cycles of Γ' is if the edge f of Λ leading to u of Γ' does not lead to a leaf. Therefore the projective indecomposable A-module P_e associated to e is uniserial with $\text{rad}^i P/\text{rad}^{i+1} P$ given by the sequence of the simple modules occurring in the cycle e belongs to. The relations of the third type ensure that the Loewy length of P_e is equal to the number of vertices in the cycle, multiplied by the exceptional multiplicity, plus 1. This proves the statement for edges e leading to a leaf.

Suppose now that e does not lead to a leaf. Then the vertex of Γ' associated to e belongs to two cycles. Hence the projective indecomposable module P_e associated to e has the property thet $\mathrm{rad}(P_e)/\mathrm{rad}^2(P_e)$ is a direct sum of two simple modules given by the edges f_1 and g_1 following e in each of the two cycles adjacent to e. Using the relations of the first type, a path in a cycle will not leave the cycle. If the vertex associated to the cycle containing e and f_1 is not exceptional, but the vertex associated to the cycle containing e and g_1 is exceptional with multiplicity μ, the paths starting at e, going to f_1, will continue in the cycle until it reaches e again. Similarly, the path starting at e, continuing to g_1, will go through the cycle e and g_1 belong to, until the path reaches e for the μ-th time again, and then the relation of second type ensures that these two paths lead to the same element in the algebra. If the path continues to f_1, then the path would have left the cycle e and g_1 belongs to, which violates the relation of the first type. Likewise for the case when the path continues with g_1. Hence $\mathrm{top}(P_e) = \mathrm{soc}(P_e)$ and $\mathrm{rad}(P_e)/\mathrm{soc}(\mathrm{P_e})$ is the direct sum of two uniserial modules given by the two cycles adjacent to e. This proves the statement.

We have shown the following

Proposition 5.10.8 *Let K be a field and let Λ be a Brauer tree with exceptional vertex of multiplicity $\mu \geq 1$, then there is a finite dimensional K-algebra A which is a Brauer tree algebra for Λ with this prescribed exceptional vertex of multiplicity $\mu \geq 1$.*

Remark 5.10.9 Observe that we could have defined a multiplicity for each vertex of a Brauer tree, instead of distinguishing only one exceptional vertex. However, these more general algebras do not occur as blocks of group algebras and hence we refrain from treating this (almost trivial) generalisation. We shall remark however that most of the subsequent properties of stable or derived equivalences treated in this chapter and Chap. 6 between these more general algebras behave in just the same way as for genuine Brauer tree algebras. These generalisations have some resonance in the literature. Actually the algebras with more than one exceptional vertex will appear in the proof of Lemma 5.10.30.

Remark 5.10.10 Let B be a Brauer tree which is a star with exceptional vertex of multiplicity $\mu \geq 1$ in the centre, i.e. the Brauer tree with e edges and $e+1$ vertices so that one of the vertices v_0 is the exceptional vertex, and this vertex is adjacent to all edges e. Then it is clear that all other vertices are adjacent to exactly one edge:

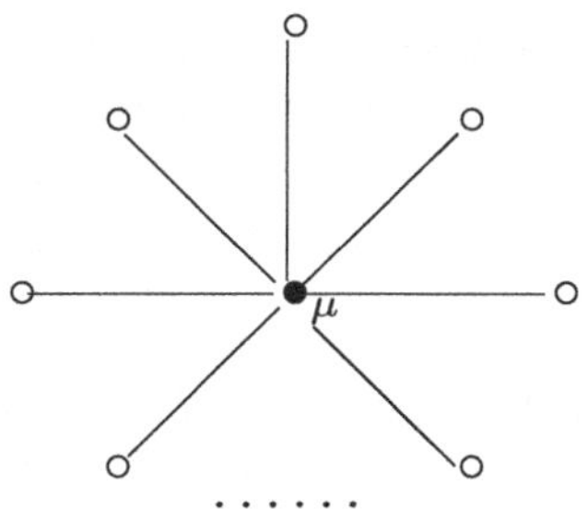

The (symmetric) Nakayama algebra $N_e^{e\mu+1}$ with e isomorphism classes of simple modules and with nilpotency degree of the radical being $e\mu + 1$ is actually a Brauer tree algebra with respect to this Brauer star. We can see this by the description of the projective indecomposable modules of a Brauer tree algebra, and its comparison with the projective indecomposables of $N_e^{e\mu+1}$. This fact can also be seen by the above description of an algebra by quiver and relation realising the Brauer tree algebra. The quiver is an oriented circle and the relation is the condition that paths of length $e\mu + 1$ are sent to 0.

5.10.2 Representation Finite Local Blocks

A simple first property is the following statement.

Proposition 5.10.11 *Let K be an infinite field and let A be a finite dimensional representation finite K-algebra which admits only one isomorphism class of indecomposable A-modules. Then every indecomposable projective A-module is uniserial.*

Remark 5.10.12 The condition on K to be infinite is actually needed in the proof.

Proof Let P be a projective indecomposable A-module. Consider $B := End_A(P)$. Then P is in fact an A–B^{op}-bimodule by putting

$$(a \otimes \varphi)(p) = \varphi(a \cdot p)$$

for every $p \in P, a \in A$ and $\varphi \in B$. Hence, we may consider for every A-module M the module $Hom_A(P, M)$, which becomes a $B = End_A(P)$-module by composing morphisms. Moreover, given a B-module N, then $Hom_B(P, N)$ is an A-module.

Then $P/\text{rad}(P) = S$ is the only simple A-module, up to isomorphism. Consider $\text{rad}(P)/\text{rad}^2(P)$. We claim that if

$$(S \oplus S) \mid (\text{rad}(P)/\text{rad}^2(P)),$$

then A is representation infinite. Indeed, in this case $K[X, Y]/(X^2, Y^2, XY)$ is a quotient ring of $B = End_A(P)$, and this ring is representation infinite by Example 2.12.10. Once this is shown, we consider the A-modules $Hom_B(P, N)$ for every

B-module N, and show that when $N \not\simeq N'$ as B-modules, then $Hom_B(P, N) \not\simeq Hom_B(P, N')$ as A-modules. Moreover, if N is indecomposable then $Hom_B(P, N)$ is indecomposable. We shall show that B maps onto the representation infinite ring $K[X, Y]/(X^2, Y^2, XY)$.

In order to prove that $K[X, Y]/(X^2, Y^2, XY)$ is a quotient ring of $B = End_A(P)$ we first observe that $\text{rad}^2(P)$, and also $\text{rad}(P)$, are characteristic submodules of P, and therefore every endomorphism α of P induces an endomorphism α_2 of $P/\text{rad}^2(P)$ and an endomorphism α_1 of $P/\text{rad}(P)$. Let

$$J_2 := \{\alpha \in B \mid \alpha_2 = 0\} \text{ and } J_1 := \{\alpha \in B \mid \alpha_1 = 0\}.$$

Then B/J_2 acts faithfully on $P/\text{rad}^2(P)$. If S^s is a direct factor of the module $\text{rad}(P)/\text{rad}^2(P)$ but S^{s+1} is not a direct factor of $\text{rad}(P)/\text{rad}^2(P)$, then

$$B/J_2 \simeq K[X_1, X_2, \ldots, X_s]/(X_i^2, X_iX_j \mid i, j \in \{1, \ldots, s\})$$

by mapping the identity to $1 \in K$ and a projection of the top of S to the j-th copy of S in $\text{rad}(P)/\text{rad}^2(P)$ to X_j. But obviously

$$K[X_1, X_2, \ldots, X_s]/(X_iX_j \mid i, j \in \{1, \ldots, s\}) \longrightarrow\!\!\!\!\rightarrow K[X_1, X_2]/(X_1^2, X_2^2, X_1X_2)$$

by mapping $X_3, \ldots, X_s$ to 0. This proves that B maps onto the representation infinite algebra $K[X, Y]/(X^2, Y^2, XY)$.

Suppose $Hom_B(P, N) \simeq Hom_B(P, N')$ as B-modules, then

$$\begin{aligned} N &\simeq Hom_B(B, N) \simeq Hom_B(End_A(P), N) \\ &\simeq Hom_B(Hom_B(P, B) \otimes_B P, N) \simeq Hom_A(Hom_B(P, B), Hom_B(P, N)) \\ &\simeq Hom_A(Hom_B(P, B), Hom_B(P, N')) \simeq Hom_B(Hom_B(P, B) \otimes_B P, N') \\ &\simeq N' \end{aligned}$$

as B-modules, since $Hom_B(P, B) \otimes_B P \simeq B$, P being B-projective. Now, observe that we constructed in Example 2.12.10 infinitely many mutually non-isomorphic $K[X, Y]/(X^2, Y^2, XY)$-modules of dimension 2. Since in a fixed dimension, an algebra of finite representation type can only have finitely many isomorphism classes of A-modules, and since the modules $Hom_B(P, N)$ have dimension bounded by

$$\dim_K(P) \cdot \dim_K(N) = 2 \cdot \dim_K(P),$$

the algebra A cannot be of finite representation type. □

5.10.3 Algebras Which Are Stably Equivalent to a Symmetric Serial Algebra

We shall give here an account of Gabriel–Riedtmann's approach to classifying algebras which are stably equivalent to symmetric serial algebras. In Lemma 2.8.2 and Remark 2.8.3 it was shown that an algebra A over a splitting field k which is serial and symmetric is Morita equivalent to the symmetric Nakayama algebra $N_m^{e\cdot m+1}$. Algebras which are stably equivalent of Morita type to one of these have a very rich structure, which we shall develop here. We mostly follow Linckelmann's thesis [23], but also use the seminal papers of Green [24] and Gabriel–Riedtmann [22] whenever appropriate.

We fix throughout the rest of this section the following notation.

Let k be a field, let A be a symmetric k-algebra and let B be the symmetric Nakayama k-algebra $N_m^{e\cdot m+1}$. Denote by $T_1, T_2, \ldots, T_m$ representatives of the isomorphism classes of simple B-modules, and let Q_i be the projective cover of T_i. Suppose that $\text{rad}^j(Q_1)/\text{rad}^{j+1}(Q_1) \simeq T_{j+1}$ for all $j \in \{0, \ldots, m-1\}$. Further, let $S_1, S_2, \ldots, S_n$ be representatives of the isomorphism classes of the simple A-modules.

Linckelmann studied the algebras which are stably equivalent to a symmetric Nakayama algebra, and he assumed that the functor inducing this equivalence comes from a functor defined between the module categories. Of course, a stable equivalence of Morita type has these properties.

We start with an easy lemma, which will be used frequently.

Lemma 5.10.13 *Let k be a field, and let A be a finite dimensional k-algebra. Let U and V be finite dimensional A-modules. Suppose $\alpha : U \longrightarrow V$ is an A-module homomorphism. Then we get:*

- *If α is injective and if α factors through an injective module, then the injective hull of U is a direct factor of V.*
- *If α is surjective and if α factors through a projective module, then the projective cover of V is a direct factor of U.*

In particular, if A is self-injective and if neither U nor V has any projective direct factors, then no injective or surjective homomorphism $U \longrightarrow V$ factors through a projective module.

Proof Let $\alpha = \beta \circ \gamma$ where $\gamma : U \longrightarrow W$ and $\beta : W \longrightarrow V$.

Suppose first that α is surjective and W is projective. Denote by $\pi : P \longrightarrow V$ the projective cover map of V. Then $\beta : W \longrightarrow V$ and $\pi : P \longrightarrow V$ are both epimorphisms of projective modules to V. Since π is the projective cover mapping, there is a $\delta : W \longrightarrow P$ such that $\beta = \pi \circ \delta$. Indeed, since P is the projective cover, we get $P/\text{rad}(P) \simeq V/\text{rad}(V)$ and α and γ are identical modulo radicals. Hence, the composition

$$U \xrightarrow{\gamma} P \longrightarrow P/\text{rad}(P)$$

is surjective, and therefore $U \longrightarrow P$ is also surjective. Hence, we may suppose that $W = P$, and then $\gamma : U \longrightarrow P$ is surjective as well. Since P is projective, γ is split, and P is a direct factor of U.

The proof of the second item is dual. Indeed, suppose α is injective and W is injective. Then $\gamma : U \longrightarrow W$ is also injective. If $\iota : U \longrightarrow I$ is the injective envelope mapping, then there is a $\delta : I \longrightarrow W$ such that $\gamma = \delta \circ \iota$. Hence we may suppose $I = W$ and $\iota = \gamma$. Moreover, $\beta : I \longrightarrow V$ needs to be injective as well, and therefore β is split, using that I is injective, proving that I is a direct factor of V. □

Let $F = M \otimes_A - : A\text{-}mod \longrightarrow B\text{-}mod$ be an exact functor mapping projectives to projectives, and let $G = N \otimes_B - : B\text{-}mod \longrightarrow A\text{-}mod$ be an exact functor mapping projectives to projectives (cf Propositions 5.3.17 and 5.5.4). Let $\underline{F} : A\text{-}\underline{mod} \longrightarrow B\text{-}\underline{mod}$ be the functor induced by F and let $\underline{G} : B\text{-}\underline{mod} \longrightarrow A\text{-}\underline{mod}$ be the functor induced by G. Suppose moreover that $\underline{G}$ is a quasi-inverse of $\underline{F}$.

Lemma 5.10.14 *A is of finite representation type (cf Definition 2.12.6).*

Proof Since $B\text{-}mod$ contains only finitely many isomorphism classes of indecomposable modules, namely the modules $Q_j/\text{rad}^\ell Q_j$ for all j and ℓ, Proposition 5.1.6 proves the statement. □

By Proposition 5.4.3 we may suppose that M and N are indecomposable, and Proposition 5.6.3 implies that $M \otimes_A V$ is indecomposable as soon as V is simple. Likewise $N \otimes_B W$ is indecomposable as soon as W is simple.

We shall start with some auxiliary lemmas in order to prove the main structure theorem in the sequel. The first lemma is a consequence of the fact that B is serial.

Lemma 5.10.15 *Let U, U_1 and U_2 be three indecomposable non-projective A-modules. Suppose $F(U)$ is simple, and suppose that the socles of U, of U_1 and of U_2 are simple. Choose the numbering so that the Loewy length of $F(U_1)$ is at most as big as the Loewy length of $F(U_2)$. Then the following holds.*

- *If U_1 and U_2 are isomorphic to submodules of U then U_2 is isomorphic to a submodule of U_1.*
- *If U is isomorphic to a submodule of U_1 and of U_2, then U_1 is isomorphic to a submodule of U_2.*

Proof Let $\alpha_1 : U_1 \hookrightarrow U$ and $\alpha_2 : U_2 \hookrightarrow U$ be the monomorphisms which exist by hypothesis. Apply F, and use that $F(U)$ is simple by hypothesis. Now, neither α_1 nor α_2 factor through a projective module, since otherwise the injective envelope of U_1 (resp. U_2) will be a direct factor of U. Hence $F\alpha_1$ and $F\alpha_2$ will not factor through a projective module, and hence are surjective since FU is simple. Since B is uniserial, $F\alpha_2$ factors through $F\alpha_1$, i.e. there is a $\varphi : FU_2 \longrightarrow FU_1$ such that $F\alpha_1 \circ \varphi = F\alpha_2$. Applying G to this equation we see that $\gamma := \alpha_1 \circ G\varphi - \alpha_2$ factors through a projective module. Hence γ is not injective, since otherwise the injective envelope of U_2 would be a direct factor of U. But the socle of U_2 is simple. Hence $\mathrm{soc}(U_2) \subseteq \ker \gamma$. If $G\varphi$ were non-injective, then $\mathrm{soc}(U_2) \subseteq \ker(G\varphi)$ and hence

$$\mathrm{soc}(U_2) \subseteq \ker(\alpha_1 \circ G\varphi - \gamma) = \ker \alpha_2 = 0.$$

Therefore $G\varphi$ is injective, which proves that U_1 is a submodule of U_2.

Let $\beta_1 : U \hookrightarrow U_1$ and $\beta_2 : U \hookrightarrow U_2$ be the monomorphisms from the hypothesis. Since U_1 and U_2 are indecomposable non-projective, Lemma 5.10.13 shows that β_1 and β_2 do not factor through a projective module. Apply F and observe again that $F\beta_1$ and $F\beta_2$ do not factor through a projective module since β_1 and β_2 do not factor through a projective module. Since FU is simple, $F\beta_1$ and $F\beta_2$ are both injective. Since FU_1 and FU_2 are uniserial with an isomorphic socle, $F\beta_2$ factors through $F\beta_1$, i.e. there is a $\psi : FU_1 \longrightarrow FU_2$ such that $F\beta_2 = \psi \circ F\beta_1$. Applying G again we see that $\beta_2 - G\psi \circ \beta_1$ factors through a projective module. Just as in the previous case we obtain that $G\psi$ is injective. □

Lemma 5.10.16 *Let U and V be indecomposable non-projective A-modules with the same socle. Suppose that FU and $\Omega^{-1}FV$ are simple and suppose that the socles of U and of V are simple. Then the following holds.*

- *$\mathrm{soc}(U)$ is the largest submodule of U which is also isomorphic to a submodule of V.*
- *Let I_U be the injective hull of U. Then the radical of I_U is the only proper submodule of I_U which has a submodule isomorphic to V.*

Proof We shall first prove the first item. Suppose U_2 is a submodule of U isomorphic to a submodule of V. If the Loewy length of $F(\mathrm{soc}(U))$ is at most as big as the Loewy length of FU_2, then the first part of Lemma 5.10.15 shows that U_2 is isomorphic to a submodule of $\mathrm{soc}(U)$. This proves the statement in this case. If the Loewy length of $F(\mathrm{soc}(U))$ is at least as big as the Loewy length of FU_2, then the Loewy length of $\Omega^{-1}F(\mathrm{soc}(U))$ is at most as big as the Loewy length of $\Omega^{-1}FU_2$ and we conclude as in the previous case, using $\Omega^{-1}F$ instead of F and V instead of U.

Denote for simplicity $\mathrm{rad}(I_U)$ by U_1 and let U_2 be a module so that $U \subseteq U_2 \subseteq U_1$ and such that V contains a submodule isomorphic to U_2. If the Loewy length of FU_1 is at most as big as the Loewy length of FU_2, then the second part of Lemma 5.10.15 shows that U_1 is isomorphic to a submodule of U_2, which implies $U_1 = U_2$. If the Loewy length of FU_1 is at least as big as the Loewy length of FU_2, then the Loewy length of $\Omega^{-1}FU_1$ is at most as big as the Loewy length of $\Omega^{-1}FU_2$, and using the second part of Lemma 5.10.15 applied to $\Omega^{-1}F$ instead of F and to V instead of U again gives the result. □

Lemma 5.10.17 *Let U and V be indecomposable non-projective A-modules and suppose U or V is simple. Let $\alpha_1 : U \longrightarrow V$ and $\alpha_2 : U \longrightarrow V$ be A-linear homomorphisms such that neither α_1 nor α_2 factors through a projective A-module. If $\ker(F\alpha_1) \subseteq \ker(F\alpha_2)$, then there is an A-linear endomorphism π of U and an A-linear endomorphism τ of V such that $\alpha_2 = \tau \circ \alpha_1 = \alpha_1 \circ \pi$.*

Proof Since FV is uniserial and since $\ker(F\alpha_1) \subseteq \ker(F\alpha_2)$ we have

$$\mathrm{im}(F\alpha_2) \subseteq \mathrm{im}(F\alpha_1) \subseteq FV.$$

Let $Q \xrightarrow{\rho} FU$ be a projective cover. Then

$$FU \xrightarrow{F\alpha_1} \mathrm{im}(F\alpha_1)$$

is an epimorphism and the composition

$$Q \xrightarrow{\rho} FU \xrightarrow{F\alpha_2} \mathrm{im}(F\alpha_2) \hookrightarrow \mathrm{im}(F\alpha_1)$$

is a morphism from a projective module to $\mathrm{im}(F\alpha_1)$. By the universal property of projective modules this morphism lifts to a morphism $Q \xrightarrow{\sigma} FU$. In other words there is a homomorphism $Q \xrightarrow{\sigma} FU$ such that $F\alpha_2 \circ \rho = F\alpha_1 \circ \sigma$. Since Q is uniserial, $\ker(\rho) \subseteq \ker(\sigma)$. Hence $F\alpha_2$ factors through $F\alpha_1$, i.e. there is an endomorphism φ of FU such that $F\alpha_2 = F\alpha_1 \circ \varphi$. Applying G to this equation we see that $\alpha_2 - \alpha_1 \circ G\varphi$ factors through a projective A-module. But since U or V is simple, no non-zero homomorphism $U \longrightarrow V$ factors through a projective homomorphism, again using Lemma 5.10.13. Hence $\alpha_2 - \alpha_1 \circ G\varphi = 0$ and we can put $\pi = G\varphi$. The proof of the existence of τ is dual. □

Corollary 5.10.18 *Let U be an indecomposable A-module, and let S be a simple A-module. Then each of the $End_A(S)$-modules $Hom_A(S, U)$ and $Hom_A(U, S)$ is either* 0 *or isomorphic to $End_A(S)$.*

Proof If U is projective, the statement is clearly true. Suppose therefore that U is not projective. Hence each non-zero homomorphism $S \longrightarrow U$ is injective and Lemma 5.10.17 applies to two such homomorphisms α_1 and α_2; indeed, if FU is uniserial then the condition on the kernel is satisfied. The statement of Lemma 5.10.17 then implies the present statement. Dually, considering non-zero homomorphisms $U \longrightarrow S$ and by the analogous argument we obtain the statement on $Hom_A(U, S)$. □

Corollary 5.10.19 *Let U be an A-module with simple socle. Then any two A-submodules are either identical or non-isomorphic.*

Proof Let V and W be two isomorphic A-submodules of U. If both are simple, then $V = W = \mathrm{soc}(U)$. We shall prove the statement by induction on the Loewy length of V (= Loewy length of W). If $V \neq W$, then each maximal submodule V' of V, and W' of W, has smaller Loewy length, and so $V' = V \cap W = W'$ is a common submodule of smaller Loewy length. Hence $(V + W)/V' = S_1 \oplus S_2$ for two simple modules $S_1 = V/V'$ and $S_2 = W/W'$. But $V/V' \simeq W/W'$ since the isomorphism taking V to W has to take V' to W' (i.e. fixes $V' = W'$). Now $V + W$ is indecomposable since $\mathrm{soc}(V + W) = \mathrm{soc}(U)$ is simple. This contradicts Corollary 5.10.18 since $Hom_A(V + W, S)$ is two-dimensional over $End_A(S)$. □

Recall that FS_i is indecomposable by Proposition 5.6.3. This fact has the following consequence.

Lemma 5.10.20 *Let $S_1, \ldots, S_n$ be representatives of the isomorphism classes of the simple A-modules, and let $T_1, \ldots, T_m$ be representatives of the isomorphism classes of the simple B-modules. Let $\gamma, \delta : \{1, \ldots, n\} \longrightarrow \{1, \ldots, m\}$ be the maps defined by*

$$T_{\gamma(i)} \simeq \mathrm{soc}(FS_i) \text{ and } T_{\delta(i)} \simeq FS_i/\mathrm{rad}(FS_i).$$

Then γ and δ are bijective.

Proof Indecomposable modules, such as FS_i, are uniserial. Hence γ and δ are well-defined. Since S_i and T_j are simple, for all indecomposable non-projective A-modules U and indecomposable non-projective B-modules V we have

$$\underline{Hom}_A(S_i, U) = Hom_A(S_i, U) \text{ and } \underline{Hom}_A(V, T_j) = Hom_A(V, T_j),$$

again using Lemma 5.10.13 and for the case of $V \longrightarrow T_j$ the dual argument applies. Hence

$$Hom_A(S_i, GT_j) = \underline{Hom}_A(S_i, GT_j) = \underline{Hom}_B(FS_i, T_j) = Hom_B(FS_i, T_j),$$

and using that each non-zero module has a socle shows that δ is surjective. Since

$$Hom_A(GT_j, S_i) = \underline{Hom}_A(GT_j, S_i) = \underline{Hom}_B(T_j, FS_i) = Hom_B(T_j, FS_i),$$

and the fact that each non-zero module has a top, γ is also surjective.

Suppose FS_{i_1} and FS_{i_2} have the same radical quotient T_j, and suppose without loss of generality that FS_{i_1} has smaller Loewy length than FS_{i_2}. Then there is an epimorphism $FS_{i_2} \twoheadrightarrow FS_{i_1}$. This epimorphism does not factor through a projective module, again using Lemma 5.10.13. Therefore

$$\underline{Hom}_B(FS_{i_2}, FS_{i_1}) = \underline{Hom}_A(S_{i_2}, S_{i_1})$$

and by Schur's lemma this is only possible if $i_1 = i_2$ since both are simple modules and the stable homomorphisms are quotients of the A-linear homomorphisms. Therefore δ is injective. The dual argument shows that γ is injective. □

Remark 5.10.21 This proves the Auslander–Reiten conjecture for algebras stably equivalent of Morita type to symmetric serial algebras. Auslander and Reiten conjectured that if A and B are finite dimensional k-algebras, and if there is an equivalence between the stable categories of A and of B, then the number of isomorphism classes of simple non-projective A-modules coincides with the number of isomorphism classes of simple non-projective B-modules.

Work by Martinez-Villa [25] shows that the conjecture is true if it is true for self-injective algebras. The conjecture is shown to hold for various very special classes of algebras. For a short overview see [18]. We will come back to this theme in Remark 5.10.26. See also Proposition 5.11.20 below for a result which sheds additional light on the conjecture.

Corollary 5.10.22 *For all $i \in \{1, \ldots, n\}$ we have*

$$S_i = \mathrm{soc}(GT_{\delta(i)}) = \mathrm{soc}(G\Omega T_{\gamma(i)})$$

and GT_j is uniserial for each j.

Proof In the proof of Lemma 5.10.20 we have seen

$$Hom_A(GT_j, S_i) = Hom_B(T_j, FS_i)$$

and

$$Hom_A(S_i, GT_j) = Hom_B(FS_i, T_j).$$

Since $Hom_B(FS_i, T_j) \neq 0$ if and only if $j = \delta(i)$ we obtain that $S_i = \mathrm{soc}(GT_{\delta(i)})$. Dually the first equation implies that $S_i = \mathrm{soc}(G\Omega T_{\gamma(i)})$.

Let U and V be two A-submodules of GT_j. Suppose that the Loewy length of FU is at least as big as the Loewy length of FV. The injection $U \longrightarrow GT_j$ does not factor through a projective module, again using Lemma 5.10.13. Lemma 5.10.15 shows that there is a monomorphism $U \overset{\iota}{\hookrightarrow} V$ and that $S_{\delta^{-1}(i)} = \mathrm{soc}(U) = \mathrm{soc}(V)$. Corollary 5.10.19 shows that the two submodules $\iota(U)$ and U of V have to be identical. Hence $U = \iota(U) \subseteq V$, which shows that GT_j is uniserial. □

Proposition 5.10.23 *Let P_i be the injective hull of S_i. Then for each $i \in \{1, \ldots, n\}$ there are unique uniserial submodules U_i and V_i such that $U_i \simeq GT_{\delta(i)}$ and $V_i \simeq G\Omega T_{\gamma(i)}$. These have the following properties:*

- $U_i + V_i = \mathrm{rad}(P_i)$
- $U_i \cap V_i = \mathrm{soc}(P_i)$
- *$U_i/(U_i \cap V_i)$ and $V_i/(U_i \cap V_i)$ do not have any common composition factors.*

Proof Lemma 5.10.20 shows that the socles of $GT_{\delta(i)}$ and of $G\Omega T_{\gamma(i)}$ are isomorphic to S_i. Since P_i is the injective hull of S_i there are submodules U_i and V_i of P_i such that $U_i \simeq GT_{\delta(i)}$ and $V_i \simeq G\Omega T_{\gamma(i)}$. Moreover, the same lemma shows that U_i and V_i are uniserial. Corollary 5.10.19 shows that U_i and V_i are the unique submodules of P_i with this property. Lemma 5.10.16 shows that $U_i + V_i = \mathrm{rad}(P_i)$ and $U_i \cap V_i = \mathrm{soc}(P_i)$. Now, U_i and V_i are uniserial. If S is a common composition factor of U_i and V_i, then taking the (indecomposable) quotient P_i/W of P_i so that S is in the socle of $U/(W \cap U)$ and in the socle of $V/(W \cap V)$, we get that $Hom_A(S, P_i/W) \simeq Hom_A(S, S^2) \neq End_A(S)$. This contradicts Corollary 5.10.18. □

Proposition 5.10.24 *Recall the mappings* $\gamma : \{1,\ldots,n\} \longrightarrow \{1,\ldots,m\}$ *and* $\delta : \{1,\ldots,n\} \longrightarrow \{1,\ldots,m\}$ *defined in Lemma 5.10.20 as well as* $\sigma : \{1,\ldots,m\} \longrightarrow \{1,\ldots,m\}$ *of Lemma 2.8.2. Put* $\rho := \gamma^{-1}\circ\delta$ *and* $\tau := \delta^{-1}\circ\sigma\circ\gamma$. *Then the following holds.*

1. $U_i/\mathrm{rad}(U_i) \simeq S_{\rho(i)}$ *and* $V_i/\mathrm{rad}(V_i) \simeq S_{\tau(i)}$.
2. *The element* $\tau\circ\rho \in \mathfrak{S}_n$ *is a transitive cycle on* $\{1,\ldots,n\}$.
3. *If* $G\Omega T_j \simeq \Omega G T_j$ *and* $G\Omega^{-1}T_j \simeq \Omega^{-1}GT_j$ *for all* $j \in \{1,\ldots,m\}$, *then for all* $i \in \{1,\ldots,n\}$ *there are short exact sequences*

$$0 \longrightarrow U_i \longrightarrow P_i \xrightarrow{\alpha_i} V_{\tau^{-1}(i)} \longrightarrow 0$$

and

$$0 \longrightarrow V_i \longrightarrow P_i \xrightarrow{\beta_i} U_{\rho^{-1}(i)} \longrightarrow 0.$$

Proof (1) The definition of γ implies

$$\begin{aligned} Hom_A(U_{i_1}, S_{i_2}) &= \underline{Hom}_A(U_{i_1}, S_{i_2}) \simeq \underline{Hom}_B(T_{\delta(i_1)}, FS_{i_2}) \\ &= Hom_B(T_{\delta(i_1)}, FS_{i_2}) \end{aligned}$$

and these spaces are non-zero if and only if $\gamma(i_2) = \delta(i_1)$. Hence

$$U_i/\mathrm{rad}(U_i) \simeq S_{(\gamma^{-1}\circ\delta)(i)} = S_{\rho(i)}.$$

Similarly,

$$\begin{aligned} Hom_A(V_{i_1}, S_{i_2}) &= \underline{Hom}_A(V_{i_1}, S_{i_2}) \simeq \underline{Hom}_B(\Omega T_{\gamma(i_1)}, FS_{i_2}) \\ &= \underline{Hom}_B(\Omega^2 T_{\gamma(i_1)}, \Omega FS_{i_2}) = \underline{Hom}_B(T_{(\sigma\circ\gamma)(i_1)}, \Omega FS_{i_2}) \\ &= Hom_B(T_{(\sigma\circ\gamma)(i_1)}, \Omega FS_{i_2}). \end{aligned}$$

This space is non-zero if and only if

$$T_{(\sigma\circ\gamma)(i_1)} \simeq \mathrm{soc}(\Omega FS_{i_2}) = FS_{i_2}/\mathrm{rad}(FS_{i_2}) = T_{\delta(i)},$$

whence if and only if $i_2 = \delta^{-1}\circ\sigma\circ\gamma(i_1) = \tau(i_1)$.

As a consequence

$$V_i/\mathrm{rad}(V_i) \simeq S_{\tau(i)}$$

as claimed.

(2) We compute

$$\tau\circ\rho = (\delta^{-1}\circ\sigma\circ\gamma)\circ(\gamma^{-1}\circ\delta) = \delta^{-1}\circ\sigma\circ\delta$$

and Lemma 2.8.2 shows that σ is a transitive cycle, which implies that its δ-conjugate $\tau \circ \rho$ is also a transitive cycle.

(3) Corollary 5.10.22 implies that we have short exact sequences

$$0 \longrightarrow GT_{\delta(i)} \longrightarrow P_i \longrightarrow \Omega^{-1}GT_{\delta(i)} \longrightarrow 0$$

and

$$0 \longrightarrow G\Omega T_{\gamma(i)} \longrightarrow P_i \longrightarrow \Omega^{-1}G\Omega T_{\gamma(i)} \longrightarrow 0.$$

Since we assume $G\Omega^{-1}T_j \simeq \Omega^{-1}GT_j$ and $G\Omega T_j \simeq \Omega GT_j$ for all $j \in \{1, \dots, m\}$, by the definition of τ and ρ we obtain

$$\Omega^{-1}G\Omega T_{\gamma(i)} = GT_{\gamma(i)} = GT_{\gamma\rho\rho^{-1}(i)} = GT_{\delta\rho^{-1}(i)} = U_{\rho^{-1}(i)}$$

using Proposition 5.10.23 for the last equality, and

$$\Omega^{-1}GT_{\delta(i)} \simeq G\Omega^{-1}T_{\delta(i)} \simeq G\Omega T_{(\sigma^{-1}\circ\delta)(i)} = V_{\tau^{-1}(i)}.$$

This proves the statement. □

The statements we have obtained so far give a good description of the structure of A.

Corollary 5.10.25 *Suppose that $G\Omega T_j \simeq \Omega GT_j$ and $G\Omega^{-1}T_j \simeq \Omega^{-1}GT_j$ for all $j \in \{1, \dots, m\}$. Then we get for all $i \in \{1, \dots, n\}$ that*

1. • *there is $a_i \in \mathbb{N}$ such that $\mathrm{rad}^\ell(U_i)/\mathrm{rad}^{\ell+1}(U_i) \simeq S_{\rho^\ell(i)}$ if $\ell \le a_i$ and $\mathrm{rad}^\ell(U_i) = 0$ if $\ell > a_i$,*
   • *there is $b_i \in \mathbb{N}$ such that $\mathrm{rad}^\ell(V_i)/\mathrm{rad}^{\ell+1}(V_i) \simeq S_{\tau^\ell(i)}$ if $\ell \le b_i$ and $\mathrm{rad}^\ell(V_i) = 0$ if $\ell > b_i$,*
2. *$a_i = a_{\rho(i)}$ and $b_i = b_{\tau(i)}$,*
3. *$\{\rho^\ell(i) \mid \ell \in \mathbb{N}\} \cap \{\tau^\ell(i) \mid \tau \in \mathbb{N}\} = \{i\}$.*

Proof The first item is an immediate consequence of the fact that the exact sequences in Proposition 5.10.24 show that $\alpha_{\tau(i)}(V_{\tau(i)}) = \mathrm{rad}(V_i)$ and that $\beta_{\rho(i)}(U_{\rho(i)}) = \mathrm{rad}(U_i)$.

The second item comes from Proposition 5.10.23, which implies the fact that the restriction of each of the mappings α_i and β_i to V_i and U_i have simple kernel and cokernel.

The third item is a consequence of Proposition 5.10.23 as well, namely that U_i and V_i only have one common composition factor, their common socle. □

Recall what we have obtained so far. The permutation $\tau \circ \rho$ is a transitive cycle. Hence we may modify the numbering so that

$$(\tau \circ \rho)(i) = i + 1 \bmod n.$$

Assume this for the moment. We then get short exact sequences

$$0 \longrightarrow U_i \longrightarrow P_i \longrightarrow V_{\tau^{-1}(i)} \longrightarrow 0$$

and since

$$(\tau \circ \rho)^{-1}(i) = \rho^{-1} \circ \tau^{-1}(i) = i - 1 \bmod n,$$

we get

$$0 \longrightarrow V_{\tau^{-1}(i)} \longrightarrow P_{\tau^{-1}(i)} \longrightarrow U_{i-1} \longrightarrow 0,$$

where the indices are taken modulo n.

We see first that we obtain a periodic projective resolution for each of the modules U_i and V_i. The length of the period is $2n$. More precisely, an injective resolution of U_1 is given by gluing and repeating the exact sequence

$$U_1 \hookrightarrow P_1 \to P_{\tau^{-1}(1)} \to P_2 \to P_{\tau^{-2}(2)} \to P_3 \to \ldots \to P_n \to P_{\tau^{-n}(n)} \longrightarrow\!\!\!\!\!\rightarrow U_1.$$

Corollary 5.10.26 *Suppose again that $G\Omega T_j \simeq \Omega G T_j$ and $G\Omega^{-1}T_j \simeq \Omega^{-1}GT_j$ for all $j \in \{1,\ldots,m\}$. Then each of the modules U_i and V_i has a periodic projective resolution in which each of the projective indecomposable modules P_i is passed exactly twice in a period. Moreover, $V_{\tau^{-1}(i+1)} = \Omega^{-(1+2i)}(U_1)$ and $U_{i+1} = \Omega^{-2i}(U_1)$ and therefore there is precisely one $(u,v) \in \{1,\ldots,2n\}^2$ such that $\Omega^u(U_i) = U_1 = \Omega^v(V_i)$.*

Proof We only need to prove the formulas $\Omega^u(U_i) = U_1 = \Omega^v(V_i)$. The unicity of (u,v) then follows. The exact sequences above imply, applying Ω^{-1} successively,

$$\begin{array}{l|l}
V_{\tau^{-1}(1)} = \Omega^{-1}(U_1) & \Omega^{-1}(V_{\tau^{-1}(1)}) = U_2 = \Omega^{-2}(U_1) \\
V_{\tau^{-1}(2)} = \Omega^{-1}(U_2) = \Omega^{-3}(U_1) & \Omega^{-1}(V_{\tau^{-1}(2)}) = U_3 = \Omega^{-4}(U_1) \\
V_{\tau^{-1}(3)} = \Omega^{-1}(U_3) = \Omega^{-5}(U_1) & \Omega^{-1}(V_{\tau^{-1}(3)}) = U_4 = \Omega^{-6}(U_1) \\
V_{\tau^{-1}(4)} = \Omega^{-1}(U_4) = \Omega^{-7}(U_1) & \Omega^{-1}(V_{\tau^{-1}(4)}) = U_5 = \Omega^{-8}(U_1) \\
 & \text{etc.}
\end{array}$$

This proves the statement. □

Remark 5.10.27 We have actually proved that A is special biserial in the sense of Sect. 1.11.2. Hence we can produce indecomposable modules as tree and as band modules.

We now use an idea of Green [24]. We draw a cyclic equioriented quiver with $2n$ edges, and label the edges successively

$$i, \tau(i), i+1, \tau(i+1), i+2, \tau(i+2), i+3 \ldots.$$

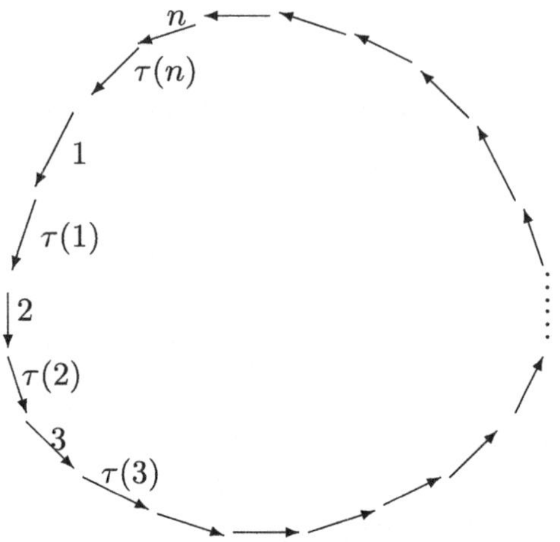

Each number i occurs exactly twice. Once labelled i, and once labelled $i = \tau(j)$ for some j. We then identify the arrow labelled i with the arrow labelled $i = \tau(j)$, but in the reverse direction, or as Green expresses the fact, "so that the directions cancel out".

Lemma 5.10.28 *The resulting topological object, which we call the Green graph Γ_A, is a tree.*

We shall prove this lemma by induction on the number of isomorphism classes of simple modules. The reduction to algebras with fewer simple modules is achieved by the existence of a uniserial projective indecomposable A-module.

The following lemma is a consequence of the fact that A is representation finite.

Lemma 5.10.29 *There is a uniserial projective indecomposable A-module.*

Proof Suppose no projective indecomposable A-module is uniserial. Then for all P_i projective indecomposable A-modules, we get that $P_i/\mathrm{rad}^2(P_i)$ has three composition factors: its simple top S_i and

$$\mathrm{soc}(P_i/\mathrm{rad}^2(P_i)) \simeq S_{\tau(i)} \oplus S_{\rho(i)}.$$

We claim that then A-*mod* is not representation finite. Indeed, for every S_i we have

$$Ext^1_A(S_i, S_{\tau(i)}) \neq 0 \neq Ext^1_A(S_i, S_{\rho(i)})$$

and $\tau(i) \neq \rho(i)$ for all i. Hence we can construct an indecomposable module, actually a string module in the sense of Sect. 1.11.2,

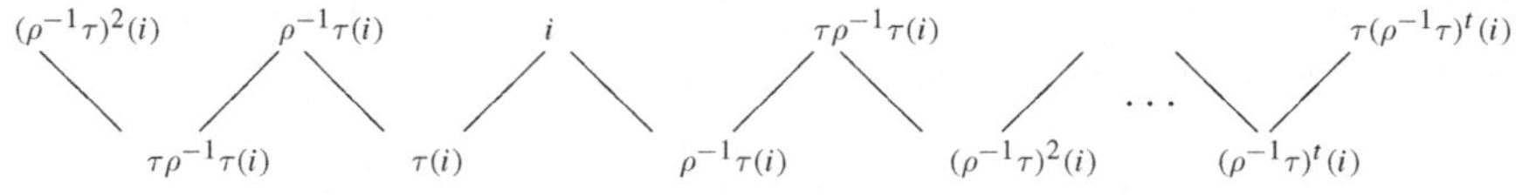

for each integer t. Here we draw a line to indicate a non-trivial extension between the simple modules corresponding to the indices. The same argument even shows that for all i there is an integer t such that either $P_{(\rho^{-1}\tau)^t(i)}$ or $P_{\tau(\rho^{-1}\tau)^t(i)}$ is uniserial.

Since the symmetric Nakayama algebra is representation finite, and since by hypothesis A is stably equivalent to a symmetric Nakayama algebra, this contradicts Proposition 5.1.6. Hence there is a uniserial projective indecomposable A-module P_i. This proves Lemma 5.10.29. □

We shall now prove Lemma 5.10.28.

Observe how a uniserial projective module occurs in Green's description by gluing the arrows in the cycle. A projective module P_i is uniserial if and only if $i = \tau(i)$ or $i + 1 = \tau(i)$ since then the corresponding module U_i is simple.

If $\tau(i) = i+1$, then we see that $(\rho\circ\tau)(i) = i+1$ implies that $\rho(i) = \tau^{-1}(i+1) = i$. Hence we can interchange ρ and τ in the Brauer tree construction.

Suppose therefore that $n = \tau(n)$. We consider the algebra $End_A(\bigoplus_{\ell=1}^{n-1} P_\ell)$ and its module category. It is clear that $End_A(\bigoplus_{\ell=1}^{n-1} P_\ell)$ is again representation finite, since

$$End_A(\bigoplus_{\ell=1}^{n-1} P_\ell) = eAe$$

for some idempotent e. Moreover, eAe satisfies exactly the analogous structure of projective indecomposable modules as described in Corollary 5.10.25. Indeed, the decomposition structure of the projective indecomposable eAe-module, which is the image of the projective indecomposable A-module P, is obtained as follows. From the decomposition structure of the indecomposable A-module P erase the simple module which is annihilated by e. This fact can be obtained by regarding the basic algebra associated to eAe as an algebra defined by a quiver with relations. Indeed, e is then a sum of pairwise disjoint "lazy paths", i.e. vertices. We say that these vertices compose e. Then eAe is given by those paths starting at a vertex composing e and ending at a vertex composing e. Hence, all projective indecomposable eAe-modules have exactly the analogous structure as those for A, except that there is one simple module less.

We now proceed by induction on the number of isomorphism classes of simple A-modules. The statement we want to prove is that a representation finite algebra A with a structure of projective indecomposable A-modules as described in Corollary 5.10.25 yields a Green-graph Γ_A being a tree.

For one simple module there is nothing to prove. For n isomorphism classes of simple modules we obtained that there must be a uniserial projective indecomposable module P_n, say, and we construct eAe as above. Then $\tau(n) = n$. The Green-graph Γ_{eAe} is a tree by induction. Let f be the idempotent corresponding to all edges adjacent to the edge n. Then fAf is a representation finite symmetric algebra, and the classification of these algebras shows that fAf is a symmetric Nakayama algebra. Hence the Green graph Γ_A is obtained from the tree Γ_{eAe} by adding a leaf at the position between 1 and $n-1$. Recall that a leaf is an edge for which one end is not attached to another edge. This gives a tree again and we have proved the statement. □

Observe that we have shown that A is a Brauer tree algebra with possibly many exceptional vertices. We shall show that there is at most one exceptional vertex for representation finite algebras A.

Suppose we have only one edge and both ends are exceptional vertices. Then $U_1 =: U$ has length bigger than 1 and likewise for $V_1 =: V$. Let P be the unique projective indecomposable A-module and let S be the unique simple. Moreover, $P_1/\mathrm{rad}^2(P_1)$ is indecomposable of Loewy length 3, where

$$\mathrm{soc}(P_1/\mathrm{rad}^2(P_1)) \simeq S \oplus S.$$

The algebra A has two generators x and y, and x maps the simple top $P/\mathrm{rad}(P)$ to the left copy of $\mathrm{soc}(P/\mathrm{rad}^2 P) = S \oplus S$ and y maps the simple top $P/\mathrm{rad}(P)$ to the right copy of $\mathrm{soc}(P/\mathrm{rad}^2 P) = S \oplus S$.

Then we can produce infinitely many indecomposable string modules M_n where the Loewy length of M_n is $2n+1$, and where the structure of M_n is given by a zig-zag

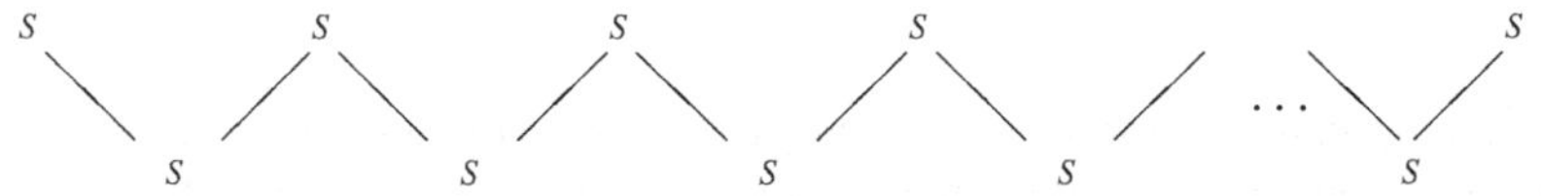

where the lines in the southwest-northeast direction indicate multiplication by x, and the lines in the northwest-southeast direction indicate multiplication by y.

The condition of having more than one exceptional vertex is equivalent to the condition in the following lemma.

Lemma 5.10.30 *Suppose that there are indices $i_1 \neq i_2$ such that one of the following holds.*

- $End_A(U_{i_1}) \neq k$ *and* $End_A(U_{i_2}) \neq k$,
- *or* $End_A(V_{i_1}) \neq k$ *and* $End_A(V_{i_2}) \neq k$,
- *or* $End_A(U_{i_1}) \neq k$ *and* $End_A(V_{i_2}) \neq k$,
- *or* $End_A(U_{i_1}) \neq k$ *and* $End_A(V_{i_1}) \neq k$,

then A is representation infinite.

Proof The proof is very similar to the proof of Lemma 5.10.29 and generalises the situation of A being local, discussed above.

Assume the first condition. Then the composition factor i_1 occurs in the composition series of U_{i_1} twice, namely in the socle and in some intermediate level. The analogous statement holds for U_{i_2}.

Let Δ_A be the full subgraph of Γ_A containing the two exceptional vertices v_1 and v_2 and all vertices v of Γ so that the geodesic from v to v_1 does not contain v_2, and so that the geodesic from v to v_1 does not contain v_2. Then v_1 and v_2 are both at the extremity of Δ_A, and are actually leaves of Δ_A. Let v_1 be adjacent to i_1 and v_2 be adjacent to i_2. Then there is a uniserial A-module T_1 of Loewy length at least 2 with top and socle isomorphic to S_{i_1}, and there is a uniserial A-module T_2 with top and socle isomorphic to S_{i_2}.

We claim that there is an indecomposable module built similarly as above by some zig-zag from T_1 to T_1. Let

be the geodesic from v_1 to v_2, where all j_s are edges of Γ_A. Then there are extensions of the following type:

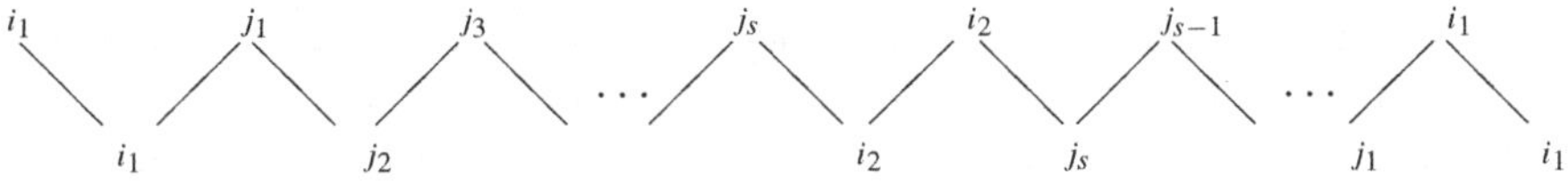

where each line indicates a uniserial module with top and socle described by the indices given in the diagram. In particular

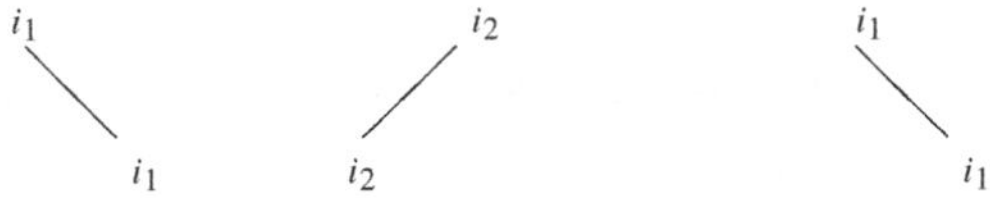

indicate the modules T_1, and T_2, and then T_1 again, respectively. We shall display this indecomposable module as follows

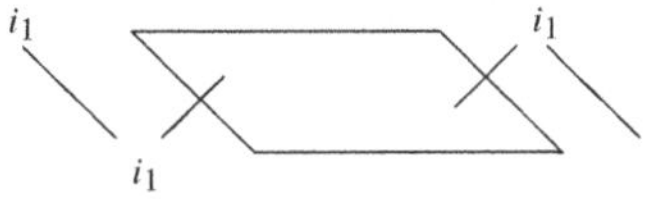

This indecomposable module can be repeated arbitrarily often, since there is an indecomposable module M with socle S_{i_1}, with top $S_{i_1} \oplus S_{j_1}$ and with $M/\mathrm{soc}(M)$ being a direct sum of two uniserial modules.

More precisely, in the scheme indicated above, we obtain

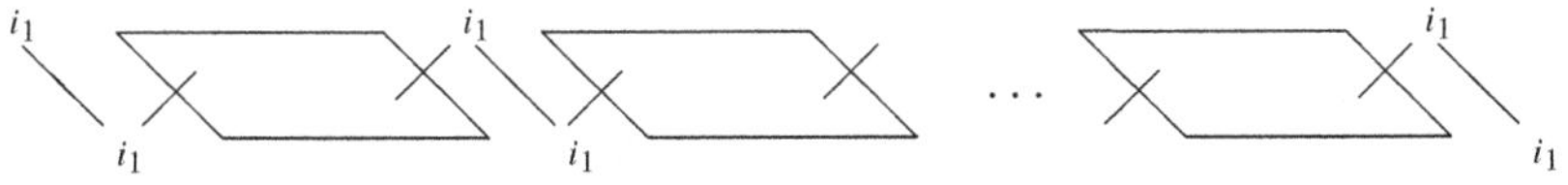

This produces an infinite series of indecomposable string modules in the sense of Sect. 1.11.2. Hence A is not of finite representation type. □

We have finally proved the very important and celebrated theorem of Gabriel–Riedtmann.

Theorem 5.10.31 (Gabriel–Riedtmann [22]) *Let A be an algebra which is stably equivalent to a symmetric Nakayama algebra B by a functor*

$$G : B\text{-}\underline{mod} \longrightarrow A\text{-}\underline{mod}$$

satisfying $G\Omega^{-1}T \simeq \Omega^{-1}GT$ and $G\Omega T \simeq \Omega GT$ for each simple B-module T. Then A is a Brauer tree algebra.

Remark 5.10.32 Gabriel and Riedtmann proved an even stronger result. They showed by somewhat different methods, the so-called covering technique, that the hypothesis on the functor G is not necessary. However, the covering technique is a quite complicated method which would need an intensive preparation.

The hypothesis on G is satisfied whenever G is a stable equivalence of Morita type, or even more generally, by Watts' theorem 3.3.16, if G is induced by an exact additive functor between the module categories.

5.10.4 Blocks with Cyclic Defect Group

Let G be a finite group, let k be a field of characteristic $p > 0$, and let D be a defect group of a block B of kG. Suppose that D is cyclic of order p^n. Then we shall show that B is a Brauer tree algebra. We proceed in several steps.

Let D_1 be the (unique!) subgroup of D of order p. We consider the following lattice of subgroups of G:

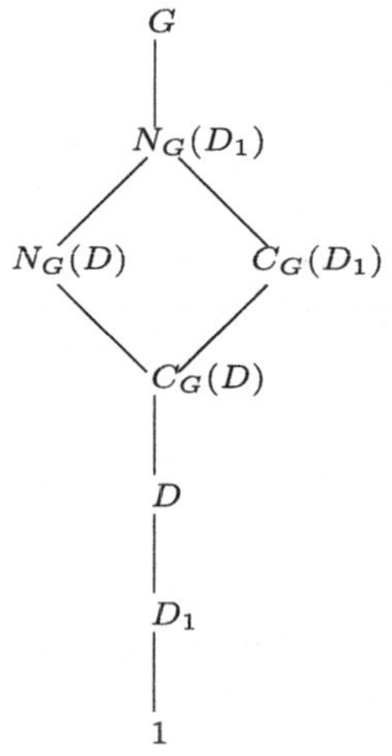

Indeed, D_1 is central in D, and hence $C_G(D_1) \geq C_G(D) \geq D$. Moreover, clearly

$$C_G(D) \leq C_G(D_1) \leq N_G(D_1)$$

and

$$D \leq C_G(D) \leq N_G(D).$$

Since D_1 is the unique subgroup of D of order p, for all $g \in N_G(D)$ the group $g \cdot D_1 \cdot g^{-1}$ is also a subgroup of D of order p, and we have the equality $g \cdot D_1 \cdot g^{-1} = D_1$ whence $N_G(D) \leq N_G(D_1)$.

We come to the corresponding blocks.

Since D is abelian, $D \leq C_G(D)$ and so $D \cdot C_G(D) = C_G(D)$. Recall that Proposition 2.11.14 shows that the projection $C_G(D) \longrightarrow C_G(D)/D$ induces a bijection between the $kC_G(D)$-blocks with defect group D and the $kC_G(D)/D$-blocks with defect group $\{1\}$. Proposition 4.4.6 shows that each $kC_G(D)$-block with defect group D is Morita equivalent to kD. Brauer correspondence associates to each block $B = kGf_B$ of G with defect group D and central primitive idempotent f_B a unique block b_D of $kN_G(D)$ with defect group D. Let $f = \beta_D(f_B)$ be the block idempotent of b_D. Recall from Remark 2.11.2 that for each subgroup H of G containing $N_G(D)$ there is a unique block b_H with defect group D so that b_H is a direct factor of the restriction of B to H. This applies in particular to $H = N_G(D_1)$. So let b be the unique block of $kN_G(D_1)$ with defect group D which is a direct factor of the restriction of B to $N_G(D_1) \times N_G(D_1)$. By Proposition 2.11.15 there is a unique $N_G(D)$-conjugacy class of blocks b' of $kC_G(D)$ with defect group D such that p does not divide $|I_{N_G(D)}(b') : C_G(D)|$ and b' is a direct factor of the restriction of b' to $C_G(D)$. Likewise there is a unique $N_G(D_1)$-conjugacy class of blocks b_1 of $kC_G(D_1)$ such that p does not divide $|I_{N_G(D_1)}(b_1) : C_G(D_1)|$ and b_1 is a direct factor of the restriction of b to $C_G(D_1)$. Let $I(b_1) := I_{N_G(D_1)}(b_1) = \{g \in N_G(D_1) |\ g \cdot b_1 \cdot g^{-1} = b_1\}$ be the inertia subgroup of b_1. Then

$$C_G(D_1) \leq I(b_1) \leq N_G(D_1)$$

and $e := |I(b_1) : C_G(D_1)|$ divides $|N_G(D_1) : C_G(D_1)|$. Since $N_G(D_1)/C_G(D_1)$ maps into $Aut(D_1) \simeq C_{p-1}$ by the conjugation action of $N_G(D_1)$ on D_1, we obtain that e divides $p-1$.

We illustrate the situation by the following scheme which opposes the subgroups to the corresponding blocks

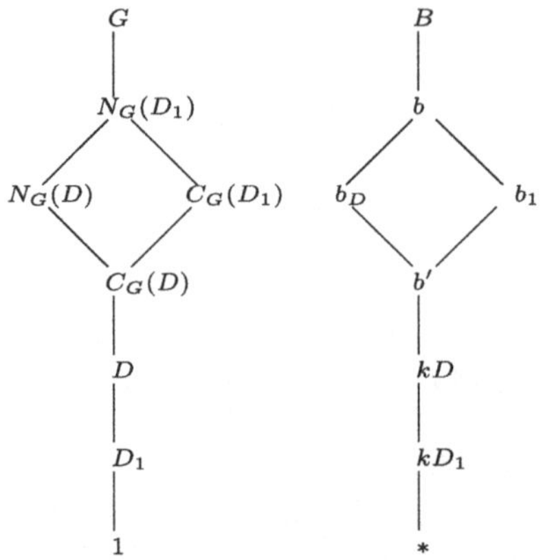

Let M be an indecomposable B-module. Proposition 2.3.6 shows that M has a vertex D_M in D. Since D is cyclic, there is a unique subgroup of D with a given order. Hence $N_G(D_1) \geq N_G(D_M) \geq N_G(D)$. Since $N_G(D) \leq N_G(D_M)$ the Green correspondence g_1 is defined between non-projective indecomposable B-modules of vertex $D_M \leq D$ and indecomposable non-projective $kN_G(D_1)$-modules of vertex D_M.

Lemma 5.10.33 *The Green correspondence g_1 produces a stable equivalence B-$\underline{mod} \longrightarrow b$-$\underline{mod}$. Moreover g_1 commutes with taking syzygies.*

Proof By Corollary 2.10.23 we see that an indecomposable B-module M belongs to B if and only if $g_1(M)$ belongs to b.

Finally, we need to determine the set $\mathfrak{Y}$ in the definition of Green correspondence. We have

$$\mathfrak{Y} = \{xDx^{-1} \mid x \notin N_G(D_1)\}$$

but if $D \cap xDx^{-1} \neq 1$, then $D_1 \leq D \cap xDx^{-1}$, since D_1 is the unique minimal subgroup of D. Hence we get $\mathfrak{Y} = \{1\}$. This is, however, the condition for g_1 to be a stable equivalence (all those indecomposable direct summands of the induced module $M \uparrow_{N_G(D_1)}^G$ which are different from $g_1(M)$ are 1-projective, hence projective.)

Finally g_1 is defined by induction, taking direct summands in the block and taking the unique non-projective direct factor. This produces an exact additive functor, and so g_1 commutes with the syzygy functor. □

Remark 5.10.34 Now, $C_G(D_1) \trianglelefteq N_G(D_1)$ and b covers b_1. By Lemma 2.11.11 and Proposition 2.11.10 applied to the normal subgroup $C_G(D_1)$ of $N_G(D_1)$ we see that b is Morita equivalent to $kI(b_1) \cdot f_1$ where f_1 is the central primitive idempotent of $kC_G(D_1)$ with $b_1 = kC_G(D_1) \cdot f_1$.

Remark 5.10.35 If $D_1 = D$, i.e. D is cyclic of order p, we have already showed at this point that blocks of cyclic defect p are Brauer tree algebras. Indeed, Lemma 5.10.33 shows that B is stably equivalent to b, and Proposition 2.12.4 together with Remark 2.12.5 show that b is Morita equivalent to a symmetric Nakayama algebra. Theorem 5.10.31 then shows that B is a Brauer tree algebra.

We shall prove in the general case that b is Morita equivalent to a symmetric Nakayama algebra with e simple modules.

Proposition 5.10.36 *Let k be algebraically closed. Then the algebra b is Morita equivalent to a symmetric Nakayama algebra $N_e^{\mu\cdot e+1}$ with e simple modules and $\mu = \frac{|D|-1}{e}$.*

Proof By Remark 5.10.34 we only need to prove that $kI(b_1)\cdot f_1$ is Morita equivalent to a Nakayama algebra $N_e^{\mu\cdot e+1}$.

For this we shall prove that b_1 is a nilpotent block. Indeed, we first observe that if Q is a subgroup of D, then Q is cyclic. Observe that $Aut(C_{p^m}) = C_{p-1}\times C_{p^{n-1}}$ for p odd, and $Aut(C_{2^n})$ is a 2-group. In particular, the automorphisms of Q of order relatively prime to p act non-trivially on D_1. Now, $N_{C_G(D_1)}(Q)/C_{C_G(D_1)}(Q)$ injects into the automorphism group of Q. Therefore, $N_{C_G(D_1)}(Q)/C_{C_G(D_1)}(Q)$ is a p-group and b_1 is a nilpotent block by Lemma 4.4.9.

Then by Theorem 4.4.10 we get that b_1 is Morita equivalent to kD. Now $\mathrm{rad}(kD)/\mathrm{rad}^2(kD) \simeq kD/\mathrm{rad}(kD) \simeq k$, the trivial module and Theorem 2.8.4 shows that kD is uniserial. Therefore b_1 is serial and has only one isomorphism class of simple modules. Let therefore S be the unique simple b_1-module, and let P_S be its uniserial projective cover.

We claim that there are precisely e ways to make S into a $kI(b_1)f_1$-module. The group $I(b_1)$ is constructed precisely in such a way that conjugation by any element of $I(b_1)$ fixes b_1. This shows that for every $g\in I(b_1)$ the $kC_G(D_1)$-module ${}^gS$ is isomorphic to S again. Recall that $N_G(D_1)/C_G(D_1)$ acts faithfully on the cyclic group D_1. Since the automorphism group of D_1 is cyclic of order $p-1$, the group $N_G(D_1)/C_G(D_1)$ is a subgroup of the cyclic group C_{p-1} of order $p-1$. Since $I(b_1)$ is isomorphic to a subgroup of $N_G(D_1)$, $I(b_1)/C_G(D_1)$ is isomorphic to a subgroup C_e of the cyclic group C_{p-1}. Let g be a generator of $I(b_1)/C_G(D_1)$. Recall that ${}^gS$ is equal to S as a vector space. Hence the isomorphism ${}^gS\simeq S$ of $kC_G(D_1)$-modules is given by an endomorphism $\theta\in End_k(S)$. Therefore

$$ {}^{g^j}S \stackrel{\theta^j}{\longrightarrow} S $$

is an isomorphism of $kC_G(D_1)$-modules for all $j\in\{1,\dots,e\}$. Since the group $I(b_1)/C_G(D_1)$ is cyclic of order e, we get that

$$\theta^e \in End_{kC_G(D_1)}(S)\simeq k.$$

Hence there is some element $\mu\in k$ such that $\theta^e(s)=\mu\cdot s$ for all $s\in S$. Since p does not divide e and k is algebraically closed, we get that there are e different roots $\lambda_1,\dots,\lambda_e$ in k of the polynomial $X^e-\mu\in k[X]$.

For each λ_i we define a $kI(b_1)$-module structure S_i on S by defining

$$g^j\cdot s := (\lambda_i^{-1}\theta)^j(s)$$

for all $s \in S$. We get that $S_i \simeq S_{i'}$ as $kI(b_1)$-modules if and only if $i = i'$. This is seen by restricting such an isomorphism to $C_G(D_1)$. Each of the modules S_i belongs to $kI(b_1)f_1$ since multiplication by f_1 is already realisable in $kC_G(D_1)$, and as modules over $C_G(D_1)$ all these modules are isomorphic to S. Frobenius reciprocity shows

$$Hom_{kI(f)}(S_i, S\uparrow_{C_G(D_1)}^{I(b_1)}) \simeq Hom_{kC_G(D_1)}(S,S) \simeq k$$

and therefore S_i is a submodule of $S\uparrow_{C_G(D_1)}^{I(b_1)}$. Hence

$$S_1 \oplus \cdots \oplus S_e \leq S\uparrow_{C_G(D_1)}^{I(b_1)} .$$

Since

$$\dim_k(S\uparrow_{C_G(D_1)}^{I(b_1)}) = \dim_k(S)\cdot|I(b_1):C_G(D_1)| = \dim_k(S)\cdot e$$

we get equality and

$$S_1 \oplus \cdots \oplus S_e \simeq S\uparrow_{C_G(D_1)}^{I(b_1)} .$$

Again by Frobenius reciprocity, if U is a simple $kI(b_1)f_1$-module,

$$Hom_{kI(b_1)}(U, S\uparrow_{C_G(D_1)}^{I(b_1)}) \simeq Hom_{C_G(D_1)}(U\downarrow_{C_G(D_1)}^{I(b_1)}, S)$$

and since b_1 is Morita equivalent to kD, all b_1-modules have non-zero homomorphisms to the unique simple module S. Therefore $U \simeq S_i$ for some i.

Now, the projective cover P_S of S is a uniserial projective b_1-module, and since b_1 is a symmetric algebra, P_S is also the injective envelope of S. Since induction $(-)\uparrow_{C_G(D_1)}^{I(b_1)}$ is exact, $P_S\uparrow_{C_G(D_1)}^{I(b_1)}$ is a projective $kI(b_1)f_1$-module with socle $S\uparrow_{C_G(D_1)}^{I(b_1)} \simeq \bigoplus_{i=1}^{e} S_i$. Hence

$$P_S\uparrow_{C_G(D_1)}^{I(b_1)} = \bigoplus_{i=1}^{e} P_i$$

for uniserial projective $kI(b_1)$-modules P_i with $\mathrm{soc}(P_i) \simeq S_i$. Hence $kI(b_1)\cdot f_1$ is a symmetric serial algebra and Lemma 2.8.2 shows that $kI(b_1)\cdot f_1$ is Morita equivalent to a symmetric Nakayama algebra $N_e^{e\cdot\mu+1}$. Since the Loewy length of each of the projective indecomposable modules P_i is equal to the Loewy length of P_S, and this is equal to the Loewy length of kD, which is $|D|$, we have proved the proposition. □

We shall prove the main theorem now. The following result is the promised structure theorem for blocks with not necessarily normal cyclic defect group.

Theorem 5.10.37 *Let k be an algebraically closed field of characteristic $p > 0$ and let G be a finite group. Let B be a block of kG with cyclic defect group D. Then B is* *a Brauer tree algebra with e isomorphism classes of simple modules and exceptional multiplicity*

$$\mu = \frac{|D|-1}{e}.$$

Proof Proposition 5.10.36 shows that b_1 is Morita equivalent to a symmetric Nakayama algebra $N_e^{\mu \cdot e+1}$. Theorem 5.10.31 then shows that B is a Brauer tree algebra with the required multiplicity of the exceptional vertex. □

Remark 5.10.38 Since the proof of Theorem 5.10.37 is quite involved we give a guideline on the different steps of the proof. Starting from G we have the block B with cyclic defect group D to study. The unique subgroup D_1 of order p of D plays a crucial role. Then Brauer correspondence gives a unique block b of $N_G(D_1)$ and covering of blocks from Sect. 2.12.2 produces a unique conjugacy class of blocks b_1 of $C_G(D_1)$. Let $I(b_1)$ be the inertia group of b_1 in $N_G(D_1)$. Then, from Sect. 2.12.2, we identify a block $kI(b_1)f_1$ of $kI(b_1)$.

As the first step we show that b_1 is nilpotent and hence, using Theorem 4.4.10, is Morita equivalent to the group ring of its defect group D. The second step is to show that $kI(b_1)f_1$ is serial symmetric which implies that this block is a symmetric Nakayama algebra, using the results from Sect. 2.8. The third step is that $kI(b_1)f_1$ is Morita equivalent to b, as is shown in Lemma 2.11.11. The fourth and final step is the fact that B and b_1 are stably equivalent by Green correspondence, and Sect. 5.10.3 then gives the theorem.

Remark 5.10.39 The structure theorem for blocks of cyclic defect group can be considered as one of the main achievements and certainly one of the most profound pieces of work of the "classical period" of the representation theory of finite groups in positive characteristic. The module category was described in its general form by Janusz and Kupisch in [26–28]. Richard Brauer described the character theoretic situation in the case $D = D_1$ in [29], and Dade generalised to cyclic defect groups in [30]. Their methods were quite different to our approach. We followed Gabriel–Riedtmann [22] and Linckelmann [23] in parts and emphasised the ring theoretic properties of the Brauer tree algebras.

The approach given here is also used in Auslander–Reiten-Smalø [31]. However, [31] does not give the link to group representations.

Remark 5.10.40 The structure of blocks of finite groups with cyclic defect can be studied further. In particular, it is possible to attach additional attributes to the Brauer trees which give additional information on the structure of the groups, knowing the attributes of the tree, or of the trees knowing the representations of the groups. In particular, there is a close link between the representation theory of characteristic p and of characteristic 0 encoded in the tree. We have not studied this at all here. An application with plenty of interesting examples of Brauer trees can be found in Hiss and Lux [32]. The latter also contains interesting information on the additional structure on Brauer trees.

Remark 5.10.41 The structure of blocks of cyclic defect group has a counterpart in orders over a complete discrete valuation ring R of characteristic 0 and residue field k of characteristic p. This theory was developed by Roggenkamp in [33] where the orders were defined, and where it is essentially proved that the orders satisfying a Green's walk around the Brauer tree [24] have a very restricted structure. Roggenkamp calls them *Green orders.* A detailed discussion of this result can be found in [8]. An alternative discussion of this subject can be found in Plesken [34].

5.11 Stable Equivalences and Self-Injectivity

We shall give an account of Reiten's proof [35] that if A and B are stably equivalent finite dimensional k-algebras, then under some conditions, A is self-injective implies that B is self-injective. More precisely,

Theorem 5.11.1 (Reiten [35, Theorem 2.4]) *Let A be a finite dimensional k-algebra over a field k and let B be an algebra such that A-$\underline{mod} \simeq B$-$\underline{mod}$. Suppose that A is self-injective and suppose that the square of the radical of each direct factor of B is non-zero. Then B is self-injective as well.*

We shall prove the theorem at the end of this Sect. 5.11.

We have already seen that the condition on the direct factors of B having a non-zero square radical is necessary since we have seen in Example 5.2.2 that

$$K[X]/X^2\text{-}\underline{mod} \simeq \begin{pmatrix} K & K \\ 0 & K \end{pmatrix}\text{-}\underline{mod}.$$

The algebra $K[X]/X^2$ is self-injective, even symmetric, whereas $\begin{pmatrix} K & K \\ 0 & K \end{pmatrix}$ is hereditary, and far from being self-injective.

In this context it is most interesting to note that Auslander–Reiten showed in [4] that if $\text{rad}^2(A) = 0$ and if B is stably equivalent to A, then $\text{rad}^2(B) = 0$ as well. Moreover, they show that each of the algebras A with $\text{rad}^2(A) = 0$ is stably equivalent to the hereditary algebra $\begin{pmatrix} A/\text{rad}(A) & 0 \\ \text{rad}(A)/\text{rad}^2(A) & A/\text{rad}(A) \end{pmatrix}$ and that two hereditary algebras are stably equivalent if and only if they are Morita equivalent, up to direct factors of semisimple algebras. We shall come back to these facts in Sect. 6.9.4.

The method used in [35] is very different from what we have done up to now. In particular the proof depends heavily on the use of functor categories. This sometimes powerful tool has not yet appeared in our presentation, so now is an ideal time to introduce the most basic results concerning this approach.

What we present here is due to Auslander–Reiten [4, 36] and Reiten [35]. Some parts of the proofs are slightly simpler and some statements are less general since we do not need them in full generality.

5.11.1 Functor Categories

Let us first give the necessary definitions. We shall specialise the definition from Example 3.1.23.

Definition 5.11.2 Let A be a finite dimensional k-algebra over a field k. Let $Fun(A\text{-}mod, \mathbb{Z}\text{-}mod)$ be the category with objects the contravariant functors $A\text{-}mod \longrightarrow \mathbb{Z}\text{-}mod$ and morphisms the natural transformations between such functors.

We remind the reader of Example 3.1.23. We note that $Fun(A\text{-}mod, \mathbb{Z}\text{-}mod)$ is an abelian category, and abstract properties of this kind do not depend on whether we take covariant or contravariant functors. Indeed, taking the opposite category interchanges covariant and contravariant. Here we need to assume that the natural transformations between two functors form a set! This is certainly the case if the isomorphism classes of objects in the category we start from form a set, which is clearly the case for $A\text{-}mod$.

Lemma 5.11.3 *Let $\mathcal{C}$ be a category whose isomorphism class of objects form a set, and let $\mathcal{A}$ be an abelian category. Then the category $Fun(\mathcal{C}, \mathcal{A})$ whose objects are functors $\mathcal{C} \longrightarrow \mathcal{A}$ and whose morphisms are natural transformations is abelian.*

Proof First, $Fun(\mathcal{C}, \mathcal{A})$ is additive by putting $(F \oplus G)(V) := F(V) \oplus G(V)$ and $(F \oplus G)(\alpha) := F(\alpha) \oplus G(\alpha)$ for all objects V, W of $\mathcal{C}$ and every morphism $\alpha \in Hom_{\mathcal{C}}(V, W)$. The set of natural transformation $Nattrans(F, G)$ is clearly an abelian group by putting $(\eta + \tau)_V := \eta_V + \tau_V$ for all objects V of $\mathcal{C}$ and all $\eta, \tau \in Nattrans(F, G)$. Moreover, if $\eta : F \longrightarrow G$ and $\tau : G \longrightarrow H$ are natural transformations then $\tau \circ \eta : F \longrightarrow H$ is a natural transformation and the composition

$$\circ : Nattrans(G, H) \times Nattrans(F, G) \longrightarrow Nattrans(F, H)$$

is $\mathbb{Z}$-bilinear, since when we evaluate at an object V of $\mathcal{C}$ the property follows from the property in $\mathcal{A}$, which is an abelian category.

Finally, let $\eta : F \longrightarrow G$ be a natural transformation. Then for all objects V of $\mathcal{C}$ we put

$$(\ker(\eta))_V := \ker(\eta_V)$$

and

$$(\text{coker}(\eta))_V := \text{coker}(\eta_V).$$

The universal properties of $\ker(\eta)$ and of $\text{coker}(\eta)$ follow from the universal properties on each evaluation. Moreover, by this definition it is also clear that if η is a monomorphism, i.e. η is a natural transformation $F \longrightarrow G$ such that η_V is a monomorphism for each V, then the natural morphism $\Gamma(\eta) : G \longrightarrow \text{coker}(\eta)$ given by $(\Gamma(\eta))_V := \text{coker}(\eta_V)$ is a natural transformation. This follows by the universal property of the cokernel. Moreover, $\ker(\Gamma(\eta)) = \eta$ since this is true on the level of evaluation on each object V. The dual argument shows that each epimorphism is the cokernel of a natural transformation, defined appropriately. □

Definition 5.11.4 Let $mod(A\text{-}mod)$ be the full subcategory of $Fun(A\text{-}mod, \mathbb{Z}\text{-}mod)$ with objects being those functors which are cokernels of natural transformations $Hom_A(-, M) \longrightarrow Hom_A(-, N)$.

Let $\underline{mod}(A\text{-}mod)$ be the full subcategory of $mod(A\text{-}mod)$ with objects being those functors F with $F(P) = 0$ for all projective A-modules P.

We should note that $mod(A\text{-}mod)$ is a rather small subcategory of $Fun(A\text{-}mod, \mathbb{Z}\text{-}mod)$. All functors in $mod(A\text{-}mod)$ commute with finite direct sums. Hence the functor mapping V to the dual of the nth tensor power of V is not in $mod(A\text{-}mod)$, for example.

Lemma 5.11.5 *The functor*

$$\begin{array}{rcl} A\text{-}mod & \xrightarrow{\rho} & mod(A\text{-}mod) \\ M & \mapsto & Hom_A(-, M) \end{array}$$

is a fully faithful embedding.

Proof It is clear that $Hom_A(-, M)$ is the cokernel of

$$Hom_A(-, 0) \longrightarrow Hom_A(-, M).$$

Yoneda's lemma shows that there is a natural isomorphism

$$Hom_{mod(A\text{-}mod)}(Hom_A(-, M), Hom_A(-, N)) \simeq Hom_A(M, N)$$

and so the functor ρ is well-defined and a fully faithful embedding. □

Lemma 5.11.6 *The functor ρ induces a natural embedding $A\text{-}\underline{mod} \xrightarrow{\underline{\rho}} \underline{mod}(A\text{-}mod)$ by setting $\underline{\rho}(M) := \underline{Hom}(-, M)$. This embedding induces an equivalence between $A\text{-}\underline{mod}$ and the full subcategory of projective objects of $\underline{mod}(A\text{-}mod)$.*

Proof If the A-linear homomorphism $\alpha : M \longrightarrow N$ factors through a projective module P, i.e.

$$(M \xrightarrow{\beta} P \xrightarrow{\gamma} N) = (M \xrightarrow{\alpha} N)$$

then applying $\underline{\rho}$ gives

$$\left(\underline{Hom}_A(-, M) \xrightarrow{\beta_*} \underline{Hom}_A(-, P) \xrightarrow{\gamma_*} \underline{Hom}_A(-, N)\right)$$
$$= \left(\underline{Hom}_A(-, M) \xrightarrow{\alpha_*} \underline{Hom}_A(-, N)\right)$$

and since P is projective, $\underline{Hom}_A(-,P) = 0$. Therefore $\underline{\rho}$ is well-defined. Yoneda's lemma gives again that $\underline{\rho}$ is a fully faithful embedding.

We need to determine the image of $\underline{\rho}$. Let $P \longrightarrow\!\!\!\!\rightarrow C$ be a projective cover mapping. Denote by (F, G) the natural transformations from the functor F to the functor G. Then

$$Hom_A(-,P) \longrightarrow Hom_A(-,C) \stackrel{\alpha}{\longrightarrow} \underline{Hom}_A(-,C) \longrightarrow 0$$

is exact. If F is a functor in $\underline{mod}(A\text{-}mod)$, then we obtain an exact sequence

$$\begin{array}{ccccccc} 0 \longrightarrow (\underline{Hom}_A(-,C), F) & \stackrel{(\alpha,F)}{\longrightarrow} & (Hom_A(-,C), F) & \longrightarrow & (Hom_A(-,P), F) \\ & & \| & & \| \\ & & F(C) & & F(P) \\ & & & & \| \\ & & & & 0 \end{array}$$

in which we may identify the first two terms. Hence $(\underline{Hom}_A(-,C), F) \simeq F(C)$. If

$$0 \longrightarrow F' \longrightarrow F \longrightarrow F'' \longrightarrow 0$$

is an exact sequence in the category $\underline{mod}(A\text{-}mod)$, we may apply the functor $Nattrans(\underline{Hom}_A(-,C), -)$, which is exact by the above, to get an exact sequence

$$0 \to (\underline{Hom}_A(-,C), F') \to (\underline{Hom}_A(-,C), F) \to (\underline{Hom}_A(-,C), F'') \to 0.$$

Hence $\underline{Hom}_A(-,C)$ is exact in $\underline{mod}(A\text{-}mod)$.

We want to prove that all projective objects in $\underline{mod}(A\text{-}mod)$ are presentable. Let F be a projective object in $\underline{mod}(A\text{-}mod)$. Since F is in $mod(A\text{-}mod)$, the functor F is a cokernel of a morphism between two representable functors. In particular there is a module C and an epimorphism $f : Hom_A(-,C) \longrightarrow F$. Since by the above

$$(\underline{Hom}_A(-,C), F) \simeq F(C) \simeq (Hom_A(-,C), F),$$

there is a $g : \underline{Hom}_A(-,C) \longrightarrow F$ making the diagram

$$\begin{array}{ccc} Hom_A(-,C) & \stackrel{f}{\longrightarrow} & F \\ \downarrow & & \uparrow g \\ \underline{Hom}_A(-,C) & = & \underline{Hom}_A(-,C) \end{array}$$

commutative. Since f is an epimorphism, g is an epimorphism as well. Since F is a projective object, g is split and hence F is a direct summand of $\underline{Hom}_A(-,C)$. Direct summands of $\underline{Hom}_A(-,C)$ are the functors $\underline{Hom}_A(-,C')$ for direct summands C' of C. Indeed, a direct summand of $\underline{Hom}_A(-,C)$ is associated with an idempotent endomorphism of $\underline{Hom}_A(-,C)$, and the endomorphisms of $\underline{Hom}_A(-,C)$ are precisely $\underline{End}_A(C)$. An idempotent of this ring corresponds with a direct summand C' of C since by Propositions 5.1.6 and 1.9.17 idempotent endomorphisms of the stable category come from idempotent endomorphisms in the module category. Hence F is presentable. □

Lemma 5.11.7 *$mod(A\text{-}mod)$ and $\underline{mod}(A\text{-}mod)$ are both abelian.*

Proof We need to show that kernels and cokernels of mappings between objects of $mod(A\text{-}mod)$ (or $\underline{mod}(A\text{-}mod)$ respectively), taken in $Fun(A\text{-}mod, \mathbb{Z}\text{-}mod)$, are in $mod(A\text{-}mod)$ (or $\underline{mod}(A\text{-}mod)$ respectively).

First $Hom_A(-, M)$ are projective objects in $Fun(A\text{-}mod, \mathbb{Z}\text{-}mod)$ by Yoneda's lemma. Hence given two objects F and G in $mod(A\text{-}mod)$ we can lift a natural transformation $\eta : F \longrightarrow G$ so that we obtain a commutative diagram

$$\begin{array}{ccccccc} Hom_A(-, M_F) & \longrightarrow & Hom_A(-, N_F) & \longrightarrow & F & \longrightarrow & 0 \\ \downarrow \xi & & \downarrow \zeta & & \downarrow \eta & & \\ Hom_A(-, M_G) & \longrightarrow & Hom_A(-, N_G) & \longrightarrow & G & \longrightarrow & 0 \end{array}$$

Since

$$Nattrans(Hom_A(-, N_F), Hom_A(-, N_G)) \simeq Hom_A(N_F, N_G)$$

again by Yoneda's lemma, and likewise for M instead of N, we see that $\xi = Hom_A(-, \alpha)$, $\zeta = Hom_A(-, \beta)$ for some morphisms $\alpha : M_F \longrightarrow M_G$ and $\beta : N_F \longrightarrow N_G$, and we can embed α and β into exact sequences

$$0 \longrightarrow K_M \longrightarrow M_F \longrightarrow M_G \longrightarrow C_M \longrightarrow 0$$

and

$$0 \longrightarrow K_N \longrightarrow N_F \longrightarrow N_G \longrightarrow C_N \longrightarrow 0.$$

These exact sequences induce a commutative diagram with exact lines and columns

$$\begin{array}{ccccccc} Hom_A(-, K_M) & \longrightarrow & Hom_A(-, K_N) & \longrightarrow & K_\eta & \longrightarrow & 0 \\ \downarrow & & \downarrow & & \downarrow & & \\ Hom_A(-, M_F) & \longrightarrow & Hom_A(-, N_F) & \longrightarrow & F & \longrightarrow & 0 \\ \downarrow \xi & & \downarrow \zeta & & \downarrow \eta & & \\ Hom_A(-, M_G) & \longrightarrow & Hom_A(-, N_G) & \longrightarrow & G & \longrightarrow & 0 \\ \downarrow & & \downarrow & & \downarrow & & \\ Hom_A(-, C_M) & \longrightarrow & Hom_A(-, C_N) & \longrightarrow & C_\eta & \longrightarrow & 0 \end{array}$$

and it is clear that $K_\eta = \ker(\eta)$ and $C_\eta = \text{coker}(\eta)$.

If now

$$0 \longrightarrow K \longrightarrow F \longrightarrow G \longrightarrow C \longrightarrow 0$$

is an exact sequence in $mod(A\text{-}mod)$, so that F, G are objects in $\underline{mod}(A\text{-}mod)$, then for every projective A-module P we get $F(P) = G(P) = 0$ and hence $K(P) = 0$ and $C(P) = 0$ since the exactness of the above sequence means that it induces exact sequences on each evaluation. □

Corollary 5.11.8 $\left(A\text{-}\underline{mod} \simeq B\text{-}\underline{mod}\right) \Longleftrightarrow \left(\underline{mod}(A\text{-}mod) \simeq \underline{mod}(B\text{-}mod)\right).$ □

Recall Remarks 1.8.22 and 3.7.11. In an abelian category $\mathcal{A}$ with enough projective objects in the sense of Remark 3.5.45 we can define $Ext^n_{\mathcal{A}}(X, Y)$ as the degree n homology of $Hom_{\mathcal{A}}(P^\bullet_X, Y)$ where $P^\bullet_X$ is a projective resolution of X. As for rings this is well-defined, and does not depend on the chosen projective resolution. Moreover, if $\mathcal{A}$ has enough injective objects in the sense of Remark 3.5.45, then we can take an injective coresolution $I^\bullet_Y$, i.e. a complex

$$0 \longrightarrow I_0 \xrightarrow{\partial_0} I_1 \longrightarrow I_2 \longrightarrow \cdots$$

with degree 0 homology $Y = \ker \partial_0$, and then

$$Ext^n_{\mathcal{A}}(X, Y) = H_n(Hom_{\mathcal{A}}(X, I^\bullet_Y)).$$

Lemma 5.11.9 *If $\mathcal{A}$ has enough projective and enough injective objects, then the two ways of defining $Ext^n_{\mathcal{A}}$ via projective resolutions or via injective co-resolutions give the same result.*

Proof This may be seen by considering $Ext^n_{\mathcal{A}}(X, Y)$ as $H_n(\mathbb{R}Hom_{\mathcal{A}}(X, Y))$ (cf Remark 3.7.9) and this can be computed by injective coresolutions in the second variable, or by projective resolutions in the first variable (cf Remark 3.7.3 and the proof of Proposition 3.7.4).

An explicit and elementary proof in the case of module categories and where $n = 1$ is given in Remark 1.8.22. The proof given there only uses category theoretical concepts of abelian categories and carries over word for word to general abelian categories. Since higher Ext-groups can be interpreted in terms of Ext^1 of syzygies (or cosyzygies, i.e. cokernels of injectives envelopes) an easy induction proof gives the statement. □

Definition 5.11.10 Let $\mathcal{A}$ be an abelian category.

- Assume that $\mathcal{A}$ has enough projective objects in the sense of Remark 3.5.45. We say that an object M has *projective dimension* at most n if there is a projective resolution

$$0 \longrightarrow P_n \longrightarrow P_{n-1} \longrightarrow \cdots \longrightarrow P_1 \longrightarrow P_0 \longrightarrow M \longrightarrow 0.$$

If there is no such n, then we say that M has infinite projective dimension. The *global dimension* of the algebra A is the supremum of the projective dimensions of all objects of $\mathcal{A}$. If $\mathcal{A}$ does not have global dimension n for any integer n, we say that the global dimension of $\mathcal{A}$ is infinite.

- Assume that $\mathcal{A}$ enough injective objects in the sense of Remark 3.5.45. We say that an object M has *injective dimension* at most n if there is an injective coresolution

$$0 \longrightarrow M \longrightarrow I_0 \longrightarrow I_1 \longrightarrow \cdots \longrightarrow I_{n-1} \longrightarrow I_n \longrightarrow 0.$$

If there is no such n, then we say that M has infinite injective dimension.

Lemma 5.11.11 *Let $\mathcal{A}$ be an abelian category. Then the global dimension of $\mathcal{A}$ is at most n if and only if $Ext^i_{\mathcal{A}}(X,Y) = 0$ for each object X and Y and each $i > n$. In particular if $\mathcal{A}$ has enough projective and enough injective objects, then the supremum of the injective dimensions equals the supremum of the projective dimensions of its objects.*

Proof Suppose $\mathcal{A}$ is an abelian category with enough projective objects. Given X, take a projective resolution $P^\bullet_X$ of X of length n. Then $Hom^\bullet_{\mathcal{A}}(P^\bullet_X, Y)$ is a complex of length n. Hence $Ext^i_{\mathcal{A}}(X,Y) = 0$ for each object X and Y and each $i > n$.

Suppose to the contrary $Ext^i_{\mathcal{A}}(X,Y) = 0$ for each object X and Y and each $i > n$. Let

$$\cdots \longrightarrow P_{n+1} \xrightarrow{\partial_{n+1}} P_n \xrightarrow{\partial_n} P_{n-1} \xrightarrow{\partial_{n-1}} \cdots \longrightarrow P_1 \longrightarrow P_0 \longrightarrow M \longrightarrow 0$$

be a projective resolution, which cannot be shorter than n. Put $Y := \mathrm{im}(\partial_n) = \ker(\partial_{n-1})$. Then the natural morphism $P_n \xrightarrow{\partial_n} Y$ gives an element in $Ext^n_{\mathcal{A}}(X,Y)$. This element is non-zero, since otherwise this natural morphism would factor through $P_n \xrightarrow{\partial_n} P_{n-1}$. This shows that ∂_n is split and the resolution can be shortened to a sequence of smaller length. Hence the projective dimension of X is at most n. Since X was arbitrary, the global dimension of $\mathcal{A}$ is at most n. Since we may compute $Ext^i_{\mathcal{A}}$ by injective coresolutions as well, we get by an analogous argument that the global dimension of $\mathcal{A}$ is the supremum of the injective dimensions of its objects. □

Remark 5.11.12 The category $\underline{mod}$(*A-mod*) is more complicated. The objects $Hom_A(-,C)$ are not in $\underline{mod}$(*A-mod*) since they do not vanish on projective objects. The objects $\underline{Hom}_A(-,C)$ are in $\underline{mod}$(*A-mod*) but their properties will differ from $Hom_A(-,C)$. This is the reason why we shall work with injective coresolutions in this case.

Lemma 5.11.13 *Let F be an object of $\underline{mod}$(A-mod). If*

$$Hom_A(-,M) \xrightarrow{Hom_A(-,f)} Hom_A(-,N) \longrightarrow F \longrightarrow 0$$

is an exact sequence of mod(A-mod), then f is an epimorphism in A-mod.

If $M \twoheadrightarrow N$ is an epimorphism of A-modules, then

$$\text{coker}(Hom_A(-, M) \longrightarrow Hom_A(-, N)) =: F$$

is an object of $\underline{mod}(A\text{-}mod)$.

Proof Indeed, the fact that $F(P) = 0$ for all projective A-modules implies that the map $Hom_A(A, f)$ is an epimorphism, and since

$$\begin{array}{ccc} Hom_A(A, M) & \stackrel{Hom_A(A,f)}{\longrightarrow} & Hom_A(A, N) \\ \downarrow\simeq & & \downarrow\simeq \\ M & \stackrel{f}{\longrightarrow} & N \end{array}$$

is commutative, f is an epimorphism and we have obtained the first statement.

If, conversely, $f : M \longrightarrow\!\!\!\!\rightarrow N$ is an epimorphism, then

$$Hom_A(-, M) \longrightarrow Hom_A(-, N) \longrightarrow F \longrightarrow 0$$

is an exact sequence in $mod(A\text{-}mod)$. Now, evaluating at A gives an exact sequence

$$Hom_A(A, M) \longrightarrow Hom_A(A, N) \longrightarrow F(A) \longrightarrow 0$$

and therefore $F(A) = 0$. Moreover, since $Hom_A(-, M)$ and $Hom_A(-, N)$ both commute with finite direct sums, F commutes with finite direct sums as well. This shows that $F(P) = 0$ for every projective A-module P and hence F is an object of $\underline{mod}(A\text{-}mod)$. □

Corollary 5.11.14 *Let F be an object of $\underline{mod}(A\text{-}mod)$. Then there is a short exact sequence*

$$0 \longrightarrow L \longrightarrow M \longrightarrow N \longrightarrow 0$$

of A-modules inducing an exact sequence

$$0 \longrightarrow Hom_A(-, L) \longrightarrow Hom_A(-, M) \longrightarrow Hom_A(-, N) \longrightarrow F \longrightarrow 0$$

in $mod(A\text{-}mod)$.

Proof Lemma 5.11.13 shows that there is an epimorphism $f : M \longrightarrow N$ such that

$$Hom_A(-, M) \longrightarrow Hom_A(-, N) \longrightarrow F \longrightarrow 0$$

is exact. Putting $L := \ker(f)$, the long exact sequence in homology (cf Proposition 3.4.11) proves the statement. □

Definition 5.11.15 A contravariant functor $F : \mathcal{A} \longrightarrow \mathcal{B}$ between exact categories is *half exact* if for each exact sequence $A \longrightarrow B \longrightarrow C$ the image $F(C) \longrightarrow F(B) \longrightarrow F(A)$ is exact.

Lemma 5.11.16 *If an object G in $\underline{mod}(A\text{-}mod)$ is half exact as a functor $A\text{-}mod \longrightarrow \mathbb{Z}\text{-}mod$, then G is injective as an object in $\underline{mod}(A\text{-}mod)$.*

Proof Let F be an object of $\underline{mod}(A\text{-}mod)$. Then by Corollary 5.11.14 we get A-modules L, M, and N and morphisms $\varphi : L \longrightarrow M$ and $\psi : M \longrightarrow N$ such that

$$0 \longrightarrow L \xrightarrow{\varphi} M \xrightarrow{\psi} N \longrightarrow 0$$

is an exact sequence and such that

$$0 \longrightarrow Hom_A(-, L) \xrightarrow{\varphi_*} Hom_A(-, M) \xrightarrow{\psi_*} Hom_A(-, N) \longrightarrow F \longrightarrow 0$$

is exact in $mod(A\text{-}mod)$. Denote by (G_1, G_2) the natural transformations from the functor G_1 to the functor G_2. Now we use the functor $Ext^1_{mod(A\text{-}mod)}(-, -)$ in the appropriate sense via Definition 3.7.8 and the remarks after Corollary 5.11.8, since $mod(A\text{-}mod)$ is abelian with sufficiently many projective objects. Apply $(-, G)$ to the exact sequence

$$Hom_A(-, L) \xrightarrow{\varphi_*} Hom_A(-, M) \xrightarrow{\psi_*} Hom_A(-, N)$$

and observe that $(Hom_A(-, K), G) = G(K)$ for all A-modules K gives that the sequence

$$G(N) \xrightarrow{G(\psi)} G(M) \xrightarrow{G(\varphi)} G(L)$$

is exact. Hence $Ext^1_{mod(A\text{-}mod)}(F, G) = 0$ whenever F is an object of $\underline{mod}(A\text{-}mod)$. Let

$$0 \longrightarrow F_1 \longrightarrow F_2 \longrightarrow F_3 \longrightarrow 0$$

be an exact sequence in $\underline{mod}(A\text{-}mod)$, then

$$0 \longrightarrow (F_1, G) \longrightarrow (F_2, G) \longrightarrow (F_3, G) \longrightarrow Ext^1_{mod(A\text{-}mod)}(F_1, G) = 0$$

is exact and therefore G is injective as claimed. □

Corollary 5.11.17 *For all A-modules C the functor $Ext^1_A(-, C)$ is injective in $\underline{mod}(A\text{-}mod)$.*

Proof Let I_C be the injective envelope of C. Then

$$Hom_A(-, I_C) \longrightarrow Hom_A(-, I_C/C) \longrightarrow Ext^1_A(-, C) \longrightarrow 0$$

is exact, and therefore $Ext^1_A(-, C)$ is an object of $\underline{mod}(A\text{-}mod)$. Moreover, $Ext^1_A(-, C)$ is half exact by Lemma 1.8.36. Lemma 5.11.16 then implies the statement. □

Lemma 5.11.18 *Let A be a finite dimensional k-algebra, let C be an indecomposable A-module, and let $P \longrightarrow C$ be its projective cover inducing a short exact sequence*

$$0 \longrightarrow K \longrightarrow P \longrightarrow C \longrightarrow 0.$$

Then the functors $Ext^1_A(-, K)$ and $Ext^1_A(-, P)$ are injective objects in the category $\underline{mod}(A\text{-}mod)$ and the above sequence induces an exact sequence

$$0 \longrightarrow \underline{Hom}_A(-, C) \longrightarrow Ext^1_A(-, K) \longrightarrow Ext^1_A(-, P)$$

so that $\underline{Hom}_A(-, C) \longrightarrow Ext^1_A(-, K)$ is an injective envelope, and the long exact sequence

$$\begin{array}{ccccccc} 0 \longrightarrow \underline{Hom}_A(-, C) \longrightarrow & Ext^1_A(-, K) & \longrightarrow & Ext^1_A(-, P) & \longrightarrow & Ext^1_A(-, C) \\ & & & & & \downarrow \\ \dots \quad \longleftarrow & Ext^2_A(-, C) & \longleftarrow & Ext^2_A(-, P) & \longleftarrow & Ext^2_A(-, K) \end{array}$$

is an injective co-resolution.

Proof It is clear that $Ext^1_A(-, K)$ and $Ext^1_A(-, P)$ vanish on projective modules.

If F is in $\underline{mod}(A\text{-}mod)$, we have seen in Corollary 5.11.14 that there is a projective resolution

$$0 \to Hom_A(-, L) \to Hom_A(-, M) \to Hom_A(-, N) \to F \to 0$$

which comes from an exact sequence

$$0 \longrightarrow L \longrightarrow M \longrightarrow N \longrightarrow 0$$

which in turn induces a long exact sequence

$$\begin{array}{ccccc} 0 \to Hom_A(-, L) & \to & Hom_A(-, M) & \to & Hom_A(-, N) \\ & & & & \downarrow \\ Ext^1_A(-, N) & \leftarrow & Ext^1_A(-, M) & \leftarrow & Ext^1_A(-, L) \end{array}$$

and since

$$0 \to Hom_A(-, L) \to Hom_A(-, M) \to Hom_A(-, N) \to F \to 0$$

is exact the map $Hom_A(-, N) \to Ext^1_A(-, L)$ factors through F, so that there is an exact sequence

$$0 \longrightarrow F \longrightarrow Ext^1_A(-, L) \longrightarrow Ext^1_A(-, M) \longrightarrow Ext^1_A(-, N).$$

Since for all K the functor $Ext_A^1(-, K)$ is half exact as a functor $A\text{-}mod \longrightarrow \mathbb{Z}\text{-}mod$, Lemma 5.11.16 shows that this is an injective resolution of F. If F itself is injective, F is a direct factor of $Ext_A^1(-, L)$ and direct factors of $Ext_A^1(-, L)$ are of the form $Ext_A^1(-, L')$ for some direct summand L' of L, as is shown using the analogous result for Hom-functors. For $Hom_A(-, M)$ the argument is the following. A direct summand of $Hom_A(-, M)$ is given by an idempotent endomorphism of $Hom_A(-, M)$. Since by Yoneda's lemma

$$Nattrans(Hom_A(-, M), Hom_A(-, M)) \simeq Hom_A(M, M) = End_A(M)$$

such an idempotent endomorphism is actually induced by an idempotent endomorphism of M, and hence a direct factor M' of M. Now, let $0 \to L \to I_L \to \mho_L \to 0$ be an exact sequence and I_L be the injective envelope of L. Then we get an exact sequence

$$0 \to Hom_A(-, L) \to Hom_A(-, I_L) \to Hom_A(-, \mho_L) \to Ext_A^1(-, L) \to 0.$$

This is a projective resolution of $Ext_A^1(-, L)$ and hence, using that the direct sum of two resolutions is the resolution of the direct sum, an idempotent endomorphism of $Ext_A^1(-, L)$ lifts to idempotent endomorphisms along the resolution. These give compatible direct sum decompositions of each of the modules L, I_L and $\mho_L$, and therefore the statement follows.

The statement on the long exact sequence as an injective coresolution is then clear by definition and since higher Ext^n can be expressed as Ext^{n-1} of appropriately defined syzygy objects. □

We even get a characterisation of the finite global dimension of A-*mod* in terms of the global dimension of $\underline{mod}(A\text{-}mod)$.

Corollary 5.11.19 *Let A be a finite dimensional algebra of global dimension at most n. Then the global dimension of $\underline{mod}(A\text{-}mod)$ is at most $3n - 1$.*

Proof Suppose that the global dimension of A is n. Then $Ext_A^n(X, Y) = 0$ for all A-modules X, Y. By Corollary 5.11.14 there is a short exact sequence

$$0 \longrightarrow L \longrightarrow M \longrightarrow N \longrightarrow 0$$

of A-modules such that

$$0 \longrightarrow Hom_A(-, L) \longrightarrow Hom_A(-, M) \longrightarrow Hom_A(-, N) \longrightarrow F \longrightarrow 0$$

is an exact sequence in $mod(A\text{-}mod)$. But continuing the long exact sequence further we get that

$$\begin{aligned} F \hookrightarrow Ext_A^1(-, L) \to Ext_A^1(-, M) \to \cdots \to Ext_A^{n-1}(-, N) \\ \to Ext_A^n(-, L) \to \cdots \end{aligned}$$

is an injective co-resolution of F in $\underline{mod}(A\text{-}mod)$. Since A is of global dimension at most n, $Ext^n_A(-, L) = 0$ and this gives an injective coresolution of length $3n$. This proves the statement. □

Proposition 5.11.20 *Every non-simple projective indecomposable A-module is injective if and only if all projective objects in* $\underline{mod}(A\text{-}mod)$ *are injective.*

Every non-simple injective indecomposable A-module is projective if and only if all injective objects in $\underline{mod}(A\text{-}mod)$ *are projective.*

Proof It is clear that the second statement is just the dual of the first. So we only need to prove the first statement.

Suppose that each non-simple projective indecomposable A-module is injective. Let $\underline{Hom}_A(-, C)$ be an indecomposable projective object of $\underline{mod}(A\text{-}mod)$, and let P be a projective cover of C with $k := \ker(P \to C)$. Then C is an indecomposable non-projective A-module and P does not have any simple direct summands. Indeed, otherwise $C/\text{rad}(C) = P/\text{rad}(P)$ has a simple direct factor S, and therefore C maps onto the projective module S. Since C is indecomposable non-projective this is a contradiction. The assumption implies therefore that P is injective. Lemma 5.11.18 shows that

$$0 \longrightarrow \underline{Hom}_A(-, C) \longrightarrow Ext^1_A(-, K) \longrightarrow Ext^1_A(-, P)$$

is exact, and the rightmost two terms are injective. But since P is injective as an A-module by hypothesis, $Ext^1_A(-, P) = 0$ and therefore

$$\underline{Hom}_A(-, C) \simeq Ext^1_A(-, K)$$

is injective.

Suppose that each projective object of $\underline{mod}(A\text{-}mod)$ is injective and let P be a non-simple projective indecomposable A-module. Then we get $\text{soc}(P) \leq \text{rad}(P)$ since P is not simple. Again by Lemma 5.11.18

$$0 \longrightarrow \underline{Hom}_A(-, P/\text{soc}(P)) \longrightarrow Ext^1_A(-, \text{soc}(P)) \longrightarrow Ext^1_A(-, P)$$

is exact, and the two rightmost terms are injective in $\underline{mod}(A\text{-}mod)$. By hypothesis $\underline{Hom}_A(-, P/\text{soc}(P))$ is injective, and therefore the sequence is split. Since

$$\underline{Hom}_A(-, P/\text{soc}(P)) \hookrightarrow Ext^1_A(-, \text{soc}(P))$$

is the injective envelope, $Ext^1_A(-, P) = 0$. This proves that P is injective.

This proves the proposition. □

Corollary 5.11.21 *Let B be a finite dimensional k-algebra which is stably equivalent to a self-injective algebra A. Then each non-simple projective B-module is injective, and each non-simple injective B-module is projective.* □

5.11.2 The Proof of Reiten's Theorem

We follow Reiten's paper [35]. In this section a finite dimensional k-algebra such that

- each non-simple indecomposable projective A-module is injective,
- each non-simple indecomposable injective A-module is projective, and
- there is at least one simple injective A-module,

is called a *counterexample algebra.*

Lemma 5.11.22 *Let K be a field, let k be a skew-field over K, let S be an algebra over K and let $M \neq 0$ be an S–k-bimodule. Suppose M, S and k are finite dimensional over K. Put $A := \begin{pmatrix} k & 0 \\ M & S \end{pmatrix}$. Then each indecomposable projective A-module is either isomorphic to $\binom{0}{P}$ for some indecomposable projective S-module P, or isomorphic to $\binom{k}{M}$.*

Proof We first observe that $I := \begin{pmatrix} 0 & 0 \\ M & 0 \end{pmatrix}$ is an ideal of A with square 0, and hence this ideal is in the radical of A. Further $A/I \simeq k \oplus S$ and since $End_A(k) = k$ is a skew-field, k is indecomposable, even simple. Hence, a simple A-module is either isomorphic to $P/\mathrm{rad}(P)$ for P a projective indecomposable S-module, or to k. However, the projective cover of k is $\binom{k}{M}$ and the projective cover of $P/\mathrm{rad}(P)$ is $\binom{0}{P}$. □

We construct A-modules in the following way. Given a k-module V and an S-module B, and given a homomorphism $f : M \otimes_k V \longrightarrow B$ then define the following action on $V \oplus B$:

$$\begin{pmatrix} \lambda & 0 \\ m & s \end{pmatrix} \cdot \begin{pmatrix} v \\ b \end{pmatrix} := \begin{pmatrix} \lambda \cdot v \\ f(m \otimes v) + s \cdot b \end{pmatrix} \quad \forall \lambda \in k, m \in M, s \in S, v \in V, b \in B.$$

This turns $V \oplus B$ into an A-module, denoted by $\binom{V}{B}_f$. A special case is given when $V = Hom_S(M, B)$ and $f(m \otimes g) = g(m)$ for all $g \in Hom_S(M, B)$ and $m \in M$. Then we denote this module by $\binom{Hom_S(M,B)}{B}_{eval}$, or just $\binom{Hom_S(M,B)}{B}$ if no confusion may arise.

Actually each A-module is isomorphic to a module $\binom{V}{B}_f$. Moreover,

$$Hom_A\left(\begin{pmatrix} V \\ B \end{pmatrix}_f, \begin{pmatrix} W \\ C \end{pmatrix}_g\right)$$
$$= \{(\alpha, \beta) \in Hom_k(V, W) \times Hom_S(B, C) \mid g \circ (id_M \otimes \alpha) = \beta \circ f\}.$$

These facts are easily verified by direct computation since any module homomorphism must commute with the A-action. In particular, the multiplication by $\begin{pmatrix} 1_k & 0 \\ 0 & 0 \end{pmatrix}$

and $\begin{pmatrix} 0 & 0 \\ 0 & 1_S \end{pmatrix}$ will give the composition as $(\alpha, \beta) \in Hom_k(V, W) \times Hom_S(W, W')$ and the condition on the maps then follows easily.

Lemma 5.11.23 *Let K be a field, let k be a skew-field over K, let S be an algebra over K and let $M \neq 0$ be an S–k-bimodule. Suppose M, S and k are finite dimensional over K. Put $A := \begin{pmatrix} k & 0 \\ M & S \end{pmatrix}$. Then each indecomposable injective A-module is isomorphic to one of the following modules:*

$$\text{coker}\left(\begin{pmatrix} 0 \\ M \end{pmatrix} \longrightarrow \begin{pmatrix} k \\ M \end{pmatrix}\right) \qquad \textit{or to} \qquad \begin{pmatrix} Hom_S(M, I) \\ I \end{pmatrix}_{eval}$$

for some indecomposable injective S-module I. Moreover $\text{coker}\left(\binom{0}{M} \longrightarrow \binom{k}{M}\right)$ *is injective.*

Proof It is clear that $\text{coker}\left(\binom{0}{M} \longrightarrow \binom{k}{M}\right)$ is an injective A-module since k is a skew-field.

A monomorphism

$$U \overset{\alpha}{\hookrightarrow} W$$

of S-modules induces a monomorphism of A-modules

$$\begin{pmatrix} Hom_S(M, U) \\ U \end{pmatrix} \hookrightarrow \begin{pmatrix} Hom_S(M, W) \\ W \end{pmatrix}.$$

If $\binom{Hom_S(M,J)}{J}$ is injective, then we need to lift any morphism

$$\begin{pmatrix} Hom_S(M, U) \\ U \end{pmatrix} \longrightarrow \begin{pmatrix} Hom_S(M, J) \\ J \end{pmatrix}$$

to a morphism

$$\begin{pmatrix} Hom_S(M, W) \\ W \end{pmatrix} \longrightarrow \begin{pmatrix} Hom_S(M, J) \\ J \end{pmatrix}.$$

Since this means any S-linear morphism $U \longrightarrow J$ has to lift along $U \hookrightarrow W$ to an S-linear morphism $W \longrightarrow J$, J has to be an injective S-module.

Let W be a k-module. Given a k-module X and a morphism $f \in Hom_S(M \otimes X, W)$, we have

$$Hom_S(M \otimes_k X, W) \simeq Hom_k(X, Hom_S(M, W))$$

and so f induces $\beta_f : X \longrightarrow Hom_S(M, W)$. Moreover,

$$eval \circ (id_M \otimes \beta_f) = f = id_W \circ f$$

and so

$$\begin{pmatrix} X \\ W \end{pmatrix}_f \stackrel{\binom{\beta_f}{id_W}}{\hookrightarrow} \begin{pmatrix} Hom_S(M,W) \\ W \end{pmatrix}$$

is an embedding. The injective hull of $\binom{X}{W}_f$ therefore has the form $\binom{Y}{I_W}_g$ for some g, and the injective hull

$$W \xrightarrow{\iota_W} I_W$$

of W. However, let W be a non-zero S-module and let I_W be its injective envelope.

$$Hom_S(M \otimes_k Y, I_W) \simeq Hom_k(Y, Hom_S(M, I_W))$$

and so a morphism $g \in Hom_S(M \otimes_k Y, I_W)$ corresponds to a well-defined morphism $\varphi_g \in Hom_k(Y, Hom_S(M, I_W))$. Then

$$\begin{array}{ccc} M \otimes_k Y & \stackrel{1_M \otimes \varphi_g}{\longrightarrow} & M \otimes_k Hom_S(M, I_W) \\ \downarrow g & & \downarrow eval \\ I_W & = & I_W \end{array}$$

is commutative, and therefore

$$\begin{pmatrix} 1 \otimes \varphi_g \\ id_{I_W} \end{pmatrix} : \begin{pmatrix} Y \\ I_W \end{pmatrix}_f \longrightarrow \begin{pmatrix} Hom_S(M, I_W) \\ I_W \end{pmatrix}$$

is a morphism between these modules. The kernel of $1 \otimes \varphi_g$ is a direct sum C of modules of type

$$\text{coker}\left(\begin{pmatrix} 0 \\ M \end{pmatrix} \longrightarrow \begin{pmatrix} k \\ M \end{pmatrix}\right)$$

and so is the cokernel D. Since k is a skew-field, C is direct factor of $\binom{Y}{I_W}_f$ and likewise D is a direct factor of $\binom{Hom_S(M,I_W)}{I_W}$. Hence if $\binom{Y}{I_W}_f$ is indecomposable injective, φ_g is an isomorphism. This proves the lemma. □

Lemma 5.11.24 *Let K be a field, let k be a skew-field over K, let S be an algebra over K and let M be an S–k-bimodule. Suppose M, S and k are finite dimensional over K. Put $A := \begin{pmatrix} k & 0 \\ M & S \end{pmatrix}$. Then A is a counterexample algebra if and only if S is a counterexample algebra, M is a simple injective S-module and $End_S(M) = k$.*

Proof Suppose first that A is a counterexample algebra.

The action of k on M which is used to define the bimodule structure of the left S-module M is denoted by $\epsilon_0 : k \longrightarrow End_S(M)$. We shall show that ϵ_0 is an isomorphism.

The homomorphism ϵ_0 is injective since k is a skew-field. Suppose ϵ is not surjective and let $\alpha \in End_S(M)$ be an endomorphism which is not in the image of ϵ_0. Since

$$\begin{pmatrix} 0 \\ M \end{pmatrix} \hookrightarrow \begin{pmatrix} k \\ M \end{pmatrix}$$

is an embedding into an injective module, and since

$$\begin{pmatrix} 0 \\ M \end{pmatrix} \xrightarrow{\begin{pmatrix} 0 & 0 \\ 0 & \alpha \end{pmatrix}} \begin{pmatrix} 0 \\ M \end{pmatrix}$$

is a homomorphism of A-modules, this endomorphism can be extended to an endomorphism of the injective module $\binom{k}{M}$. Hence there is a matrix $\begin{pmatrix} u & v \\ w & \alpha \end{pmatrix}$ such that the diagram

$$\begin{array}{ccc} \begin{pmatrix} 0 \\ M \end{pmatrix} & \xrightarrow{\begin{pmatrix} 0 & 0 \\ 0 & \alpha \end{pmatrix}} & \begin{pmatrix} 0 \\ M \end{pmatrix} \\ \downarrow & & \downarrow \\ \begin{pmatrix} k \\ M \end{pmatrix} & \xrightarrow{\begin{pmatrix} u & v \\ w & \alpha \end{pmatrix}} & \begin{pmatrix} k \\ M \end{pmatrix} \end{array}$$

commutes. Since $\begin{pmatrix} u & v \\ w & \alpha \end{pmatrix}$ is A-linear, it commutes with $\begin{pmatrix} 1 & 0 \\ 0 & 0 \end{pmatrix}$ in particular, and therefore $v = 0$ and $w = 0$. But this morphism is not A-linear since it does not commute with $\begin{pmatrix} 1 & 0 \\ m & s \end{pmatrix}$. This contradiction shows that $End_S(M) \overset{\epsilon_0}{\simeq} k$ and so M is indecomposable since k is a skew field.

Let P be a non-simple projective S-module. Then $\binom{0}{P}$ is a non-simple projective A-module, and hence injective since A is a counterexample algebra. Lemma 5.11.23 implies that P is an injective S-module and $Hom_S(M, P) = 0$.

We now show that M is an injective S-module: The module

$$\begin{pmatrix} k \\ M \end{pmatrix} = \begin{pmatrix} k \\ M \end{pmatrix}_{id}$$

is projective, non-simple since it contains the submodule $\binom{0}{M}$, and therefore injective since A is a counterexample algebra. Lemma 5.11.23 shows that M is an injective S-module and $Hom_S(M, M) \simeq k$, as a k-module.

We shall show that M is simple as an S-module. Suppose otherwise let N be a simple S-submodule of M and let T be a simple quotient module of M. Let I_T be the injective envelope of T. Then $Hom_S(M, I_T) \neq 0$ by construction. If T is not injective, then the A-module $\binom{Hom_S(M,I_T)}{I_T}$ is indecomposable non-simple injective, and hence projective by the same argument as above. Again we deduce in the very same way that $I_T \simeq M$, and therefore $N \simeq T$. If this isomorphism is not just an identity, the composition

$$M \longrightarrow\!\!\!\!\rightarrow T \simeq N \hookrightarrow M$$

is nilpotent and non-zero. This contradicts the fact that $k \simeq End_S(M)$, which we proved above. Hence M is simple injective as an S-module.

We need to show that each non-simple injective S-module I is projective. Indeed, if I is non-simple indecomposable injective, then the injective hull of $\binom{0}{I}$ as an A-module must be isomorphic to $\binom{Hom_S(M,I)}{I}$, which is non-simple injective and therefore projective since A is a counterexample algebra. But the only two indecomposable projective A-modules of this type have one of the properties $Hom_S(M, I) = k$ or $Hom_S(M, I) = 0$. In the first case we get $M \simeq I$, which is excluded since M is simple. In the second case $\binom{0}{I}$ is projective as an A-module since A is a counterexample algebra and I is non-simple injective. Therefore I is projective as an S-module, and we are done.

If $S = S_1 \times S_2$ is decomposable, then A is decomposable. Indeed, since M is indecomposable, either S_1 or S_2 acts as 0 on M, say S_2, and then $A \simeq A_1 \times S_2$ for $A_1 = \begin{pmatrix} k & 0 \\ M & S_1 \end{pmatrix}$.

Suppose now that S is a counterexample algebra, that the S–k-bimodule M is a simple injective S-module and that the natural map $k \stackrel{\epsilon_0}{\to} End_S(M)$ is an isomorphism.

Each non-simple indecomposable projective A-module $\binom{0}{P}$ is given by a non-simple indecomposable projective S-module P, which is by hypothesis injective as an S-module. Since M is simple injective, $Hom_S(M, P) = 0$, and therefore its injective hull is isomorphic to

$$\begin{pmatrix} Hom_S(M, P) \\ P \end{pmatrix} = \begin{pmatrix} 0 \\ P \end{pmatrix}$$

which is therefore an injective A-module.

The indecomposable projective A-module $\binom{k}{M}$ is injective as well, since M is a simple injective S-module and since $k \stackrel{\epsilon_0}{\to} End_S(M)$ is an isomorphism.

Let $\binom{Hom_S(M,I)}{I}$ be a non-simple indecomposable injective A-module. Since M is simple injective, either $M \simeq I$ or $Hom_S(M, I) = 0$. If $M \simeq I$ we are done. In the second case $\binom{0}{I}$ is a non-simple indecomposable injective A-module if and only if I is a non-simple indecomposable injective S-module, I is projective by hypothesis, and therefore $\binom{0}{I}$ is a projective A-module.

The simple A-module

$$\operatorname{coker}\left(\begin{pmatrix}0\\M\end{pmatrix}\hookrightarrow\begin{pmatrix}k\\M\end{pmatrix}\right)$$

is injective and not projective. Therefore A is a counterexample algebra as well. □

Lemma 5.11.25 *Let A be a counterexample algebra over some field K. Then the global dimension of A is finite.*

Proof We may suppose that A is basic since the global dimension is a Morita invariant. We shall prove the lemma by induction on the number n of isomorphism classes of simple A-modules.

If $n = 1$, then the only simple module must be projective and hence $\text{gldim}(A) = 0$.

Suppose $n > 1$. Since A is assumed to be a counterexample algebra, there is a simple injective A-module T. Then $Hom_K(T, K)$ is a projective A^{op}-module. Let P be a projective A^{op}-module such that $P \oplus Hom_K(T, K)$ is a progenerator of A^{op} and $End_{A^{op}}(P \oplus Hom_K(T, K))$ is basic. (Note that in order to define P one has to take one copy of each isomorphism class of a projective indecomposable A^{op}-module not isomorphic to $Hom_K(T, K)$ and form the direct sum of these modules.) Since T is simple injective,

$$Hom_{A^{op}}(P, Hom_K(T, K)) = Hom_K(P \otimes_A T, K) = Hom_A(T, Hom_K(P, K))$$

is 0 since any non zero homomorphism has to be injective, and splits therefore since T is injective. Hence there is an indecomposable direct factor of P isomorphic to the dual of T. This was excluded. Therefore

$$A^{op} \simeq \begin{pmatrix} End_{A^{op}}(Hom_K(T, K)) & 0 \\ Hom_{A^{op}}(P, Hom_K(T, K)) & End_{A^{op}}(P) \end{pmatrix}$$

and putting

$$\begin{aligned} M &:= Hom_{A^{op}}(P, Hom_K(T, K)),\\ k &:= End_A(T) = End_{A^{op}}(Hom_K(T, K)),\\ S &:= End_{A^{op}}(P), \end{aligned}$$

this is exactly the situation of Lemma 5.11.24. Hence $S := End_{A^{op}}(P)$ is a counterexample algebra as well, and by construction has $n - 1$ isomorphism classes of simple modules. The induction hypothesis then shows that the global dimension of S is finite.

The simple A-module

$$L := \operatorname{coker}\left(\begin{pmatrix}0\\M\end{pmatrix}\hookrightarrow\begin{pmatrix}k\\M\end{pmatrix}\right)$$

has a projective resolution

$$\hat{Q}^\bullet \longrightarrow \binom{k}{M} \longrightarrow L \longrightarrow 0$$

where

$$Q^\bullet : \cdots \longrightarrow Q^2 \longrightarrow Q^1 \longrightarrow Q^0 \longrightarrow M$$

is a projective resolution of M as an S-module, and

$$\hat{Q}^\bullet : \cdots \longrightarrow \binom{0}{Q^2} \longrightarrow \binom{0}{Q^1} \longrightarrow \binom{0}{Q^0} \longrightarrow \binom{0}{M} \longrightarrow 0$$

is a projective resolution of $\binom{0}{M}$ as an A-module. Hence if the projective dimension of each simple S-module is at most m, then the projective dimension of each simple A-module is at most $m+1$. This shows that the global dimension of A is at most $m+1$. Indeed, if $U^\bullet$ is a projective resolution of a module U_0 and $V^\bullet$ is a projective resolution of V_0, and if

$$0 \longrightarrow U_0 \longrightarrow W_0 \longrightarrow V_0 \longrightarrow 0$$

is an exact sequence, by the Horseshoe Lemma 3.5.50 we can get a projective resolution of the module W_0 whose homogeneous degrees are the direct sum of the homogeneous degrees of $U^\bullet$ and of $V^\bullet$. This finishes the proof. □

We can now prove Theorem 5.11.1 First we may suppose without loss of generality that A is indecomposable and suppose that each non-simple indecomposable projective A-module is injective and that each non-simple indecomposable injective A-module is projective. Proposition 5.11.20 then shows that each projective object of $\underline{mod}(A\text{-}mod)$ is injective and each injective object of $\underline{mod}(A\text{-}mod)$ is projective. By Corollary 5.11.8 we get that $\underline{mod}(A\text{-}mod) \simeq \underline{mod}(B\text{-}mod)$ and hence each projective object of $\underline{mod}(B\text{-}mod)$ is injective and each injective object of $\underline{mod}(B\text{-}mod)$ is projective. Again by Proposition 5.11.20 this means that each non-simple indecomposable injective B-module is projective and each non-simple indecomposable projective B-module is injective. If B is not self-injective, B is a counterexample algebra and by Lemma 5.11.25, of finite global dimension. By Corollary 5.11.19 the global dimension of $\underline{mod}(B\text{-}mod)$ is finite as well. But since every projective object in $\underline{mod}(B\text{-}mod)$ is also injective, the global dimension of $\underline{mod}(B\text{-}mod)$ is finite if and only if the global dimension of $\underline{mod}(B\text{-}mod)$ is 0.

By Corollary 5.11.21 each non simple projective B-module is injective, and each non-simple injective B-module is projective.

Let T be a simple B-module which is not injective and let I_T be the injective envelope of T as a B-module. Since T is simple, I_T is indecomposable. Since T is assumed to be non-injective, I_T properly contains T, which implies that I_T is projective by Corollary 5.11.21. Moreover, $Ext^1_B(-,T)$ is an injective object in $\underline{mod}(B\text{-}mod)$. Since the global dimension of $\underline{mod}(B\text{-}mod)$ is 0 the functor $Ext^1_B(-,T)$ is projective as an object of $\underline{mod}(B\text{-}mod)$ as well, and by Lemma 5.11.5 there is a B-module

M with

$$Ext^1_B(-,T) \simeq \underline{Hom}_B(-,M).$$

From the exact sequence

$$0 \longrightarrow T \longrightarrow I_T \longrightarrow I_T/T \longrightarrow 0$$

we obtain an exact sequence

$$Hom_B(-,I_T) \to Hom_B(-,I_T/T) \to Ext^1_B(-,T) \to Ext^1_B(-,I_T)$$

and where I_T is projective and injective. Hence $Ext^1_B(-,I_T) = 0$. But this shows that

$$Ext^1_B(-,T) \simeq \underline{Hom}(-,I_T/T)$$

and we get $M = I_T/T$.

Suppose that $\text{rad}(M) \neq 0$. Since I_T is an indecomposable projective-injective B-module, $I_T/\text{rad}(I_T) = M/\text{rad}(M) =: S$ is simple. Then there is a non-split short exact sequence

$$0 \longrightarrow \text{rad}(M) \longrightarrow M \longrightarrow S \longrightarrow 0.$$

But this implies that we get a non-zero natural transformation

$$\underline{Hom}_B(-,M) \longrightarrow \underline{Hom}_B(-,S)$$

which is left and right split since the global dimension of $\underline{mod}(B\text{-}mod)$ is 0. However,

$$Nattrans(\underline{Hom}_B(-,S), \underline{Hom}_B(-,M)) \simeq \underline{Hom}_B(S,M)$$

and likewise

$$Nattrans(\underline{Hom}_B(-,M), \underline{Hom}_B(-,S)) \simeq \underline{Hom}_B(M,S).$$

This can be seen using a projective resolution

$$Hom_B(-,P_M) \to Hom_B(-,M) \to \underline{Hom}_B(-,M) \to 0$$

of M in $mod(B\text{-}mod)$ and likewise for S. A morphism in $mod(B\text{-}mod)$ then lifts to the projective resolution, and there we have to apply Yoneda's lemma to get the statement. This shows that $M \twoheadrightarrow S$ is split in the stable category, which is absurd. Hence $\text{rad}(M) = 0$, and therefore all projective B-modules have radical squared 0. But we have excluded this case. This proves the Theorem. □

Remark 5.11.26 Auslander and Reiten proved many more of these correspondences using the technique of functor categories. This was the first appearance of questions culminating in the famous Auslander–Reiten conjecture [31, Conjecture (5) p. 409] which says that if A and B are stably equivalent algebras then the number of isomorphism classes of non-projective simple modules of A and of B should coincide.

There has been quite some progress on the Auslander–Reiten conjecture. As already announced in Remark 5.10.21 Martinez-Villa showed in [25] that the conjecture is true if it can be proved for all self-injective algebras. Indeed, he showed that the conjecture is true for a pair of two algebras A and B if and only if it is true for the pair of self-injective algebras $\Gamma(A)$ and $\Gamma(B)$ where $\Gamma(A)$ is constructed from A by a quite involved process.

This and further developments on the Auslander–Reiten conjecture was proved using Auslander–Reiten quivers and almost split sequences.

References

1. Auslander, M., Kleiner, M.: Adjoint functors and an extension of the Green correspondence for group representations. Adv. Math. **104**, 297–314 (1994)
2. Krause, H.: Stable equivalence preserves representation type. Commentarii Mathematici Helvetici **72**, 266–284 (1997)
3. Krause, H., Zwara, G.: Stable equivalence and generic modules. Bull. Lond. Math. Soc. **32**, 615–618 (2000)
4. Auslander, M., Reiten, I.: Stable equivalence of Artin algebras. In: Proceedings of the Conference on Orders, Group Rings and Related Topics. Springer Lecture Notes in Mathematics (Ohio State University, Columbus, 1972), pp. 8–71, vol. 353 (1973)
5. Broué, M.: Equivalences of blocks of group algebras. Finite dimensional algebras and related topics (Ottawa, 1992), pp. 1–26, NATO Advanced Science Institute Series C Mathematics Physics Science, 424. Kluwer Academic Publisher, Dordrecht (1994)
6. Auslander, M., Reiten, I.: On a theorem of Ed Green on the dual of the transpose. In: Canadian Mathematical Society Conference Proceedings, vol. 11, pp. 53–65 (1991) (Selected works of M. Auslander, vol. 1 (877–889), American Mathematical Society 1999)
7. Fröhlich, A., Taylor, M.: Algebraic Number Theory. Cambridge Studies in Advanced Mathematics, p. 27. Cambridge University Press, Cambridge (1993)
8. König, S., Zimmermann, A.: Derived Equivalences for Group Rings (with contributions by Keller, B., Linckelmann, M., Rickard, J., Rouquier, R.) Lecture Notes in Mathematics 1685. Springer, Berlin (1998)
9. Rickard, J.: Some recent advances in modular representation theory. Algebras and Modules, CMS Conference Proceedings, vol. 23, pp. 157–178 (1998)
10. Liu, Y.: On stable equivalences of Morita type for finite dimensional algebras. Proc. Am. Math. Soc. **131**, 2657–2662 (2003)
11. Dugas, A.S., Martinez-Villa, R.: A note on stable equivalence of Morita type. J. Pure Appl. Algebra **208**, 421–433 (2007)
12. Liu, Y.: Summands of stable equivalences of Morita type. Commun. Algebra **36**(10), 3778–3782 (2008)
13. König, S., Liu, Y.: Simple-minded systems in stable module categories. Q. J. Math. **63**, 653–674 (2012)
14. Tashikawa, H., Wakamatsu, T.: Cartan matrices and Grothendieck groups of stable catgegories. J. Algebra **144**, 390–398 (1991)
15. Xi, C.C.: Stable equivalences of adjoint type. Forum Mathematicum **20**, 81–97 (2008)

16. Bouc, S.: Bimodules, trace généralisée et transferts en homologie de Hochschild, preprint (1999)
17. Roggenkamp, K.W., Huber-Dyson, V.: Lattices over Orders 1. Lecture Notes in Mathematics, vol. 115. Springer, Berlin (1970)
18. Liu, Y., Zhou, G., Zimmermann, A.: Higman ideal, stable Hochschild homology and Auslander-Reiten conjecture. Mathematische Zeitschrift **270**, 759–781 (2012)
19. Héthelyi, L., Horváth, E., Külshammer, B., Murray, J.: Central ideals and Cartan invariants of symmetric algebras. J. Algebra **293**, 243–260 (2005)
20. König, S., Liu, Y., Zhou, G.: Transfer maps in Hochschild (co)homology and applications to stable and derived invariants and to the Auslander-Reiten conjecture. Trans. Am. Math. Soc. **364**, 195–232 (2012)
21. Zhou, G., Zimmermann, A.: Classifying tame blocks and related algebras up to stable equivalences of Morita type. J. Pure Appl. Algebra **215**, 2969–2986 (2011)
22. Gabriel, P., Riedtmann, C.: Group representations without groups. Commentarii Mathematici Helvetici **54**, 240–287 (1979)
23. Linckelmann, M.: Variations sur les blocs à groupes de défaut cyclique, in Séminaire sur les groupes finis (Séminaire Claude Chevalley) Tôme IV, pp. 1–83. Publications mathématiques de l'université Paris VII (1989)
24. Green, J.A.: Walking around the Brauer tree. J. Aust. Math. Soc. **17**, 197–213 (1974)
25. Martinez-Villa, R.: Properties that are left invariant under stable equivalence. Commun. Algebra **18**(12), 4141–4169 (1990)
26. Janusz, G.J.: Indecomposable modules for finite groups. Ann. Math. **89**, 209–241 (1969)
27. Kupisch, H.: Projektive Moduln endlicher Gruppen mit zyklischer p-Sylow-Gruppe. J. Algebra **10**, 1–7 (1968)
28. Kupisch, H.: Unzerlegbare Moduln endlicher Gruppen mit zyklischer p-Sylow-Gruppe. Math. Z. **108**, 77–104 (1969)
29. Brauer, R.: Investigations on group characters. Ann. Math. **42**, 936–958 (1941)
30. Dade, E.: Blocks with cyclic defect groups. Ann. Math. **84**, 20–48 (1966)
31. Auslander, M., Reiten, I., Smalø, S.O.: Representation Theory of Artin Algebras. Cambridge Studies in Advanced Mathematics, Cambridge (1995)
32. Hiss, G., Lux, K.: Brauer Trees of Sporadic Groups. Oxford Science Publications, Clarendon Press, Oxford (1989)
33. Roggenkamp, K.W.: Blocks of cyclic defect and Green orders. Commun. Algebra **20**, 1715–1734 (1992)
34. Plesken, W.: Group Rings of Finite Groups over p-adic Integers. Lecture Notes in Mathematics 1026. Springer, Berlin (1983)
35. Reiten, I.: Stable equivalence of selfinjective algebras. J. Algebra **40**, 63–74 (1976)
36. Auslander, M., Reiten, I.: Stable equivalence of dualizing R-varieties. Adv. Math. **12**, 306–366 (1974) (Representation theory of artin algebras, Cambridge Studies in Advanced Mathematics, Cambridge (1995))

Chapter 6
Derived Equivalences

In Chap. 4 we gave a necessary and sufficient criterion telling us when, for two algebras A and B, one has $A\text{-}Mod \simeq B\text{-}Mod$ as abelian categories. In Chap. 5 we studied the question of when $A\text{-}\underline{Mod} \simeq B\text{-}\underline{Mod}$.

We shall give an analogous theory now describing when $D^b(A\text{-}Mod) \simeq D^b(B\text{-}Mod)$ for two algebras A and B. There are basically two approaches, one given by Rickard, the other given by Keller. We shall use both sources and switch from one to the other when appropriate. Throughout this chapter we shall use the abbreviation $D^*(A)$ for $D^*(A\text{-}Mod)$ to shorten the notation.

6.1 Tilting Complexes

6.1.1 Historical and Methodological Remarks

Let K be a commutative ring and let A and B be K-algebras. Recall from Theorem 4.2.8 that if $F : B\text{-}Mod \longrightarrow A\text{-}Mod$ is a functor, then $F(B) =: M$ is an A–B-bimodule and $End_A(M) \simeq B$. In this case, the Morita Theorem 4.2.8 tells us that

$$M \otimes_B - : B\text{-}Mod \longrightarrow A\text{-}Mod$$

is an equivalence. Further, whenever there is a bimodule M which is projective as an A-module, projective as a B-module and such that $End_A(M) = B$ and $End_B(M) = A$, then

$$M \otimes_B - : B\text{-}Mod \longrightarrow A\text{-}Mod$$

is an equivalence.

Similar characterisations exist for equivalences between derived categories as well. This is Rickard's famous theorem [1] which holds in a very general context. Essentially the result is that $D^b(A) \simeq D^b(B)$ as triangulated categories if and only if

Algebra and Applications 19, DOI: 10.1007/978-3-319-07968-4_6,

there is an object T of $D^b(A)$, called a tilting complex, satisfying certain properties. The disadvantage of Rickard's result [1] is that it just gives a necessary and sufficient criterion for the existence of an equivalence $D^b(A) \simeq D^b(B)$. However it does not give an actual functor $D^b(A) \longrightarrow D^b(B)$ and in general the criterion does not describe such a functor. If A and B are algebras over a commutative ring k such that *A is flat as a k-module*, then Rickard proves in [2] that there is an equivalence

$$X \otimes_B^{\mathbb{L}} - : D^b(B) \longrightarrow D^b(A)$$

for a complex X in $D^b(A \otimes_k B^{op}\text{-}mod)$. Rickard does not give an explicit method how to actually *construct* X once T is known. For algebras A and B over a commutative ring k such that *A is projective as a k-module,* there is a constructive proof due to Keller [3] which gives an explicit construction of X once T is known. This approach will be presented here. The slight loss of generality which comes from the additional assumption that A and B should be projective as k-modules should not pose any serious problem since we are mainly interested in finite dimensional algebras or classical orders.

6.1.2 An Equivalence Gives a Tilting Complex

Let $F : D^-(A) \longrightarrow D^-(B)$ be a functor. Recall from Lemma 3.5.49 that $A\text{-}Mod \hookrightarrow D^-(A)$ is an embedding of A-*Mod* into $D^-(A)$ as a full subcategory. Then the image of the regular module A in $D^-(A)$ is also denoted by A and we may consider $F(A) \in D^-(B)$. Then we get an algebra homomorphism

$$A^{op} \simeq End_A(A) \simeq End_{D^-(A)}(A) \longrightarrow End_{D^-(B)}(F(A))$$

and if F is an equivalence, then $End_{D^-(B)}(F(A)) \simeq A^{op}$ as algebras. Moreover, since A is a projective A-module $Ext_A^i(A, A) = 0$ for all $i \neq 0$ and if F is an equivalence commuting with the shift functor [1], we get

$$\begin{aligned} Hom_{D^-(B)}(F(A), F(A)[i]) &\simeq Hom_{D^-(B)}(F(A), F(A[i])) \\ &\simeq Hom_{D^-(A)}(A, A[i]) \simeq Ext_A^i(A, A) = 0 \end{aligned}$$

for all $i \neq 0$. Finally, the smallest triangulated subcategory of $D^-(A)$ containing all direct summands of finite direct sums of A is $K^b(A\text{-}proj)$. Here, as usual, we denote by A-*proj* the category of finitely generated projective A-modules. Indeed, this is a triangulated full subcategory of $D^-(A) \simeq K^-(A\text{-}Mod)$ (cf Proposition 3.5.43). Moreover, all direct summands of finite direct factors of A give precisely the category A-*proj* by definition of a finitely generated projective A-module. Hence, given a bounded complex X

$$\cdots \longrightarrow 0 \longrightarrow X_n \xrightarrow{d_n} X_{n-1} \xrightarrow{d_{n-1}} \cdots \longrightarrow X_0 \longrightarrow 0 \longrightarrow \cdots$$

of finitely generated projective modules, define $\check{X}$ to be the complex

$$\cdots \longrightarrow 0 \longrightarrow X_{n-1} \xrightarrow{d_{n-1}} X_{n-2} \longrightarrow \cdots \longrightarrow X_0 \longrightarrow 0 \longrightarrow \cdots .$$

Then the morphism of complexes $X_n[n-1] \longrightarrow \check{X}$

$$\begin{array}{c} X_n \\ \downarrow d_n \\ \cdots \longrightarrow 0 \longrightarrow X_{n-1} \longrightarrow X_{n-2} \longrightarrow \cdots \longrightarrow X_0 \longrightarrow 0 \longrightarrow \cdots \end{array}$$

has cone X. By induction on the length of the complex, we know that $\check{X}$ is in the smallest triangulated full subcategory of $D^-(A)$ containing A-*proj*. Since shift in degrees can be realised by triangles as well, we know that $X_n[n-1]$ is in the smallest triangulated full subcategory of $D^-(A)$ containing A-*proj*. Therefore, X is also in the smallest triangulated full subcategory of $D^-(A)$ containing A-*proj*.

Now, we need a method to characterise $K^b(A\text{-}proj)$ in an intrinsic way inside $D^-(A)$. We first use Proposition 3.5.43 so that we may replace $D^-(A)$ by $K^-(A\text{-}Proj)$.

Lemma 6.1.1 *$K^b(A\text{-}Proj)$ is the subcategory of $K^-(A\text{-}Proj)$ consisting of those objects X such that for all objects Y of $K^-(A\text{-}Proj)$ there is an $i(Y)$ with $Hom_{K^-(A\text{-}Proj)}(Y, X[i]) = 0$ for all $i < i(Y)$.*

Proof Indeed, if X is in $K^b(A\text{-}Proj)$ the complex X is concentrated in a finite number of degrees. Hence all homogeneous components of X are 0 in degrees superior to some $n_+(X) \in \mathbb{N}$. Hence, since a given Y has homogeneous components in degrees at least $n_-(Y)$ only, the complex being right bounded, there is no non-zero morphism complex $Y \longrightarrow X[n_-(Y) - n_+(X) - j]$ for all $j \geq 0$ since in each degree either $Y_i = 0$ or $(X[n_-(Y) - n_+(X) - j])_i = 0$ for all $i \in \mathbb{Z}$.

If the homology of X is not bounded to the left, then

$$Hom_{K^-(A\text{-}Proj)}(A, X[-k]) = Hom_{K^-(A\text{-}Proj)}(A[k], X) = H_k(X)$$

may be non-zero for arbitrary large k. If X is not in $K^b(A\text{-}Proj)$, then the kernel of the differential d_k^X for high enough degrees k is not projective. Then the embedding induces a non-zero element in $Hom_{K^-(A\text{-}Proj)}(\ker(d_k^X)[k], X)$. This proves the statement. □

Recall from Definition 3.3.14 the concept of a *compact object* in an additive category admitting arbitrary coproducts. An object is compact if the covariant *Hom*-functor commutes with arbitrary coproducts.

Proposition 6.1.2 *The compact objects in $D^-(A)$ are precisely the objects isomorphic to objects in $K^b(A\text{-}proj)$.*

Proof Using Proposition 3.5.43 we need to show that the compact objects in $K^-(A\text{-}Proj)$ are precisely $K^b(A\text{-}proj)$.

First, we shall prove that compact objects in $K^b(A\text{-}Proj)$ are in $K^b(A\text{-}proj)$. The first step is to observe that the compact objects in A-*Proj* are A-*proj*. But this follows from Lemma 3.3.13. Let

$$X : \cdots \longrightarrow 0 \longrightarrow X_n \longrightarrow X_{n-1} \longrightarrow \cdots \longrightarrow X_{m+1} \longrightarrow X_m \longrightarrow 0 \longrightarrow \cdots$$

be a bounded complex of finitely generated projective modules. Then stupid truncation gives a distinguished triangle

$$X_n[n-1] \longrightarrow X' \longrightarrow X \longrightarrow X_n[n]$$

where

$$X' : \cdots \longrightarrow 0 \longrightarrow X_{n-1} \longrightarrow \cdots \longrightarrow X_{m+1} \longrightarrow X_m \longrightarrow 0 \longrightarrow \cdots$$

is the complex obtained from X by erasing the leftmost term and the leftmost mapping in the triangle is given by the differential in degree n. Let $Y = \coprod_{i\in I} Y_i$. There is a natural morphism

$$\coprod Hom_{K^-(A\text{-}Proj)}(X, Y_i) \longrightarrow Hom_{K^-(A\text{-}Proj)}(X, \coprod_{i\in I} Y_i)$$

given by the universal property of the coproduct. Use induction on $|m-n|$ and observe that by the induction hypothesis $Hom(X', -)$ commutes with coproducts. We may apply $Hom_{K^-(A\text{-}Proj)}(-, Y)$ to this triangle and obtain a long exact sequence (abbreviating $(U, V) = Hom_{K^-(A\text{-}Proj)}(U, V)$ for all U, V) to get a commutative diagram with exact lines

$$\begin{array}{ccccccccc}
(X_n[k], Y) & \leftarrow & (X', Y) & \leftarrow & (X, Y) & \leftarrow & (X_n[n], Y) & \leftarrow & (X'[1], Y) \\
\| & & \alpha \uparrow & & \beta \uparrow & & \| & & \alpha[1] \uparrow \\
\coprod(X_n[k], Y_i) & \leftarrow & \coprod(X', Y_i) & \leftarrow & \coprod(X, Y_i) & \leftarrow & \coprod(X_n[n], Y_i) & \leftarrow & \coprod(X'[1], Y_i)
\end{array}$$

where we have used the abbreviation $\coprod$ for $\coprod_{i\in I}$ and $k := n-1$. The second left and rightmost vertical mappings α and $\alpha[1]$ are isomorphisms by the induction hypothesis. Therefore the middle vertical arrow β is also an isomorphism.

Suppose to the contrary that

$$X : \cdots \longrightarrow 0 \longrightarrow X_n \longrightarrow X_{n-1} \longrightarrow \cdots \longrightarrow X_{m+1} \longrightarrow X_m \longrightarrow 0 \longrightarrow \cdots$$

is a bounded complex of projective modules and suppose $Hom_{K^-(A)}(X, -)$ commutes with direct sums. Suppose that X does not belong to $K^b(A\text{-}proj)$ and choose $n - m$ to be minimal with respect to this property and the property that its covariant Hom-functor commutes with arbitrary coproducts. If $n = m$, then the question is actually a question regarding modules. The compact objects in A-*Proj* are precisely the objects in A-*proj* by Lemma 3.3.13. Therefore, $n > m$.

Again we get a distinguished triangle

$$X_n[n-1] \longrightarrow X' \longrightarrow X \longrightarrow X_n[n]$$

where

$$X' : \cdots \longrightarrow 0 \longrightarrow X_{n-1} \longrightarrow \cdots \longrightarrow X_{m+1} \longrightarrow X_m \longrightarrow 0 \longrightarrow \cdots$$

is the complex obtained from X by erasing the left-most term. Again, for $Y = \coprod_{i \in I} Y_i$ we obtain a natural transformation

$$\coprod_{i \in I} Hom_{K^-(A\text{-}Proj)}(-, Y_i) \longrightarrow Hom_{K^-(A\text{-}Proj)}\left(-, \coprod_{i \in I} Y_i\right)$$

so that there is a commutative diagram

$$\begin{array}{ccccccccc}
(X, Y) & \leftarrow & (X_n[n], Y) & \leftarrow & (X'[1], Y) & \leftarrow & (X[1], Y) & \leftarrow & (X_n[k], Y) \\
\uparrow & & \uparrow & & \uparrow & & \uparrow & & \uparrow \\
\coprod(X, Y_i) & \leftarrow & \coprod(X_n[n], Y_i) & \leftarrow & \coprod(X'[1], Y_i) & \leftarrow & \coprod(X[1], Y_i) & \leftarrow & \coprod(X_n[k], Y_i)
\end{array}$$

with exact lines, where again we use the abbreviation $\coprod$ for $\coprod_{i \in I}$ and $k := n + 1$. The transformation evaluated on X and on $X[1]$ is an isomorphism by assumption.

If X_n is finitely generated, then the transformation evaluated on $X_n[n]$ and $X_n[n+1]$ is also an isomorphism. Hence, the transformation evaluated on X' is also an isomorphism. This contradicts to the minimality of $n - m$.

Therefore we may assume that X_n is not finitely generated. Then X_n is a direct factor of $\coprod_{i \in I} A$ for an infinite set I and the morphism of complexes $X \longrightarrow \coprod_{i \in I} A[n]$ has to commute with the above sum decomposition. But this is the case only if it is homotopy equivalent to a mapping with image in a finitely generated direct factor Y of X_n.

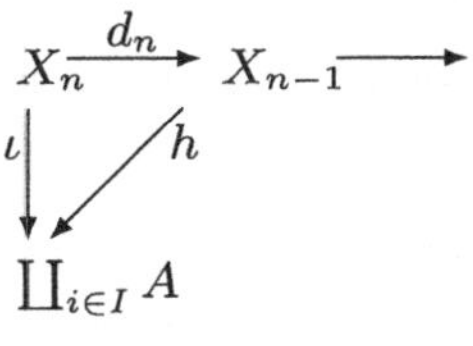

where the homotopy is displayed as h. Let $X_n = Y \oplus Z$ so that the restriction of d_n to Z is split by h. Hence the complex is isomorphic, as a complex, to

$$\cdots \longrightarrow 0 \longrightarrow Y \oplus Z \longrightarrow X'_{n-1} \oplus Z \longrightarrow X_{n-2} \longrightarrow \cdots \longrightarrow X_m \longrightarrow 0 \longrightarrow \cdots$$

for some X'_{n-1}, and the differential being the identity on Z. In $K^-(A\text{-}Proj)$ the complex X is isomorphic to

$$\cdots \longrightarrow 0 \longrightarrow Y \longrightarrow X'_{n-1} \longrightarrow X_{n-2} \longrightarrow \cdots \longrightarrow X_m \longrightarrow 0 \longrightarrow \cdots$$

with a finitely generated Y.

We now show that a compact object in $K^-(A\text{-}Proj)$ has homology in only finitely many degrees. Let X be a right bounded compact complex of projective modules. Suppose X is not isomorphic in the homotopy category to a complex with bounded homology. Then X has homology in arbitrary high degrees, that is, there is a strictly increasing sequence $(n_k)_{k\in\mathbb{N}}$ of integers with $H_{n_k}(X) \neq 0$. But

$$\begin{array}{ccccccc} X_{n_k+1} & \stackrel{d_{n_k+1}}{\longrightarrow} & \ker(d_{n_k}) & \longrightarrow & H_{n_k}(X) & \longrightarrow & 0 \\ \| & & \downarrow & & \downarrow & & \\ X_{n_k+1} & \stackrel{d_{n_k+1}}{\longrightarrow} & X_{n_k} & \longrightarrow & \operatorname{coker}(d_{n_k+1}) & \longrightarrow & 0 \end{array}$$

is commutative with vertical monomorphisms and exact rows. Hence there is a non-zero homomorphism of X to $\operatorname{coker}(d_{n_k+1})[n_k]$ so that we may consider $Y := \coprod_{k\in\mathbb{N}} \operatorname{coker}(d_{n_k+1})[n_k]$. We get that

$$Hom_{K^-(A\text{-}Proj)}(X, Y) \neq \coprod_{k\in\mathbb{N}} Hom_{K^-(A\text{-}Proj)}(X, \operatorname{coker}(d_{n_k+1})[n_k]).$$

Indeed, the left-hand side is strictly bigger.

We now show that a compact object in $K^{-,b}(A\text{-}Proj)$ is actually in $K^b(A\text{-}Proj)$. Given a compact object X in $K^{-,b}(A\text{-}Proj)$. If X is not in $K^b(A\text{-}Proj)$, then the kernel of the differential d_n of X is not projective for all n. Then $Hom_{K^-(A\text{-}Proj)}(X, \ker(d_n)[n]) \neq 0$ for for arbitrary high n. Let $Y := \bigoplus_{n\in\mathbb{N}} \ker(d_n)[n]$ for all n. Then

$$\begin{aligned} Hom_{K^-(A\text{-}Proj)}(X, Y) &= Hom_{K^-(A\text{-}Proj)}(X, \bigoplus_{n\in\mathbb{N}} \ker(d_n)[n]) \\ &= \coprod_{n\in\mathbb{N}} Hom_{K^-(A\text{-}Proj)}(X, \ker(d_n)[n]) \end{aligned}$$

since X is compact. However, there are non-zero morphisms from X to $\ker(d_n)[n]$ for arbitrary high n. This contradiction proves the statement. □

Definition 6.1.3 Let B be an algebra over a commutative ring K. A *tilting complex* T over B is a complex isomorphic to a complex in $K^b(B\text{-}proj)$ satisfying

- $Hom_{D^-(B)}(T, T[i]) = 0$ for all $i \in \mathbb{Z} \setminus \{0\}$

- the smallest triangulated full subcategory of $D^-(B)$ containing all direct factors of finite direct sums of T is $K^b(B\text{-}proj)$.

Remark 6.1.4 The concept of a tilting complex is crucial for equivalences between derived categories. These axioms were first stated in this form by Rickard [1]. The concept in a more rudimentary form first appeared in work of Brenner-Butler and of Happel.There mainly tilting modules (or more precisely quasi-tilting modules), that is tilting complexes which are modules, were used. The notation also comes from there, as their action in a certain sense can be visualised by a "tilting" of the Auslander-Reiten quiver in the case of tame hereditary algebras. However, even in this case the tilting modules only tell a small part of the story. The breakthrough was obtained by Rickard with the concept of tilting complexes.

Proposition 6.1.5 *Let $F : D^-(A) \longrightarrow D^-(B)$ be an equivalence of triangulated categories. Then $F(A) =: T$ is a tilting complex over B with endomorphism ring A^{op}.*

Proof Since A is compact in $D^-(A)$, by Lemma 3.2.10 T is also compact in $D^-(B)$. By Proposition 6.1.2 we get that T is in $K^b(B\text{-}proj)$. We have already seen that since $Ext^i_A(A, A) = 0$ for all $i \neq 0$, also $Hom_{D^-(A)}(T, T[i]) = 0$ for all $i \in \mathbb{Z} \setminus \{0\}$. Since $End_A(A) = A^{op}$, also $End_{D^-(B)}(T) = A^{op}$. This proves the statement. □

Corollary 6.1.6 *Let K be a commutative ring, let A and B be K-algebras which are projective as K-modules, and let $X \in D^b(A \otimes_K B^{op})$ and $Y \in D^b(B \otimes_K A^{op})$ be objects such that $X \otimes^{\mathbb{L}}_B Y \simeq A$ in $D^b(A \otimes_K A^{op})$ and $Y \otimes^{\mathbb{L}}_A X \simeq B$ in $D^b(B \otimes_K B^{op})$. Then the image $X \otimes^{\mathbb{L}}_B B$ of X in $D^b(A)$ is isomorphic to a tilting complex.*

Proof If A and B are projective as K-modules then any projective A–B-bimodule is projective as an A-module and projective as a B-module. Hence, we may replace X and Y by their projective resolutions and we see that the left derived tensor product can be replaced by the ordinary tensor product. This is associative. Therefore, $X \otimes^{\mathbb{L}}_B -$ is an equivalence and $X \otimes^{\mathbb{L}}_B B$ is the image of X in $D^b(A)$. □

6.2 Some Background on Co-algebras

We shall study tilting complexes in more detail here. Using Keller's approach [3] given a tilting complex T over A with endomorphism ring B we shall construct a complex X of A–B-bimodules such that X is isomorphic to T in $D^-(A)$.

In the case of module categories this is an easy thing to do. Given an A-module M the module M is by definition an A–B-bimodule for $B := End_A(M)$. In the case of derived categories this is more complicated. If T is a complex in $D^-(A)$, then $End_{C^-(A\text{-}mod)}(T)$ acts on each of the homogeneous components of T. However $End_{D^-(A)}(T)$ does not act on each of the homogeneous components of T. Indeed, we already have that $End_{K^-(A\text{-}mod)}(T)$ does not act, since this is a proper quotient of $End_{C^-(A\text{-}mod)}(T)$ and two morphisms of complexes which differ by a homotopy will not act in the same way on the homogeneous components of T.

There are different ways to circumvent this problem. Perhaps the most elementary solution is that given by Keller in [3]. In order to give Keller's proof we need some theory.

Throughout this Sect. 6.2 we shall use the convention that the symbol $\otimes$ without subscript will mean $\otimes_K$.

6.2.1 Co-algebras, Bi-algebras, Hopf-Algebras

The first notion we shall need is the notion of coalgebras. There is an extensive literature around this concept. We recommend Montgomery's book [4] and Brzezinski-Wisbauer's monograph [5].

Let K be a commutative ring.A K-algebra A is given by a multiplication mapping $\mu : A \otimes_K A \longrightarrow A$ and a unit mapping $\epsilon : K \longrightarrow A$.The associativity of the multiplication is given by the fact that the diagram

$$\begin{array}{ccc} A \otimes A \otimes A & \stackrel{id\otimes\mu}{\longrightarrow} & A \otimes A \\ \downarrow \mu \otimes id & & \downarrow \mu \\ A \otimes A & \stackrel{\mu}{\longrightarrow} & A \end{array}$$

is commutative. The fact that ϵ describes a unit is given by the commutativity of the diagrams

$$\begin{array}{ccc} K \otimes A & \stackrel{\epsilon\otimes id}{\longrightarrow} & A \otimes A \\ \downarrow id \otimes id & & \downarrow \mu \\ K \otimes_K A & \stackrel{nat}{\simeq} & A \end{array}$$

and

$$\begin{array}{ccc} A \otimes K & \stackrel{id\otimes\epsilon}{\longrightarrow} & A \otimes A \\ \downarrow id \otimes id & & \downarrow \mu \\ A \otimes_K K & \stackrel{nat}{\simeq} & A \end{array}$$

We reverse all arrows in the above diagrams to get the concept of a coalgebra. More precisely

Definition 6.2.1 A *coalgebra* (C, Δ, η) is given by a K-module C together with a K-linear mapping $\Delta : C \longrightarrow C \otimes_K C$ and a K-linear mapping $\eta : C \longrightarrow K$ such that the diagrams

$$\begin{array}{ccc} C\otimes C\otimes C & \stackrel{id\otimes\Delta}{\longleftarrow} & C\otimes C \\ \uparrow \Delta\otimes id & & \uparrow \Delta \\ C\otimes C & \stackrel{\Delta}{\longleftarrow} & C \end{array}$$

and

$$\begin{array}{ccc} K\otimes C & \stackrel{\eta\otimes id}{\longleftarrow} & C\otimes C \\ \uparrow id\otimes id & & \uparrow \Delta \\ K\otimes_K C & \stackrel{nat}{\simeq} & C \end{array}$$

as well as

$$\begin{array}{ccc} C\otimes K & \stackrel{id\otimes\eta}{\longleftarrow} & C\otimes C \\ \uparrow id\otimes id & & \uparrow \Delta \\ C\otimes_K K & \stackrel{nat}{\simeq} & C \end{array}$$

are commutative.

The mapping Δ is called a *coproduct* and the mapping η is called a *counit.*

Do not confuse the notion of a coproduct of a coalgebra with the notion of a coproduct in an additive category. Of course the same coincidence occurs for a product in an algebra and the (direct) product in a category. The term is the same in both cases, the meaning is completely different.

Example 6.2.2 We mention some examples for coalgebra structures.

1. A group ring is a coalgebra. Indeed, $\Delta(g) = g\otimes g$ for all $g \in G$ and $\eta(g) = 1$ for all $g \in G$ gives the structure of a coalgebra on KG. The counit is the *augmentation map*.
2. The most important example we shall consider in this section is the tensor coalgebra. Let A be a K-algebra and let

$$B := K \oplus \bigoplus_{i=1}^{\infty} A^{\otimes i}.$$

Then

$$\begin{aligned} &\Delta(a_1 \otimes a_2\otimes \cdots \otimes a_n) \\ &\quad := 1 \otimes (a_1 \otimes \cdots \otimes a_n) + \sum_{i=2}^{n} (a_1 \otimes \cdots \otimes a_{i-1}) \otimes (a_i \otimes \cdots \otimes a_n) \\ &\qquad + (a_1 \otimes \cdots \otimes a_n) \otimes 1 \end{aligned}$$

and η being the identification of K with the degree 0 component gives the structure of a coalgebra on B.
3. If (C, Δ, η) is a coalgebra, then $(C \otimes C, \Delta \otimes \Delta, \eta \otimes \eta)$ is also a coalgebra.

The concept of a morphism of algebras can be expressed in terms of diagrams as well: Let (A, μ, ϵ) and (A', μ', ϵ') be algebras. Then a K-linear mapping $\varphi : A \longrightarrow A'$ is a morphism of K-algebras if

$$\begin{array}{ccc} A \otimes A & \stackrel{\mu}{\longrightarrow} & A \\ \varphi \otimes \varphi \downarrow & & \downarrow \varphi \\ A' \otimes A' & \stackrel{\mu'}{\longrightarrow} & A' \end{array}$$

and

$$\begin{array}{ccc} K & \stackrel{\epsilon}{\longrightarrow} & A \\ \| & & \downarrow \varphi \\ K & \stackrel{\epsilon'}{\longrightarrow} & A' \end{array}$$

are commutative. Likewise a K-linear mapping $\varphi : C \longrightarrow C'$ is a morphism of coalgebras if

$$\begin{array}{ccc} C \otimes C & \stackrel{\Delta}{\longleftarrow} & C \\ \varphi \otimes \varphi \uparrow & & \uparrow \varphi \\ C' \otimes C' & \stackrel{\Delta'}{\longleftarrow} & C' \end{array}$$

and

$$\begin{array}{ccc} K & \stackrel{\eta}{\longleftarrow} & C \\ \| & & \uparrow \varphi \\ K & \stackrel{\eta'}{\longleftarrow} & C' \end{array}$$

are commutative.

We can even combine both structures and obtain in this way the concept of a bi-algebra.

Definition 6.2.3 Let K be a commutative ring and let A be a K-vector space. Then $(A, \mu, \epsilon, \Delta, \eta)$ is a *bi-algebra* if (A, μ, ϵ) is an algebra, if (A, Δ, η) is a coalgebra, and if μ and ϵ are morphisms of coalgebras, and if Δ and η are morphisms of algebras.

We observe that a group ring *KG* is a bi-algebra. The tensor coalgebra is also a bi-algebra, with the unit mapping being the projection onto the degree 0 component, and the multiplication being

$$\mu(a_1 \otimes \cdots \otimes a_n) = \sum_{i=0}^{n} (-1)^{i+1} (a_1 \otimes \cdots \otimes a_{i-1} \otimes a_i a_{i+1} \otimes a_{i+2} \otimes \cdots \otimes a_n) .$$

We shall not need or use the following concept of a Hopf algebra. Nevertheless, we give the definition for completeness, since it is not much of an effort to do so. Let C be a co-algebra and A be an algebra. Then $Hom_K(C, A)$ carries the *convolution product*

$$f * g := \mu \circ (f \otimes g) \circ \Delta$$

for all $f, g \in Hom_K(C, A)$.

Definition 6.2.4 A *Hopf-algebra* is a bi-algebra $(A, \mu, \epsilon, \Delta, \eta)$ together with a map $S \in Hom_K(A, A)$ which is inverse to id_A with respect to the convolution product. Such a map S is then called an *antipode.*

Example 6.2.5 Group algebras of finite groups are Hopf algebras with antipode $G \ni g \mapsto g^{-1} \in G$. We mention that there are Hopf algebras which are not group algebras.

The axioms of a (left) module over an algebra (A, μ, ϵ) can also be formulated in the form of a diagram as well: A module over an algebra (A, μ, ϵ) is a K-module M together with a K-linear map $\gamma : A \otimes M \longrightarrow M$ making the diagrams

$$\begin{array}{rcc} A \otimes A \otimes M & \stackrel{\mu \otimes id}{\longrightarrow} & A \otimes M \\ id \otimes \gamma \downarrow & & \downarrow \gamma \\ A \otimes M & \stackrel{\gamma}{\longrightarrow} & M \end{array}$$

and

$$\begin{array}{ccc} K \otimes M & \stackrel{\epsilon \otimes id}{\longrightarrow} & A \otimes M \\ \downarrow nat & & \downarrow \gamma \\ M & = & M \end{array}$$

commutative. Likewise one can formulate the axioms of a right module.

Definition 6.2.6 Let K be a field and let (C, Δ, η) be a coalgebra. A left *comodule* M is a K-vector space together with a K-linear map $\rho : M \longrightarrow C \otimes M$ such that the diagrams

$$\begin{array}{rcc} C \otimes C \otimes M & \stackrel{\Delta \otimes id}{\longleftarrow} & C \otimes M \\ id \otimes \rho \uparrow & & \uparrow \rho \\ C \otimes M & \stackrel{\rho}{\longleftarrow} & M \end{array}$$

and

$$\begin{array}{ccc} K \otimes M & \stackrel{\eta \otimes id}{\longleftarrow} & C \otimes M \\ \uparrow nat & & \uparrow \gamma \\ M & = & M \end{array}$$

are commutative.

Likewise a right C-comodule is a K-module M together with a K-linear map $M \longrightarrow M \otimes C$ satisfying the analogous diagrams.

A K-linear map $\lambda : M \longrightarrow N$ of A-modules is a morphism of modules if the diagram

$$\begin{array}{ccc} A\otimes M & \stackrel{id\otimes\lambda}{\longrightarrow} & A\otimes N \\ \downarrow \mu_M & & \downarrow \mu_N \\ M & \stackrel{\lambda}{\longrightarrow} & N \end{array}$$

is commutative. Morphisms of comodules are defined as morphisms of modules.

Definition 6.2.7 Let C be a coalgebra and let M and N be comodules with structure mappings $\rho_M : M \longrightarrow C\otimes M$ and $\rho_N : N \longrightarrow C\otimes N$. Then a K-linear mapping $\nu : M \longrightarrow N$ is a *morphism of comodules* if the diagram

$$\begin{array}{ccc} C\otimes M & \stackrel{id\otimes\nu}{\longrightarrow} & C\otimes N \\ \uparrow \rho_M & & \uparrow \rho_N \\ M & \stackrel{\nu}{\longrightarrow} & N \end{array}$$

is commutative.

Remark 6.2.8 Let (C, Δ, η) be a K-coalgebra, and let L be a K-module. Then $(C\otimes L, \Delta\otimes 1_L)$ is trivially a C-comodule. Moreover, for every C-comodule (U, Δ_U) we get a morphism $Hom_{C\text{-}comod}(U, C\otimes L) \to Hom_K(U, L)$ by $\varphi \mapsto (\eta\otimes 1_L)\circ\varphi$. Inversely we get a morphism

$$\begin{array}{rcl} Hom_K(U, L) & \longrightarrow & Hom_{C\text{-}comod}(U, C\otimes L) \\ \alpha & \mapsto & (1\otimes\alpha)\circ\Delta_U \end{array}$$

Indeed, this is dual to the well-known fact that

$$\begin{array}{rl} Hom_K(L, V) \simeq & Hom_A(A\otimes_K L, V) \\ & \alpha \mapsto (a\otimes\ell \mapsto \mu_V(a, \alpha(\ell))) = \mu_V\circ(1\otimes\alpha) \end{array}$$

for a K-algebra (A, μ, ϵ) and an A-module (V, μ_V).

We say that $C\otimes L$ is a C-cofree C-comodule.

6.2.2 Derivations, Coderivations

This section follows a mixture of the presentation of Keller [3] and Stasheff [6].

Definition 6.2.9 Let A be a K-algebra for a commutative ring K. Then a *derivation* on A is a K-linear map $\delta : A \longrightarrow A$ satisfying

$$\delta(ab) = \delta(a)b + a\delta(b)$$

for all $a, b \in A$. A derivation δ is *inner* if there is an $a \in A$ such that

$$\delta(b) = ab - ba$$

for all $b \in A$.

Note that for every $b \in A$ the mapping $\delta_a(b) := ab - ba$ is a derivation. Of course, the notion of derivation comes from the usual rule for the derivative of a product of functions. It has further implications and applications in Lie algebras, where the notion is central. In our context the notion of derivation occurs primarily in the setting of Hochschild cohomology.

Remark 6.2.10 The set $Der(A, A)$ of derivations is trivially a K-module and the set $IDer(A, A)$ of inner derivations is a submodule. Let K be a field and let A be a finite dimensional K-algebra. Then the degree 1 Hochschild cohomology $HH^1(A)$ is isomorphic as a K-vector space to $Der(A, A)/IDer(A, A)$. This follows trivially from the bar resolution in Proposition 3.6.4.

Observe that the notion of a derivation can be written in terms of the multiplication mapping $\mu : A \otimes A \to A$. The map δ is a derivation if

$$\delta \circ \mu = \mu \circ (\delta \otimes id) + \mu \circ (id \otimes \delta).$$

We can easily define the coproperty of the derivation using this description to get a coderivation. In this way we have a dual concept for coalgebras.

Definition 6.2.11 Let K be a commutative ring and let C be a coalgebra with coproduct $\Delta : C \to C \otimes C$. A K-linear map $b : C \to C$ is a *coderivation* if

$$\Delta \circ b = (b \otimes id + id \otimes b) \circ \Delta.$$

Let (C, Δ, η) be a coalgebra with coderivation b_C and let (M, ρ) be a comodule for C. A coderivation for M with respect to b_C is a K-linear endomorphism b of M such that

$$\rho \circ b = (id_C \otimes b + b_C \otimes id_M) \circ \rho.$$

Example 6.2.12 We have seen in Example 6.2.2 item 2 that for every K-algebra V the $\mathbb{N}$-graded vector space $T(V) := K \oplus \bigoplus_{i=1}^{\infty} V^{\otimes i}$ is a coalgebra, where the grading is defined so that K is in degree 0 and $V^{\otimes i}$ is in degree i.

We continue to discuss this example with respect to coderivations in several steps.

1. If $V = A$ is an algebra then define the homogeneous endomorphism $\gamma_{T(A)}$ of $T(A)$ of degree -1 by

$$\gamma_{T(A)}(a_1 \otimes \cdots \otimes a_n) = \sum_{i=1}^{n-1} (-1)^{i+1} a_1 \otimes \cdots \otimes a_{i-1} \otimes a_i a_{i+1} \otimes a_{i+2} \otimes \cdots \otimes a_n.$$

   We observe that $\gamma_{T(A)}$ is a coderivation. In order to verify this property observe that we use the graded tensor product of mappings: If f and g are homogeneous morphisms of degree n and m between $\mathbb{Z}$-graded K-modules, and if x and y are homogeneous elements of degree u and v, then

$$(f \otimes g)(x \otimes y) := (-1)^{m \cdot u} f(x) \otimes g(y).$$

2. Now, $\gamma_{T(A)}^2 = 0$ since in the composition the term

$$a_1 \otimes \cdots \otimes a_{i-2} \otimes a_{i-1}a_i a_{i+1} \otimes a_{i+2} \otimes \cdots \otimes a_n$$

occurs twice with different signs, and since the term

$$a_1 \otimes \cdots \otimes a_{j-1} \otimes a_j a_{j+1} \otimes a_{j+2} \otimes \cdots \otimes a_{i-1} \otimes a_i a_{i+1} \otimes a_{i+2} \otimes \cdots \otimes a_n$$

again occurs in two different ways, with different signs as well.
Therefore $(T(A), \gamma_{T(A)})$ is a complex.
3. Now, given a second K-algebra B and a $\mathbb{Z}$-graded B-module L, the space $X := T(A) \otimes L$ can be written as

$$X = L \oplus (A \otimes L) \oplus (A^{\otimes 2} \otimes L) \oplus (A^{\otimes 3} \otimes L) \oplus \cdots$$

and one obtains two gradings. It has a grading since $T(A)$ is $\mathbb{Z}$-graded, but since L is $\mathbb{Z}$-graded, the module $T(A) \otimes L$ has a second grading coming from L. Hence X is bigraded, or in other words it is $\mathbb{Z} \times \mathbb{Z}$-graded.
4. Moreover by Remark 6.2.8 X is a comodule over $T(A)$. The comodule structure on X is given by $\Delta_X = \Delta \otimes id_L$ where Δ is the codiagonal mapping of the coalgebra $T(A)$. More explicitly

$$\begin{aligned} \Delta_X : X &\longrightarrow T(A) \otimes X \\ a_1 \otimes \cdots \otimes a_{n-1} \otimes x &\mapsto 1 \otimes (a_1 \otimes \cdots \otimes a_{n-1} \otimes x) \\ &\quad + \sum_{i=2}^{n-2} (a_1 \otimes \cdots \otimes a_{i-1}) \otimes (a_i \otimes \cdots \otimes a_{n-1} \otimes x) \\ &\quad + (a_1 \otimes \cdots \otimes a_{n-1}) \otimes x \end{aligned}$$

5. Let $\varepsilon : X \longrightarrow L$ be the projection onto degree 0. Then the space of coderivations $Coder(X, X)$ on X (with respect to the coderivation $\gamma_{T(A)}$) is $\mathbb{Z}$-graded and we get a graded vector space homomorphism

$$\begin{aligned} Coder(X, X) &\xrightarrow{\Psi} Hom_K(X, L) \\ \gamma &\mapsto \varepsilon \circ \gamma \end{aligned}$$

We claim that this homomorphism is an isomorphism.
Indeed, we can give an inverse map. Let $b : X \longrightarrow L$ be a K-linear map of degree e, where the degree on X is given by the degree inherited by L. Hence,

$b = \sum_{\ell=0}^{\infty} b_\ell$ where b_ℓ is the degree ℓ homogeneous component of b, where the degree ℓ refers to the the grading obtained by $T(A)$. Put

$$\begin{aligned}
&\gamma_b(a_1 \otimes \cdots \otimes a_{n-1} \otimes x) \\
&:= \sum_{i=1}^{n-2} (-1)^{e(i-1)}(a_1 \otimes \cdots \otimes a_{i-1} \otimes a_i a_{i+1} \otimes a_{i+2} \otimes \cdots \otimes a_{n-1} \otimes x) \\
&\quad + \sum_{\ell=1}^{n} (-1)^{e(n-\ell)}(a_1 \otimes \cdots \otimes a_{n-\ell} \otimes b_\ell(a_{n-\ell+1} \otimes \cdots \otimes a_{n-1} \otimes x))
\end{aligned}$$

for all $a_i \in A, i \in \{1, 2, \ldots, n-1\}, x \in L$.
We define

$$\begin{aligned}
Hom_K(X, L) &\xrightarrow{\Phi} Coder(X, X) \\
b &\mapsto \gamma_b
\end{aligned}$$

by sending b to γ_b. It is clear that $\Psi \circ \Phi = id_{Hom_K(X,L)}$ since the projection of γ_b to L is $b = \sum b_\ell$. The fact that $\Phi \circ \Psi = id_{Coder(X,X)}$ is more technical: Given a coderivation γ' of X of degree e we see by definition $\Phi(\Psi(\gamma')) = \Phi(\varepsilon \circ \gamma')$. We obtain that $\varepsilon \circ \gamma' =: b' \in Hom_K(X, L)$ is a homomorphism of degree e. Let $b'_i : A^{\otimes i-1} \otimes L \longrightarrow L$ be the homogeneous component of b'. We obtain $\Phi(\Psi(\gamma')) = \Phi(\varepsilon \circ \gamma') = \Phi(b') =: \gamma$ where

$$\begin{aligned}
&\gamma(a_1 \otimes \cdots \otimes a_{n-1} \otimes x) \\
&= \sum_{i=1}^{n-2} (-1)^{e(i-1)}(a_1 \otimes \cdots \otimes a_i a_{i+1} \otimes \cdots \otimes a_{n-1} \otimes x) \\
&\quad + \sum_{\ell=1}^{n} (-1)^{n-\ell} a_1 \otimes \cdots \otimes a_{n-\ell} \otimes b'_\ell(a_{n-\ell+1} \otimes \cdots \otimes a_{n-1} \otimes x)
\end{aligned}$$

by the definition of b' and of $\Phi(b') = \gamma_{b'}$ above.

We need to show that $\gamma = \gamma'$.
The case $n = 1$ follows from $\gamma'(x) = b'_1(x) = \gamma(x)$.
The case $n = 2$ follows from the argument below.

$$\gamma'(a_1 \otimes x) = b'_2(a_1 \otimes x) + \sum_j a_{1,j} \otimes x_j$$

and we shall show that

$$\gamma'(a_1 \otimes x) = a_1 \otimes b'_1(x) + b'_2(a_1 \otimes x).$$

We shall use the defining property of a coderivation γ' on X, namely

$$\Delta_X \circ \gamma' = (\gamma_{T(A)} \otimes 1_X + 1_{T(A)} \otimes \gamma') \circ \Delta_X.$$

The right-hand side gives

$$\begin{aligned}
&\Delta_X(\gamma'(a_1 \otimes x)) \\
&\quad = \Delta_X(b_2'(a_1 \otimes x) + \sum_j a_{1,j} \otimes x_j) \\
&\quad = 1_{T(A)} \otimes b_2'(a_1 \otimes x) + 1_{T(A)} \otimes \sum_j a_{1,j} \otimes x_j + \sum_j a_{1,j} \otimes x_j.
\end{aligned}$$

But

$$\begin{aligned}
&(\Delta_X \circ \gamma')(a_1 \otimes x) \\
&\quad = (\gamma_{T(A)} \otimes 1_X + 1 \otimes \gamma')(\Delta_X(a_1 \otimes x)) \\
&\quad = (\gamma_{T(A)} \otimes 1_X + 1 \otimes \gamma')(1 \otimes (a_1 \otimes x) + a_1 \otimes x) \\
&\quad = 1_{T(A)} \otimes \gamma'(a_1 \otimes x) + a_1 \otimes \gamma'(x) \\
&\quad = 1_{T(A)} \otimes b_2'(a_1 \otimes x) + 1_{T(A)} \otimes \sum_j a_{1,j} \otimes x_j + a_1 \otimes b_1'(x)
\end{aligned}$$

and so $\gamma(a_1 \otimes x) = a_1 \otimes b_1'(x) + b_2'(a_1 \otimes x)$ as claimed.
For $n = 3$ we put

$$\gamma'(a_1 \otimes a_2 \otimes x) = b_3'(a_1 \otimes a_2 \otimes x) + \sum_j (a_{1,j} \otimes x) + \sum_k (a_{1,k} \otimes a_{2,k} \otimes x)$$

for certain elements $a_{1,j}, a_{1,k}, a_{2,k}, x_j, x_k$ (where we hope that the slight abuse of notation will not cause additional problems) and compute

$$\begin{aligned}
&(\Delta_X \circ \gamma')(a_1 \otimes a_2 \otimes x) \\
&\quad = ((\gamma_{T(A)} \otimes 1_X + 1_{TA} \otimes \gamma') \circ \Delta_X)(a_1 \otimes a_2 \otimes x) \\
&\quad = (\gamma_{T(A)} \otimes 1_X + 1_{TA} \otimes \gamma') \\
&\qquad\qquad (1 \otimes (a_1 \otimes a_2 \otimes x) + a_1 \otimes (a_2 \otimes x) + (a_1 \otimes a_2) \otimes x) \\
&\quad = a_1a_2 \otimes x + 1 \otimes \gamma'(a_1 \otimes a_2 \otimes x) + a_1 \otimes \gamma'(a_2 \otimes x) + (a_1 \otimes a_2) \otimes \gamma'(x) \\
&\quad = a_1a_2 \otimes x + 1 \otimes b_3'(a_1 \otimes a_2 \otimes x) \\
&\qquad + 1 \otimes \sum_j (a_{1,j} \otimes x_j) + 1 \otimes \sum_k (a_{1,k} \otimes a_{2,k} \otimes x_k) \\
&\qquad + a_1 \otimes (a_1 \otimes b_1'(x)) + a_1 \otimes b_2'(a_1 \otimes x) + (a_1 \otimes a_2) \otimes b_1'(x).
\end{aligned}$$

Computing directly

$$\begin{aligned}
&(\Delta_X \circ \gamma')(a_1 \otimes a_2 \otimes x) \\
&= \Delta_X(b_3'(a_1 \otimes a_2 \otimes x) + \sum_j (a_{1,j} \otimes x_j) + \sum_k (a_{1,k} \otimes a_{2,k} \otimes x_k)) \\
&= 1 \otimes b_3'(a_1 \otimes a_2 \otimes x) + 1 \otimes \sum_j (a_{1,j} \otimes x_j) \\
&\quad + 1 \otimes \sum_k (a_{1,k} \otimes a_{2,k} \otimes x_k) + \sum_k a_{1,k} \otimes (a_{2,k} \otimes x_k) \\
&\quad + \sum_k (a_{1,k} \otimes a_{2,k}) \otimes x_k
\end{aligned}$$

and comparing the homogeneous components with respect to the still present grading we get

$$\sum_j (a_{1,j} \otimes x_j) = a_1 a_2 \otimes x + a_1 \otimes b_2'(a_1 \otimes x)$$

and

$$\sum_k (a_{1,k} \otimes a_{2,k} \otimes x) = (a_1 \otimes a_2) \otimes b_1'(x).$$

This shows that $\gamma = \gamma'$.

In the same way, one proceeds for $n \geq 4$, using repeatedly the defining property of a coderivation. The defining equation for a coderivation defines maps $A^{\otimes k} \otimes L \to A^{\otimes m} \otimes L$ for $0 < m \leq k$ in a unique way. I wish to thank Intan Muchtadi-Alamsyah who performed these computations explicitly.

6. Given a coderivation γ of $X = T(A) \otimes L$, we get

$$(id_{T(A)} \otimes \gamma) \circ (\gamma_{T(A)} \otimes id_X) = -(\gamma_{T(A)} \otimes id_X) \circ (id_{T(A)} \otimes \gamma)$$

as a K-linear endomorphism of $T(A) \otimes X$ using our sign rule for the tensor products of mappings. Now $\gamma_{T(A)}^2 = 0$, as we have seen above, and we obtain

$$\begin{aligned}
\Delta_X \circ \gamma^2 &= (id_{T(A)} \otimes \gamma + \gamma_{T(A)} \otimes id_X)^2 \circ \Delta_X \\
&= (id_{T(A)} \otimes \gamma^2 + \gamma_{T(A)} \otimes \gamma - \gamma_{T(A)} \otimes \gamma + \gamma_{T(A)}^2 \otimes id_X) \circ \Delta_X \\
&= (id_{T(A)} \otimes \gamma^2) \circ \Delta_X.
\end{aligned}$$

Hence $\gamma^2 : X \longrightarrow X$ is a morphism of $T(A)$-comodules.

7. Let $\gamma \in Coder(X, X)$. Then $\gamma^2 : X \longrightarrow X$ is a homomorphism of $T(A)$-comodules by Item 6. Remark 6.2.8 implies that $\gamma^2 = 0$ if and only if $\varepsilon \circ \gamma^2 = 0$. We need to write down the explicit formula for $\varepsilon \circ \gamma^2$, and in order to do so, we

set $\gamma = \sum_{\ell=0}^{\infty} b_\ell$ for homogeneous components b_ℓ of degree ℓ. Hence $\gamma^2 = 0$ if and only if

$$0 = \sum_{i=1}^{n-2} (-1)^{i+1} b_{n-1}(a_1 \otimes \cdots \otimes a_{i-1} \otimes a_i a_{i+1} \otimes a_{i+2} \otimes \cdots \otimes a_{n-1} \otimes x)$$
$$+ \sum_{\ell=1}^{n} (-1)^{n-\ell} b_{n-\ell+1}(a_1 \otimes \cdots \otimes a_{n-\ell} \otimes b_\ell(a_{n-\ell+1} \otimes \cdots \otimes a_{n-1} \otimes x))$$

for all $a_i \in A, i \in \{1, 2, \ldots, n-1\}, x \in L, n \geq 1$.

6.3 Strong Homotopy Action

We describe a version of a strong homotopy action due to J. D. Stasheff which is presented in [7].

6.3.1 Strong Homotopy Actions for Complexes Without Self-Extensions

Remark 6.3.1 The technique of strong homotopy actions essentially comes from the theory of A_∞-algebras. This theory is highly sophisticated and since we will not need the theory of A_∞-algebras in the sequel, we will refrain from giving a complete and systematic treatment. The origins of these methods come from algebraic topology.

Let A and B be K-algebras. By definition, a complex of B-modules is a $\mathbb{Z}$-graded B-module endowed with a differential of degree -1. Let L be a $\mathbb{Z}$-graded B-module.

Definition 6.3.2 A *strong homotopy action* of A is the datum of B-linear homomorphisms

$$m_n : A^{\otimes n-1} \otimes_K L \longrightarrow L$$

for each $n \geq 1$ where m_n is graded of degree $n-2$ satisfying the following equation:

$$\sum_{i=1}^{n-2} (-1)^i m_{n-1}(a_1 \otimes \cdots \otimes a_{i-1} \otimes a_i a_{i+1} \otimes a_{i+2} \otimes \cdots \otimes a_{n-1} \otimes x)$$
$$= \sum_{\ell=1}^{n-2} (-1)^{n-\ell} m_{n-\ell+1}(a_1 \otimes \cdots \otimes a_{n-\ell} \otimes m_\ell(a_{n-\ell+1} \otimes \cdots \otimes a_{n-1} \otimes x)).$$

Remark 6.3.3 In particular for $n = 1$ we get that m_1 is a homomorphism of degree -1 satisfying

$$m_1 \circ m_1 = 0$$

and for $n = 2$ we get that m_2 is a homomorphism of degree 0 such that

$$m_2(a \otimes m_1(x)) = m_1(m_2(a \otimes x))$$

for each $a \in A$ and $x \in L$. Hence, (L, m_1) is a complex and $m_2(a \otimes -)$ is a morphism of complexes for all $a \in A$.

For $n = 3$ we get a B-linear morphism of degree 1 satisfying

$$m_3(a \otimes b \otimes m_1(x)) + m_2(ab \otimes x) - m_2(a \otimes m_2(b \otimes x)) + m_1(m_3(a \otimes b \otimes x)) = 0$$

for all $a, b \in A$ and $x \in L$. This equation is equivalent to

$$m_2(ab \otimes x) = m_2(a \otimes m_2(b \otimes x)) - (m_1(m_3(a \otimes b \otimes x)) - m_3(a \otimes b \otimes m_1(x)))\,.$$

Recall that m_1 was a differential. Then m_3 can be interpreted as a homotopy map, depending on a and b, and the "multiplication" m_2 is associative up to this homotopy map m_3. Hence, m_2 is associative, up to a homotopy map m_3.

Remark 6.3.4 We continue with Example 6.2.2 and Example 6.2.12.

Given an algebra A and a $\mathbb{Z}$-graded B-module L, then, by Example 6.2.12 Item 6, for each degree -1 coderivation γ of $T(A) \otimes L$ we get that γ^2 is an endomorphism of $T(A) \otimes L$ as $T(A)$-comodules. Now if $\gamma^2 = 0$ then γ actually satisfies the defining equation for a strong homotopy action by Item 7 in Example 6.2.12. In other words the degree -1 coderivation differentials on $T(A) \otimes L$ are precisely the strong homotopy actions on $T(A) \otimes L$.

Proposition 6.3.5 *Let A be an algebra, let B be an algebra and let L be a $\mathbb{Z}$-graded B-module. Suppose there are homogeneous B-linear mappings*

$$m_1 : L \longrightarrow L, \text{ of degree } -1$$

$$m_2 : A \otimes L \longrightarrow L \text{ of degree } 0$$

and

$$m_3 : A \otimes A \otimes L \longrightarrow L \text{ of degree } 1$$

such that $m_1^2 = 0$, $m_1(m_2(a \otimes x)) = m_2(a \otimes m_1(x))$ and

$$m_3(a \otimes b \otimes m_1(x)) + m_2(ab \otimes x) - m_2(a \otimes m_2(b \otimes x)) + m_1(m_3(a \otimes b \otimes x)) = 0$$

for all $a, b \in A$ and $x \in L$. Suppose moreover that

$$Hom_{K(B\text{-}Mod)}(A^{\otimes n} \otimes L, L[n-2]) = 0$$

for all $n \geq 3$. Then there are B-linear morphisms $m_n : A^{\otimes n-1} \otimes L \longrightarrow L$ such that $(m_i)_{i \in \mathbb{N}}$ is a strong homotopy action of A on L.

Proof The mappings $m_1 : L \longrightarrow L$, $m_2 : A \otimes L \to L$, and $m_3 : A \otimes A \otimes L \to L$ are the images of the coderivations γ_i under ε^*. Hence we have coderivations γ_1, γ_2 and γ_3 of $T(A) \otimes L$ such that

$$\gamma_1\gamma_1 = 0, \quad \gamma_1\gamma_2 + \gamma_2\gamma_1 = 0, \quad \gamma_1\gamma_3 + \gamma_2\gamma_2 + \gamma_3\gamma_1 = 0$$

and we need to find coderivations γ_i for $i \geq 4$ of $T(A) \otimes L$ such that

$$(\dagger): \quad \gamma_1\gamma_n + \gamma_2\gamma_{n-1} + \cdots + \gamma_{n-1}\gamma_2 + \gamma_n\gamma_1 = 0$$

for all $n \in \mathbb{N}$. We shall construct γ_n by induction. Suppose we have already constructed γ_i for $i \leq N-1$ satisfying ($\dagger$). Then we need to find γ_N such that

$$\gamma_1\gamma_N + \gamma_N\gamma_1 + (\gamma_2\gamma_{N-1} + \cdots + \gamma_{N-1}\gamma_2) = 0.$$

Let

$$\delta := (\gamma_2\gamma_{N-1} + \cdots + \gamma_{N-1}\gamma_2)$$

and let

$$X_N := L \oplus (A \otimes L) \oplus (A^{\otimes 2} \otimes L) \oplus \cdots \oplus (A^{\otimes N-1} \otimes L).$$

Since $\gamma_i(X_N) \subseteq X_{N-i+1}$ we get $\delta(X_N) \subseteq X_1 = L$ and $\delta(X_{N-1}) = 0$. Hence δ induces a degree -2 morphism

$$\delta : X_N/X_{N-1} \longrightarrow L.$$

Observe that

$$X_N/X_{N-1} = A^{\otimes N-1} \otimes L$$

and if we can show that δ commutes with the differential γ_1, then

$$\delta \in Hom_{K(B\text{-}Mod)}(A^{\otimes N-1} \otimes L, L[N-3]),$$

which is 0 by hypothesis. Hence, in this case δ is 0-homotopic, which gives a homotopy b_N, and this homotopy induces a coderivation γ_N as requested. Let $\lambda := \gamma_1 + \gamma_2 + \cdots + \gamma_{N-1}$ and observe that by the induction hypothesis ($\dagger$) holds for $N-1$, and therefore the restriction of λ^2 and δ to X_N coincide (as usual, for every mapping $f : Y \longrightarrow Z$ we denote by $f|_X$ the restriction of f to the subset $X \leq Y$):

$$\lambda^2|_{X_N} = \delta|_{X_N}.$$

Hence

$$\lambda^2(X_{N-1}) = \delta(X_{N-1}) = 0$$

and

$$\lambda^2(X_N) = \delta(X_N) \subseteq L.$$

Therefore, since $\gamma_i(L) = 0$ if $i > 1$ and therefore also $\delta(L) = \gamma_1(L)$, we obtain

$$\gamma_1\delta|_{X_N} = \gamma_1\lambda^2|_{X_N} = \lambda^3|_{X_N} = \lambda^2\gamma_1|_{X_N} = \delta\gamma_1|_{X_N}.$$

This implies that δ is a morphism of complexes, as we needed to show. □

Remark 6.3.6 Observe how the morphisms b_N and γ_N were constructed. A first observation was that they were constructed recursively. Then, as a second observation, one should note that the maps γ_N and b_N are homotopies for a well-defined morphism of complexes which can be explicitly computed once the maps γ_i for $i < N$ are known. We see that in some sense the maps γ_n are constructed by first modifying by a homotopy so that the action up to homotopy becomes an action. Then one needs to modify the modification by a further homotopy, etc., using that the modifications are of higher and higher degrees. This fact allows us to explicitly construct a complex X of B–A-bimodules with an explicit action of A, and which restricts to a fixed complex T of B-modules without self-extensions and with endomorphism ring A in the homotopy category.

We should mention that the explicit construction of such a complex X was an open problem for quite some time and only in some very specific settings was a construction known (e.g. [8, 9]).

6.3.2 Morphisms Between Strong Homotopy Actions

Let A and B be K-algebras and recall that a strong homotopy action of A on $\mathbb{Z}$-graded B-modules L and M is given as a sequence of mappings

$$m_n^L : A^{\otimes n-1} \otimes L \longrightarrow L$$

and

$$m_n^M : A^{\otimes n-1} \otimes M \longrightarrow M$$

of degree $n-2$ such that

$$\sum_{i=1}^{n-2}(-1)^i m_{n-1}^L(a_1 \otimes \cdots \otimes a_{i-1} \otimes a_i a_{i+1} \otimes a_{i+2} \otimes \cdots \otimes a_{n-1} \otimes x)$$

$$= \sum_{\ell=1}^{n-2}(-1)^{n-\ell} m^L_{n-\ell+1}(a_1 \otimes \cdots \otimes a_{n-\ell} \otimes m^L_\ell(a_{n-\ell+1} \otimes \cdots \otimes a_{n-1} \otimes x))$$

and

$$\sum_{i=1}^{n-2}(-1)^i m^M_{n-1}(a_1 \otimes \cdots \otimes a_{i-1} \otimes a_i a_{i+1} \otimes a_{i+2} \otimes \cdots \otimes a_{n-1} \otimes x)$$
$$= \sum_{\ell=1}^{n-2}(-1)^{n-\ell} m^M_{n-\ell+1}(a_1 \otimes \cdots \otimes a_{n-\ell} \otimes m^M_\ell(a_{n-\ell+1} \otimes \cdots \otimes a_{n-1} \otimes x)).$$

Definition 6.3.7 Let A and B be K-algebras and let L and M be $\mathbb{Z}$-graded B-modules with a strong homotopy action of A on L and on M. A *morphism* $f : L \longrightarrow M$ *between modules* M *and* N *with strong homotopy action* of A is a sequence of B-linear morphisms $f_n : A^{\otimes n-1} \otimes L \longrightarrow M$ of degree $n-1$ such that for all $n \in \mathbb{N}$ we have

$$\sum_{\ell=1}^{n} m^M_\ell(a_1 \otimes \ldots a_{\ell-1} \otimes f_{n-\ell+1}(a_\ell \otimes \cdots \otimes a_{n-1} \otimes x))$$
$$= \sum_{i=1}^{n-2}(-1)^{i-1} f_{n-1}(a_1 \otimes \cdots \otimes a_{i-1} \otimes a_i a_{i+1} \otimes a_{i+2} \otimes \cdots \otimes a_{n-1} \otimes x)$$
$$+ \sum_{\ell=1}^{n}(-1)^{n-\ell} f_{n-\ell+1}(a_1 \otimes \cdots \otimes a_{n-\ell} \otimes m^L_\ell(a_{n-\ell+1} \otimes \cdots \otimes a_{n-1} \otimes x))$$

for all $a_i \in A, i \in \{1, \ldots, n-1\}$ and $x \in L$.

Remark 6.3.8 Recall that the degree 1 condition of a strong homotopy action on L (or M resp.) is just the condition that m^L_1 (or m^M_1 resp.) is a differential turning $T(A)\otimes L$ (or $T(A)\otimes M$ resp.) into a complex. Consider the condition for morphisms of strong homotopy actions in degree 1, i.e. the condition on f_1. The defining formula for f being a morphism of strong homotopy actions is just the condition

$$f_1 m^L_1 = m^M_1 f_1.$$

This is equivalent to saying that f_1 is a morphism of complexes

$$f_1 : (L, m^L_1) \longrightarrow (M, m^M_1).$$

The case of f_2 reveals another classical situation. Indeed, the equation becomes

$$m_1^M(f_2(a \otimes x)) + f_2(a \otimes m_1^L(x)) = f_1(m_2^L(a \otimes x)) - m_2^M(a \otimes f_1(x))$$

for all $a \in A$ and $x \in L$. Recall that we could interpret m_2 as multiplication by A, which was associative up to some homotopy m_3. This just means that now f_1 is A-linear up to a homotopy $x \mapsto f_2(a \otimes x)$, a homotopy which depends on $a \in A$.

Definition 6.3.9 Let A and B be K-algebras and let L and M be $\mathbb{Z}$-graded B-modules with a strong homotopy action $(m_i^L)_{i\in\mathbb{N}}$ on L and $(m_i^M)_{i\in\mathbb{N}}$ on M. A morphism of strong homotopy actions $f : L \longrightarrow M$ is called *nullhomotopic* if there is a sequence $(h_i)_{i\in\mathbb{N}}$ of graded B-linear morphisms $h_i : A^{\otimes i-1} \otimes L \longrightarrow M$ of degree i such that for each $n \in \mathbb{N}$ we have

$$\begin{aligned} f_n &= \sum_{\ell=1}^{n}(-1)^{\ell-1} m_\ell^M(a_1 \otimes \cdots \otimes a_{\ell-1} \otimes h_{n-\ell+1}(a_\ell \otimes \cdots \otimes a_{n-1} \otimes x)) \\ &\quad + \sum_{i=1}^{n-2}(-1)^{i-1} h_{n-1}(a_1 \otimes \cdots \otimes a_{i-1} \otimes a_i a_{i+1} \otimes a_{i+2} \otimes \cdots \otimes a_{n-1} \otimes x) \\ &\quad + \sum_{\ell=1}^{n}(-1)^{n-\ell} h_{n-\ell+1}(a_1 \otimes \cdots \otimes a_{n-\ell} \otimes m_\ell^L(a_{n-\ell+1} \otimes \cdots \otimes a_{n-1} \otimes x)). \end{aligned}$$

Two strong homotopy actions are *homotopy equivalent* if and only if their difference is nullhomotopic.

Again, the case $n = 1$ leads to the usual concept of being nullhomotopic.

Recall from Example 6.2.12 Item 8 that the structure of a strong homotopy action of A on the $\mathbb{Z}$-graded B-module L is equivalent to a degree -1 differential on the $T(A)$-comodule structure on $T(A) \otimes L$, and likewise the structure of a strong homotopy action of A on the $\mathbb{Z}$-graded B-module M is equivalent to a degree -1 differential on the $T(A)$-comodule structure on $T(A) \otimes M$.

As before let $\varepsilon^M : T(A) \otimes M \longrightarrow M$ be the projection on the degree 0 component of $T(A) \otimes M$. Let $f : T(A) \otimes L \longrightarrow T(A) \otimes M$ be a homomorphism of $T(A)$-comodules. Then $\varepsilon^M \circ f : T(A) \otimes L \longrightarrow M$ is a morphism of graded modules. Then, similar to Example 6.2.12 Item 5, the mapping $f \mapsto \varepsilon^M \circ f$ is a bijection between the $T(A)$-comodule morphisms $f : T(A) \otimes L \longrightarrow T(A) \otimes M$ and the graded homomorphisms $T(A) \otimes L \longrightarrow M$. Under this bijection the morphisms between graded $T(A)$-comodules with a differential of degree -1 correspond precisely to the morphisms between strong homotopy actions. Nullhomotopic strong homotopy actions correspond precisely to nullhomotopic morphisms of $T(A)$ comodules with degree -1 differential. The proofs of these facts are a straightforward computation using the above formulas.

6.3.3 Constructing an Algebra Action Out of a Strong Homotopy Action

Let L and M be $\mathbb{Z}$-graded B-modules. We have seen that a $T(A)$-comodule structure on $T(A) \otimes L$ with differential is equivalent to the structure of a strong homotopy action of A on L, and likewise for M. Moreover, we have seen that the vector space of graded $T(A)$-comodule homomorphisms

$$T(A) \otimes L \longrightarrow T(A) \otimes M$$

of degree p of graded $T(A)$-comodules with differential is in bijection with the morphisms of strong homotopy action $L \longrightarrow M$ by sending

$$f : T(A) \otimes L \longrightarrow T(A) \otimes M$$

to

$$\varepsilon^M \circ f : T(A) \otimes L \longrightarrow M.$$

Since the space of graded $T(A)$-comodule homomorphisms $T(A) \otimes L \longrightarrow T(A) \otimes M$ of graded $T(A)$-comodules with differential b^L on L and differential b^M on M is $\mathbb{Z}$-graded, we may impose a differential ∂ on this space by

$$f \overset{\partial}{\mapsto} (b^M \circ f - (-1)^p f \circ b^L).$$

This is indeed a differential, since

$$\begin{aligned}
\partial^2(f) &= \partial(b^M \circ f + (-1)^p f \circ b^L) \\
&= b^M \circ (b^M \circ f + (-1)^p f \circ b^L) + (-1)^{p-1}(b^M \circ f + (-1)^p f \circ b^L) \circ b^L \\
&= (-1)^p b^M \circ f \circ b^L + (-1)^{p-1} b^M \circ f \circ b^L \\
&= 0.
\end{aligned}$$

It is clear that $\partial(f)$ is a morphism of $T(A)$-comodules if f is such a morphism.

By the bijection of morphisms of $T(A)$-comodules $T(A) \otimes L$ to $T(A) \otimes M$ with differential and the morphisms of strong homotopy actions on L to M we obtain a complex $Hom^{\bullet}_{stronghomotopy}(L, M)$ whose degree p component is the degree p morphisms of strong homotopy actions $L \longrightarrow M$ and whose differential, which we denote by ∂^{∞}, is the image of ∂ under ε^M_*.

Lemma 6.3.10 *The complex* $(Hom^{\bullet}_{stronghomotopy}(L, M), \partial^{\infty})$ *has the property that* $\ker(\partial_0^{\infty})$ *contains precisely the morphisms of strong homotopy actions* $L \longrightarrow M$ *of* A. *Moreover* $\mathrm{im}(\partial_1^{\infty})$ *contains precisely the nullhomotopic morphisms of strong homotopy actions.*

Proof Let f be a $T(A)$-comodule morphism $T(A) \otimes L \longrightarrow T(A) \otimes M$ of degree p. Then $\partial^\infty f$ induces a sequence of mappings $(\varepsilon^M \circ \partial^\infty f)_n : A^{\otimes n-1} \otimes L \longrightarrow M$ of degree $p+n-1$ given by

$$\begin{aligned}
&(\varepsilon^M \circ \partial^\infty f)_n(a_1 \otimes \cdots \otimes a_{n-1} \otimes x) \\
&= \sum_{\ell=1}^{n} (-1)^{p(\ell-1)} m_\ell^M(a_1 \otimes \cdots \otimes a_{\ell-1} \otimes f_{n-\ell+1}(a_\ell \otimes \cdots \otimes a_{n-1} \otimes x)) \\
&\quad - \sum_{i=1}^{n-2} (-1)^{p+i-1} f_{n-1}(a_1 \otimes \cdots \otimes a_{i-1} \otimes a_i a_{i+1} \otimes a_{i+2} \otimes \cdots \otimes a_{n-1} \otimes x) \\
&\quad - \sum_{\ell=1}^{n} (-1)^{p+n-\ell} f_{n-\ell+1}(a_1 \otimes \cdots \otimes a_{n-\ell} \otimes m_\ell^L(a_{n-\ell+1} \otimes \cdots \otimes a_{n-1} \otimes x)).
\end{aligned}$$

This proves the statement. □

Let K be a commutative ring, let A and B be K-algebras and suppose that A is flat over K. Then denote by $SHaction(A-B)$ the category whose objects are strong homotopy actions of A on $\mathbb{Z}$-graded B-modules and morphisms the morphisms of strong homotopy actions of A.

Now, every complex X of A–B-bimodules induces a strong homotopy action on the restriction of X to B by mapping the differential to m_1, the action of A to m_2 and putting $m_i^X = 0$ for all $i \geq 3$. A morphism of complexes of A–B-bimodules $X \longrightarrow Y$ induces a morphism of strong homotopy actions $X \longrightarrow Y$. Hence, we obtain a functor "weakening"

$$W : C^-(A \otimes B\text{-}Mod) \longrightarrow SHaction(A-B)$$

which associates to a strict action its strong homotopy action with $m_i = 0$ for all $i \geq 3$. We shall show that W has a left adjoint R, "rigidifying" the homotopy action to an actual action. Moreover, we shall be able to give an explicit construction of RL for a strong homotopy action of A on L.

Given a $\mathbb{Z}$-graded B-module L with a strong homotopy action of A on L, let

$$RL := A \otimes T(A) \otimes L$$

as a graded module and equip it with the differential

$$\begin{aligned}
&d(a_0 \otimes \cdots \otimes a_{n-1} \otimes x) \\
&:= -a_0 a_1 \otimes a_2 \otimes \cdots \otimes a_{n-1} \otimes x \\
&\quad + \sum_{i=1}^{n-2} (-1)^{i-1} a_0 \otimes \cdots \otimes a_{i-1} \otimes a_i a_{i+1} \otimes a_{i+2} \ldots a_{n-1} \otimes x
\end{aligned}$$

$$+\sum_{\ell=1}^{n}(-1)^{n-\ell}a_0\otimes\cdots\otimes a_{n-\ell}\otimes m_\ell^X(a_{n-\ell+1}\otimes\cdots\otimes a_{n-1}\otimes x).$$

Observe that we consider the grading of $A\otimes T(A)$ as the grading coming from $T(A)$, i.e. since

$$T(A)=K\oplus A\oplus A^{\otimes 2}\oplus A^{\otimes 3}\oplus\cdots$$

with K in degree 0, A in degree 1, $A^{\otimes n}$ in degree n, we have

$$A\otimes T(A)=A\oplus A^{\otimes 2}\oplus A^{\otimes 3}\oplus A^{\otimes 4}\oplus\cdots$$

with A in degree 0, $A^{\otimes 2}$ in degree 1, and $A^{\otimes n+1}$ in degree n. We need to show that d is indeed a differential. The fact that d is homogeneous of degree -1 is immediate. We need to compute that $d^2=0$. Let

$$\begin{aligned}&d_{A\otimes T(A)}(a_0\otimes\cdots\otimes a_{n-1})\\&\quad=-a_0a_1\otimes a_2\otimes\cdots\otimes a_{n-1}+a_0\otimes\gamma_{T(A)}(a_1\otimes a_2\otimes\cdots\otimes a_{n-1})\end{aligned}$$

and abbreviate

$$\alpha(a_0\otimes\cdots\otimes a_{n-1})=-a_0a_1\otimes a_2\otimes\cdots\otimes a_{n-1}$$

as well as

$$\beta(a_0\otimes\cdots\otimes a_{n-1})=a_0\otimes\gamma_{T(A)}(a_1\otimes a_2\otimes\cdots\otimes a_{n-1}).$$

Then

$$(d_{A\otimes T(A)})^2=\alpha^2+\alpha\beta+\beta\alpha+\beta^2$$

and since we can compute

$$\begin{aligned}\alpha^2(a_0\otimes\cdots\otimes a_{n-1})&=a_0a_1a_2\otimes a_3\otimes\cdots\otimes a_{n-1}\\ \alpha\beta(a_0\otimes\cdots\otimes a_{n-1})&=-a_0a_1a_2\otimes a_3\otimes\cdots\otimes a_{n-1}\\&\quad+a_0a_1\otimes a_2a_3\otimes\cdots\otimes a_{n-1}\\&\quad-a_0a_1\otimes a_2\otimes a_3a_4\otimes\cdots\otimes a_{n-1}\\&\quad-\cdots\\&\quad-(-1)^{n-1}a_0a_1\otimes a_2\otimes\cdots\otimes a_{n-2}a_{n-1}\\ \beta\alpha(a_0\otimes\cdots\otimes a_{n-1})&=-a_0a_1\otimes a_2a_3\otimes\cdots\otimes a_{n-1}\\&\quad+a_0a_1\otimes a_2\otimes a_3a_4\otimes\cdots\otimes a_{n-1}\\&\quad+\cdots\\&\quad+(-1)^{n-1}a_0a_1\otimes a_2\otimes\cdots\otimes a_{n-2}a_{n-1}\end{aligned}$$

$$\beta^2 = 0$$

we get $(d_{A\otimes T(A)})^2 = 0$. The differential on $T(A) \otimes L$ is given by

$$\begin{aligned}
&\gamma(a_1 \otimes \cdots \otimes a_{n-1} \otimes x) \\
&:= \sum_{i=1}^{n-2} (-1)^{i+1} (a_1 \otimes \cdots \otimes a_{i-1} \otimes a_i a_{i+1} \otimes a_{i+2} \otimes \cdots \otimes a_{n-1} \otimes x) \\
&+ \sum_{\ell=1}^{n-1} (-1)^{n-\ell} (a_1 \otimes \cdots \otimes a_{n-\ell} \otimes m_\ell^L (a_{n-\ell+1} \otimes \cdots \otimes a_{n-1} \otimes x))
\end{aligned}$$

as was defined in Example 6.2.12. The morphism

$$1_A \otimes \Delta_{T(A)} \otimes 1_L : A \otimes T(A) \otimes L \longrightarrow A \otimes T(A) \otimes T(A) \otimes L$$

defines an isomorphism on a graded submodule of $A \otimes T(A) \otimes T(A) \otimes L$ and a differential on this graded submodule is induced by the differential on $A \otimes T(A) \otimes T(A) \otimes L$ so that

$$\begin{array}{ccc}
A \otimes T(A) \otimes L & \stackrel{1_A \otimes \Delta_{T(A)} \otimes 1_L}{\longrightarrow} & A \otimes T(A) \otimes T(A) \otimes L \\
d_{A\otimes T(A)\otimes L} \downarrow & & d_{A\otimes T(A)} \otimes 1_{T(A)\otimes L} + \downarrow 1_{A\otimes T(A)} \otimes d_{T(A)\otimes L} \\
A \otimes T(A) \otimes L & \stackrel{1_A \otimes \Delta_{T(A)} \otimes 1_L}{\longrightarrow} & A \otimes T(A) \otimes T(A) \otimes L
\end{array}$$

is commutative. Since we know already that

$$d_{A\otimes T(A)} \otimes 1_{T(A)\otimes L} + 1_{A\otimes T(A)} \otimes d_{T(A)\otimes L}$$

is a differential on $A \otimes T(A) \otimes T(A) \otimes L$, $d_{A\otimes T(A)\otimes L}$ is also a differential on $A \otimes T(A) \otimes L$.

Let X be a $\mathbb{Z}$-graded B-module with a strong homotopy action of A on X. Suppose Y is a complex of A–B-bimodules. Since $m_i^Y = 0$ for all $i \geq 3$ we see that

$$Hom^\bullet_{stronghomotopy}(X, WY) \simeq Hom^\bullet_K(T(A) \otimes X, Y)$$

as complexes. But now,

$$Hom^\bullet_K(T(A) \otimes X, Y) \simeq Hom^\bullet_A(A \otimes T(A) \otimes X, Y)$$

by Frobenius reciprocity (or in other words $(A \otimes_K -, Hom_A(A, -))$ is an adjoint pair of functors between K-*Mod* and A-*Mod*). But finally

$$Hom^\bullet_A(A \otimes T(A) \otimes X, Y) = Hom^\bullet_A(RX, Y).$$

More explicitly the composition of the isomorphisms gives explicit mappings in each direction:

$$\begin{aligned} Hom^\bullet_{stronghomotopy}(X, WY) &\xrightarrow{\varphi} Hom^\bullet_A(RX, Y) \\ f &\mapsto [(a \otimes ta \otimes x) \mapsto af(ta \otimes x)] \\ Hom^\bullet_A(RX, Y) &\longrightarrow Hom^\bullet_{stronghomotopy}(X, WY) \\ g &\mapsto [(ta \otimes x) \mapsto g(1 \otimes ta \otimes x)] \end{aligned}$$

The fact that φ commutes with the differentials is an easy verification using the explicit formulas:

$$d(\varphi(f)) = a_0 d(f(a_1 \otimes \cdots \otimes a_{n-1} \otimes x))$$

and then if f is an element in the degree p homogeneous component of the $Hom^\bullet_{stronghomotopy}$-complex,

$$\begin{aligned} &\varphi(f)(d(a_0 \otimes \cdots \otimes a_{n-1} \otimes x)) \\ &\quad = -(a_0 a_1) f(a_2 \otimes \cdots \otimes a_{n-1} \otimes x) \\ &\qquad + \sum_{i=1}^{n-2} (-1)^{i-1} a_0 f(a_2 \otimes \cdots \otimes a_i a_{i+1} \otimes \cdots \otimes a_{n-1} \otimes x) \\ &\qquad + \sum_{\ell=1}^{n} (-1)^{n-\ell} a_0 f(a_1 \otimes \cdots \otimes a_{n-\ell} \otimes m_\ell(a_{n-\ell+1 \otimes \cdots \otimes a_{n-1} \otimes x})), \\ &\varphi(d(f))(a_0 \otimes \cdots \otimes a_{n-1} \otimes x) \\ &\quad = a_0 d(f(a_1 \otimes \cdots \otimes a_{n-1} \otimes x) - (-1)^p a_0(a_1 f(a_2 \otimes \cdots \otimes a_{n-1} \otimes x)) \\ &\qquad - \sum_{i=1}^{n-2} (-1)^{p+i-1} a_0 f(a_1 \otimes \cdots \otimes a_i a_{i+1} \otimes \ldots a_{n-1} \otimes x) \\ &\qquad - \sum_{\ell=1}^{n} (-1)^{p+n-\ell} a_0 f(a_1 \otimes \cdots \otimes a_{n-\ell} \otimes m_\ell(a_{n-\ell+1} \otimes \cdots \otimes a_{n-1} \otimes x)). \end{aligned}$$

Hence $\varphi(d(f)) = d(\varphi(f)) - (-1)^p \varphi(f) d$, which is precisely the image of the differential in degree p.

We have shown the following

Lemma 6.3.11 *The functors (R, W) form a pair of adjoint functors*

$$W : K^-(A \otimes B\text{-}Mod) \longrightarrow SHaction(A - B)/homotopy$$

and

$$R : SHaction(A - B)/homotopy \longrightarrow K^-(A \otimes B\text{-}Mod).$$

Proof The only thing to verify is that R and W respect nullhomotopic morphisms. But this is clear. □

A morphism $f = (f_n)_{n\in\mathbb{N}}$ in $SHaction(A - B)$ is called a *quasi-isomorphism* if f_1 is a quasi-isomorphism of complexes. Since W does not change anything in the structure of complexes, it is clear that W preserves quasi-isomorphisms. Since A is K-flat, $A \otimes_K -$ preserves exact sequences and hence R also preserves quasi-isomorphisms. In the same way as we constructed a derived category of an additive category out of a homotopy category by formally inverting quasi-isomorphisms, we may consider the derived category $D(SHaction(A - B))$ of the additive category $SHaction(A - B)$.

Lemma 6.3.12 *The functors* (R, W) *form a pair of adjoint functors*

$$W : D^-(A \otimes B\text{-}Mod) \longrightarrow D(SHaction(A - B))$$

and

$$R : D(SHaction(A - B)) \longrightarrow D^-(A \otimes B\text{-}Mod).$$

Proof This follows immediately from the fact that W and R preserve quasi-isomorphisms and from Lemma 6.3.11. □

Considering the counit $\eta : RW \longrightarrow id$ of the adjunction, and we see that

$$RWY = A \otimes TA \otimes Y$$

with differential given by

$$\begin{aligned}
&d(a_0 \otimes \cdots \otimes a_{n-1} \otimes y)\\
&\quad := -a_0a_1 \otimes a_2 \otimes \cdots \otimes a_{n-1} \otimes y\\
&\qquad + \sum_{i=1}^{n-2}(-1)^{i-1} a_0 \otimes \cdots \otimes a_{i-1} \otimes a_i a_{i+1} \otimes a_{i+2} \otimes \cdots \otimes a_{n-1} \otimes y\\
&\qquad + \sum_{\ell=1}^{2}(-1)^{n-\ell} a_0 \otimes \cdots \otimes a_{n-\ell} \otimes m_\ell^Y(a_{n-\ell+1} \otimes \cdots \otimes a_{n-1} \otimes y)\\
&\quad = -a_0a_1 \otimes a_2 \otimes \cdots \otimes a_{n-1} \otimes y\\
&\qquad + \sum_{i=1}^{n-2}(-1)^{i-1} a_0 \otimes \cdots \otimes a_{i-1} \otimes a_i a_{i+1} \otimes a_{i+2} \otimes \cdots \otimes a_{n-1} \otimes y\\
&\qquad + (-1)^{n-1} a_0 \otimes \cdots \otimes a_{n-1} \otimes d^Y(y)\\
&\qquad + (-1)^{n-2} a_0 \otimes \cdots \otimes a_{n-2} \otimes a_{n-1} y
\end{aligned}$$

$$= \sum_{i=0}^{n-2}(-1)^{i-1}a_0 \otimes \cdots \otimes a_{i-1} \otimes a_i a_{i+1} \otimes a_{i+2} \otimes \cdots \otimes a_{n-1} \otimes y$$
$$+ (-1)^{n-2}a_0 \otimes \cdots \otimes a_{n-2} \otimes a_{n-1}y$$
$$+ (-1)^{n-1}a_0 \otimes \cdots \otimes a_{n-1} \otimes d^Y(y).$$

Hence the evaluation $RWY \xrightarrow{\eta_Y} Y$ of the counit on Y is the bar resolution of Y. We know by Proposition 3.6.4 and Remark 3.6.5 that this resolution is a quasi-isomorphism. Therefore $RW \xrightarrow{\eta} id$ is an isomorphism.

Proposition 6.3.13 *The functor*

$$W : D^-(A \otimes B\text{-}Mod) \longrightarrow D(SHaction(A - B))$$

induces an equivalence between the category of A–B-bimodules and the full subcategory of $D(SHaction(A - B))$ given by those strict homotopy actions of A for which $m_2^L(1 \otimes -) : L \longrightarrow L$ is the identity morphism on homology.

Proof Since the counit $RW \longrightarrow id$ is an isomorphism we get that W induces an embedding into a full subcategory of $D(SHaction(A - B))$. The full subcategory is obtained as those objects for which the unit $id \longrightarrow WR$ is an isomorphism.

Let L be a $\mathbb{Z}$-graded B-module with a strong homotopy action of A such that the unit $L \longrightarrow WR(L)$ is an isomorphism in the derived category. By Remark 6.3.8 we see that

$$\begin{array}{ccc} L & \longrightarrow & WR(L) \\ m_2^L(1,-)\downarrow & & \downarrow m_2^{WR(L)}(1,-) \\ L & \longrightarrow & WR(L) \end{array}$$

commutes up to homotopy. The morphism $m_2^{WR(L)}(1,-)$ is multiplication by 1 on the complex $WR(L)$, since the A-module structure is given by multiplication on the left term A of $A \otimes T(A) \otimes L$. Since the unit $L \longrightarrow WR(L)$ is a quasi-isomorphism, the action of $m_2^L(1,-)$ on the homology of L is the identity as well.

Suppose now that

$$m_2^L(1,-) : L \longrightarrow L$$

is the identity morphism on homology. Then we may filter $A \otimes T(A) \otimes L$ in the following way.

$$F^p(A \otimes T(A) \otimes L) := (A \otimes L) \oplus (A \otimes A \otimes L) \oplus \cdots \oplus (A \otimes A^{\otimes p} \otimes L)$$

for all $p \geq 0$. We take the trivial filtration $F^p L := L$ for all $p \geq 0$ on L, and then the morphism

$$\begin{aligned} f : L &\longrightarrow A \otimes T(A) \otimes L \\ x &\mapsto 1 \otimes x \end{aligned}$$

is compatible with the filtrations, in the sense of Sect. 3.9. Both filtrations are regular. Then the first page E_1 of the spectral sequence of the filtration of $A \otimes TA \otimes L$ is the bar resolution of the graded module $H^*(L)$ (cf Definition 3.6.4). The bar resolution is quasi-isomorphic to $H^*(L)$ (cf Corollary 3.6.2) via the map f. Hence f induces isomorphisms on the E_2-terms of the spectral sequences. Proposition 3.9.11 then shows that f is a quasi-isomorphism. This shows the proposition. □

6.4 Constructing a Functor from a Tilting Complex; Keller's Theorem

Theorem 6.4.1 (Keller [3]) *Let K be a commutative ring and let A and B be two K-algebras. Suppose that A is projective as a K-module. Let T be a right bounded complex of projective B-modules and suppose that a homomorphism $\alpha : A \longrightarrow End_{K^-(B\text{-}Proj)}(T)$ of K-algebras is given. Suppose moreover that $Hom_{K^-(B\text{-}Proj)}(T, T[n]) = 0$ for each $n > 0$.*

Then there is a right bounded complex X in $K^-(B\text{-}Proj)$, a quasi-isomorphism $\varphi : T \longrightarrow X$ in $K^-(B\text{-}Proj)$, and a homomorphism of complexes $\beta : A \longrightarrow End_{C^-(B\text{-}Proj)}(X)$ such that for all $a \in A$ the diagram

$$\begin{array}{ccc} T & \stackrel{\varphi}{\longrightarrow} & X \\ \alpha(a) \downarrow & & \downarrow \beta(a) \\ T & \stackrel{\varphi}{\longrightarrow} & X \end{array}$$

is commutative.

Proof Since T is a complex of B-modules with a morphism $A \longrightarrow End_{D^-(B)}(T)$ we may put m_1 the differential and m_2 the action of A on T, which is associative up to some homotopy m_3. This verifies the defining conditions of the first terms of a strong homotopy action by Remark 6.3.3. Since A is projective as a K-module, the vanishing conditions on homomorphisms to shifted copies of T of Proposition 6.3.5 are satisfied. Hence, by Proposition 6.3.5 the mappings m_1, m_2 and m_3 can be completed to a strong homotopy action of A on L. Since $A \longrightarrow End_{K^-(B)}(T)$ is a ring homomorphism, $1 \in A$ acts as the identity on T, and therefore $m_2(1 \otimes -)$ induces the identity on the homology of the strong homotopy action complex of T. Therefore, by Proposition 6.3.13, we get that the strong homotopy action of A on T comes from a complex of A–B-bimodules X in the sense that the unit $f : T \longrightarrow WRT$ induces an isomorphism on the homology. Hence f_1 is a quasi-isomorphism and we may put $\varphi := f_1$. □

Remark 6.4.2 The complex X can be constructed explicitly. Actually $X = RT$, which is $A \otimes T(A) \otimes T$ as a graded module, and whose differentials are defined by the mappings m_i.

Remark 6.4.3 We obtained a complex X in $K^-(B\text{-}Proj)$ and a homomorphism $A \longrightarrow End_{C^-(B\text{-}Proj)}(X)$. Since A induces endomorphisms on the graded B-module X, we see that A acts on each of the homogeneous components of X in a way compatible with the action of B. Therefore X is a complex of B–A-bimodules and we may consider X in $D^-(A \otimes B)$.

Observe that in order to compute $\mathbb{R}Hom(X, -)$ and $X \otimes^{\mathbb{L}}_A -$ we only need to consider X up to isomorphism in $D^-(A \otimes B)$. Hence, we may replace X by its projective resolution and obtain a complex X of A–B-bimodules which is projective in each degree.

Proposition 6.4.4 *Let K be a commutative ring, let A and B be K-algebras and suppose that A is projective as a K-module. Let X be a right bounded complex of B–A-bimodules. Then there is a complex $\hat{X}$ of B–A-bimodules such that $\hat{X} \simeq X$ in $D^-(B \otimes A^{op})$ and such that the image of $\hat{X}$ in $D^-(B)$ is a complex of right bounded projective B-modules. If $B \otimes_K A^{op}$ is Noetherian, and if all homogeneous components of X are finitely generated, then all homogeneous components of $\hat{X}$ can be chosen finitely generated.*

Let T be a bounded complex of projective B-modules and suppose that $X \simeq T$ in $D^-(B)$. Then there is a bounded complex X' isomorphic to X in $D^-(B \otimes A^{op})$ such that each homogeneous component of X' is projective as a $B \otimes A^{op}$-module except the component in the highest degree m, where X'_m is projective as a B-module. Moreover, if $T_v = 0$, then $X'_v = 0$.

If B is also projective as a K-module, and if there is a bounded complex T' of projective A-modules which is isomorphic to X in the derived category of A-modules, then X' can be chosen so that all homogeneous components of X' are projective as A-modules and as B-modules.

Proof Since A is projective over K, we observe that the free module $B \otimes A^{op}$ is projective as a B-module. If B is projective as a K-module then $B \otimes_K A^{op}$ is projective as an A^{op}-module. Replace X by its projective resolution $\hat{X}$ using Proposition 3.5.43. Now, each homogeneous component of $\hat{X}$ is projective as a $B \otimes A^{op}$-module. If $B \otimes_K A^{op}$ is Noetherian and if X is a complex of finitely generated bimodules, then also $\hat{X}$ is a complex of finitely generated bimodules. However, $\hat{X}$ is not necessarily bounded.

Let T be concentrated in degrees $m, m-1, \dots, n$ for $m \geq n$ and let X' be the intelligent truncation of $\hat{X}$ at degree m, so that X' is concentrated in degrees at most m. All homogeneous components of X' are projective, except possibly X'_m. Moreover, X' is isomorphic to $\hat{X}$ in $D^-(B)$. The restriction of $\hat{X}$ to B is in $K^-(B\text{-}Proj)$, and by Proposition 3.5.43 we have an isomorphism $T \xrightarrow{\alpha} \hat{X}$ in $K^-(B\text{-}Proj)$. Hence, the cone of α is zero in $K^-(B\text{-}Proj)$, which implies that the cone of α is homotopic to 0. Moreover, since there is a quasi-isomorphism $\hat{X} \xrightarrow{\pi} X'$ the homomorphism of complexes $\pi \circ \alpha : T \to X'$ is a quasi-isomorphism. Let $\beta := \pi \circ \alpha$. By Lemma 3.5.32 the cone $C(\beta)$ of β is acyclic. The construction of $C(\beta)$ in Definition 3.5.26 shows that the rightmost homogeneous component of $C(\beta)$ is X_s, for some s, the homogeneous component of degree u for $u \in \{s+1, \dots, m-1\}$ is $T_{u-1} \oplus X_u$, the degree m homogeneous component of $C(\beta)$ is $T_{m-1} \oplus X'_m$ and the degree $m+1$ homogeneous component of $C(\beta)$ is T_m. Since $C(\beta)$ is acyclic, and since only the degree m homogeneous component of $C(\beta)$ is possibly non-projective, the differential of $C(\beta)$ is split up to degree $m-1$. Hence there is a short exact sequence

$$0 \to T_m \to T_{m-1} \oplus X'_m \to U_{m-1} \to 0$$

for a projective module U_{m-1}. Since U_{m-1} is projective, the sequence splits and therefore X'_m is also projective.

If B is also projective as a K-module, choosing m such that also the homogeneous components of T' are zero in degrees higher than m, the same argument shows that the complex X' has the required property. This proves the statement. □

Remark 6.4.5 The technique of coderivations can also be used to show the invariance of further structures under equivalences of derived categories. We shall show in Proposition 6.7.10 that the Hochschild cohomology is invariant in some sense under an equivalence of derived categories. Hochschild cohomology carries a rich structure, in particular the structure of a Lie algebra, the so-called Gerstenhaber structure. Keller showed that the Gerstenhaber structure is in a certain sense invariant under equivalences of derived categories (see Remark 6.12.33 below). An account of parts of this approach is given in [10]. We have already mentioned the Gerstenhaber structure in Remark 5.9.25.

6.5 Rickard's Morita Theorem for Derived Equivalences

Morita equivalences are very nicely behaved, as we have seen. Derived equivalences are much more complicated, although there is something similar to Morita's theorem for derived categories. This result was proved by Rickard [1] in a slightly more general fashion than the proof below. The original proof is discussed in detail in [11, Chap. 1].

6.5.1 The Theorem and Its Proof

Most astonishing is that the converse to Proposition 6.1.5 is true as well. This is the statement of Rickard's Morita theorem for derived equivalences.

Theorem 6.5.1 [1] *Let A and B be two algebras over a commutative ring K and suppose that A is projective over K. Then*

- *$D^-(A\text{-}Mod) \simeq D^-(B\text{-}Mod)$ as triangulated categories if and only if there is a tilting complex T over B with endomorphism algebra A^{op} and if there is an equivalence $D^-(A\text{-}Mod) \simeq D^-(B\text{-}Mod)$ as triangulated categories, then T is the image of the rank one regular module under the equivalence.*
- *Moreover,*

$$\begin{aligned} D^-(A\text{-}Mod) \simeq D^-(B\text{-}Mod) &\Leftrightarrow D^b(A\text{-}Mod) \simeq D^b(B\text{-}Mod) \\ &\Leftrightarrow K^b(A\text{-}proj) \simeq K^b(B\text{-}proj) \\ &\Leftrightarrow K^b(A\text{-}Proj) \simeq K^b(B\text{-}Proj) \\ &\Leftrightarrow K^-(A\text{-}Proj) \simeq K^-(B\text{-}Proj) \end{aligned}$$

- *and if A and B are Noetherian and finitely generated as R-modules, then*

$$D^-(A\text{-}Mod) \simeq D^-(B\text{-}Mod) \Leftrightarrow D^b(A\text{-}mod) \simeq D^b(B\text{-}mod).$$

Remark 6.5.2 Parts of the proof of the first part come from Keller's approach in [11, Chap. 8] and the second and third part of the proof partly follows Rickard [1].

Proof If $F : D^-(A\text{-}Mod) \longrightarrow D^-(B\text{-}Mod)$ is an equivalence of triangulated categories, then $F(A) =: T$ is a tilting complex in $D^-(B\text{-}Mod)$ with endomorphism ring A^{op} by Proposition 6.1.5.

Now, let T be a tilting complex T in $D^-(B\text{-}Mod)$ with endomorphism ring A^{op}. By Theorem 6.4.1 there is a complex X of B–A-bimodules such that $X \simeq T$ in $D^b(B\text{-}Mod)$. Therefore we may replace T by X and obtain a functor

$$\mathbb{R}Hom_B(X, -) : D^-(B\text{-}Mod) \longrightarrow D^-(A\text{-}Mod).$$

By Proposition 3.7.16 the functor $\mathbb{R}Hom_B(X, -)$ has a left adjoint

$$X \otimes^{\mathbb{L}}_A - : D^-(A\text{-}Mod) \longrightarrow D^-(B\text{-}Mod).$$

Let

$$\varepsilon : id_{D^-(A\text{-}Mod)} \longrightarrow \mathbb{R}Hom_B(X, X \otimes^{\mathbb{L}}_A -)$$

be the unit of this adjoint pair and let $\mathcal{T}$ be the full subcategory generated by those objects M of $D^-(A\text{-}Mod)$ such that ε_M is an isomorphism.

We claim that the regular A-module A is in $\mathcal{T}$. Indeed,

$$\varepsilon_A : A \longrightarrow \mathbb{R}Hom_B(X, X \otimes^{\mathbb{L}}_A A) = \mathbb{R}Hom_B(X, X) = End_{D^-(B\text{-}Mod)}(T)$$

is an isomorphism by Theorem 6.4.1 and the hypothesis that T is a tilting complex with endomorphism ring A^{op}.

We claim that $\mathcal{T}$ is triangulated. Indeed, by Proposition 3.7.4 and Definition 3.7.8 we obtain that $\mathbb{R}Hom(X, -)$ sends a distinguished triangle to a distinguished triangle, and $X \otimes_A^{\mathbb{L}} -$ sends a triangle to a triangle. Moreover, both functors commute with the shift functor. Hence, denoting by C the cone of a morphism, for any $f : X \longrightarrow Y$ we get

$$C(\mathbb{R}Hom_B(X, X \otimes_A^{\mathbb{L}} f)) \simeq C(f)$$

if X and Y are in $\mathcal{T}$, where the isomorphism is induced by ε. But then axiom TR3 of a triangulated category together with the long exact sequence on homology, Proposition 3.5.29 shows that $C(f)$ is also in $\mathcal{T}$. Since T is a tilting complex, the smallest triangulated category containing the direct summands of finite direct sums of T is $K^b(B\text{-}proj)$. The smallest triangulated category closed under direct summands and containing A is $K^b(A\text{-}proj)$ and so $\mathbb{R}Hom_B(X, -)$ induces an equivalence

$$K^b(B\text{-}proj) \longrightarrow K^b(A\text{-}proj).$$

Now, X is compact as a complex of B-modules and hence $\mathbb{R}Hom_B(X, -)$ commutes with arbitrary direct sums.

Hence ϵ_P is an isomorphism for any projective A-module P. Moreover, the category $K^b(A\text{-}Proj)$ is the triangulated category generated by projective A-modules. Moreover, $K^b(B\text{-}Proj)$ is the smallest triangulated category containing all direct summands of arbitrary sums of T and since T is mapped to A under $\mathbb{R}Hom_B(X, -)$, the functor $\mathbb{R}Hom_B(X, -)$ induces an equivalence between $K^b(A\text{-}Proj)$ and $K^b(B\text{-}Proj)$.

The smallest triangulated subcategory of $K^-(A\text{-}Proj)$ containing A and which is closed under direct summands of arbitrary direct sums is $K^-(A\text{-}Proj) \simeq D^-(A\text{-}Mod)$. Likewise the smallest triangulated subcategory of $K^-(B\text{-}Proj)$ containing T and which is closed under direct summands of arbitrary direct sums is $K^-(B\text{-}Proj) \simeq D^-(B\text{-}Mod)$. Hence $\mathbb{R}Hom_B(X, -)$ induces an equivalence between $D^-(A\text{-}Mod)$ and $D^-(B\text{-}Mod)$.

We claim that

$$K^b(B\text{-}proj) \simeq K^b(A\text{-}proj) \Rightarrow K^-(B\text{-}proj) \simeq K^-(A\text{-}proj)$$

and

$$K^b(B\text{-}proj) \simeq K^b(A\text{-}proj) \Rightarrow K^-(B\text{-}Proj) \simeq K^-(A\text{-}Proj).$$

The image of the regular A-module under an equivalence $K^b(A\text{-}proj) \simeq K^b(B\text{-}proj)$ is a tilting complex T over B with endomorphism ring A^{op}. Let X be a complex constructed by Keller's theorem. Let U be a right bounded complex of projective A-modules. We need to show that U is in the image of $\mathbb{R}Hom_B(X, -)$. We know that bounded complexes of projective A-modules are in the image of $\mathbb{R}Hom_B(X, -)$. Let $\kappa_{\leq n}U$ be the complex which has the same modules and differentials as U in all degrees less than or equal to n, and which is 0 in degrees strictly larger than n. Then we get an injective morphism of complexes

$$\kappa_{\leq i}U \overset{\iota_i}{\hookrightarrow} \kappa_{\leq i+1}U$$

for all $i \in \mathbb{N}$ and

$$U = \mathrm{colim}_{n\in\mathbb{N}}\kappa_{\leq n}U$$

is the inductive limit. We know by Proposition 3.1.18 that we get an exact sequence of complexes

$$0 \longrightarrow \coprod_{n\in\mathbb{N}} \kappa_{\leq n}U \longrightarrow \coprod_{n\in\mathbb{N}} \kappa_{\leq n}U \longrightarrow U \longrightarrow 0$$

and we observe that by Proposition 3.5.51 this means that in $K^-(A\text{-}proj)$ we get a distinguished triangle

$$\coprod_{n\in\mathbb{N}} \kappa_{\leq n}U \longrightarrow \coprod_{n\in\mathbb{N}} \kappa_{\leq n}U \longrightarrow U \longrightarrow \left(\coprod_{n\in\mathbb{N}} \kappa_{\leq n}U\right)[1].$$

Now, for each $n \in \mathbb{N}$ there is an object V_n of $K^-(B\text{-}proj)$ such that

$$\kappa_{\leq n}U \simeq \mathbb{R}Hom_B(X, V_n)$$

and such that $\iota_n = Hom_B(X, \alpha_n)$ for some $\alpha_n : V_n \longrightarrow V_{n+1}$. Since the functor $\mathbb{R}Hom_B(X, -)$ preserves triangles and since, X being compact, this functor preserves direct sums, we get that $\mathrm{colim}_{n\in\mathbb{N}} V_n =: V$ maps to U under $\mathbb{R}Hom_B(X, -)$.

Therefore $\mathbb{R}Hom_B(X, -)$ also induces an equivalence

$$K^b(B\text{-}Proj) \longrightarrow K^b(A\text{-}Proj)$$

and an equivalence

$$K^-(B\text{-}Proj) \longrightarrow K^-(A\text{-}Proj)$$

as well as an equivalence

$$K^-(B\text{-}proj) \longrightarrow K^-(A\text{-}proj).$$

Since $K^-(A\text{-}Proj) \simeq D^-(A\text{-}Mod)$ and $K^-(B\text{-}Proj) \simeq D^-(B\text{-}Mod)$ we have shown that the existence of a tilting complex induces an equivalence of each of the triangulated categories in the second part of the theorem, except $D^b(A\text{-}Mod) \simeq D^b(B\text{-}Mod)$.

We get that $K^b(A\text{-}proj) \simeq K^b(B\text{-}proj)$ implies $D^b(A\text{-}Mod) \simeq D^b(B\text{-}Mod)$.

We already know that we may produce an equivalence of triangulated categories $K^-(A\text{-}Proj) \simeq K^-(B\text{-}Proj)$ from an equivalence $K^b(A\text{-}proj) \simeq K^b(B\text{-}proj)$. First $D^b(A\text{-}Mod)$ is characterised in $K^-(A\text{-}proj)$ as those objects C in $K^-(A\text{-}proj)$ such that for all D in $K^-(A\text{-}Proj)$ there is an integer $m(C, D)$ such that $Hom_{K^-(A\text{-}Proj)}$

$(D, C[m]) = 0$ for all $m < m(C, D)$. Indeed, if C is in $D^b(A\text{-}Mod)$, since we are only dealing with D in $K^-(A\text{-}Proj)$, we do not need to invert quasi-isomorphisms and we may work in the homotopy category only. Hence, C has 0 components only in high degrees. D however is right bounded. If we shift C far to the right so that all the non-zero homogeneous components of C are in positions where there are no non-zero components of D, we obtain only the zero morphism. On the other hand, Corollary 3.5.52 implies that $Hom_{K^-(A\text{-}Proj)}(A, C[n]) = H_n(C)$ for all $n \in \mathbb{N}$ and the analogous characterisation holds for B. So, homomorphisms from A to shifted copies of C determine when if C has bounded homology or not. Therefore an equivalence of $K^-(A\text{-}Proj) \simeq K^-(B\text{-}Proj)$ induces an equivalence $K^{-,b}(A\text{-}Proj) \simeq K^{-,b}(B\text{-}Proj)$ and hence an equivalence $D^b(A\text{-}Mod) \simeq D^b(B\text{-}Mod)$.

An equivalence $D^b(A\text{-}Mod) \simeq D^b(B\text{-}Mod)$ of triangulated categories induces an equivalence $K^b(A\text{-}proj) \simeq K^b(B\text{-}proj)$.

Now, C in $D^b(A\text{-}Mod)$ is in $K^b(A\text{-}Proj)$ if and only if for all M in $D^b(A\text{-}Mod)$ there is an $n(C, M)$ so that $Hom_{D^b(A\text{-}Mod)}(C, M[n]) = 0$ if $n > n(C, M)$. Indeed, replace C by its projective resolution using Proposition 3.5.43 and denote the resulting complex by C again. If C is not bounded, then there exist infinitely many indices $n_i > 0$ such that $\ker(d^C_{n_i})$ is not projective. Otherwise, one could split off a zero-homotopic complex so that C is bounded and has projective components. But then $\bigoplus_{i\in\mathbb{N}} \ker(d^C_{n_i}) =: M$ has the property that $Hom_{K^-(A\text{-}Proj)}(C, M[n_i]) \neq 0$ for all i. Hence an equivalence $D^b(A\text{-}Mod) \simeq D^b(B\text{-}Mod)$ induces an equivalence $K^b(A\text{-}proj) \simeq K^b(B\text{-}proj)$.

Suppose now that A and B are Noetherian. Then $D^b(A\text{-}mod) \simeq K^{-,b}(A\text{-}proj)$ and $D^-(A\text{-}mod) \simeq K^-(A\text{-}proj)$ and likewise for B. But Keller's Theorem 6.4.1 and the second step implies that T induces a complex of finitely generated bimodules X such that $\mathbb{R}Hom_B(X, -) : D^-(B\text{-}Mod) \to D^-(A\text{-}Mod)$ is an equivalence with quasi-inverse $X \otimes^{\mathbb{L}}_A -$. Both functors restrict to functors between $D^-(A\text{-}mod)$ and $D^-(B\text{-}mod)$ and the arguments of the second step showing why bounded homology is preserved apply here as well. If $D^b(A\text{-}mod) \simeq D^b(B\text{-}mod)$, then the image of the regular A-module A is a tilting complex with endomorphism ring A^{op}, and the first part of the theorem shows that $D^-(A\text{-}Mod) \simeq D^-(B\text{-}Mod)$. □

Remark 6.5.3 Let us give two remarks on the statement of Theorem 6.5.1.

- Let K be a field and let A and B be finite dimensional K-algebras. Then we have seen combining Proposition 6.4.4 and Theorem 6.5.1 that if $D^b(A\text{-}Mod) \simeq D^b(B\text{-}Mod)$, then there is a complex X of A–B-bimodules such that

$$X \otimes^{\mathbb{L}}_B - : D^b(B\text{-}Mod) \longrightarrow D^b(A\text{-}Mod)$$

  is an equivalence. It is not known at present if this implies that every equivalence of triangulated categories $D^b(A\text{-}Mod) \simeq D^b(B\text{-}Mod)$ is of this form. Indeed, Theorem 6.5.1 does not state that *every* equivalence of triangulated categories $D^b(A\text{-}Mod) \simeq D^b(B\text{-}Mod)$ is of this form. However no equivalence is known which is not given by a tensor product with a complex of bimodules. Equivalences given by a tensor product with a complex of bimodules will be called *of standard type*.

- The proof of Theorem 6.5.1 gives slightly stronger statements. In some steps we show that every equivalence between the concerned triangulated categories implies an equivalence of some other linked triangulated categories. In particular any equivalence $D^-(A\text{-}Mod) \simeq D^-(B\text{-}Mod)$ of triangulated categories restricts to an equivalence $K^b(A\text{-}proj) \simeq K^b(B\text{-}proj)$ since an equivalence preserves the full subcategory of compact objects.

6.5.2 Derived Equivalences of Standard Type

Definition 6.5.4 Let K be a commutative ring and let A and B be two K-algebras. An equivalence of triangulated categories $F : D^b(A) \longrightarrow D^b(B)$ is said to be *of standard type* if there is a complex X of A-B-bimodules such that

$$F \simeq X \otimes_B^{\mathbb{L}} -.$$

Proposition 6.5.5 *Let K be a commutative ring and let A and B be K-algebras such that B is projective as a K-module. If $D^b(A) \simeq D^b(B)$, then there is an equivalence $X \otimes_A^{\mathbb{L}} - : D^b(A) \longrightarrow D^b(B)$ of standard type.*

Proof This follows from Proposition 6.4.4 and Theorem 6.5.1 and the fact that $X \otimes_A^{\mathbb{L}} -$ is left adjoint to $\mathbb{R}Hom_B(X, -)$ by Proposition 3.7.16. Finally Proposition 3.2.8 proves that $X \otimes_A^{\mathbb{L}} -$ is an equivalence. □

Equivalences of standard type have many nice properties which we shall present in the sequel. The first one is that the inverse is standard as well.

Proposition 6.5.6 *Let K be a commutative ring, let A and B be Noetherian K-algebras, suppose that A and B are finitely generated projective as K-modules, and let*

$$X \otimes_B^{\mathbb{L}} - : D^b(B) \longrightarrow D^b(A)$$

be an equivalence of standard type. Then

$$\mathbb{R}Hom_A(X, -) : D^b(A) \longrightarrow D^b(B)$$

is a quasi-inverse of $X \otimes_B^{\mathbb{L}} -$ and

$$\mathbb{R}Hom_A(X, -) \simeq \mathbb{R}Hom_A(X, A) \otimes_A^{\mathbb{L}} -.$$

Proof We may assume that X is a complex of A–B-bimodules, its restriction being finitely generated projective on either side by Proposition 6.4.4. Hence

$$\mathbb{R}Hom_A(X,-) \simeq Hom_A(X,-)$$

where we understand that we use for $Hom_A(X,-)$ the complex defined for the right derived Hom-functor, except that we no longer need to replace the argument X by its projective resolution. Lemma 4.2.5 shows that

$$\mathbb{R}Hom_A(X,-) \simeq Hom_A(X,-) \simeq Hom_A(X,A) \otimes_A -.$$

Proposition 3.2.8 shows that the right adjoint $Hom_A(X,A) \otimes_A -$ is an equivalence if and only if $X \otimes_B -$ is an equivalence. □

Corollary 6.5.7 *Let K be a commutative ring, let A and B be Noetherian K-algebras such that A and B are both finitely generated projective as K-modules, and let*

$$X \otimes_A^{\mathbb{L}} - : D^b(A) \longrightarrow D^b(B)$$

be an equivalence of standard type. Then there is a bounded complex Y of A-B-bimodules, projective on either side, such that

$$X \otimes_A^{\mathbb{L}} Y \simeq B \text{ in } D^b(B \otimes B^{op}) \text{ and } Y \otimes_B^{\mathbb{L}} X \simeq A \text{ in } D^b(A \otimes A^{op}).$$

Proof Indeed, Proposition 6.5.6 shows that a quasi-inverse to $X \otimes_A^{\mathbb{L}} -$ is standard as well, whence of the form $Y \otimes_B^{\mathbb{L}} -$. Since $Y \otimes_B^{\mathbb{L}} -$ is quasi-inverse to $X \otimes_A^{\mathbb{L}} -$, we get that

$$\left(X \otimes_A^{\mathbb{L}} -\right) \circ \left(Y \otimes_B^{\mathbb{L}} -\right) \simeq id \text{ and } \left(Y \otimes_B^{\mathbb{L}} -\right) \circ \left(X \otimes_A^{\mathbb{L}} -\right) \simeq id.$$

Applying the composition of functors to the regular module, and using functoriality to get the bimodule structure, this proves the statement. □

Definition 6.5.8 Let K be a commutative ring and let A and B be K-algebras such that A and B are projective as K-modules. Then a bounded complex X in $D^b(B \otimes A^{op})$ is a *two-sided tilting complex* if there is another complex Y in $D^b(A \otimes B^{op})$ such that

$$X \otimes_A^{\mathbb{L}} Y \simeq B \text{ in } D^b(B \otimes B^{op})$$

and

$$Y \otimes_B^{\mathbb{L}} X \simeq A \text{ in } D^b(A \otimes A^{op}).$$

Remark 6.5.9 Observe that the situation of equivalences induced by two-sided tilting complexes is very similar to that of Morita equivalences.

We obtain a unicity result.

Proposition 6.5.10 *Let K be a commutative ring and let A and B be Noetherian K-algebras which are finitely generated projective as K-modules. Let T be a tilting complex in $D^b(B)$ with endomorphism ring isomorphic to A. If X and X' are complexes of B–A-bimodules with a quasi-isomorphism $\varphi' : T \longrightarrow X'$ and a quasi-isomorphism $\varphi : T \longrightarrow X$ then there is a unique invertible A–A bimodule M such that*

$$X' \simeq X \otimes_A M.$$

In particular if K is a field and A is basic and finite dimensional over K, then there is an automorphism α of A such that

$$X' \simeq X \otimes_A {}_1A_\alpha.$$

Proof Indeed, we have seen that X and X' are both two-sided tilting complexes. We may even assume that X and X' are both complexes of modules which are projective on either side. Let Y be a quasi-inverse of X and let Y' be a quasi-inverse of X', and again we may assume that both are formed by modules which are projective on either side. Since T is the image of the regular A-module A, we get that

$$X \otimes_A Y' \simeq A$$

as complexes of left modules. Moreover

$$(X \otimes_A Y') \otimes_A (X' \otimes Y) \simeq A$$

and

$$(X' \otimes Y) \otimes_A (X \otimes_A Y') \simeq A$$

as complexes of A–A-bimodules. Furthermore, since X, Y, X', Y' are all complexes of projective A-modules, so is $X \otimes_A Y'$ and we get that $X \otimes_A Y'$ is an invertible A-A-bimodule in the sense of Definition 4.6.1, whence it is an element in the Picard group of A. If A is a basic finite dimensional algebra over a field we know by Corollary 4.6.4 that $Pic_K(A) \simeq Out_K(A)$ and hence the statement follows. □

6.6 The Case of Two-Term Complexes

In practice tilting complexes with only two non-zero homogeneous components are particularly widely used. They are moreover the main tool in Okuyama's method for proving that certain blocks of group rings are derived equivalent. This method will be described in more detail in Sect. 6.10.2.

6.6.1 The Two-Sided Tilting Complex of a Two-Term Tilting Complex

We shall give an explicit formula for the two-sided tilting complex associated to a two-term tilting complex. In general Keller's Theorem 6.4.1 gives the general formula, and the formula is as explicit as the homotopy can be made explicit for morphisms from T to a shifted copy. However, even though the formula is explicit, it is not easy to exploit in specific examples. This is why it is instructive to see what the formula means for two-term tilting complexes, in which case it is quite easy to give. Moreover, two-term tilting complexes will be of quite some importance in examples, and many of the classical approaches to using equivalences of derived categories use two-term tilting complexes. Another approach to give an explicit formula was given in [8, 9] in some very specific cases, using only Proposition 6.6.2 below, and no knowledge of homotopy of morphisms of complexes at all. Note that Proposition 6.6.2 is a general fact which does not use Keller's theorem 6.4.1 or Rickard's Theorem 6.5.1, and so the formulas in [8, 9] are completely independent facts.

The presentation we choose here again closely follows Keller's original article [3].

Let K be a commutative ring, let A and B be two K-algebras and suppose that A is projective as a K-module. Let

$$T: \quad \cdots \longrightarrow 0 \longrightarrow T_1 \stackrel{d}{\longrightarrow} T_0 \longrightarrow 0 \longrightarrow \cdots$$

be a complex of projective B-modules and suppose that there is a ring homomorphism

$$A \stackrel{\alpha}{\longrightarrow} End_{K^b(B\text{-}proj)}(T).$$

Suppose moreover that

$$Hom_{K^b(B\text{-}proj)}(T, T[n]) = 0 \ \ \forall n < 0.$$

Now,

$$End_{C^b(B\text{-}proj)}(T) \stackrel{\pi}{\longrightarrow\!\!\!\!\to} End_{K^b(B\text{-}proj)}(T)$$

is surjective and K-linear. Since A is K-projective, we can find a K-linear map

$$A \stackrel{\tilde{\alpha}}{\longrightarrow} End_{C^b(B\text{-}proj)}(T)$$

such that

$$\alpha = \pi \circ \tilde{\alpha}.$$

Put

$$\begin{array}{rcl} A \otimes_K T & \xrightarrow{m_2} & T \\ a \otimes x & \mapsto & (\tilde{\alpha}(a))(x) \end{array}$$

There is no reason why m_2 should be an action of A on T. Indeed, usually it will not be associative in the sense that, for all $a, b \in A$ and $x \in T$, in general

$$m_2(ab \otimes x) \neq m_2(a \otimes m_2(b \otimes x)).$$

If we had equality, and $m_2(1 \otimes x) = x$ for all x, then m_2 would be an action of A on the complex T. Denote by

$$\begin{array}{rcl} A \otimes_K A & \xrightarrow{\mu_A} & A \\ a \otimes b & \mapsto & ab \end{array}$$

the multiplication in A. If the square

$$\begin{array}{ccc} A \otimes_K A \otimes_K T & \xrightarrow{id_A \otimes m_2} & A \otimes_K T \\ \mu_A \otimes id_T \downarrow & & \downarrow m_2 \\ A \otimes_K T & \xrightarrow{m_2} & T \end{array}$$

would be commutative, the construction would be finished. However, it is not commutative in general, but since α is a ring homomorphism it is commutative in the homotopy category. In other words, there is a morphism of B-modules, homogeneous of degree -1,

$$m_3 : A \otimes_K A \otimes_K T \longrightarrow T$$

which is our homotopy, such that for the difference between the two mappings in the square we get

$$m_2(a \otimes m_2(b \otimes x)) - m_2(ab \otimes x) = m_3(a \otimes b \otimes d(x)) + d(m_3(a \otimes b \otimes x))$$

for all $a, b \in A$; $x \in T$.

We now define a bi-graded module $\tilde{X}$ as follows.

degrees	$(*, 1)$	$(*, 0)$
$(2, *)$	$A \otimes_K A \otimes_K A \otimes_K T_1$	$A \otimes_K A \otimes_K A \otimes_K T_0$
$(1, *)$	$A \otimes_K A \otimes_K T_1$	$A \otimes_K A \otimes_K T_0$
$(0, *)$	$A \otimes_K T_1$	$A \otimes_K T_0$

Then we define maps $d_{(i,j)}$ between the homogeneous components for all $(i, j) \in \{(2, 1), (2, 0), (1, 1), (1, 0), (0, 1), (0, 0)\}$ as follows. For all $a, b, c \in A$ and $x \in T$

we put

$$\begin{aligned} d_{(2,*)}(a \otimes b \otimes c \otimes x) &:= -ab \otimes c \otimes x + a \otimes bc \otimes x + a \otimes m_3(b \otimes c \otimes x) \\ &\quad - a \otimes b \otimes m_2(c \otimes x) + a \otimes b \otimes c \otimes d(x) \\ d_{(1,*)}(a \otimes b \otimes x) &:= -ab \otimes x + a \otimes m_2(b \otimes x) - a \otimes b \otimes d(x) \\ d_{(0,*)}(a \otimes x) &:= a \otimes d(x). \end{aligned}$$

The whole picture can be displayed in the following scheme:

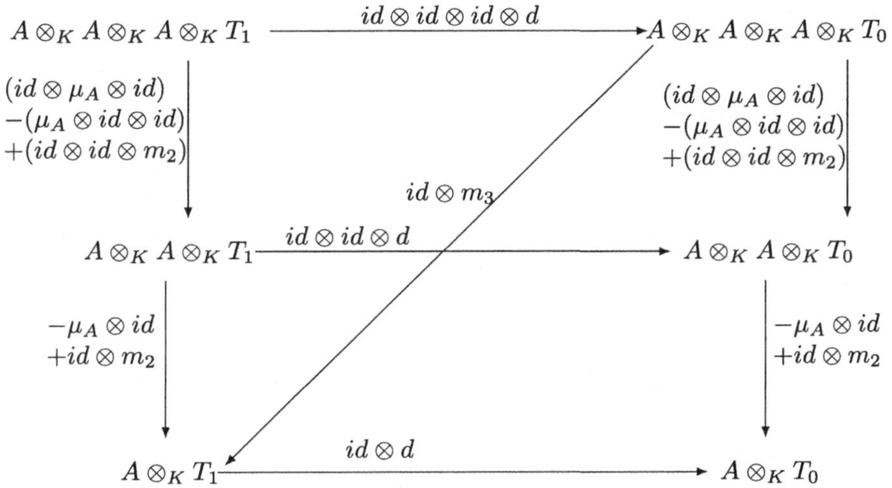

Observe that all the mappings are of total degree 1.

Then form the total complex (!) of this scheme, that is consider

$$\begin{aligned} \tilde{X}_3 &:= A \otimes_K A \otimes_K A \otimes_K T_1 \\ \tilde{X}_2 &:= (A \otimes_K A \otimes_K A \otimes_K T_0) \oplus (A \otimes_K A \otimes_K T_1) \\ \tilde{X}_1 &:= (A \otimes_K A \otimes_K T_0) \oplus (A \otimes_K T_1) \\ \tilde{X}_0 &:= A \otimes_K T_0 \end{aligned}$$

with the corresponding mappings $d_{(i,j)}$ and form

$$X: \quad \cdots \longrightarrow 0 \longrightarrow \tilde{X}_1/\mathrm{im}(d_{(2,0)} + d_{(1,1)}) \longrightarrow \tilde{X}_0 \longrightarrow 0 \longrightarrow \cdots .$$

Define $X_1 := \tilde{X}_1/\mathrm{im}(d_{(2,0)} + d_{(1,1)})$ and $X_0 := \tilde{X}_0$.

Corollary 6.6.1 *The complex X is a two-sided tilting complex corresponding to the one-sided tilting complex T.*

We define a mapping

$$\begin{array}{ccc} T & \xrightarrow{f} & X \\ x & \mapsto & 1 \otimes x \end{array}$$

where 1 represents the unit element of A in the various tensor product components. The action of A on X and the homomorphism $A \longrightarrow End_{K^b(B\text{-}proj)}(T)$ can be compared via f. They coincide up to some homotopy mapping f_2!

Indeed, the map

$$\begin{array}{ccc} A \otimes_K T & \xrightarrow{f_2} & \tilde{X} \\ a \otimes x & \mapsto & 1 \otimes a \otimes x \in A \otimes_K A \otimes_K T \end{array}$$

is of degree -1, and we verify

$$d(f_2(a \otimes x)) + f_2(a \otimes d(x)) = f(m_2(a \otimes x)) - af(x).$$

Since there is a natural morphism of complexes $\tilde{X} \longrightarrow X$ we obtain the comparison of the action of A as $End_{K^b(B\text{-}proj)}(T)$ and the action of A on X as required.

6.6.2 Endomorphism Rings of Two-Term Tilting Complexes

It is particularly easy to describe the endomorphism rings of two-term tilting complexes. We shall present a method for computing the endomorphism ring of such a complex in terms of its homology. This result is due to Muchtadi-Alamsyah [12, 13] and the proof we present uses spectral sequences and is due to Keller. Muchtadi originally gave a purely elementary proof, however the argument is slightly longer than the one we give here. Keller's approach is published in [12, Appendix].

Let K be a commutative ring, let A be a K-algebra and let X and Y be two objects in $D^b(A)$. Suppose moreover that $H_k(X) = 0$ and $H_k(Y) = 0$ whenever $k \notin \{0, 1\}$.

We shall now use Proposition 3.9.17. This shows that there is a spectral sequence

$$E_1^{p,q} = \prod_{\ell \in \mathbb{Z}} Ext_A^{2p+q}(H_\ell(X), H_{\ell+p}(Y)) \Rightarrow Hom_{D^b(A)}(X, Y[p+q]).$$

The spectral sequence converges since X and Y are both bounded complexes and so the sequence is limited in a finite array. The differential on this page is induced by the differential on X.

Obviously, $E_1^{p,q} = 0$ if $2p + q < 0$. Moreover, since $H_m(X) = 0 = H_m(Y)$ if $m \notin \{0, 1\}$, we get

$$E_1^{p,q} = Ext_A^{2p+q}(H_0(X), H_p(Y)) \oplus Ext_A^{2p+q}(H_1(X), H_{1+p}(Y)).$$

This implies $p \in \{1, 0, -1\}$. The only possibly non-zero terms in the first page are

$$\begin{aligned} E_1^{-1,q} &= Ext_A^{q-2}(H_1(X), H_0(Y)) \\ E_1^{0,q} &= Ext_A^q(H_0(X), H_0(X)) \oplus Ext_A^q(H_1(X), H_1(X)) \\ E_1^{1,q} &= Ext_A^{q+2}(H_0(X), H_1(Y)). \end{aligned}$$

The differential d_1 on E_1 is induced by the differential on X. We consider the term $p+q=0$, which will give the homomorphism space $Hom_{D^b(A)}(X, Y)$ in the infinite page. The lowest degrees of interest are

$$\begin{aligned} E_1^{1,0} &= Ext_A^2(H_0(X), H_1(Y)) \\ E_1^{0,0} &= Hom_A(H_0(X), H_0(X)) \oplus Hom_A(H_1(X), H_1(X)) \\ E_1^{1,-1} &= Ext_A^1(H_0(X), H_1(Y)). \end{aligned}$$

We now argue as in Proposition 3.9.15. In the lowest degrees we already get $E_1^{1,-1} = E_\infty^{1,-1}$. Moreover,

$$0 \longrightarrow E_\infty^{1,-1} \longrightarrow H^0(\mathbb{R}Hom(X, Y)) \longrightarrow E_\infty^{0,1} \longrightarrow 0$$

is exact. The page $E_\infty^{0,1}$ can be obtained as

$$E_\infty^{0,1} = E_2^{0,1} = \ker(E_1^{0,0} \longrightarrow E_1^{1,0}).$$

But therefore the sequence

$$0 \longrightarrow E_\infty^{1,-1} \longrightarrow H^0(\mathbb{R}Hom(X, Y)) \longrightarrow E_1^{0,0} \longrightarrow E_1^{1,0}$$

is exact.

Proposition 6.6.2 *Let K be a commutative ring, let A be a K-algebra and let X and Y be two objects in $D^b(A)$. Suppose moreover that $H_k(X) = 0$ and $H_k(Y) = 0$ whenever $k \notin \{0, 1\}$. If $Ext_A^1(H_0(X), H_1(Y)) = 0$, then there is a pullback*

$$\begin{array}{ccc} Hom_{D^b(A)}(X, Y) & \longrightarrow & Hom_A(H_0(X), H_0(X)) \\ \downarrow & & \downarrow \\ Hom_A(H_1(X), H_1(X)) & \longrightarrow & Ext_A^2(H_0(X), H_1(Y)) \end{array}$$

Proof The exact sequence from the observation above transforms precisely to this pullback. Indeed, this is a consequence of the explicit description of the pullback in Proposition 1.8.25. This proves the statement. □

Remark 6.6.3 Let us give some interpretation of Proposition 6.6.2.

- It can easily be seen that the map

$$Hom_{D^b(A)}(X, Y) \longrightarrow Hom_A(H_i(X), H_i(X))$$

  is the map H_i that maps a homomorphism between complexes to the one that it induces on homology.
- Let A be a self-injective K-algebra and suppose that X and Y are complexes of projective modules with homology concentrated in degree 0 and 1. Then

$$Ext_A^2(H_0(X), H_1(Y)) \simeq \underline{Hom}_A(H_1(X), H_1(Y)).$$

  Indeed, let

$$X : \cdots \longrightarrow 0 \longrightarrow X_1 \xrightarrow{d_1^X} X_0 \longrightarrow 0 \longrightarrow \cdots$$

  and

$$Y : \cdots \longrightarrow 0 \longrightarrow Y_1 \xrightarrow{d_1^Y} Y_0 \longrightarrow 0 \longrightarrow \cdots,$$

  then

$$H_1(X) = \ker(d_1^X)\ ,\ H_1(Y) = \ker(d_1^Y)$$

  and

$$H_0(X) = \mathrm{coker}(d_1^X)\ ,\ H_0(Y) = \mathrm{coker}(d_1^Y).$$

  Hence, X and Y are the first terms of a projective resolution of $H_0(X)$, resp. $H_0(Y)$. We get $H_1(X) = \Omega^2(H_0(X))$ and $H_1(Y) = \Omega^2(H_0(Y))$. This shows

$$\begin{aligned} Ext_A^2(H_0(X), H_1(Y)) &\simeq Ext_A^2(H_0(X), \Omega^2(H_0(Y))) \\ &\simeq \underline{Hom}_A(H_0(X), \Omega^{-2}\Omega^2(H_0(Y))) \\ &\simeq \underline{Hom}_A(H_0(X), H_0(Y)) \\ &\simeq \underline{Hom}_A(H_1(X), H_1(Y)) \end{aligned}$$

where the second isomorphism comes from Definition 1.8.19.

Corollary 6.6.4 *Let A be a self-injective algebra and let T be a tilting complex over A with homology concentrated in degree 0 and 1. Then*

$$\begin{array}{ccc} End_{D^b(A)}(T) & \longrightarrow & End_A(H_0(T)) \\ \downarrow & & \downarrow \\ End_A(H_1(T)) & \longrightarrow & \underline{End}_A(H_0(T)) \end{array}$$

is a pullback diagram of algebras.

Proof We need to verify the hypotheses of Proposition 6.6.2. Once this is done, the pullback over a second extension group follows from Proposition 6.6.2. The second second extension group is just the stable endomorphism group of the homology by Remark 6.6.3. Let

$$T = (\cdots \longrightarrow T_1 \stackrel{d}{\longrightarrow} T_0 \longrightarrow 0 \longrightarrow \cdots)$$

and let $\alpha \in Hom_A(H_1(T), H_0(T))$ by non-zero. Denote by $\iota : H_1(T) \longrightarrow T_1$ the natural embedding and by $\pi : T_0 \longrightarrow H_0(T)$ the natural projection. Then T_1 is projective, and since A is self-injective, T_1 is also an injective module. Hence the map $\alpha : H_1(T) \longrightarrow H_0(T)$ factorises $H_1(T) \longrightarrow T_1 \longrightarrow H_0(T)$, i.e. there is a morphism $\beta : T_1 \longrightarrow H_0(T)$ so that $\alpha = \beta \circ \iota$. Since T_1 is projective, there is a $\gamma : T_1 \longrightarrow T_0$ such that $\beta = \pi \circ \gamma$. The map γ induces a morphism of complexes $\delta : T \longrightarrow T[1]$ such that $H_1(\delta) = \alpha \neq 0$:

$$\begin{array}{ccccccccccc} \cdots \longrightarrow & 0 & \longrightarrow & 0 & \longrightarrow & T_1 & \longrightarrow & T_0 & \longrightarrow & 0 & \longrightarrow \cdots \\ & & & \downarrow & & \downarrow \delta & & \downarrow & & & \\ \cdots \longrightarrow & 0 & \longrightarrow & T_1 & \longrightarrow & T_0 & \longrightarrow & 0 & \longrightarrow & 0 & \longrightarrow \cdots \end{array}$$

Since T is a tilting complex, $Hom_{D^b(A)}(T, T[1]) = 0$. This contradiction proves the first statement. The morphisms are indeed ring homomorphisms, as is seen by Remark 6.6.3. □

Two-term tilting complexes are actually easily constructed.

Lemma 6.6.5 (Rickard (1990), appears in [14]) *Let K be a field and let A be a finite dimensional symmetric K-algebra. Let $P_1, P_2, \ldots, P_n$ be a set of representatives for the isomorphism classes of projective indecomposable A-modules, and put $S_i := P_i/\mathrm{rad}(P_i)$ for all $i \in \{1, \cdots, n\}$. Let*

$$I_0 := \{1, \ldots, m\} \subset \{1, \cdots, n\} =: I$$

be a proper subset. For each $j \notin I_0$ let M_j be the biggest quotient of P_j such that for all composition factors S_i of M_j we have $i \notin I_0$. Suppose R_j is the projective cover of $\ker(P_j \longrightarrow M_j)$, and denote by $\delta_j : R_j \longrightarrow P_j$ the composition

$$R_j \twoheadrightarrow \ker(P_j \longrightarrow M_j) \hookrightarrow P_j.$$

Put $R_* := \bigoplus_{j=m+1}^n R_j$ and $P_* := \bigoplus_{j=m+1}^n P_j$, as well as

$$R_* \stackrel{\delta}{\longrightarrow} P_*$$

where

$$\delta := \begin{pmatrix} \delta_{m+1} & 0 & \dots\dots\dots & & 0 \\ 0 & \delta_{m+2} & 0 & & \vdots \\ \vdots & \ddots & \ddots & \ddots & \vdots \\ \vdots & & \ddots & \ddots & \ddots \; \vdots \\ \vdots & & & \ddots & \ddots \; 0 \\ 0 & \dots & \dots\dots & 0 & \delta_n \end{pmatrix},$$

and $P^* := \bigoplus_{i \in I_0} P_i$. *We construct the complex* T

$$\dots \longrightarrow 0 \longrightarrow P^* \oplus R_* \xrightarrow{\binom{0}{\delta}} P_* \longrightarrow 0 \longrightarrow \dots$$

where $P^* \oplus R_*$ *is in degree* 1 *and* P_* *is in degree* 0.

Then T *is a tilting complex.*

Proof Since $P_i/\mathrm{rad}(P_i) = S_i$ is always a composition factor of M_i, for all $i \in I_0$ we get that for each direct factor P_i of P^* the module P_i is never a direct factor of P_*. Put

$$C_j : \dots \longrightarrow 0 \longrightarrow R_j \xrightarrow{\delta_j} P_j \longrightarrow 0 \longrightarrow \dots$$

and $C := \bigoplus_{j=m+1}^n C_j$.

Observe that

$$Hom_{D^b(A)}(C_j, P^*[1]) = Hom_A(M_j, P^*).$$

Let $\varphi \in Hom_A(M_j, P^*)$. The socle of P^* is a direct sum of simple modules S_i where $i \in I_0$. All composition factors of M_j, and hence of $\mathrm{im}(\varphi)$ are S_k for $k \notin I_0$. Since $\mathrm{soc}(\mathrm{im}(\varphi)) \subseteq \mathrm{soc}(P^*)$, we get $\varphi = 0$.

Now, any morphism $\psi : P^* \longrightarrow C_j$ is presented by a mapping $\psi_0 : P^* \longrightarrow P_j$ and we get that the composition $P^* \longrightarrow P_j \longrightarrow M_j$ is zero, since no composition factor of $P^*/\mathrm{rad}(P^*)$ is a composition factor of M_j. Therefore, ψ_0 has image in $\ker(P_j \longrightarrow M_j)$, and since P^* is projective, ψ_0 factors through δ_j. This shows that ψ is homotopy equivalent to 0.

Now, $Hom_{D^b(A)}(C_j, C_\ell[1]) = Ext_A^1(M_j, M_\ell)$. But an extension representing a non-zero element in $Ext_A^1(M_j, M_\ell)$ gives a module with composition factors only in $\{S_{m+1}, \dots, S_n\}$. This contradicts the maximality of M_j and of M_ℓ.

Consider $Hom_{D^b(A)}(C_j[1], C_\ell) = Hom_A(M_j, \ker(\delta_\ell))$. Since $M_j/\mathrm{rad}(M_j) = S_j$ is simple, and S_j is not a composition factor of $\ker(\delta_\ell)$, this shows that $Hom_{D^b(A)}(C_j[1], C_\ell) = 0$.

Therefore $Hom_{D^b(A)}(T, T[n]) = 0$ for $n \neq 0$. Moreover, $\ker(P_j \longrightarrow M_j)/\mathrm{rad}$ $(\ker(P_j \longrightarrow M_j))$ has composition factors S_ℓ for $\ell \in I_0$, by the choice of maximality of M_j, and therefore R_j is a direct factor of a direct sum of s copies of P^* for some s. This shows that we may define a morphism $C_j \longrightarrow (P^*)^s$ with cone having a direct summand $P_j[1]$. This gives that each projective indecomposable A-module is in the triangulated category generated by $add(T)$. We have shown that T is a tilting complex. □

Remark 6.6.6 We will come back to the construction of Lemma 6.6.5 in Sects. 6.10.1 and 6.10.2, where it shall be used.

Using Lemma 6.6.5 Rickard constructed in [14] an infinite family of algebras all derived equivalent to each other, but belonging to infinitely many Morita equivalence classes.

6.7 First Properties of Derived Equivalences

When looking at examples, we try certain classes of rings at first. More complicated rings, admitting more sophisticated derived equivalences, are discussed in detail later.

6.7.1 A First Non-trivial Example

Let $K = \mathbb{Z}$, let p be a prime number and let

$$A = \begin{pmatrix} \mathbb{Z} & \mathbb{Z} \\ p\mathbb{Z} & \mathbb{Z} \end{pmatrix}.$$

There are two immediate indecomposable projective A-modules, namely the two columns

$$P_1 := \begin{pmatrix} \mathbb{Z} \\ p\mathbb{Z} \end{pmatrix} \text{ and } P_2 := \begin{pmatrix} \mathbb{Z} \\ \mathbb{Z} \end{pmatrix},$$

where $P_1 \leq P_2$ with quotient being $\mathbb{Z}/p\mathbb{Z}$ as an abelian group. Now, let

$$T_1 := \cdots \longrightarrow 0 \longrightarrow P_1 \stackrel{\iota}{\longrightarrow} P_2 \longrightarrow 0 \longrightarrow \cdots$$

be the complex with homology $\mathbb{Z}/p\mathbb{Z}$ concentrated in degree 0. Then $T := T_1 \oplus P_2$ is a tilting complex. Indeed, P_2 is a direct factor, and there is a morphism $P_2 \longrightarrow T_1$ with cone

$$\cdots \longrightarrow 0 \longrightarrow P_1 \oplus P_2 \stackrel{(\iota, id_{P_2})}{\longrightarrow} P_2 \longrightarrow 0 \longrightarrow \cdots$$

which is isomorphic in the homotopy category to $P_1[1]$. Hence $P_1[1]$ and P_2 are in the category generated by direct summands of direct sums of T and hence $A = P_1 \oplus P_2$.

Moreover $Ext^i_A(P_2, P_2) = 0$ if $i \neq 0$, and

$$Hom_{K(A\text{-}proj)}(T_1, P_2[n]) = 0 = Hom_{K(A\text{-}proj)}(P_2[n], T_1)$$

if $n \notin \{0, 1\}$ already by the fact that the complexes are 0 elsewhere. Moreover

$$Hom_{K(A\text{-}proj)}(T_1, P_2[1]) = 0$$

since $Hom_{K(A\text{-}proj)}(P_1, P_2) = \mathbb{Z} \cdot \iota$ and each of these morphisms factors through ι (which corresponds to 1 in $\mathbb{Z}$). Hence this factorisation corresponds to a homotopy $h = \lambda \cdot id$:

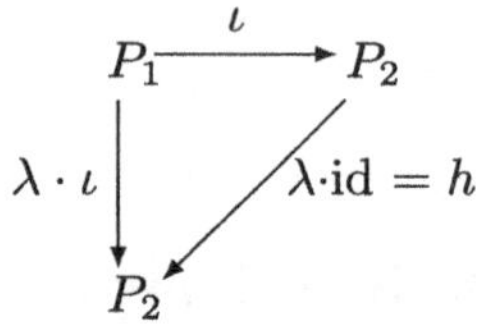

Similarly, $Hom_{K(A\text{-}proj)}(P_2[1], T_1) = 0$ since ι is injective, and a possible morphism has to compose to 0 when followed by ι:

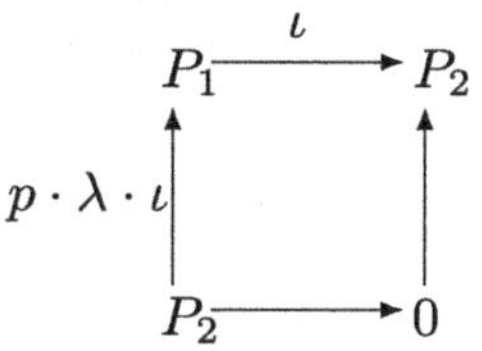

is commutative only for $\lambda = 0$.

We compute the endomorphism ring.

$End_A(P_2) = \mathbb{Z}$ and $Hom_{K^-(A\text{-}proj)}(T_1, P_2) = 0$ since there is no non-zero morphism $P_2 \to P_2$ which precomposes with ι to 0:

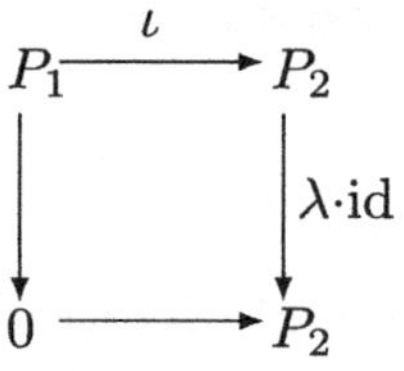

is commutative only for $\lambda = 0$.

Now, T_1 is isomorphic to P_2/P_1 in the derived category, and hence

$$End_{K^-(A\text{-}proj)}(T_1) = End_A(P_2/P_1) \simeq \mathbb{Z}/p\mathbb{Z}.$$

Finally, $Hom_{K^-(A\text{-}proj)}(P_2, T_1)$ can be computed by the diagrams above. $Hom_{C^-(A\text{-}proj)}(P_2, T_1) = \mathbb{Z} \cdot id_{P_2}$ and the homotopies are precisely the morphisms factoring through ι, whence $p \cdot \iota \cdot \mathbb{Z}$:

$$
\begin{array}{ccc}
P_1 & \xrightarrow{\ \iota\ } & P_2 \\
 & {\scriptstyle p\cdot\lambda\cdot\,\mathrm{id}}\ \nwarrow & \uparrow {\scriptstyle p\cdot\lambda\cdot\iota} \\
 & & P_2
\end{array}
$$

Hence

$$Hom_{K^-(A\text{-}proj)}(P_2, T_1) = \mathbb{Z}/p\mathbb{Z}$$

and

$$End_{K^-(A\text{-}proj)}(T_1 \oplus P_2) \simeq \begin{pmatrix} \mathbb{Z} & 0 \\ \mathbb{Z}/p\mathbb{Z} & \mathbb{Z}/p\mathbb{Z} \end{pmatrix}.$$

Corollary 6.7.1 *The algebra*

$$\begin{pmatrix} \mathbb{Z} & \mathbb{Z}/p\mathbb{Z} \\ 0 & \mathbb{Z}/p\mathbb{Z} \end{pmatrix}$$

is derived equivalent to the algebra

$$\begin{pmatrix} \mathbb{Z} & \mathbb{Z} \\ p\mathbb{Z} & \mathbb{Z} \end{pmatrix}.$$

Proof The opposite algebra of this matrix algebra is given by the matrix transpose. □

Remark 6.7.2 A similar argument gives that another tilting complex T' for A is given by $T' := T_1 \oplus P_1[1]$. Then it is completely analogous to see that

$$End_{K^-(A\text{-}proj)}(T_1 \oplus P_1[1]) \simeq \begin{pmatrix} \mathbb{Z} & \mathbb{Z}/p\mathbb{Z} \\ 0 & \mathbb{Z}/p\mathbb{Z} \end{pmatrix}$$

and hence we get equivalences of triangulated categories

$$D^b\left(\begin{pmatrix} \mathbb{Z} & \mathbb{Z} \\ p\mathbb{Z} & \mathbb{Z} \end{pmatrix}\right) \simeq D^b\left(\begin{pmatrix} \mathbb{Z} & 0 \\ \mathbb{Z}/p\mathbb{Z} & \mathbb{Z}/p\mathbb{Z} \end{pmatrix}\right) \simeq D^b\left(\begin{pmatrix} \mathbb{Z} & \mathbb{Z}/p\mathbb{Z} \\ 0 & \mathbb{Z}/p\mathbb{Z} \end{pmatrix}\right)$$

between the derived categories of these three algebras.

Remark 6.7.3 We mention that apparently a derived equivalence of a $\mathbb{Z}$-order will not in general preserve the property of being $\mathbb{Z}$ torsion-free, nor will it preserve the property of being an order. A detailed study of this type of example is given in [15].

We also mention that the tilting complex used in the above example is isomorphic to a module. Actually, the homology of this tilting complex is non-zero only in degree 0.

In contrast to this phenomenon we recall that a ring which is Morita equivalent to an order is an order again. Hence the tilting complex, which is a module, cannot be a Morita bimodule. We remark that a tilting complex may be isomorphic to a module, without being a Morita bimodule. The module is then called a *quasi-tilting module.*

6.7.2 Local Rings and Commutative Rings

We know a self-equivalence of $D^b(A)$ for all rings A, namely the functor $[1]$, the shift in degrees. Moreover, any Morita equivalence A-*Mod* $\longrightarrow$ B-*Mod* gives rise to a derived equivalence $D^b(A) \longrightarrow D^b(B)$ since a Morita bimodule is actually also a two-sided tilting complex when considered as a complex using Lemma 3.5.49. In some cases all derived equivalences are compositions of these two special types.

Let A be a local ring. We shall show that any derived equivalence is actually the composition of a shift with a Morita equivalence. Recall that a ring is local if the set of non-invertible elements is an ideal.

Proposition 6.7.4 *Let K be a commutative ring and let A be a local K-algebra such that the Krull-Schmidt theorem is valid for A-proj. Let B be a K-algebra and let $F : D^b(A) \longrightarrow D^b(B)$ be an equivalence of triangulated categories. Then there is a Morita equivalence $M : A$-Mod $\longrightarrow B$-Mod and $n \in \mathbb{Z}$ such that*

$$F \simeq M \circ [n].$$

Proof Let (T, d^T) be a tilting complex over A with endomorphism ring B^{op}. We first observe that if T is concentrated in only one degree, then any homotopy $h : T \longrightarrow T$ of degree 1 is actually 0 in the category of complexes. Hence there is an integer $n \in \mathbb{Z}$ such that $T = M[n]$ and

$$End_{D^b(A)}(T) = End_{A\text{-}mod}(M) \simeq B^{op}.$$

As a consequence T is an A–B-bimodule. The fact that T is a tilting complex ($add(T)$ generates $K^b(A\text{-}proj)$) shows in addition that M is a progenerator of A-*mod* with endomorphism ring B^{op}. By the Morita theorem B is Morita equivalent to A and T is as it was claimed.

Suppose that there is a tilting complex (T, d^T) which is not of the form claimed and hence not isomorphic in $K^b(A\text{-}proj)$ to a complex concentrated in a single degree. We may suppose that T is a complex without zero homotopic direct factor. Recall that T is bounded and let

$$\longrightarrow 0 \longrightarrow T_n \longrightarrow T_{n-1} \longrightarrow \cdots \longrightarrow T_{m+1} \longrightarrow T_m \longrightarrow 0 \longrightarrow \cdots .$$

Since A is local, the only indecomposable projective A-module is the rank one free module, and any projective A-module is therefore free. We choose

$$rank_A(T) = rank_A(T_n \oplus T_{n+1} \oplus \cdots \oplus T_m)$$

to be minimal amongst all the tilting complexes in $K^b(A\text{-}proj)$ not isomorphic to a shifted copy of a Morita bimodule. Now $T^m = A^{k_m}$ and $T_n = A^{k_n}$ for integers k_m and k_n. Hence there is an A-linear morphism

$$T_n = A^{k_n} \xrightarrow{\varphi_0} A^{k_m} = T_m$$

by projection onto the first component of A^{k_n} followed by injection into the first component of A^{k_m}. Then define a morphism $\varphi \in Hom_{D^b(A)}(T, T[n-m])$ by

$$\begin{array}{ccccccc} \cdots \longrightarrow & 0 & \longrightarrow & T_n & \xrightarrow{d_n^T} & T_{n-1} & \\ & \downarrow & & \downarrow \varphi_0 & & \downarrow & \\ \cdots \longrightarrow & T_{m+1} & \xrightarrow{d_{m+1}^T} & T_m & \longrightarrow & 0 & \longrightarrow \cdots \end{array}$$

Since T is a tilting complex

$$Hom_{D^b(A)}(T, T[n-m]) = 0$$

unless $n = m$, which was excluded. Hence we get a homotopy $h = (h_k)_{k \in \mathbb{Z}}$ with $h_k : T_k \longrightarrow T_{n-m+1}$ such that

$$\varphi = h_n \circ d_n^T + d_{m+1}^T \circ h_{n+1}.$$

Consider the projection π onto the first component of $T_m = A^{k_m}$. Now

$$\pi \circ \varphi_0 = \pi \circ (h_n \circ d_n^T + d_{m+1}^T \circ h_{n+1}) = \pi \circ h_n \circ d_n^T + \pi \circ d_{m+1}^T \circ h_{n+1}$$

is surjective by definition. Since the image of $\pi \circ h_n \circ d_n^T$ is an ideal of A, as is the image of $\pi \circ d_{m+1}^T \circ h_{n+1}$, and since A is local, one of these ideals $\mathrm{im}(\pi \circ h_n \circ d_n^T)$ or $\mathrm{im}(\pi \circ d_{m+1}^T \circ h_{n+1})$ has to be all of A. Hence either the restriction of d_n^T to the first component is split injective, or the restriction to the last component of d_{m+1}^T is split surjective. In any case T is isomorphic in the homotopy category to a complex with lower A-rank. This contradicts the choice of T and proves the proposition. □

Corollary 6.7.5 *Let K be a field of characteristic $p > 0$ and let P be a finite p-group. If A is a K-algebra and $D^b(A) \simeq D^b(KP)$ as triangulated categories, then A is Morita equivalent to KP.*

Proof We know that KP is local by Proposition 1.6.22 or by Theorem 2.9.7. Proposition 6.7.4 shows that every derived equivalence of standard type is a composition of a shift with a Morita equivalence, and Proposition 6.5.5 shows that there is an equivalence of standard type. □

Remark 6.7.6 We mention explicitly that this is in contrast to the situation for stable equivalences of Morita type. For group rings of dihedral 2-groups over fields of characteristic 2 there are endotrivial modules which are not syzygies of the trivial module [16–18]. Since syzygies in the stable category correspond to the shift functor in the derived category we obtain that the statement which is analogous to Proposition 6.7.4 for the stable category does not hold.

Lemma 6.7.7 *Let A be a commutative k-algebra and let T be a tilting complex in $K^b(A\text{-}proj)$. Then T is isomorphic to its homology $T \simeq H(T)$ in $K^b(A\text{-}proj)$.*

Proof Let $\mathfrak{m}$ be a maximal ideal of A. Then it is a classical fact, and actually easy to prove, that the localisation $A_\mathfrak{m}$ is a flat extension of A. Indeed, let $M \leq N$ be an inclusion of A-modules. If $m \in M$ which becomes 0 in $A_\mathfrak{m} \otimes_A N$, then there is an $a \in A \setminus \mathfrak{m}$ such that $am = 0$ in N. But then this relation already holds in M, and hence m is 0 in $A_\mathfrak{m} \otimes_A M$.

Then, by Lemma 3.8.6, we get for all bounded complexes of finitely generated A-modules U, V

$$Hom_{D^b(A\text{-}mod)}(U, V) \otimes_A A_\mathfrak{m} \simeq Hom_{D^b(A_\mathfrak{m}\text{-}mod)}(U \otimes_A A_\mathfrak{m}, V \otimes_A A_\mathfrak{m})$$

and therefore for all $n \neq 0$

$$\begin{aligned} 0 &= A_\mathfrak{m} \otimes_A Hom_{K^b(A\text{-}proj)}(T, T[n]) \\ &\simeq Hom_{K^b(A_\mathfrak{m}\text{-}proj)}(A_\mathfrak{m} \otimes_A T, A_\mathfrak{m} \otimes_A T[n]). \end{aligned}$$

Moreover, if A is in the triangulated category generated by $add(T)$, then $A_\mathfrak{m}$ is in the triangulated category generated by $add(T \otimes_A A_\mathfrak{m}))$. Hence $T \otimes_A A_\mathfrak{m}$ is a tilting complex with

$$End_{K^b(A_\mathfrak{m}\text{-}proj)}(T \otimes_A A_\mathfrak{m}) \simeq End_{K^b(A\text{-}proj)}(T) \otimes_A A_\mathfrak{m}.$$

Moreover, since $A_\mathfrak{m}$ is flat as an A-module,

$$H(T \otimes_A A_\mathfrak{m}) \simeq H(T) \otimes_A A_\mathfrak{m}.$$

We may therefore suppose that $A = A_\mathfrak{m}$ is local. From Proposition 6.7.4 we then obtain Morita bimodules $M(\mathfrak{m})$ such that

$$T \otimes_A A_\mathfrak{m} \simeq M(\mathfrak{m})[n_\mathfrak{m}] \simeq H(T \otimes_A A_\mathfrak{m})$$

for each maximal ideal $\mathfrak{m}$ of A.

The above isomorphism is an isomorphism in $K^b(A_{\mathfrak{m}}\text{-}proj)$. Hence $H(T)\otimes_A A_{\mathfrak{m}}$ is projective for each maximal ideal $\mathfrak{m}$ of A. Hence

$$Ext^1_{A_{\mathfrak{m}}}(H(T)\otimes_A A_{\mathfrak{m}}, -) = 0.$$

But by Lemma 3.8.6 this implies that

$$Ext^1_{A_{\mathfrak{m}}}(H(T)\otimes_A A_{\mathfrak{m}}, M\otimes_A A_{\mathfrak{m}}) \simeq Ext^1_A(H(T), M)\otimes_A A_{\mathfrak{m}}$$

for all A-modules M.

Let N be an A-module such that

$$N\otimes_A A_{\mathfrak{m}} = 0$$

for each maximal ideal $\mathfrak{m}$ and let $n \in N$. Then there is an $a \notin \mathfrak{m}$ such that $a\cdot n = 0$. Hence the annihilator

$$ann(n) := \{a \in A \mid a\cdot n = 0\}$$

is not contained in $\mathfrak{m}$. But $ann(n)$ is an ideal of A, which has to be contained in a maximal ideal, if $ann(n) \neq A$. Therefore $ann(n) = A$ and this implies that $1 \in ann(n)$, which shows that $1\cdot n = n = 0$. Hence $N = 0$.

This implies that $Ext^1_A(H(T), M) = 0$, and by consequence each short exact sequence

$$0 \longrightarrow M \longrightarrow E \longrightarrow H(T) \longrightarrow 0$$

splits. This shows that $H(T)$ is a projective A-module. We will have finished as soon as we have proven the following lemma.

Lemma 6.7.8 *Let T be a right bounded complex of A-modules with finitely generated homology $H(T)$ which is projective as an A-module. Then $T \simeq H(T)$ in $K^{-,b}(A\text{-}proj)$.*

Proof Let s be the lowest degree such that $H_s(T) \neq 0$. We may therefore suppose that $T_{s-1} = 0$, and therefore the differential $T_{s+1} \longrightarrow T_s$ has cokernel $H_s(T)$. The module $H_s(T)$ is projective by hypothesis, and hence the projection $T_s \twoheadrightarrow H_s(T)$ splits. Therefore

$$T \simeq T' \oplus H_s(T)$$

for some complex T' with

$$H(T') \oplus H_s(T) = H(T),$$

whence T' has homology concentrated in degrees strictly bigger than s. By induction on the number of degrees in which the homology is non-zero we see that $T \simeq H(T)$. This proves the statement. □

6.7.3 Hochschild (Co-)Homology and Derived Equivalences

As we have seen in Proposition 4.3.1, a Morita equivalence between the rings Λ and Γ induces an isomorphism between the centres $Z(\Lambda)$ and $Z(\Gamma)$ of Λ and Γ. The proof there was relatively elementary and simple. The analogous argument will not work for derived equivalences, due to the problem that the endomorphism ring of a tilting complex has to be taken *up to homotopy*.

For derived equivalences, however, a more sophisticated approach is possible through Hochschild cohomology.

Proposition 6.7.9 *Let K be a commutative ring and let A and B be two Noetherian K-algebras. Suppose that A and B are finitely generated projective as K-modules. Let*

$$F := X \otimes_A^{\mathbb{L}} - : D^b(A) \longrightarrow D^b(B)$$

be a derived equivalence of standard type with quasi-inverse

$$G := Y \otimes_B^{\mathbb{L}} - : D^b(B) \longrightarrow D^b(A).$$

Then

$$\hat{F} := \left(X \otimes_A^{\mathbb{L}} -\right) \otimes_A^{\mathbb{L}} Y : D^b(A \otimes_K A^{op}) \longrightarrow D^b(B \otimes_K B^{op})$$

is an equivalence of triangulated categories of standard type.

Proof Since A and B are both finitely generated projective over K and Noetherian, using Proposition 6.4.4 we may choose X and Y so that the restriction of X to the left and to the right are complexes of finitely generated projective modules, using that if A is projective as a K-module, then $A \otimes_K B^{op}$ is a finitely generated free B-right module, and likewise for A. Therefore we may replace $D^b(A \otimes_K A^{op})$ by $K^{-,b}(A \otimes A^{op}\text{-}Proj)$ and in this way we can replace the derived tensor product by the ordinary tensor product, which is associative. We define

$$\hat{G} := \left(Y \otimes_B^{\mathbb{L}} -\right) \otimes_B^{\mathbb{L}} X : D^b(B \otimes_K B^{op}) \longrightarrow D^b(A \otimes_K A^{op})$$

and compute

$$(\hat{F} \circ \hat{G} \simeq id) \text{ as well as } (\hat{G} \circ \hat{F} \simeq id)$$

since $X \otimes_A^{\mathbb{L}} Y \simeq B$ in $D^b(B \otimes_K B^{op})$ and $Y \otimes_B^{\mathbb{L}} X \simeq A$ in $D^b(A \otimes_K A^{op})$. This shows that $\hat{F}$ is an equivalence. Hence

$$X \otimes_A - \otimes_A Y \simeq (X \otimes_K Y) \otimes_{A \otimes A^{op}} -$$

and we obtain the statement. □

Proposition 6.7.10 (Rickard [2]) *Let K be a commutative ring and let A and B be two Noetherian K-algebras. Suppose that A and B are both finitely generated projective as K-modules. Then $D^b(A) \simeq D^b(B)$ implies that*

$$HH^*(A) \simeq HH^*(B)$$

as K-algebras.

Proof By Definition 3.7.8 we have

$$HH^n(A) = Ext^n_{A\otimes_K A^{op}}(A, A)$$

and by Definition 3.7.8 we have

$$Ext^n_{A\otimes_K A^{op}}(A, A) \simeq Hom_{D^b(A\otimes_K A^{op})}(A, A[n]).$$

Let X be a two-sided tilting complex in $D^b(B\otimes_K A^{op})$ inducing a derived equivalence of standard type. Proposition 6.7.9 shows that in the notation used there

$$\hat{F}(A) = X \otimes_A A \otimes_A Y = X \otimes_A Y = B$$

in $D^b(B \otimes_K B^{op})$. Hence

$$\begin{aligned} HH^n(A) &= Hom_{D^b(A\otimes_K A^{op})}(A, A[n]) \\ &\simeq Hom_{D^b(B\otimes_K B^{op})}(\hat{F}(A), \hat{F}(A[n])) \text{ applying the equivalence } \hat{F} \\ &\simeq Hom_{D^b(B\otimes_K B^{op})}(\hat{F}(A), \hat{F}(A)[n]) \text{ since } \hat{F} \text{ is triangulated} \\ &\simeq Hom_{D^b(B\otimes_K B^{op})}(B, B[n]) \\ &= HH^n(B). \end{aligned}$$

Observe that the isomorphism is basically given by the effect of $\hat{F}$ on morphisms. Since for each

$$\alpha \in Hom_{D^b(A\otimes_K A^{op})}(A, A[n])$$

and

$$\beta \in Hom_{D^b(A\otimes_K A^{op})}(A, A[m])$$

we get

$$\hat{F}(\alpha[m] \circ \beta) = \hat{F}(\alpha[m]) \circ \hat{F}(\beta) = \hat{F}(\alpha)[m] \circ \hat{F}(\beta)$$

which shows that $HH^*(A) \simeq HH^*(B)$ is compatible with the multiplication. Additivity of the isomorphism is clear. □

Corollary 6.7.11 *Let K be a commutative ring and let A and B be two Noetherian K-algebras. Suppose that A and B are finitely generated projective as K-modules. Then $Z(A) \simeq Z(B)$.*

Proof We get that
$$HH^*(A) \simeq HH^*(B)$$
as K-algebras by Proposition 6.7.10. Since
$$HH^0(A) = End_{A\otimes_K A^{op}}(A) = Z(A)$$
and likewise $HH^0(B) = Z(B)$, we obtain the statement. □

As an application we get another interesting property of derived equivalences of direct products, and as a consequence, of semisimple algebras.

Lemma 6.7.12 *Let K be a commutative ring and let A_1 and A_2 be two Noetherian K-algebras. Suppose A_1 and A_2 are finitely generated projective as a K-modules. Let B be a K-algebra, projective as K-module, and suppose $D^b(A_1 \times A_2) \simeq D^b(B)$. Then $B \simeq B_1 \times B_2$ with $D^b(A_1) \simeq D^b(B_1)$ and $D^b(A_2) \simeq D^b(B_2)$.*

Proof Put $A := A_1 \times A_2$. Let $X \in D^b(A \otimes_K B^{op})$ be a two-sided tilting complex inducing the derived equivalence
$$X \otimes_B^{\mathbb{L}} - : D^b(B) \longrightarrow D^b(A).$$
Let $e_1 \in Z(A)$ be the idempotent with $A_1 = A \cdot e_1$. By Corollary 6.7.11 we get that X induces an idempotent $f_1 \in Z(B)$ which is the image of e_1 under the derived equivalence. We obtain that $e_1 X f_1$ is a two-sided tilting complex in $D^b(A_1 \otimes_R (Bf_1)^{op})$ and $(1-e_1)X(1-f_1)$ is a two-sided tilting complex in $D^b(A_2 \otimes_R (B(1-f_1))^{op})$. This proves the statement. □

Corollary 6.7.13 *Let A be a semisimple artinian K-algebra over a field K. Then any algebra B which is derived equivalent to A is Morita equivalent to A, and hence is semisimple as well.*

Proof By Lemma 6.7.12 we may suppose that A is actually simple. Since A is simple, A is Morita equivalent to a skew-field by Wedderburn's Theorem 1.4.16. Hence A is local in this case and we are done by Proposition 6.7.4. □

The case of Hochschild homology is slightly more difficult and actually less well-documented in the literature. A treatment is given in [10], however the statement that Hochschild homology is a derived invariant was considered by the specialist as a known fact much before the proof in [10] appeared.

Proposition 6.7.14 *Let K be a commutative ring and let A and B be two Noetherian K-algebras. Suppose that A and B are both finitely generated projective as K-modules. Then $D^b(A) \simeq D^b(B)$ implies that*

$$HH_*(A) \simeq HH_*(B)$$

as graded K-modules.

Proof We know by Definition 3.6.7 that

$$HH_n(A) = Tor_n^{A\otimes_K A^{op}}(A, A).$$

Let

$$\mathbb{B}A^\bullet \longrightarrow\!\!\!\!\rightarrow A$$

be the bar resolution of A (cf Definition 3.6.4), which is a free resolution of A as $A \otimes_K A^{op}$-modules. Hence

$$\mathbb{B}A^\bullet \longrightarrow\!\!\!\!\rightarrow A$$

is a quasi-isomorphism in $D^b(A \otimes_K A^{op})$.

Let X be a two-sided tilting complex in $D^b(B\otimes_K A^{op})$ with inverse $Y \in D^b(A\otimes_K B^{op})$. Since A and B are Noetherian and projective over K by Proposition 6.4.4 we may assume that X is a complex of projective A^{op}-modules, of projective B-modules and that Y is a complex of projective A-modules and projective B^{op}-modules. Hence

$$X \otimes_A \mathbb{B}A \otimes_A Y \longrightarrow\!\!\!\!\rightarrow X \otimes_A Y$$

is a surjective morphism of complexes of $B \otimes B^{op}$-modules which is actually a quasi-isomorphism in $D^b(B \otimes_K B^{op})$ since the bar resolution is also a quasi-isomorphism. Observe that $X \otimes_A \mathbb{B}A \otimes_A Y$ is a complex of projective $B \otimes_K B^{op}$-modules. Indeed, $\mathbb{B}A$ is a complex of free $A \otimes_K A^{op}$-modules. Hence, we need to show that

$$X \otimes_A (A \otimes_K A^{op})^n \otimes_A Y \simeq (X \otimes_K Y)^n$$

is a complex of projective $B \otimes_K B^{op}$-modules. But if X_a is a projective B-module, and if Y_b is a projective B^{op}-module, then

$$X_a \oplus X'_a \simeq B^{n_a} \text{ and } Y_b \oplus Y'_b \simeq (B^{op})^{n_b}$$

and therefore $X_a \otimes Y_b$ is a direct factor of

$$(X_a \oplus X'_a) \otimes (Y_b \oplus Y'_b) \simeq (B \otimes_K B^{op})^{n_a \cdot n_b}$$

which is free.

Now

$$Y \otimes_B X \simeq A$$

in $D^b(A \otimes_K A^{op})$ and hence

$$\begin{aligned}\mathbb{B}A \otimes_{A\otimes A^{op}} A &\simeq \mathbb{B}A \otimes_{A\otimes A^{op}} (Y \otimes_B X)\\ &\simeq (X \otimes_A \mathbb{B}A \otimes_A Y) \otimes_{B\otimes B^{op}} B\end{aligned}$$

where the last isomorphism is given by

$$\begin{aligned}\mathbb{B}A \otimes_{A\otimes A^{op}} (Y \otimes_B X) &\longrightarrow (X \otimes_A \mathbb{B}A \otimes_A Y) \otimes_{B\otimes B^{op}} B\\ u \otimes x \otimes y &\mapsto (y \otimes u \otimes x) \otimes 1\end{aligned}$$

This is clearly well-defined, and is an isomorphism since

$$\begin{aligned}(X \otimes_A \mathbb{B}A \otimes_A Y) \otimes_{B\otimes B^{op}} B &\longrightarrow \mathbb{B}A \otimes_{A\otimes A^{op}} (Y \otimes_B X)\\ (x \otimes u \otimes y) \otimes b &\mapsto (u \otimes y \otimes bx) = (u \otimes yb \otimes x)\end{aligned}$$

is an inverse morphism. Now taking homology

$$\begin{aligned}HH_m(A) &= Tor_m^{A\otimes A^{op}}(A, A)\\ &\simeq H_m(\mathbb{B}A \otimes_{A\otimes A^{op}} A)\\ &\simeq H_m(\mathbb{B}A \otimes_{A\otimes A^{op}} (Y \otimes_B X))\\ &\simeq H_m((X \otimes_A \mathbb{B}A \otimes_A Y) \otimes_{B\otimes B^{op}} B)\\ &\simeq Tor_m^{B\otimes B^{op}}(B, B)\\ &= HH_m(B).\end{aligned}$$

This proves the proposition. □

Corollary 6.7.15 *Let K be a commutative ring and let A and B be Noetherian K-algebras. Suppose A and B are projective as K-modules. Then $D^b(A) \simeq D^b(B)$ as triangulated categories implies $A/[A, A] \simeq B/[B, B]$ as K-modules.*

Proof Indeed, $HH_0(A) = A/[A, A]$ and likewise for B. □

Remark 6.7.16 Keller showed [19] that cyclic homology and cyclic cohomology is invariant under derived equivalences.

6.7.4 Symmetric Algebras

Recall from Sect. 1.10.2 that if K is a commutative ring, then a K-algebra A is symmetric if

$$A \simeq Hom_K(A, K)$$

as an $A \otimes_K A^{op}$-module. Rickard showed that if A is a symmetric algebra over a field K and if B is a K-algebra such that $D^b(A) \simeq D^b(B)$, then B is symmetric as well.

Proposition 6.7.17 [2, 20] *Let K be a commutative ring and let A and B be Noetherian K-algebras. Suppose A and B are projective as K-modules. If there is an equivalence $D^b(A) \simeq D^b(B)$ given by an equivalence F of standard type, then F induces an equivalence $F^e : D^b(A \otimes_K A^{op}) \longrightarrow D^b(B \otimes_K B^{op})$ satisfying $F^e(Hom_K(A, K)) \simeq Hom_K(B, K)$ and $F^e(A) = B$. In particular, if A is symmetric, then B is symmetric as well.*

Proof Once we have proved that $F^e(Hom_K(A, K)) \simeq Hom_K(B, K)$ and $F^e(A) = B$, then we have proved that B is symmetric if A is symmetric. Indeed,

$$A \simeq Hom_K(A, K)$$

as bimodules implies

$$B \simeq F^e(A) \simeq F^e(Hom_K(A, K)) \simeq Hom_K(B, K)$$

as bimodules.

Let T be a tilting complex in $K^b(B\text{-}proj)$ with endomorphism ring A^{op}. Construct a two-sided tilting complex $X \in D^b(A \otimes_K B^{op})$ by Keller's construction Theorem 6.4.1 so that

$$X \otimes_B^{\mathbb{L}} - : D^b(B) \longrightarrow D^b(A)$$

is an equivalence of triangulated categories. We apply Proposition 6.4.4 to replace X by a quasi-isomorphic complex which has finitely generated projective homogeneous components (as bimodules) in each degree, except in the highest non-zero degree, where the homogeneous component is projective as an A-module and projective as a B-module. By Proposition 6.5.6, the functor $\mathbb{R}Hom_A(X, -)$ is a quasi-inverse of $X \otimes_B^{\mathbb{L}} -$, and $\mathbb{R}Hom_A(X, -) \simeq Hom_A(X, A) \otimes_A^{\mathbb{L}} -$. Put $X' := \mathbb{R}Hom_A(X, A)$.

Hence we can assume that X and X' are complexes of projective B-modules and we may replace the left derived tensor product $- \otimes_B^{\mathbb{L}} -$ by the ordinary tensor product over B. In this case the tensor product is associative:

$$\begin{aligned}(X \otimes_B^{\mathbb{L}} -) \otimes_B^{\mathbb{L}} X' &= (X \otimes_B -) \otimes_B X' \\ &= X \otimes_B (- \otimes_B X') = X \otimes_B^{\mathbb{L}} (- \otimes_B^{\mathbb{L}} X')\end{aligned}$$

so that we can write $X \otimes_B^{\mathbb{L}} - \otimes_B^{\mathbb{L}} X'$ without specifying the parentheses.

Since $X \otimes_B X' \simeq A$ in the derived category of A–A-bimodules, and since $X' \otimes_A X \simeq B$ in the derived category of B–B-bimodules, the functor $X \otimes_B^{\mathbb{L}} - \otimes_B^{\mathbb{L}} X' : D^b(B \otimes B^{op}) \to D^b(A \otimes A^{op})$ is an equivalence. Denote by G this functor. We have

$$
\begin{aligned}
&Hom_{D^b(B\otimes_K B^{op})}(B \otimes_K B^{op}, G^{-1}(Hom_K(A, K)[n])) \\
&\quad \simeq Hom_{D^b(A\otimes_K A^{op})}(X \otimes_K \mathbb{R}Hom_A(X, A), Hom_K(A, K)[n]) \\
&\quad \simeq Hom_{D^b(A)}(X, Hom_{A^{op}}(\mathbb{R}Hom_A(X, A), Hom_K(A, K))[n]) \\
&\quad \simeq Hom_{D^b(A)}(X[-n], Hom_K(\mathbb{R}Hom_A(X, A) \otimes_A A, K)) \\
&\quad \simeq Hom_{D^b(A)}(X[-n], Hom_K(\mathbb{R}Hom_A(X, A), K)) \\
&\quad \simeq Hom_K(\mathbb{R}Hom_A(X, A) \otimes^{\mathbb{L}}_A X[-n], K) \\
&\quad \simeq Hom_K(X' \otimes^{\mathbb{L}}_A X[-n], K) \\
&\quad \simeq Hom_K(Hom_{D^b(A)}(X, X[-n]), K) \\
&\quad \simeq \begin{cases} Hom_K(B, K) & \text{if } n = 0 \\ 0 & \text{if } n \neq 0 \end{cases}
\end{aligned}
$$

where we apply various times diverse adjointness formulas. This means that the homology of $G^{-1}(Hom_K(A, K))$ is concentrated in one degree and moreover

$$G^{-1}(Hom_K(A, K)) \simeq Hom_K(B, K)$$

in $D^b(B \otimes_K B^{op})$. With $F^e := G^{-1}$ we obtain the statement. □

Remark 6.7.18 There is an approach for self-injective algebras as well.

- Let K be an algebraically closed field and let A and B be finite dimensional K-algebras. Suppose $D^b(A) \simeq D^b(B)$. Then A is self-injective if and only if B is self-injective. This is a result due to Al-Nofayeeh [21], which uses most surprisingly, geometric methods.
- Recall that Corollary 6.7.1 shows that there are examples of two algebras which have equivalent derived categories. One of these algebras is an order, whereas the other is not an order.

6.8 Grothendieck Groups and Cartan-Brauer Triangle

We shall continue with the Grothendieck group defined in Definition 2.6.1. Our purpose will be to prove that the Grothendieck group of a module category is invariant under derived equivalences. This is not immediate and we shall need to define first a Grothendieck group of a derived category, following Hartshorne, in turn following Grothendieck. Then we shall show that the Grothendieck group of a derived category is actually isomorphic to the Grothendieck group of the algebra. Finally we shall develop the theory for group algebras where we study Broué's abelian defect conjecture.

6.8.1 The Grothendieck Group of a Triangulated Category

We loosely follow Hartshorne [22] in this section.

Let $\mathcal{T}$ be a triangulated category with suspension functor T. We introduced the Grothendieck group of a triangulated category in Sect. 5.7.1. Recall the construction we gave there. We define the Grothendieck group $G_0(\mathcal{T})$ of $\mathcal{T}$ as the quotient of the free abelian group on isomorphism classes $[X]$ of objects X of $\mathcal{T}$ modulo the subgroup generated by the elements $[Y]-[X]-[Z]$ whenever there is a distinguished triangle

$$X \to Y \to Z \to TX$$

in $\mathcal{T}$.

Remark 6.8.1 Recall that if $X \to Y \to Z \to TX$ is a distinguished triangle, then $Y \to Z \to TX \to TY$ is also a distinguished triangle. Moreover, $X \to X \to 0 \to TX$ is a distinguished triangle, and hence so is $X \to 0 \to TX \to TX$. Therefore $[0] = [X] + [TX]$ in the Grothendieck group, which is equivalent to

$$[TX] = -[X]$$

in the Grothendieck group. Hence the relation $[Y] = [X] + [Z]$ in the Grothendieck group coming from a distinguished triangle $X \to Y \to Z \to TX$ is the same relation as the one coming from the shifted distinguished triangle $Y \to Z \to TX \to TY$.

We now specialise to the case $\mathcal{T} = D^b(A)$ for a Noetherian K-algebra A.

Lemma 6.8.2 *Let $X = (\bigoplus_{i\in\mathbb{Z}} X_i, \partial_i)$ be a complex in $D^b(A\text{-mod})$. Then $[X] = \sum_{i\in\mathbb{Z}}(-1)^i[X_i]$ in $G_0(D^b(A\text{-mod}))$. The same holds for a complex in $K^b(A\text{-proj})$ and its value in $G_0(K^b(A\text{-proj}))$.*

Proof We define a homomorphism of abelian groups by

$$\begin{array}{ccc} G_0(D^b(A\text{-}mod)) & \stackrel{\Psi}{\longrightarrow} & G_0(A\text{-}mod) \\ [(\bigoplus_{i\in\mathbb{Z}} X_i, \partial_i)] & \mapsto & \sum_{i=-\infty}^{\infty}(-1)^i[X_i] \end{array}$$

Here we denote by $X = (\bigoplus_{i\in\mathbb{Z}} X_i, \partial_i)$ a complex where X_i is the degree i homogeneous component and $\partial_i : X_i \to X_{i-1}$ is the degree i homogeneous part of the differential. The sum on the right-hand side is finite since X is bounded. Let $X_i = 0$ for $i > i_0$. Then stupid truncation yields a distinguished triangle

$$X_{i_0}[i_0 - 1] \stackrel{\partial_{i_0}}{\longrightarrow} X_{\leq i_0 - 1} \longrightarrow X \longrightarrow X_{i_0}[i_0]$$

where the morphisms are displayed in the following diagram:

$$
\begin{array}{rccccccccc}
X_{i_0}[i_0-1]: & 0 & \longrightarrow & 0 & \longrightarrow & X_{i_0} & \longrightarrow & 0 & & \\
& \downarrow & & \downarrow & & \downarrow \partial_{i_0} & & \downarrow & & \\
X_{\leq i_0-1}: & 0 & \longrightarrow & 0 & \longrightarrow & X_{i_0-1} & \stackrel{\partial_{i_0-1}}{\longrightarrow} & X_{i_0-2} & \longrightarrow & \dots \\
& \downarrow & & \downarrow & & \downarrow id & & \downarrow id & & \\
X: & 0 & \longrightarrow & X_{i_0} & \stackrel{\partial_{i_0}}{\longrightarrow} & X_{i_0-1} & \stackrel{\partial_{i_0-1}}{\longrightarrow} & X_{i_0-2} & \longrightarrow & \dots \\
& \downarrow & & \downarrow id & & \downarrow & & & & \\
X_{i_0}[i_0]: & 0 & \longrightarrow & X_{i_0} & \longrightarrow & 0 & & & &
\end{array}
$$

Hence,

$$[X] + [X_{i_0}[i_0 - 1]] = [X_{\leq i_0-1}]$$

and since $[Y[1]] = -[Y]$ we get

$$[X] + (-1)^{i_0-1}[X_{i_0}] = [X_{\leq i_0-1}]$$

or equivalently

$$[X] = [X_{\leq i_0-1}] + (-1)^{i_0}[X_{i_0}].$$

Induction on i_0 gives

$$[X] = \sum_{i\in\mathbb{Z}}(-1)^i[X_i]$$

where we remind the reader again that X is a bounded complex and hence the sum is finite. The case of a complex in $K^b(A\text{-}proj)$ is identical. □

Lemma 6.8.3 *Let $X \in D^b(A\text{-}mod)$. Then*

$$[X] = \sum_{i\in\mathbb{Z}}(-1)^i[H_i(X)]$$

in $G_0(D^b(A\text{-}mod))$.

Proof Let again $X_i = 0$ for $i > i_0$. Intelligent truncation yields a distinguished triangle

$$\ker(\partial_{i_0})[i_0] \longrightarrow X \longrightarrow \tau_{\leq i_0}X \longrightarrow \ker(\partial_{i_0})[i_0 + 1]$$

displayed by the diagram

$$
\begin{array}{rccccccccc}
\ker(\partial_{i_0})[i_0]: & 0 & \longrightarrow & \ker(\partial_{i_0}) & \longrightarrow & 0 & & & & \\
& \downarrow & & \downarrow incl & & \downarrow & & & & \\
X: & 0 & \longrightarrow & X_{i_0} & \stackrel{\partial_{i_0}}{\longrightarrow} & X_{i_0-1} & \longrightarrow & \dots & & \\
& \downarrow & & \downarrow & & \downarrow id & & & & \\
\tau_{\leq i_0}X: & 0 & \longrightarrow & \mathrm{im}(\partial_{i_0}) & \stackrel{incl}{\longrightarrow} & X_{i_0-1} & \stackrel{\partial_{i_0-1}}{\longrightarrow} & X_{i_0-2} & \longrightarrow & \dots
\end{array}
$$

The fact that the cone of the map *incl* is isomorphic to $\tau_{\leq i_0}X$ is an easy exercise. Since

$$H_{i_0}(X) = \ker(\partial_{i_0})$$

we get

$$[X] = (-1)^{i_0}[H_{i_0}(X)] + [\tau_{\leq i_0}X]$$

and by induction on i_0 we obtain the statement. □

Given a complex X with bounded homology, the complex

$$H := \left(\cdots \longrightarrow 0 \longrightarrow \bigoplus_{i\in\mathbb{Z}} H_{2i+1}(X) \xrightarrow{0} \bigoplus_{i\in\mathbb{Z}} H_{2i}(X) \longrightarrow 0 \longrightarrow \cdots\right)$$

with homology concentrated in degrees 0 and 1 has exactly the same image in the Grothendieck group of $D^b(A\text{-}mod)$ as X:

$$[H] = [X] \in G_0(D^b(A\text{-}mod)).$$

Given an element E in $G_0(A\text{-}mod)$, we can write $E = [M]-[N]$ for two A-modules M and N and see that this element is the image of

$$\cdots \longrightarrow 0 \longrightarrow N \xrightarrow{0} M \longrightarrow 0 \longrightarrow \cdots$$

with homology in degrees 0 and 1 in the Grothendieck group of $D^b(A\text{-}mod)$. Hence

Proposition 6.8.4 *Let K be a field and let A be a finite dimensional K-algebra. Then the embedding A-mod $\longrightarrow D^b(A\text{-}mod)$ induces an isomorphism of abelian groups*

$$G_0(A\text{-}mod) \simeq G_0(D^b(A\text{-}mod)).$$

Proof The embedding induces a map in one direction. The inverse mapping is given above. □

A somewhat easier argument applies to $K_0(A)$, the Grothendieck group of finitely generated projective modules introduced in Definition 2.6.3. By Lemma 6.8.2 we see that the image of a complex $X = (\bigoplus_{i\in\mathbb{Z}} X_i, \partial_i)$ in $G_0(K^b(A\text{-}proj))$ is

$$[X] = \sum_{i\in\mathbb{Z}} (-1)^i [X_i]$$

so that we obtain the following proposition in a manner completely analogous to Proposition 6.8.4.

Proposition 6.8.5 *Let K be a splitting field for the finite dimensional K-algebra A. Then*

$$G_0(K^b(A\text{-}proj)) \simeq K_0(A)$$

is induced by the embedding A-proj $\longrightarrow K^b(A\text{-}proj)$.

Proof Analogous to the proof of Proposition 6.8.4. □

6.8.2 Derived Equivalences and Cartan-Brauer Triangles

Recall that Remark 2.6.10 showed that there is a non-degenerate bilinear pairing

$$\langle\ ,\ \rangle : K_0(A) \times G_0(A\text{-}mod) \longrightarrow \mathbb{Z}$$

with a pair of mutually dual bases

$$\{[P_L] \mid L \text{ simple } \Gamma\text{-module}\}$$

for $K_0(A)$ and

$$\{[L] \mid L \text{ simple } \Gamma\text{-module}\}$$

for $G_0(A\text{-}mod)$. Here we denote as usual by P_L the projective cover of the module L.

Corollary 6.8.6 *Let K be a field and let A and B be two finite dimensional K-algebras. Let $F : D^b(A\text{-}mod) \longrightarrow D^b(B\text{-}mod)$ be an equivalence of triangulated categories of standard type. Then F induces isomorphisms $F : K_0(K^b(A\text{-}proj)) \longrightarrow K_0(K^b(B\text{-}proj))$ and $\overline{F} : G_0(D^b(A\text{-}mod)) \longrightarrow G_0(D^b(B\text{-}mod))$ of abelian groups.*

Proof By Theorem 6.5.1 we get that F induces the isomorphisms on the Grothendieck groups as claimed. □

Let R be a complete discrete valuation ring with residue field k and field of fractions K and suppose that Λ and Γ are R-orders. Put $A := k \otimes_R \Lambda$ and $B := k \otimes_R \Gamma$, which are then finite dimensional k-algebras and let $M_\Lambda := K \otimes_R \Lambda$ and $M_\Gamma := K \otimes_R \Gamma$, which are finite dimensional semisimple K-algebras. Let now X be a two-sided tilting complex in $D^b(\Gamma \otimes_R \Lambda^{op})$, all of whose homogeneous components are projective if restricted to either side, so that $X \otimes_\Lambda -$ is an equivalence

$$X \otimes_\Lambda - : D^b(\Lambda) \longrightarrow D^b(\Gamma).$$

Then X is invertible by Corollary 6.5.7, and hence there is a complex Y in $D^b(\Lambda \otimes_R \Gamma^{op})$ which is inverse to X with respect to $- \otimes^{\mathbb{L}}_\Lambda -$ and $- \otimes^{\mathbb{L}}_\Gamma -$. But then $k \otimes_R X$ is invertible with inverse $k \otimes_R Y$ and $K \otimes_R X$ is invertible with inverse $K \otimes_R Y$.

We have proved the following.

Proposition 6.8.7 *Let R be a complete discrete valuation ring with residue field k and field of fractions K and suppose that Λ and Γ are R-orders. Put $A := k \otimes_R \Lambda$ and $B := k \otimes_R \Gamma$ and let $M_\Lambda := K \otimes_R \Lambda$ and $M_\Gamma := K \otimes_R \Gamma$. Let X be a two-sided tilting complex in $D^b(\Gamma \otimes_R \Lambda^{op})$, and suppose that all homogeneous components of X are projective as Λ-modules and as Γ-modules. Then*

$$\begin{array}{rcl} X \otimes^{\mathbb{L}}_\Lambda - : & D^b(\Lambda) & \longrightarrow D^b(\Gamma) \\ (k \otimes_R X) \otimes^{\mathbb{L}}_A - : & D^b(A) & \longrightarrow D^b(B) \\ (K \otimes_R X) \otimes^{\mathbb{L}}_{M_\Lambda} - : & D^b(M_\Lambda) & \longrightarrow D^b(M_\Gamma) \end{array}$$

are equivalences of triangulated categories. □

We consider now the Cartan-Brauer triangle from Proposition 2.6.7 and combine it with Proposition 6.8.7 to obtain the following diagram.

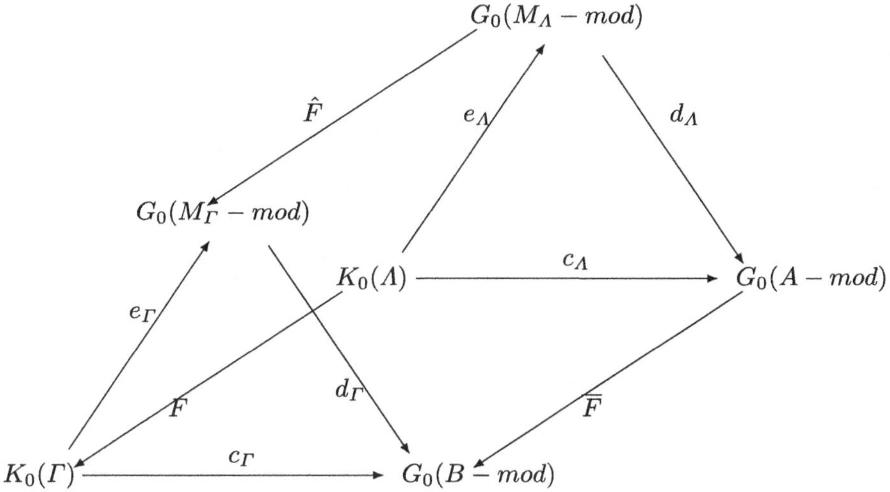

Here we denote by

$$\begin{aligned} F([L]) &:= [X \otimes^{\mathbb{L}}_\Lambda L] \\ \hat{F}([V]) &:= [(K \otimes_R X) \otimes^{\mathbb{L}}_{M_\Lambda} V] \\ \overline{F}([M]) &:= [(k \otimes_R X) \otimes^{\mathbb{L}}_A M] \end{aligned}$$

the corresponding isomorphisms on the Grothendieck groups.

Theorem 6.8.8 *Let R be a complete discrete valuation ring with residue field k and field of fractions K and suppose that Λ and Γ are R-orders. Put $A := k \otimes_R \Lambda$ and $B := k \otimes_R \Gamma$ and let $M_\Lambda := K \otimes_R \Lambda$ and $M_\Gamma := K \otimes_R \Gamma$. Let X be a two-sided tilting complex in $D^b(\Gamma \otimes_R \Lambda^{op})$, and suppose that all homogeneous components of X are projective as Λ-modules and as Γ-modules. Then the squares in the above diagram are commutative. More precisely*

$$\overline{F} \circ c_\Lambda = c_\Gamma \circ F,\quad \hat{F} \circ e_\Lambda = e_\Gamma \circ F \;\; and \;\; \overline{F} \circ d_\Lambda = d_\Gamma \circ \hat{F}$$

Proof Let P be a projective Λ-module. The hypotheses on X imply that we may replace the left derived tensor product by the ordinary tensor product. Then

$$\overline{F} \circ c_\Lambda([P]) = [(k \otimes_R X) \otimes_{\Lambda} (k \otimes_\Lambda P)] = [(k \otimes_R (X \otimes_{\Lambda} P)] = c_\Gamma \circ F([P]).$$

This shows the first equation.

Moreover,

$$\hat{F} \circ e_\Lambda([P]) = [(K \otimes_R X) \otimes_{M_\Lambda} (K \otimes_R P)] = [K \otimes_R (X \otimes_{\Lambda} P)] = e_\Gamma \circ F([P]).$$

This shows the second equation.

Let V be an M_Λ-module and let L_V be a full Λ-lattice in V (i.e. L_V is a Λ-lattice and $KL_V = V$). Then

$$(K \otimes_R X) \otimes_{M_\Lambda} V \simeq K \otimes_R (X \otimes_{\Lambda} L_V)$$

and $\bigoplus_{i\in\mathbb{Z}} \left((X \otimes_{\Lambda} L_V)\right)_{2i}$ is a full lattice in $\bigoplus_{i\in\mathbb{Z}} \left((K \otimes_R X) \otimes_{M_\Lambda} V\right)_{2i}$. Likewise, $\bigoplus_{i\in\mathbb{Z}} \left((X \otimes_{\Lambda} L_V)\right)_{2i+1}$ is a full lattice in $\bigoplus_{i\in\mathbb{Z}} \left((K \otimes_R X) \otimes_{M_\Lambda} V\right)_{2i+1}$. Hence

$$d_\Gamma([(K \otimes_R X) \otimes_{M_\Lambda} V] = [k \otimes_R (X \otimes_\Lambda L_\Lambda)]$$

and likewise for d_Λ. Therefore

$$\begin{aligned} d_\Gamma \circ \hat{F}([V]) &= d_\Gamma([(K \otimes_R X) \otimes_{M_\Lambda} V]) = [k \otimes_R (X \otimes_{\Lambda} L_V)] \\ &= [(k \otimes_R X) \otimes_{\Lambda} (k \otimes_R L_V)] = \overline{F} \circ d_\Lambda([V]). \end{aligned}$$

This proves the statement. □

Proposition 6.8.9 *Let k be a field and let A and B be finite dimensional k-algebras. Suppose that k is a splitting field for A and for B. If $D^b(A) \simeq D^b(B)$, then there is a matrix Φ in $Gl_{rank_{\mathbb{Z}}(G_0(A))}(\mathbb{Z})$ such that, if we denote by C_A the Cartan matrix of A and by C_B the Cartan matrix of B, then $\Phi \cdot C_B \cdot \Phi^{tr} = C_A$, where Φ^{tr} denotes the transpose of Φ. In particular the elementary divisors of the two Cartan matrices coincide.*

Proof Let $F : D^b(A) \to D^b(B)$ be an equivalence of triangulated categories. For two complexes X, Y in $K^b(A\text{-}proj)$ let

$$\langle X, Y\rangle_A^e := \sum_{j\in\mathbb{Z}} (-1)^j \dim_k Hom_{K^b(A-proj)}(X, Y[j])$$

and define $\langle U, V\rangle_B^e$ analogously. Since X and Y are bounded complexes of projective modules, only finitely many terms on the right-hand side are non-zero. This fact and Proposition 3.5.23 prove that this value is well-defined. Finally, $\langle X, Y\rangle_A^e = \langle F(X), F(Y)\rangle_B^e$, since F is an equivalence. Moreover, denoting by $P_1, \dots, P_n$ representatives of the isomorphism classes of the indecomposable projective A-modules, and by $Q_1, \dots, Q_n$ representatives of the isomorphism classes of the indecomposable projective B-modules, we have that $\langle P_i, P_j\rangle_A^e = (C_A)_{i,j}$ is the (i, j)-th coefficient of the Cartan matrix of A. Likewise $\langle Q_i, Q_j\rangle_B^e = (C_B)_{i,j}$ is the (i, j)-th coefficient of the Cartan matrix of B. Let X be a bounded complex of finitely generated projective A-modules, and suppose that $X_m = 0$ whenever $m > n$ or $m < k$. Stupid truncation of X gives a distinguished triangle

$$\sigma^{<n}X \to X \to X_n \to \sigma^{<n}X[1]$$

and hence, denoting $(X, Y) := Hom_{K^b(A\text{-}proj)}(X, Y)$, we get a long exact sequence

$$\cdots \to (\sigma^{<n}X[1], Y) \to (X_n, Y) \to (X, Y) \to (\sigma^{<n}X, Y) \to (X_n[-1], Y) \cdots .$$

Since X and Y are both in $K^b(A\text{-}proj)$, this long exact sequence is actually bounded below and above. An easy induction shows that $\langle X, Y\rangle_A^e$ only depends on the value of X in $K_0(A\text{-}proj)$. The dual argument, using covariant Hom functors, shows that $\langle X, Y\rangle_A^e$ only depends on the value of Y in $K_0(A\text{-}proj)$ as well. The analogous statement holds for $\langle\ ,\ \rangle_B^e$. Now, $F(P_i) =: T_i$ and let $[T_i] = \sum_{s=1}^n f_{i,s}[Q_s]$ in $K^b(B\text{-}proj)$. Since F is an equivalence, $T := \bigoplus_{i=1}^n T_i$ is a tilting complex, hence $add(T)$ generates $K^b(B\text{-}proj)$. Therefore the matrix $\Phi := (f_{i,s})_{1\le i,s\le n}$ is invertible in $\mathbb{Z}$. Moreover,

$$\langle P_i, P_j\rangle_A^e = \langle T_i, T_j\rangle_B^e = \sum_{s,t=1}^n f_{i,s} f_{j,t} \langle Q_s, Q_t\rangle_B^e.$$

Hence $\Phi \cdot C_B \cdot \Phi^{tr} = C_A$. This proves the statement. □

Remark 6.8.10 We remark that the elementary divisors of the Cartan matrix usually encode a lot of information about the algebra. A consequence is, for example, that the absolute value of the determinant of the Cartan matrix, the so-called Cartan determinant, is just the product of the elementary divisors, and hence the Cartan determinant is invariant under derived equivalences.

Remark 6.8.11 Let R be a complete discrete valuation ring of characteristic 0, let K be its field of fractions and let k be its residue field of characteristic $p > 0$. Let G and H be finite groups such that K and k are splitting fields for G and for H. By Corollary 2.5.18 the central primitive idempotents of kG are in bijection with the central primitive idempotents of RG by means of the residue map $RG \longrightarrow kG$. Hence let e_G (resp. e_H) be a central primitive idempotent of RG and let $\overline{e_G}$ (resp $\overline{e}_H$) be its image in kG (resp kH). Let

$$X \otimes^{\mathbb{L}}_{RG} - : D^b(RGe_G) \to D^b(RHe_H)$$

be an equivalence of standard type, and we may and will suppose that each homogeneous component of X is projective as an RG-module and as an RH-module. We consider the isomorphism $G_0(KGe_G\text{-}mod) \to G_0(KHe_H\text{-}mod)$ induced by the equivalence

$$(K \otimes_R X) \otimes_{KG} - : D^b(KGe_G) \to D^b(KHe_H).$$

Since KG and KH are both semisimple, $K \otimes_R X \simeq H(K \otimes_R X)$ in $D^b(KHe_H \otimes_K (KGe_G)^{op})$.

We have a $\mathbb{Z}$-bilinear inner product on the Grothendieck group $G_0(KGe_G\text{-}mod)$ given by

$$\langle [M_1], [M_2] \rangle_G := \dim_K(Hom_{KG}(M_1, M_2))$$

for all KGe_G-modules M_1 and M_2. Likewise we have the inner product $\langle [N_1], [N_2] \rangle_H := \dim_K(Hom_{KH}(N_1, N_2))$ for all KHe_H-modules N_1 and N_2. Observe that the inner product on the Grothendieck group of the module category transforms under the isomorphisms $G_0(KGe_G\text{-}mod) \simeq G_0(D^b(KGe_G\text{-}mod))$ into the inner product

$$\langle U, V \rangle := \sum_{i \in \mathbb{Z}} (-1)^i \dim_K(Hom_{D^b(KGe_G)}(U, V[i]))$$

in the Grothendieck group of the derived category.

We claim that $(K \otimes_R X) \otimes_{KG} - : D^b(KGe_G) \to D^b(KHe_H)$ induces an isometry $G_0(KGe_G\text{-}mod) \simeq G_0(KHe_H\text{-}mod)$. We first observe that X is a two-sided tilting complex, and hence KX is also a two-sided tilting complex. Therefore for each simple KGe_G-module S there is exactly one degree n_S in which $H^{n_S}((K \otimes_R X) \otimes_{KG} S) \neq 0$. Defining $\hat{F}_X([-]) := [(K \otimes_R X) \otimes_{KG} -]$ we get for two KGe_G-modules M_1 and M_2

$$\begin{aligned}
&\langle \hat{F}_X(M_1), \hat{F}_X(M_2) \rangle_H \\
&\quad = \sum_{i \in \mathbb{Z}} (-1)^i \dim_K(Hom_{D^b(KH)}(KX \otimes_{KG} M_1, KX \otimes_{KG} M_2[i])) \\
&\quad = \sum_{i \in \mathbb{Z}} (-1)^i \dim_K(Hom_{D^b(KG)}(M_1, M_2[i])) = \dim_K(Hom_{KG}(M_1, M_2)) \\
&\quad = \langle M_1, M_2 \rangle_G.
\end{aligned}$$

For every simple KGe_G-module S we get $\langle [S], [S] \rangle = 1$ and hence $[S]$ is mapped by $\hat{F}_X$ to an element U of $KHe_H\text{-}mod$ with $\langle U, U \rangle = 1$. This shows that $U = [T]$ or $U = -[T]$ for some simple KHe_H-module T.

The isomorphism $\hat{F}_X$ has many interesting properties which can be expressed by arithmetic properties of character values. Broué [23, 24] calls this correspondence a *perfect isometry*.

6.9 Singularity and Stable Categories as Quotients of Derived Categories

Let K be a commutative ring and let A be a Noetherian K-algebra. Recall that $K^b(A\text{-}proj)$ is a full subcategory of $D^b(A\text{-}mod)$ since $D^b(A\text{-}mod) \simeq K^{-,b}(A\text{-}proj)$. We shall imitate the quotient construction of a stable category in a sense to be made precise.

6.9.1 Definition and First Properties of the Singularity Category

We shall "localise" $K^{-,b}(A\text{-}proj)$ at morphisms $\alpha : X \longrightarrow Y$ so that $C(\alpha)$ is an object in $K^b(A\text{-}proj)$ in the same way as we obtained the derived category $D^-(A)$ from $K^-(A\text{-}Mod)$. We shall invert in this way all morphisms in $K^{-,b}(A\text{-}proj)$ with cone being in $K^b(A\text{-}proj)$. The localised category will be called the singularity category. In other words, the singularity category has the same objects as $K^{-,b}(A\text{-}proj)$ and (analogously to Definition 3.5.35) a morphism from X to Y in the singularity category is an equivalence class of triples

$$X \stackrel{\nu}{\longleftarrow} Z \stackrel{\alpha}{\longrightarrow} Y$$

where $C(\nu)$ is isomorphic to an object of $K^b(A\text{-}proj)$. Two such triples are equivalent if they are both covered by a third triple in the sense described in Definition 3.5.35. Composition of two such triples is defined as in Lemma 3.5.33 whose proof carries over word by word to this situation, changing the property "acyclic complex" to "complex belonging to $K^b(A\text{-}proj)$".

Definition 6.9.1 We define the *singularity category* $D_{sg}(A)$ of A to be the category with the same objects as $K^{-,b}(A\text{-}proj)$, and morphisms from X to Y are equivalence classes of triples

$$X \stackrel{\nu}{\longleftarrow} Z \stackrel{\alpha}{\longrightarrow} Y$$

where $C(\nu)$ is isomorphic to an object of $K^b(A\text{-}proj)$. Two triples $X \stackrel{\nu}{\longleftarrow} Z \stackrel{\alpha}{\longrightarrow} Y$ and $X \stackrel{\nu'}{\longleftarrow} Z' \stackrel{\alpha'}{\longrightarrow} Y$ are equivalent if there is a triple $X \stackrel{\nu''}{\longleftarrow} Z'' \stackrel{\alpha''}{\longrightarrow} Y$ and there are morphisms $Z \stackrel{\beta}{\longleftarrow} Z'' \stackrel{\beta'}{\longrightarrow} Z'$ in $K^{-,b}(A\text{-}proj)$ such that the diagram

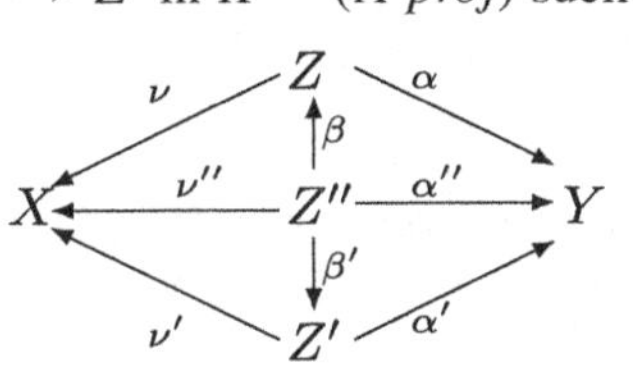

is commutative.

Remark 6.9.2 We now give some historical remarks on the development of singularity categories.

- The notion and concept of the singularity category goes back to an unpublished manuscript of Buchweitz [25]. There Buchweitz discovered that the category $D_{sg}(A)$ is related to the study of certain singularities in algebraic varieties and maximal Cohen-Macaulay modules. Later, Orlov [26, 27] independently rediscovered the category $D_{sg}(A)$ in connection with problems in theoretical physics.
- The notion of the singularity category comes from the following nice result due to Orlov [26]. For the statement we will need some elements of algebraic geometry. We will not be using these considerations again, but those readers who know about this part of algebraic geometry might find the link useful. Let us start with a Noetherian scheme X over a field K. The category of coherent sheaves over X is an abelian category, and it is of great interest to form the derived category $D^b(X)$ of this abelian category; actually this was the original motivation for studying derived categories. The singularity category $D_{sg}(X)$ is then produced as in Definition 6.9.1. Let X be such a scheme and let U be an open neighbourhood of the singular locus $Sing(X)$ of X. Then Orlov shows that $D_{sg}(X) \simeq D_{sg}(U)$. For more details on this result we refer to [28].

There is a natural functor $K^{-,b}(A\text{-}proj) \longrightarrow D_{sg}(A)$ defined to be the identity on objects and a morphism $\alpha : X \longrightarrow Y$ is mapped to the triple $X \overset{id}{\longleftarrow} X \overset{\alpha}{\longrightarrow} Y$. Composing with the equivalence $D^b(A\text{-}mod) \simeq K^{-,b}(A\text{-}mod)$ we obtain a natural functor $D^b(A\text{-}mod) \longrightarrow D_{sg}(A)$. We define distinguished triangles of $D_{sg}(A)$ to be the images of the distinguished triangles of $D^b(A\text{-}mod)$ and the suspension functor of $D_{sg}(A)$ to be the shift in degrees, as in $D^b(A\text{-}mod)$.

Proposition 6.9.3 *Let K be a commutative ring and let A be a Noetherian K-algebra. Then the singularity category $D_{sg}(A)$ is triangulated and the functor $D^b(A\text{-}mod) \longrightarrow D_{sg}(A)$ is a functor of triangulated categories in the sense that it commutes with the shift functor and that it sends distinguished triangles to distinguished triangles.*

Proof The proof is word for word the same as the proof of Proposition 3.5.40, changing "the cone is an acyclic complex" to "the cone is a complex in $K^b(A\text{-}proj)$". The details are left to the reader. □

We can compose the embedding $A\text{-}mod \to D^b(A)$ with the functor $D^b(A\text{-}mod) \to D_{sg}(A)$ to get a functor $A\text{-}mod \to D_{sg}(A)$.

Proposition 6.9.4 *Let K be a commutative ring and let A be a Noetherian K-algebra. The natural functor $A\text{-}mod \longrightarrow D_{sg}(A)$ factors through the functor $A\text{-}mod \longrightarrow A\text{-}\underline{mod}$. More precisely, there is a functor*

$$A\text{-}\underline{mod} \longrightarrow D_{sg}(A)$$

making the diagram

$$\begin{array}{ccc} A\text{-}mod & \longrightarrow & D^b(A\text{-}mod) \\ \downarrow & & \downarrow \\ A\text{-}\underline{mod} & \longrightarrow & D_{sg}(A) \end{array}$$

commutative.

Proof From Proposition 3.5.18 we know that there is a full embedding

$$A\text{-}mod \longrightarrow K^{-,b}(A\text{-}proj)$$

which sends a module M to a projective resolution P_M, and a morphism to its lift along the projective resolution.

Now

$$K^{-,b}(A\text{-}proj) \longrightarrow D_{sg}(A)$$

is defined to be the identity on objects and hence the composition

$$A\text{-}mod \longrightarrow D_{sg}(A)$$

is again the identity on objects.

Suppose now $\alpha : M \longrightarrow N$ is a morphism in A-*mod* factoring through a projective module P:

$$\begin{array}{ccc} M & \stackrel{\beta}{\longrightarrow} & P \\ \| & & \downarrow\gamma \\ M & \stackrel{\alpha}{\longrightarrow} & N \end{array}$$

Since $D^b(A\text{-}mod)$ is triangulated, we may form distinguished triangles

$$M \longrightarrow P \longrightarrow C(\beta) \longrightarrow M[1],$$

$$M \longrightarrow N \longrightarrow C(\alpha) \longrightarrow M[1]$$

and

$$P \longrightarrow N \longrightarrow C(\gamma) \longrightarrow P[1].$$

Considering these triangles in $D_{sg}(A)$ we get that P is in $K^b(A\text{-}proj)$ and hence $C(\beta)[-1] \longrightarrow M$ is an isomorphism, as is $N \longrightarrow C(\gamma)$. Therefore the cone P of $C(\beta)[-1] \longrightarrow M$ is isomorphic to 0 in $D_{sg}(A)$ using Lemma 3.4.9. This shows that the image of α in $D_{sg}(A)$ factors through the image of P, which is 0. This shows that $\alpha = 0$ in $D_{sg}(A)$. Hence the natural functor $A\text{-}mod \longrightarrow D_{sg}(A)$ factors through $A\text{-}\underline{mod}$. This proves the statement. □

Remark 6.9.5 Let A be a hereditary K-algebra over a field K. Then every finite dimensional A-module M has a projective resolution of length 2: By Lemma 1.11.3 there is an exact sequence

$$0 \longrightarrow P_1 \longrightarrow P_0 \longrightarrow M \longrightarrow 0$$

for two projective modules P_1 and P_0. Hence, $A\text{-}mod \longrightarrow D^b(A\text{-}mod)$ actually has image in $K^b(A\text{-}proj)$ and so the functor

$$A\text{-}\underline{mod} \longrightarrow D_{sg}(A)$$

is equivalent to the 0-functor.

As in Corollary 3.5.41 we get the following universal property.

Lemma 6.9.6 *Let A be a finite dimensional K-algebra and denote by $N : D^b(A\text{-}mod) \longrightarrow D_{sg}(A)$ the natural functor. Let $\mathcal{C}$ be a triangulated category and let $F : D^b(A\text{-}mod) \longrightarrow \mathcal{C}$ be a functor of triangulated categories. Suppose that for all $\alpha : X \longrightarrow Y$ with cone in $K^b(A\text{-}proj)$ we have that $F(\alpha)$ is invertible in $\mathcal{C}$. Then there is a functor $G : D_{sg}(A) \longrightarrow \mathcal{C}$ such that $F = G \circ N$.*

Proof We define $G(X) := F(X)$ for all objects X of $D_{sg}(A)$. For any morphism $X \longrightarrow Y$ in $D_{sg}(A)$ represented by (ν, Z, α) we define

$$G((\nu, Z, \alpha)) := F(\alpha) \circ F(\nu)^{-1}$$

where ν is a morphism with cone in $K^b(A\text{-}proj)$ and hence $F(\nu)$ is invertible by hypothesis. This obviously gives a functor G and $F = G \circ N$ is clear from the definition. □

Recall that a functor $F : \mathcal{C} \longrightarrow \mathcal{D}$ is dense if for every object D in $\mathcal{D}$ there is an object C_D in $\mathcal{C}$ such that there is an isomorphism $F(C_D) \simeq D$ in $\mathcal{D}$.

Proposition 6.9.7 *Let A be a finite dimensional K-algebra for a field K. Consider the functor $F : A\text{-}\underline{mod} \to D_{sg}(A)$. Then for every object X in $D_{sg}(A)$ there is an integer $n \in \mathbb{N}$ and an object M in $A\text{-}\underline{mod}$ such that $X \simeq FM[n]$. If A is self-injective, then F is dense.*

Proof Let X be an object of $D_{sg}(A)$. Then X is a right bounded complex of projective A-modules. Suppose that $H_m(X) = 0$ if $|m| \geq n_0$ and let $\sigma_{\leq n_0}X$ be the stupidly truncated complex at degree n_0. Recall that $(\sigma_{\leq n_0}X)_n = X_n$ if $n \leq n_0$ and $(\sigma_{\leq n_0}X)_n = 0$ if $n > n_0$. The differential on $\sigma_{\leq n_0}X$ is the same as the differential on X in all degrees $n \leq n_0$. These complexes fit into an exact sequence of morphisms of complexes

$$\begin{array}{cccccccccccc}
\to & X_{n+2} & \to & X_{n_0+1} & \to & 0 & \to & 0 & \to & \cdots & \to & 0 & & \\
 & \uparrow id & & \uparrow id & & \uparrow & & \uparrow & & & & \uparrow & & \\
\to & X_{n+2} & \to & X_{n_0+1} & \stackrel{\partial_{n_0+1}}{\to} & X_{n_0} & \stackrel{\partial_{n_0}}{\to} & X_{n_0-1} & \stackrel{\partial_{n_0-1}}{\to} & \cdots & \stackrel{\partial_{m+1}}{\to} & X_m & \to 0 \to & \cdots \\
 & \uparrow & & \uparrow & & \uparrow id & & \uparrow id & & & & \uparrow id & & \\
\to & 0 & \to & 0 & \to & X_{n_0} & \stackrel{\partial_{n_0}}{\to} & X_{n_0-1} & \stackrel{\partial_{n_0-1}}{\to} & \cdots & \stackrel{\partial_{m+1}}{\to} & X_m & \to 0 \to & \cdots
\end{array}$$

The choice of n_0 implies that this induces a distinguished triangle

$$\sigma_{\leq n_0}X \longrightarrow X \longrightarrow F(C_{n_0}(X))[n_0+1] \longrightarrow \sigma_{\leq n_0}X[1]$$

where $C_{n_0}(X) = \ker(\partial_{n_0})$. Observe that $\sigma_{\leq n_0}X$ is an object of $K^b(A\text{-}proj)$. Lemma 3.4.9 implies that $X \longrightarrow F(C_{n_0}(X))[n_0+1]$ is an isomorphism in $D_{sg}(A)$. This proves the first part of the statement. If A is self-injective, then $A\text{-}\underline{mod}$ is triangulated and $F(C_{n_0}(X))[n_0+1] \simeq F(\Omega^{-n_0-1}(C_{n_0}(X)))$ since the functor $A\text{-}\underline{mod} \longrightarrow D_{sg}(A)$ is triangulated. This proves the second statement. □

6.9.2 Singularity Categories and Self-Injective Algebras; Link to the Stable Case

As we shall now see the situation is simpler in case of self-injective algebras.

Proposition 6.9.8 (Rickard [29]) *Let K be a commutative ring and let A be a Noetherian K-algebra. Then the functor*

$$F : A\text{-}\underline{mod} \longrightarrow D_{sg}(A)$$

has the property $F \circ \Omega \simeq [-1] \circ F$. If K is a field and A is finite dimensional self-injective, then F is an equivalence between triangulated categories.

Proof We replace, as usual, $D^b(A\text{-}mod)$ by $K^{-,b}(A\text{-}proj)$ to simplify the discussion.
We first show $F \circ \Omega \simeq [-1] \circ F$. Let X be an A-module with projective resolution

$$\cdots \longrightarrow P_2 \stackrel{\partial_2}{\longrightarrow} P_1 \stackrel{\partial_1}{\longrightarrow} P_0 \longrightarrow X \longrightarrow 0.$$

Then F sends X to the complex

$$\cdots \longrightarrow P_2 \stackrel{\partial_2}{\longrightarrow} P_1 \stackrel{\partial_1}{\longrightarrow} P_0 \longrightarrow 0 \longrightarrow \cdots$$

where P_i is placed in degree i. Now, $\Omega(X)$ is mapped by F to the complex

$$\cdots \longrightarrow P_2 \stackrel{\partial_2}{\longrightarrow} P_1 \longrightarrow 0 \longrightarrow 0 \longrightarrow \cdots$$

where P_i is placed in degree $i-1$. This complex is the image of X under $F \circ \Omega$.

On the other hand, $F(X)[-1]$ is the complex

$$\cdots \longrightarrow P_2 \stackrel{\partial_2}{\longrightarrow} P_1 \stackrel{\partial_1}{\longrightarrow} P_0 \longrightarrow 0 \longrightarrow 0$$

where P_i is placed in position $i-1$. However,

$$\begin{array}{rccccccccc} F(\Omega X): & \cdots \longrightarrow & P_2 & \stackrel{\partial_2}{\longrightarrow} & P_1 & \longrightarrow & 0 & \longrightarrow & 0 & \longrightarrow \cdots \\ & & \uparrow & & \uparrow & & \uparrow & & \uparrow & \\ F(X)[-1]: & \cdots \longrightarrow & P_2 & \stackrel{\partial_2}{\longrightarrow} & P_1 & \stackrel{\partial_1}{\longrightarrow} & P_0 & \longrightarrow & 0 & \longrightarrow \cdots \\ & & \uparrow & & \uparrow & & \uparrow & & \uparrow & \\ P_0[-1]: & \longrightarrow & 0 & \longrightarrow & 0 & \longrightarrow & P_0 & \longrightarrow & 0 & \longrightarrow \cdots \end{array}$$

is a sequence of locally split morphisms of complexes, which induces a distinguished triangle

$$P_0[-1] \longrightarrow F(X)[-1] \longrightarrow F(\Omega X) \longrightarrow P_0$$

and since P_0 is in $K^b(A\text{-}proj)$, we get that $F(X)[-1] \longrightarrow F(\Omega X)$ is an isomorphism in $D_{sg}(A)$.

Suppose now for the rest of the proof that A is self-injective.

Proposition 6.9.7 shows that the natural functor $F : A\text{-}\underline{mod} \longrightarrow D_{sg}(A)$ is dense. The first step of the proof shows that $F \circ \Omega \simeq [-1] \circ F$.

If A is self-injective, then F is a functor of triangulated categories. Indeed, a distinguished triangle in $A\text{-}\underline{mod}$ is obtained by the following construction: Let $X \stackrel{\alpha}{\longrightarrow} Y$ be a homomorphism. Then let $X \longrightarrow I$ be an injective hull of X and form the pushout

$$\begin{array}{ccccccccc} 0 \longrightarrow & X & \stackrel{\iota}{\longrightarrow} & I & \stackrel{\pi\circ\beta}{\longrightarrow} & \Omega^{-1}X & \longrightarrow 0 \\ & \downarrow \alpha & & \downarrow \beta & & \| & \\ 0 \longrightarrow & Y & \stackrel{\lambda}{\longrightarrow} & U & \stackrel{\pi}{\longrightarrow} & \Omega^{-1}X & \longrightarrow 0 \end{array}$$

to obtain the distinguished triangle

$$X \stackrel{\alpha}{\longrightarrow} Y \stackrel{\lambda}{\longrightarrow} U \stackrel{\pi}{\longrightarrow} \Omega^{-1}X.$$

But this pushout implies that

$$0 \longrightarrow X \stackrel{(\alpha,-\iota)}{\longrightarrow} Y \oplus I \stackrel{\binom{\lambda}{\beta}}{\longrightarrow} U \longrightarrow 0$$

is an exact sequence of A-modules. Since exact sequences in $A\text{-}mod$ become distinguished triangles in $D^b(A\text{-}mod)$, we get that the embedding $A\text{-}mod \longrightarrow D^b(A\text{-}mod)$ maps the above exact sequence to the distinguished triangle

$$X \stackrel{(\alpha,-\iota)}{\longrightarrow} Y \oplus I \stackrel{\binom{\lambda}{\beta}}{\longrightarrow} U \longrightarrow X[1]$$

and via the functor

$$D^b(A\text{-}mod) \longrightarrow D_{sg}(A)$$

this distinguished triangle is mapped to the distinguished triangle

$$X \stackrel{\alpha}{\longrightarrow} Y \stackrel{\lambda}{\longrightarrow} U \longrightarrow X[1]$$

since I is an injective module, hence a projective module, A being self-injective, hence in $K^b(A\text{-}proj)$.

For self-injective algebras there is no finite projective resolution of non-projective modules by Lemma 1.10.8 and so the functor

$$F : A\text{-}\underline{mod} \longrightarrow D_{sg}(A)$$

has the property that only the 0-module is sent to 0.

We shall now prove that F is full. Since F is dense, we only need to show that a triple (ν, Z, α) for a module Z, a module homomorphism $\alpha : Z \longrightarrow Y$ and a module homomorphism $\nu : Z \longrightarrow X$ with cone in $K^b(A\text{-}proj)$ is in the image of F. By the fact that F is dense and that $A\text{-}mod$ is a full subcategory of $D^b(A\text{-}mod)$ any morphism α is in the image of F. We need to show that a module homomorphism $\nu : Z \longrightarrow X$ with cone in $K^b(A\text{-}proj)$ is actually invertible in $A\text{-}\underline{mod}$. Indeed, the cone of ν is given by the above pushout construction

$$\begin{array}{ccccccccc} 0 \longrightarrow & Z & \stackrel{\iota}{\longrightarrow} & I & \stackrel{\pi\circ\beta}{\longrightarrow} & \Omega^{-1}Z \longrightarrow 0 \\ & \downarrow \nu & & \downarrow \mu & & \| \\ 0 \longrightarrow & X & \stackrel{\lambda}{\longrightarrow} & U & \stackrel{\pi}{\longrightarrow} & \Omega^{-1}Z \longrightarrow 0 \end{array}$$

and hence the cone U is a module again. (Note that we defined distinguished triangles in the stable categories slightly differently, using projective covers instead of injective hulls. It is straightforward to see that this definition here is equivalent to the one given in Sect. 5.1.4.) Now, the only module in $K^b(A\text{-}proj)$ is actually a projective module. Therefore U is projective and the pushout construction gives the exact sequence

$$0 \longrightarrow Z \stackrel{(\nu,-\iota)}{\longrightarrow} X \oplus I \stackrel{\binom{\lambda}{\beta}}{\longrightarrow} U \longrightarrow 0$$

of A-modules. Since now U is projective, the sequence splits, and hence, using that I is projective as well, ν is an isomorphism in the stable category. This shows that F is full.

Moreover, the fact that only projective modules are sent to 0 by F also shows that F is faithful. Indeed, let $\alpha : X \longrightarrow Y$ in $A\text{-}\underline{mod}$ be a morphism such that $F(\alpha) = 0$.

Then the distinguished triangle

$$X \xrightarrow{\alpha} Y \xrightarrow{\lambda} U \xrightarrow{\pi} \Omega^{-1}X$$

is mapped to the distinguished triangle

$$FX \xrightarrow{0} FY \xrightarrow{F\lambda} FU \longrightarrow (FX)[1].$$

By Lemma 3.4.9 $F\lambda$ is a split monomorphism, and so there is a $\hat{\sigma} : FU \longrightarrow FY$ such that $\hat{\sigma} \circ F\lambda = id_{FY}$. But since F is full, there is a $\sigma : U \longrightarrow Y$ such that $F\sigma = \hat{\sigma}$. Hence $F(\sigma \circ \lambda) = id_{FY} = F(id_Y)$. Therefore the cone $C(\sigma \circ \lambda)$ of $\sigma \circ \lambda$ is mapped by F to

$$F(C(\sigma \circ \lambda)) \simeq C(F(\sigma \circ \lambda)) \simeq C(id_{FY}) \simeq 0.$$

We have seen that $FM \simeq 0$ implies $M \simeq 0$, and so $C(\sigma \circ \lambda) \simeq 0$. This shows that $\sigma \circ \lambda$ is an automorphism, again using Lemma 3.4.9, and therefore λ is a split monomorphism. Lemma 3.4.9 then shows that $\alpha = 0$. □

Remark 6.9.9 Observe that we used at various places in the second part of the proof of Proposition 6.9.8 that A is self-injective. If A is not self-injective, then F is neither full nor faithful. Actually $D_{sg}(A)$ is quite complicated for non-self-injective A.

Lemma 6.9.10 *Let A be a (not necessarily selfinjective) finite dimensional K-algebra and let $F : A\text{-}\underline{mod} \longrightarrow D_{sg}(A)$ be the natural functor. Let M and N be two A-modules such that there is an $n \in \mathbb{N}$ with $\Omega^n(M) \simeq \Omega^n(N)$, then $F(M) \simeq F(N)$.*

Proof We get

$$F(M)[-n] \simeq F(\Omega^n(M)) \simeq F(\Omega^n(N)) \simeq F(N)[-n]$$

using Proposition 6.9.8. Hence $F(M) \simeq F(N)$ since the shift functor is invertible in $D_{sg}(A)$. □

Example 6.9.11 Let K be a field and let A be the algebra given by the quiver

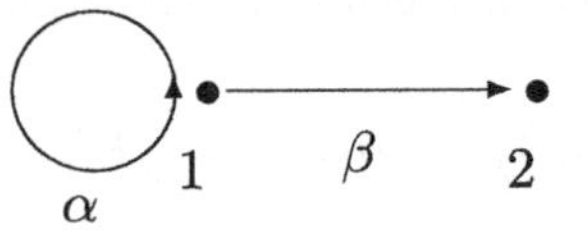

with relation $\alpha^2 = 0$, $\beta\alpha = 0$. Then the projective module associated to the vertex 2 is uniserial with top the simple A-module 2 and socle the simple A-module 1. The projective module associated to the vertex 1 is uniserial of length 2 with both composition factors isomorphic to the simple 1. Hence the simple module S_2 has a projective resolution

$$\cdots \longrightarrow P_1 \longrightarrow \cdots \longrightarrow P_1 \longrightarrow P_1 \longrightarrow P_1 \longrightarrow P_2 \longrightarrow S_2 \longrightarrow 0$$

and the simple module S_1 has a projective resolution

$$\cdots \longrightarrow P_1 \longrightarrow \cdots \longrightarrow P_1 \longrightarrow P_1 \longrightarrow P_1 \longrightarrow P_1 \longrightarrow S_1 \longrightarrow 0$$

and in particular $\Omega(S_2) = S_1$. Since A is serial there are only four indecomposable A-modules, P_1, P_2, S_2, S_1 and therefore A-$\underline{mod}$ has two indecomposable objects S_1 and S_2 with no morphisms between them, and endomorphisms K for each of them. Observe that $S_1[1] \simeq S_1$ in $D_{sg}(A)$. Lemma 6.9.10 then shows that $D_{sg}(A)$ has only one indecomposable object. It can be shown that this object has endomorphism ring K in $D_{sg}(A)$.

Theorem 6.9.12 (Rickard [2], Keller-Vossieck [30]) *Let K be a field and let A and B be self-injective finite dimensional K-algebras. Then $D^b(A) \simeq D^b(B)$ implies that A-$\underline{mod} \simeq B$-$\underline{mod}$ and A and B are stably equivalent of Morita type.*

Proof By Rickard's Morita theorem 6.5.1 for derived categories there is a tilting complex T over A with endomorphism ring B^{op} and by Keller's Theorem 6.4.1 we know that there is a two-sided tilting complex X in $D^b(A \otimes_K B^{op})$ realising an equivalence of triangulated categories

$$F := X \otimes_B^{\mathbb{L}} - : D^b(B\text{-}mod) \longrightarrow D^b(A\text{-}mod).$$

Again by Rickard's Theorem 6.5.1 and by the second item of Remark 6.5.3 we know that F maps $K^b(B\text{-}proj)$ to $K^b(A\text{-}proj)$ and the same holds true for a quasi-inverse G to F. So F induces an equivalence

$$\underline{F} : B\text{-}\underline{mod} \simeq D_{sg}(B) \longrightarrow D_{sg}(A) \simeq A\text{-}\underline{mod}.$$

By Proposition 6.4.4 we may replace X by a quasi-isomorphic bounded complex with all homogeneous components being projective bimodules, except the homogeneous component in highest degree. This component is nevertheless projective as an A-module, and projective as a B-module. Let therefore X be the complex

$$\cdots \to 0 \to X_k \to X_{k-1} \to \cdots \to X_n \to 0 \to \cdots$$

where $X_{k-1}, \ldots, X_n$ are projective as A-B-bimodules, whereas X_k is projective as an A-module and projective as a B-module. Therefore $X \otimes_B^{\mathbb{L}} - = X \otimes_B -$. The image of a B-module C under $X \otimes_B^{\mathbb{L}} -$ is the complex

$$\cdots \to 0 \to X_k \otimes_B C \to X_{k-1} \otimes_B C \to \cdots \to X_n \otimes_B C \to 0 \to \cdots$$

Stupid truncation at degree k yields a distinguished triangle

$$\tau_{\leq k-1} X \otimes_B C \to X \otimes_B C \to (X_k \otimes_B C)[k] \to \tau_{\leq k-1} X \otimes_B C[1].$$

Now, $P \otimes_B C$ is a projective A-module if P is a projective A–B-bimodule. Indeed, $(A \otimes_K B)^s \otimes_B C \simeq A^s \otimes_K C \simeq A^{s \cdot \dim_K C}$ and tensor products of direct summands of $(A \otimes_K B)^s$ with C yield direct summands of $A^{s \cdot \dim_K C}$, whence projective modules. Hence $\tau_{\leq k-1} X \otimes_B C$ is in $K^b(A\text{-}proj)$, so that the image of $X \otimes_B C$ in $D_{sg}(A)$ is $(X_k \otimes_B C)[k] = \Omega^{-k}(X_k \otimes_B C)$. We may replace X by $X[-k]$ from the beginning and may hence assume that $k = 0$. Since X_k is projective as an A-module and as a B-module, $X_k \otimes_B -$ satisfies the hypotheses of Proposition 5.3.17. This proposition now shows that $X \otimes_B -$ induces a stable equivalence of Morita type between A and B. □

Example 6.9.13 Let k be a field of characteristic 2 and let P be a dihedral 2-group. Then there are non-trivial stable self-equivalences of kP-$\underline{mod}$. The group of self-equivalences of kP-$\underline{mod}$ contains the subgroup given by tensoring with $\Omega^n(k)$ over k for $n \in \mathbb{Z}$ and the group $Out_k(kP)$ of Morita self-equivalences. But the group generated by these two subgroups is strictly smaller than the group of self-equivalences of kP-$\underline{mod}$. This is a result due to Carlson and Thévenaz [16–18].

As we have seen in Proposition 6.7.4, any derived self-equivalence of $D^b(kP)$ is a composition of a Morita self-equivalence and a shift. We recall Remark 6.7.6.

Example 6.9.14 Let K be a field. We recall Example 5.2.2 which shows that the stable category of the symmetric local algebra $K[X]/X^2$ is stably equivalent as K-linear category to the stable category of the hereditary algebra $\begin{pmatrix} K & K \\ 0 & K \end{pmatrix}$. Of course these two algebras cannot have equivalent derived categories since $K[X]/X^2$ is local and any algebra with equivalent derived category to $D^b(K[X]/X^2)$ is Morita equivalent to $K[X]/X^2$ by Proposition 6.7.4.

Remark 6.9.15 The reader should be aware of the following two facts.

- Theorem 6.9.12 was the motivation for Broué to define stable equivalences of Morita type.
- Of course Theorem 6.9.12 implies that all invariants of self-injective algebras under stable equivalences of Morita type are invariants under derived equivalences. For example
  - the stable Grothendieck group from Proposition 5.7.4
  - and its consequence the classification of Nakayama algebras which was shown in Lemma 5.7.6,
  - positive degree Hochschild homology from Theorem 5.8.12,
  - positive degree Hochschild cohomology from Theorem 5.8.17,
  - the stable centres from Proposition 5.9.5,
  - the stable cocentre from Theorem 5.9.20 and
  - the Külshammer ideals by Proposition 5.9.21.
  - in particular if B_G and B_H are blocks of the groups G and H, and if $D^b(B_G) \simeq D^b(B_H)$, then the defect groups of B_G and of B_H have the same exponent (cf Lemma 2.9.18).

Of course, Hochschild cohomology is more immediately seen to be an invariant by Proposition 6.7.10. This also includes the invariance of the centre. For symmetric algebras this also implies the invariance of the Hochschild homology. Indeed, Proposition 3.7.16 shows that the functor

$$A \otimes^{\mathbb{L}}_{A\otimes_K A^{op}} - : D^-(A \otimes_K A^{op}\text{-}Mod) \longrightarrow D^-(K\text{-}Mod)$$

admits a right adjoint

$$\mathbb{R}Hom_K(A, -) : D^-(K\text{-}Mod) \longrightarrow D^-(A \otimes_K A^{op}\text{-}Mod).$$

Hence

$$\begin{aligned} Hom_K(HH_n(A), K) &= Hom_{D^-(K\text{-}Mod)}(A \otimes^{\mathbb{L}}_{A\otimes_K A^{op}} A, K[n]) \\ &\simeq Hom_{D^-(A\otimes_K A^{op}\text{-}Mod)}(A, \mathbb{R}Hom_K(A, K[n])) \\ &= Hom_{D^-(A\otimes_K A^{op}\text{-}Mod)}(A, A[n]) \text{ since } A \text{ is symmetric} \\ &= Ext^n_{A\otimes_K A^{op}}(A, A) \\ &= HH^n(A). \end{aligned}$$

Proposition 6.7.17 now shows that a derived equivalence of standard type between A and B sends $Hom_K(A, K)$ to $Hom_K(B, K)$.

This shows

Lemma 6.9.16 *Let A be a symmetric K-algebra, then the Hochschild cohomology of A is the K-linear dual of its Hochschild homology.*

Nevertheless, we have already seen that Hochschild homology is invariant under derived equivalences. This was the subject of Proposition 6.7.14.

6.9.3 Homomorphisms in Singularity Categories as Limits

Example 6.9.17 Let K be a field and let $A = K[X, Y]/(X^2, Y^2, XY)$. This algebra is local, and the simple module S satisfies

$$\Omega^n(S) = S^{2^{n+1}}.$$

Indeed,

$$0 \to \text{rad}A \to A \to S \to 0$$

is exact, and $\text{rad}(A) \simeq S^2$ is semisimple of dimension 2. Let $F : A\text{-}\underline{mod} \longrightarrow D_{sg}(A)$ be the canonical functor. But then, by Lemma 6.9.10,

$$FS \simeq (F(\Omega^n S))[n] \simeq (F(S^{2^n}))[n] \simeq ((FS)^{2^n})[n]$$

for all $n \in \mathbb{N}$ in $D_{sg}(A)$. Hence

$$\begin{aligned} End_{D_{sg}(A)}(FS) &\simeq End_{D_{sg}(A)}(((FS)^{2^n})[n]) \\ &\simeq End_{D_{sg}(A)}(((FS)^{2^n})) \text{ (since [1] is a self-equivalence)} \\ &\simeq Mat_{2^n \times 2^n}(End_{D_{sg}(A)}(FS)). \end{aligned}$$

Since S is not of finite global dimension in A-*mod*, we get $FS \neq 0$. As a consequence $End_{D_{sg}(A)}(FS)$ cannot be a finite dimensional K-vector space.

This example already gives a rather good image of what happens in the singularity category. If A is an algebra and if M and N are A-modules, then taking syzygies is a well-defined functor

$$\Omega : A\text{-}\underline{mod} \longrightarrow A\text{-}\underline{mod}$$

and hence induces

$$\underline{Hom}_A(M, N) \to \underline{Hom}_A(\Omega M, \Omega N) \to \underline{Hom}_A(\Omega^2 M, \Omega^2 N) \to \cdots$$

for all A-modules M and N. These maps form a directed system, so that we can consider the inductive limit of this system.

Proposition 6.9.18 (Beligiannis [31, Corollary 3.3, Proposition 3.4]; [30] in a special case) *Let K be a field and let A be a finite dimensional K-algebra. Let M and N be finite dimensional A-modules and let $F : A\text{-}\underline{mod} \longrightarrow D_{sg}(A)$ be the canonical functor. Then for all $k \in \mathbb{Z}$*

$$Hom_{D_{sg}(A)}(FM, FN[-k]) \simeq \text{colim}_n \underline{Hom}_A(\Omega^n M, \Omega^{n+k} N).$$

Proof Since F induces a morphism

$$\underline{Hom}_A(M, \Omega^k N) \longrightarrow Hom_{D_{sg}(A)}(FM, F\Omega^k N)$$

and therefore a commutative diagram

$$\begin{array}{ccc} \underline{Hom}_A(\Omega M, \Omega^{k+1} N) & \longrightarrow & Hom_{D_{sg}(A)}(F\Omega M, F\Omega^{k+1} N) \\ & & \| \\ \uparrow & & Hom_{D_{sg}(A)}((FM)[-1], (FN)[-k-1]) \\ & & \| \\ \underline{Hom}_A(M, \Omega^k N) & \longrightarrow & Hom_{D_{sg}(A)}(FM, (FN)[-k]) \end{array}$$

we get by the universal property of the inductive limit a morphism

$$\Lambda : \text{colim}_n \underline{Hom}_A(\Omega^n M, \Omega^{n+k} N) \longrightarrow Hom_{D_{sg}(A)}(FM, FN[-k]).$$

In order to prove that this is an isomorphism we need to show that a homomorphism $\hat{\nu} : Z \longrightarrow M$ in $D^b(A)$ with cone in $K^b(A\text{-}proj)$ comes from a sequence of homomorphisms $\nu^n : \Omega^n Z \longrightarrow \Omega^n M$ which are invertible in the inductive limit.

Since Z and M are both shifted copies of A-modules, working in $K^{-,b}(A\text{-}proj)$ we may replace Z by a projective resolution P^Z and M by a projective resolution P^M. Further, $\hat{\nu}$ is a morphism of complexes up to homotopy, and the cone P_C is a bounded complex of projective A-modules. Hence there is an $n_0 \in \mathbb{N}$ such that $P_C^n = 0$ for $n \geq n_0$ and therefore $\Omega^n(\hat{\nu})$ is an isomorphism if $n > n_0$. This shows that the natural map Λ is surjective. The fact that Λ is injective follows immediately from the construction in Proposition 6.9.7. This proves the proposition. □

Definition 6.9.19 We say that in an additive category $\mathcal{A}$ *idempotents split* if for every idempotent endomorphism e of an object X there is a unique object Y and $u \in Mor_{\mathcal{A}}(X, Y)$ as well as $v \in Mor_{\mathcal{A}}(Y, X)$ such that $e = v \circ u$ and $id_Y = u \circ v$.

Remark 6.9.20 Let A be a finite dimensional K-algebra. Then idempotents split in A-*mod*, since $Y = e(X)$, and u and v being the natural maps give the result. Also idempotents split in A-$\underline{mod}$ since we may assume that the object X does not have a projective direct factor, and then endomorphisms factoring through projective modules are nilpotent and so endomorphisms which are idempotent modulo endomorphisms factoring through projective modules lift to idempotent endomorphisms (Proposition 1.9.17).

Corollary 6.9.21 *Let K be a field and let A be a finite dimensional K-algebra. Then idempotents split in $D_{sg}(A)$.*

Proof Let X be an object of $D_{sg}(A)$ and let $e^2 = e \in End_{D_{sg}(A)}(X)$. We may assume that $W \simeq M[n]$ for some finite dimensional A-module M. By Proposition 6.9.18 we can represent e as a sequence $e^{(n)}$ of coherent stable endomorphisms of $\Omega^n(M)$. For large n we see that $e^{(n)}$ is an idempotent endomorphism. We need to show that for large enough n the endomorphism $e^{(n)} \in \underline{End}_A(\Omega^n M)$ is a split idempotent. But since $\Omega^n M$ is a finite dimensional A-module and since idempotents split in A-$\underline{mod}$ we are done. □

6.9.4 Algebras with Radical Squared 0; an Example

We shall give a rather detailed description of the singularity category of finite dimensional algebras A with $\text{rad}^2(A) = 0$. In particular the question of when the singularity category has finite dimensional homomorphism spaces is answered for these algebras. This is work due to Xiao-Wu Chen [32] and we shall closely follow his presentation.

Let A be a finite dimensional algebra over a field with $rad^2(A) = 0$. Observe that Example 6.9.17 is of this type. In order to simplify the notation we put $\rho := rad(A)$ and $\overline{A} := A/rad(A)$. Then ρ has a natural $\overline{A} \otimes \overline{A}$-module module structure. We put

$$\rho^{\otimes n+1} := \rho^{\otimes n} \otimes_{\overline{A}} \rho \text{ for all } n \in \mathbb{N} \text{ and } \rho^{\otimes 0} := \overline{A}.$$

Since the functor $\rho \otimes - : \overline{A}\text{-}mod \longrightarrow \overline{A}\text{-}mod$ maps $\rho^{\otimes n}$ to $\rho^{\otimes n+1}$, the functor $\rho \otimes -$ induces a sequence of ring homomorphisms

$$End_{\overline{A}}(\rho^{\otimes 0}) \stackrel{\lambda_0}{\to} End_{\overline{A}}(\rho^{\otimes 1}) \stackrel{\lambda_1}{\to} End_{\overline{A}}(\rho^{\otimes 2}) \stackrel{\lambda_2}{\to} \cdots \stackrel{\lambda_{n-1}}{\to} End_{\overline{A}}(\rho^{\otimes n}) \stackrel{\lambda_n}{\to} \cdots$$

Let $\Gamma(A)$ be the colimit of this sequence.

We first show that $\Gamma(A)$ is a ring.

Fix $n \in \mathbb{N}$ and consider $Hom_{\overline{A}}(\rho^{\otimes i}, \rho^{\otimes i-n})$ for each $i \geq n$. Then again these vector spaces form an inductive system, and consider the inductive limit of these. The object one obtains is denoted by $K^n(A)$, and carries a natural structure of a $\Gamma(A)$–$\Gamma(A)$-bimodule. Of course, $K^0(A) = \Gamma(A)$ as $\Gamma(A)$–$\Gamma(A)$-bimodules. Composition of mappings gives $\Gamma(A)$–$\Gamma(A)$-bimodule morphisms

$$K^n(A) \otimes_{\Gamma(A)} K^m(A) \longrightarrow K^{n+m}(A).$$

Since A is an algebra with $rad^2(A) = 0$ the syzygy (removing projective direct factors) of every simple module is semisimple. Indeed, let S be simple. Then S is cyclic (cf Remark 1.4.25), and the projective indecomposable module P mapping onto S has radical length at most 2. Therefore $\ker(P \longrightarrow S) \subseteq soc(P)$ and this module is semisimple. Moreover, if S is projective simple, then

$$\Omega(S) = 0 = \rho \otimes_{\overline{A}} S$$

and therefore Ω and $\rho \otimes_{\overline{A}}$ - are endofunctors of $\overline{A}\text{-}mod$, and both have value 0 on projective simple A-modules. This shows that Ω and $\rho \otimes_{\overline{A}} -$ are both endofunctors of the full subcategory $\overline{A}\text{-}mod \cap A\text{-}\underline{mod}$ of the stable category $A\text{-}\underline{mod}$ which is generated by $\overline{A}$-modules.

Lemma 6.9.22 *The endofunctors Ω and $\rho \otimes_{\overline{A}} -$ of $\overline{A}\text{-}mod \cap A\text{-}\underline{mod}$ coincide.*

Proof Let S be a semisimple A-module and let P be its projective cover. Then

$$\Omega(S) = \ker(P \to S) = rad(P) \simeq \rho \otimes_A P \simeq \rho \otimes_{\overline{A}} P/rad(P) \simeq \rho \otimes_{\overline{A}} S$$

and all the isomorphisms are easily verified to be functorial. □

Lemma 6.9.23 *Let $F : A\text{-}\underline{mod} \longrightarrow D_{sg}(A)$ be the canonical functor. Then for each $k \in \mathbb{Z}$ we get a natural isomorphism*

$$K^k(A) \simeq Hom_{D_{sg}(A)}(F(\overline{A}), F(\overline{A})[k]).$$

In particular, $K^0(A) \simeq End_{D_{sg}(A)}(F\overline{A}) = \Gamma(A)$.

Proof If $k \geq 0$, we get by Proposition 6.9.18

$$\begin{aligned} &Hom_{D_{sg}(A)}(F(\overline{A}), F(\overline{A})[k]) \\ &\quad = Hom_{D_{sg}(A)}(F(\overline{A})[-k], F(\overline{A})) = Hom_{D_{sg}(A)}(F(\overline{A})[-k], F(\overline{A})) \\ &\quad = Hom_{D_{sg}(A)}(F(\Omega^k(\overline{A})), F(\overline{A})) \overset{F}{\simeq} \mathrm{colim}_n \underline{Hom}_A(\Omega^{k+n}\overline{A}, \Omega^n\overline{A}) \\ &\quad = \mathrm{colim}_n \underline{Hom}_A(\rho^{\otimes k+n}, \rho^{\otimes n}) \ \text{ (by Lemma 6.9.22).} \end{aligned}$$

Therefore we get an epimorphism $K^n(A) \overset{\psi}{\twoheadrightarrow} Hom_{D_{sg}(A)}(F(\overline{A}), F(\overline{A})[k])$. Let $f \in \ker(Hom_{\overline{A}\text{-}mod}(\rho^{\otimes n}, \rho^{\otimes m}) \to \underline{Hom}_{A\text{-}mod}(\rho^{\otimes n}, \rho^{\otimes m}))$. But now,

$$Hom_{A\text{-}\underline{mod} \cap \overline{A}\text{-}mod}(\rho^{\otimes n}, \rho^{\otimes m}) \simeq \underline{Hom}_A(\rho^{\otimes n}, \rho^{\otimes m})$$

for all $n, m \in \mathbb{Z}$ since every morphism factoring through a projective module actually factors through a semisimple projective module. The functor $\rho \otimes -$ annihilates semisimple projective modules, and therefore $\rho \otimes f = 0$. Lemma 6.9.22 then implies ψ is injective.

If $k < 0$, then

$$\begin{aligned} Hom_{D_{sg}(A)}(\overline{A}, \overline{A}[k]) &= Hom_{D_{sg}(A)}(F(\overline{A}), F(\overline{A})[k]) \\ &= Hom_{D_{sg}(A)}(F(\overline{A}), F(\Omega^{-k}(\overline{A}))) \\ &\overset{F}{\simeq} \mathrm{colim}_n \underline{Hom}_A(\Omega^n\overline{A}, \Omega^{n-k}\overline{A}) \\ &= \mathrm{colim}_n \underline{Hom}_A(\rho^{\otimes n}, \rho^{\otimes n-k}) \ \text{ (by Lemma 6.9.22)} \\ &= K^k(A) \end{aligned}$$

by the same argument as above. Putting all of this together, we have proved the statement. □

Lemma 6.9.24 *Let $\mathcal{A}$ be a semisimple abelian category, and let Σ be a self-equivalence of $\mathcal{A}$. Then there is a unique choice of distinguished triangles such that $(\mathcal{A}, \Sigma)$ is a triangulated category with this choice of distinguished triangles.*

Proof Indeed, the axioms of a triangulated category imply that the triangles

$$C \xrightarrow{id} C \longrightarrow 0 \longrightarrow \Sigma C$$

$$0 \longrightarrow C \xrightarrow{id} C \longrightarrow \Sigma 0$$

$$C \longrightarrow 0 \longrightarrow \Sigma C \xrightarrow{id} \Sigma C$$

for all possible indecomposable objects C of $\mathcal{A}$ are distinguished triangles, triangles isomorphic to these triangles, and direct sums of these triangles have to be distinguished triangles as well. It is trivial to check that defining distinguished triangles to be this set of triangles, including direct sums of triangles isomorphic to these, satisfies the axioms of a triangulated category. In order to verify this fact it is sufficient to note that each morphism in $\mathcal{A}$ is a direct sum of morphisms of one of the following types:

$$0 \longrightarrow C, \quad C \longrightarrow 0, \quad C \xrightarrow{\alpha} C$$

for objects C of $\mathcal{A}$ and isomorphisms α. □

As in Sect. 6.9.4 let A be a finite dimensional algebra over a field and suppose $\mathrm{rad}^2(A) = 0$. Then the category $\Gamma(A) - fp\text{-}mod$ of finitely presented $\Gamma(A)$-modules is semisimple. Indeed, let M be a finitely presented $\Gamma(A)$-module. Then M is actually presented by elements in $End_{\overline{A}}(\rho^{\otimes n})$ for some large n. If $N \twoheadrightarrow M$ is then an epimorphism between finitely presented $\Gamma(A)$-modules, then this is an epimorphism of $End_{\overline{A}}(\rho^{\otimes n})$ for some large n, and since $End_{\overline{A}}(\rho^{\otimes n})$ is semisimple, the epimorphism is split. Hence $N \twoheadrightarrow M$ is split as $\Gamma(A)$-modules, and therefore $\Gamma(A) - fp\text{-}mod$ is semisimple. In particular $\Gamma(A) - fp\text{-}mod = \Gamma(A)\text{-}proj$. Denote as before by $F : A\text{-}\underline{mod} \to D_{sg}(A)$ the natural functor.

Proposition 6.9.25 *Let A be a finite dimensional algebra over a field with* $\mathrm{rad}^2(A) = 0$. *Then*

$$D_{sg}(A) = add(\overline{A}).$$

Moreover there is a self-equivalence Σ of $\Gamma(A)$-proj such that the triangulated category $(\Gamma(A)\text{-}proj, \Sigma)$ defined in Lemma 6.9.24 is equivalent to $D_{sg}(A)$:

$$D_{sg}(A) \simeq (\Gamma(A)\text{-}proj, \Sigma_A).$$

Proof Since $\mathrm{rad}^2(A) = 0$, we get that any finitely generated A-module is a direct sum of a projective and a semisimple module:

$$\Omega(A\text{-}mod) \subseteq add(A \oplus \overline{A}).$$

We claim that then $D_{sg}(A) = add(\overline{A})$. Indeed, by Proposition 6.9.7 we know that for all objects X of $D_{sg}(A)$ there is an A-module M and an integer n_1 such that $X \simeq FM[n_1]$. But now, for $\mathcal{C} := A\text{-}mod$, we have

$$add(A \oplus \overline{A}) \supseteq add(\Omega(\mathcal{C})) \supseteq add(\Omega^2(\mathcal{C})) \supseteq add(\Omega^3(\mathcal{C})) \supseteq \cdots .$$

We see that there is an n_0 such that

$$add(\Omega^n(A\text{-}mod)) = add(\Omega^{n+1}(A\text{-}mod))$$

for each $n \geq n_0$. Indeed, there are only finitely many indecomposable direct summands of $\overline{A}$. Hence the descending sequence

$$add(A \oplus \overline{A}) \supseteq add(\Omega(\mathcal{C})) \supseteq add(\Omega^2(\mathcal{C})) \supseteq add(\Omega^3(\mathcal{C})) \supseteq \dots$$

of subcategories cannot be infinite, where again $\mathcal{C} := A\text{-}mod$. Therefore

$$add(\Omega^{n_0+n_1}(A\text{-}mod)) = add(\Omega^{n_0}(A\text{-}mod)).$$

In Proposition 6.9.7 we constructed M so $X \simeq FM[n]$ for some $n \in \mathbb{N}$. Since $F(\Omega^m M)[m] \simeq FM$, for any $m \in \mathbb{N}$ we may replace n by $n + m$ and replace M by $\Omega^m M$ such that $X \simeq FM[n_1]$ for some $M \in add(\Omega^{n_0}(A\text{-}mod))$. Therefore, since $add(\Omega^{n_0+n_1}(A\text{-}mod)) = add(\Omega^{n_0}(A\text{-}mod))$, there is an A-module N such that

$$M \oplus N \in \Omega^{n_0+n_1}(A\text{-}mod).$$

Hence

$$M \oplus N = \Omega^{n_1}(L)$$

for some $L \in \Omega^{n_0}(A\text{-}mod)$.

But this shows that $FL \simeq F(M \oplus N)[n_1]$ in $D_{sg}(A)$. Since $X \simeq FM[n_1]$ we get that X is a direct summand of L and L is isomorphic to an object in

$$add(\Omega^{n_0}(A\text{-}mod)) \subseteq add(A \oplus \overline{A}).$$

Since projective A-modules are 0 in $D_{sg}(A)$ we get $D_{sg}(A) = add(\overline{A})$.

For the second statement we now apply the functor

$$Hom_{D_{sg}(A)}(\overline{A}, -) : D_{sg}(A) \longrightarrow End_{D_{sg}(A)}(\overline{A})\text{-}mod$$

and observe that

$$Hom_{D_{sg}(A)}(\overline{A}, U) \to Hom_{End_{D_{sg}(A)}(\overline{A})}(Hom_{D_{sg}(A)}(\overline{A}, \overline{A}), Hom_{D_{sg}(A)}(\overline{A}, U))$$

is an isomorphism. Indeed

$$Hom_{End_{D_{sg}(A)}(\overline{A})}(Hom_{D_{sg}(A)}(\overline{A}, \overline{A}), Hom_{D_{sg}(A)}(\overline{A}, V)) \simeq Hom_{D_{sg}(A)}(\overline{A}, V))$$

for all V and hence, by the additivity of Hom-functors,

$$Hom_{D_{sg}(A)}(S, U) \to Hom_{End_{D_{sg}(A)}(\overline{A})}(Hom_{D_{sg}(A)}(\overline{A}, S), Hom_{D_{sg}(A)}(\overline{A}, U))$$

is an isomorphism for all direct factors S of $\overline{A}$. Since $D_{sg}(A) = add(\overline{A})$ we obtain that the functor $Hom_{D_{sg}(A)}(\overline{A}, -)$ is fully faithful. Again $D_{sg}(A) = add(\overline{A})$ implies that the image of the functor $Hom_{D_{sg}(A)}(\overline{A}, -)$ is precisely the category $End_{D_{sg}(A)}(\overline{A})$-*proj*.

We have shown that

$$D_{sg}(A) \simeq (End_{D_{sg}(A)}(\overline{A}))\text{-}proj.$$

Lemma 6.9.23 now shows that $End_{D_{sg}(A)}(\overline{A}) \simeq \Gamma(A)$. The shift functor [1] on $D_{sg}(A)$ is mapped by the equivalence

$$D_{sg}(A) \simeq \Gamma(A)\text{-}proj$$

to some self-equivalence Σ_A of $\Gamma(A)$-*proj*. Since finitely generated $\Gamma(A)$-modules are semisimple, Lemma 6.9.24 implies that there is a unique triangulated structure on $\Gamma(A)$-*proj* with shift functor Σ_A. Unicity then proves the statement. □

Lemma 6.9.26 *In* $(\Gamma(A)\text{-}proj, \Sigma_A)$ *we have* $\Sigma_A^n \simeq K^n(A) \otimes_{\Gamma(A)} -$.

Proof Indeed, the equivalence

$$D_{sg}(A) \xrightarrow{\Psi} (\Gamma(A)\text{-}proj, \Sigma)$$

is induced by $Hom_{D_{sg}(A)}(\overline{A}, -)$ and maps $\overline{A}$ to $\Gamma(A)$. Since Σ_A is defined by the equation

$$\Sigma_A \circ Hom_{D_{sg}(A)}(\overline{A}, -) = Hom_{D_{sg}(A)}(\overline{A}, -) \circ [1],$$

and therefore

$$\Sigma_A^n \circ Hom_{D_{sg}(A)}(\overline{A}, -) = Hom_{D_{sg}(A)}(\overline{A}, -) \circ [n],$$

and since $D_{sg}(A) = add(\overline{A})$, the (additive) self-equivalence is completely determined by the image of $\overline{A}$. Now the image of $\overline{A}$ is

$$\Sigma_A^n(\Gamma(A)) = \Sigma_A^n(Hom_{D_{sg}(A)}(\overline{A}, \overline{A}) = Hom_{D_{sg}(A)}(\overline{A}, A[n]) = K^n(A),$$

and by the fact that Σ_A commutes with direct sums, an argument as in Watts' Theorem 3.3.16 shows that

$$\Sigma_A^n(-) = K^n(A) \otimes_{\Gamma(A)} -$$

as claimed. □

Proposition 6.9.27 $K^n(A) \in Pic(\Gamma(A))$.

Proof Indeed, Σ_A^n is a self-equivalence and therefore $K^n(A) \otimes_{\Gamma(A)} -$ is a self-equivalence as well. Hence, composing $\Sigma_A^n \circ \Sigma_A^m = \Sigma_A^{n+m}$ implies that

$$K^n(A) \otimes_{\Gamma(A)} K^m(A) \longrightarrow K^{n+m}(A)$$

is an isomorphism and therefore the inverse bimodule to $K^n(A)$ is $K^{-n}(A)$. This proves the statement. □

6.10 Examples

6.10.1 Blocks of Cyclic Defect

Recall from Definition 5.10.4 the definition of a Brauer tree algebra and from Theorem 5.10.37 that blocks of finite groups over an algebraically closed field and with cyclic defect group are Brauer tree algebras.

We shall prove a result due to Rickard that two Brauer tree algebras over the same field k and with the same number of isomorphism classes of simple modules and the same exceptional multiplicity have equivalent derived categories. More precisely:

Theorem 6.10.1 *Let k be a field and let A be a Brauer tree algebra over k with exceptional multiplicity μ and associated to a Brauer tree with e edges. Then $D^b(A) \simeq D^b(N_e^{\mu e+1})$, where $N_e^{e\mu+1}$ is the self-injective Nakayama algebra with e simple modules and nilpotency degree of the radical being $e\mu + 1$.*

Two Brauer tree algebras over the same base field k have equivalent derived categories if and only if they satisfy both of the following two conditions:

- *they have the same number of isomorphism classes e of simple modules, and*
- *they have the same exceptional multiplicity.*

Remark 6.10.2 Recall that Remark 5.10.10 shows that the Nakayama algebra with e isomorphism classes of simple modules and nilpotency degree of the radical being $\mu e + 1$ is actually a Brauer tree algebra associated to a star with exceptional multiplicity concentrated in the centre.

Proof Brauer tree algebras are symmetric, and hence an equivalence between derived categories between two Brauer tree algebras imply a stable equivalence between the two Brauer tree algebras (cf Theorem 6.9.12). Suppose we have proved that a Brauer tree algebra has a derived category equivalent to a Nakayama algebra $N_e^{e\mu+1}$. Then e is an invariant of a derived equivalence by Proposition 6.8.9 using Proposition 2.5.9. The exceptional multiplicity of Brauer tree algebras is an invariant under stable equivalences of Morita type by Lemma 5.7.6. Therefore we only need to show that a Brauer tree algebra has derived category equivalent to the derived category of a Nakayama algebra $N_e^{e\mu+1}$.

Let A be a Brauer tree algebra associated to a Brauer graph Γ and let v_0 be its exceptional vertex (if there is no exceptional vertex, choose an arbitrary vertex as the exceptional vertex). For each vertex v of Γ let d_v be the distance between v and v_0 in the graph. Denote by Γ_0 the set of vertices of Γ and by Γ_1 the set of edges of Γ. Concerning the distance in a graph we use the following definition of distances between vertices in a connected and finite graph.

Since Γ is connected, there is a chain of edges $e_1, e_2, \ldots, e_s$ such that $e_i \neq e_j$ for $i \neq j$ and such that

- e_1 is adjacent with the vertices v_0 and v_1, e_2 is adjacent with the vertices v_1 and v_2, and e_i is adjacent with v_{i-1} and v_i for all $i \in \{1, \ldots, s\}$
- $v_{s+1} = v$
- s is minimal with respect to these properties.

We then say that the distance $d(v, v_0)$ between v_0 and v is s. We put $d(v_0, v_0) = 0$.

We shall prove the result by induction on $d_\sum := \sum_{v\in\Gamma_0} d(v, v_0)$. It is clear that for a tree with e edges we have $d_\sum \geq e$, and that $d_\sum = e$ if and only if Γ is a star with exceptional vertex in the centre. This proves the result for $d_\sum = e$. In this case $\max(\{d(v, v_0) \mid v \in \Gamma_0\}) = 1$.

Let

$$d_{\max} := \max(\{d(v, v_0) \mid v \in \Gamma_0\}).$$

Since we may assume that $d_\Sigma \geq e + 1$, we get that $d_{\max} \geq 2$. All vertices v in $\{v \in \Gamma_0 | \; d(v, v_0) = d_{\max}\}$ have the property that there is only one edge e_v adjacent to v. Indeed, otherwise one of the edges leads to v_0 and the other edge would lead to a vertex of strictly greater distance. This shows by the definition of a Brauer tree algebra that the projective indecomposable module associated to e_v is uniserial.

Let $\hat{v} \in \{v \in \Gamma_0 | \; d(v, v_0) = d_{\max}\}$ be a vertex, with edge $\hat{e}$ adjacent to $\hat{v}$ and another vertex $\check{v}$. Within the set of all these vertices there is at least one such that in addition in the cyclic ordering of the edges at the vertex $\check{v}$ the edge $\check{e}$ immediately following $\hat{e}$ is adjacent to $\check{v}$ and some other vertex v' and that $d(v', v_0) = d_{\max} - 2$. Let S_f be the simple module associated to the edge f and denote by P_f the indecomposable projective A-module associated to the edge $f \in \Gamma_1$.

Then the beginning of a minimal projective resolution of $S_{\hat{e}}$ is

$$\cdots \longrightarrow P_{\check{e}} \stackrel{d}{\longrightarrow} P_{\hat{e}} \longrightarrow S_{\hat{e}} \longrightarrow 0.$$

Let

$$E_0 := \{f_1, f_2, \ldots, f_s, \check{e}, \hat{e}\}$$

be the edges of Γ which are adjacent with $\check{v}$ and let

$$E_1 := \Gamma_1 \setminus E_0.$$

Suppose that the edges adjacent to v' are

$$\check{e}, g_1, g_2, \dots, g_t$$

written in the cyclic ordering at v'.

Then the projective module $P_{\hat{e}}$ has composition series

$$\begin{pmatrix} S_{\hat{e}} \\ S_{\check{e}} \\ S_{f_1} \\ \vdots \\ S_{f_s} \\ S_{\hat{e}} \end{pmatrix}$$

and the module $\ker(d)$ is uniserial with composition series

$$\begin{pmatrix} S_{g_1} \\ S_{g_2} \\ \vdots \\ S_{g_t} \\ S_{\check{e}} \end{pmatrix}.$$

Let

$$P_1 := \bigoplus_{f \in \Gamma_1 \setminus \{\hat{e}\}} P_f.$$

Consider the two-term complex T

$$\cdots \longrightarrow 0 \longrightarrow (P_{\check{e}} \oplus P_1) \stackrel{(d\ 0)}{\longrightarrow} P_{\hat{e}} \longrightarrow 0 \longrightarrow \cdots$$

Lemma 6.6.5 shows that T is a tilting complex.

Let us consider $B := (End_{D^b(A)}(T))^{op}$. We see that the projective indecomposable B-modules are indecomposable direct factors of T. By Proposition 6.7.17, since A is symmetric, B is symmetric as well.

We hence consider the indecomposable direct factors of T. These are all the indecomposable projective A-modules except $P_{\hat{e}}$, plus $\hat{T}$, which replaces $P_{\hat{e}}$.

Associated to the vertex $\check{v}$ are the uniserial projective indecomposable modules

$$P_{\hat{e}}, P_{f_1}, P_{f_2}, \dots, P_{f_s}$$

and as A-modules the homomorphisms with maximal image between these modules are

$$P_{\hat{e}} \longleftarrow P_{f_1} \longleftarrow P_{f_2} \longleftarrow \cdots \longleftarrow P_{f_s} \longleftarrow P_{\check{e}} \longleftarrow P_{\hat{e}}$$

as is seen from the composition series of $P_{\hat{e}}$ displayed above. Now, $P_{\hat{e}}$ is not a direct factor of T. Any morphism $P_{f_i}[1] \longrightarrow \check{T}$ is 0, for every i. Any morphism $P_{\check{e}}[1] \longrightarrow \hat{T}$ has image in the socle of $\ker(d)$ and hence factors

$$P_{\check{e}}[1] \longrightarrow P_{g_1}[1] \longrightarrow \hat{T}.$$

Moreover, any morphism $\hat{T} \longrightarrow P_{f_i}[1]$ is (homotopic to) zero, since any morphism $P_{\check{e}} \longrightarrow P_{f_i}$ factors through $P_{\hat{e}}$.

Associated to the vertex v' we obtain the projective indecomposable A-modules

$$P_{\check{e}}, P_{g_1}, P_{g_2}, \ldots, P_{g_t}$$

and we assume they have been ordered so that we obtain morphisms with maximal image as follows:

$$P_{\check{e}} \longleftarrow P_{g_1} \longleftarrow P_{g_2} \longleftarrow \cdots \longleftarrow P_{g_t} \longleftarrow P_{\check{e}},$$

as is seen by the composition series of $\ker(d)$.

The morphism

$$\begin{array}{ccccccc} 0 \longrightarrow & P_{\check{e}} & \longrightarrow & P_{\hat{e}} & \longrightarrow 0 \\ & \downarrow \text{id} & & & \\ & P_{\check{e}} & & & \end{array}$$

does not factor through any other direct factor of T. Further, since $\ker(d)$ is uniserial with head S_{g_1}, there is a morphism

$$\begin{array}{ccccc} & P_{g_1} & & & \\ & \downarrow & & & \\ 0 \longrightarrow & P_{\check{e}} & \longrightarrow & P_{\hat{e}} & \longrightarrow 0 \end{array}$$

which does not factor through any other direct factor of T, since for any projective module P we get

$$Hom_{D^b(A)}(P[1], \hat{T}) = Hom_A(P[1], \ker(d)).$$

We obtain that the projective indecomposable B-module $\hat{Q}$ corresponding to the direct factor $\hat{T}$ is uniserial, has simple top denoted $S_{\hat{T}}$ and simple socle $S_{\hat{T}}$ and composition series

$$\begin{pmatrix} S_{\hat{T}} \\ S_{\check{e}} \\ S_{g_1} \\ S_{g_2} \\ \vdots \\ S_{g_t} \\ S_{\hat{T}} \end{pmatrix}.$$

The indecomposable projective B-module Q_{f_i} corresponding to the direct factors $P_{f_i}[1]$ of T are still uniserial with one composition factor less, namely $S_{\hat{e}}$ is no longer a composition factor, and except for this modification, the composition series remain identical.

The indecomposable projective B-module Q_{g_j} corresponding to the indecomposable factor P_{g_j} is exactly the same as for A, with one exception: The simple factor $S_{\hat{T}}$ is introduced below S_{g_t} and above $S_{\check{e}}$:

$$\mathrm{rad}(Q_{g_j})/\mathrm{soc}(Q_{g_j}) = \hat{U}_{g_j} \oplus \hat{V}_{g_j}$$

where

$$\mathrm{rad}(P_{g_j})/\mathrm{soc}(P_{g_j}) = U_{g_j} \oplus V_{g_j}$$

and where U_{g_j}, V_{g_j}, $\hat{U}_{g_j}$ and $\hat{V}_{g_j}$ are all uniserial. The composition series of V_{g_j} and of $\hat{V}_{g_j}$ are identical (identifying the simple A-modules and the simple B-modules in the canonical fashion). The composition series of U_{g_j} and $\hat{U}_{g_j}$ are

$$U_{g_j} \leftrightarrow \begin{pmatrix} S_{g_{j+1}} \\ S_{g_{j+2}} \\ \vdots \\ S_{g_t} \\ S_{\check{e}} \\ S_{g_1} \\ \vdots \\ S_{g_{j-1}} \end{pmatrix}, \quad \hat{U}_{g_j} \leftrightarrow \begin{pmatrix} S_{g_{j+1}} \\ S_{g_{j+2}} \\ \vdots \\ S_{g_t} \\ S_{\hat{T}} \\ S_{\check{e}} \\ S_{g_1} \\ \vdots \\ S_{g_{j-1}} \end{pmatrix}.$$

Similar considerations as for Q_{g_j} hold for the projective indecomposable B-module $Q_{\check{e}}$ associated to the direct factor $P_{\check{e}}$. All other projective indecomposable modules remain unchanged.

We have proved the following lemma which is interesting in its own right.

Proposition 6.10.3 (König and Zimmermann [33]) *Let A be a Brauer tree algebra, let S be a simple A-module with uniserial projective cover, and let T be the complex*

$$0 \longrightarrow P \oplus \hat{P} \stackrel{(d\ 0)}{\longrightarrow} Q \longrightarrow 0$$

so that

$$\cdots \longrightarrow P \stackrel{d}{\longrightarrow} Q \longrightarrow S \longrightarrow 0$$

is a minimal projective resolution of S, where $\hat{P}$ is the direct sum of all indecomposable projective A-modules (one copy for each isomorphism class) different from Q. Then T is a tilting complex with endomorphism ring being a Brauer tree algebra for the Brauer tree indicated as below.

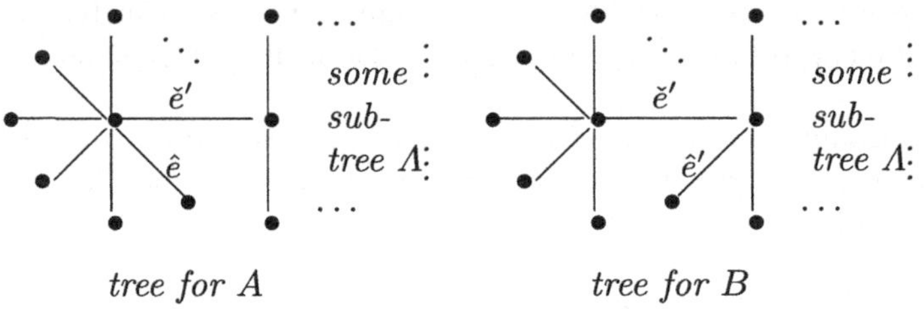

Here $\hat{e}$ is the edge corresponding to the simple module S, and $\hat{e}'$ its image in B under the derived equivalence and Λ is the same sub-tree for A and for B.

The proof of Proposition 6.10.3 has been given above. Lemma 6.10.3 now shows that the value $d_{\sum}$ for B is one smaller than the value $d_{\sum}$ for the algebra A:

$$d_{\sum}(B) = d_{\sum}(A) - 1.$$

Since $d_{\sum}(B)$ is smaller than $d_{\sum}(A)$, the induction hypothesis shows that $D^b(B) \simeq D^b(N_e^{e\mu+1})$. Since T is a tilting complex, Rickard's Theorem 6.5.1 shows that $D^b(A) \simeq D^b(B)$. This proves Theorem 6.10.1. □

Proposition 6.10.4 *Let A be an algebra and let B be a Brauer tree algebra. Then $D^b(A) \simeq D^b(B)$ implies that A is a Brauer tree algebra as well.*

Proof Indeed, since B is symmetric, A is symmetric as well by Proposition 6.7.17. Hence the existence of an equivalence $D^b(A) \simeq D^b(B)$ implies the existence of an equivalence of standard type, and this implies the existence of a stable equivalence of Morita type between A and B (cf Theorem 6.9.12). Now, Theorem 5.10.31 shows that A is a Brauer tree algebra. □

Remark 6.10.5 Proposition 6.10.3 was first proved in [33]. Many different approaches have been taken to prove Theorem 6.10.1. We single out only some of the different proofs.

The first proof was given by Rickard [1]. Later [33] gave the proof we have given here but for an "order" version of Brauer trees, called Green orders, developed by Roggenkamp. This approach, including a complete development of Green orders, is also given in detail in [11]. Note that in [11, page 76 line 15ff] a short argument using the uniqueness of a maximal order in a skew-field [34, Theorem 12.8] is missing. Further, Schaps and Zakay-Illouz gave an independent proof in [35].

Example 6.10.6 Let K be a field and let A and B be two finite dimensional K-algebras. Recall that a K-algebra D is called split semisimple if K is a splitting field for the semisimple K-algebra D. The derived category of a simple K-algebra D for which K is a splitting field is nothing but $\bigoplus_{i\in\mathbb{Z}} K\text{-}mod$ since any complex is isomorphic to its homology. Since $D^b(A \times B\text{-}mod) \simeq D^b(A\text{-}mod) \times D^b(B\text{-}mod)$, we get an analogous result for split semisimple K-algebras. Suppose K is a splitting field for A and for B. If $D^b(A) \simeq D^b(B)$ as triangulated categories, then we see that $D^b(A/\text{rad}(A)) \simeq D^b(B/\text{rad}(B))$ since $G_0(A) \simeq G_0(B)$ and the Grothendieck group completely determines the derived category of a split semisimple K-algebra.

We might ask if $D^b(A) \simeq D^b(B)$ implies $D^b(A/\text{rad}^s(A)) \simeq D^b(B/\text{rad}^s(B))$ for every $s \in \mathbb{Z}$. This is false in general and Brauer tree algebras give a counterexample.

Indeed, let A be the Brauer tree algebra associated to the graph

and without exceptional vertex.

Let B be the Brauer tree algebra associated to the graph

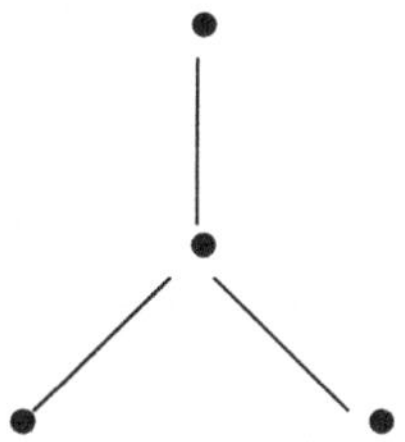

and without exceptional vertex.

The algebra A has the property $\text{rad}^3(A) = 0$ whereas $\text{rad}^4(B) = 0 \neq \text{rad}^3(B)$. Moreover $B/\text{rad}^3(B)$ is isomorphic to the Nakayama algebra N_3^3, which is non-symmetric, whereas $A/\text{rad}^3(A) = A$ is symmetric, as is any Brauer tree algebra. Proposition 6.7.17 then shows that

$$D^b(A/\text{rad}^3(A)) \not\simeq D^b(B/\text{rad}^3(B))$$

but nevertheless we know from Theorem 6.7.17 that

$$D^b(A) \simeq D^b(B).$$

6.10.2 Okuyama's Method

Perhaps the most striking and up to now most successful approach to producing derived equivalences between blocks of group rings is due to Okuyama [36].

Unfortunately the paper remains unpublished, however a description in Japanese appeared in [37].[1]

The approach was used by a group of authors to prove Broué's abelian defect conjecture in various cases. We cite here Okuyama, Koshitani, Kunugi, Müller and Waki [36, 38–45].

We follow Okuyama's paper [37] in the presentation here. Recall the construction of two-term tilting complexes from Lemma 6.6.5: Suppose A is a symmetric k-algebra for some algebraically closed field k. Let $I = \{1, \ldots, n\}$ and suppose that $S_1, \ldots, S_n$ is a set of representatives of the isomorphism classes of simple A-modules, then for each proper subset I_0 of I we constructed a two-term tilting complex $T(I_0)$.

The Method

Let G be a finite group, let k be an algebraically closed field of characteristic $p > 0$, and let B be a block of kG with defect group D. Let $H := N_G(D)$ and let b be the Brauer correspondent of B in kH. We want to show that $D^b(B) \simeq D^b(b)$. As above let $S_1, \ldots, S_n$ be a set of representatives of the isomorphism classes of simple A-modules.

The method uses the fact that in many cases it is possible to find a stable equivalence of Morita type

$$M_0 \otimes_B - : B\text{-}\underline{mod} \simeq b\text{-}\underline{mod}.$$

If the defect group is a trivial intersection group (see the introductory remarks of Sect. 5.1.1), a stable equivalence of Morita type is simply given by the restriction functor, hence just the Green correspondence. Let $g := M_0 \otimes_B -$ be a stable equivalence of Morita type between B and b.

If g is really the Green correspondence the Green correspondents $g(S_i)$ of S_i in kH for all $i \in \{1, \ldots, n\}$ can be determined, admittedly sometimes with quite some effort, but it is a classical task. The Green correspondent $g(S_i)$ is the unique direct factor of the restriction $S_i \downarrow_H^G$ with the same vertex D as S_i, and by Proposition 2.4.3 the Green correspondent $g(S_i)$ is a b-module. Frequently much is known about the Green correspondents in concrete examples. The corresponding work is usually quite technical.

Then Okuyama chooses some $I_0^{(1)} \subseteq I$, which, by Lemma 6.6.5, induces an equivalence $F_1 : D^b(b) \simeq D^b(A_1)$ for $A_1 := (End_{D^b(b)}(T(I_0^{(1)})))^{op}$. Moreover, Theorem 6.9.12 implies that F_1 induces a stable equivalence of Morita type given by the pair of bimodules $({}_{A_1}M_b^{(1)}, {}_bN_{A_1}^{(1)})$. We observe that since $T(I_0^{(1)})$ is a two-term tilting complex, the module $M^{(1)}$ is given by the bimodule X_1 in Corollary 6.6.1.

Compute $\tilde{L}_i^{(1)} := M^{(1)} \otimes_b g(S_i)$ and observe that by Lemma 5.6.3 and Proposition 5.4.3 for each $i \in \{1, \ldots, n\}$ there is a unique indecomposable non-projective direct factor $L_i^{(1)}$ of $\tilde{L}_i^{(1)}$.

[1] S. Koshitani kindly translated Okuyama's paper [37] for me. I am very grateful to him.

The aim is to obtain $L_i^{(1)}$ of a simpler Loewy structure than S_i, at least for one i_0, and not to get more complicated Loewy structure for S_i for any of the $i \in I \setminus \{i_0\}$.

Then, choosing $I_0^{(2)} \subset I$, we construct the two-term tilting complex $T(I_0^{(2)})$ with endomorphism ring A_2 and the A_2–A_1-bimodule $M^{(2)}$. We again compute the module $L_i^{(2)}$ as the unique non-projective direct summand of $M^{(2)} \otimes_{A_1} L_i^{(1)}$, again with the goal of simplifying the Loewy structure of $L_i^{(2)}$ with respect to the Loewy structure of $L_i^{(1)}$.

Iteratively, choosing subsets $I_0^{(1)}, I_0^{(2)}, \dots, I_0^{(s)}$ for some s, we obtain algebras $A_1, A_2, \dots, A_s$ such that

$$D^b(b) \simeq D^b(A_1) \simeq \cdots \simeq D^b(A_s).$$

Moreover, $L_i^{(s)}$ is the image of $g(S_i)$ under the stable equivalence of Morita type induced by the composition $D^b(b) \xrightarrow{F} D^b(A_s)$ of these equivalences of triangulated categories. Recall the natural functor $D^b(A\text{-}mod) \longrightarrow A\text{-}\underline{mod} = D_{sg}(A)$ for every self-injective algebra A (cf Proposition 6.9.3 and Proposition 6.9.8).

We obtain the following commutative diagram of functors between categories.

$$\begin{array}{ccccc} & & D^b(b\text{-}mod) & \xrightarrow{F} & D^b(A_s\text{-}mod) \\ & & \downarrow & & \downarrow \\ B\text{-}\underline{mod} & \xrightarrow{g} & b\text{-}\underline{mod} & \xrightarrow{\underline{F}} & A_s\text{-}\underline{mod} \end{array}$$

Suppose now we managed to get that $L_i^{(s)}$ is a simple A_s-module. Then, by Proposition 5.6.5, we get that $\underline{F} \circ g$ is a Morita equivalence. Hence B and A_s are Morita equivalent, and $D^b(b) \simeq D^b(A_s)$ by construction of the tilting complexes $T^{(\ell)}$ for $\ell \in \{1, \dots, s\}$. Hence

$$D^b(b) \simeq D^b(B)$$

in this case. We summarise Okuyama's criterion in the following proposition.

Proposition 6.10.7 (Okuyama [36]). *Let B and b be two symmetric k-algebras over an algebraically closed field k. Suppose*

$$B\text{-}\underline{mod} \xrightarrow{g} b\text{-}\underline{mod}$$

is a stable equivalence of Morita type, and suppose that there are algebras $b = A_0, A_1, \dots, A_s$ and standard equivalences

$$F_\ell : D^b(A_{\ell-1}) \longrightarrow D^b(A_\ell)$$

for each $\ell \in \{1, \dots, s\}$ which induce stable equivalences of Morita type $\underline{F_\ell} : A_{\ell-1}\text{-}\underline{mod} \longrightarrow A_\ell\text{-}\underline{mod}$. If now for each simple B-module S the indecompos-

able non-projective module $(\underline{F}_\ell \circ \underline{F}_{\ell-1} \circ \dots \underline{F}_1 \circ g)(S)$ *is again simple, then* $D^b(B) \simeq D^b(b)$.

Remark 6.10.8 We assumed that the field is algebraically closed. A weaker assumption is sufficient, for example we could assume instead that the base field is perfect.

Okuyama did not use Keller's construction of Corollary 6.6.1 to obtain the image of the modules $g(S_i)$ under a stable equivalence induced by some tilting complex. Instead another computation was proposed.

Lemma 6.10.9 (Okuyama [36]) *Let* k *be an algebraically closed field and let* A_1 *and* A_2 *be symmetric* k*-algebras. Let*

$$F = Y \otimes^{\mathbb{L}}_{A_2} - : D^b(A_2) \longrightarrow D^b(A_1)$$

be an equivalence of standard type between triangulated categories, and let $T = F(A_2)$ *be a tilting complex in* $K^b(A_1\text{-}proj)$ *with endomorphism ring* A_2^{op}. *Let* ${}_{A_2}M_{A_1}$ *be an* A_2–A_1*-bimodule such that* F^{-1} *induces the stable equivalence*

$$M \otimes_{A_1} - : A_1\text{-}\underline{mod} \longrightarrow A_2\text{-}\underline{mod}.$$

Then for all A_1*-modules* X *with projective resolution* P_X *we get that*

$$H^m(Hom_{A_1}(T, X)) = 0 \;\forall\, m \neq n_0$$

implies

$$\Omega^{n_0}(Hom_{K^{-,b}(A_1\text{-}proj)}(T, P_X[n_0])) \simeq M \otimes_{A_1} X$$

in the stable category of A_1*-modules.*

Proof By Proposition 6.4.4 and Proposition 6.5.5 we may first suppose that Y is a two-sided tilting complex in $D^b(A_1 \otimes_K A_2^{op})$ isomorphic to T in $K^{-,b}(A_2\text{-}proj)$, with all homogeneous components projective as bimodules, except the highest degree in which the homogeneous component is projective as a left and as a right module. Then we may replace $\mathbb{R}Hom_{A_2}(Y, -)$ by $Hom_{A_2}(Y, -)$ and $Y \otimes^{\mathbb{L}}_{A_1} -$ by $Y \otimes_{A_1} -$ and

$$Hom_{K^{-,b}(A_1\text{-}proj)}(T, X[n_0]) \simeq Hom_{K^{-,b}(A_1\text{-}proj)}(Y, X[n_0]).$$

Moreover, $Hom_{K^{-,b}(A_1\text{-}proj)}(Y, -)$ is right adjoint to $Y \otimes_{A_2} -$ by Proposition 3.7.16.

Since $Y \otimes_{A_2} -$ is an equivalence, its right adjoint $Hom_{K^{-,b}(A_1\text{-}proj)}(Y, -)$ is a quasi-inverse equivalence. Therefore

$$Hom_{K^{-,b}(A_1\text{-}proj)}(T, X[n_0]) = F^{-1}(X[n_0]).$$

Now, denote by

$$\nu_{A_2} : D^b(A_2\text{-}mod) \longrightarrow A_2\text{-}\underline{mod}$$

and

$$\nu_{A_1} : D^b(A_1\text{-}mod) \longrightarrow A_1\text{-}\underline{mod}$$

the natural functors of triangulated categories. We get

$$\nu_{A_2} \circ F^{-1} = \underline{F}^{-1} \circ \nu_{A_1} \text{ and } \Omega^{-1} \circ \nu_{A_2} = \nu_{A_2} \circ [1].$$

Therefore

$$\begin{aligned}\nu_{A_2}(Hom_{K^{-,b}(A_1\text{-}proj)}(T, X[n_0])) &\simeq \nu_{A_2}(Hom_{K^{-,b}(A_1\text{-}proj)}(Y, X[n_0])) \\ &= \nu_{A_2}(F^{-1}(X[n_0])) \\ &= \Omega^{-n_0}(\nu_{A_2}(F^{-1}(X))).\end{aligned}$$

By hypothesis, $H^m(Hom_A(T, X)) = 0$ if $m \neq n_0$. This shows that the complex $Hom_A(T, X)$ is isomorphic to its homology, and therefore

$$\begin{aligned}\nu_{A_2}(Hom_A(T, X)) &\simeq \nu_{A_2}(H^{n_0}(Hom_A(T, X))) \\ &= \nu_{A_2}(H^{n_0}(\mathbb{R}Hom_A(T, X))) \\ &\simeq \nu_{A_2}(Hom_{D^b(A_1)}(T, X[n_0])) \\ &= \nu_{A_2}(F^{-1}(X[n_0])) \\ &\simeq M \otimes_{A_1} \nu_{A_1}(X[n_0]) \\ &\simeq \Omega^{-n_0}(M \otimes_{A_1} \nu_{A_1}(X)).\end{aligned}$$

This proves the statement. □

An Example of Okuyama's Method

Okuyama tried the method himself in various cases. In all cases a quite detailed knowledge of the structure of the involved group rings is necessary. Since we haven't developed these rather specific examples in such detail, in this section we refer to the original treatments, simply citing how the sets are chosen.

Example 6.10.10 The alternating group A_7 of degree 7 and A_6 of degree 6 both have Sylow 3 subgroups isomorphic to $P := C_3 \times C_3$. Let k be an algebraically closed field of characteristic 3. All Sylow subgroups are trivial intersection subgroups.

Moreover,

$$N_{A_6}(P)/C_{A_6}(P) \simeq C_4 \simeq N_{A_7}(P)/C_{A_7}(P).$$

We observe that the conjugacy classes of A_6 are represented by the elements

3-singular	3-regular
(123), (123)(456),	1, (12)(34),
	(1234)(56), (12345), (12346)

and the conjugacy classes of A_7 are represented by the elements

3-singular	3-regular
(123), (123)(45)(67), (123)(456),	1, (12)(34), (1234)(56),
	(12345), (1234567), (1234576)

where we call an element 3-regular if its order is not divisible by 3, and 3-singular otherwise. Therefore A_6 has five 3-regular conjugacy classes and A_7 has six 3-regular conjugacy classes and so

$$K_0(kA_7) \simeq \mathbb{Z}^6 \text{ and } K_0(kA_6) \simeq \mathbb{Z}^5.$$

It is known that the principal blocks of kA_6 and of kA_7 both have four simple modules. We start with kA_7. The Green correspondents $g(S_i)$ of the simple modules S_i for $i \in \{1, 2, 3, 4\}$ are simple for $i \neq 2$ and

$$g(S_1) = S_1' \ , \ g(S_2) = \begin{matrix} & S_2' & \\ S_1' & & S_3' \\ & S_2' & \end{matrix} \ , \ g(S_3) = S_3' \ , \ g(S_4) = S_4'$$

which means that $g(S_2)/\mathrm{rad}(g(S_2)) \simeq S_2'$, $\mathrm{rad}(g(S_2))/\mathrm{rad}^2(g(S_2)) \simeq S_1' \oplus S_3'$, $\mathrm{rad}^2(g(S_2))/\mathrm{rad}^3(g(S_2)) \simeq S_2'$, and $\mathrm{rad}^3(g(S_2)) = 0$. Then choosing $I_0 = \{2\}$ in Okuyama's method gives that the images of each of the $g(S_i)$ under the stable equivalence induced by this complex are simple.

The case of kA_6 is slightly more complicated. The Green correspondents of the four simple kA_6-modules T_1, T_2, T_3, T_4 are

$$g(T_1) = T_1' \ , \ g(T_2) = \begin{matrix} T_2' \\ T_1' \\ T_3' \end{matrix} \ , \ g(T_3) = \begin{matrix} T_3' \\ T_1' \\ T_2' \end{matrix} \ , \ g(T_4) = \begin{matrix} & T_4' & \\ T_2' & & T_3' \\ & T_4' & \end{matrix}$$

Taking first $I_0 = \{4\}$ in Okuyama's method, the stable equivalence of Morita type maps the Green correspondents $g(T_i)$ to $X^{(1)}(T_i)$ where

$$X^{(1)}(T_1) = T_1'' \ , \ X^{(1)}(T_2) = \begin{matrix} T_2'' \\ T_1'' \\ T_3' \end{matrix} \ , \ X^{(1)}(T_3) = \begin{matrix} T_3'' \\ T_1'' \\ T_2'' \end{matrix} \ , \ X^{(1)}(T_4) = T_4''.$$

Now, in a second step, taking $I_0 = \{1, 3\}$ we get that the stable equivalence induced by this new derived equivalence applied to these images gives modules $X^{(2)}(T_i)$ where

$$X^{(2)}(T_1) = T_1''' \ , \ X^{(2)}(T_2) = T_2''' \ , \ X^{(2)}(T_3) = T_3''' \ , \ X^{(2)}(T_4) = T_4''' .$$

Hence the Broué conjecture for principal blocks is true for A_6 and A_7 in characteristic 3 and also, since

$$N_{A_6}(P)/C_{A_6}(P) \simeq C_4 \simeq N_{A_7}(P)/C_{A_7}(P),$$

the principal blocks of kA_6 and kA_7 are derived equivalent.

6.10.3 Derived Equivalences Defined by Simple Modules

As we have seen in Sect. 6.10.2 the images of simple modules under a stable equivalence, and a fortiori also under an equivalence between derived categories, provide important information. In [46] Rickard gave a criterion for the existence of an equivalence between the derived categories of two symmetric algebras A and B in terms of the possible images of simple B-modules under such an equivalence. The images of indecomposable projective modules are the indecomposable direct factors of a tilting complex, as was shown in Proposition 6.1.5. We shall follow the ideas of [46] in this section.

Throughout let k be a field and let A and B be finite dimensional symmetric k-algebras. We collect some properties of simple B-modules. Let Σ be a set of representatives of the isomorphism classes of the simple B-modules. Then

- $End_B(S)$ is a skew-field for each $S \in \Sigma$.
- $Hom_B(S_1, S_2) = 0$ for each $S_1, S_2 \in \Sigma$ whenever $S_1 \not\simeq S_2$.
- The smallest triangulated subcategory of $D^b(B)$ containing Σ is $D^b(B)$.
- $Hom_{D^b(B)}(S_1, S_2[m]) = 0$ for each $S_1, S_2 \in \Sigma$ and any integer $m < 0$.

The first two items are just Schur's lemma, the third item comes from the fact that any module has a finite composition series, and the last item is just the observation that negative Ext-groups between modules are 0.

Now, each of these items remains invariant under an equivalence $F : D^b(B) \to D^b(A)$ of triangulated categories. Let $\mathcal{X} := \{F(S) \mid S \in \Sigma\}$, then

1. $End_{D^b(A)}(X)$ is a skew-field for each $X \in \mathcal{X}$.
2. $Hom_{D^b(A)}(X_1, X_2) = 0$ for each $X_1, X_2 \in \mathcal{X}$ whenever $X_1 \not\simeq X_2$.
3. The smallest triangulated subcategory of $D^b(A)$ containing $\mathcal{X}$ is $D^b(A)$.
4. $Hom_{D^b(A)}(X_1, X_2[m]) = 0$ for each $X_1, X_2 \in \mathcal{X}$ and any integer $m < 0$.

Remark 6.10.11 König and Yang [47] call systems $\mathcal{X}$ of objects in a triangulated category satisfying the properties 1–4 above a simple minded collection. They show that this concept is essentially the same as what was called by Aihara and Iyama a silting object [48]. Further, these correspond to t-structures on the derived category. In 1988 a connection between silting objects and t-structures was discovered by

Keller and Vossieck [49]. Aihara and Iyama were mainly interested in links to cluster categories, whereas König and Yang developed their concept with the stable category in mind and in particular the Auslander-Reiten conjecture.

The main result of this section is the following.

Proposition 6.10.12 (Rickard [46]) *Let K be a field, and let A be a finite dimensional symmetric K-algebra. Let $\mathcal{X}$ be a set of objects in $D^b(A)$ satisfying the properties* 1, 2, 3, 4, *and suppose that the cardinality of $\mathcal{X}$ is equal to the $\mathbb{Z}$-rank of the Grothendieck group of A.*

Then there is a finite dimensional symmetric K-algebra B and an equivalence $F : D^b(B) \simeq D^b(A)$ of triangulated categories such that, for each simple B-module S, we get $F(S)$ is isomorphic to an element in $\mathcal{X}$ and each element in $\mathcal{X}$ is isomorphic to the image of a simple B-module under F.

Proof We shall divide the proof into several steps. We shall first recursively construct a tilting complex T in $D^b(A)$. For each $X \in \mathcal{X}$ put $X^{(0)} := X$.

Suppose we have constructed $X^{(n-1)}$ for each $X \in \mathcal{X}$. For each $Y \in \mathcal{X}$ choose an $End_{D^b(A)}(X)$-basis $B_X^{(n-1)}(Y,t)$ of $Hom_{D^b(A)}(Y[t], X^{(n-1)})$. Put

$$Z_X^{(n-1)}(Y,t) := \bigoplus_{\alpha \in B_X^{(n-1)}(Y,t)} Y[t]$$

and since for each $\alpha \in B_X^{(n-1)}(Y,t)$ we canonically obtain a morphism $\alpha : Y[t] \longrightarrow X^{(n-1)}$, by the universal property of the direct sum we get a morphism

$$\alpha_X^{(n-1)}(Y,t) : Z_X^{(n-1)}(Y,t) \longrightarrow X^{(n-1)},$$

which is just α on the component indexed by α.

Put

$$Z_X^{n-1} := \bigoplus_{t<0;\, Y \in \mathcal{X}} Z_X^{(n-1)}(Y,t).$$

Again by the universal property of the direct sum, this in turn induces a homomorphism

$$\alpha_X^{(n-1)} : Z_X^{n-1} \longrightarrow X^{(n-1)}.$$

Define

$$X^{(n)} := cone(\alpha_X^{(n-1)})$$

and hence obtain a distinguished triangle

$$Z_X^{(n-1)} \stackrel{\alpha_X^{(n-1)}}{\longrightarrow} X^{(n-1)} \stackrel{\beta_X^{(n-1)}}{\longrightarrow} X^{(n)} \longrightarrow Z_X^{(n-1)}[1].$$

The following properties of the construction are crucial and will be used in the sequel.

- We observe that by construction $Z_X^{(n-1)}$ is a possibly infinite direct sum of objects $Y[t]$ for $Y \in \mathcal{X}$ and $t < 0$.
- Moreover, also by construction,

$$Hom_{D(A)}(Y[t], Z_X^{(n-1)}) \stackrel{Hom_{D(A)}(Y[t],\alpha_X^{(n-1)})}{\longrightarrow} Hom_{D(A)}(Y[t], X^{(n-1)})$$

is surjective for all $t < 0$ and $Y \in \mathcal{X}$.
- We claim that

$$Hom_{D(A)}(Y[-1], Z_X^{(n-1)}) \stackrel{Hom_{D(A)}(Y[-1],\alpha_X^{(n-1)})}{\longrightarrow} Hom_{D(A)}(Y[-1], X^{(n-1)})$$

is an isomorphism for all $Y \in \mathcal{X}$.

We shall need to prove the third property. By hypothesis we get for all $Y, Y' \in \mathcal{X}$

$$Hom_{D(A)}(Y[-1], Y'[t]) = Hom_{D(A)}(Y, Y'[t+1]) = 0$$

whenever $Y \not\simeq Y'$ or $t \neq -1$. Hence,

$$\begin{aligned} Hom_{D(A)}(Y[-1], Z_X^{(n-1)}) &= Hom_{D(A)}(Y[-1], \bigoplus Y[-1]) \\ &= \bigoplus End_{D(A)}(Y[-1]) \\ &= \bigoplus End_{D(A)}(Y) \end{aligned}$$

where the sums run over a basis of $Hom_{D(A)}(Y[-1], X^{(n-1)})$ as a vector space over the skew-field $End_{D(A)}(Y)$. Each of the copies of $End_{D(A)}(Y)$ is mapped to the basis element of $Hom_{D(A)}(Y[-1], X^{(n-1)})$ to which it corresponds. Hence, the map $Hom_{D(A)}(Y[-1],\alpha_X^{(n-1)})$ identifies the $End_{D(A)}(Y)$-vector space $Hom_{D(A)}(Y[-1], Z_X^{(n-1)})$ with tuples of $End_{D(A)}(Y)$ after having chosen a basis. This proves the third item.

Now, define the endomorphism

$$\beta_X := \bigoplus_{n=0}^{\infty} \beta_X^{(n)}$$

of $\bigoplus_{n=0}^{\infty} X^{(n)}$. Then $T_X := cone(id - \beta_X)$ so that we get a distinguished triangle

$$\bigoplus_{n=0}^{\infty} X^{(n)} \stackrel{id-\beta_X}{\longrightarrow} \bigoplus_{n=0}^{\infty} X^{(n)} \longrightarrow T_X \longrightarrow \bigoplus_{n=0}^{\infty} X^{(n)}[1].$$

Recall our conventions for limits in a category from Definition 3.1.16 and Proposition 3.1.18.

Claim 6.10.13 For each object U of $D^-(A\text{-}mod)$ the space $Hom_{D(A)}(U, T_X)$ is canonically isomorphic to the colimit of the direct system given by the modules $Hom_{D(A)}(U, X^{(n)})$ and the maps $Hom_{D^-(A)}(U, \beta_X^{(n)})$.

Proof of Claim 6.10.13 By hypothesis $\mathcal{X}$ is a finite set of objects in $D^b(A)$. Hence there is an integer n_0 such that $H_m(X) = 0$ for each m with $|m| > n_0$. But this shows that $H_m(Z_X^{(n)}) = 0$ whenever $m > n_0$ since $Z_X^{(n)}$ is a direct sum of copies of $Y[t]$ for $Y \in \mathcal{X}$ and $t < 0$. Further, we can replace each object $X^{(n)}$ by its injective coresolution by Proposition 3.5.43. Since A is symmetric, injective and projective modules coincide, so that each $X^{(n)}$ can be presented by a left bounded complex of projective modules. Again by Proposition 3.5.43 we may replace U by its projective resolution, so that U can be assumed to be a right bounded complex of finitely generated projective A-modules. We observe in a first step that the natural morphism gives an isomorphism

$$Hom_{D(A)}(U, \bigoplus_{n\in\mathbb{N}} X^{(n)}) \simeq \bigoplus_{n\in\mathbb{N}} Hom_{D(A)}(U, X^{(n)}).$$

Indeed, in order to compute the homomorphisms from U to $\bigoplus_{n\in\mathbb{N}} X^{(n)}$ we consider the stupidly truncated U at degree n_0, to obtain the complex $\sigma_{\geq n_0+1}U$ (cf Remark 3.5.22). We see that

$$Hom_{D(A)}(U, \bigoplus_{n\in\mathbb{N}} X^{(n)}) \simeq Hom_{D(A)}(\sigma_{\geq n_0+1}U, \bigoplus_{n\in\mathbb{N}} X^{(n)})$$

since in degrees greater than n_0 all homogeneous components of $X^{(n)}$ are 0. But, $\sigma_{\geq n_0+1}U$ is a compact object by Proposition 6.1.2 and so

$$\begin{aligned} Hom_{D(A)}(U, \bigoplus_{n\in\mathbb{N}} X^{(n)}) &\simeq Hom_{D(A)}(\sigma_{\geq n_0+1}U, \bigoplus_{n\in\mathbb{N}} X^{(n)}) \\ &\simeq \bigoplus_{n\in\mathbb{N}} Hom_{D(A)}(\sigma_{\geq n_0+1}U, X^{(n)}) \\ &\simeq \bigoplus_{n\in\mathbb{N}} Hom_{D(A)}(U, X^{(n)}). \end{aligned}$$

Now, we apply $Hom_{D(A)}(U, -)$ to the distinguished triangle

$$\bigoplus_{n=0}^{\infty} X^{(n)} \stackrel{id-\beta_X}{\longrightarrow} \bigoplus_{n=0}^{\infty} X^{(n)} \longrightarrow T_X \longrightarrow \bigoplus_{n=0}^{\infty} X^{(n)}[1]$$

to obtain a long exact sequence

$$(U, \bigoplus_n X^{(n)}[m]) \to (U, \bigoplus_n X^{(n)}[m]) \to (U, T_X[m]) \to (U, \bigoplus_n X^{(n)}[m+1])$$

where we abbreviate $(-,-) := Hom_{D(A)}(-,-)$. However, in three of the four terms shown above, taking morphism spaces $(U,-)$ commutes with the direct sums, and it is clear that the leftmost morphism is injective for all m. In particular, since the case $m = 1$ is injective, this implies that the long exact sequence gives rise to a short exact sequence

$$\begin{array}{ccccccccc} 0 \to & (U, \bigoplus_{n=0}^{\infty} X^{(n)}) & \to & (U, \bigoplus_{n=0}^{\infty} X^{(n)}) & \to & (U, T_X) & \to 0 \\ & \| & & \| & & & \\ 0 \to & \bigoplus_{n=0}^{\infty}(U, X^{(n)}) & \to & \bigoplus_{n=0}^{\infty}(U, X^{(n)}) & \to & \mathrm{colim}_n (U, X^{(n)}) & \to 0 \end{array}$$

where the rightmost lower entry is the colimit of the direct system by Proposition 3.1.18. Unicity of the cokernel proves the claim. □

Claim 6.10.14 For each integer $m \in \mathbb{Z}$ and all $X, Y \in \mathcal{X}$ we get that $Hom_{D(A)}(Y, T_X[m]) = 0$ if $X \not\simeq Y$ or $m \neq 0$. Moreover $Hom_{D(A)}(X, T_X) = End_{D^b(A)}(X)$.

Proof of Claim 6.10.14 We apply $Hom_{D(A)}(Y,-)$ to the distinguished triangle

$$Z_X^{(n-1)} \stackrel{\alpha_X^{(n-1)}}{\longrightarrow} X^{(n-1)} \stackrel{\beta_X^{(n-1)}}{\longrightarrow} X^{(n)} \longrightarrow Z_X^{(n-1)}[1].$$

We use further that

$$Hom_{D(A)}(Y, Z_X^{(n-1)}[m]) = 0$$

for all $m \le 0$, by hypothesis and by the fact that $Z_X^{(n-1)}[m]$ is a direct sum of right-shifted copies of Y' for $Y' \in \mathcal{X}$. Therefore

$$Hom_{D(A)}(Y, X^{(n-1)}) \longrightarrow Hom_{D(A)}(Y, X^{(n)})$$

is injective and

$$Hom_{D(A)}(Y, X^{(n-1)}[m]) \longrightarrow Hom_{D(A)}(Y, X^{(n)}[m])$$

is an isomorphism for $m < 0$.

Since

$$Hom_{D(A)}(Y[-1], Z_X^{(n-1)}) \stackrel{Hom_{D(A)}(Y[-1], \alpha_X^{(n-1)})}{\longrightarrow} Hom_{D(A)}(Y[-1], X^{(n-1)})$$

is an isomorphism for all $Y \in \mathcal{X}$,

$$Hom_{D(A)}(Y, X^{(n-1)}) \longrightarrow Hom_{D(A)}(Y, X^{(n)})$$

is also surjective. Now, for $m \le 0$ we get

$$Hom_{D^b(A)}(Y, X^{(0)}[m]) = Hom_{D^b(A)}(Y, X[m]) = 0$$

unless $X \simeq Y$ and $m = 0$. In this case

$$Hom_{D^b(A)}(Y, X^{(0)}[m]) = End_{D(A)}(X).$$

By Claim 6.10.14 we obtain the statement for $m \leq 0$. If $m > 0$, then by the second item of the three properties preceding Claim 3.1.18,

$$Hom_{D(A)}(Y[t], Z_X^{(n-1)}) \overset{Hom_{D(A)}(Y[t],\alpha_X^{(n-1)})}{\longrightarrow} Hom_{D(A)}(Y[t], X^{(n-1)})$$

is surjective for all $t = -m < 0$ and $Y \in \mathcal{X}$. This implies the claim. □

Claim 6.10.15 For all $X \in \mathcal{X}$ we get that T_X is a compact object in $D(A)$. In particular T_X is isomorphic to a bounded complex of finitely generated projective A-modules.

Proof of Claim 6.10.15 By Claim 6.10.14 we get that the infinite direct sum $\bigoplus_{m\in\mathbb{Z}} Hom_{D(A)}(Y, T_X[m])$ is a finite dimensional $End_{D(A)}(Y)$-vector space for each $Y \in \mathcal{X}$. Since $Y \in \mathcal{X}$ is an object of $D^b(A\text{-}mod)$, the skew-field $End_{D(A)}(Y)$ is finite dimensional over K. Hence $\bigoplus_{m\in\mathbb{Z}} Hom_{D(A)}(Y, T_X[m])$ is a finite dimensional K-vector space for each $Y \in \mathcal{X}$. The class of objects Y in $D(A)$ for which $\bigoplus_{m\in\mathbb{Z}} Hom_{D(A)}(Y, T_X[m])$ is a finite dimensional K-vector space forms a full triangulated subcategory of $D(A)$. Since $\mathcal{X}$ generates $D^b(A\text{-}mod)$ as a triangulated category for all Y in $D^b(A\text{-}mod)$ we get that $\bigoplus_{m\in\mathbb{Z}} Hom_{D(A)}(Y, T_X[m])$ is a finite dimensional K-vector space.

In particular, we may choose $Y = A$, and obtain that

$$\bigoplus_{m\in\mathbb{Z}} Hom_{D(A)}(A, T_X[m]) = \bigoplus_{m\in\mathbb{Z}} H_m(T_X)$$

is a finite dimensional K-vector space. Therefore T_X is an object of the bounded derived category $D^b(A\text{-}mod)$ of finite dimensional A-modules. We may choose for Y a simple A-module S, and denote by I_S the injective hull of S. Then $\bigoplus_{m\in\mathbb{Z}} Hom_{D(A)}(S, T_X[m])$ is a finite dimensional K-vector space implies that I_S occurs only a finite number of times in an injective coresolution of T_X. Therefore, the injective coresolution of T_X is a bounded complex of finitely generated injective modules, hence a bounded complex of finitely generated projective modules since A is symmetric, whence self-injective. This proves the claim. □

Using Lemma 1.10.30 for Claim 6.10.14 we obtain

Claim 6.10.16 For each integer $m \in \mathbb{Z}$ and all $X, Y \in \mathcal{X}$ we get that $Hom_{D(A)}(T_X[m], Y) = 0$ unless $X \simeq Y$ and $m = 0$. If $m = 0$ and $X \simeq Y$ we get that $Hom_{D(A)}(T_X[m], Y) = End_{D^b(A)}(X)$ is a skew-field. □

Claim 6.10.17 $Hom_{D(A)}(T_X, T_Y[m]) = 0$ for all $X, Y \in \mathcal{X}$ and $m \in \mathbb{Z} \setminus \{0\}$.

Proof of Claim 6.10.17 By Claim 6.10.16 we get $Hom_{D(A)}(T_X, Y'[m]) = 0$ for each $m < 0$ and $X, Y' \in \mathcal{X}$. Since $Z_Y^{(n-1)}$ is a direct sum of negatively shifted copies of the object $Y' \in \mathcal{X}$, we get $Hom_{D(A)}(T_X, Z_Y^{(n-1)}) = 0$ for all $X, Y \in \mathcal{X}$. We now apply the functor $Hom_{D(A)}(T_X, -)$ to the distinguished triangle

$$Z_Y^{(n-1)} \xrightarrow{\alpha_Y^{(n-1)}} Y^{(n-1)} \xrightarrow{\beta_Y^{(n-1)}} Y^{(n)} \longrightarrow Z_Y^{(n-1)}[1].$$

Denote $Hom_{D(A)}(T_X, -) =: (T_X, -)$ for the moment. For every $m < 0$ this distinguished triangle yields an exact sequence

$$(T_X, Y^{(n-1)}[m]) \to (T_X, Y^{(n)}[m]) \to (T_X, Z_Y^{(n)}[m+1])$$

where the rightmost term is 0, as we have seen. Induction on n shows that $Hom_{D(A)}(T_X, Y^{(n)}[m]) = 0$ for all $m < 0$. Claim 6.10.13 then shows that $Hom_{D(A)}(T_X, T_Y[m]) = 0$ whenever $m < 0$. Using again that A is symmetric and Lemma 1.10.13, $Hom_{D(A)}(T_Y[m], T_X) = 0$ for $m < 0$, and therefore $Hom_{D(A)}(T_X, T_Y[m]) = 0$ for all $m \neq 0$. This proves the claim. □

Claim 6.10.18 For each non-zero object C of $D^-(A)$ there exists $m \in \mathbb{Z}$ and $X \in \mathcal{X}$ such that $Hom_{D^-(A)}(C, T_X[m]) \neq 0$.

Proof of Claim 6.10.18 First, each $X \in \mathcal{X}$ is in $D^b(A\text{-}mod)$ and therefore X is bounded. Then C is right bounded, and therefore $Hom_{D^-(A)}(C, X[m]) = 0$ for each $m << 0$ sufficiently small.

We know that $\mathcal{X}$ generates $D^b(A)$. If therefore $Hom_{D^-(A)}(C, X[m]) = 0$ for all X and all m, we get $Hom_{D^-(A)}(C, A[m]) = 0$ for all m, and hence, using again that A is symmetric and Lemma 1.10.30,

$$H_m(C) = Hom_{D^-(A)}(A[m], C) = 0$$

for all m, and therefore $C = 0$.

This shows that there exist $m \geq 0$ and $X \in \mathcal{X}$ such that

$$Hom_{D^-(A)}(C, X[m]) \neq 0.$$

Suppose that m is minimal with this property. In other words, we may assume $Hom_{D^-(A)}(C, Y[m']) = 0$ for all $m' < m$ and $Y \in \mathcal{X}$. Such a minimal m exists since each $X \in \mathcal{X}$ is bounded and C is right bounded. We shall now apply $Hom_{D(A)}(C, -)$ to the distinguished triangle

$$Z_X^{(n-1)} \xrightarrow{\alpha_X^{(n-1)}} X^{(n-1)} \xrightarrow{\beta_X^{(n-1)}} X^{(n)} \longrightarrow Z_X^{(n-1)}[1].$$

The first part of the proof of Claim 6.10.13 shows that

$$Hom_{D^-(A)}(C, Z_X^{(n-1)}[m]) = 0$$

and we hence obtain an exact sequence

$$0 \to Hom_{(A)}(C, X^{(n-1)}[m]) \to Hom_{(A)}(C, X^{(n)}[\text{-}modm]).$$

Therefore $Hom_{(A)}(C, X^{(n-1)}[m]) \to Hom_{(A)}(C, X^{(n)}[m])$ is injective for all $n \geq 1$, and by Claim 6.10.13 we get $Hom_{D^-(A)}(C, T_X[m]) \neq 0$. This proves the claim. □

We claim that $T := \bigoplus_{X \in \mathcal{X}} T_X$ is a tilting complex. Indeed, T is in $K^b(A\text{-}proj)$ by Claim 6.10.15. Claim 6.10.17 shows that $Hom_{D(A)}(T, T[m]) = 0$ for $m \neq 0$. Let $B := (End_{D^b(A)}(T))^{op}$. We may apply Keller's Theorem 6.4.1 and we may replace T by a complex $\tilde{T}$, so that $T \simeq \tilde{T}$ in $D^-(A)$, and so that $\tilde{T}$ is a complex of B–A-bimodules, projective as an A-module, and projective as a B-module. Hence, the functor

$$F := \mathbb{R}Hom_A(\tilde{T}, -) : D^-(A) \longrightarrow D^-(B)$$

has a left adjoint

$$G := \tilde{T} \otimes_B^{\mathbb{L}} - : D^-(B) \longrightarrow D^-(A).$$

We claim that F is an equivalence. Assume to the contrary that F is not an equivalence. Then there is some object U of $D^-(A)$ such that the counit $GFU \xrightarrow{\eta_U} U$ of the adjunction evaluated at U is not an isomorphism. However, since (G, F) is an adjoint pair, by Lemma 3.2.6 we obtain that $FGFU \xrightarrow{F\eta_U} FU$ is an isomorphism. If we denote by C_U the mapping cone of η_U, then

$$Hom_{D^-(A)}(T, C_U[m]) = H_m(FC_U) = 0.$$

Since A is symmetric, using Lemma 1.10.30, we get $Hom_{D^-(A)}(C_U[m], T) = 0$ for each $m \in \mathbb{Z}$. This contradicts Claim 6.10.18. By Theorem 6.5.1 we get that $add(T)$ generates $K^b(A\text{-}proj)$ as a triangulated category. Hence T is a tilting complex.

Only the following argument remains.

By Theorem 6.4.1 and Theorem 6.5.1 there is an equivalence $G : D^b(B) \to D^b(A)$ of standard type for which $G({}_BB) = T$. By Proposition 6.7.17 B is also a finite dimensional symmetric k-algebra. Since $G({}_BB) = T$, each indecomposable projective B-module P is mapped to an indecomposable direct factor T_P of T. Let $S_P := P/\text{rad}(P)$ and let $\mathcal{P}$ be a complete set of representatives of indecomposable projective B-modules. Then, by the introductory remarks of this section, $\mathcal{Y} := \{G(S_P) \mid P \in \mathcal{P}\}$ satisfies the properties 1, 2, 3, 4 from the beginning of the section.

We claim that $\mathcal{X} = \mathcal{Y}$. Indeed, applying $F := G^{-1}$ we may assume that $\mathcal{X}$ is a set of objects in $D^b(A\text{-}mod)$ so that for each projective indecomposable A-module P there is a unique $X_P \in \mathcal{X}$ such that $Hom_{D^b(A)}(P, X_P[m])$ is a skew-field for $m = 0$, is 0 for $m \neq 0$ and so that for all other $X' \in \mathcal{X}$ with $X' \not\simeq X_P$ we have $Hom_{D^b(A)}(P, X'[m]) = 0$. We need to show that under these hypotheses, $\mathcal{X}$ consists of simple A-modules only.

First, we need to show that each element X of $F(\mathcal{X})$ is a module. Indeed, using Corollary 3.5.52, any simple module S in the socle of the degree m homology of X will induce a non zero morphism $S[m] \longrightarrow X$. Combining with the projective cover map $P \longrightarrow S$ we get a non-zero morphism $P[m] \longrightarrow X$. Since $Hom_{D^b(A)}(P, X[m]) = 0$ if $m \neq 0$, we get that X is a module. We hence have to deal with properties in the module category. There it is clear that the above properties ensure that $\mathcal{X}$ consists only of simples modules. Since $\mathcal{X}$ generates $D^b(A\text{-}mod)$, all simple modules occur in $F(\mathcal{X})$. Since for each P there is a unique X_P with non-zero homomorphism set, $\mathcal{X} = \mathcal{Y}$. □

6.11 Broué's Abelian Defect Group Conjecture; Historical Remarks

In 1978, Alperin, Broué and Puig created theories which aimed to predict the representation theory of a finite group G over a field k of characteristic p from the representation theory of subgroups of G linked to the structure of the most eminent p-subgroups of G, such as the Sylow subgroups, the centralisers of p-subgroups, the normalisers of p-subgroups, and so on. This grew out of a natural desire to understand the fusion of Sylow subgroups, i.e. if two p-subgroups are conjugate in G, may they already be conjugate over smaller subgroups. First, in the spirit of the time, the main focus was on character theory, but soon Broué felt that character theory is just a shadow of more profound correspondences that hold in structures derived from the structure of the group, and actually in the module categories.

At around the same time Brenner and Butler [50] gave a nice interpretation of some phenomena discovered by Bernstein-Gel'fand-Ponomarev on the Auslander-Reiten quiver, and defined by certain A-modules M with $Ext^1_A(M, M) = 0$ the procedure they called tilting. The correspondence described by Bernstein-Gel'fand-Ponomarev was then a functorial correspondence "tilting" parts of the Auslander-Reiten quiver.

Happel then discovered the importance of derived categories for the representation theory of artinian algebras [51, 52] in interpreting the Brenner–Butler tilting as an equivalence between derived categories and explained in this way some strange behaviour of the Brenner–Butler tilting.

After Happel's discovery Rickard [1] and Keller [53] developed in their respective theses a Morita theory for derived categories. The result is Theorem 6.5.1 above. When it became clear how derived equivalences work, Broué realised that this is exactly what is needed to explain the correspondences between character tables of blocks of finite groups and their Brauer correspondents. In particular this explains the signs occurring mysteriously in the correspondences of the character tables (cf Remark 6.8.11).

This and Rickard's proof that Brauer tree algebras with the same numerical data are derived equivalent motivated Broué to formulate his most famous Abelian Defect Conjecture.

Conjecture 6.11.1 (Broué [54, Conjecture 4.9]) *Let k be an algebraically closed field of characteristic $p > 0$, let G be a finite group, let P be an abelian Sylow p-subgroup of G and let $H = N_G(P)$. Let B_G be the principal block of kG and let B_H be the principal block of kH. Then there should be an equivalence of triangulated categories*

$$D^b(B_G) \simeq D^b(B_H).$$

It should be noted that the hypothesis of P being abelian is really necessary. In [54] Broué states that Gerald Cliff observed that for the group $Sz(8)$ and $p = 2$ the principal blocks B_G and B_H do not have isomorphic centres and hence cannot be derived equivalent (cf Corollary 6.7.11).

For non-principal blocks Broué was more hesitant. However, by now it seems to be commonly accepted that the conjecture should be generalised to non-principal blocks in a natural way, replacing the two principal blocks by Brauer correspondent blocks, and the Sylow group by the defect group of the blocks. Broué only formulated it once and not as a conjecture, but as a question.

Conjecture 6.11.2 (Broué [23, Question 6.2]) *Let k be an algebraically closed field of characteristic $p > 0$ and let G be a finite group. Let B_G be a block of kG with abelian defect group D, let $H := N_G(D)$ and let B_H be the Brauer correspondent of B_G in kH. Then there should be an equivalence of triangulated categories*

$$D^b(B_G) \simeq D^b(B_H).$$

As we have seen, an equivalence between derived categories of finite dimensional algebras implies an isomorphism of the Grothendieck groups and actually, if the algebras are blocks of group rings of finite groups, or more generally residue algebras of classical orders over complete discrete valuation rings, even isomorphic Cartan-Brauer triangles (cf Theorem 6.8.8), and in the case of group rings additional properties of character values. The existence of such a correspondence of the Grothendieck group of ordinary representations satisfying further arithmetic properties which can be shown to hold in the case of an isometry derived from an equivalence between derived categories is called a perfect isometry by Broué. Broué conjectures in [23] that for a not necessarily principal block B_G of kG with abelian defect group D and the Brauer correspondent B_H of B_G in kH, for $H = N_G(D)$, the blocks B_G and B_H should be perfectly isometric.

Much more is known to hold for perfect isometries than for derived equivalences. In fact, a perfect isometry can be detected by examination of the character tables and is therefore much easier to verify. However, just to recognise that two blocks are perfectly isometric, once the character tables are known, is not very satisfactory. A structural explanation is desirable, and the existence of an equivalence between the derived categories is such a structural explanation. Furthermore, Broué's abelian defect conjecture implies various other conjectures which were a source of research in the representation theory of groups for many years. We mention Alperin's conjecture, Alperin-McKay conjecture and others. For a very nice overview of these links we refer to Külshammer [55].

Moreover, derived equivalences caught up in the meantime. More and more techniques were developed and eventually most people became interested in equivalences of derived categories, and not quite as much in perfect isometries only.

The first breakthrough was the fact that blocks with cyclic defect group satisfy the Broué's conjecture 6.11.2. This was first proved by Rickard, and then later reproved in various ways by many people. For a proof, see our Theorem 6.10.1.

Then Okuyama gave a method to prove the conjecture by examining Green correspondence and a very tricky application of the fact that derived equivalent self-injective algebras are stably equivalent of Morita type (cf Theorem 6.9.12) as well as Proposition 5.6.5, which states that a stable equivalence of Morita type which maps every simple module to a simple module is actually a Morita equivalence. Details are given in Sect. 6.10.2.

Broué's abelian defect conjecture was proved for many groups using this method. Mainly finite simple groups were considered, and in the meantime Marcus developed in [56] a method to prove Broué's abelian defect conjecture for a group G if one knows it for all groups involved in G (i.e. subgroups, quotients, and quotients of subgroups), and if one is ready to check several highly technical conditions on these equivalences and on the structure of the group. In this way it was possible to prove Conjecture 6.11.1 for $p = 2$. The method is just to check explicitly all possible counterexamples. This is an (almost) finite list since only simple groups have to be considered. The exceptional groups are examined with quite some effort using Okuyama's method. The remaining task, although very technical, is possible.

A completely different approach was used by Chuang and Rouquier for the symmetric groups. Chuang and Rouquier [57] gave a very specific argument, using the fact that one has a tower of groups nicely embedded one in the other, and the very well-known representation theory of symmetric groups, to prove Broué's conjecture 6.11.2 for symmetric groups, and actually much more. Their paper was one of the origins of what is now called the categorification technique.

Further developments concerned non-abelian defect groups. Erdmann classified all possible blocks with dihedral, semidihedral or quaternion defect groups up to Morita equivalences. Using this, and a detailed case by case analysis, a derived equivalence classification was obtained by Holm [58]. Of course, this is not covered by Broué's abelian defect conjecture, but it nevertheless provides a lot of information on these algebras which was previously unknown.

Perhaps the most interesting effect of Broué's abelian defect conjecture is that ever since people first started working on it, and in increasing intensity since then, the representation theory of finite groups has become closely linked to other subjects in mathematics. Since Broué's abelian defect conjecture needs quite a deep knowledge of homological algebra methods, specialists having started working in mathematics after 1990 became familiar with these techniques, became interested in parallel developments in other branches of mathematics, and since then an interconnection between algebraic topology, algebraic geometry, mathematical physics and the representation theory of groups and algebras has been established. Furthermore, although they are closely related subjects, the representation theory of algebras and the representation theory of finite groups had difficulty interacting before 1990 (with some exceptions, of course). Now, the subjects often borrow ideas and methods from each other, to their mutual benefit.

6.12 Picard Groups of Derived Module Categories

As an application of Morita's theorem we studied in Sect. 4.6 the group of self-equivalences of the module category of a k-algebra A. Morita's Theorem 4.2.8 showed that a self-equivalence A-*Mod* $\longrightarrow$ A-*Mod* which restricts to an equivalence A-*mod* $\longrightarrow$ A-*mod* for an algebra A is always given by taking the tensor product with an A–A-bimodule M and hence we can just study the isomorphism classes of A–A-bimodules, which are invertible under the operation $- \otimes_A -$.

6.12.1 The General Definition

In principle we might as well study self-equivalences of $D^b(A)$ and their behaviour under composition. The first question is, of course, will this be a group? Composition is associative, but taking all self-equivalences, even up to isomorphism, it is not absolutely clear that this will provide a set of objects. Moreover, we do not really have a tool to study them, unless we can describe them as tensor product with something.

Here, a problem is that we do not know if all self-equivalences are actually of standard type, given by a tensor product with a complex of bimodules, i.e. a two-sided tilting complex (cf Remark 6.5.3). We therefore need to restrict to this case. The proof for the existence of two-sided tilting complexes needs that A is projective as k-module. It therefore seems natural to assume this condition. However, there is a more important reason to assume that A is projective, and hence projective as a k-module. Let X, Y be two-sided tilting complexes in $D^b(A \otimes_k A^{op})$. Then the composition of the self-equivalence given by $X \otimes_A^{\mathbb{L}} -$ and $Y \otimes_A^{\mathbb{L}} -$ is

$$X \otimes_A^{\mathbb{L}} (Y \otimes_A^{\mathbb{L}} -) : D^b(A) \longrightarrow D^b(A),$$

which means that we first replace Y by its projective resolution, and then replace X by its projective resolution. There is no a priori reason why this should be the same as

$$(X \otimes^{\mathbb{L}}_A Y) \otimes^{\mathbb{L}}_A - : D^b(A) \longrightarrow D^b(A).$$

Here we need to compute the projective resolution of $X \otimes^{\mathbb{L}}_A Y$, and this will not in general be the tensor product of the projective resolutions of each term, unless A is projective as a k-module. In general the left derived tensor product will not be associative!

If A is projective as a k-module, then we may take a projective resolution P_X of X, and then observe that an A–A-bimodule M which is projective as a bimodule will be projective when restricted to A and also when restricted to A^{op}. Hence we can replace the complex X by P_X, truncate intelligently and then the left derived tensor product is just the total complex of the ordinary tensor product and this is an associative operation.

Definition 6.12.1 (Rouquier and Zimmermann [59]) Let k be a commutative ring and let A be a k-algebra which is projective as a k-module. Then let $DPic_k(A)$ be the set which is formed by the isomorphism classes $[X]$ of bounded complexes of finitely generated A–A-bimodules X so that there is a bounded complex of finitely generated A–A-bimodules Y satisfying

$$X \otimes^{\mathbb{L}}_A Y \simeq A \simeq Y \otimes^{\mathbb{L}}_A X$$

in $D^b(A \otimes_k A^{op})$. Then this forms a group under the group law

$$[X_1] \cdot [X_2] := [X_1 \otimes^{\mathbb{L}}_A X_2].$$

The group $(DPic_k(A), - \otimes^{\mathbb{L}}_A -)$ is the *derived Picard group* of A.

Remark 6.12.2 Note that the derived Picard group $DPic_k(A)$ of A is not the derived group (i.e. the commutator group) $[Pic_k(A), Pic_k(A)]$ of the usual Picard group $Pic_k(A)$ of A. To avoid this possible confusion the notation $TrPic_k(A)$ was used in [59]. However, it seems that the notation $DPic_k(A)$ is now commonly used.

Remark 6.12.3 By the very definition of a triangulated category $\mathcal{T}$, we always have a self-equivalence of $\mathcal{T}$ by the suspension functor, i.e. the degree shift [1] in the case of derived categories. In the case of the derived category $D^b(A)$ the shift in degree will be of infinite order.

Hence the group $\langle [1] \rangle$ generated by the shift in degree is a cyclic group of infinite order, and is a subgroup of $DPic_k(A)$. Since

$$X \otimes_A C[1] \simeq (X \otimes_A C)[1]$$

for all objects C of $D^b(A)$ and all X with $[X] \in DPic_k(A)$ we get that $\langle [1] \rangle$ is a subgroup of the centre of $DPic_k(A)$.

A first immediate consequence is

Lemma 6.12.4 *Let k be a commutative ring and let A and B be Noetherian k-algebras which are assumed to be projective as k-modules. If $D^b(A) \simeq D^b(B)$ then $DPic_k(A) \simeq DPic_k(B)$.*

Proof By Proposition 6.5.5 and Proposition 6.5.6 there is a two-sided tilting complex Z in $D^b(A \otimes_k B^{op})$ with inverse $\hat{Z}$. Then

$$\begin{array}{rcl} DPic_k(A) & \longrightarrow & DPic_k(B) \\ [X] & \mapsto & [\hat{Z} \otimes^{\mathbb{L}}_A (X \otimes^{\mathbb{L}}_A Z)] \end{array}$$

is a well-defined group isomorphism, as is readily verified. Note that since A and B are assumed to be projective as k-modules, we have

$$(\hat{Z} \otimes^{\mathbb{L}}_A X) \otimes^{\mathbb{L}}_A Z \simeq \hat{Z} \otimes^{\mathbb{L}}_A (X \otimes^{\mathbb{L}}_A Z).$$

This proves the statement. □

Denote by $Aut_k(B)$ the automorphism group of the k-algebra B. For ordinary Picard groups we obtained a homomorphism to the automorphism group of the centre of the algebra by comparison of left and right multiplication of central elements. This was the object of Proposition 4.6.7. For *DPic* this approach is not really appropriate. In the derived category setting it is more sensible to apply homological algebra methods.

Lemma 6.12.5 *Let k be a commutative ring and let A be a k-algebra which is projective as a k-module. Then there is a natural group homomorphism $DPic_k(A) \longrightarrow Aut_k(Z(A))$.*

Proof Let X be a two-sided tilting complex giving $[X] \in DPic_k(A)$, and let $[X]^{-1} = [\hat{X}]$ in $DPic_k(A)$. By Proposition 6.4.4 we may assume that X and $\hat{X}$ are complexes with projective homogeneous components. By Proposition 6.7.9 we get that $X \otimes_A - \otimes_A \hat{X} : D^b(A \otimes_k A^{op}) \to D^b(A \otimes_k A^{op})$ is a self-equivalence. This induces an isomorphism

$$Hom_{D^b(A \otimes_k A^{op})}(A, A) \xrightarrow{\varphi_X} Hom_{D^b(A \otimes_k A^{op})}(A, A)$$

satisfying $\varphi_X \circ \varphi_Y = \varphi_{X \otimes_A Y}$. Since

$$Hom_{D^b(A \otimes_k A^{op})}(A, A) = Hom_{A \otimes_k A^{op}}(A, A) \simeq Z(A)$$

this gives a group homomorphism $DPic_k(A) \longrightarrow Aut_k(Z(A))$ as claimed. □

Remark 6.12.6 The attentive reader will recognise that the above arguments are very analogous to those used in Proposition 6.7.10 showing that Hochschild cohomology is invariant under derived equivalences. It is straightforward to extend the group homomorphism

$$DPic_k(A) \longrightarrow Aut_k(Z(A))$$

to a group homomorphism

$$DPic_k(A)) \longrightarrow Aut_k(HH^*(A))$$

of k-automorphisms of the Hochschild cohomology k-algebra. The proof is identical.

As in Definition 4.6.8 for the ordinary Picard group of an algebra we obtain the following

Definition 6.12.7 Let A be a k-algebra which is projective as a k-module. Then put $DPicent(A) := \ker(DPic_k(A) \longrightarrow Aut_k(Z(A))$.

Lemma 6.12.8 *Let A be a local k-algebra which is projective as a k-module. Then*

$$DPic_k(A) \simeq Pic_k(A) \times \langle[1]\rangle.$$

Let A be a commutative indecomposable k-algebra which is projective as a k-module. Then

$$DPic_k(A) \simeq Pic_k(A) \times \langle[1]\rangle.$$

Proof The first statement is an immediate consequence of Proposition 6.7.4 and the second statement follows from Lemma 6.7.7. □

Lemma 6.12.9 *Let k be a commutative ring and let S be a commutative ring with a ring homomorphism $k \to S$. Let A be a k-algebra. Then for all $[X] \in DPic_k(A)$ we have $[S \otimes_k X] \in DPic_S(S \otimes_k A)$. Moreover, $S \otimes_k -$ induces a group homomorphism $DPic_k(A) \to DPic_S(S \otimes_k A)$.*

Proof Let $[Y] \in DPic_k(A)$ such that $[X] \cdot [Y] = [A]$ and by Proposition 6.4.4 we may assume that X and Y are complexes of projective modules, when restricted to one side. Then

$$(S \otimes_k X) \otimes_{S \otimes_k A} (S \otimes_k Y) \simeq S \otimes_k (X \otimes_A Y) \simeq S \otimes_k A$$

and likewise

$$(S \otimes_k Y) \otimes_{S \otimes_k A} (S \otimes_k X) \simeq S \otimes_k A.$$

Since by the same arguments

$$(S \otimes_k X) \otimes_{S \otimes_k A} (S \otimes_k Z) \simeq S \otimes_k (X \otimes_A Z)$$

we get that $S \otimes_k -$ induces a group homomorphism, as claimed. □

6.12.2 Fröhlich's Localisation Sequence

We shall briefly state some results on Picard groups in case when Λ is an R-order in a semisimple K-algebra A in the sense of Definition 2.5.15. Recall that this means that R is an integral domain with field of fractions K, that Λ is an R-algebra such that Λ is finitely generated projective as R-module and such that $K \otimes_R \Lambda = A$. A *Dedekind domain* R is a commutative ring without zero divisors such that each ideal is a unique product of prime ideals (up to permutation of factors). We shall use the following number theoretical property of R-orders. We shall not give the (number theoretical) proof here and instead refer the reader to [34].

Proposition 6.12.10 (cf e.g. Reiner [34, Theorem 5.3]) *Let R be a Dedekind domain with field of fractions K and let Λ be an R-order in a semisimple K-algebra A. Let $Specmax(R)$ be the set of maximal ideals of R. Each maximal ideal of R induces a discrete valuation v on K and*

$$M = K \otimes_R M \cap \bigcap_{v \in Specmax(R)} \left(\hat{R}_v \otimes_R M\right).$$

Moreover, for each $v \in Specmax(R)$ let there be an $\hat{R}_v \otimes_R \Lambda$-lattice $M(v) \subseteq \hat{K}_v \otimes_R \Lambda$ such that there is a finite subset $V \subseteq Specmax(R)$ with $M(v) = \hat{R}_v \otimes_R \Lambda$ for all $v \in Specmax(R) \setminus V$. Then

$$\widetilde{M} := K \otimes_R M \cap \bigcap_{v \in Specmax(R)} M(v)$$

is a Λ-lattice with

$$\hat{R}_v \otimes_R \widetilde{M} \simeq M(v).$$

The main result of this section is that there is an exact sequence, due to Fröhlich [60], linking $Picent(\Lambda)$ and $Picent(\hat{R}_\wp \otimes_R \Lambda)$ for completions $\hat{R}_\wp$ of R at the prime $\wp$.

Theorem 6.12.11 ([60] for *Picent*, [59] for *DPicent*) *Let R be a Dedekind domain with field of fractions K and let Λ be an R-order in a semisimple K-algebra A. Let $Specmax(R)$ be the maximal ideals of R, and for each maximal ideal let R_v be the corresponding discrete valuation ring of the valuation. Then there is an exact sequence*

$$1 \to DPicent(Z(\Lambda)) \xrightarrow{\tau} DPicent(\Lambda) \xrightarrow{\sigma} \prod_{v \in Specmax(R)} DPicent(\hat{R}_v \otimes_R \Lambda)$$

where σ is induced by $M \mapsto \hat{R}_v \otimes_R M$ and τ is given by $[L] \mapsto [L \otimes_{Z(\Lambda)} \Lambda]$.

Any element in the subset $\widetilde{\prod}_{v\in Specmax(R)} Picent(\hat{R}_v \otimes_R \Lambda)$*, which is formed by sequences of invertible bimodules, indexed by* v*, such that all but a finite number of them are trivial, is in the image of* σ*.*

Proof ([59, 60]) By Lemma 6.12.8 we get for indecomposable k-algebras Λ

$$DPic_k(Z(\Lambda)) = Pic_k(\Lambda)\times < [1] > .$$

Hence, we may assume that $[L]$ is an element in $Picent(Z(\Lambda))$ rather than in $DPicent(Z(\Lambda))$. Since $[L] \in Picent(Z(\Lambda))$, we have that L is an invertible $Z(\Lambda)$-$Z(\Lambda)$-bimodule such that $Z(\Lambda)$ acts the same way on the left as on the right. Hence we get that $L \otimes_{Z(\Lambda)} \Lambda$ is a Λ–Λ-bimodule by putting

$$\lambda_1 \cdot (\ell \otimes x) \cdot \lambda_2 = \ell \otimes (\lambda_1 x \lambda_2).$$

Now L is invertible, that is there is a $[\Gamma] \in Picent(Z(\Lambda))$ such that

$$L \otimes_{Z(\Lambda)} \Gamma \simeq Z(\Lambda) \simeq \Gamma \otimes_{Z(\Lambda)} L$$

and therefore

$$\begin{aligned}(L \otimes_{Z(\Lambda)} \Lambda) \otimes_\Lambda (\Gamma \otimes_{Z(\Lambda)} \Lambda) &\simeq (L \otimes_{Z(\Lambda)} (\Lambda \otimes_\Lambda \Gamma)) \otimes_{Z(\Lambda)} \Lambda\\ &\simeq (L \otimes_{Z(\Lambda)} \Gamma) \otimes_{Z(\Lambda)} \Lambda\\ &\simeq Z(\Lambda) \otimes_{Z(\Lambda)} \Lambda\\ &\simeq \Lambda\end{aligned}$$

as Λ–Λ-bimodules. Likewise

$$(\Gamma \otimes_{Z(\Lambda)} \Lambda) \otimes_\Lambda (L \otimes_{Z(\Lambda)} \Lambda) \simeq \Lambda$$

and we get a well-defined mapping. Now, we see that for $[L] \in Picent(Z(\Lambda))$ we obtain

$$L \otimes_{Z(\Lambda)} \Lambda \simeq \Lambda \otimes_{Z(\Lambda)} L$$

as Λ–Λ-bimodules. This also shows that τ is a group homomorphism by a direct computation similar to the computation for the inverse.

For any Λ–Λ-bimodule M we put

$$M^\Lambda := \{m \in M \mid \lambda m = m\lambda \ \ \forall \lambda \in \Lambda\} \simeq Hom_{\Lambda\otimes\Lambda^{op}}(\Lambda, M).$$

This is obviously a $Z(\Lambda) = End_{\Lambda\otimes\Lambda^{op}}(\Lambda)$-module. Then $M \simeq N$ as Λ–Λ-bimodules implies $M^\Lambda \simeq N^\Lambda$ as $Z(\Lambda)$-modules. In order to show the injectivity we see that if $L \otimes_{Z(\Lambda)} \Lambda \simeq \Lambda$ as bimodules then

$$(L \otimes_{Z(\Lambda)} \Lambda)^{\Lambda} = L \otimes_{Z(\Lambda)} (\Lambda^{\Lambda}) = L \otimes_{Z(\Lambda)} Z(\Lambda) = L$$

by definition of the action of Λ on the induced module $L \otimes_{Z(\Lambda)} \Lambda$. Hence

$$L \otimes_{Z(\Lambda)} \Lambda \simeq \Lambda \Rightarrow L = (L \otimes_{Z(\Lambda)} \Lambda)^{\Lambda} \simeq \Lambda^{\Lambda} = Z(\Lambda)$$

which implies the injectivity of τ.

The mapping σ is well-defined by Lemma 6.12.9. We observe that

$$Z(\hat{R}_v \otimes_R \Lambda) = \hat{R}_v \otimes_R Z(\Lambda)$$

since R is in the centre of Λ, and Λ is R-projective. The fact that

$$\begin{array}{rcl} Picent(\Lambda) & \longrightarrow & Picent(\hat{R}_v \otimes_R \Lambda) \\ [M] & \mapsto & [\hat{R}_v \otimes_R M] \end{array}$$

is a group homomorphism is clear from this. Hence σ is a group homomorphism as well.

We need to show that $\sigma \circ \tau = 0$. Let L be an invertible $Z(\Lambda)$-module. We need to show that

$$\hat{R}_v \otimes_R (\Lambda \otimes_{Z(\Lambda)} L) \simeq \hat{R}_v \otimes_R \Lambda$$

as $(\hat{R}_v \otimes_R \Lambda)$–$(\hat{R}_v \otimes_R \Lambda)$-bimodules. Since $Z(\Lambda)$ is a commutative R-order, $Z(\Lambda)$ is in particular basic, and since $\hat{R}_v$ is complete Proposition 2.5.17 shows that $\hat{R}_v \otimes_R Z(\Lambda) = Z(\hat{R}_v \otimes_R \Lambda)$ satisfies the Krull-Schmidt property on projective modules. Hence Corollary 4.6.4 shows that

$$Pic_{\hat{R}_v}(Z(\hat{R}_v \otimes_R \Lambda)) \simeq Out_{\hat{R}_v}(Z(\hat{R}_v \otimes_R \Lambda))$$

and in particular

$$Picent(Z(\hat{R}_v \otimes_R \Lambda)) = 1.$$

Therefore

$$\hat{R}_v \otimes_R L \simeq Z(\hat{R}_v \otimes_R \Lambda) = \hat{R}_v \otimes_R Z(\Lambda)$$

which implies that

$$\hat{R}_v \otimes_R (\Lambda \otimes_{Z(\Lambda)} L) \simeq (\hat{R}_v \otimes_R L) \otimes_{Z(\Lambda)} \Lambda \simeq \hat{R}_v \otimes_R \Lambda$$

as $(\hat{R}_v \otimes_R \Lambda)$–$(\hat{R}_v \otimes_R \Lambda)$-bimodules.

We need to show that if X is an object of $DPicent(\Lambda)$ with $\hat{R}_v \otimes_R X \simeq \hat{R}_v \otimes_R \Lambda$ for all v, then X is in the image of τ. First, since $\hat{R}_v \otimes_R X \simeq \hat{R}_v \otimes_R \Lambda$ for all v, the homology of X is concentrated in degree 0, and actually projective (cf Lemma 3.8.6). By Lemma 3.8.6 we get that X is an invertible bimodule. Therefore we can suppose

$[X] \in Picent(\Lambda)$. Moreover, $K \otimes_R X \simeq K \otimes_R \Lambda$ implies that we can choose an isomorphism so that $X \subseteq \Lambda$. Indeed, let $\{b_1, \dots, b_n\} \subseteq K\Lambda$ be a generating set of X, then we may choose $r \in R \setminus \{0\}$ so that $\{rb_1, \dots, rb_n\} \subseteq \Lambda$. Moreover, let again $\{b_1, \dots, b_n\} \subseteq \Lambda$ be a generating set of X, and let $\{c_1, \dots, c_m\}$ be an R-generating set of Λ. Then express $b_i = \sum_{j=1}^m d_{i,j} c_j$. We consider the valuations of the (finite number of) coefficients $d_{i,j}$ in the expression of b_i in the R-basis of Λ, and obtain that there is an $r \in R \setminus \{0\}$ such that

$$r \cdot \Lambda \subseteq X \subseteq \Lambda.$$

This shows that there are only a finite set V of valuations v with $\hat{R}_v \otimes_R X \neq \hat{R}_v \otimes_R \Lambda$. Hence

$$\hat{R}_v \otimes_R X = (\hat{R}_v \otimes_R \Lambda) \cdot u_v$$

for some $u_v \in Z(\hat{R}_v \otimes_R \Lambda)$, in order to get a bimodule. We may put $u_v = 1$ if $v \in Specmax(R) \setminus V$ and for all valuations v we get that u_v is a unit since $K \otimes_R X = K \otimes_R \Lambda$. Put

$$C := \bigcap_v (R_v \otimes_R Z(\Lambda)) \cdot u_v$$

and

$$\check{C} := \bigcap_v (R_v \otimes_R Z(\Lambda)) \cdot u_v^{-1}.$$

Then both C and $\check{C}$ are $Z(\Lambda)$-modules. Moreover,

$$K \otimes_R C = K \otimes_R Z(\Lambda) = Z(K \otimes_R \Lambda) = K \otimes_R \check{C}.$$

Since

$$\hat{R}_v \otimes_R C = \left(\hat{R}_v \otimes_R Z(\Lambda)\right) \cdot u_v \text{ and } \hat{R}_v \otimes_R \check{C} = \left(\hat{R}_v \otimes_R Z(\Lambda)\right) \cdot u_v^{-1}$$

for each valuation v, we get

$$\begin{aligned}
&\hat{R}_v \otimes_R \left(C \otimes_{Z(\Lambda)} \check{C}\right) \\
&= \left(\left(\hat{R}_v \otimes_R Z(\Lambda)\right) \cdot u_v\right) \otimes_{\hat{R}_v \otimes_R Z(\Lambda)} \left(\left(\hat{R}_v \otimes_R Z(\Lambda)\right) \cdot u_v^{-1}\right) \\
&= \left(\hat{R}_v \otimes_R Z(\Lambda)\right) \cdot u_v \cdot u_v^{-1} \\
&= \hat{R}_v \otimes_R Z(\Lambda)
\end{aligned}$$

and likewise

$$\hat{R}_v \otimes_R \left(\check{C} \otimes_{Z(\Lambda)} C\right) = \hat{R}_v \otimes_R Z(\Lambda)$$

we obtain that $[C] \in Picent(Z(\Lambda))$ by Proposition 6.1210. Moreover,

$$\hat{R}_v \otimes_R (C \otimes_{Z(\Lambda)} \Lambda) = \left(\hat{R}_v \otimes_R \Lambda\right) \cdot u_v$$

for all v, and therefore, by Proposition 6.12.10

$$C \otimes_{Z(\Lambda)} \Lambda \simeq X.$$

We need to show that any element in $\widetilde{\prod}_{v \in Specmax(R)} Picent(\hat{R}_v \otimes_R \Lambda)$ is in the image of σ. We start with a collection $[X(v)] \in Picent(\hat{R}_v \otimes_R \Lambda)$ such that there is a finite subset $V \subseteq Specmax(R)$ so that $X(v) = \hat{R}_v \otimes_R \Lambda$ whenever $v \in Specmax(R) \setminus V$ and such that $X(v) \subseteq \hat{R}_v \otimes_R \Lambda$ for all $v \in Specmax(R)$. For each $v \in Specmax(R)$ there is an $\check{X}(v) \in Picent(\hat{R}_v \otimes_R \Lambda)$ such that

$$[X(v)] \cdot [\check{X}(v)] = [\hat{R}_v \otimes_R \Lambda].$$

Again $\check{X}(v) = \hat{R}_v \otimes_R \Lambda$ for all $v \in Specmax(R) \setminus V$ and $\check{X}(v) \subseteq \hat{R}_v \otimes_R \Lambda$ for all $v \in Specmax(R)$.

Put $X := \bigcap_{v \in Specmax(R)} X(v)$ and $\check{X} := \bigcap_{v \in Specmax(R)} \check{X}(v)$. Since

$$\hat{R}_v \otimes_R X \otimes_\Lambda \check{X} = X(v) \otimes_{\hat{R}_v \otimes_R \Lambda} \check{X}(v) = \hat{R}_v \otimes_R \Lambda$$

and

$$\hat{R}_v \otimes_R \check{X} \otimes_\Lambda X = \check{X}(v) \otimes_{\hat{R}_v \otimes_R \Lambda} X(v) = \hat{R}_v \otimes_R \Lambda$$

we get that $[X] \in Picent(\Lambda)$ by Proposition 6.12.10. Moreover,

$$\hat{R}_v \otimes_R X_v = X(v)$$

and therefore σ has the claimed images. □

Remark 6.12.12 It can be shown (cf e.g. [34, Theorem 37.28]) that we always get $Picent(\hat{R}_v \otimes_R \Lambda) = 1$ for all but a finite number of valuations.

Remark 6.12.13 The theorem can be used to lift automorphisms of $\hat{R}_v \otimes_R \Lambda$ for all v to some invertible bimodule X of $Picent(\Lambda)$. Local-global principles are very useful, and the argument was applied in [61] where a group G and an automorphism α of G was constructed so that α is not inner, but α becomes inner in RG where R is a finite extension of $\mathbb{Z}$.

6.12.3 An Example: The Brauer Tree Algebras

The group $DPic_k(A)$ is very big in general. We already have seen that it is always infinite since it contains the infinite cyclic group $\langle[1]\rangle \simeq C_\infty$ given by shift in degrees. However, most astonishing at first sight, the degree shift is not a direct factor in general. We shall first give an example to illustrate this fact.

By Lemma 6.12.4 we get that for every Brauer tree algebra A without exceptional vertex the group $DPic_k(A)$ is independent of the shape of the Brauer tree. We suppose therefore that A is a Brauer tree algebra without exceptional vertex associated to a stem, i.e. to the Brauer tree

$$\bullet_1 \text{———} \bullet_2 \text{———} \cdots\cdots \text{———} \bullet_n \text{———} \bullet_{n+1}$$

with projective indecomposable modules $P_1, \dots, P_n$, so that P_i is associated to the edge adjacent to the vertices $\{i, i+1\}$. We then consider the complex X_i given by

$$\cdots \longrightarrow 0 \longrightarrow P_i \otimes_k Hom_A(P_i, A) \stackrel{\epsilon}{\longrightarrow} A \longrightarrow 0 \longrightarrow \cdots$$

where $\epsilon(x \otimes f) := f(x)$ is the evaluation map. Now, there is a primitive idempotent $e_i \in A$ such that $P_i = Ae_i$ and $Hom_A(P_i, A) = e_iA$. Hence $\mathrm{im}(\epsilon) = Ae_iA$. In particular, Ae_i and e_iA both belong to the image of ϵ.

We consider the images $X_i \otimes_A P_j$. Since

$$(Ae_i \otimes_k e_iA) \otimes_A Ae_j = Ae_i \otimes_k e_iAe_j = Ae_i \otimes_k Hom_A(Ae_i, Ae_j)$$

and since

$$\dim_k(Hom_A(Ae_i, Ae_j)) = \begin{cases} 0 & \text{if } |i-j| > 1 \\ 1 & \text{if } |i-j| = 1 \\ 2 & \text{if } |i-j| = 0 \end{cases}$$

and since $A \otimes_A Ae_j = Ae_j$, we get the following isomorphisms of two-term complexes

$$X \otimes_A P_i = (P_i^2 \stackrel{\epsilon\otimes id_{P_i}}{\longrightarrow} P_i) \simeq (P_i \longrightarrow 0).$$

The right-hand isomorphism comes from the fact that the differential $\epsilon \otimes id_{P_i}$ is surjective onto the projective module, whence split. Overall we get

$$(*) \quad X_i \otimes_A P_j \simeq \begin{cases} 0 \longrightarrow P_j & \text{if } |i-j| > 1 \\ P_i \longrightarrow P_j & \text{if } |i-j| = 1 \\ P_i \longrightarrow 0 & \text{if } i = j \end{cases}.$$

Lemma 6.12.14 *The complex X_i is a two-sided tilting complex.*

Proof Since all homogeneous components are projective regarded as a left A-modules or as right A-modules, we get the following isomorphism of bicomplexes

$$X_i \otimes_A Hom_A(X_i, A) \simeq Hom_A(X_i, X_i).$$

The total complex of this bicomplex is then $X_i \otimes_A^{\mathbb{L}} Hom_A(X_i, A)$. However, by Lemma 3.7.10

$$H_0(Hom_A(X_i, X_i)) = H_0(\mathbb{R}Hom_A(X_i, X_i)) = End_{D^b(A)}(X_i)$$

and

$$H_s(Hom_A(X_i, X_i)) = Hom_{D^b(A)}(X_i, X_i[s])$$

for $s \neq 0$.

Now, X_i regarded as a complex of A-left modules is isomorphic to

$$\cdots \longrightarrow 0 \longrightarrow P_i^4 \longrightarrow A \longrightarrow 0 \longrightarrow \cdots.$$

The direct factor P_i of A on the right of the differential is in the image, and hence this complex is homotopy equivalent to

$$\cdots \longrightarrow 0 \longrightarrow P_i^3 \longrightarrow A/P_i \longrightarrow 0 \longrightarrow \cdots.$$

We now adopt the convention that the rightmost degree written is the degree 0 component and direct factors are just written vertically above one another. With this notational simplification we observe that this complex is actually

$$\begin{array}{ccc} & & P_1 \\ & & \vdots \\ & & P_{i-2} \\ P_i & \longrightarrow & P_{i-1} \\ P_i & & \\ P_i & \longrightarrow & P_{i+1} \\ & & P_{i+2} \\ & & \vdots \\ & & P_n \end{array}$$

as was actually shown in the remarks just before the lemma. By Corollary 6.6.4 and Lemma 6.6.5 this is a tilting complex with endomorphism ring A^{op}. Hence, by the functoriality of the construction

$$X_i \otimes_A Hom_A(X_i, A) \simeq A$$

in the derived category of complexes of A–A-bimodules. By the analogous argument, using right modules instead of left modules

$$Hom_A(X_i, A) \otimes_A X_i \simeq A$$

in the derived category of complexes of A–A-bimodules. □

Remark 6.12.15 If A and B are symmetric algebras, and if X is a bounded complex of A–B-bimodules such that $X \otimes_B B$ is a tilting complex in $D^b(A)$ with endomorphism ring B^{op} and such that $A \otimes_A X$ is a tilting complex in $D^b(B^{op})$ with endomorphism ring A^{op}, then X is a two-sided tilting complex. This fact is due to Rickard and can be found in [8].

We claim that for all i we have

$$\begin{array}{l}(X_i \otimes_A X_{i-1} \otimes_A \cdots \otimes_A X_1) \otimes_A P_j \\ \simeq \begin{cases} P_i \to P_{i-1} \to \cdots \to P_1 \to 0 & \text{if } j = 1 \\ P_{j-1} \to 0 & \text{if } 1 \neq j \leq i \\ P_i \to P_{i+1} & \text{if } i+1 = j \\ 0 \to P_j & \text{if } j > i+1 \end{cases}.\end{array}$$

We already have this equation for $i = 1$ by the above equation $(*)$. We suppose this is true for $i > 1$, and prove it for $i + 1$. We compute immediately

$$X_{i+1} \otimes_A P_j = P_j$$

if $j > i + 1$ and

$$X_{i+1} \otimes_A P_j[1] = P_j[1]$$

whenever $j < i$ by $(*)$. By the same argument,

$$X_{i+1} \otimes_A P_i = (P_{i+1} \to P_i).$$

Moreover,

$$\begin{aligned} X_{i+1} \otimes_A (P_i \to P_{i+1}) &\simeq X_{i+1} \otimes_A cone(P_i \to P_{i+1}) \\ &\simeq cone(X_{i+1} \otimes_A P_i \to X_{i+1} \otimes_A P_{i+1}) \\ &\simeq cone((P_{i+1} \to P_i) \to (P_{i+1} \to 0)) \\ &\simeq P_i \to 0 \end{aligned}$$

since the mappings are all with maximal image, using that the complex X_{i+1} is a two-sided tilting complex, and hence $X_{i+1} \otimes_A T$ is a direct summand of a two-sided tilting complex for all direct factors T of a tilting complex. Similarly,

$$X_{i+1} \otimes_A (P_{i-1} \to \cdots \to P_1 \to 0) \simeq (P_{i-1} \to \cdots \to P_1 \to 0)$$

since $X_{i+1} \otimes_A P_j = P_j$ whenever $j < i$ again by $(*)$. Hence

$$\begin{aligned}X_{i+1}\otimes_A(P_i \to \cdots \to P_1 \to 0)\\ &\simeq X_{i+1} \otimes_A cone(P_i[i-1] \to (P_{i-1} \to \cdots \to P_1 \to 0)\\ &\simeq cone((P_{i+1} \to P_i)[i-1] \to (P_{i-1} \to \cdots \to P_1 \to 0))\\ &\simeq (P_{i+1} \to P_i \to \cdots \to P_1 \to 0).\end{aligned}$$

This proves the statement.

We claim that

$$X_1 \otimes_A (X_2 \otimes_A X_1) \otimes_A \cdots \otimes_A (X_n \otimes_A \cdots \otimes_A X_1) \otimes_A P_j \simeq P_{n+1-j}[n]$$

for all j. Indeed, by the first step we get

$$(X_n \otimes_A X_{n-1} \otimes_A \cdots \otimes_A X_1) \otimes_A P_j = \begin{cases} P_n \to P_{n-1} \to \cdots \to \quad P_1 \quad \to 0 & \text{if } j = 1 \\ \qquad\qquad\qquad\qquad P_{j-1} \to 0 & \text{if } 1 \neq j \end{cases}.$$

Then

$$(X_r \otimes_A \cdots \otimes_A X_1) \otimes_A P_j[1] \simeq P_{j-1}[2]$$

if $1 \neq j < r$, and for all r. Therefore we just need to examine the initial term. Here we get

$$(X_i \otimes_A \cdots \otimes_A X_1) \otimes_A (P_r \to P_{r-1} \to \cdots \to P_1) \simeq (P_r \to P_{r-1} \to \cdots \to P_{i+1})[i]$$

for all $i < r$. Indeed, this can be shown by induction on i. For $i = 1$ we compute

$$\begin{aligned}X_1 \otimes_A (P_r \to P_{r-1} \to \cdots \to P_1)\\ &= X_1 \otimes_A cone((P_r \to \cdots \to P_3 \to 0) \to (P_2 \to P_1))\\ &= cone((P_r \to \cdots \to P_3 \to 0) \to (cone(P_1 \to P_2) \to (P_1 \to 0))\\ &= cone((P_r \to \cdots \to P_3 \to 0) \to (P_2 \to 0))\\ &= (P_r \to \cdots \to P_3 \to P_2 \to 0).\end{aligned}$$

Suppose the formula is true for $i-1$. Then we get

$$\begin{aligned}X_i \otimes_A (P_{i+1} \to P_i) &\simeq cone(X_i \otimes_A P_{i+1} \to X_i \otimes_A P_i)\\ &\simeq cone((P_i \to P_{i+1}) \to (P_i \to 0))\\ &\simeq P_{i+1} \to 0.\end{aligned}$$

Hence

$$\begin{aligned}(X_i\otimes_A \cdots \otimes_A X_1) \otimes_A (P_r \to P_{r-1} \to \cdots \to P_1)\\ &= X_i \otimes_A (P_r \to P_{r-1} \to \cdots \to P_i)[i-1]\\ &= X_i \otimes_A cone((P_r \to P_{r-1} \to \cdots \to P_{i+2} \to 0) \to (P_{i+1} \to P_i))[i-1]\end{aligned}$$

$$= cone((P_r \to P_{r-1} \to \cdots \to P_{i+2} \to 0) \to (P_{i+1} \to 0))[i-1]$$
$$= (P_r \to P_{r-1} \to \cdots \to P_{i+2} \to P_{i+1})[i].$$

This proves the statement. In particular, we have

$$(X_{r-1} \otimes_A \cdots \otimes_A X_1) \otimes_A (P_r \to P_{r-1} \to \cdots \to P_1) \simeq P_r[r-1].$$

Concluding we obtain

$$X_1 \otimes_A (X_2 \otimes_A X_1) \otimes_A \cdots \otimes_A (X_n \otimes_A \cdots \otimes_A X_1) \otimes_A P_j \simeq P_{n+1-j}[n]$$

for all j, as claimed.

We observe that if A is basic, then there is an automorphism σ of A which corresponds to the symmetry of the tree, i.e.

$${}^{\sigma}P_j \simeq P_{n+1-j} \quad \forall j \in \{1, \ldots, n\}.$$

This automorphism induces a self-equivalence M of the module category $A-mod$ given by ${}_{\sigma}A_1 \otimes_A -$. If A is not basic, then the self-equivalence M still exists and induces the automorphism σ on the basic algebra.

Corollary 6.12.16 *Let A be the Brauer tree algebra associated to a stem without exceptional vertex. Then there is an automorphism σ of A and self-equivalences $F_i := X_i \otimes_A - : D^b(A) \longrightarrow D^b(A)$ such that, denoting ${}_{\sigma}A_1 =: M$,*

$$(M \cdot F_1 \cdot (F_2F_1) \cdot \cdots \cdot (F_nF_{n-1} \ldots F_1))(P) \simeq P[n]$$

for all projective A-modules P. □

By Proposition 6.5.10 we get that for any two two-sided tilting complexes X and Y we obtain that $X \otimes_A A \simeq Y \otimes_A A$ in $D^b(A)$ if and only if there is an automorphism α of A with ${}_{\alpha}A_1 \otimes_A X \simeq Y$ in $D^b(A \otimes_k A^{op})$.

Lemma 6.12.17 *Let A be the Brauer tree algebra associated to a stem without exceptional vertex. Let α be a Morita self-equivalence of A. If there is a projective indecomposable A-module P such that ${}^{\alpha}P \not\simeq P$, then ${}^{\alpha}P \simeq {}^{\sigma}P$ for each projective A-module P.*

Proof Without loss of generality the Brauer tree algebra A will be assumed to be basic. Then α and σ are both automorphisms.

For every primitive idempotent e of A, also $\alpha(e)$ is also a primitive idempotent of A. Since

$$\dim_k(Ae) = 3 \Rightarrow \dim_k(A\alpha(e)) = 3$$

and since the only two primitive idempotents e of A with three dimensional Ae are $e = e_1$ with $Ae_1 = P_1$ or $e = e_n$ with $Ae_n = P_n$. Modifying α with σ if necessary we may suppose that $\alpha(e_1) = e_1$, and then also $\alpha(e_n) = e_n$. If f is a primitive idempotent of A with $\dim_k(fAe_1) = 1$, then $f = e_2$. We apply α to this equation and obtain that $\alpha(e_2) = e_2$ as well. Suppose that we have shown already that $\alpha(e_j) = e_j$ for all $j < i$, then there is a unique primitive idempotent $e \notin \{e_1, \dots, e_{i-1}\}$, namely $e = e_i$, with $\dim_k(eAe_{i-1}) = 1$. Hence, again applying α to this equation we obtain $\alpha(e_i) = e_i$.

This shows that either α or $\sigma \circ \alpha$ fixes all projective indecomposable A-modules. □

Corollary 6.12.18 *Let A be the Brauer tree k-algebra associated to a stem without exceptional vertex. Let $Out_0(A)$ be the group of those elements of $DPic_k(A)$ such that $F(P) \simeq P$ for all projective A-modules P. Then there is an $M_0 \in Out_0(A)$ such that we have an isomorphism*

$$(M_0 \cdot M \cdot F_1 \cdot (F_2F_1) \cdot \dots \cdot (F_nF_{n-1} \dots F_1)) \simeq [n]$$

of elements in $DPic_k(A)$.

Proof Indeed, Corollary 6.12.16 shows that the action of the two terms $(F_1 \cdot (F_2F_1) \cdot \dots \cdot (F_nF_{n-1} \dots F_1))$ and $M[n]$ coincide on projectives. Hence Lemma 6.12.17 shows that after modifying with an element which is trivial on all projectives we have equality in $DPic_k(A)$. □

Remark 6.12.19 By Rickard's theorem 6.5.1 and Proposition 6.1.2 we get for any $[X] \in DPic_k(A)$ that the self-equivalence $X \otimes_A^{\mathbb{L}} -$ of $D^b(A)$ restricts to a self-equivalence of $K^b(A\text{-}proj)$. Lemma 6.8.2 then shows that the self-equivalence $X \otimes_A^{\mathbb{L}} -$ of $K^b(A\text{-}proj)$ induces an automorphism of $G_0(K^b(A\text{-}proj))$. Hence we get a group homomorphism

$$DPic_k(A) \longrightarrow Aut_{\mathbb{Z}}(G_0(A\text{-}proj))$$

and we observe that $Out_0(A)$ is in the kernel of this group homomorphism. However, $Out_0(A)$ is strictly smaller than this kernel since, for example, [2] acts trivially on $G_0(K^b(A\text{-}proj))$ but does not act trivially on any projective module.

Lemma 6.12.20 $Out_0(A) \le Pic_k(A)$.

Proof Let $[X] \in Out_0(A)$. Then $X \otimes_A P \simeq P$ for all projective A-modules P. This implies that $X \otimes_A P[i] \simeq P[i]$ for all integers i, and therefore $X \otimes_A A[i] \simeq A[i]$ for all i. The proof of Corollary 3.5.52 shows that

$$H_i(X) \simeq Hom_{D^b(A)}(A[i], X) \simeq Hom_{D^b(A)}(X[i], X)$$

which is 0 unless $i = 0$, using that X is a tilting complex in $D^b(A)$. Therefore the homology of X is concentrated in degree 0 and X is isomorphic to an A–A bimodule. Since the same argument also holds for the inverse $[Y]$ of $[X]$ in $DPic_k(A)$, X is an invertible bimodule. This is precisely the condition for $[X] \in Pic_k(A)$. □

We shall need to compute $Pic_k(A)$ for a basic Brauer tree algebra A without exceptional vertex and associated to a Brauer tree which is just a line.

Lemma 6.12.21 *Let A be a basic Brauer tree algebra A without exceptional vertex and associated to a Brauer tree which is just a line. Then*

$$Out_0(A) = k^\times.$$

Proof By Corollary 4.6.4 $Out_0(A)$ is equal to the automorphism group of A. Any automorphism α is determined by the images of the vertices and of the arrows of the corresponding quiver. We have seen in Lemma 6.12.17 that we may assume that $\alpha(e) = e$ for all idempotents e of A, or what is equivalent, that α induces an element in $Out_0(A)$. The quiver and relations of A were obtained in Sect. 5.10.1. We have n vertices $e_1, \dots, e_n$ and arrows $e_i \xrightarrow{\alpha_i} e_{i+1}$ as well as $e_{i+1} \xrightarrow{\beta_i} e_i$ for all $i \in \{1, \dots, n-1\}$. The relations are:

$$\begin{array}{lll} \alpha_i\alpha_{i+1} = 0 & \beta_{i+1}\beta_i = 0 & \alpha_{i+1}\beta_{i+1} = \beta_i\alpha_i \\ \alpha_1\beta_1\alpha_1 = 0 & & \beta_{n-1}\alpha_{n-1}\beta_{n-1} = 0. \end{array}$$

Hence

$$\begin{aligned} \alpha_i &\mapsto \lambda_i\alpha_i \\ \beta_i &\mapsto \mu_i\beta_i \end{aligned}$$

for $\lambda_i, \mu_i \in k \setminus \{0\}$ are the only possibilities for an automorphism $\alpha \in Out_0(A)$. The relations

$$\alpha_{i+1}\beta_{i+1} = \beta_i\alpha_i$$

imply that

$$\lambda_i\mu_i = \lambda_1\mu_1 =: \kappa$$

for all $i \in \{2, \dots, n-1\}$. Fix now $\gamma_1, \dots, \gamma_n \in k \setminus \{0\}$. Then the element

$$u_\gamma := \sum_{i=1}^{n} \gamma_i e_i$$

is invertible in A, with inverse

$$u_\gamma^{-1} = \sum_{i=1}^{n} \gamma_i^{-1} e_i.$$

We obtain

$$u_\gamma \cdot \alpha_i \cdot u_\gamma^{-1} = \frac{\gamma_i}{\gamma_{i+1}}\alpha_i$$
$$u_\gamma \cdot \beta_i \cdot u_\gamma^{-1} = \frac{\gamma_{i+1}}{\gamma_i}\beta_i$$

for all $i \in \{1, \ldots, n-1\}$. Hence conjugation by different u produce all automorphisms given by $(\lambda_1, \ldots, \lambda_{n-1}; \kappa)$ with the property that $\kappa = 1$. Inner automorphisms give the identity as self-Morita equivalence and hence we get that $Out_0(A)$ is a quotient of $k^\times$, that is, any automorphism is determined by κ, and conversely any automorphism determines a specific κ. On the other hand, if $v = u_\gamma + n$ for some $n \in \mathrm{rad}(A)$, we get $v^{-1} = u^{-1} + m$ for $m \in \mathrm{rad}(A)$ and hence

$$v \cdot \alpha_i \cdot v^{-1} \in u_\gamma \cdot \alpha_i \cdot u_\gamma^{-1} + \mathrm{rad}^2(A)$$

and likewise

$$v \cdot \beta_i \cdot v^{-1} \in u_\gamma \cdot \beta_i \cdot u_\gamma^{-1} + \mathrm{rad}^2(A).$$

Since

$$e_i \cdot rad^2(A) \cdot e_{i+1} = 0 = e_{i+1} \cdot \mathrm{rad}^2(\mathrm{A}) \cdot \mathrm{e_i}$$

for all $i \in \{1, \ldots, n-1\}$, we get

$$v \cdot \alpha_i \cdot v^{-1} = u_\gamma \cdot \alpha_i \cdot u_\gamma^{-1}$$

and

$$v \cdot \beta_i \cdot v^{-1} = u_\gamma \cdot \beta_i \cdot u_\gamma^{-1}.$$

This proves the lemma. □

Remark 6.12.22 The attentive reader will recognise that the result announced in [59] is different. The proof given there contains a mistake which we correct here. The mistake was mentioned to us independently by Dusko Bogdanic and by Alexandra Zvonareva.

Proposition 6.12.23 *Let A be a Brauer tree algebra associated to a stem and without exceptional vertex and let G be the subgroup of $DPic_k(A)$ generated by $F_1, \ldots, F_n$. Then*

$$Out_0(A) \times G \le DPic_k(A)$$

and $Out_0(A) \times \langle\sigma\rangle = Out(A)$.

In particular,

$$(F_1 \cdot (F_2F_1) \cdot \cdots \cdot (F_nF_{n-1} \ldots F_1)) \simeq M[n]$$

in $DPic_k(A)$.

Proof We will first show that ${}_{\alpha}(X_i)_{\alpha^{-1}} \simeq (X_i)$ for every $\alpha \in Out_0(A)$. Indeed

$$\bigoplus_{i=1}^{n} P_i \otimes_A Hom_A(P_i, A) \xrightarrow{\epsilon} A$$

is the projective cover map of A as A–A-bimodules. Since α preserves all projective indecomposable A-modules,

$${}_{\alpha}A_1 \otimes_A \left(\bigoplus_{i=1}^{n} P_i \otimes_A Hom_A(P_i, A) \xrightarrow{\epsilon} A \right) \otimes_A {}_{\alpha}A_1$$

is also a projective cover of A as an A–A-bimodule. Hence there is an isomorphism

$${}_{\alpha}A_1 \otimes_A X_i \otimes_A {}_{\alpha}A_1 \simeq X_i.$$

We shall show now that $Out(A) \cap G = 1$. By construction of the stable equivalence of Morita type induced by a two-sided tilting complex, we see that X_i induces the identity on A-$\underline{mod}$ since the stable equivalence of Morita type induced by X_i is induced by tensoring with A (the term $P_i \otimes_k Hom_A(P_i, A)$ is projective as a bimodule). However, a bimodule ${}_{\alpha}A_1$ acts trivially on each object in A-$\underline{mod}$ if and only if it acts trivially on each object in A-*mod*, and this is the case if and only if α is inner. Indeed, if α acts trivially on each of the A-modules, then it acts trivially on each of the simple modules, hence also on their projective covers. Hence ${}_{\alpha}A_1 \simeq A$ as A–A-bimodules, and therefore α is inner (cf Lemma 1.10.9). Hence

$$Out_0(A) \cap \langle F_1, \ldots, F_n \rangle = 1.$$

We have already seen in Lemma 6.12.17 that

$$\langle \sigma \rangle \times Out_0(A) = Out(A)$$

and therefore

$$\langle F_1, \ldots, F_n, \sigma \rangle \times Out_0(A) \leq DPic_k(A).$$

This proves the proposition. □

Observe that Corollary 6.12.18 for $n = 2$ tells us

$$F_1 F_2 F_1 = M[2].$$

Since $M F_2 M = F_1$ we obtain from this equation

$$(F_1 M)^3 = [-1] \text{ and of course } M^2 = id.$$

We shall need the following result, which is well-known and can be found e.g. in Serre [62].

Proposition 6.12.24 *We have $PSL_2(\mathbb{Z}) \simeq \langle a, b \mid a^3 = b^2 = 1\rangle$, i.e. the group $PSL_2(\mathbb{Z})$ is generated by the symbols a and b, subject to the relations $a^3 = 1$ and $b^2 = 1$.*

The proof is quite sophisticated and uses the classification of the structure of groups acting on trees. We just mention that the two generators correspond to the matrices

$$a \leftrightarrow \begin{pmatrix} 1 & -1 \\ 1 & 0 \end{pmatrix} =: S \text{ and } b \leftrightarrow \begin{pmatrix} 0 & 1 \\ -1 & 0 \end{pmatrix} =: T.$$

We therefore obtain a group homomorphism

$$PSL_2(\mathbb{Z}) \xrightarrow{\chi} DPic_k(A)/\langle[1]\rangle.$$

Theorem 6.12.25 [59] *Let k be an algebraically closed field and let A be a Brauer tree algebra with 2 edges and without an exceptional vertex. Then the map $\chi : PSL_2(\mathbb{Z}) \longrightarrow DPic_k(A)/\langle[1]\rangle$ which maps S to F_1M and T to M is surjective and the kernel coincides with the direct factor $Out_0(A) \simeq k^\times$ of $DPic_k(A)$.*

Proof We shall closely follow [59]. For every $[C]$ in $DPic_k(A)$ let $C_1 := C \otimes_A P_1$ and $C_2 := C \otimes_A P_2$. We shall always replace C_1 and C_2 by the unique complexes C_1' and C_2' in $K^b(A\text{-}proj)$ which satisfy $C_1' \simeq C_1$ and $C_2' \simeq C_2$ in $K^b(A\text{-}proj)$ and so that C_1' and C_2' are of smallest possible k-dimension. These complexes are unique up to isomorphism of complexes since the spaces of non-isomorphisms between indecomposable projective A-modules are at most 1-dimensional (cf Proposition 3.5.23). We call these complexes the reduced parts of C_i, and denote them again by C_i.

We shall proceed with various auxiliary results. The following result proves the injectivity.

Proposition 6.12.26 *If $\chi\left(\begin{pmatrix} a & b \\ c & d \end{pmatrix}\right) =: C$, then $C_1' \simeq P_1^{|a|} \oplus P_2^{|b|}$ and $C_2' \simeq P_1^{|c|} \oplus P_2^{|d|}$ as A-modules (i.e. forgetting the differential).*

Assume ab and cd are not both zero. Let $C_{12} := cone(C_1 \to C_2)$ and $C_{21} := cone(C_2 \to C_1)$.

If $ab \leq 0$ and $cd \leq 0$, then $\dim_k C_{12} = |\dim_k C_1 - \dim_k C_2|$ and $\dim_k C_{21} = \dim_k C_1 + \dim_k C_2$.

If $ab \geq 0$ and $cd \geq 0$, then $\dim_k C_{12} = \dim_k C_1 + \dim_k C_2$ and $\dim_k C_{21} = |\dim_k C_1 - \dim_k C_2|$.

Proof Note first that an element $x = \begin{pmatrix} a & b \\ c & d \end{pmatrix}$ of $PSL_2(\mathbb{Z})$ is determined by $|a|$, $|b|$, $|c|$, $|d|$ and by the signs of ab and cd. Note that if both ab and cd are non-zero, then these signs are equal.

The proposition is clear when $ab = cd = 0$, since then C is isomorphic, up to shift, to A or to M.

So, we assume $(ab, cd) \neq (0, 0)$. We will prove the proposition by induction on $|a| + |b| + |c| + |d|$.

Conjugating if necessary x by the matrix T, we may assume that $ab \leq 0$ and $cd \leq 0$. Let us assume that $|b| \leq |a|$ and $|d| \leq |c|$. The other case can be dealt with following the same proof as below, conjugating all matrices by $\begin{pmatrix} 0 & 1 \\ 1 & 0 \end{pmatrix}$.

We have

$$x = \begin{pmatrix} -b & a+b \\ -d & c+d \end{pmatrix} S.$$

When $b|a+b| = d|c+d| = 0$, we have two cases:

If $x = \begin{pmatrix} 1 & 0 \\ -1 & 1 \end{pmatrix}$ then $C \simeq X_2$, up to a shift.

If $x = \begin{pmatrix} 1 & -1 \\ 1 & 0 \end{pmatrix}$ then $C \simeq X_1 \otimes_A M$, up to a shift.

In the first case, we have

$$\begin{aligned} &C_1 \simeq P_2 \to P_1, \quad C_{12} \simeq P_1[1], \\ &C_2 \simeq P_2 \to 0, \quad C_{21} \simeq P_2 \to P_2 \to P_1 \end{aligned}$$

and we are done.

In the second case, we have

$$\begin{aligned} &C_1 \simeq P_1 \to P_2, \quad C_{12} \simeq P_2[1], \\ &C_2 \simeq P_1 \to 0, \quad C_{21} \simeq P_1 \to P_1 \to P_2 \end{aligned}$$

and we are also done.

Assume now $b|a+b|$ and $d|c+d|$ are not both zero. Denote by C' a two-sided tilting complex such that the class of C in $DPic(A)$ is the product of the class of C' by $F_1 M$. Let C_1' be the reduced part of $C' \otimes_A P_1$ and let C_2' be the reduced part of $C' \otimes_A P_2$. The image of C' in $G/\langle[1]\rangle$ is equal to $\chi\begin{pmatrix} -b & a+b \\ -d & c+d \end{pmatrix}$.

Note that our assumptions on the sign of ab and cd, on the absolute value of a compared to that of b and on the absolute value of c compared to that of d imply $|a+b| + |c+d| < |a| + |c|$. Hence by induction we have $\dim_k \operatorname{cone}(C_1' \to C_2') = \dim_k C_1' + \dim_k C_2'$ and $\dim_k \operatorname{cone}(C_2' \to C_1') = |\dim_k C_1' - \dim_k C_2'|$. We have $C_1 \simeq \operatorname{cone}(C_1' \to C_2')$ and $C_2 \simeq C_1'[1]$. Consider the canonical map $Hom(C_1'[1], C_1'[1]) \to Hom(C_1, C_2)$. Then the morphism $C_1 \to C_2$, which is the image of the identity morphism on $C_1'[1]$ under this canonical map, is not homotopy equivalent to zero. So, $\dim_k \operatorname{cone}(C_1 \to C_2) = \dim_k C_2' = \dim_k C_1 - \dim_k C_2$.

We need to prove now that $\dim_k \operatorname{cone}(C_2 \to C_1) = \dim_k C_1 + \dim_k C_2$.

Since $End_{A\otimes A^{op}}(A) \simeq Z(A)$ and the head and the socle of A as an $(A \otimes A^{op})$-module are isomorphic to $S_1 \otimes Hom_A(S_1, A) \oplus S_2 \otimes Hom_A(S_2, A)$, let $z \in Z(A)$ be an $(A \otimes A^{op})$-endomorphism of A with image isomorphic to $S_1 \otimes Hom_A(S_1, A)$. Then multiplication by z induces a non-zero but non-invertible endomorphism of P_1.

Let z' be the image of z under the automorphism of $Z(A)$ induced by C'. Then under the isomorphism $End(P_1) \simeq End(C_1')$ induced by C', the image of the endomorphism given by multiplication by z is the endomorphism given by multiplication by z'.

Multiplication by z' on a projective module has image in the socle of this module. Hence, the morphism $f : C_1'[1] \to C_1'[1]$ given by multiplication by z' extends to a morphism $g : C_1'[1] \to \text{cone}(C_1' \to C_2')$. Now, the identity map $C_1'[1] \to C_1'[1]$ extends to a map $h : \text{cone}(C_1' \to C_2') \to C_1'[1]$ and we have $f = hg$. As f is not zero, g is not zero either. The reduced part of the cone of g has dimension $\dim C_1' + \dim_k \text{cone}(C_1' \to C_2')$. Now, a non-zero morphism $C_2 \to C_1$ is equal to g up to a scalar. Hence, its cone has dimension $\dim_k C_2 + \dim_k C_1$. So, the second part of the proposition holds for x.

Finally, we know by induction that C_1' is isomorphic to $P_1^{|b|} \oplus P_2^{|d|}$ and C_2' is isomorphic to $P_1^{|a+b|} \oplus P_2^{|c+d|}$, when the differential and the grading are omitted. As $\dim_k C_1 = \dim_k C_1' + \dim_k C_2'$ and $C_1 = \text{cone}(C_1' \to C_2')$, we deduce that C_1 is isomorphic to $P_1^{|a|} \oplus P_2^{|c|}$ when the differential and the grading are omitted. Since $C_2 \simeq C_1'[1]$, when omitting the differential and the grading, C_2 becomes isomorphic to $P_1^{|b|} \oplus P_2^{|d|}$. So, the first part of the proposition holds for x. □

In order to prove surjectivity of χ we prove that, for A a Brauer tree algebra with two edges and no exceptional vertex, the group $DPic_k(A)$ is generated by F_1, F_2 and $Pic_k(A) \times \langle [1] \rangle$.

Let us start with some general properties of the image of simple modules by a derived equivalence.

For any finite dimensional algebra A over a field k and for any bounded complex X over A with non-zero homology, we denote by $\text{rb}(X)$ the smallest integer i with $H_i(X) \neq 0$. Similarly, we denote by $\text{lb}(X)$ the largest integer i with $H_i(X) \neq 0$. We define the *amplitude* of X as

$$\Lambda(X) = \{\text{rb}(X), \text{rb}(X) + 1, \ldots, \text{lb}(X)\}.$$

Finally, the length $\ell(X)$ of X is the cardinality of its amplitude.

Lemma 6.12.27 *Let* $U \longrightarrow V \longrightarrow W \longrightarrow U[1]$ *be a distinguished triangle in* $D^b(A)$.

Then $\Lambda(V) \subseteq \Lambda(U) \cup \Lambda(W)$.

If $rb(U) \neq rb(W) - 1$ *and* $lb(U) \neq lb(W) - 1$, *then we have*

$$\Lambda(V) = \Lambda(U) \cup \Lambda(W).$$

Proof This follows immediately from the long exact sequence

$$\cdots \to H_i(U) \to H_i(V) \to H_i(W) \to H_{i-1}(U) \to \cdots$$

from Corollary 3.5.31. □

Let C be a two-sided tilting complex in $D^b(A \otimes A^{op})$. Replacing C by an isomorphic complex, we may and will assume that the homogeneous components $C_i = 0$ for $i \notin \Lambda(C)$. To avoid trivialities, we assume furthermore that $\Lambda(C)$ has more than one element, or in other words that C is not a shifted module. Denote by $F : D^b(A) \to D^b(A)$ the functor $C \otimes_A -$.

Lemma 6.12.28 *If M is an A-module, then $\Lambda(F(M)) \subseteq \Lambda(C)$.*

Proof: Clear. □

Lemma 6.12.29 *We have $\Lambda(C) = \bigcup_V \Lambda(F(V))$ where V runs over the simple A-modules.*

Proof Since, as an A-module, $C \simeq F(A)$, and as A has a composition series of simple modules, Lemma 6.12.27 gives the inclusion $\Lambda(C) \subseteq \bigcup_V \Lambda(F(V))$. The reverse inclusion follows from Lemma 6.12.28. □

Lemma 6.12.30 *If V is a simple module, then $\Lambda(F(V)) \neq \Lambda(C)$.*

Proof Let V be a simple module with $\Lambda(F(V)) = \Lambda(C)$ and let T be the restriction of $Hom_A(C, A)$ to A. Then we get that the complex of k-modules $\mathbb{R}Hom_A(T, V) \simeq C \otimes_A V$ has amplitude $\Lambda(C) = \Lambda(T^*)$. Let $n = \mathrm{rb}(T)$ and $m = \mathrm{lb}(T)$. There is a non-zero morphism $T \to V[n]$, hence a non-zero morphism $f : P[n] \to T$, which is injective, where P is a projective cover of V. There is also a non-zero morphism $T \to V[m]$. This means that T_m has a direct summand isomorphic to P whose intersection with $H_m(T)$ is non-zero. So, there is a non-zero morphism $g : T \to P[m]$ which is surjective. Now, the morphism $f[m-n] \circ g : T \to T[m-n]$ is non-zero:

$$\begin{array}{ccccccc} & & 0 & \to T_m & \to T_{m-1} & \to \cdots \\ & & & \downarrow & & \\ & & & P & & \\ & & & \downarrow & & \\ \cdots \to & T_{n+1} & \to & T_n & \to & 0 \end{array}$$

Since T is a tilting complex, this implies $m = n$, which has been excluded. This contradiction proves the lemma. □

From now, A is a Brauer tree algebra with two edges and no exceptional vertex. The algebra A has two simple modules S_1 and S_2 and we may assume the indexing is chosen so that

$$(*)\ \mathrm{rb}(F(S_1)) > \mathrm{rb}(F(S_2)) \text{ and } \mathrm{lb}(F(S_1)) > \mathrm{lb}(F(S_2))$$

since by Lemma 6.12.29 and by Lemma 6.12.30, there is no inclusion between the sets $\Lambda(F(S_1))$ and $\Lambda(F(S_2))$. Note that, we then have $\mathrm{rb}(C) = \mathrm{rb}(F(S_2))$ and $\mathrm{lb}(C) = \mathrm{lb}(F(S_1))$.

Denote by $X = P_1 \to P_2$ a complex with P_2 in degree 0 and where the differential $P_1 \to P_2$ is non-zero. We have $H_0(X) \simeq S_2$ and $H_1(X) \simeq S_1$, hence we have a distinguished triangle

$$S_2[-1] \to S_1[1] \to X \to S_2$$

and applying F, a distinguished triangle

$$F(S_1)[1] \to F(X) \to F(S_2) \to F(S_1)[2].$$

By Lemma 6.12.27, this implies

$$\Lambda(F(X)) \subseteq \{\mathrm{rb}(F(S_2)), \ldots, \mathrm{lb}(F(S_1)) + 1\},$$

using $(*)$. In particular, $\ell(F(X)) < \ell(C)$.

Let L be the kernel of a surjective map $P_2 \to S_2$. We have an exact sequence

$$0 \longrightarrow S_2 \longrightarrow L \longrightarrow S_1 \longrightarrow 0,$$

hence a distinguished triangle

$$F(S_2) \longrightarrow F(L) \longrightarrow F(S_1) \longrightarrow F(S_2)[1].$$

By Lemma 6.12.27, we obtain $\Lambda(F(L)) = \Lambda(C)$. The distinguished triangle

$$F(L) \longrightarrow F(P_2) \longrightarrow F(S_2) \longrightarrow F(L)[1]$$

shows that $\mathrm{rb}(F(P_2)) = \mathrm{rb}(C)$. The distinguished triangle

$$F(S_1) \longrightarrow F(P_1) \longrightarrow F(L) \longrightarrow F(S_1)[1]$$

shows that $\mathrm{lb}(F(P_1)) = \mathrm{lb}(C)$.

If $\mathrm{lb}(F(P_2)) < \mathrm{lb}(C)$, then $\Lambda(F(X) \oplus F(P_2))$ is strictly contained in $\Lambda(C)$. Let $C' = C \otimes_A Hom_A(X_2, A)[1]$. Then

$$C' \otimes_A P_1 \simeq F(X)$$

and

$$C' \otimes_A P_2 \simeq F(P_2).$$

Consequently, $\Lambda(C')$ is strictly contained in $\Lambda(C)$.

If $\text{lb}(F(P_2)) = \text{lb}(C)$, then

$$\Lambda(F(X)[1] \oplus F(P_1)) \subseteq \Lambda(C)$$

and

$$\ell(F(X)) + \ell(F(P_1)) < \ell(F(P_2)) + \ell(F(P_1)).$$

Let $C' = C \otimes_A X_1[1]$. Then $C' \otimes_A P_1 \simeq F(P_1)$ and $C' \otimes_A P_2 \simeq F(X)[1]$. So,

$$\Lambda(C') \subseteq \Lambda(C)$$

and

$$\ell(C' \otimes_A P_1) + \ell(C' \otimes_A P_2) < \ell(C \otimes_A P_1) + \ell(C \otimes_A P_2).$$

It follows by induction first on $\ell(C)$, then on $\ell(C \otimes_A P_1) + \ell(C \otimes_A P_2)$, that, modulo the subgroup generated by F_1 and F_2, every element of $DPic_k(A)$ is in $Pic_k(A) \times \langle [1] \rangle$. This proves Theorem 6.12.25. □

Remark 6.12.31 It is known that $PSL_2(\mathbb{Z})$ is isomorphic to the quotient of the braid group on 3 strings modulo its centre. In [59] it is shown that actually there is a homomorphism of the braid group on $n+1$ strings to $DPic_k(A_n)$ where A_n is the Brauer tree algebra with n edges and without exceptional vertex. Furthermore, if $n = 2$ then $DPic_k(A_2)$ is the direct product of a central extension of B_3 of degree 2 and $Out_0(A_2)$.

This result attracted much interest. In a completely independent approach [63] Seidel and Thomas, as well as Khovanov and Seidel [64], discovered the homomorphism of the braid group on $n+1$ strings to $DPic_k(A_n)$ in the context of mirror symmetry, mathematical physics and differential geometry. They show by very sophisticated methods that the homomorphism of the braid group to $DPic_k(A_n)$ is injective for all n.

Remark 6.12.32 Proposition 6.12.26 has various refinements. In [65] the following interesting results are proved.

Recall the short exact sequence

$$1 \longrightarrow k^\times \longrightarrow PSL_2(\mathbb{Z}) \xrightarrow{\chi} DPic_k(A)/\langle [1] \rangle \longrightarrow 1$$

of groups from Theorem 6.12.25.

The group ring $\mathbb{F}_3\mathfrak{S}_3$ of the symmetric group of order 6 in characteristic 3 is a Brauer tree algebra with two simple modules and no exceptional vertex. Since the projective indecomposable $\mathbb{F}_3\mathfrak{S}_3$-modules are reductions modulo 3 of projective indecomposable $\hat{\mathbb{Z}}_3\mathfrak{S}_3$-modules, we can actually lift the two two-sided tilting complexes X_1 and X_2 to two-sided tilting complexes $\hat{X}_1$ and $\hat{X}_2$ over $\hat{\mathbb{Z}}_3\mathfrak{S}_3$, giving elements in $Pic_{\hat{\mathbb{Z}}_3}(\hat{\mathbb{Z}}_3\mathfrak{S}_3)$. We can then try to find the kernel of the mapping

$$DPic_{\hat{\mathbb{Z}}_3}(\hat{\mathbb{Z}}_3\mathfrak{S}_3) \longrightarrow DPic_{\hat{\mathbb{Q}}_3}(\hat{\mathbb{Q}}_3\mathfrak{S}_3)$$

given by tensor product with $\hat{\mathbb{Q}}_3$ over $\hat{\mathbb{Z}}_3$. Then this subgroup is mapped by χ^{-1} to the derived subgroup of the level 2 congruence subgroup $\Gamma(2)$ of $PSL_2(\mathbb{Z})$. Here

$$\Gamma(2) = \ker(PSL_2(\mathbb{Z}) \longrightarrow PSL_2(\mathbb{F}_2)).$$

Moreover, $Picent(\mathbb{F}_3\mathfrak{S}_3)$ is mapped by χ^{-1} to $\Gamma(2)$. It is known that $\Gamma(2)$ is a free group on 2 generators, and actually

$$\Gamma(2) = < \begin{pmatrix} 1 & 2 \\ 0 & 1 \end{pmatrix}, \begin{pmatrix} 1 & 0 \\ -2 & 1 \end{pmatrix} > .$$

These two matrices correspond to F_1^2 and F_2^2 in the above proof. Then the commutator of F_1^2 with F_2^2 corresponds to the commutator matrix, which is

$$\begin{pmatrix} 13 & 8 \\ 8 & 5 \end{pmatrix}.$$

This means that the smallest one-sided tilting complex realising this derived equivalence is isomorphic to

$$(P_1^{13} \oplus P_2^8) \oplus (P_1^8 \oplus P_2^5)$$

as an $\mathbb{F}_3\mathfrak{S}_3$-module. For details on these and other matrix groups the reader may consult Newman [66].

Remark 6.12.33 Keller used in [67] the group *DPic* to show that the Gerstenhaber structure on the Hochschild cohomology is invariant under derived equivalences. More precisely, Keller considered the functor $S \mapsto DPic_S(A \otimes_k S)$ from the category of commutative k-free k-algebras S to the category of groups. In the same way as a Lie algebra is obtained naturally from an algebraic group Keller constructed a Lie algebra from this functor. He showed that this Lie algebra is precisely the Gerstenhaber structure on the Hochschild cohomology of A.

References

1. Rickard, J.: Morita theory for derived categories. J. Lond. Math. Soc. **39**(2), 436–456 (1989)
2. Rickard, J.: Derived equivalences as derived functors. J. Lond. Math. Soc. **43**, 37–48 (1991)
3. Keller, B.: Bimodule complexes via strong homotopy actions. Algebras Represent. Theory **3**, 357–376 (2000)
4. Montgomery, S.: Hopf algebras and their actions on rings. CBMS Regional Conference Series in Mathematics, vol. 82. American Mathematical Society, Providence, R.I. (1993)

5. Brzezinski, T., Wisbauer, R.: Corings and comodules. London Mathematical Society Lecture Notes, vol. 309. Cambridge University Press, Cambridge (2003)
6. Stasheff, J.D.: The intrinsic bracket on the deformation complex of an associative algebra. J. Pure Appl. Algebra **89**, 231–235 (1993)
7. Stasheff, J.D.: Homotopy associativity of H-spaces. I, II. Trans. Am. Math. Soc. **108**, 275–292 (1963) (ibid. 108 (1963) 293–312)
8. Zimmermann, A.: Twosided tilting complexes for Green orders and Brauer tree algebras. J. Algebra **187**, 446–473 (1997)
9. Zimmermann, A.: Twosided tilting complexes for Gorenstein orders. J. Algebra **209**, 585–621 (1998)
10. Zimmermann, A.: Fine Hochschild invariants of derived categories for symmetric algebras. J. Algebra **308**, 350–367 (2007)
11. König, S., Zimmermann, A.: Derived equivalences for group rings, with contributions by Bernhard Keller, Markus Linckelmann, Jeremy Rickard and Raphaël Rouquier. Lecture Notes in Mathematics, vol. 1685. Springer, Berlin (1998)
12. Muchtadi-Alamsyah, I.: Endomorphismes de complexes déterminés par leurs homologies, thèse de doctorat Université de Picardie (2004)
13. Muchtadi-Alamsyah, I.: Homomorphisms of complexes via homologies. J. Algebra **294**, 321–345 (2005)
14. Rickard, J.: Infinitely many algebras derived equivalent to a block (preprint 2013), http://arxiv.org/abs/1310.2403v1arXiv:1310.2403v1
15. König, S., Zimmermann, A.: Tilting hereditary orders. Commun. Algebra **24**(6), 1897–1913 (1996)
16. Carlson, J., Thévenaz, J.: Torsion endo-trivial modules. Algebras Represent. Theory **3**, 303–335 (2000)
17. Carlson, J., Thévenaz, J.: The classification of endo-trivial modules. Inventiones Math. **158**, 389–411 (2004)
18. Carlson, J., Thévenaz, J.: The classification of torsion endo-trivial modules. Ann. Math. **162**, 823–883 (2005)
19. Keller, B.: Invariance and localization for cyclic homology of DG algebras. J. Pure Appl. Algebra **123**, 223–273 (1998)
20. Zimmermann, A.: Tilted symmetric orders are symmetric orders. Archiv der Mathematik **73**, 15–17 (1999)
21. Al-Nofayeeh, S.: Equivalences of derived categories for selfinjective algebras. J. Algebra **313**, 897–904 (2007)
22. Hartshorne, R.: Residues and duality, Lecture Notes of a Seminar on the Work of A. Grothendieck, Given at Harvard 1963/64. Lecture Notes in Mathematics, vol. 20. Springer, Berlin (1966)
23. Broué, M.: Isométries parfaites, types de blocs, catégories dérivées. Astérisque **181–182**, 61–92 (1990)
24. Broué, M.: Isométries parfaites, types de blocs, catégories dérivées, Représentations Linéaires des Groupes Finis, Luminy 16–21 mai 1988, ed: Marc Cabanes; pages 61–92, Astérisque **181–182** (1990)
25. Buchweitz, R.O.: Maximal Cohen-Macaulay modules and Tate cohomology, manuscript Universität Hannover (1986)
26. Orlov, D.: Triangulated categories of singularities and D-branes in Ginzburg-Landau models. Trudy Steklov Math. Inst. **246**, 240–262 (2004)
27. Orlov, D.: Derived categories of coherent sheaves and triangulated categories of singularities, Algebra, arithmetic, and geometry: in honor of Yurij I. Manin. Volume II, 503–531. Progress in Mathematics, vol. 270. Birkhäuser Boston, Boston (2009)
28. Orlov, D.: Formal completions and idempotent completions of triangulated categories of singularities. Adv. Math. **226**, 206–217 (2011)
29. Rickard, J.: Derived categories and stable equivalences. J. Pure Appl. Algebra **61**, 303–317 (1989)

30. Keller, B., Vossieck, D.: Sous les catégories dérivées. Comptes Rendus de l'Académie des Sciences Paris, Série I Mathématique **305**(6), 225–228 (1987)
31. Beligiannis, A.: The homological theory of contravariantly finite subcategories: Auslander-Reiten contexts, Gorenstein categories and (co-)stabilizations. Commun. Algebra **28**(10), 4547–4596 (2000)
32. Chen, X.-W.: The singularity category of an algebra with radical square zero. Documenta Mathematica **16**, 921–936 (2011)
33. König, S., Zimmermann, A.: Tilting selfinjective algebras and Gorenstein orders, Oxford. Q. J. Math. **48**(2), 351–361 (1997)
34. Reiner, I.: Maximal Orders. Academic Press, London (1975)
35. Schaps, M., Zakay-Illouz, E.: Pointed Brauer trees. J. Algebra **246**, 647–672 (2001)
36. Okuyama, T.: Some examples of derived equivalent blocks of finite groups (unpublished preprint 1997)
37. Okuyama, T.: Remarks on splendid tilting complexes, (Japanese). In: Koshitani, S. (ed.) Representations of finite groups and related topics RIMS Kokyuroku. Proceedings of the Research Institute of Mathematical Sciences, vol. 1149, pp. 53–59. Kyoto University (2000)
38. Kunugi, N.: Derived equivalences in symmetric groups. Surikaisekikenkyusho Kokyuroku **1140**, 131–135 (2000) (Research on the cohomology theory of finite groups (Japanese) (Kyoto, 1999))
39. Kunugi, N.: Morita equivalent 3-blocks of the 3-dimensional projective special linear groups. Proc. Lond. Math. Soc. **80**, 575–589 (2000)
40. Kunugi, N., Waki, K.: Derived equivalences for the 3-dimensional special unitary groups in non-defining characteristic. J. Algebra **240**, 251–267 (2001)
41. Koshitani, S., Kunugi, N.: Broué's conjecture holds for principal 3-blocks with elementary abelian defect group of order 9. J. Algebra **248**, 575–604 (2002)
42. Koshitani, S., Kunugi, N., Waki, K.: Broué's conjecture for non-principal 3-blocks of finite groups. J. Pure Appl. Algebra **173**, 177–211 (2002)
43. Koshitani, S., Kunugi, N., Waki, K.: Broué's Abelian defect group conjecture for the Held group and the sporadic Suzuki group. J. Algebra **279**, 638–666 (2004)
44. Koshitani, S., Kunugi, N., Waki, K.: Broué's abelian defect conjecture holds for the Janko simple group J_4. J. Pure Appl. Algebra **212**, 1438–1456 (2008)
45. Koshitani, S., Müller, J.: Broué's abelian defect group conjecture holds for the Harada-Norton sporadic simple group HN. J. Algebra **324**, 394–429 (2010)
46. Rickard, J.: Equivalences of derived categories for symmeric algebras. J. Algebra **257**, 460–481 (2002)
47. König, S., Yang, D.: Silting objects, simple-minded collections, t-structures and co-t-structures for finite-dimensional algebras. Doc. Math. **19**, 403–438 (2014)
48. Aihara, T., Iyama, O.: Silting mutation in triangulated categories. J. Lond. Math. Soc. **85**, 633–668 (2012)
49. Keller, B., Vossieck, D.: Aisles in derived categories. Bulletin de la Société mathématique de la Belgique, Série A **40**, 239–253 (1988)
50. Brenner, S., Butler, M.: Generalizations of the Bernstein-Gel'fand-Ponomarev reflection functors. Representation theory, II (Proceedings of the Second International Conference, Carleton University, Ottawa, Ontario, 1979), pp. 103–169. Lecture Notes in Mathematics, vol. 832. Springer, Berlin-New York (1980)
51. Happel, D.: On the derived category of a finite-dimensional algebra. Commentarii Mathematici Helvetici **62**, 339–389 (1987)
52. Happel, D.: Triangulated categories in the representation theory of finite-dimensional algebras. London Mathematical Society Lecture Note Series, vol. 119. Cambridge University Press, Cambridge (1988)
53. Keller, B.: Deriving DG categories. Annales Scientifiques de l'École Normale Supérieure **27**(4), 63–102 (1994)
54. Broué, M.: Equivalences of blocks of group algebras. Finite dimensional algebras and related topics (Ottawa, ON, 1992), 126, NATO Advanced Science Institute Series C Mathematics Physics Science, vol. 42. Kluwer Academic Publisher, Dordrecht (1994)

55. Külshammer, B.: Offene Probleme in der Darstellungstheorie endlicher Gruppen. Jahresbericht der Deutschen Mathematiker Vereinigung **94**, 98–104 (1992)
56. Marcus, A.: On equivalences between blocks of group algebras: reduction to the simple components. J. Algebra **184**, 372–396 (1996)
57. Chuang, J., Rouquier, R.: Derived equivalences for symmetric groups and $\mathfrak{sl}_2$-categorification. Ann. Math. **167**(2), 245–298 (2008)
58. Holm, T.: Derived equivalence classification of algebras of dihedral, semidihedral, and quaternion type. J. Algebra **211**, 159–205 (1999)
59. Rouquier, R., Zimmermann, A.: Picard groups for derived module categories. Proc. Lond. Math. Soc. **87**(3), 197–225 (2003)
60. Fröhlich, A.: The Picard group of noncommutative rings, in particular of orders. Trans. Am. Math. Soc. **180**, 1–45 (1973)
61. Roggenkamp, K.W., Zimmermann, A.: Outer group automorphisms can become inner in the integral group ring. J. Pure Appl. Algebra **103**, 91–99 (1995)
62. Serre, J.P.: Arbres, amalgames, SL_2, Société Mathématique de France, Astérisque 46, Paris (1977)
63. Seidel, P., Thomas, R.: Braid group actions on derived categories of coherent sheaves. Duke Math. J. **108**, 37–108 (2001)
64. Khovanov, M., Seidel, P.: Quivers, and braid group actions. J. Am. Math. Soc. **15**, 203–271 (2002)
65. Zimmermann, A.: Automorphisms of green orders and their derived categories. Algebras Represent. Theory **7**(1), 19–34 (2004)
66. Newman, M.: Integral Matrices. Academic Press, New York (1972)
67. Keller, B.: Hochschild cohomology and derived Picard groups. J. Pure Appl. Algebra **190**, 177–196 (2004)

Index

Symbols
$- \otimes^{\mathbb{L}} -$, 353
A_∞-algebra, 574
A-Mod, 260
A-$Proj$, 280
A-mod, 260
A-$proj$, 280
B covers b, 246
$D(\mathcal{A})$, 328
$DPicent(A)$, 671
$D_{sg}(A)$, 627
$D_{\mathcal{A}'}(\mathcal{A})$, 328
Ext, 80, 87
  via injectives, 539
Ext-quiver, 206
$Ext_A^1(M, N)$, 87
$Ext_A^i(M, N)$, 94
$Ext_A^i(X, Y)$, 355
$HH^0(A)$, 343
HH^n, 342
$HH_0(A)$, 214, 343
HH_n, 342
$Hom_A(L, \alpha)$, 76
$Hom_A(\alpha, L)$, 76
I-adic completion, 189
K_0 of a ring, 199
$Pic_K(A)$, 419
$Picent(A)$, 424
R-linear category, 273
R-linear functor, 273
R-order in a semisimple algebra, 189
Tor-sequence, 361
$Tor_k^A(N, M)$, 342, 353
$Z^{st}(A)$, 486
$[A, A]$, 214
rb, 688
lb, 688
$DPic_k(A)$, 669
b^G, 246
cde-triangle, 202
n-th syzygy, 79
p-adic integers, 187
p-adic numbers, 188
p-group, 40
p-permutation module, 70
t-structures, 657
$\mathcal{E}ns$, 260
$\mathbb{R}Hom_A(X, Y)$, 355
$\mathcal{A}b$, 260
$K^{x,y}(\mathcal{A})$, 304

A
Abelian
  category, 266, 279
  defect conjecture, 666
  group, 3
    Baer sum, 88
    divisible, 77
    finitely generated, 77
    free, 48
Absolutely simple, 32
Acyclic complex, 309
Additive category, 279
Adjacent, 505
Adjoint
  functor, 274
  pair, 274, 358, 445, 584
Admissible
  ideal, 146, 414
  relation, 146
Algebra, 1, 564
  A_∞, 574

A. Zimmermann, *Representation Theory*,
Algebra and Applications 19, DOI: 10.1007/978-3-319-07968-4,

basic, 402, 418, 421
bi-, 566
Brauer tree, 506, 528, 532, 683
co-, 564
epimorphism, 2
extension, 160
free-, 29
Frobenius, 126, 417, 418
graded, 173
hereditary, 140, 415
homomorphism, 2
Hopf, 567
indecomposable, 177
monomorphism, 2
Nakayama, 128, 472
opposite, 3
Picard group of, 419
quaternion, 22, 55
quiver, 142, 414
Kronecker, 42
self-injective, 126
semisimple, 24, 28
separable, 346
serial, 47, 209, 210
simple, 27
special biserial, 146
symmetric, 134
weakly symmetric, 133
not symmetric, 135
Weyl, 30
Almost split sequences, x, 553
Amplitude, 688
Annihilator, 27
Antipode, 567
Arrow of a quiver, 142
Artinian, 14
Associated graded module, 372
Associative bilinear form, 127
Augmentation
ideal, 107
map, 107, 109, 565
Auslander–Kleiner lemma, 445
Auslander–Reiten
conjecture, 519, 553, 657
quivers, x
Automorphism of modules, 4

B

Baer sum, 88, 89
Balanced mapping, 51
Band modules, 151
Bar resolution, 342
Base field extension, 31
Basic algebra, 402, 418, 421
Belongs to a block, 177
Bi-algebra, 566
Bicomplex, 378
Bilinear form, 127
Bimodule, 51
Biproduct in categories, 261, 262
Blichfeldt's theorem, 172
Block
belongs to, 177
Brauer pair belongs to, 238
class, 205
covers, 246
nilpotent, 406
of a group ring, 176
of cyclic defect group, structure theorem, 532
principal, 177
Bouc's trace map, 476, 479, 497
Boundaries of a complex, 301
Bounded
above complex, 301
below complex, 301
complex, 301
Braid group, 691
Brauer
correspondence, 176
generalised sense, 241
correspondent, 181
first main theorem, 180
group, 23
homomorphism, 228
pair, 234, 406
belongs to block, 238
quotient, 234
star algebra, 509
third main theorem, 183
tree, 505
algebra, 506, 528, 532, 683
Brauer tree algebras
derived equivalence, 645
Brenner–Butler tilting, 563
Broué's
abelian defect conjecture, viii, 652, 666
stable equivalence of Morita type, 446

C

Cartan
determinant, 625
mapping, 199, 623
matrix, 199, 625

Cartan Brauer triangle, 202
 and derived equivalence, 623
Category, 259
 R-linear, 273
 abelian, 266, 279
 additive, 279
 colimit in, 265
 comma, 260
 derived, 328
  isomorphism in, 328
 equivalence of -, 268
 exact, 280
 full sub-, 273
 generator, 387
 homotopy, 304
 isomorphism of -, 268
 limit in, 265
 morphisms in, 259
 objects of, 259
 of abelian groups, 260
 of complexes, 300
 of functors, 267
 of sets, 260
 opposite, 260
 progenerator, 394
 singularity, 627
 stable, 428
 triangulated, 289
Cauchy sequence in valuation rings, 186
Cayley-Hamilton theorem, 26
Centre
 projective, 486
 stable, 486
Centre of an algebra, 1, 33
Chain complex, 300
Change of rings, 59, 194, 364
Circulant matrix, 472
Classical order, 189
Clifford's theorem, 170
Closed path, 142
Cluster categories, 657
Coalgebra, 564
Coboundary
 group, 105
 Hochschild, 342
Cochain complex, 300
Cocycles
 group, 105
 Hochschild, 342
Coderivation, 569
Codirect system, 265
Coequaliser, 263
Cofree comodule, 568
Cogenerator, 387
Cohomological functor, 299
Cohomology
 cyclic, 616
 Hochschild, 342, 343
 of groups, 101
Cokernel
 in a category, 263
 of a module homomorphism, 4
Colimit in a category, 265
Comma category, 260
Commutator quotient, 214
Commutator subspace, 214
Comodule, 567
Compact object, 283, 559
Compatible grading and filtration, 373
Complete
 discrete valuation ring, 187
 set of primitive idempotents, 124
Completion, 187
 I-adic, 189
Complex
 boundaries of, 301
 acyclic, 309
 bi, 378
 category, 300
 chain, 300
 cochain, 300
 cycles of, 301
 differential of, 300
 double, 378
 filtered, 372
 homogeneous component, 300
 homology of, 301
 of modules, 300
 total, 353
Composition factors, 43
Composition series, 43
Concatenation of paths, 142
Cone of a morphism, 313
Connecting morphism, 281
Contravariant, 263
Convergent
 sequence in valuation rings, 186
 spectral sequence, 381
Convolution product, 566
Coproduct, 565
 in categories, 261
 of modules, 7
Coresolution, 120
Counit
 of a coalgebra, 565
 of the adjunction, 274

Covariant, 263
Covering of blocks, 246
Covering technique, 511, 528
Cross effect, 392
Cup product, 357
Cutting, 309
Cycles of a complex, 301
Cyclic
  group, 10
  homology, 616
  module, 27
  normal defect, 252
  ordering, 505

D

Decomposable module, 17
Decomposition
  map, 200, 623
  matrix, 201
  number, 201
Dedekind domain, 672
Defect, 178
  abelian d. conjecture, 666
  group, 178
    cyclic, 254
    cyclic normal, 252
  of principal block, 180
Dense functor, 273
Derivation, 102, 345, 568
Derived
  category, 328
    isomorphism in, 328
  functor, 350
Derived equivalence, 589
  p-groups, 609
  and being a symmetric algebra, 617
  and Brauer tree algebras, 645
  and Cartan Brauer triangle, 623
  and centre, 614
  and Grothendieck group, 621
  and Hochschild cohomology, 613
  and Hochschild homology, 615
  and indecomposability, 614
  and stable invariants, 636
  commutative algebras, 610
  for local rings, 608
  of standard type, 594
Differential of a complex, 300
Dihedral group, 66
Dimension
  global, 80, 539
  injective, 539
  projective, 539
Direct product of modules, 6
Direct sum of modules, 7
Direct system, 265
Discrete valuation, 185
  ring, 187
Divisible abelian group, 77
Double
  class, 66
  complex, 378
  coset, 66
Dual numbers, 17
  stable category of, 440

E

Eilenberg Watts theorem, *see* Watts' theorem
Elementary divisors, 471
Endomorphism of modules, 4
Enough
  injectives, 337
  projectives, 337
Epimorphism
  in a category, 261
  of algebras, 2
  of modules, 4
Equaliser, 263
Equivalence
  derived categories, 594
  Morita, 399
  of categories, 268
  of short exact sequences, 82
  of standard type, 593, 594
  stable, 440
    of Morita type, 446
    of Morita type generalised sense, 452
Erdmann classification of tame blocks, 221, 667
Essentially surjective functor, 273
Exact
  category, 280
  functor, 284
  half, 541
  sequence
    equivalence, 82
    long, 79
    short, 74
Exceptional
  multiplicity, 505, 531, 532
  vertex, 505
Exponent of a group, 225
Ext-groups, 87, 94, 355
Extension

of algebras, 160
separable, 160, 346

F
Faithful functor, 273
Faithful module, 172
Faithfully flat, 193, 361
Feit-Thompson theorem, 115
Field, 1
Filtered
complex, 372
differential module, 372
graded module, 373
Filtration, 372
regular, 373
Finite representation type, 252, 254
Finitely generated
abelian groups, 77
module, 11, 14
Finitely presented module, 283
Fitting's lemma, 17
for orders, 193
Five lemma for derived categories, 291
Five lemma for triangulated categories, 291
Flat, 193, 361, 362, 449
and localisation, 610
and projective, 363
faithfully, 193
Forgetful functor, 264
Fröhlich's localisation sequence, 672
Frattini quotient, 115
Free
abelian group, 48
algebra, 29
module, 73
resolution, 79
Frobenius
algebra, 125, 126, 417, 418
reciprocity, 64, 97, 358
Full functor, 273
Full subcategory, 273
Functor, 263
R-linear, 273
adjoint, 274
category, 267
between modules, 535
projective object in, 270
cohomological, 299
dense, 273
derived, 350
essentially surjective, 273
exact, 284
faithful, 273
forgetful, 264
full, 273
homological, 299, 339
Nakayama, 416
of triangulated categories, 299
polynomial, 392
preserves direct sums, 391
quasi-inverse, 268
representable, 76, 264, 269, 285, 298
shift, 289
suspension, 289, 669

G
Gabriel's theorem, 414
Gabriel–Riedtmann theorem, 511, 528
Galois module structure, 197
Generalised quaternion group, 22
Generator, 387
Gerstenhaber structure, 504, 589, 692
Global dimension, 80, 539
Graded
algebra, 173
filtered module, 373
module, 300
morphism, 300
Graph, 505
Green
correspondence, 166
correspondent, 167
indecomposability theorem, 176
order, 534
walk around the Brauer tree, 523
Grothendieck
group
for general algebras, 198
invariance under derived equivalences, 622
invariance under stable equivalence, 470
of derived category, 622
of triangulated category, 619
stable, 469
universe, 264
Group
abelian
divisible, 77
finitely generated, 77
algebra, 8
braid, 691
cohomology, 101
2-coboundary, 105

2-cocycle, 105
dihedral, 66
graded algebra, 173
inertia, 170
modular, 686
quaternion, 22, 55
generalised, 22
representation, 8
ring, 8
symmetric, 62

H

Half exact, 541
Hattori Stallings trace map, 498
Head, 37
Hereditary algebra, 140, 415
Hereditary order, 605
Higher torsion groups, 354
Higman
criterion, 232
ideal, 489
and projective centre, 492
and Reynolds ideal, 495
Hochschild
1-coboundary, 569
1-cocycle, 345, 569
2-coboundary, 344
2-cocycle, 344
Bouc's trace
on complex, 476
on homology, 479
coboundary, 342
cocycle, 342
cohomology, 342, 343, 569
derived equivalence, 613
stable equivalence of Morita type, 485
homology, 214, 342
derived equivalence, 615
stable equivalence of Morita type, 483
values in a bimodule, 479
Homogeneous component of a complex, 300
Homological functor, 299, 339
Homology
as a functor, 301
cyclic, 616
of a complex, 301
Homomorphism
of algebras, 2
of modules, 4
Homotopic to zero, 304
Homotopy, 304
category of complexes, 304
equivalent, 304
Hopf algebra, 59, 567
Hopkins' theorem, 36, 40
Horseshoe lemma, 338
Hypercohomology Ext, 355

I

Idempotent
element, 13, 37, 121
endomorphism, 13
split, 639
Image of module homomorphism, 4
Indecomposable
algebra, 177
module, 17
Index of a subgroup, 60
Induced module, 59
Inductive limit, 265
Inertia group, 170
of a block, 245
of an idempotent, 245
Infinite page of a spectral sequence, 376
Infinite representation type, 252
Initial object, 262
Injective
coresolution, 120
dimension, 539
enough objects, 337
envelope, 118
hull, 118
module, 76
object in a category, 264
sufficiently many, 337
Inner
automorphism, 67
derivation, 102, 568
Intelligent cutting, 309
Intelligent truncation, 309
Inverse limit, 188
Invertible bimodule, 400
Isomorphism
in a category, 261
in derived category, 328
of modules, 4

J

Jacobson radical, 35

K
Külshammer
approach to Brauer pairs, 234
approach to nilpotent blocks, 407
definition of separable extension, 160
ideal, 221
and stable equivalence of Morita type, 500
spaces, 216
Keller's theorem
on derived equivalences, 557
on strictifying homotopy actions, 587
Kernel
in a category, 263
of module homomorphism, 4
Kronecker quiver, 41, 42, 255
Krull Schmidt theorem, 17, 370
for lattices over orders, 194

L
Lattice, 192
Lazy path in a quiver, 142
Left
adjoint, 274
derived
functor, 350
tensor product, 353
exact functor, 284
module, 23
Left bound, 688
Lifting of idempotents, 124
Limit
inductive, 265
inverse, 188
projective, 188, 265
Limit in a category, 265
Linkage class, 206
Local ring, 18
derived equivalence between, 608
is basic, 402
Localisation, 332
Loewy
factors, 40
length, 40
series, 40
Long exact sequence, 79, 298, 361
Loop in a quiver, 142

M
Mackey's formula, 66
Malcev Wedderburn theorem, *see* Wedderburn Malcev theorem
Maschke's theorem, 11
Maximal element, 16
May's lemma, 291
Minimal element, 16
Mitchell's theorem, 280
Modular group, 686
Module, 3
associated graded, 372
cyclic, 27
decomposable, 17
faithful, 172, 401
finitely presented, 283
free, 73
Galois m., 197
graded, 300
homomorphism, 4
indecomposable, 17
induced, 59
left, 3, 23
permutation, 71
quasi-tilting, 608
right, 3, 23
source, 164
string, 149
trivial source, 71
uniserial, 46, 209, 210
vertex, 161
Monoid, 260
Monomorphism
in a category, 261
of algebras, 2
of modules, 4
Morita
bimodule, 399
equivalence, 399
and centres, 401
centraliser of defect groups, 404
principal blocks and normal subgroups, 403
stable equivalence of M. type, 446
Morita theorem, 398
Morphism
cone of m., 313
of comodules, 568
Morphisms
in a category, 259
of graded modules, 300
of strong homotopy actions, 578
Multiplicity
exceptional, 505

N
Nakayama

algebra, 128, 472
stable equivalence of Morita type, 473
automorphism, 132
functor, 416
lemma, 36
permutation
for Frobenius algebra, 132
for self-injective algebra, 140
Natural isomorphism, 266
Natural transformation, 266
Nil-radical, 38
Nilpotent
block, 406
element, 17, 37
Noether Deuring theorem
for derived categories, 368, 371
for modules, 195
Noether Skolem theorem, *see* Skolem Noether theorem
Noetherian, 14
Non-degenerate bilinear form, 127
Nullhomotopic strong homotopy action, 579

O

Object
in a category, 259
initial, 262
terminal, 262
zero, 262
Odd order theorem, 115
Okuyama's method, 468, 651
Okuyama's theorem, 653
One-sided tilting complex, 562
Opposite
algebra, 3
category, 260
Order in a semisimple algebra, 189
Orthogonal idempotent, 121
Osima's theorem, 230

P

Page of the spectral sequence, 375
Path in a quiver, 142
Perfect isometry, 625, 666
Permutation basis, 70
Permutation module, 70, 71
Picard group
Fröhlich's sequence, 672
of an algebra, 419
of derived category, 669
and centre, 670
of a commutative algebra, 671
of a local algebra, 671
of Brauer tree algebra, 686
Polynomial functor, 392
Primitive idempotent, 121, 123
Principal block, 177
Product in categories, 261
Progenerator, 394
Projective
and flat, 363
centre, 486
and Cartan mapping, 496
and Higman ideal, 492
cover, 118
dimension, 539
enough objects, 337
limit, 188, 265
module, 74
object in a category, 264
relative to a
subalgebra, 155
subgroup, 156
resolution, 79
sufficiently many, 337
Proper submodule, 4
Puig's theorem, 407
Pullback, 79
in categories, 263
of groups, 91
of modules, 89
of sets, 91
Pushout
in categories, 263
of modules, 81, 89

Q

Quasi-inverse functor, 268
Quasi-isomorphism, 585
Quasi-tilting module, 563, 608
Quaternion
algebra, 22, 55
group, 22, 55
generalised, 22
Quiver, 41, 128, 142
Ext, 206
admissible ideal, 146
algebra, 142, 412, 414
arrow, 142
Kronecker, 42
lazy path, 142
loop, 142
path, 142

tree, 505
vertex, 142
walk, 142

R

Radical, 35
of a p-group algebra, 40
square zero algebra, 442, 639
Regular filtration, 373
Regular module, 3
Reiten's theorem, 534
Relation
admissible, 146
Relatively projective, 155
Representable functor, 76, 264, 269, 285, 298
Representation
of a group, 8
of cyclic group, 10
regular, 3
trivial, 8
type, 252
stable equivalence, 430
Resolution, 79
Restriction of a module, 64
Reynolds ideal, 221, 495
Rickard's theorem
on derived equivalences, 557, 589
defined by simples, 657
on two-term tilting complexes, 603
Right
adjoint, 274
derived functor, 350
derived hom functor, 355
exact functor, 284
module, 23
Right bound, 688
Ring, 1
Rosenberg's lemma, 124

S

Schanuel's lemma, 78
Schur
group, 23
lemma, 22
Zassenhaus theorem, 114
Self-injective algebra, 126
invariance under
derived equivalence, 618
stable equivalence, 534
stable category is triangulated, 433
Semilinear map, 224
Semilocal, 189, 371
Semisimple algebra, 11, 24, 26, 28
Semisimple module, 11
Separable algebra, 346
Separable extension, 160, 346, 406, 448
Serial algebra, 47, 209, 210
Sheaves, 280
Short exact sequence, 74, 343, 381
equivalence, 82
of groups, 107
split, 113
splits, 74
Silting object, 657
Simple algebra, 27
Simple minded collection, 657
Simple module, 11
Singularity category, 627
and stable category, 631
and stable equivalence, 635
homomorphisms via syzygy, 638
Skolem Noether theorem, 421
Snake lemma, 281
Socle, 44
Source
algebra, 406
idempotent, 406
of a module, 164
Special biserial algebra, 146
Spectral sequence, 371
convergence, 381
induced short exact sequences, 381
infinite page, 376
of a differential filtered module, 375
of a double complex, 378
of a filtered complex, 376
page, 375
Split
epimorphism, 261
monomorphism, 261
semisimple, 32
short exact sequence, 74
Splitting field, 31, 32
Stabiliser, 62
Stable
category, 428
self-injective algebra, 433
centre, 486
equivalence, 440
elementary divisors of Cartan mapping, 471
finite representation type, 430
indecomposable module, 430
equivalence of Morita type, 446

and Bouc's trace map, 503
and Hochschild cohomology, 485
and Hochschild homology, 483
and indecomposability, 461
and Külshammer ideals, 500
and Nakayama algebras, 473
and simple modules, 467, 468
and stable centres, 488, 503
and stable cocentre, 499
in the generalised sense, 452
Grothendieck group, 469
Standard type, 594
Star, 509
String module, 149
Strong homotopy action, 574
homotopy equivalent, 579
morphism, 578
nullhomotopic, 579
Strongly graded algebra, 173
Structure constants, 31
Stupid truncation, 310
Submodule, 4
Sufficiently many
injectives, 337
projectives, 337, 351
Suspension functor, 669
Symmetric
algebra, 134
invariance under derived equivalence, 617
invariance under stable equivalence, 465
group, 13, 62
Symmetrising bilinear form, 134, 136
Syzygy, 78, 79, 429
for self-injective algebra, 432

T
Tensor co-algebra, 565, 569
Tensor product, 48, 49
Terminal object, 262
Tilting
complex, 557, 589
one-sided, 562
two-sided, 595
two-term, 597
module, 563
Top, 37
Torsion groups, 353
Total complex, 353
Trace map, 163, 231
Bouc's, 479
Hattori Stallings, 498
Track, 147
Transitive permutation module, 70, 71
Tree, 505
Triangulated category, 289, 433
Trivial
intersection group, 427, 652
module, 8
representation, 8
source module, 71
submodule, 63
Truncation, 309
intelligent, 309
stupid, 310
Twist of a module by automorphism, 67
Twisted bimodule, 131
Two-sided tilting complex, 595
Two-term tilting complex, 597

U
Underlying graph of a quiver, 505
Uniformiser, 186
Uniserial module, 46, 209, 210, 513
Unit
of an algebra, 564
of the adjunction, 274

V
Valuation, 185
Verdier
axioms, 289
localisation, 332
Vertex
adjacent to edge, 505
of a module, 161
of a quiver, 142

W
Walk
around the Brauer tree, 523
in a quiver, 142
Watts' theorem, 285
Weakly symmetric algebras, 133
Wedderburn
Malcev theorem, 345
theorem, 24
Weyl algebra, 30

Y
Yoneda's lemma, 269

Z
Zassenhaus Schur theorem, *see* Schur Zassenhaus theorem
Zero object, 262
Zorn's lemma, 16

The manufacturer's authorised representative in the EU is Springer Nature Customer Service Centre GmbH, Europaplatz 3, 69115 Heidelberg, Germany. If you have any concerns regarding our products, please contact ProductSafety@springernature.com

Printed and bound by CPI Group (UK) Ltd, Croydon, CR0 4YY
30/09/2026
02225589-0020